P9-AOI-927

Fluid Mechanics Through Problems

R.J. Garde
Pro Vice-chancellor
Indira Gandhi National Open University
New Delhi, India
(Formerly, Professor of Hydraulic Engineering
University of Roorkee
Roorkee, India)

JOHN WILEY & SONS
New York Chichester Brisbane Toronto Singapore

First published in 1989 by
WILEY EASTERN LIMITED
4835/24 Ansari Road, Daryaganj
New Delhi 110 002, India

Distributors:

Australia and New Zealand:
JACARANDA-WILEY LTD.
GPO Box 859, Brisbane, Queensland 4001, Australia

Canada:
JOHN WILEY & SONS CANADA LIMITED
22 Worcester Road, Rexdale, Ontario, Canada

Europe and Africa:
JOHN WILEY & SONS LIMITED
Baffins Lane, Chichester, West Sussex, England

South East Asia:
JOHN WILEY & SONS, INC.
05–04 Block B, Union Industrial Building
37 Jalan Pemimpin, Singapore 2057

Africa and South Asia:
WILEY EXPORTS LIMITED
4835/24 Ansari Road, Daryaganj
New Delhi 110 002, India

North and South America and rest of the world:
JOHN WILEY & SONS, INC.
605 Third Avenue, New York, N.Y. 10158, USA

Library of Congress Cataloging in Publication Data

Garde, R.J.
Fluid mechanics through problems/R.J. Garde.
p. cm.

1. Fluid mechanics—Problems, exercises, etc. I. Title.
TA357.G29 1989 88-29770
620.1′06′076—dc 19 CIP

ISBN 0-470-21332-9 John Wiley & Sons, Inc.
ISBN 81-224-0106-6 Wiley Eastern Limited

Printed in India at Prabhat Press, Meerut.

To

VIDYA

Preface

A large number of text books in Fluid Mechanics at Undergraduate level are available at present. These books deal with the basic concepts, describe certain flow phenomena and give the relevant derivations. However, because of space limitation these books do not do justice to amply illustrate the applications of the theory to a number of problems encountered in practice. Experience in the application of principles to numerical problems helps the student in grasping full meaning of the theory.

Therefore, a need is felt for a book in fluid mechanics in SI units which will discuss the basic theory in brief, then give a large number of graded solved problems illustrating the wide applications of fluid mechanics, followed by graded problems with answers, and a set of descriptive questions. The above need is fulfilled by this book. Answers to the descriptive questions are not given and student is expected to refer to a standard text in fluid mechanics and search the answer. The author is convinced that this arrangement will be found extremely beneficial by the students.

In seventeen chapters the book covers the syllabus of Fluid Mechanics and Machinery for undergraduate students of Civil, Mechanical and Chemical Engineering. The topics covered include fluid properties, fluid statics, kinematics, fluid dynamics including energy and momentum equations, dimensional analysis, laminar flow, turbulent flow and its applications, forces on immersed bodies, compressible flow, open channel flow, pumps and turbines, and unsteady flow. With this coverage the book will be extremely useful to diploma students and students preparing for AMIE and other competitive examinations.

The author acknowledges the assistance given by Prof. K.G. Ranga Raju and Mr. U.C. Kothyari of the Civil Engineering Department, University of Roorkee in carefully going through the manuscript and making valuable suggestions, and to Rashmi Garde for carefully going through the proofs.

R.J. GARDE

Contents

CHAPTER I

Properties of Fluids

1.1 INTRODUCTION

A fluid can be defined as a substance which deforms or yields continuously when shear stress is applied to it, no matter how small it is.

Fluids can be subdivided into liquids and gases. Liquids occupy a certain volume and have a free surface. Gases have a tendency to expand and fill container in which they are kept; they do not have free surface. Gases, when subjected to normal stress change their volume considerably. Liquids can be compressed to a small extent. Solids are least compressible. Solids when subjected to shear stress deform until internal resistance to deformation equals the externally applied stress. Some of the examples of fluids are water, air, hydrogen gas, oils, paint, blood, glycerine, brine, honey, etc.

In this book SI system (Systeme Internationale d'unites) of units is adopted, in which the following are the units of basic quantities that are used in fluid mechanics.

Length	metre (m)
Mass	kilogramme (kg)
Time	second (s)
Thermodynamic temperature	kelvin (K)
Temperature	celsius (°C)

A unit of force is newton (N) which is the force that produces 1 m/s^2 acceleration in a mass of 1 kg. Unit of work is joule (J) which is the work done when 1.0 N force acts through a distance of 1.0 m. Unit of power is watt (W) which is the power necessary for 1 J of work in 1 s.

$$1\,\text{W} = \frac{1\,\text{J}}{1\,\text{s}} = \frac{1\,\text{N m}}{1\,\text{s}}$$

Unit of frequency is hertz (H) having dimension of s^{-1}.

1.2 MASS DENSITY, SPECIFIC WEIGHT, SPECIFIC VOLUME, RELATIVE DENSITY, PRESSURE

Mass density, ρ (Rho) is the mass per unit volume

$$\rho = \lim_{\Delta V \to 0} \frac{\Delta M}{\Delta V} \tag{1.1}$$

$$\text{and} \quad \rho = f(x, y, z, T)$$

where x, y, z are co-ordinates of the point in flow field and T is temperature. At 20 °C and atmospheric pressure

water: $\rho = 998$ kg/m³

air: $\rho = 1.208$ kg/m³

Specific weight γ (Gama) is weight per unit volume;

$$\gamma = \rho g \ \text{N/m}^3 \tag{1.2}$$

Specific volume is volume per unit weight and hence

$$v = \frac{1}{\gamma} \ \text{m}^3/\text{N} \tag{1.3}$$

Relative density is the ratio of mass density to mass density of pure water at standard pressure of 101.325 N/m² and temperature of 4°C.

Pressure is force acting on unit area normal to it and has units of N/m²;

$$p = \lim_{\Delta A \to 0} \frac{\Delta F}{\Delta A} \tag{1.4}$$

1.3 VISCOSITY

Viscosity is that property of fluid by which it offers resistance to shear acting on it. According to Newton's law of viscosity the shear F acting between two layers of fluid is proportional to difference in their velocities Δu and area A, and inversely proportional to the distance Δy between them (see Fig. 1.1).

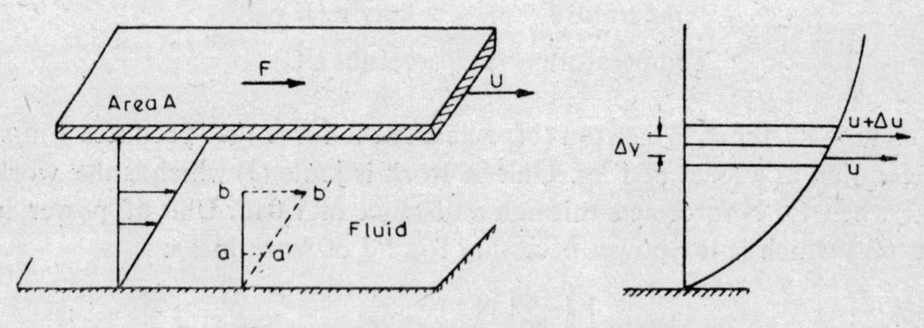

Fig. 1.1. Shear and velocity distribution.

Therefore

$$F = \mu A \frac{\Delta u}{\Delta y}$$

τ is the shear stress; or

$$\tau = \frac{F}{A} = \mu \frac{du}{dy} \tag{1.5}$$

where μ (Mu) is the constant of proportionality with dimensions of

$$[\mu] = \frac{[F/A]}{[du/dy]} = \frac{\text{N}}{\text{m}^2}\frac{\text{s m}}{\text{m}}, \text{ i.e., } \frac{\text{N s}}{\text{m}^2} \text{ or } \frac{\text{kg}}{\text{m s}}$$

One gm/cm s dynamic viscosity is known as poise (*P*).

du/*dy* gives the angular velocity of line *ab* or it is the rate of angular deformation.

$$\text{Coefficient of kinematic viscosity, } \nu = \frac{\mu}{\rho} \tag{1.6}$$

ν (Nu) has dimensions of m^2/s. One cm^2/s kinematic viscosity is known as stoke (S).

$$\mu \text{ or } \nu = f(p, T)$$

Variation of viscosity of liquids with pressure is very small and can be neglected. The variation with temperature is given by

$$\mu_T = Ae^{B/T} \tag{1.7}$$

where μ_T is dynamic viscosity at absolute temperature T and *A* and *B* are constants. Viscosity of liquids decreases with increase in temperature.
In the presence of suspended matter the viscosity of liquids increases according to the law

$$\mu_m/\mu = 1 + aC_v \tag{1.8}$$

where μ_m is the viscosity of liquid when suspended matter concentration in absolute volume is C_v and μ is the viscosity of clear liquid; *a* is 2.5 according to Einstein for C_v upto 0.30 and 4.5 for higher concentration as recommended by Ward. Viscosity of gases increases with increase in temperature and can be given by the formula proposed by Sutherland

$$\mu_T = \frac{bT^{1/2}}{1 + a/T} \tag{1.9}$$

where *a* and *b* are constants for a given gas.

Fluids are classified according to the relation between shear τ and rate of angular deformation:

$$\tau = 0 \qquad \text{Ideal fluids}$$

$$\tau = \mu \frac{du}{dy} \qquad \text{Newtonian fluids}$$

$$\tau = \text{const} + \mu \left(\frac{du}{dy}\right) \qquad \text{Ideal plastics or Bingham plastics}$$

$$\tau = \text{const} + \mu \left(\frac{du}{dy}\right)^n \qquad \text{Thyxotropic fluids}$$

$$\tau = \mu \left(\frac{du}{dy}\right)^n \qquad \text{Non Newtonian fluids}$$

For non Newtonian fluids, if n is less than unity they are called pseudo-plastics while fluids in which n is greater than unity are known as dilatants. These are shown in Fig. 1.2.

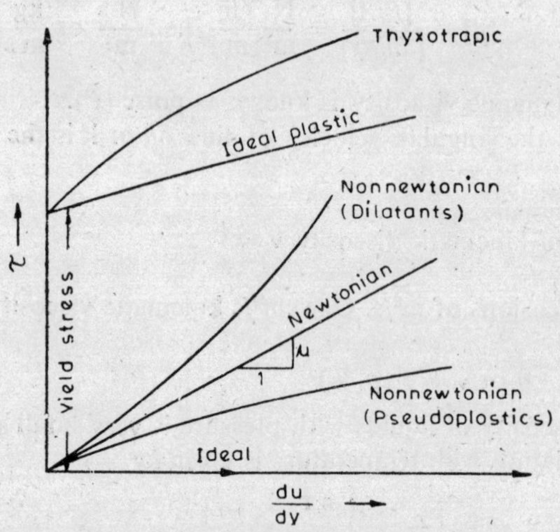

Fig. 1.2. Rheological classification of fluids.

1.4 BULK MODULUS OF ELASTICITY

Bulk modulus of elasticity of fluid, E, is defined as follows:

$$E = -v\frac{dp}{dv} \tag{1.10}$$

where v is the specific volume and dv is the change in v due to increase in pressure dp. E is expressed in N/m² or kN/m² (kilonewtons per m²).

1.5 GAS LAWS

Compression or expansion of gases can occur either under isothermal condition (i.e. at constant temperature) or under adiabatic condition (i.e. at constant heat content).

Isothermal condition is governed by Boyle's law according to which, for a given weight of gas at constant temperature,

$$pV = \text{const.} \tag{1.11}$$

where p is the pressure in N/m² or kN/m² and V is the volume of given weight of gas. According to Charles' law, for a constant volume of a given weight of gas

$$p/T = \text{const.} \tag{1.12}$$

where T is the absolute temperature of gas in kelvins.

$$T = t°\text{C} + 273.16$$

Equation of state of gas is

$$pv/T = R \tag{1.13}$$

where R is known as the universal gas constant having the units of

$$\frac{\text{N m}}{\text{N K}} \quad \text{or} \quad \text{m/K}.$$

In adiabatic process

$$pv^k = \text{const.} \tag{1.14}$$

If adiabatic process is reversible it is known as isentropic. Value of k varies with different gases.

1.6 SURFACE TENSION AND CAPILLARITY

Coefficient of surface tension σ is force per unit length (N/m) which is caused when liquid-gas interface meets a solid wall.

$$\text{Capillary rise or depression,} \quad h = \frac{2\sigma \cos \theta}{\gamma R} \tag{1.15}$$

where θ is the angle of contact between solid surface and liquid. If θ is greater than 90, there will be depression of surface in the tube (e.g. mercury in glass tube). If θ is less than 90° there will be a capillary rise (e.g. water in glass tube); (Fig. 1.3). For clean glass and water, θ is nearly equal to zero.

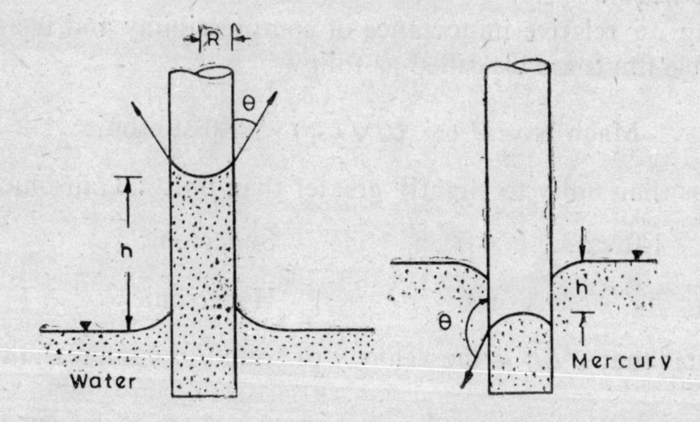

Fig. 1.3. Capillary rise and depression.

Difference of pressure Δp between inside and outside of a droplet of radius R

$$= \frac{2\sigma}{R} \tag{1.16}$$

Appendix A gives the basic properties of common fluids at 20°C, whereas Appendix B gives properties of water at different temperatures. Appendix C lists the properties of air at different temperatures and at atmospheric pressure. Appendix D lists the properties of common gases while Appendix E gives properties of standard atmosphere.

1.7 FLOW REGIMES

Depending on relative effects of viscosity and inertia, flows can be classified into laminar and turbulent flows.

In pipes: Reynolds No. $\text{Re}\left(=\frac{UD\rho}{\mu}\right) < 2100$ laminar flow

$\text{Re} > 3000$ turbulent flow

Here U is average velocity of flow in a pipe of diameter D.

In open channels: $\text{Re}\left(=\frac{UR\rho}{\mu}\right) > 500$ laminar flow

$\text{Re} > 2000$ turbulent flow

Depending on the relative importance of gravity and inertial forces, flow in open channels can be classified into sub-critical and supercritical flows.

Froude No. $\text{Fr}\left(=\frac{U}{\sqrt{gD}}\right) < 1.0$ subcritical flow

$\text{Fr} = 1.0$ critical flow

$\text{Fr} > 1.0$ supercritical flow

Here R is the hydraulic radius which equals area of flow divided by wetted perimeter; D is hydraulic mean depth which equals area of flow divided by water surface width.

Depending on relative importance of compressibility and inertia, flows of compressible fluids are classified as follows:

Mach No. $M\ (= U/\sqrt{E/\rho}) < 1.0$ subsonic

M slightly less than unity to slightly greater than unity: Transonic

$1.0 < M < 6.0$ Supersonic

$M > 6.0$ Hypersonic

It may be noted that $\sqrt{E/\rho}$ is the velocity of sound or compression wave in the medium.

Depending on distance between different molecules of the fluid relative to characteristic length of flow, one can have either continuum flow or free molecular flow.

$$\text{Knudsen Number } N = \frac{\text{Mean free path of molecules}}{\text{Characteristic length of flow}}$$

If $N < 0.01$ Continuum flow

$N > 10$ Free molecular flow

In all discussions in this book, the flow is treated as continuum.

ILLUSTRATIVE EXAMPLES

1.1 If 3.5 m³ of oil weighs 32.95 kN, calculate its specific weight, specific volume, mass density and relative density.

$$\text{Specific weight } \gamma = \frac{\text{weight}}{\text{volume}} = \frac{32.95}{3.5}$$

$$= 9.414 \text{ kN/m}^3$$

$$\text{Specific volume } v = \frac{1}{\gamma} = \frac{1}{9.414} = 0.106 \text{ m}^3\text{/kN}$$

$$\text{Mass density } \rho = \frac{\gamma}{g} = \frac{9.414 \times 10^3}{9.806} = 960.0 \text{ kg/m}^3$$

$$\text{Relative density} = \frac{960}{1000} = 0.960$$

1.2 If the kinematic viscosity of benzene is 7.42×10^{-3} stokes and its mass density is 860 kg/m³, determine its dynamic viscosity in kg/m s.

$$\nu = 7.42 \times 10^{-3} \text{ S}$$

$$= 7.42 \times 10^{-3} \times 10^{-4} \text{ m}^2\text{/s} = 7.42 \times 10^{-7} \text{ m}^2\text{/s}$$

$$\therefore \quad \mu = \rho\nu = 860 \times 7.42 \times 10^{-7}$$

$$= 6\,381 \times 10^{-4} \text{ kg/ms}$$

1.3 If the mass density of a fluid is 789 kg/m³, determine its specific weight and specific volume.

$$\gamma = \rho g = 789 \times 9.806$$

$$= 7736.934 \text{ N/m}^3 \quad \text{or} \quad 7.737 \text{ kN/m}^3$$

$$\text{Specific volume } v = \frac{1}{\gamma} = \frac{1}{7.737} = 0.129 \text{ m}^3\text{/kN}$$

1.4 Determine the kinematic viscosity of air at 20°C if its dynamic viscosity is 1.85×10^{-4} poise and its mass density is 1.208 kg/m³.

$$\mu = 1.85 \times 10^{-4} \text{ poise}$$

$$= 1.85 \times 10^{-5} \text{ kg/m s}$$

$$\therefore \quad \nu = \frac{\mu}{\rho} = \frac{1.85 \times 10^{-5}}{1.208} = 1.531 \times 10^{-5} \text{ m}^2\text{/s}$$

1.5 The space between two parallel horizontal plates kept 5 mm apart is filled with crude oil of dynamic viscosity 2.5 kg/m s. If the lower plate is stationary and upper plate is pulled with a velocity of 1.75 m/s, determine the shear stress on the lower plate.

$$\frac{du}{dy} = \frac{1.75}{5 \times 10^{-3}} = 0.35 \times 10^3 \text{ s}^{-1}$$

$$\tau = \mu \frac{du}{dy} = 2.5 \times 0.35 \times 10^3$$

$$= 875.0 \text{ N/m}^2$$

1.6 A 50 mm diameter and 0.10 m long cylindrical body slides vertically down in a 52 mm diameter cylindrical tube. The space between the cylindrical body and tube wall is filled with oil of dynamic viscosity 1.9 N s/m². Determine its velocity of fall if its weight is 16 N (Fig. 1.4).

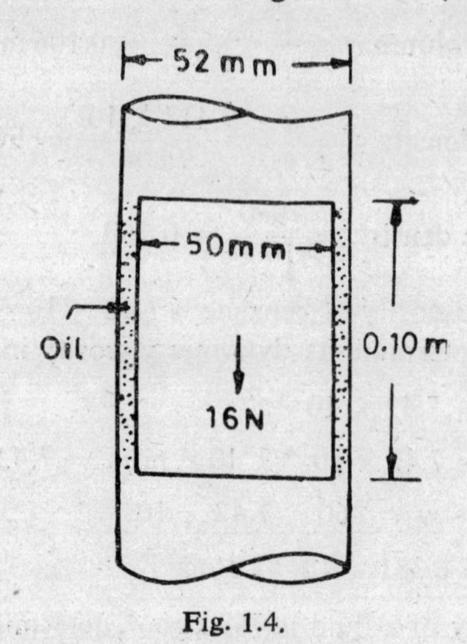

Fig. 1.4.

Let U be its terminal velocity of fall. Shear stress τ will be

$$\tau = \mu \frac{du}{dy} = 1.9 \times \frac{U}{1 \times 10^{-3}}$$

$$= 1.9 \times 10^3 \, U \text{ N/m}^2$$

The shear stress will act on the surface of the cylinder. Hence

$$\text{Total force } F = \tau \times A$$

$$= 1.9 \times 10^3 \times U \times 3.142 \times 50 \times 10^{-3} \times 0.10$$

$$= 29.849 \text{ U}$$

Under equilibrium condition, the weight will be balanced by the total shear force. Hence

$$16.0 = 29.849 \, U$$

$$\text{or} \quad U = 0.536 \text{ m/s}$$

Here it is assumed that velocity variation across the gap of 1 mm is linear.

1.7 A rectangular plate of 0.50 m × 0.50 m dimensions weighing 500 N slides down an inclined plane making 30° angle with horizontal, at a velocity of 1.75 m/s. If the 2 mm gap between the plate and inclined surface is filled with a lubricating oil, find its viscosity and express it in poise as well as in Ns/m^2 (Fig. 1.5).

$$F = 0.5 \times 0.5 \times \mu \times \frac{1.75}{2 \times 10^{-3}}$$

$$= 0.219 \times 10^3 \mu$$

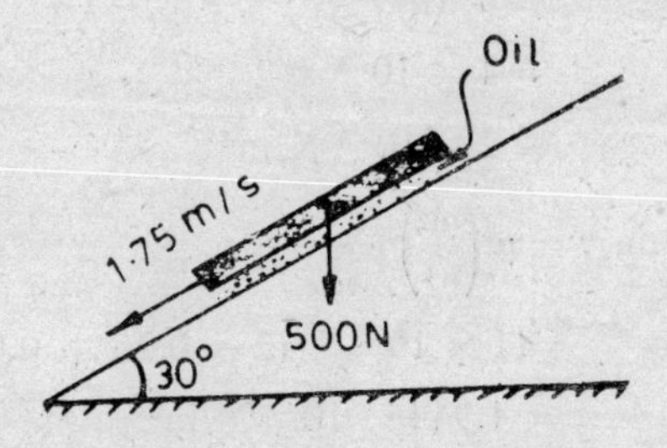

Fig. 1.5.

For equilibrium, this must be equal to component of weight along the inclined surface.

$$500 \sin 30 = 0.219 \times 10^3 \mu$$

$$\therefore \quad \mu = \frac{500 \times 0.50}{0.219} \times 10^{-3}$$

$$= 1.142 \text{ N s/m}^2$$

The viscosity in poise units will be 11.42 poise.

1.8 The velocity distribution, for small values of y, in laminar boundary layer on a flat plate is given by the equation

$$u = 5y - 2y^3$$

in which u is the velocity in m/s at a distance y m above the plate. Determine shear stress at $y = 0$ and $y = 0.25$ m if $\mu = 1.85 \times 10^{-5}$ kg/m s.

$$\frac{du}{dy} = 5 - 6y^2$$

$$\therefore \quad \left(\frac{du}{dy}\right)_{y=0} = 5 \text{ s}^{-1} \quad \text{and} \quad \left(\frac{du}{dy}\right)_{y=0\cdot25} = 5.0 - 0.375 = 4.625 \text{ s}^{-1}$$

$$\therefore \quad \tau \text{ at the boundary} = 5 \times 1.85 \times 10^{-5}$$

$$= 9.25 \times 10^{-5} \text{ N/m}^2$$

$$\text{At } y = 0.25 \text{ m}, \tau = 4.625 \times 1.85 \times 10^{-5}$$

$$= 8.556 \times 10^{-5} \text{ N/m}^2$$

1.9 If the velocity distribution in laminar boundary layer over a flat plate is given by

$$u = 2y - 2y^3 + y^4$$

determine the shear stress on the plate and at $y = 0.10$ m if dynamic viscosity of the fluid is 1×10^{-3} kg/m s.

$$du/dy = 2 - 6y^2 + 4y^3$$

$$\therefore \text{ at } y = 0, \tau = \mu \left(\frac{du}{dy}\right)_{y=0}$$

$$= 1 \times 10^{-3} \times 2$$

$$= 2 \times 10^{-3} \text{ N/m}^2$$

$$\text{At } y = 0.10 \text{ m}, \tau = \mu \left(\frac{du}{dy}\right)_{y=0\cdot 10}$$

$$= 1 \times 10^{-3} \times (2 - 0.06 + 0.004)$$

$$= 1.944 \times 10^{-3} \text{ N/m}^2$$

1.10 A closed vessel of volume 80 litres contains 0.50 N of gas at a pressure of 150 kN/m². If the gas is compressed isothermally to half its volume, determine the resulting pressure.

For isothermal process $\quad p_1 v_1 = p_2 v_2$

$$\therefore \quad p_2 = \frac{p_1 v_1}{v_2} = \frac{150 \times 80}{40} = 300 \text{ kN/m}^2$$

1.11 When the pressure of an enclosed gas is doubled, its new volume is 0.591 times the initial volume. Determine the value of k assuming the process to be adiabatic.

For adiabatic process $p_1 v_1^k = p_2 v_2^k$

$$\therefore \quad \frac{p_2}{p_1} = \left(\frac{v_1}{v_2}\right)^k \quad \therefore \quad 2 = \left(\frac{v_1}{0.591\, v_1}\right)^k = 1.691^k$$

$$\therefore \quad k = 1.319$$

1.12 What will be the change in pressure of a gas enclosed in a container at 300 kN/m² pressure if its temperature is changed from 30°C to 65°C.

$$\text{For constant volume } \frac{p_1}{T_1} = \frac{p_2}{T_2} \quad \text{or} \quad p_2 = \frac{T_2}{T_1} p_1$$

$$\therefore \quad p_2 = \frac{(273.16 + 65.00)}{(273.16 + 30.00)} \times 300$$

$$= 334.635 \text{ kN/m}^2$$

$$\therefore \quad \text{Change in pressure} = (334.635 - 300.000)$$

$$= 34.635 \text{ kN/m}^2$$

1.13 Determine the specific weight of hydrogen at 40°C at a pressure of two atmospheres absolute.

$$\text{One atmospheric pressure} = 101.325 \text{ kN/m}^2$$

$$\therefore \quad p = 2 \times 101.325 = 202.650 \text{ kN/m}^2$$

$$= 202.65 \times 10^3 \text{ N/m}^2$$

From Appendix D, $R = 420.3$ and

$$T = 273.16 + 40$$

$$= 313.16° \text{ K}$$

$$\frac{pv}{T} = R \quad \text{or} \quad \gamma = \frac{1}{v} = \frac{p}{RT}$$

$$\therefore \quad \gamma = \frac{202.65 \times 10^3}{420.3 \times 313.16} = 1.540 \text{ N/m}^3$$

1.14 What should be the internal diameter of a glass tube if capillary rise in it is not to exceed 2.0 mm?

$$\text{Since } h = \frac{2\sigma \cos \theta}{\gamma R} \quad \text{and} \quad \theta = 0°$$

$$R = \frac{2 \times 0.0735}{2 \times 10^{-3} \times 9.787} = 7.5 \times 10^{-3} \text{ m}$$

$$= 7.5 \text{ mm}$$

$$\therefore \quad \text{diameter} = 2 \times 7.5 = 15 \text{ mm}$$

1.15 Assuming that sap in trees has the same characteristics as water and that it rises purely due to capillary phenomenon, what will be the average diameter of capillary tubes in a tree if the sap is carried to a height of 10 m?

$$\text{Taking } \theta = 0°, \quad h = 2\sigma/\gamma R$$

$$R = \frac{2 \times 0.0735}{10 \times 9787} = 1.502 \times 10^{-6} \text{ m}$$

$$\therefore \quad \text{diameter} = 0.004 \times 10^{-6} \text{ m}$$

$$= 3 \times 10^{-3} \text{ mm}$$

1.16 What will be the difference of pressure between inside and outside of a droplet of water 1 mm in diameter.

$$\text{Since } p = 2\sigma/R = 4\sigma/d$$

$$= \frac{4 \times 0.0735}{1 \times 10^{-3}} = 0.2940 \times 10^3 \text{ N/m}^2$$

$$= 294 \text{ N/m}^2$$

1.17 **A glass tube of 2 mm internal diameter is immersed in an oil of mass density 960 kg/m³ to a depth of 10 mm. If a pressure of 172 N/m² is needed to form a bubble which is just released, determine the surface tension of the oil (Fig. 1.6).**

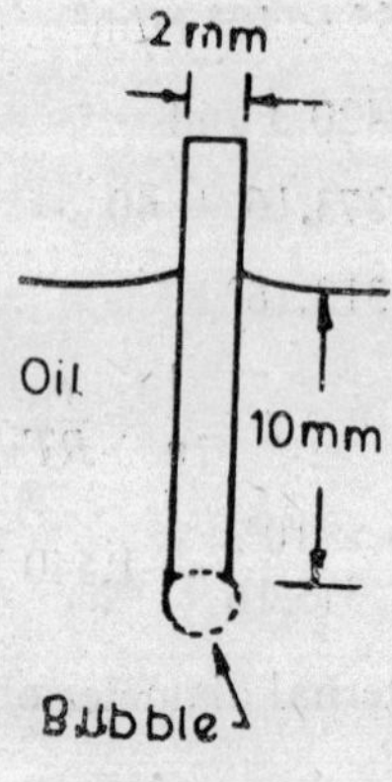

Fig. 1.6

$$(p_i - p_o) = 2\sigma/r; \text{ but } p_0 = 960 \times 9.806 \times 10 \times 10^{-3}$$

$$= 94.138 \text{ N/m}^2$$

$$\text{Inside pressure } p_i = 172.00 \text{ N/m}^2$$

$$\therefore \quad (p_i - p_o) = (172.000 - 94.138) = 77.862 \text{ N/m}^2$$

$$\therefore \quad \sigma = \frac{77.862 \times 1 \times 10^{-3}}{2} = 0.0389 \text{ N/m}$$

1.18 **If velocity distribution for laminar flow in a pipe is given by**

$$u/u_{max} = [1 - (r^2/R^2)]$$

where u is the velocity at a distance r from the centre line, u_{max} is the centre line velocity and R is the pipe radius, determine expression for shear stress τ.

$$\text{Since } r = R - y, \; 1 = -\frac{dy}{dr} \quad \text{or} \quad dy = -dr$$

$$\therefore \quad \tau = \mu \frac{du}{dy} = -\mu \frac{du}{dr}$$

$$\text{However,} \quad \frac{du}{dr} = u_{max}\left(\frac{-2r}{R^2}\right) = \frac{-2r\,u_{max}}{R^2}$$

$$\therefore \quad \tau = \frac{2\mu r\, u_{max}}{R^2}$$

1.19 **Calculate the force required to lift a thin ring of wire 20 mm in diameter from water surface. Neglect the weight of the ring.**

$$\text{Perimeter of ring} = \pi D$$
$$= 3.142 \times 20 \times 10^{-3} \text{ m}$$
$$= 0.06284 \text{ m}$$

Surface tension force will be due to surface tension on both sides of wire

$$F = 2 \times 0.06284 \times 0.0735$$
$$= 9.238 \times 10^{-3} \text{ N}$$

1.20 In the flow conditions given in Fig. 1.7 determine the velocity at which the central plate of area 5.0m² will move if a force of 150 N is applied to it. The dynamic viscosities of the two oils are in the ratio of 1:3 and viscocity of top oil 0.10 Ns/m².

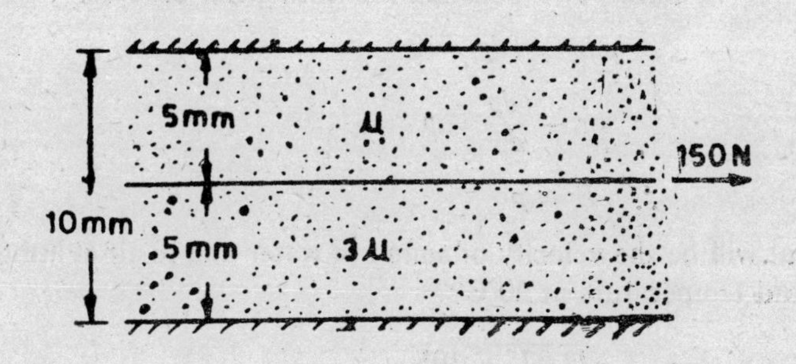

Fig. 1.7.

$$F = F_1 + F_2$$
$$= A\,\mu_1\left(\frac{du}{dy}\right) + A\,\mu_2\left(\frac{du}{dy}\right)$$
$$\therefore\quad 150 = 5 \times 0.1 \times \frac{U}{5 \times 10^{-3}} + 5 \times 0.3 \times \frac{U}{5 \times 10^{-3}}$$
$$= U\,(100 + 300) = 400\,U$$
$$\therefore\quad U = 150/400 = 0.375 \text{ m/s}$$

1.21 Determine whether the flow in a 1.5 m wide rectangular channel with 0.75 m depth of flow and 1.25 m/s velocity is (1) subcritical or supercritical, (2) laminar or turbulent.

$$R = \frac{By}{B + 2y} = \frac{(1.5 \times 0.75)}{(1.5 + 0.75 \times 2)} = 0.375\text{m}$$

$\therefore$ Assuming water temperature to be 20°C, $\mu = 1 \times 10^{-3}$ kg/m s, and

$$\rho = 998 \text{ kg/m}^3$$

$$\therefore\quad Re = \frac{UR\rho}{\mu} = \frac{1.25 \times 0.375 \times 998}{1 \times 10^{-3}}$$

$$= 4.678 \times 10^5$$

Hence the flow is turbulent.

$$\text{Hydraulic mean deapth} = \frac{A}{T} = \frac{1.5 \times 0.75}{1.5} = 0.75 \text{ m}$$

$$\therefore \quad Fr = \frac{U}{\sqrt{gA/T}} = \frac{1.25}{\sqrt{9.806 \times 0.75}} = 0.461$$

Hence flow is subcritical.

1.22 Obtain an expression for volume modulus of elasticity of atmosphere assuming the process is isothermal.

$$E = -v\frac{dp}{dv}$$

but $pv =$ constant for isothermal process.

$$\therefore \quad pdv + vdp = 0$$

$$\text{or } -\frac{vdp}{dv} = p$$

$$\therefore \quad E = p$$

1.23 What will be the velocity of sound in water and in air at atmospheric pressure and temperature at 20°C?

$$\text{For water } \sqrt{E/\rho} = \sqrt{\frac{2.075 \times 10^9}{998}} = 1441.93 \text{ m/s}$$

For air at 20°C and assuming isothermal process

$$\sqrt{E/\rho} = \sqrt{p/\rho} = \sqrt{\frac{101.325 \times 10^3}{1.208}} = 289.61 \text{ m/s}$$

If the process is assumed to be adiadatic $E = kp$ (see problem 1.31 for solution). Taking $k = 1.4$,

$$\sqrt{E/\rho} = \sqrt{kp/\rho} = \sqrt{\frac{1.4 \times 101.325 \times 10^3}{1.208}} = 342.68 \text{ m/s}$$

1.24 Air is introduced through a nozzle into a tank of water to form a stream of bubbles. If the process requires 2.5 mm diameter bubbles to be formed, by how much the air pressure at the nozzle must exceed that of the surrounding water?

$$\Delta p = \frac{2\sigma}{R}$$

Assuming water temperature to be 20°, $\sigma = 0.0735$ N/m

$$\therefore \quad \Delta p = 2 \times 0.0735/1.25 \times 10^{-3} = 117.6 \text{ N/m}^2.$$

PROBLEMS

1.1 If the relative density of a fluid is 1.59, calculate its mass density, specific weight and specific volume.

(1590 kg/m^3, 15.591 kN/m^3, 0.064 m^3/kN)

1.2 If relative density of a liquid is 13.6, find its specific volume.

(0.00750 m^3/kN)

1.3 If the dynamic viscosity of a liquid is 0.012 poise and its relative density is 0.79, obtain its kinematic viscosity in m^2/s.

(1.522×10^{-6} m^2/s)

1.4 Two horizontal plates are kept 12.5 mm apart and the space between them is filled with oil of dynamic viscosity of 14 poise. If the top plate is moved at a constant velocity of 2.5 m/s, determine the shear stress on the lower plate. (280.0 N/m^2)

1.5 When a real fluid flows past a plate held parallel to flow, the velocity distribution near the plate is given by

$$u/U = \tfrac{3}{2}(y/\delta) - \tfrac{1}{2}(y/\delta)^2$$

where $u = U$ when $y = \delta$. Determine the shear stress at $y = 0$ and when $y/\delta = 0.50$. ($3\mu U/2\delta$, $\mu U/\delta$)

1.6 A cylindrical body of 75 mm diameter and 0.15 m length falls freely in a 80 mm diameter circular tube kept vertically. If the space between the cylindrical body and tube is filled with oil of dynamic viscosity 0.9 poise, determine the weight of the body when it falls at a speed of 1.5 m/s. (1.909 N)

1.7 A circular disc of 0.30 m diameter and weighing 50 N is kept on an inclined surface with a slope of 45°. The space of 2 mm between disc and inclined surface is filled with oil of dynamic viscosity 1.0 N s/m^2. What force will be required to pull the disc up the inclined plane at velocity at 0.50 m/s? (53.012 N)

1.8 What will be the dynamic viscosity of muddy water at 20°C if it contains 30 per cent fine sediment by volume. (2.35×10^{-3} kg/m s)

1.9 Express the bulk modulus of elasticity in terms of mass density ρ of the fluid and pressure.

$$\left(E = \rho \frac{dp}{d\rho}\right)$$

1.10 Find the increase in the pressure required to reduce the volume of water by 0.8 percent if its bulk modulus of elasticity is 2.075×10^9 N/m^2.

(1.66×10^9 N/m^2)

1.11 Assuming depth of sea to be 10000 m, determine the mass density of sea water at this depth if its relative density at sea level is 1.026. Take a constant value of E as 2.113×10^9 N/m^2. (107.60 kg/m^3)

1.12 Determine the bulk modulus of elasticity of a fluid that has a density increase of 0.002 percent for a pressure increase of 44.540 kN/m^2.

(2.227×10^9 N/m^2)

1.13 What change in pressure is required to compress a given mass of gas to one third its volume under isothermal condition? $(2p_1)$

1.14 If 1.0 m^3 of gas is compressed to 0.20 m^3 volume when its initial pressure was 150 kN/m^2 determine the final pressure if the compression takes place under (i) isothermal condition, (ii) adiabatic condition. Take $k = 1.30$. (750 kN/m^2, 1215.492 kN/m^2)

1.15 In an adiabatic compression when final pressure was four times the initial pressure, the ratio of final to initial volume was 0.315. Determine the value of adiabatic constant. (1.20)

1.16 Determine the temperature and pressure of oxygen if from its original p, v and t of 1 bar, 1.5 m^3 and 20°C it is compressed to 0.30 m^3 volume under adiabatic condition. Take $k = 1.29$.

(7.974 bars, 194.37°C)

1.17 A cylinder of 0.30 m length and 0.10 m diameter rotates about a vertical axis inside a fixed cylindrical tube of 105 mm diameter and 0.30 m length. If the space between the tube and the cylinder is filled with liquid of dynamic viscosity 0.125 N s/m^2, determine the speed of rotation of the cylinder which will be obtained if an external torque of 1.0 N m is applied to it. (81.03 rpm)

1.18 A circular disc of radius R is held parallel to a large plane and stationary surface at a small distance from it. If the space t between the two is filled with oil of dynamic viscosity μ and the disc is rotated at N rpm, determine the torque required to maintain this rotation.

$$\left(\frac{\pi^2 N \mu R^4}{60t}\right)$$

1.19 A circular cylinder of radius R_1 and height h rotates at N rpm in a cylindrical container of radius R_2 with their axes vertical and coinciding. If the spacing between the bottoms of the cylindrical container and cylinder is t which is small and if the space between the cylinder and container is filled with oil of dynamic viscosity μ, obtain an expression for the total torque T required to maintain the motion. Assume R_2 to be slightly larger than R_1.

$$\left(T = \frac{\pi^2 N \mu R_1^3 h}{15(R_2 - R_1)} + \frac{\pi^2 N \mu R_1^4}{60t}\right)$$

1.20 What will be the temperature of oxygen in °C if at 4.0 kN/m^2 pressure its unit weight is 0.50 N/m^3? (Use value of $R = 26.8$ m/°K).

(25.35°C)

1.21 What percent error will be made in the viscosity of blood if clear water viscosity at 20°C is assumed when it contains 50 percent by volume of red blood cells? Use Ward's equation. (225 percent)

1.22 Determine the capillary rise in a clean glass tube of internal diameter 2.5 mm if the liquid is carbon tetrachloride. (2.71 mm)

1.23 Two vertical parallel glass plates distance t apart are partially submerged in a liquid of specific weight γ and surface tension σ. Show that the capillary rise is given by

$$h = \frac{2\sigma \cos\theta}{t\gamma}$$

1.24 Determine the diameter of a droplet of water in mm if the pressure inside is to be greater than the outside pressure by 130 N/m².

(2.262 mm)

1.25 A soap bubble of 50 mm diameter has a pressure difference of 20 N/m² between its inside and outside. Determine the coefficient of surface tension of the solution. (2.5×10^{-1} N/m)

1.26 For flow in a pipe take

$$\tau = \mu \frac{\partial u}{\partial y}$$

and shear stress at the wall τ_0 to be related to shear stress at distance y by the relation $\tau = \tau_0 \left(1 - \frac{y}{R}\right)$. Determine the velocity distribution by integration of the equation for shear assuming $u = u_{max}$ at the centre and $u = 0$ at the wall. Here R is the radius of pipe.

$$\left(\frac{u}{u_{max}} = \left(1 - \frac{r^2}{R^2}\right)\right)$$

1.27 Three cylindrical tubes of 0.50 m length are placed co-axially and the central tube is rotated at 5 rpm applying a torque of 6.0 N m. Determine the viscosity of oil which fills the space between the tubes. Take R_1, R_2 and R_3 as 0.150, 0.152 and 0.154 m.

($2.076 \times 7 \times 10^{-3}$ Ns/m²)

1.28 Determine whether the flow in the following cases is laminar or turbulent:

(i) 0.30 m diameter pipe carrying water at 20°C at 3.0 m/s velocity

(ii) Flow of oil at 0.50 m/s velocity in 25 mm diameter tube. The dynamic viscosity of oil is 1.0 N s/m² and its mass density is 970 kg/m³.

(iii) Water flowing at 1.5 m/s velocity in a rectangular channel of 4.0 m width and 1.0 m depth. Water temperature is 20°C. Also note the Reynolds numbers in each case.

(8.982×10^5 turbulent, 12.125 laminar, 9.985×10^5 turbulent)

1.29 Determine whether the flow in the following cases is subcritical, critical or supercritical (list the Fr number in each case).

(i) Water flowing at 0.75 m/s velocity in a rectangular channel of width 5.0 m and depth 1.0 m.

(ii) Water flowing in a triangular channel of central angle 90° when the velocity is 0.99 m/s and depth of 0.20 m.

(iii) Water flowing at 3.0 m/s velocity in a trapezoidal channel of bottom width 1.0 m and side slope of 1: 1, at a depth of 0.8 m. (0.239 subcritical, 1.0 critical, 1.287 supercritical)

1.30 If an aeroplane is moving through air at 800 km/hr, determine the flow regime if velocity of sound in air is 340 m/s. (Note the Mach number) (0.654 subsonic)

1.31 Obtain expression for E when the flow is adiabatic. ($E = kp$)

1.32 A long rigid pipe of 0.30 m diameter is used for pumping oil across the country. The pipe becomes plugged at some unknown point so that no fluid can flow. A piston inserted from one end of the pipe slides without leakage through 0.50 m causing an increase in pressure of 200 kN/m². Assume E for oil to be 1.8×10^9 N/m², determine the approximate location of obstruction. (4.5 km)

DESCRIPTIVE QUESTIONS

1.1 Does definition of fluid include only substances in liquid phase? Explain.

1.2 List the differences between liquids, solids and granular material.

1.3 List the differences between liquids and gases.

1.4 Give five examples of fluid flow phenomena encountered in every day life.

1.5 What is continuum? Is air a continuum? Does it always remain so?

1.6 Classify the fluids A and B for which following values of deformation rate and shear stress are obtained experimentally.

τ N/m²	0	100	200	300	400
Fluid A $\dfrac{du}{dy}$ s^{-1}	0	0.30	0.60	0.90	1.2
Fluid B $\dfrac{du}{dy}$ s^{-1}	0	54.8	77.5	94.9	109.5

1.7 How small should be the value of ΔV in the definition of mass density $\rho = \lim\limits_{\Delta \forall \to 0} \dfrac{\Delta M}{\Delta \forall}$?

1.8 Give one example each of Newtonian fluid, pseudoplastic, dilatant, ideal plastic and thyxotropic fluid.

1.9 Classify the following fluids: Water, sugar solution, printer's ink, air, glycerine, and molten metal.

1.10 Under what conditions is the elasticity of the fluid important?

1.11 Differentiate between adiabatic and isontropic processes.

1.12 Even though needle is heavier than water, it can float on it if it is placed lengthwise on the water surface. Why?

1.13 Why does oil spread when it is poured on water surface?

1.14 Why does the viscosity of a liquid decrease with increase in temperature whereas it increases with increase in temperature in the case of a gas?

1.15 Give two examples each of (i) laminar flow, (ii) turbulent flow (iii) supersonic flow.

1.16 Show that the velocity gradient can be interpreted as rate of angular deformation.

1.17 Which fluid property/properties are important in the following physical phenomena:

Lubrication
Rise of sap in trees
Ground water flow
Settling of a sediment particle in water
Force acting on the bottom of water tank
Formation of droplets
Energy loss in pipelines
Water hammer phenomenon in pipes
Echo sounders
Formation of waves on water surface
Cavitation
Pumping of blood through arteries to capillaries.

1.18 Arrange the following fluids according to increasing values of dynamic viscosity and also according to decreasing values of mass density: air, water, alcohol, glycerine, castor oil.

1.19 Give one example each where air can be treated as an incompressible fluid and water has to be treated as compressible fluid.

1.20 Mention different ways in which the fluid can be set in motion; give one example each.

1.21 Draw a sketch of smoke issuing from the cigarette and sketch its path. Indicate to what regimes of flow the portion near the cigarette, away from it and in between the two belong.

1.22 A fluid is a substance that

(i) has to be kept in a closed container
(ii) is almost incompressible
(iii) has zero shear stress
(iv) flows when even a small shear is applied to it.

1.23 A Nowtonian fluid is that

(i) which follows Newton's laws of motion
(ii) which needs a minimum shear before it starts deforming
(iii) for which shear and deformation are related as $\tau = \mu \partial u/\partial y$.

1.24 Dimensions of dynamic viscosity are

(i) L^2/T (ii) M/LT (iii) MT/L (iv) T/L^2

1.25 Dynamic viscosity of a gas

(i) increases as temperature decreases

(ii) increases as temperature increases
(iii) is independent of temperature
(iv) may increase or decrease with increase in temperature, depending on nature of gas.

1.26 An isentropic process is one in which
(i) pv = constant (ii) pv^k = constant (iii) pv^k = constant, and process is reversible (iv) none of the above.

1.27 Bulk modulus of elasticity of liquids has the dimensions
(i) MT^2/L (ii) ML^2/T (iii) M/LT^2

CHAPTER II

Kinematics of Fluid Flow

2.1 VELOCITY

Kinematics of fluid flow deals with fluid motion in terms of displacements, velocities, accelerations, rotations etc., without regard to forces responsible for motion.

Velocity of a fluid particle, V_s, is defined as

$$V_s = \lim_{\Delta t \to 0} \frac{\Delta S}{\Delta t} = \frac{dS}{dt} \tag{2.1}$$

where ΔS is the distance travelled in time Δt. Velocity has both magnitude and direction and hence it is a vector quantity. It can be represented in cartesian co-ordinate system as

$$\mathbf{V}_s = u\mathbf{i} + v\mathbf{j} + w\mathbf{k} \tag{2.2}$$

where $u = dx/dt$, $v = dy/dt$ and $w = dz/dt$ are the velocity components in x, y, z directions respectively, and $\mathbf{i}$, $\mathbf{j}$ and $\mathbf{k}$ are unit vectors in those directions (Fig. 2.1).

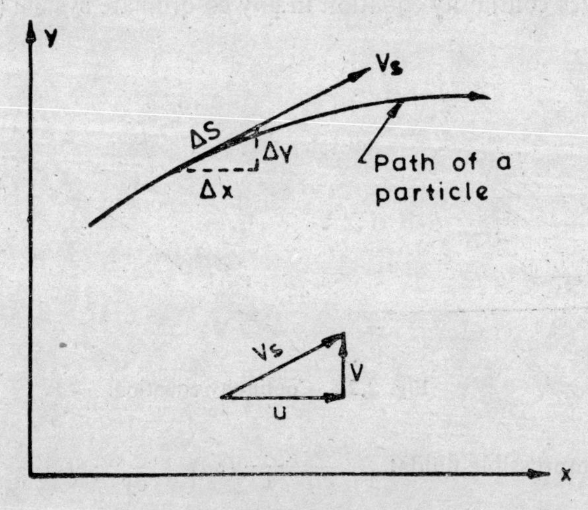

Fig. 2.1 Definition sketch.

2.2 TYPES OF FLOW

If F is a flow or fluid property such as velocity, pressure, mass density or temperature, then the following types of flows can be defined

$$\left.\begin{array}{ll} \text{Steady flow:} & \left(\dfrac{\partial F}{\partial t}\right)_{\text{at a point or section}} = 0 \\ \text{Unsteady flow:} & \left(\dfrac{\partial F}{\partial t}\right)_{\text{at a point or section}} \neq 0 \end{array}\right| \quad (2.3)$$

$$\left.\begin{array}{ll} \text{Uniform flow:} & \left(\dfrac{\partial F}{\partial S}\right)_{t=t_0} = 0 \\ \text{Nonuniform flow:} & \left(\dfrac{\partial F}{\partial S}\right)_{t=t_0} \neq 0 \end{array}\right| \quad (2.4)$$

$$\left.\begin{array}{ll} \text{One dimensional:} & F = F(x, t) \quad \text{or} \quad F(s, t) \\ \text{Two dimensional:} & F = F(x, y, t) \\ \text{Three dimensional:} & F = F(x, y, z, t) \end{array}\right| \quad (2.5)$$

2.3 STREAM LINES, PATH LINES, STREAK LINES

An imaginary line in the flow field such that at every point along it the velocity vector is tangential to it is known as a stream line. Equation of stream line is

$$\frac{u}{dx} = \frac{v}{dy} = \frac{w}{dz} \quad (2.6)$$

A path line is the path followed by a fluid particle during its travel. A streak line is the path followed by all the fluid particles passing through a given point in space. A stream tube consists of a group of stream lines.

2.4 CONTINUITY EQUATION

Application of the principle of conservation of mass to an elementary volume gives continuity equation in any co-ordinate system (Fig. 2.2)

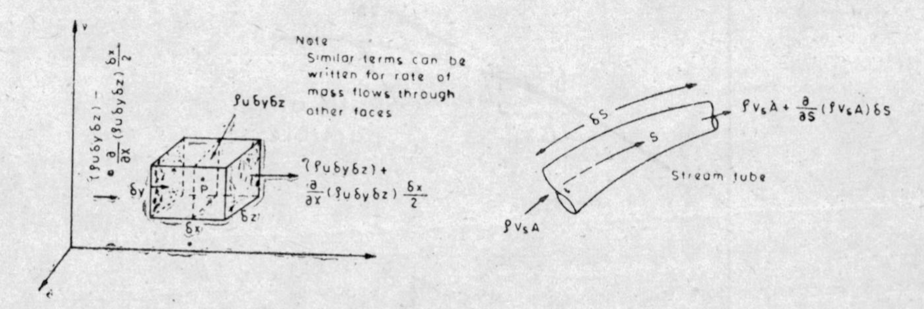

Fig. 2.2. Continuity equation.

For compressible fluids: $\dfrac{\partial \rho}{\partial t} + \dfrac{\partial}{\partial x}(\rho u) + \dfrac{\partial}{\partial y}(\rho v) + \dfrac{\partial}{\partial z}(\rho w) = 0$ (2.7) or in vector form

$$\frac{\partial \rho}{\partial t} + \text{div}\,(\rho \mathbf{V}) = 0 \tag{2.8}$$

For incompressible (homogeneous or nonhomogeneous) fluids:

$$\frac{\partial u}{\partial x} + \frac{\partial v}{\partial y} + \frac{\partial w}{\partial t} = 0 \tag{2.9}$$

In cylindrical-polar coordinate system the equation for compressible fluid is

$$\frac{\partial \rho}{\partial t} + \frac{\partial}{\partial x}(\rho V_x) + \frac{1}{r}\frac{\partial}{\partial r}(\rho V_r.r) + \frac{1}{r}\frac{\partial}{\partial \theta}(\rho V_\theta) = 0 \tag{2.10}$$

For incompressible (homogeneous or nonhomogeneous) fluids the equation is

$$\frac{\partial V_x}{\partial x} + \frac{1}{r}\frac{\partial (V_r.r)}{\partial r} + \frac{1}{r}\frac{\partial}{\partial \theta}(V_\theta) = 0 \tag{2.11}$$

In the same manner continuity equation for a stream tube can be written:

Compressible fluids: $$\frac{\partial \rho}{\partial t} + \frac{1}{A}\frac{\partial}{\partial S}(\rho V_s A) = 0 \tag{2.12}$$

For steady flow of compressible fluids: $$\rho V_s A = \text{const.} \tag{2.13}$$

For incompressible fluids: $$\left.\begin{aligned} &\frac{\partial}{\partial S}(V_s A) = \text{const.} \\ &\text{or}\quad V_s.A = \text{const.} \end{aligned}\right| \tag{2.14}$$

Average velocity U over the area $A = \int_A u\,dA$

Discharge Q in m³/s $= UA$

Here dA and A are perpendicular to velocity vector u or U, A is in m² and U in m/s. Discharge can also be expressed in litres/s

$$1 \text{ l/s} = 10^{-3} \text{ m}^3/\text{s}$$

2.5 ACCELERATION

Acceleration is the rate of change of velocity with time. It is a vector quantity, it can be caused due to change in magnitude or direction (or both) of the velocity vector.

Convective acceleration is due to nonuniformity of flow whereas local acceleration is due to unsteadiness of flow. Convective tangential acceleration $a_s = V_s \frac{\partial V_s}{\partial S}$. It is along the streamline and it is due to change in magnitude of velocity. Local tangential acceleration is given by $\frac{\partial V_s}{\partial t}$.

Total tangential acceleration: $$\frac{dV_s}{dt} = \frac{\partial V_s}{\partial t} + V_s \frac{\partial V_s}{\partial S} \tag{2.15}$$

Convective normal acceleration due to change in direction of flow along a streamline is equal to $\frac{V_s^2}{R}$ (Fig. 2.3), where R is radius of curvature of streamline. Local normal acceleration $= \frac{\partial V_n}{\partial t}$, where V_n is the normal component of velocity generated due to change in direction.

$$\frac{dV_n}{dt} = \frac{\partial V_n}{\partial t} + \frac{V_s^2}{R} \tag{2.16}$$

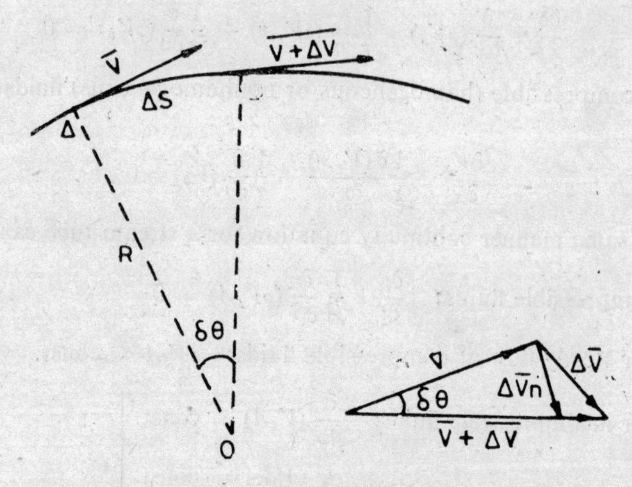

Fig. 2.3 Normal acceleration,

In the Cartesian co-ordinate system the acceleration vector **a** can be written as

$$\mathbf{a} = a_x\mathbf{i} + a_y\mathbf{j} + a_z\mathbf{k} \tag{2.17}$$

where the three acceleration components are given by

$$\begin{aligned} a_x &= \frac{\partial u}{\partial t} + u\frac{\partial u}{\partial x} + v\frac{\partial u}{\partial y} + w\frac{\partial u}{\partial z} \\ a_y &= \frac{\partial v}{\partial t} + u\frac{\partial v}{\partial x} + v\frac{\partial v}{\partial y} + w\frac{\partial v}{\partial z} \\ a_z &= \frac{\partial w}{\partial t} + u\frac{\partial w}{\partial x} + v\frac{\partial w}{\partial y} + w\frac{\partial w}{\partial z} \end{aligned} \tag{2.18}$$

local acceleration : convective acceleration

2.6 ROTATION, VORTICITY, CIRCULATION

Rotation ω about any axis is defined as the average of angular velocities of two elements, such as OA and OB in a plane perpendicular to the axis. Thus (Fig. 2.4),

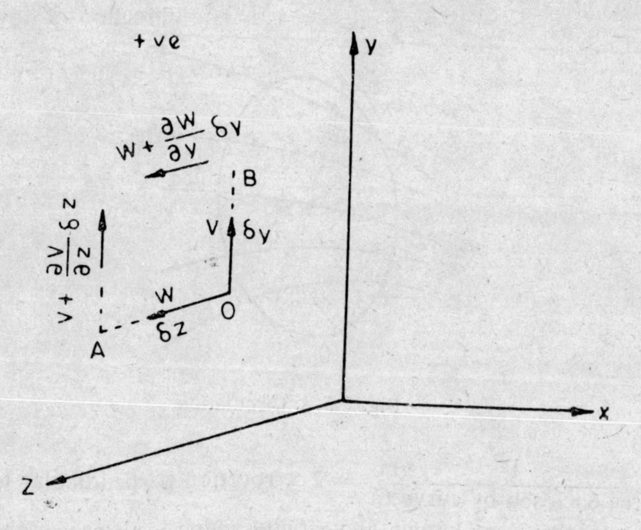

Fig. 2.4. Rotation.

$$\left.\begin{aligned}\omega_x &= \frac{1}{2}\left(\frac{\partial w}{\partial y}-\frac{\partial v}{\partial z}\right)\\ \omega_y &= \frac{1}{2}\left(\frac{\partial u}{\partial z}-\frac{\partial w}{\partial x}\right)\\ \omega_z &= \frac{1}{2}\left(\frac{\partial v}{\partial x}-\frac{\partial u}{\partial y}\right)\end{aligned}\right\} \tag{2.19}$$

Rotation vector

$$\begin{aligned}\boldsymbol{\omega} &= \omega_x\mathbf{i}+\omega_y\mathbf{j}+\omega_z\mathbf{k}\\ &= \frac{1}{2}\,\text{Curl}\,\mathbf{V}\end{aligned} \tag{2.20}$$

In cylindrical-polar coordinate system

$$\left.\begin{aligned}\omega_x &= \frac{1}{2}\left[\frac{\partial}{r\partial r}(rV_\theta)-\frac{\partial}{r\partial\theta}(V_r)\right]\\ \omega_r &= \frac{1}{2}\left[\frac{\partial V_x}{r\partial\theta}-\frac{1}{r}\frac{\partial}{\partial x}(V_\theta . r)\right]\\ \omega_\theta &= \frac{1}{2}\left[\frac{\partial V_r}{\partial x}-\frac{\partial V_x}{\partial r}\right]\end{aligned}\right\} \tag{2.21}$$

Circulation Γ around a closed curve C is defined as the line integral of **V.ds** along the curve C, taken positive in anticlockwise direction (Fig. 2.5).

$$\Gamma = \int_C \mathbf{V}\cdot\mathbf{ds} = \int_C (u\,dx+v\,dy+w\,dz) \tag{2.22}$$

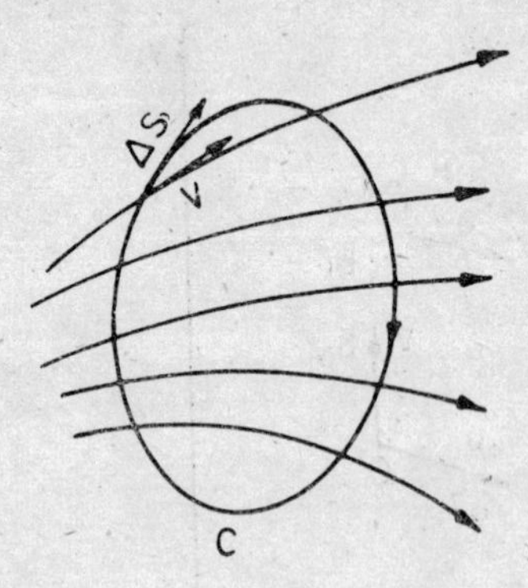

Fig. 2.5. Circulation.

$$\frac{\Gamma}{\text{area enclosed by curve } C} = 2 \times \text{rotation perpendicular to area}$$

The quantity 2ω is known as vorticity which is also a vector quantity. If at every point in the flow field $\omega_x = \omega_y = \omega_z$ is equal to zero, flow is called irrotational; otherwise it is rotational flow.

2.7 VELOCITY POTENTIAL AND STREAM FUNCTION

Velocity potential function ϕ is a scalar function of x, y, z and t such that its negative derivative with respect to any of x, y, z gives the velocity component in that direction. Thus $\phi = \phi\ (x, y, z, t)$, and

$$-\frac{\partial \phi}{\partial x} = u, -\frac{\partial \phi}{\partial y} = v, -\frac{\partial \phi}{\partial z} = w \tag{2.23}$$

Equations (2.9) and (2.23) show that for incompressible fluids, ϕ satisfies Laplace equation, i.e.

$$\frac{\partial^2 \phi}{\partial x^2} + \frac{\partial^2 \phi}{\partial y^2} + \frac{\partial^2 \phi}{\partial z^2} = 0 \tag{2.24}$$

Further ϕ satisfies the condition of irrotationality i.e.

$$\omega_x = \omega_y = \omega_z = 0.$$

In cylindrical polar co-ordinate system $\phi = \phi\ (x, y, \theta, t)$ and

$$-\frac{\partial \phi}{\partial x} = V_x, -\frac{\partial \phi}{\partial r} = V_r \text{ and } -\frac{1}{r}\frac{\partial \phi}{\partial \theta} = V_\theta \tag{2.25}$$

Laplace equation then becomes

$$\frac{\partial^2 \phi}{\partial x^2} + \frac{1}{r}\frac{\partial \phi}{\partial r} + \frac{\partial^2 \phi}{\partial r^2} + \frac{1}{r^2}\frac{\partial^2 \phi}{\partial \theta^2} = 0 \tag{2.26}$$

For two dimensional rotational or irrotational flows of incompressible fluids, a scalar function $\psi\ (x, y, t)$ can be defined such that

$$-\frac{\partial \psi}{\partial y} = u, \quad +\frac{\partial \psi}{\partial x} = v \text{ (Cartesian)} \tag{2.27}$$

$$-\frac{1 \partial \psi}{r \partial \theta} = V_r, + \frac{\partial \psi}{r} = V_\theta \text{ (Polar)} \tag{2.28}$$

For irrotational flow ψ satisfies Laplace's equation, viz.

$$\frac{\partial^2\psi}{\partial x^2}+\frac{\partial^2\psi}{\partial y^2}=0 \tag{2.29}$$

otherwise
$$\nabla^2\psi = 2\omega_z \tag{2.30}$$

For two-dimensional irrotational flow of incompressible fluids, a set of equipotential lines (i.e. lines with ϕ = constant) and streamlines (i.e. ψ = constant) can be drawn for given boundary configuration. This forms a flow net. Equipotential lines and streamlines intersect orthogonally.

Stagnation point is characterised by $u = v = 0$, i.e. $\frac{\partial\psi}{\partial y}=\frac{\partial\psi}{\partial x}=0$. At such a point streamlines intersect orthogonally.

2.8 FLOW-NET

For two dimensional irrotational flow of incompressible fluids, a set of stream lines and equipotential lines can be drawn using ϕ = constant and ψ = constant and $\Delta\phi = \Delta\psi$. This constitutes a flow-net at any given instant. Typical flow-net for a two dimensional sharp bend is shown in Fig. 2.6. In drawing flow-net by graphical method the following points be kept in mind.

1. The boundaries act as limiting stream lines.
2. In uniform flow region, streamlines will be parallel and equidistant. (Note: $\psi_2 - \psi_1 = \Delta\psi$ represents discharge passing between the two successive stream lines).

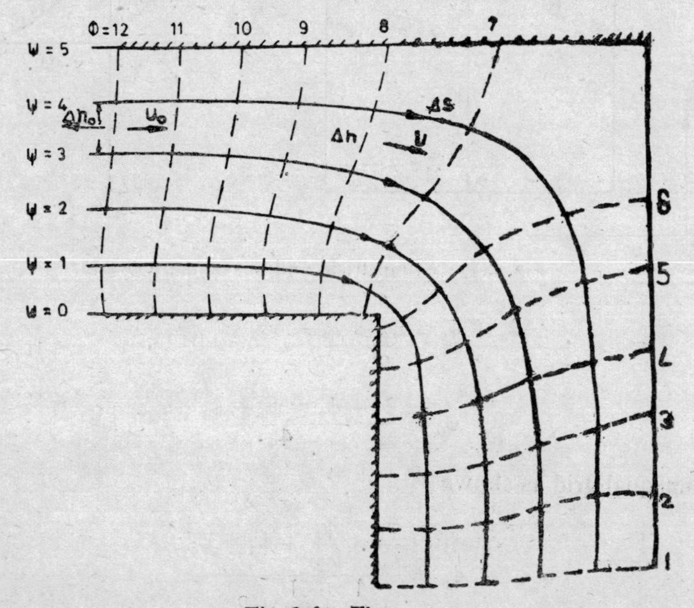

Fig. 2.6. Flow-net.

3. In nonuniform flow region, sketch the streamlines approximately.
4. Since $\delta\phi = \delta\psi$, $\Delta S = \Delta n$ where Δn is the spacing between stream lines and ΔS that between successive equipotential lines at a given point. Keeping this in mind and that equipotential lines will intersect streamlines normally, these lines can be drawn.
5. In the limit, the flow-net should consist of small squares whose diagonals should be equal.
6. For boundary curving sharply towards the flow and away from flow, a five cornered figure will be formed.

For a stream tube, since no flow can go out through boundary stream lines, $\Delta n_0 U_0 = \Delta n U$, or

$$\frac{U}{U_0} = \frac{\Delta n_0}{\Delta n}$$

Thus, spacing between streamlines is inversely proportional to velocity.

2.9 RELAXATION METHOD OF DRAWING FLOW-NET

Since for two dimensional irrotational flow ψ satisfies Laplace equation

$$\frac{\partial^2\psi}{\partial x^2} + \frac{\partial^2\psi}{\partial y^2} = 0$$

which can be written in finite difference form as (Fig. 2.7).

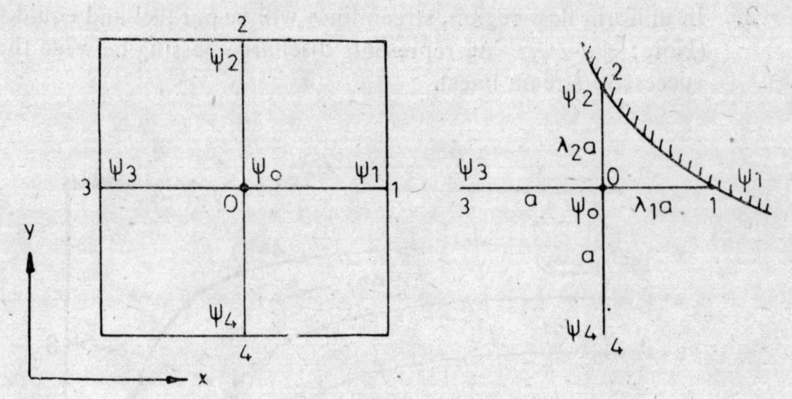

Fig. 2.7. Symmetrical and unsymmetrical grid.

$$\frac{\psi_1 + \psi_2 + \psi_3 + \psi_4 - 4\psi_0}{a^2} = 0$$

or

$$\psi_0 = \frac{\psi_1 + \psi_2 + \psi_3 + \psi_4}{4}$$

For unequal grid as shown

$$\psi_0 = \frac{\dfrac{\psi_1}{\lambda_1} + \dfrac{\psi_2}{\lambda_2} + \psi_3 + \psi_4}{\dfrac{1}{\lambda_1} + \dfrac{1}{\lambda_2} + 2}$$

Hence in the flow field, a grid can be formed and values of ψ at the boundary and at nodes can be given by inspection. These are corrected successively using the above relations. When the above equations are reasonably satisfied at all points, stream lines can be sketched for constant ψ values.

ILLUSTRATIVE EXAMPLES

2.1 During a flood, surface velocity in a stream is measured with the help of a float. A float in the form of an empty sealed tin container travels a distance of 52.50 m in 30 s on the surface of a stream. Determine the surface velocity.

$$\text{Velocity} = \frac{\text{distance travelled}}{\text{time}}$$
$$= 52.50/30$$
$$= 1.75 \text{ m/s}$$

2.2 The velocity field in a fluid is given by

$$\mathbf{V}_s = (3x + 2y)\,\mathbf{i} + (2z + 3x^2)\,\mathbf{j} + (2t - 3z)\,\mathbf{k}.$$

(i) What are the velocity components u, v, w at any point in the flow field?

(ii) Determine the speed at the point (1, 1, 1).

(iii) Determine the speed at time $t = 2$ s at point (0, 0, 2).

The velocity components at any point (x, y, z) are

(i) $u = (3x + 2y)$, $v = (2z + 3x^2)$, $w = (2t - 3z)$

(ii) Substitute $x = 1$, $y = 1$, $z = 1$ in the expressions for u, v and w.

$$\therefore \quad u = (3 + 2) = 5, \; v = (2 + 3) = 5, \; w = (2t - 3)$$
$$\therefore \quad V^2 = u^2 + v^2 + w^2$$
$$= 5^2 + 5^2 + (2t - 3)^2$$
$$= 25 + 25 + 4t^2 - 12t + 9$$
$$= 4t^2 - 12t + 59$$
$$\therefore \quad V_{(1, 1, 1)} = \sqrt{(4t^2 - 12t + 59)}$$

(iii) Substitute $t = 2$, $x = 0$, $y = 0$, $z = 2$ in the expressions for u, v, w.

$$\therefore \quad u = 0, \; v = (4 + 0) = 4, \; w = (4 - 6) = -2$$
$$\therefore \quad V^2_{(0, 0, 2, 2)} = (0 + 16 + 4) = 20$$
$$\therefore \quad V = \sqrt{20}$$

2.3 Classify the velocity field in Ex. 2.2 as steady or unsteady, uniform or nonuniform and one, two or three dimensional.

(i) Since V_s at given (x, y, z) depends on t, it is unsteady flow.

(ii) Since at given t velocity changes in the x direction it is nonuniform.

(iii) Since V_s depends on x, y, z, it is three dimensional flow.

2.4 Determine the discharge of water flowing through a large pipe of 1.5 m diameter if the average velocity of flow is 3.0 m/s.

$$Q = UA = 3.0 \times \frac{\pi}{4}(1.5)^2$$

$$= 5.299 \text{ m}^3/\text{s}$$

2.5 In a wide rectangular open channel, the velocity distribution in the vertical is given by $u = 1.25\, y^{1/6}$. Assuming depth of flow to be 2m determine the discharge flowing per unit width of channel (Fig. 2.8).

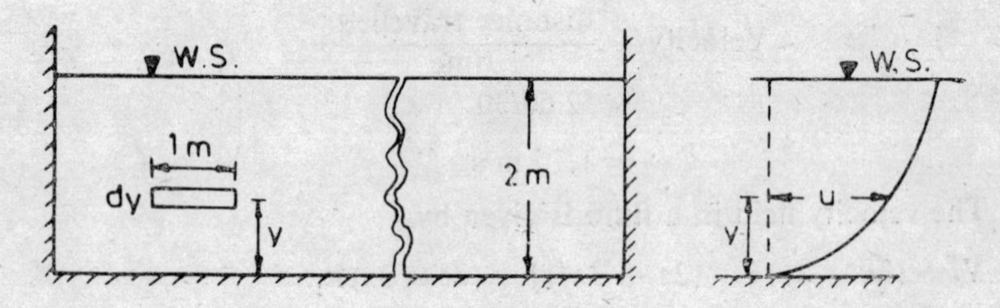

Fig. 2.8. Flow in a channel.

For unit width $dA = 1 \times dy$

$$dq = u\,dA = 1.25\, y^{1/6}\, dy$$

$$\therefore \quad q = \int dq = \int_0^2 1.25\, y^{1/6}\, dy$$

$$= 1.25 \times \frac{6}{7}\left(y^{7/6}\right)_0^2 = 2.405 \text{ m}^3/\text{s m}$$

2.6 Determine whether the continuity equation is satisfied by the following velocity components for incompressible fluid.

$$u = x^3 - y^3 - z^2x, \quad v = y^3 - z^3$$

$$w = -3x^2z - 3y^2z + \frac{z^3}{3}$$

$$\frac{\partial u}{\partial x} = 3x^2 - z^2, \quad \frac{\partial v}{\partial y} = 3y^2$$

$$\frac{\partial w}{\partial z} = -3x^2 - 3y^2 + z^2$$

$$\therefore \quad \frac{\partial u}{\partial x} + \frac{\partial v}{\partial y} + \frac{\partial w}{\partial z} = 3x^2 - z^2 + 3y^2 - 3x^2 - 3y^2 + z^2 = 0$$

Therefore, these velocity components satisfy the continuity equation.

2.7 Determine the missing component of velocity distribution such that they satisfy continuity equation

$$u = ?$$

$$v = ax^3 - by^2 + cz^2$$

$$w = bx^3 - cy^2 + az^2x$$

Since $$\frac{\partial u}{\partial x} + \frac{\partial v}{\partial y} + \frac{\partial w}{\partial z} = 0$$

$$\frac{\partial u}{\partial x} = -\left(\frac{\partial v}{\partial y} + \frac{\partial w}{\partial z}\right)$$

$$= -(-2by + 2azx)$$

$$= 2by - 2azx$$

$\therefore$ $$u = \int \frac{\partial u}{\partial x}\, dx = \int (2by - 2azx)\, dx$$

$$= 2byx - 2az\frac{x^2}{2} + f(y, z)$$

The exact nature of $f(y, z)$ will be known if the boundary conditions are known.

2.8 A 0.30 m diameter pipe carrying oil at 1.5 m/s velocity suddenly expands to 0.60 m diameter pipe. Determine the discharge and velocity in 0.60 m diameter pipe.

$$Q = A_1 U_1 = A_2 U_2$$

$$\therefore \quad Q = \frac{\pi}{4} D_1^2\, U_1 = 0.785 \times 0.30^2 \times 1.50$$

$$= 0.106 \text{ m}^3/\text{s} \quad \text{or} \quad 106 \text{ l/s}$$

$$U_2 = \frac{Q}{A_2} = \frac{0.106}{0.785 \times 0.60^2}$$

$$= 0.375 \text{ m/s.}$$

2.9 A pipe diameter changes from 0.50 m to 1.0 m in a length of 1 m giving a pipe diffuser or transition. If a discharge Q flows from the 0.50 m diameter section towards 1.0 m diameter section, obtain a general expression for velocity (Fig. 2.9). From geometry of the figure

$$\frac{L-1}{L} = \frac{0.25}{0.50}$$

$$\therefore \quad 0\,50\,L - 0.50 = 0.25\,L$$

$$L = 2 \text{ m}$$

Pipe radius R_x at a distance x from the beginning is given by

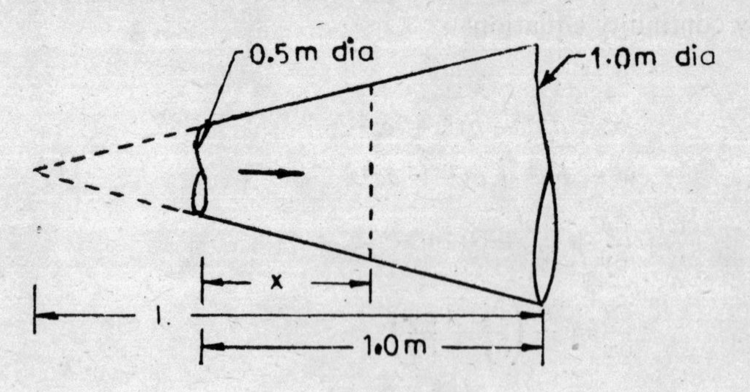

Fig. 2.9 Diffuser

$$\frac{R_x}{0.25} = \frac{1+x}{1} \qquad \therefore \quad R_x = (0.25 + 0.25\,x)$$

$$\therefore \quad A_x = \pi R_x^2 = \pi\,(0.25 + 0.25x)^2$$

$$U_x = \frac{Q}{A_x} = \frac{Q}{\pi\,(0.25 + 0.25\,x)^2} = \frac{Q}{0.0625\,\pi\,(1 + x)^2}$$

2.10 Apply the principle of conservation of mass principle to a balloon which is being inflated.

Let R be the radius of balloon and ρ be the mass density of air. In time dt mass entering the balloon $= \rho AV dt$ where A is area of entrance to balloon and V velocity through it. As a result, balloon radius will increase to $(R + dR)$ and density of air inside to $(\rho + d\rho)$.

Therefore

$$\text{Increase in the mass in the balloon} = \frac{4}{3}\pi\,(R + dR)^3\,(\rho + d\rho) - \frac{4}{3}\pi R^3\rho$$

Neglecting higher order terms of dR, this reduces to

$$= \frac{4}{3}\pi R^3 d\rho + 4\pi R^2 dR\rho$$

Equating the two, we get

$$\rho VAdt = \frac{4}{3}\pi R^3 d\rho + 4\pi R^2 dR\rho$$

$$\therefore \quad \rho AV = \frac{4}{3}\pi R^3\,\frac{d\rho}{dt} + 4\pi R^2\rho\,\frac{dR}{dt}$$

2.11 Determine the equation of a streamline passing through point (2, 3) if velocity components for two dimensional flow are given by $u = a$ and $v = a$ where a is non-zero constant. The equation of stream line is

$$\frac{u}{dx} = \frac{v}{dy} \quad \therefore \quad \frac{a}{dx} = \frac{a}{dy}$$

or $dy = dx$ which on integration gives

$y = x + c$; the value of the constant of integration can be found from the condition that the stream line must pass through

$$x = 2, y = 3$$

$$\therefore \quad c = 3 - 2 = 1$$

$\therefore$ The equation of stream line is $y = x + 1$.

Thus it is a straight line with slope of 45° and intersecting the y axis at 1.

2.12 Determine the equation of a stream line passing through the point $(a, 0)$ if the two dimensional flow is described by

$$u = \frac{-y}{b^2}, v = \frac{x}{a^2}$$

Since
$$\frac{dx}{u} = \frac{dy}{v}$$

$$\frac{b^2 dx}{-y} = \frac{a^2 dy}{x} \quad \text{or} \quad \frac{xdx}{a^2} + \frac{ydy}{b^2} = 0$$

$\therefore \quad \frac{x^2}{a^2} + \frac{y^2}{b^2} = 2c$, where c is constant of integration. Since the stream line passes through $(a, 0)$, $2c = \frac{a^2}{a^2} + 0 = 1$, or $c = \frac{1}{2}$. Hence, equation of stream line is $\frac{x^2}{a^2} + \frac{y^2}{b^2} = 1$.

2.13 In example 2.9 if the pipe carries a discharge of 0.785 m³/s determine the convective tangential acceleration at 0.5 m from entrance.

$$a_x = V_x \frac{\partial V_x}{\partial x} \quad \text{but} \quad V_x = \frac{Q}{0.0625\,\pi\,(1 + x)^2}$$

$$\therefore \quad a_x = \frac{Q^2}{(0.0625\,\pi)^2} \cdot \frac{1}{(1 + x)^2} \left(\frac{-2}{(1 + x)^3}\right)$$

Substituting values of Q and x, we get

$$a_x = \frac{-0.785^2 \quad 2}{(0.0625\,\pi)^2\,(1.5)^5} = -4.209 \text{ m/s}^2$$

The negative sign shows that a_x is deceleration.

2.14 If in Example 2.11 the discharge increases from 0.785 m³/s to 1.570 m³/s in 10 seconds, determine total acceleration.

$$\text{Local acceleration} = \frac{\partial V_s}{\partial t}$$

At $x = 0.5$ m

$R = (0.25 + 0.125) = 0.375$ m

$\therefore$ Area $= 3.142 \times 0.375^2 = 0.442$ m²

$\therefore$ Initial velocity $V_{s1} = \dfrac{0.785}{0.442} = 1.776$ m/s

Final velocity $V_{s2} = \dfrac{1.570}{0.442} = 3.552$ m/s

$$\therefore \quad \frac{\partial V_s}{\partial t} = \frac{(3.552 - 1.776)}{10} = 0.178 \text{ m/s}^2$$

$\therefore$ Total acceleration $= 0.178 - 4.209 = -4.031$ m/s^2

2.15 If the velocity components for two dimensional flow are given by

$$u = \frac{x}{x^2 + y^2} \quad \text{and} \quad v = \frac{y}{x^2 + y^2}$$

determine the acceleration components a_x and a_y.

$$a_x = u\frac{\partial u}{\partial x} + v\frac{\partial u}{\partial y}$$

$$= \frac{x}{(x^2 + y^2)}\frac{(y^2 - x^2)}{(x^2 + y^2)^2} - \frac{2y^2x}{(x^2 + y^2)^3}$$

$$= \frac{xy^2 - x^3 - 2xy^2}{(x^2 + y^2)^3} = -\frac{x}{(x^2 + y^2)^2}$$

$$a_y = u\frac{\partial v}{\partial x} + v\frac{\partial v}{\partial y}$$

$$= \frac{x(-2xy)}{(x^2 + y^2)^3} + \frac{y(x^2 - y^2)}{(x^2 + y^2)^3}$$

$$= \frac{-2x^2y + x^2y - y^3}{(x^2 + y^2)^3}$$

$$= \frac{-y(x^2 + y^2)}{(x^2 + y^2)^3}$$

$$= \frac{-y}{(x^2 + y^2)^2}$$

2.16 For flow described by velocity components in Example 2.15, determine the rotation ω_z.

Since

$$u = x/(x^2 + y^2) \quad \text{and} \quad v = y/(x^2 + y^2)$$

$$\frac{\partial v}{\partial x} = \frac{(x^2 + y^2) \times 0 - 2xy}{(x^2 + y^2)^2} = -2xy/(x^2 + y^2)^2$$

$$\frac{\partial u}{\partial y} = \frac{(x^2 + y^2) \times 0 - 2xy}{(x^2 + y^2)^2} = -2xy/(x^2 + y^2)^2$$

$$\omega_z = \frac{1}{2}\left(\frac{\partial v}{\partial x} - \frac{\partial u}{\partial y}\right) = \frac{1}{2}\left[\frac{-2xy + 2xy}{(x^2 + y^2)^2}\right] = 0$$

Hence the flow is irrotational.

2.17 If the velocity field is given by $V_x = 0$, $V_r = 0$ and $V_\theta = a.r$ show that the flow is rotational.

$$\text{Since } \omega_x = \frac{1}{2}\left[\frac{\partial}{r\partial r}(rV_\theta) - \frac{1}{r}\frac{\partial}{\partial\theta}(V_r)\right]$$

$$= \frac{1}{2}\left[\frac{\partial V_\theta}{\partial r} + \frac{V_\theta}{r}\right]$$

$$= \tfrac{1}{2}[a + a] = a$$

Flow is rotational.

2.18 Find the circulation around the closed curve defined by

$$y = 1,\ x = 2,\ y = 4,\ x = 4$$

when the velocity field is given by $u = (16y - 8x)$, $v = (8y - 7x)$ (Fig. 2.10).

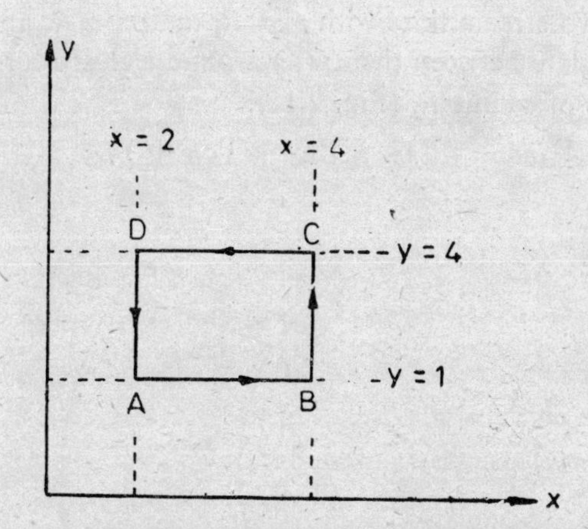

Fig. 2.10. Circulation around a curve.

$$\Gamma_{ABCD} = \int_{ABCD} (udx + vdy)$$

$$= \int_{AB} (udx + vdy) + \int_{BC} (udx + vdy) + \int_{CD} (udx + vdy)$$

$$+ \int_{DA} (udx + vdy)$$

$$= \int_2^4 (16y - 8x)\, dx + \int_1^4 (8y - 7x)\, dy + \int_4^2 (16y - 8x)\, dx$$

$$+ \int_4^1 (8y - 7x)\, dy$$

In the 1st integral $y = 1$, in 2nd integral $x = 4$, in the third integral $y = 4$ and in 4th integral $x = 2$.

$$\Gamma_{ABCD} = [16yx - 4x^2]_2^4 + [4y^2 - 7xy]_1^4 + [16yx - 4x^2]_4^2 + [4y^2 - 7xy]_4^1$$

$$= (64 - 64 - 32 + 16) + (64 - 112 - 4 + 28)$$

$$+ (128 - 16 - 256 + 64) + (4 - 14 - 64 + 56)$$

$$= -138$$

Area of curve $ABCD = 6$

$$\therefore \quad \text{Circulation per unit area} = -\frac{138}{6} = -23$$

It can be confirmed that this is equal to $2\,\omega_z$. (See problem 2.19).

2.19 If $V = \frac{c}{r}$ gives variation of tangential velocity with radius r in a vortex formed in a wash basin, determine the circulation (*i*) around a closed curve formed by two stream lines with $r = R_1$ and $r = R_2$ and two radius vectors with an angle θ between them; (*ii*) around a closed curve in the form of concentric circle of radius R_1 (Fig. 2.11).

$$\Gamma_{ABCD} = \Gamma_{AB} + \Gamma_{BC} + \Gamma_{CD} + \Gamma_{DA}$$

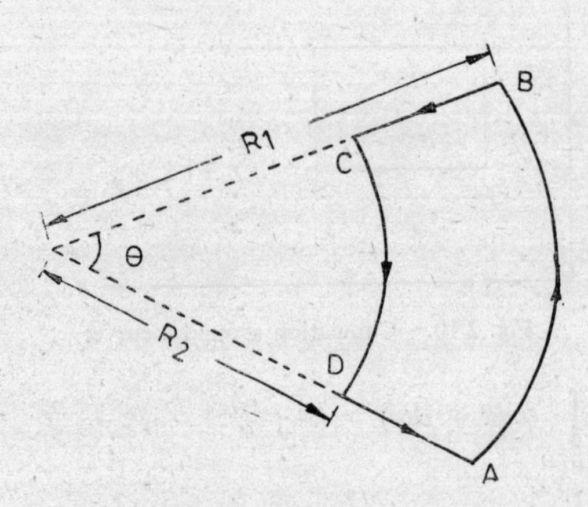

Fig. 2.11. Circulation in frec vortex.

Since there is no velocity along radial directions $\Gamma_{BC} = \Gamma_{DA} = 0$.

$$\therefore \quad \Gamma_{ABCD} = \frac{c}{R_1} R_1\theta - \frac{c}{R_2} . R_2\theta = 0$$

$\therefore$ Circulation per unit area $= 0$

Circulation around a closed curve in the form of a circle of radius R_1 will be

$$\Gamma = \frac{c}{R_1} \times 2\pi R_1 = 2\pi C.$$

Thus, it seems that if centre is included in the closed curve, one gets rotational flow, otherwise for any other curve it gives irrotational flows. Here the centre is a **singular point**. The flow represented by $V = c/r$ is known as free vortex flow or irrotational vortex.

2.20 Check whether the following functions represent possible flow phenomena of irrotational type

(i) $\phi = x^2 - y^2 + y$

(ii) $\phi = \sin(x + y + z)$

(iii) $\phi = \dfrac{Ax}{x^2 + y^2}$

(iv) $\phi = Ur\cos\theta$

(i) $\dfrac{\partial\phi}{\partial x} = 2x \quad \therefore \quad \dfrac{\partial^2\phi}{\partial x^2} = 2$

$\dfrac{\partial\phi}{\partial y} = -2y + 1 \quad \therefore \quad \dfrac{\partial^2\phi}{\partial y^2} = -2$

$\therefore \quad \dfrac{\partial^2\phi}{\partial x^2} + \dfrac{\partial^2\phi}{\partial y^2} = 2 - 2 = 0.$

Hence $\phi = x^2 - y^2 + y$ satisfies Laplace's equation and can represent irrotational flow.

(ii) $\phi = \sin(x + y + z)$

$\therefore \quad \dfrac{\partial\phi}{\partial x} = \cos(x + y + z), \quad \dfrac{\partial^2\phi}{\partial x^2} = -\sin(x + y + z)$

Similarly $\dfrac{\partial^2\phi}{\partial y^2} = -\sin(x + y + z)$

and $\dfrac{\partial^2\phi}{\partial z^2} = -\sin(x + y + z)$

$\therefore \quad \dfrac{\partial^2\phi}{\partial x^2} + \dfrac{\partial^2\phi}{\partial y^2} + \dfrac{\partial^2\phi}{\partial z^2} = -3\sin(x + y + z) \neq 0$

$\therefore$ $\phi = \sin(x + y + z)$ does not satisfy Laplace's equation and hence it cannot represent irrotational flow phenomena.

(iii) $\phi = Ax/(x^2 + y^2)$ where A is constant.

$$\frac{\partial\phi}{\partial x} = \frac{A(x^2 + y^2) - 2Ax^2}{(x^2 + y^2)^2} = \frac{A(y^2 - x^2)}{(x^2 + y^2)^2}$$

$$\frac{\partial^2\phi}{\partial x^2} = \frac{-(x^2 + y^2)^2\, 2Ax - 2A(y^2 - x^2)(x^2 + y^2) \times 2x}{(x^2 + y^2)^4}$$

$$\therefore \quad \frac{\partial^2\phi}{\partial x^2} = \frac{2Ax^3 - 6Axy^2}{(x^2 + y^2)^3}$$

Similarly $\dfrac{\partial\phi}{\partial y} = \dfrac{(x^2 + y^2) \times 0 - 2Axy}{(x^2 + y^2)^2} = \dfrac{-2Axy}{(x^2 + y^2)^2}$

and $$\frac{\partial^2\phi}{\partial y^2} = \frac{-2Ax^3 + 6Axy^2}{(x^2 + y^2)^3}$$

$\therefore$ $\frac{\partial^2\phi}{\partial x^2} + \frac{2\phi}{y^2} = 0$. Hence $\phi = Ax/(x^2 + y^2)$ can represent irrotational flow.

(iv) $$\phi = U \cos \theta / r$$

Laplace's equation in polar co-ordinates is

$$\frac{1}{r}\frac{\partial\phi}{\partial r} + \frac{\partial^2\phi}{\partial r^2} + \frac{1}{r^2}\frac{\partial^2\phi}{\partial\theta^2} = 0. \text{ Hence}$$

$$\frac{\partial\phi}{\partial r} = -\frac{U \cos\theta}{r^2}, \quad \frac{\partial^2\phi}{\partial r^2} = \frac{2U\cos\theta}{r^3}$$

$$\frac{\partial\phi}{\partial\theta} = -U\sin\theta/r, \quad \frac{\partial^2\phi}{\partial\theta^2} = -U\cos\theta/r$$

Substituting these values in right hand side of Laplace's equation, we get

$$\text{LHS} = \frac{1}{r}\left(-\frac{U\cos\theta}{r^2}\right) + \frac{2U\cos\theta}{r^3} + \frac{1}{r^2}\left(\frac{U\cos\theta}{r}\right)$$

$$= -\frac{U\cos\theta}{r^2} + \frac{2U\cos\theta}{r^3} - \frac{U\cos\theta}{r^3}$$

$$= 0$$

Hence $\phi = \frac{U\cos\theta}{r}$ satisfies Laplace's equation and can represent irrotational flow.

2.21 If $\phi = 2xy$ determine ψ

It is known that

$$\frac{-\partial\phi}{\partial x} = u = -\frac{\partial\psi}{\partial y} \text{ and } \frac{-\partial\phi}{\partial y} = v = +\frac{\partial\psi}{\partial x}$$

$\therefore$ $$\frac{\partial\phi}{\partial x} = 2y = \frac{\partial\psi}{\partial y}$$

On integration, $$\psi = y^2 + f(x)$$

$\therefore$ $$\frac{\partial\psi}{\partial x} = f'(x) = -\frac{\partial\phi}{\partial y} = -2x$$

$\therefore$ $$\frac{\partial f}{\partial x} = -2x \quad \therefore \quad f = -x^2 + C$$

$\therefore$ $\psi = (y^2 - x^2) + C$, where C is a numerical constant.

2.22 If $\phi = \frac{-A}{2\pi} \ln r$, where A is a positive constant, determine ψ and plot typical equipotential lines and stream lines. Identify the flow pattern.

$$\phi = \frac{-A}{2\pi} \ln r \qquad \therefore \quad \frac{\partial \phi}{\partial r} = \frac{-A}{2\pi r} = \frac{\partial \psi}{r \partial \theta}$$

$\therefore$ On integration, $\psi = -\int \frac{A}{2\pi}\, d\theta + f(r) = -\frac{A\theta}{2\pi} + f(r)$

$$\therefore \qquad \frac{-d\psi}{\partial r} = f'(r) \quad \text{and} \quad -\frac{\partial \psi}{\partial r} = \frac{1}{r}\frac{\partial \phi}{\partial \theta} \quad \text{and} \quad \frac{\partial \phi}{\partial \theta} = 0$$

$$\therefore \qquad \frac{-\partial \psi}{\partial r} = 0 = f'(r)$$

$\therefore$ $f(r) = C$, where C is a numerical constant.

$$\therefore \qquad \psi = -\frac{A\theta}{2\pi} + C \quad \text{or} \quad -\frac{A\theta}{2\pi} \text{ if } C \text{ is included in } \psi.$$

$$\psi = \text{constant implies } \frac{A\theta}{2\pi} = \text{const. i.e. } \theta = \text{constant.}$$

This will give radial lines as stream lines (Fig. 2.12).

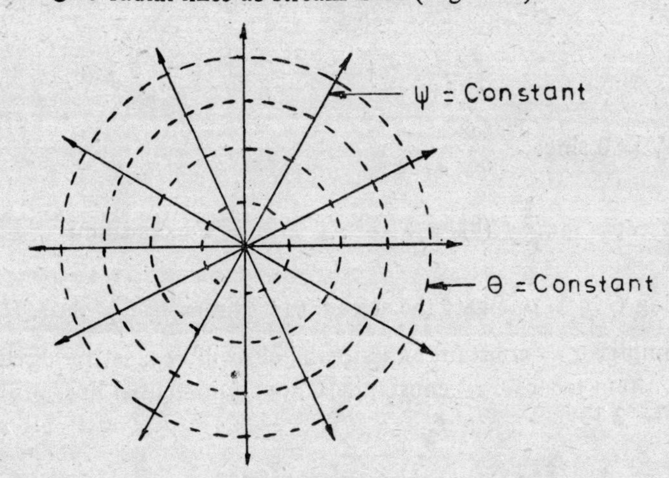

Fig. 2.12. Source.

Since

$$V_r = -\frac{\partial \psi}{r \partial \theta} = \frac{A}{2\pi r}$$

is positive, flow is outwards. It is known as a source. $\phi =$ constant means $r =$ constant. Hence equipotential lines will be circles.

$$Q = V_r \times 2\pi r \times 1 = \frac{A}{2\pi r} \cdot 2\pi r = A$$

Therefore, $A =$ discharge per unit depth. It is also known as the source strength. Further since at $r = 0$, V_r is infinite, the origin is a singular point.

2.23 Examine the flow given by $\phi = \frac{B}{2\pi}\,\theta$, where B is a constant.

First it can be checked whether ϕ satisfies Laplace's equation in polar co-ordinate system

$$\frac{1}{r}\frac{\partial\phi}{\partial r}+\frac{\partial^2\phi}{\partial^2 r^2}+\frac{1}{r^2}\frac{\partial^2\phi}{\partial\theta^2}=0$$

$$\frac{\partial\phi}{\partial r}=0=\frac{\partial^2\phi}{\partial r^2}\ ;\ \frac{\partial\phi}{r\theta}=\frac{B}{2\pi}\ \text{and}\ \frac{\partial^2\phi}{\partial\theta^2}=0$$

Hence $\phi=\frac{B}{2\pi}\theta$ satisfies Laplace's equation and will represent irrotational flow.

$$-\frac{1}{r}\frac{\partial\phi}{\partial\theta}=-\frac{1}{r}\frac{B}{2\pi}=V_\theta=\frac{\partial\psi}{\partial r}$$

$\therefore$ We can integrate $\frac{\partial\psi}{\partial r}=-\frac{1}{r}\frac{B}{2\pi}$ to get ψ.

$$\therefore\qquad \psi=-\frac{B}{2\pi}\ln r+f(\theta)$$

$$\therefore\qquad \frac{\partial\psi}{\partial\theta}=f'(\theta)\ \text{and}\ -\frac{1}{r}\frac{\partial\psi}{\partial\theta}=-\frac{1}{r}f'(\theta)=V_r$$

but $V_r=0$ since $\frac{\partial\phi}{\partial r}=0$

$$\therefore\qquad \frac{1}{r}f'(\theta)=0\qquad\therefore\ f(\theta)=C\ \text{constant.}$$

Including C in ψ, one gets the stream function as $\psi=\frac{B}{2\pi}$ ln r. Hence $\psi=$ const. implies $r=$ constant and streamlines will be a set of circles. While $\phi=$ constant means $\theta=$ constant. Hence equipotential lines will be radial lines (Fig 2.13).

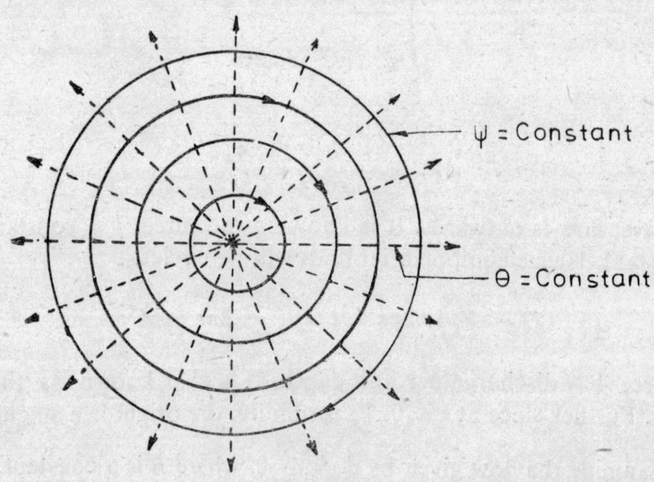

Fig. 2.13. Free Vortex.

Since $V_\theta = -\frac{B}{r2\pi}$, tangential velocity increases as r decreases and is infinite when $r = 0$. Negative value of V_θ indicates that it is in clockwise direction.

Also $V_\theta \cdot r =$ constant $\left(-\frac{B}{2\pi}\right)$. This flow pattern is known as *irrotational or free vortex* and it is found to occur while emptying wash basins and bath tubs. B is known as the vortex strength and has the dimensions of L^2/T.

2.24 Examine the irrotational flow represented by $\phi = \frac{-A\cos\theta}{2\pi r}$, where A is a numerical constant. It can be checked that ϕ satisfies Laplace equation.

$$-\frac{\partial\phi}{\partial r} = +\frac{A\cos\theta}{2\pi r^2} = \frac{1}{r}\frac{\partial\psi}{\partial\theta}$$

$$\therefore \qquad \psi = \frac{A\sin\theta}{2\pi r^2} + f(r)$$

$$\frac{\partial\psi}{\partial r} = -\frac{A\sin\theta}{2\pi r^2} + f'(r) \quad \therefore \quad V_\theta = -\frac{\partial\phi}{r\partial\theta}$$

$$= -\frac{A\sin\theta}{2\pi r^2} \quad \therefore \quad f'(r) = 0 \text{ or } f(r) = \text{constant.}$$

$\therefore \quad \psi = \frac{A\sin\theta}{2\pi r}$. For constant value of ψ, say C, we can write

$r = \frac{A}{2\pi C}\sin\theta$, or $r = r_0\sin\theta$, where $r_0 = \frac{A}{2\pi C}$.

For various r_0 values $r = r_0 \sin \theta$ will plot as a circle passing through origin and diameter r_0 along y axis. The x axis will be tangent to these circles. In the same way $\phi =$ constant, one can write $r = r_0\cos\theta$ where $r_0 = \frac{-A}{2\pi C}$. These represent circles passing through origin and with diameter r_0 along x axis. The y axis will be tangent to these circles. This is known as doublet. It can be verified by the student that doublet is formed when a source is kept at $-a$ and sink at $+a$ along x axis and letting a tend to zero. Hence the flow in Fig. 2.14 is shown from left to right. If source is at $+a$ and sink at $-a$ and then $a \to 0$ the direction of flow will be from right to left. Doublet is used in developing complicated flow patterns by the principle of superposition and in aerodynamics.

2.25 Examine the two dimensional flow obtained by superposition of rectilinear flow on a source.

Rectilinear flow is represented by $\phi = Ux$, $\psi = -Uy$ and source by $\phi = -\frac{A}{2\pi}\ln r$ and $\psi = -\frac{A\theta}{2\pi}$. The velocity potential and stream function for tqe resulting flow are given by

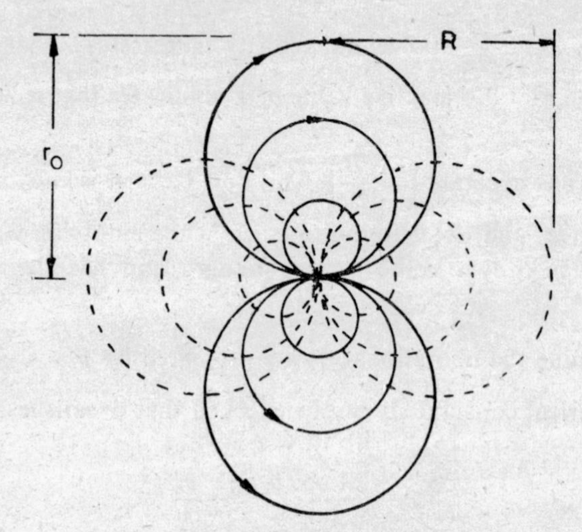

Fig. 2.14. Doublet.

$$\phi = -\frac{A}{2\pi}\ln r - Ux$$

$$\psi = -\frac{A\theta}{2\pi} - Uy \quad \text{and } y = r\sin\theta \text{ and } x = r\cos\theta$$

At the stagnation point $V_r = V_\theta = 0$. Hence

$$-\frac{1}{r}-\frac{\partial\psi}{\partial\theta} = \frac{A}{2\pi r} + U\cos\theta = 0$$

$$\frac{\partial\psi}{\partial r} = -U\sin\theta = 0$$

These two conditions are satisfied if $\theta = \pi$ and $r = A/2\pi U$. Substitution of these values in the expression for ψ yields

$$\psi = -\frac{A}{2} - \frac{UA}{2\pi U}\times 0 = -\frac{A}{2}$$

Hence equation of stream line passing through the stagnation point is given by

$$-\frac{A}{2} = -\frac{A\theta}{2\pi} - Ur\sin\theta \quad \text{or} \quad \frac{A}{2}\left(1-\frac{\theta}{\pi}\right) = Ur\sin\theta$$

The distance of the point from x axis is given by $r\sin\theta$

$$y = r\sin\theta = \frac{A}{2U}\left(1-\frac{\theta}{2\pi}\right)$$

The maximum value of y occurs at $\theta = 0$ and $x \to \alpha$.

Therefore, $y_{max} = \frac{A}{2U}$. This height is approached asymptotically. Since there is no flow across the stream line,

$$\frac{A}{2}\left(1-\frac{\theta}{2\pi}\right) = U\, r \sin\theta$$

gives the surface of a long bluff body. The source will be inside the body. Figure 2.15 shows flows past such a body.

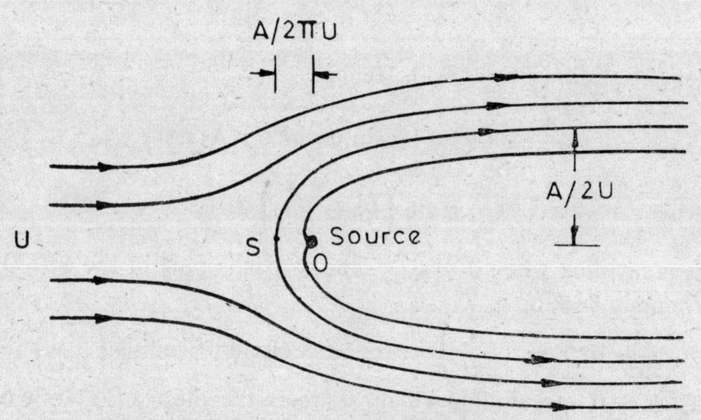

Fig. 2.15. Flow past half body.

2.26 Examine the flow pattern produced by superposition of uniform flow over the doublet.

It can be shown that when a source and a sink pair is kept in a uniform flow, it represents flow past a two dimensional oval shaped body (Fig. 2.16). As the spacing between them is reduced, the length of the oval is gradually reduced and in the limit when that tends to zero, flow past a circle or two dimensional cylinder kept in uniform stream is produced. Hence superposition of doublet and uniform flow produces flow past a two-dimensional circular cylinder.

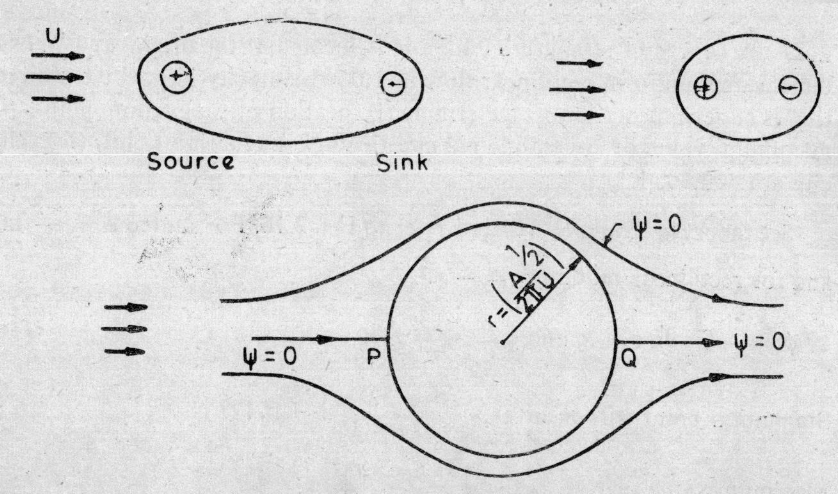

Fig. 2.16. Flow past a cylinder.

Superposition of uniform flow and doublet yields the velocity potential and stream function as

$$\phi = -Ux - \frac{A}{2\pi r}\cos\theta$$

$$\psi = -Uy + \frac{A}{2\pi r}\sin\theta$$

Consider the stream line $\psi = 0$. Hence

$$\frac{A}{2\pi r}\sin\theta = Ur\sin\theta \quad \text{since } y = r\sin\theta$$

$$\therefore \qquad \sin\theta\left(Ur - \frac{A}{2\pi r}\right) = 0$$

This is satisfied when $\theta = 0, \pi$, and x axis is part of the $\psi = 0$ streamline. When $\sin\theta \neq 0$,

$Ur = \frac{A}{2\pi r}$ or $r = \left(\frac{A}{2\pi U}\right)^{1/2}$. Hence a circle of radius $\left(\frac{A}{2\pi U}\right)^{1/2}$ also represents $\psi = 0$ streamline which assumes the shape of a circle or a two-dimensional cylinder. Velocities at points P and Q whose co-ordinates are $\left(\frac{A}{2\pi U}\right)^{1/2}$, π and $\left(\frac{A}{2\pi U}\right)^{1/2}$, 0 will yield $V_r = V_\theta = 0$ at both the points. Hence P and Q are stagnation points. Since there is no flow across the stream line, the resulting flow is flow past a two-dimensional circular cylinder.

For any value of θ other than 0 or π and $r = \sqrt{\frac{A}{2\pi U}}$, $V_r = 0$ and $V_\theta = -U\sin\theta - U\sin\theta = -2U\sin\theta$.

The negative sign implies that V_θ is in clockwise direction. V_θ is maximum when $\theta = \frac{\pi}{2}$ or $\frac{3\pi}{2}$ and its value is $V_\theta = -2U$.

2.27 A source of strength 3.0 m^3/s m is located at the origin and a two-dimensional uniform rectilinear flow at 1.0 m/s velocity in positive x direction is superimposed on it. Determine (i) location of stagnation point, (ii) maximum value of streamline passing through stagnation point, (iii) velocity at $(-0.30, 1.5)$.

See theory in Example 2.24 and refer to Fig. 2.16. For source $\phi = \frac{-3}{2\pi}\ln r$ and for rectilinear motion $\phi = -x$.

$$\therefore \phi = -\frac{3}{2\pi}\ln r - x \text{ and } \psi = \frac{-3\theta}{2\pi} - y$$

Stagnation point will occur at $x = \frac{-3.0}{2\pi \times 1} = -0.477$ m

Maximum value of $y = \frac{A}{2U} = \frac{3}{2 \times 1} = 1.5$ m

Also at the point (– 0.30, 1.50), $r = \sqrt{0.3^2 + 1.5^2} = 1.530$ m

and $\theta = \cos^{-1} - \dfrac{0.3}{1.53} = 101.31°$.

Since

$$\phi = \frac{-3}{2\pi} \ln r - r \cos \theta \quad (r \cos \theta = x)$$

$$V_r = \frac{-\partial \phi}{\partial r} = \frac{3}{2\pi r} + \cos \theta$$

$$= \frac{3.0}{2 \times 3.142 \times 1.53} - 0.1961 = 0.116 \text{ m/s}$$

$$V_\theta = -\frac{1}{r}\frac{\partial \phi}{\partial \theta} = -\sin \theta = -0.981 \text{ m/s}$$

$$V = \sqrt{V_r^2 + V_\theta^2} = \sqrt{0.116^2 + 0.981^2} = 0.988 \text{ m/s}$$

2.28 Superposition of a circulation of strength – Γ, on uniform flow of velocity U, normal to a circular cylinder of radius, a, produces a streamline pattern represented by the stream function

$$\psi = U\left(r - \frac{a^2}{r}\right) \sin \theta + \frac{\Gamma}{2\pi} \ln r$$

Sketch neatly flow patterns for different values of Γ clearly marking the positions of stagnation points.

It can be seen that the stream function ψ is obtained by superposition of uniform flow ($\psi_1 = Ur \sin \theta$) doublet $\left(\psi_2 = \dfrac{Ua^2}{r} \sin \theta\right)$ and an irrotational vortex $\left(U_3 = \dfrac{\Gamma}{2\pi} \ln r\right)$.

$$V_\theta = \frac{\partial \psi}{\partial r} = U \sin \theta + \frac{Ua^2}{r^2} \sin \theta + \frac{\Gamma}{2\pi r}$$

Therefore on the surface on the cylinder $r = a$ and

$$V_\theta = 2\, U \sin \theta + \frac{\Gamma}{2\pi a}$$

Positive sign of V_θ shows that it is in anticlockwise direction. At stagnation point $V_\theta = V_r = 0$.

$$\therefore \qquad \sin \theta = -\frac{\Gamma}{4\pi U a}$$

The position of stagnation points for different values of Γ are shown in Fig. 2.17. The condition Γ = 0 corresponds to flow perpendicular to the cylinder.

2.29 Draw typical streamlines for two-dimensional sudden contraction using relaxation method.

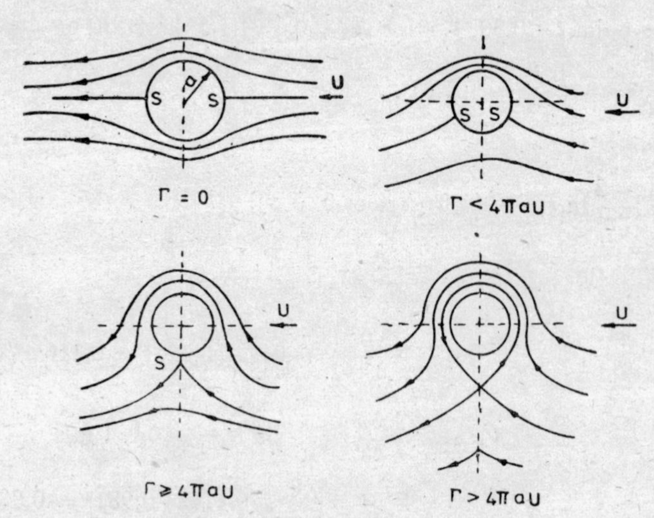

Fig. 2.17. Rectilinear flow past a cylinder without and with circulation.

The flow field is divided into grids as shown in Fig. 2.18. One boundary, which is a limiting streamline, is given $\psi = 0$. The other boundary is given $\psi = 100$ arbitrarily. Values between zero and 100 are arbitrarily given at nodes. Starting from left each value at node is crossed and the average of four nearest surrounding values is written. When this is complete, the process is repeated until changes in ψ values due to relaxation become insignificant. Now lines of constant ψ, say $\psi = 25, 50, 75$ can be drawn by interpolation. A similar procedure can be followed for drawing equipotential lines.

PROBLEMS

2.1 In each of the following cases state, giving reasons, whether the flow is steady, unsteady, uniform or nonuniform.

(i) $u = 10xt + 15x^2$

(ii) $u = 20$

(iii) Flow in a pipe bend with constant discharge.

(iv) Flow in a converging pipe in which discharge is gradually increased.

(v) Flow in a constant diameter pipe in which discharge is continuously increasing.

(unsteady nonuniform, steady uniform, steady nonuniform, unsteady nonuniform, unsteady uniform)

2.2 Check whether the velocity components $u = 3x$, $v = (2z + 3x^2)$ and $w = -(3z + 2t)$, satisfy the continuity equation. (yes)

2.3 The upper end of a 100 mm diameter vertical pipe is connected

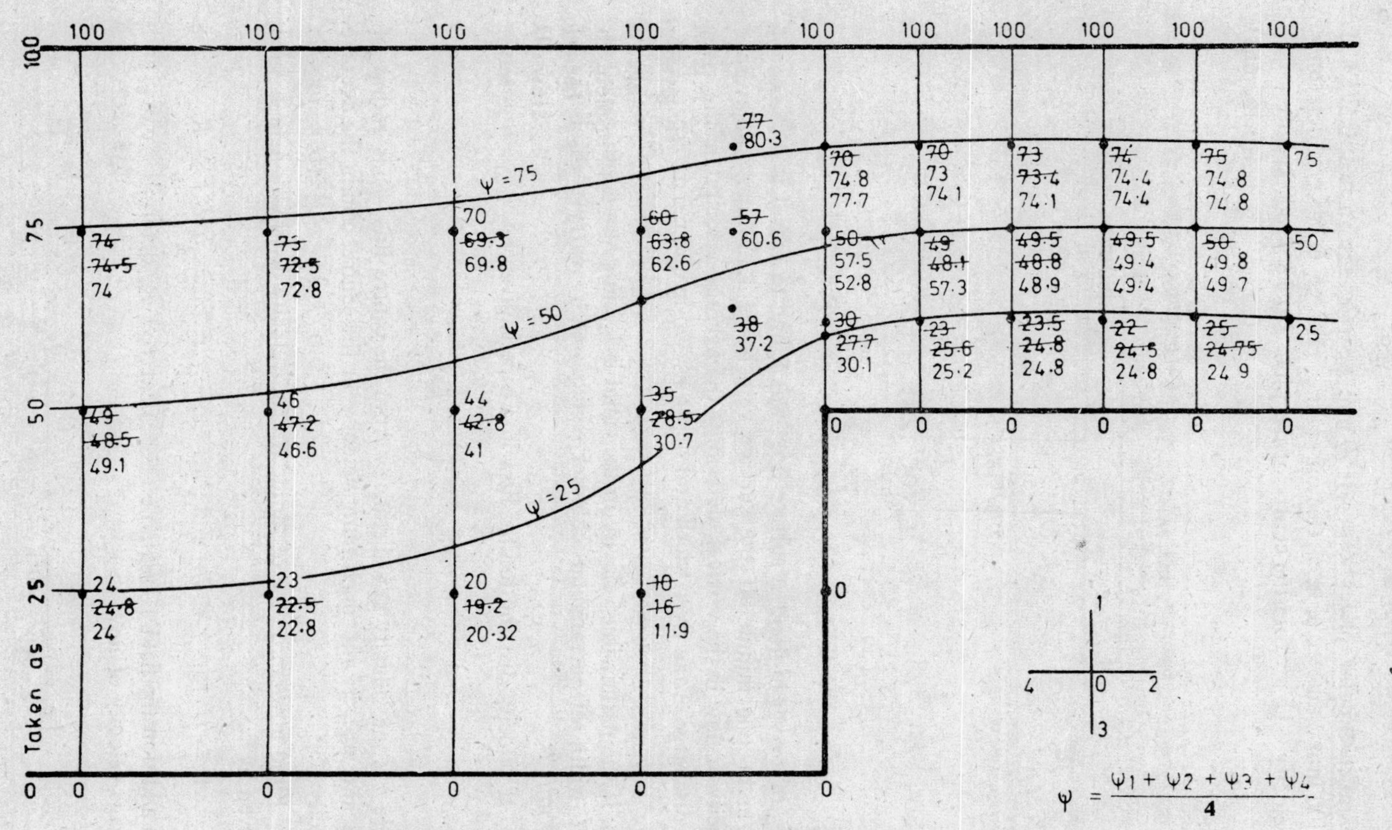

Fig. 2.18. Plotting of streamlines by **relaxation method.**

to parallel circular plates kept at 50 mm apart as shown in Fig. 2.19. If the 100 mm diameter pipe carries oil at the rate of 6.28 kg/s, determine its velocity 0.15 m from the centre of the pipe. Take $\rho = 800\ \text{kg/m}^3$.

(0.167 m/s)

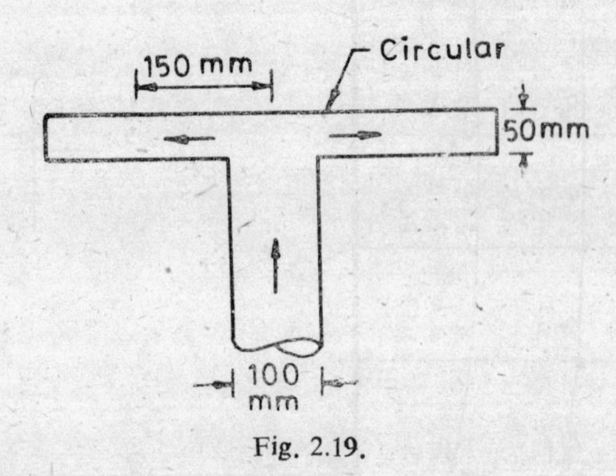

Fig. 2.19.

2.4 In a river in flood, a surface float was found to travel a distance of 150 m in one minute. If the surface velocity is 1.13 times the average velocity of flow in the vertical, determine the average velocity in the river and discharge. Take river width and depth as 300 m and 6.0 m respectively.

(2.212 m/s, 3982.3 m³/s)

2.5 If the permissible velocity in the canal is 1.5 m/s at a depth of 2.0 m, what width of rectangular channel is required to carry a discharge of 45.0 m³/s ?

(15.0 m)

2.6 If velocity distribution for flow in a pipe is given by

$$u/u_{max} = \left(1 - \frac{r^2}{R^2}\right)$$

where u is the velocity at a distance r from the centre line, R is radius of pipe and u_{max} is the maximum velocity at the centre, obtain the average velocity U.

$(U = u_{max}/2)$

2.7 Fluid flows in a pipe of radius R. The velocity distribution in a pipe is given by

$$u = U \text{ for } r \leqslant b \quad \text{and} \quad u = c_1 - c_2 r^2 \text{ for } b \leqslant r \leqslant R.$$

If in addition the field is real, i.e. $u = 0$ at $r = R$, determine c_1, c_2 and average velocity of flow.

$$\left(c_1 = \frac{UR^2}{(R^2 - b^2)},\ c_2 = \frac{U}{(R^2 - b^2)},\ U_{av} = \frac{U(R^2 + b^2)}{2R^2}\right)$$

2.8 In a pipe line system two pipes of 100 mm and 300 mm diameter join into a 450 mm diameter pipe. If the average velocities in 100 mm and 300 mm diameter pipes are 1.5 m/s and 2.0 m/s respectively, determine the average velocity in 450 mm diameter pipe and the total flow rate.

(0.963 m/s, 0.153 m³/s)

2.9 Show that the velocity components

$$u = -x/(x^2 + y^2) \text{ and } v = -y/(x^2 + y^2)$$

satisfy the continuity equation for two-dimensional flow of an incompressible fluid.

2.10 Consider an elementary volume in cylindrical polar co-ordinate system with δx, δr and $r\delta\theta$ as sides (Fig. 2.20). Let V_x, V_r, V_θ be the velocity components in the x, r and θ directions respectively. Obtain continuity equation for compressible fluid.

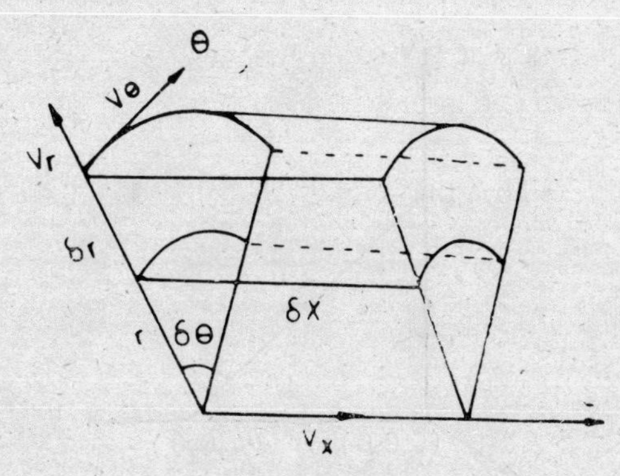

Fig. 2.20.

$$\left(\frac{\partial \rho}{\partial t} + \frac{\partial}{\partial x}(\rho V_x) + \frac{\partial}{r\partial r}(\rho\ V_r . r) + \frac{\partial}{r\partial\theta}(\rho V_\theta) = 0\right)$$

2.11 If $u = -ay$ and $v = ax$ are the velocity components for two dimensional flow, determine the equation of streamline passing through (3, 2).

$(x^2 + y^2 = 13)$

2.12 Obtain the equation of streamline passing through the point (2, 3) if the velocity components for two-dimensional flow are given by

$$u = -y^2 \text{ and } v = -3x. \qquad (2y^2 - 9x^2 = 18)$$

2.13 Find the equation of streamline for two-dimensional flow described by $u = xe^{-kt}$ and $v = y$.

$(y = cx^{ekt})$

2.14 In a converging section of a pipe the velocity at a given section is 1.50 m/s, where the diameter is 1.0 m. In a length of 2.0 m, its diameter is reduced to 0.50 m. Determine the convective tangential accelerations at two sections.

(3.375 m/s², 13.500 m/s²)

2.15 Water flows at a velocity of 3.0 m/s in a bend of radius 30 m in an open channel. What is the convective normal acceleration ? (0.30 m/s²)

2.16 If the velocity field in three dimensional fluid flow is given by

$$\mathbf{V} = 12xy\,\mathbf{i} + 6x^3\,\mathbf{j} + (3t^2x + z)\,\mathbf{k},$$

find the local acceleration at the point (3, 4, 4) when $t = 3$ s and also the total acceleration at point (1, 1, 1) when $t = 3$ s.

(54.0 m/s² in z direction, 479.804 m/s²)

2.17 Velocity field for two-dimensional steady flow is given by

$$\mathbf{V} = \left(10 + \frac{20x}{x^2 + y^2}\right)\mathbf{i} + \frac{20y}{x^2 + y^2}\mathbf{j}$$

Find acceleration at (10, 10, 5). (–0.10 m/s², –1.10 m/s² and 0)

2.18 Determine the circulation of $\mathbf{V} = 4x^2\mathbf{i} + 2yz\,\mathbf{j} - (4y^2 + z^2)\mathbf{k}$, about a closed triangle which has its vertices at (0, 0, 0), (0, 4, 0) and (4, 0, 0) (Fig. 2.21). (0)

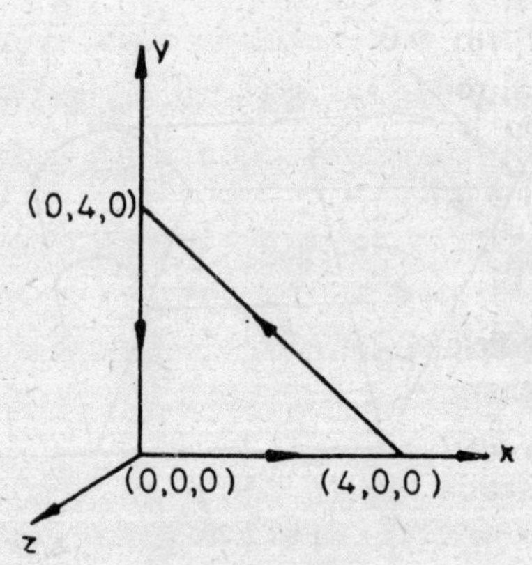

Fig. 2.21.

2.19 If $\mathbf{V} = (16y - 8x)\mathbf{i} + (8y - 7x)\,\mathbf{j}$ is the flow field for two-dimensional flow, confirm that the velocity components satisfy the continuity equation. Then find the total convective acceleration at (3, 4). Is the flow rotational or irrotational? ($\mathbf{a} = -144\mathbf{i} - 192\mathbf{j}$, rotational)

2.20 If $u = ax$ and $v = -ay$ give velocity distribution for two dimensional flow, determine the equation of streamline passing through the point (3, 1). Is the flow irrotational or rotational? Identify the flow pattern.

($xy = 3$, irrotational, flow around a corner)

2.21 If velocity components for three dimensional flow are given by $u = yzt$, $v = xzt$ and $w = xyt$, check whether the velocity components satisfy continuity equation. If it does, determine the rotation vector.

(yes, 0)

2.22 In problem 2.21 obtain the equation of streamline passing through the point (1, 2, 1). $\left(\frac{x}{yz} - \frac{y}{xz} = -3 \text{ and } \frac{x}{yz} = \frac{z}{xy}\right)$

2.23 If $u = ax$ and $v = ay$ and $w = -2az$ are the velocity components for a fluid flow in a particular case, check whether they satisfy the continuity equation. If they do, is the flow rotational or irrotational? Also obtain equation of streamline passing through the point (2, 2, 4).

(yes, irrotational, $x = y$, $xz^{1/2} = 4$)

2.24 Which of the following flow fields satisfy continuity and when they do, is the flow irrotational or rotational?

(i) $u = 0, v = a$ (ii) $u = ax, v = -ay$

(iii) $u = a, v = 0$ (iv) $u = ay, v = ax$

(v) $u = x^2y + y^2, v = x^2 - xy^2$

(vi) $u = x^3 \sin y, v = 3x^2 \cos y$

(vii) $V_r = 2r \sin \theta \cos \theta, V_\theta = -2r \sin^2 \theta$

(viii) $u = x^2 + 2xy, v = -(y^2 + 2xy)$

[(i) yes, irrotational, (ii) yes, irrotational, (iii) yes, irrotational (iv) yes irrotational, (v) yes, rotational, (vi) yes, rotational (vii) yes, rotational (viii) yes, rotational)].

2.25 If the equation of streamlines for a given fluid flow problem is $x^2 - y^2 =$ const., determine the magnitude and direction of velocity vector at (3, 4).

(5, 36.37° with x axis)

2.26 If the velocity field is given by $u = y, v = -x$, find the circulation around the closed curve defined by $x = \pm 1$ and $y = \pm 2$.

(−16)

2.27 If the velocity field is given by $u = x^2 + 2xy$ and $v = -(y^2 + 2xy)$, determine the circulation around a closed curve defined by $x = 1, x = 3$, $y = 1, y = 4$.

(−86)

2.28 If $V_s = cr$ gives the tangential velocity in a vortex flow at a distance r from the centre, determine the circulation around a curve formed by two streamlines with $r = R_1$ and $r = R_2$ and the two radius vectors with an angle θ between them, and express it as circulation per unit area.

($2c$)

2.29 Check whether the following functions represent the possible irrotational flow phenomenon

(i) $\phi = x + y + z$ (ii) $\phi = A(x^2 - y^2)$

(iii) $\phi = zx^2 - y^2 - z^2$ (iv) $\phi = \ln x$

(v) $\phi = \mu \ln x$ (vi) $\phi = \sin x$

(vii) $\phi = Ur \cos \theta + \frac{U}{r} \cos \theta$ (viii) $\phi = 3x^2 - 3x + 3y^2 + 16t + 2zt$

Here A, μ, U are constants.

[(i) yes, (ii) yes, (iii) no, (iv) no, (v) yes, (vi) no, (vii) yes (viii) no]

2.30 Which of the following stream functions are possible irrotational flow fields.

(i) $\psi = 2Axy$ (ii) $\psi = A(x^2 - y^2)$

(iii) $\psi = Ax + By$ (iv) $\psi = 2xy + y$

(v) $\psi = A \sin \frac{\pi x}{L} + B \sin \frac{\pi y}{L}$ (vi) $\psi = a\theta$

Here A, B, a and L are constants.

[(i) yes, (ii) yes, (iii) yes, (iv) yes, (v) no (vi) yes)]

2.31 If the velocity field is given by $u = x^2 - y^2 + x$ and $v = -(2xy + y)$ determine ϕ and ψ.

$$\left(\phi = c + \frac{y^2}{2} - (2x-1) - \frac{x^2}{2} - \frac{x^3}{3},\quad \psi = c - (2xy + y)\right)$$

2.32 Examine the flow pattern given by $\phi = \frac{A}{2\pi} \ln r$, where A is the +ve constant. Determine ψ.

$$\left(\text{Sin } k, \frac{A\theta}{2\pi}\right)$$

2.33 If $\phi = Ay$ where A is a positive constant, determine ψ and draw typical streamlines and equipotential lines. Identify the flow.

($\psi = -Ax$, Flow parallel to y axis with velocity $-A$)

2.34 If ϕ_1 and ϕ_2 are the velocity potentials, check whether (i) $\phi_1 + \phi_2$ (ii) $C_1\phi_1 + C_2\phi_2$ (iii) $\phi_1 - \phi_2$ (iv) $\frac{\phi_1}{\phi_2}$ and (v) $\phi_1\phi_2$ satisfy Laplace equation.

[(i) yes, (ii) yes, (iii) yes, (iv) no (v) No)]

2.35 Identify the two flows given by $\phi_1 = -Ay$ and $\phi_2 = -Bx$ where A and B are constants. Superimpose the two flows and describe the resulting flow.

(Rectilinear flow making an angle $\theta = \tan^{-1} \frac{A}{B}$ with +ve x-axis).

2.36 If $\phi = A(x^2 - y^2)$ is the velocity potential, determine ψ. Draw typical streamlines and equipotential lines and identify the flow. Determine the discharge passing between streamlines passing through the points (3, 1) and (3, 3). ($\psi = 2Axy$, flow around a corner, $12A$)

2.37 A source of strength 0.2 m³/s m and a vortex of strength 0.30 m³/s m are located at the origin. Determine the velocity potential and stream function for the resulting flow. Also determine the velocity components at point (2, 2).

$$\left(\phi = \frac{-0.20}{2\pi} \ln r + \frac{0.30}{2\pi}\theta,\ \psi = \frac{0.20}{2\pi}\theta - \frac{0.30}{2\pi} \ln r,\ V_\theta = -0.0169 \text{ m/s},\right.$$

$$\left. V_r = 0.0113 \text{ m/s}\right)$$

2.38 Show that, for two-dimensional irrotational flow, streamlines and equipotential lines intersect orthogonally.

2.39 Show that for two-dimensional rotational flow

$$\frac{\partial^2\psi}{\partial x^2} + \frac{\partial^2\psi}{\partial y^2} = 2\omega_z.$$

2.40 Approximate flow pattern for two-dimensional steady flow around a cylinder kept between parallel plates is shown in Fig. 2.22. Indicate

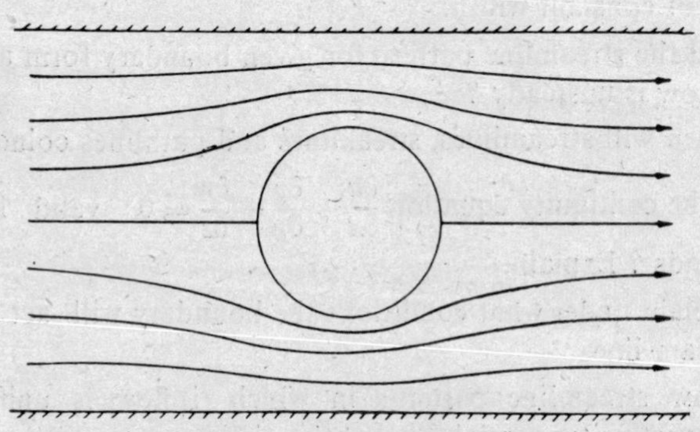

Fig. 2.22. Streamline pattern around cylinder kept between parallel plates.

points of stagnation, maximum velocity, location where normal acceleration is maximum and normal acceleration is small. Also plot U/U_0 variation along the plate as well as along the cylinder.

2.41 Draw a flow net for the boundary form shown in Fig. 2.23 and

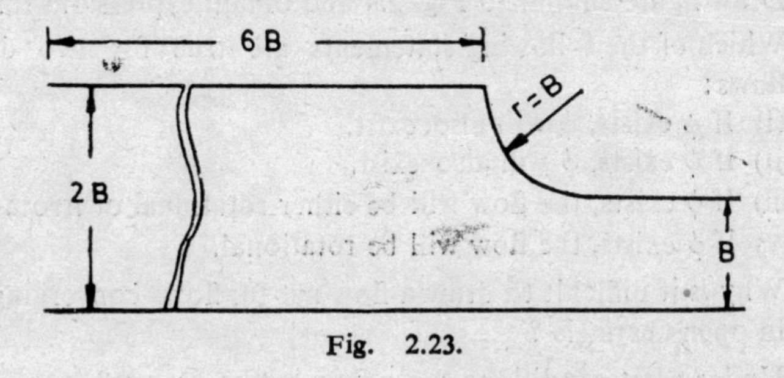

Fig. 2.23.

show points of stagnation and regions for uniform flow, predominantly normal acceleration and predominantly convective tangential accelerations.

2.42 Suggest an electrical analogy set up for obtaining flow net for geometry shown in Prob. 2.40.

2.43 For geometry shown in Fig. 2.23, draw streamlines using relaxation technique.

DESCRIPTIVE QUESTIONS

2.1 Classify the following flows as steady, unsteady, uniform, nonuniform:

(i) Constant discharge flowing through a converging pipe

(ii) Constant discharge flowing in a pipe bend of constant diameter
(iii) Increasing discharge in a straight pipe of constant diameter
(iv) Flow of constant water discharge in a long rectangular channel of constant width.

2.2 Will the streamline pattern for given boundary form always change if flow is unsteady ?

2.3 When will streamlines, streaklines and pathlines coincide ?

2.4 Is the continuity equation $\frac{\partial u}{\partial x}+\frac{\partial v}{\partial y}+\frac{\partial w}{\partial z}=0$ valid for stratified liquids ? Explain.

2.5 Explain under what conditions the boundary will act as a limiting stream line.

2.6 Draw streamline patterns in which (i) flow is uniform, (ii) only convective tangential acceleration is present, (iii) only convective normal acceleration is present, and (iv) both convective tangential and convective normal accelerations are present.

2.7 If the velocity distribution in an open channel is given by $\frac{u}{u_m}=\frac{y}{D}$, where u_m is the velocity at the water surface and D is depth of flow, is the flow rotational or irrotational ?

2.8 Draw figure similar to Fig. 2.4 and obtain expressions for ω_y and ω_z.

2.9 Which of the following statements are true for two dimensional flows:
(i) If ϕ exists, ψ will also exist.
(ii) If ψ exists, ϕ will also exist.
(iii) If ψ exists, the flow will be either rotational or irrotational.
(iv) If ϕ exists, the flow will be rotational.

2.10 Why is it difficult to draw a flow net for flows converging in section in open channels ?

2.11 Under what conditions can a flow net be drawn ?

2.12 Is it true that flow net can be drawn only for ideal fluids ?

2.13 When nothing such as 'ideal fluid' exists, what is the utility of studying flow of ideal fluids ?

2.14 Given below are some velocity distributions (Fig. 2.24). Which of them represent irrotational flow ? Give reasons.

2.15 A sphere is moving at a constant velocity in still air. Is the flow at a given point steady or unsteady ? Draw typical stream line, streak line and path line in front of the ball.

2.16 Give few instances where streamlining can improve the performance of a hydraulic structure.

2.17 In what ways can the separation of flow be avoided ?

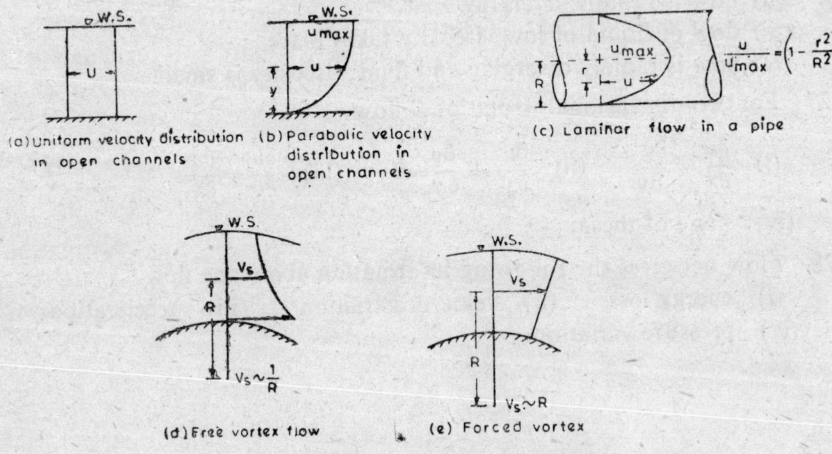

Fig. 2.24. Velocity distribution.

2.18 Draw typical flow pattern for flow past a Maruti-Suzuki car and a fully loaded truck as they travel at 50 km/hr. Comment on the two flows.

2.19 Is separation always harmful ? If your answer is in the negative, give an example or two where occurrence of separation is beneficial.

2.20 What are the characteristics of flow where stagnation and separation would occur ?

2.21 If $\phi = \phi_1$ and $\phi = \phi_2$ are two velocity potentials which satisfy Laplace equation, then $\phi = \phi_1 + \phi_2$ will also satisfy Laplace equation. What is its significance in terms of developing complex flow patterns ?

2.22 List the ways in which flow patterns for a particular flow phenomenon can be obtained.

2.23 A stream line is a line in flow field
(i) that is traced by all the fluid particles passing through a given point
(ii) along which a fluid particle travels
(iii) such that at every point on it the velocity is tangential to it.

2.24 If for steady flow, stream lines are converging straight lines
(i) only convective normal acceleration is present
(ii) only convective tangential acceleration is present
(iii) both convective normal and tangential accelerations are present
(iv) there is no acceleration.

2.25 Dimensions of ϕ are
(i) FL (ii) L/T^2 (iii) L^2/T^2 (iv) L^2/T

22.6 Flow net can be drawn for steady flows if
(i) flow is rapidly convergi ng and fluid viscosity is small

(ii) flow is rapidly diverging
(iii) flow of liquid of low viscosity takes place
(iv) flow is rapidly diverging and fluid viscosity is small.

2.27 For two dimensional irrotational flow

(i) $\frac{\partial u}{\partial x}=\frac{\partial v}{\partial y}$ (ii) $\frac{\partial u}{\partial y}+\frac{\partial v}{\partial x}=0$ (iii) $\frac{\partial u}{\partial y}=\frac{\partial v}{\partial x}$
(iv) none of these.

2.28 Flow net gives the following information about the flow
(i) energy loss (ii) velocity variation (iii) accelerations
(iv) pressure variation.

CHAPTER III

Equations of Motion and Energy Theorem

3.1 EQUATIONS OF MOTION

For flow of an ideal incompressible fluid along a stream line, where the forces governing the motion are due to gravity and pressure, Newton's second law yields

$$\frac{\partial V_s}{\partial t}+V_s\frac{\partial V_s}{\partial S}=-g\frac{\partial Z}{\partial S}-\frac{1}{\rho}\frac{\partial p}{\partial S} \tag{3.1}$$

where V_s is tangential velocity at any point, Z is the elevation and p is the

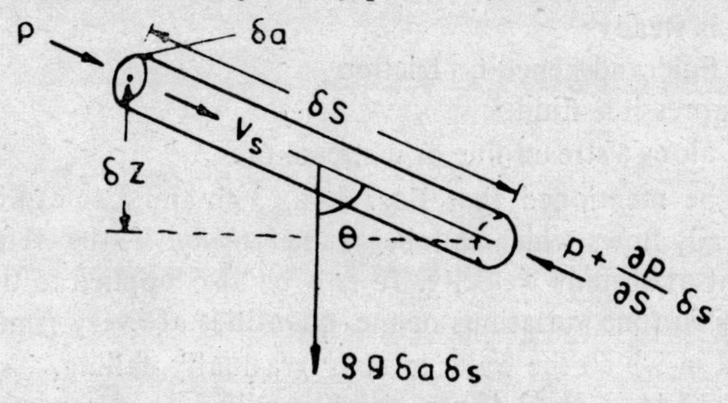

Fig 3.1 Definition sketch.

pressure (Fig. 3.1). Corresponding equations in cartesian co-ordinate system are known as Euler's equations of motion. These are,

$$\left.\begin{aligned}\frac{\partial u}{\partial t}+u\frac{\partial u}{\partial x}+v\frac{\partial u}{\partial y}+w\frac{\partial u}{\partial z}&=X-\frac{1}{\rho}\frac{\partial p}{\partial x}\\ \frac{\partial v}{\partial t}+u\frac{\partial v}{\partial x}+v\frac{\partial v}{\partial y}+w\frac{\partial v}{\partial z}&=Y-\frac{1}{\rho}\frac{\partial p}{\partial y}\\ \frac{\partial w}{\partial t}+u\frac{\partial w}{\partial x}+v\frac{\partial w}{\partial y}+w\frac{\partial w}{\partial z}&=Z-\frac{1}{\rho}\frac{\partial p}{\partial z}\end{aligned}\right\} \tag{3.2}$$

where u, v and w are velocity components in x, y and z directions respectively and X, Y, Z are components of fluid weight per unit mass, in x, y, z directions respectively. Also

$$V_s = ui + vj + wk. \tag{3.3}$$

3.2 INTEGRATION OF EULER'S EQUATION OF MOTION ALONG A STREAM LINE

Integration of Eq. 3.1 along a stream line yields

$$\frac{V_S^2}{2g} + \frac{p}{\gamma} + Z = \text{constant } C_1 \tag{3.4a}$$

$$\frac{\rho V_S^2}{2} + p + \gamma z = \text{constant } C_2 \tag{3.4b}$$

$$\frac{V_S^2}{2} + \frac{p}{\rho} + gZ = \text{constant } C_3 \tag{3.4c}$$

Equation 3.4 is known as Bernoulli's equation. The terms in Eqs. 3.4a, 3.4b and 3.4c represent energy per unit weight, energy per unit volume, and energy per unit mass of fluid respectively.

It can be written in the form

$$\frac{U_1^2}{2} + \frac{p_1}{\rho} + gZ_1 = \frac{U_2^2}{2} + \frac{p_2}{\rho} + gZ_2 \tag{3.4d}$$

Equations 3.4a, 3.4b and 3.4c are applicable under following conditions:

1 Flow is steady
2 Ideal fluid and hence no friction
3 Incompressible fluid
4 Flow along a stream line or a stream tube.

It may be mentioned that Eqs. 3.4a, 3.4b and 3.4c can be applied to those unsteady flows which can be made steady if the reference frame translates at a constant velocity. It can be also applied to those unsteady flows in which time variations of the quantities are very small; e.g. flow from a tank in which the water level is gradually falling. As an approximation Eqs. 3.4a, 3.4b, 3.4c can also be applied to compressible flows if temperature change does not occur and Mach number M is less than 0.20, or large changes in velocity or elevation do not occur, e.g. wind forces on structures, flow in ventilation systems, etc.

Bernoulli's equation for compressible fluid under adiabatic condition is

$$\frac{U^2}{2} + gZ + \frac{k}{k-1}\frac{p}{\rho} = \text{const.} \tag{3.5}$$

where k is equal to c_p/c_v i.e, ratio of specific heat at constant pressure and specific heat at constant volume.

3.3 PHYSICAL MEANING OF TERMS IN EQUATION (3.4a) AND APPLICATIONS

(i) $U^2/2g$ is kinetic energy per unit weight of fluid with dimensions Nm/N or m.

(ii) $\frac{p}{\gamma}$ represents work done by pressure force per unit weight of fluid: it also has dimension of length.

(iii) Z is potential energy per unit weight of fluid measured above an arbitrary datum.

(iv) The term $\left(\frac{p}{\gamma} + Z\right)$ is known as piezometric head since liquid under pressure will rise to this level if a vertical glass tube, called piezometer, is connected to the boundary.

Bernoulli's equation needs the following corrections when it is applied as an energy equation.

(i) If there are losses due to boundary shear or change of shape, a term h_L must be added to the right hand side; h_L is energy loss per unit weight of fluid.

$$\frac{U_1^2}{2g} + \left(\frac{p_1}{\gamma} + Z_1\right) = \frac{U_2^2}{2g} + \left(\frac{p_2}{\gamma} + Z_2\right) + h_L \tag{3.6}$$

(ii) When the equation is applied to a stream tube of finite dimensions, velocity distribution over the section may be nonuniform needing correction to kinetic energy term. If u is the velocity over area δA and U is average velocity over area A, kinetic energy correction factor α is defined as

$$\frac{1}{2}\alpha U^3 A\rho = \int_A \frac{1}{2}\rho u^3 \, dA$$

or

$$\alpha = \frac{1}{AU^3}\int_A u^3 dA \tag{3.7}$$

Hence Eq. 3.4(a) can be written as

$$\frac{\alpha_1 U_1^2}{2g} + \left(\frac{p_1}{\gamma} + Z_1\right) = \frac{\alpha_2 U_2^2}{2g} + \left(\frac{p_2}{\gamma} + Z_2\right) \tag{3.8}$$

(iii) If mechanical energy E_m per unit weight of fluid is added to or removed from the system, Eq. 3.4a becomes

$$\frac{U_1^2}{2g} + \frac{p_1}{\gamma} + Z_1 = \frac{U_2^2}{2g} + \frac{p_2}{\gamma} + Z_2 \pm E_m \tag{3.9}$$

The negative sign (w.r.t. E_m) is used in case of a pump and positive sign in case of a turbine). Power developed by turbine or given by pump to the fluid is equal to

$$P = \frac{\gamma Q E_m}{1000} \text{ kW} \tag{3.10}$$

3.4 ENERGY EQUATION WITH HEAT TRANSFER

When dealing with compressible fluid involving changes in temperature, addition or removal of heat, and shaft work (work done or extracted from fluid by a machine), one can write the following equation for steady flow.

$$\frac{U_1^2}{2g} + Z_1 + \frac{p_1}{g\rho_1} + e_1 + H = \frac{U_2^2}{2g} + Z_2 + \frac{p_2}{g\rho_2} + e_2 \pm E_m \quad (3.11)$$

Here e_1 and e_2 are internal energy of fluid at the two states and

$$(e_2 - e_1) = C_v J\,(T_2 - T_1)$$

where T_1 and T_2 are the absolute temperatures, C_v is specific heat at constant volume and $J = 4.187$ the conversion factor to change calories into joules. H is the heat added per unit weight of fluid and E_m is external work done per unit weight of fluid, both being expressed in joules. The term E_m is sometimes known as the shaft work, its sign being $-$ve for pumps and $+$ve for turbines as mentioned earlier.

It may be noted that energy equation for steady frictionless incompressible fiuid reduces to Bernoulli's equation provided there is no shaft work or heat transfer.

ILLUSTRATIVE EXAMPLES

3.1 A pitot tube inserted at the centre of 0.20 m diameter pipe is connected to the wall static pressure tap through an inverted U-tube manometer. When the pipe carries water, U-tube shows a difference of 10 cm. Determine the velocity of flow (Fig. 3.2).

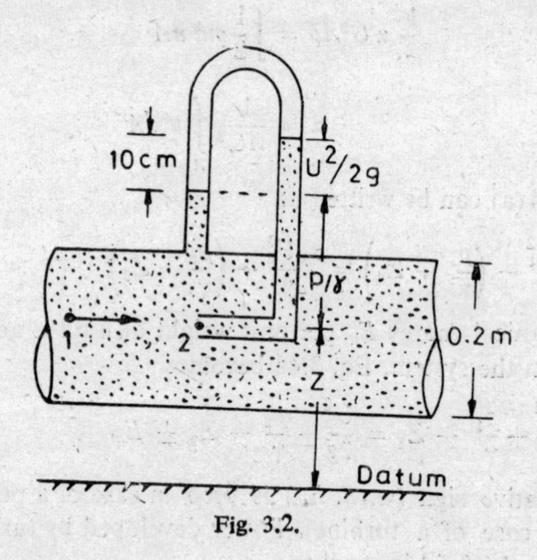

Fig. 3.2.

Applying Bernoulli's theorem between points 1 and 2 (which is a stagnation pont) one gets

$$\frac{U_1^2}{2g}+\frac{p_1}{\gamma}=\frac{p_2}{\gamma}$$

Therefore

$$\frac{U_1^2}{2g}=\frac{p_2-p_1}{\gamma}=0.10\,\text{m}$$

$$U_1=\sqrt{2\times 9.806\times 0.1}=1.400\text{ m/s}$$

3.2 Draw a neat sketch of Prandtl's pitot tube and explain its working.

To determine the velocity of flow by a pitot tube, one needs to measure the static pressure head in the section and the stagnation pression head which is equal to $\left(\frac{p}{\gamma}+\frac{U^2}{2g}\right)$. These two are measured through a single combination of two tubes in case of Prandtl tube (Fig. 3.3). The front portion of the tube is so rounded that there is no separation on the shaft of tube

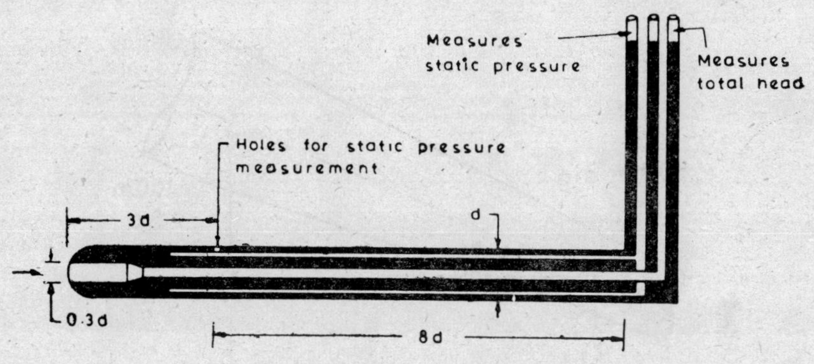

Fig. 3.3 Prandtl's pitot tube.

and at a distance of about $3d$ the pressure on the tube surface is static. Holes at this location are connected to another tube inside which, along with the central total head tube are connected to a manometer. If ΔH represents the difference of head between these two tubes, the velocity is given by

$$U=C_d\sqrt{2g\Delta H} \tag{3.12}$$

The coefficient C_d = 0.95 to 0.99 and needs to be determined through experimental calibration. Prandtl pitot tube is relatively insensitive to direction of flow. Even if flow makes an angle of $\pm$ 20° with the axis of tube, the error in velocity is only about $\pm$ 2 to 3 per cent.

3.3 Water flows through a pipe of 200 mm diameter at a discharge of 62.8 l/s. At a section which is 5.0 m above the datum the pipe pressure is 150 kN/m². Determine the total energy of the flow.

$$U=\frac{Q}{A}=\frac{0.0628}{0.785\times 0.20^2}=2.0\text{ m/s}$$

Total energy per unit weight of fluid

$$= \frac{U^2}{2g} + \frac{p}{\gamma} + Z$$

$$= \frac{2^2}{2 \times 9.806} + \frac{150}{9.797} + 5.0$$

$$= 0.204 + 15.326 + 5.00$$

$$= 20.530 \text{ m}$$

3.4 A 50 mm diameter tube gradually expands to 100 mm diameter tube in a length of 10 m. If the tube makes an angle of 20° in upward direction with the horizontal (Fig. 3.4), determine the pressure p_2 at the exit if the tube carries a discharge of 3.925 l/s and the inlet pressure p_1 is 60 kN/m², assuming (i) no energy loss, (ii) a loss of 0.20 m.

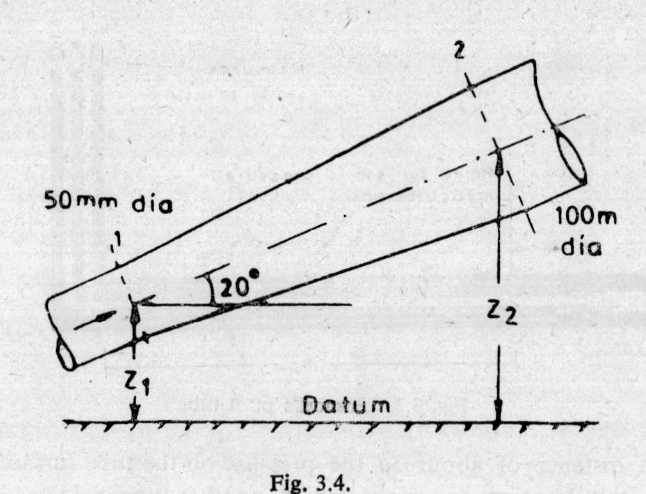

Fig. 3.4.

$$U_1 = \frac{0.003925}{0.785 \times 0.05^2} = 2.0 \text{ m/s}$$

$$U_2 = \frac{0.003925}{0.785 \times 0.10^2} = 0.50 \text{ m/s}$$

$$(Z_2 - Z_1) = 10 \sin 20°$$

$$= 3.42 \text{ m}$$

(i) $$Z_1 + \frac{60}{9.787} + \frac{2^2}{2 \times 9.806} = Z_2 + \frac{p_2}{9.787} + \frac{0.5^2}{2 \times 9.806}$$

$$\therefore \quad \frac{p_2}{9.787} = -3.42 + 6.132 + 0.204 - 0.0127$$

$$= 2.902$$

$$\therefore \quad p_2 = 2.902 \times 9.787 = 28.40 \text{ kN/m}^2$$

(ii) $$Z_1 + \frac{60}{9.78} + \frac{2^2}{2 \times 9.806} = Z_2 + \frac{p_2}{9.787} + \frac{0.5^2}{2 \times 9.806} + 0.20$$

$$\therefore \quad \frac{p_2}{9.787} = -3.42 + 6.132 + 0.204 - 0.0127 - 0.20$$

$$= 2.702$$

$$\therefore \quad p_2 = 2.702 \times 9.787 = 26.44 \text{ kN/m}^2.$$

3.5 Calculate the average velocity of flow and kinetic energy correction factor α for laminar flow in a pipe for which velocity distribution is given by

$$\frac{u}{u_m} = \left(1 - \frac{r^2}{R^2}\right)$$

where u is the velocity of a distance r from the centre line, u_m is the centre line velocity and R is the pipe radius.

Take an annular strip of length $2\pi r$ and width dr.

$$dQ = 2\pi r . dr \cdot u$$

$$\therefore \quad Q = \int_0^R 2\pi r dr u = 2\pi u_m \int_0^R \left(1 - \frac{r^2}{R^2}\right) r dr$$

$$= 2\pi \, u_m \left(\frac{R^2}{2} - \frac{R^2}{4}\right)$$

$$= \pi R^2 u_m / 2$$

$$\therefore \quad U = \frac{Q}{A} = \frac{u_m}{2} \frac{\pi R^2}{\pi R^2} = \frac{u_m}{2}$$

$$\alpha = \frac{\int_0^R \frac{\rho}{2} u_m^3 \left(1 - \frac{r^2}{R^2}\right)^3 2\pi r dr}{\frac{\rho}{2} \pi R^2 U^3}$$

$$= \frac{8 \times 2\pi u_m^3}{\pi R^2 u_m^3} \int_0^R \left(r - \frac{3r^3}{R^2} + \frac{3r^5}{R^4} - \frac{r^7}{R^6}\right) dr$$

$$= \frac{16}{R^2} \left(\frac{R^2}{2} - \frac{3}{4} R^2 + \frac{3}{6} R^2 - \frac{1}{8} R^2\right)$$

$$= \frac{16}{R^2} \cdot \frac{3R^2}{24}$$

$$= 2.0$$

3.6 A jet of water issues vertically upwards from 0.20 m high nozzle whose inlet and outlet diameters are 100 mm and 40 mm respectively. If the pressure at the inlet is 20 kN/m² above the atmospheric pressure, determine the discharge and the height to which the jet will rise (Fig. 3.5). Mention all the assumptions made.

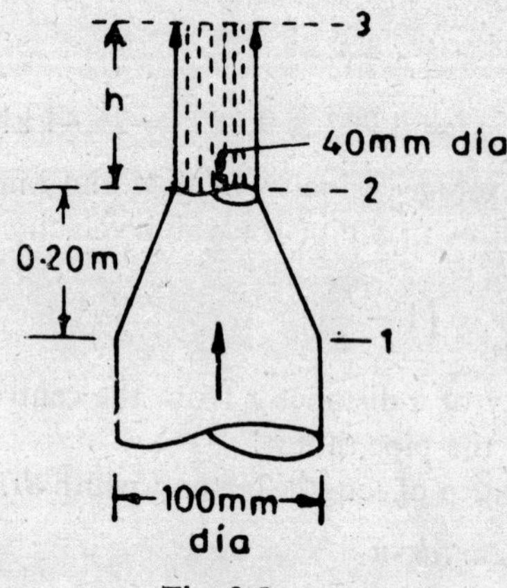

Fig. 3.5.

Assumptions made are:
During its travel the jet does not disintegrate.
Velocity distribution is uniform.
There is no frictional resistance.
Apply Bernoulli's equation between section (1) and (2),

$$Z_1 + \frac{p_1}{\gamma} + \frac{U_1^2}{2g} = Z_2 + \frac{p_2}{\gamma} + \frac{U_2^2}{2g}$$

$$Z_1 = 0,\ Z_2 = 0.2\ \text{m},\ p_1 = 20\ \text{kN/m}^2,\ p_2 = 0$$

$$\therefore\quad 0 + \frac{20}{9.787} + \frac{U_1^2}{2g} = 0.20 + 0 + \frac{U_2^2}{2 \times 9.806}$$

and continuity equation yields

$$0.785 \times 0.10^2\ U_1 = 0.785 \times 0.040^2\ U_2$$

Therefore, $U_2 = 6.25\ U_1$. Substituting this value in the above equation one gets,

$$\frac{38.062\ U_1^2}{2 \times 9.806} = 2.0436 - 0.2000$$

$$= 1.8436$$

$$\therefore\quad U_1 = 0.975\ \text{m/s and}\ U_2 = 6.25 \times 0.975 = 6.092\ \text{m/s}$$

$$Q = 0.785 \times 0.1^2 + 0.975 = 7.654 \times 10^{-3}\ \text{m}^3/\text{s or}\ 7.654\ \text{l/s}$$

If h is the height to which the jet rises, application of Bernoulli's equation between sections 2 and 3 yields

$$\frac{(6.092)^2}{2 \times 9.806} + 0 + 0 = 0 + 0 + h$$

$$\therefore\quad h = 1.892\ \text{m}$$

3.7 For the study of free (irrotational) vortex flow that occurs in bath tubs or wash basins, consider Z axis to be vertically upwards, r axis radially outwards and θ direction perpendicular to r axis (Fig. 3 6). Assuming no

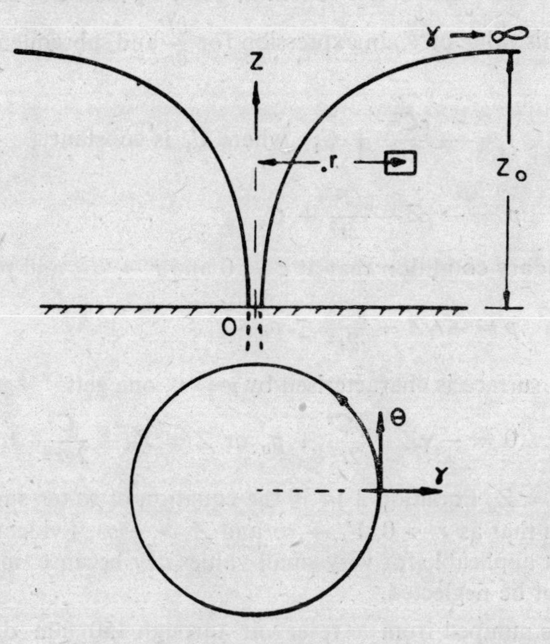

Fig. 3.6. Free-vortex.

accelerations in θ and Z directions and V_s^2/r as the acceleration in r direction towards the centre, write Euler's equations of motion and integrate them to get the equation for water surface profile.

Consider a unit volume of fluid at a distance r from centre line. The three equations of motion will be

$$-\frac{\partial p}{\partial r} = -\rho\frac{V_s^2}{r}$$

$$-\frac{\partial p}{r\partial\theta} = 0$$

$$-\frac{\partial p}{\partial Z} - \gamma = 0$$

Integration of second equation yields

$$p = p(r, Z)$$

Integration of third equation yields

$$p = -\gamma Z + f(r)$$

Hence, $\frac{\partial p}{\partial r} = \frac{\partial f}{\partial r}$, and from first equation $\frac{\partial p}{\partial r} = \rho\frac{V_s^2}{r}$

Therefore, $$\frac{\partial f}{\partial r} = \rho\frac{V_s^2}{r}$$

Since irrotational motion is maintained without any external torque the rate of change of angular momentum must be zero (see Chapter VI).

$$\frac{\partial}{\partial t}(MV_s.\, r) = 0 \quad \text{or} \quad V_s r = C \text{ i.e. } V_s = \frac{C}{r} \tag{3.13}$$

Here M is fluid mass and C is a constant. Thus V_s increases as r decreases. Substituting this value of V_s in expression for $\frac{\partial f}{\partial r}$ and subsequent integration yields

$$f = -\frac{\rho C^2}{2r^2} + C_1 \quad \text{where } C_1 \text{ is constant.}$$

$$\therefore \qquad p = -\gamma Z - \frac{\rho C^2}{2r^2} + C_1$$

with the boundary condition that at $z = 0$ and $r \to \infty$; and $p = p_0$

$$p = -\gamma Z - \frac{\rho C^2}{2r^2} + p_0$$

Since the free surface is characterised by $p = 0$, one gets

$$0 = -\gamma Z - \frac{\rho C^2}{2r^2} + p_0 \text{ or } Z = Z_0 - \frac{C^2}{2gr^2} \tag{3.14}$$

Since $p_0/\gamma = Z_0$ Equation 3.14 is the equation of water surface profile. It can be seen that as $r \to 0$, $V_s \to \infty$ and $Z \to -\infty$. Evidently the above analysis is not applicable for very small values of r because in this region friction cannot be neglected.

3.8 Water is pumped from a reservoir through 150 mm diameter pipe and is delivered at a height of 15 m from centre line of pump through a 100 mm nozzle connected to 150 mm discharge line as shown in Fig. 3.7.

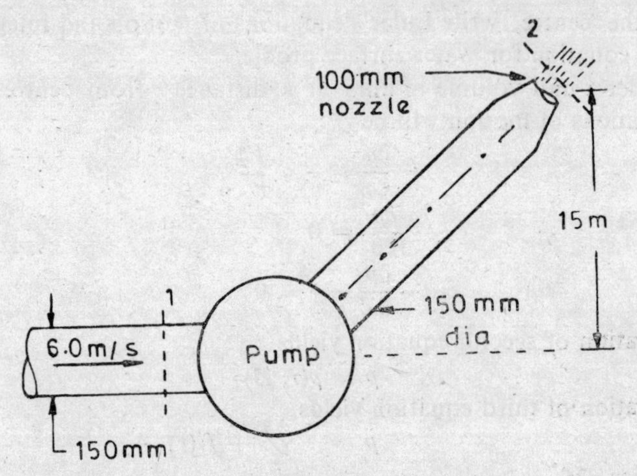

Fig. 3.7.

If the pressure at pump inlet is 210.0 kN/m² absolute, inlet velocity of 6.0 m/s and the jet is discharged into atomsphere, determine the energy supplied by the pump. Assume atomspheric pressure as 101.3 kN/m² and no friction.

Assuming water temperature to remain same, apply Bernoulli's equation between sections 1 and 2.

$$\frac{U_1^2}{2g} + 0 + \frac{210}{9.787} = \frac{U_2^2}{2g} + 15 + \frac{101.3}{9.787} - E_m$$

and continuity equation yields $U_2 = \left(\frac{150}{100}\right)^2 U_1$

$$= 13.5 \text{ m/s}$$

$$\therefore \quad E_m = 9.293 + 15.0 + 10.350 - 1.836 - 21.457$$

$$= 11.35 \text{ m}$$

$$\text{Energy supplied} = \frac{Q\gamma E_m}{1000} = \frac{(0.785 \times 0.15^2 \times 6.0) \times 9787 \times 11.35}{1000}$$

$$= 11.772 \text{ kW}$$

3.9 The suction and delivery pipe diameters of a pump are 200 mm and 100 mm respectively. If the inlet and outlet pressures are 70 kN/m² and 210 kN/m² respectively and pump delivers 4.0 kW power to the fluid, find the discharge of water flowing through the pump. Assume exit to be 0.50 m above the inlet.

Writing $U_1 = \frac{Q}{0.785 \times 0.2^2}$ and $U_2 = \frac{Q}{0.785 \times 0.1^2}$

Bernoulli's equation can be expressed as

$$\frac{Q^2}{(2 \times 9.806)(0.785 \times 0.2^2)^2} + \frac{70}{9.787} + 0 = \frac{Q^2}{(2 \times 9.806)(0.785 \times 0.1^2)^2}$$

$$+ \frac{210}{9.787} + 0.50 - \frac{4.0 \times 1000}{Q \times 9787}$$

$$51.715Q^2 + 7.152 = 827.44Q^2 + 21.457 + 0.500 - \frac{4.087}{Q}$$

$$775.725Q^3 + 14.805Q - 4.087 = 0$$

or $\quad Q^3 + 0.0191Q - 0.00527 = 0$

$Q = 0.10 \qquad \text{RHS} = 0.001 + .00191 - 0.00527$

$$= -0.00236$$

$Q = 0.13 \qquad \text{RHS} = 0.00220 + 0.00243 - 0.00527$

$$= -0.000587$$

$Q = 0.14 \qquad \text{RHS} = 0.00274 + 0.00267 - 0.00527$

$$= +0.000144$$

$Q = 0.135 \qquad \text{RHS} = 0.00246 + 0.00258 - 0.00527$

$$= -0.000231 \approx 0$$

Hence Q will lie between 0.140 and 0.135; or $Q = 0.137$ m²/s.

3.10 Water flows upwards through a vertical pipe of 300 mm diameter and then flows out radially into atmosphere between two parallel circular plates of 750 mm diameter kept 10 mm apart (Fig. 3.8). If presure at B is 25 kN/m² gauge, find the discharge through the system.

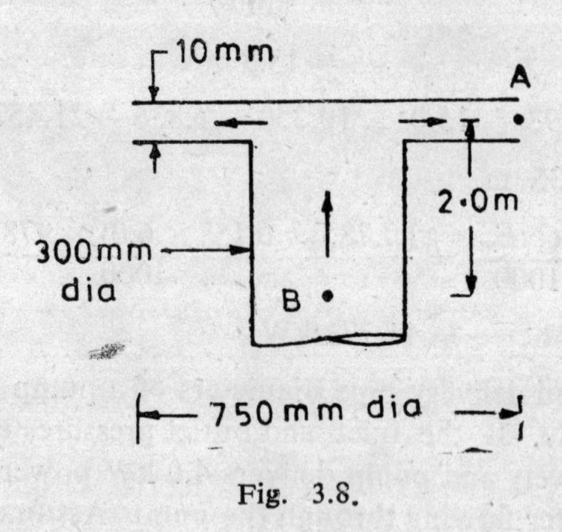

Fig. 3.8.

Apply Bernoulli's equation between points B and A, assuming no loss.

$$\frac{U_B^2}{2g} + \frac{p_B}{\gamma} + 0 = \frac{U_A^2}{2g} + 0 + 2.0$$

and continuity equation gives

$$0.785 \times 0.30^2\, U_B = 2 \times 3.142 \times 0.375 \times 0.01 U_A$$

or
$$U_B = 0.333\, U_A$$

$$\frac{(0.333 U_A)^2}{2 \times 9.806} + \frac{25}{9.787} + 0 = \frac{U_A^2}{2 \times 9.806} + 0 + 2.0$$

$$\frac{0.889\, U_A^2}{2 \times 9.806} = 0.554 \quad \text{or} \quad U_A = 3.497 \text{ m/s}$$

$\therefore$
$$U_B = 0.333 \times 3.497 = 1.166 \text{ m/s}$$

$$Q = 0.785 \times 0.3^2 \times 1.166 = 0.0824 \text{ m}^3\text{/s} \quad \text{or } 82.4 \text{ l/s.}$$

3.11 Applying Bernoulli's equation and the equation of continuity, show that the discharge in the venturimeter is given by

$$Q = \frac{(\pi d^2/4)\sqrt{2g\,(h_1 - h_2)}}{\sqrt{1 - (d/D)^4}}$$

A venturimeter is made up of a rapid contraction followed by gradual expansion in a pipe as shown in Fig. 3.9. It is used as a discharge measuring device. Let D and d be the pipe diameter and throat diameter respectively. For given diameter Q

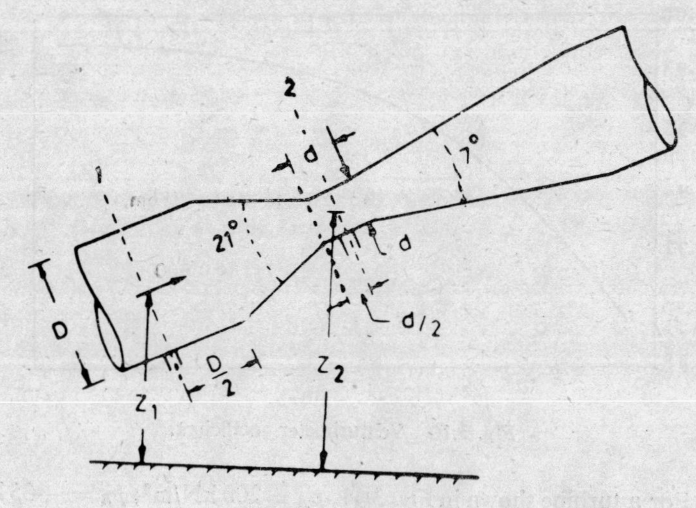

Fig. 3.9. Venturimeter.

$$\frac{\pi}{4} D^2 U_1 = \frac{\pi}{4} d^2 U_2$$

or
$$U_1 = (d/D)^2 U_2$$

Bernoulli's equation gives (assume no loss between sections 1 and 2),

$$\frac{U_1^2}{2g} + \frac{p_1}{\gamma} + Z_1 = \frac{U_2^2}{2g} + \frac{p_2}{\gamma} + Z_2$$

Letting $\left(\frac{p}{\gamma} + Z\right) = h$ the piezometric head,

$$(d/D)^4 \frac{U_2^2}{2g} + h_1 = h_2 + \frac{U_2^2}{2g}$$

or
$$\frac{U_2^2}{2g} [1 - (d/D)^4] = (h_1 - h_2)$$

and
$$U_2 = \frac{1}{\sqrt{1 - \left(\frac{d}{D}\right)^4}} \sqrt{2g(h_1 - h_2)}$$

Hence
$$Q = \frac{\pi d^2}{4} U_2 = \frac{\pi d^2/4}{\sqrt{1-(d/D)^4}} \sqrt{2g\,(h_1 - h_2)} \tag{3.15}$$

(To take into account the effect of viscosity, the discharge equation is written as $Q = \frac{C\pi\, d^2/4}{\sqrt{1 - (d/D)^4}} \sqrt{2g\,(h_1 - h_2)}$ where C varies from 0.93–0.99 and is a function of Reynolds number $U_2 d\rho/\mu$. Figure 3.10 shows variation of C with $\frac{U_2 d\rho}{\mu}$.

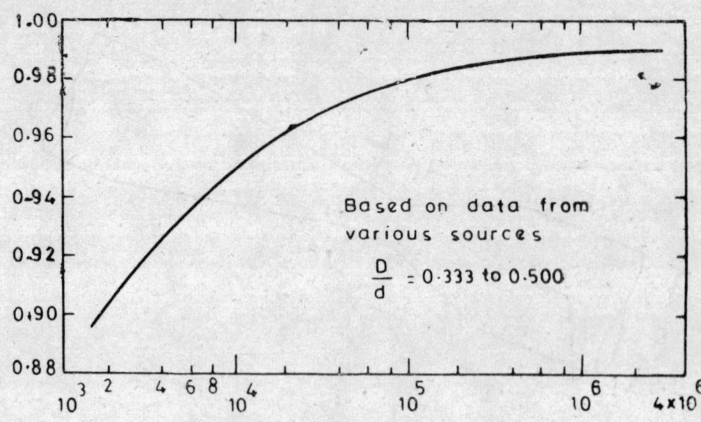

Fig. 3.10 Venturimeter coefficient.

3.12 For a turbine shown in Fig. 3.11, $p_A = 200$ kN/m²; $p_B = -35$ kN/m², $Q = 0.40$ m³/s. Determine the energy output at the machine, if its efficiency is 80 per cent.

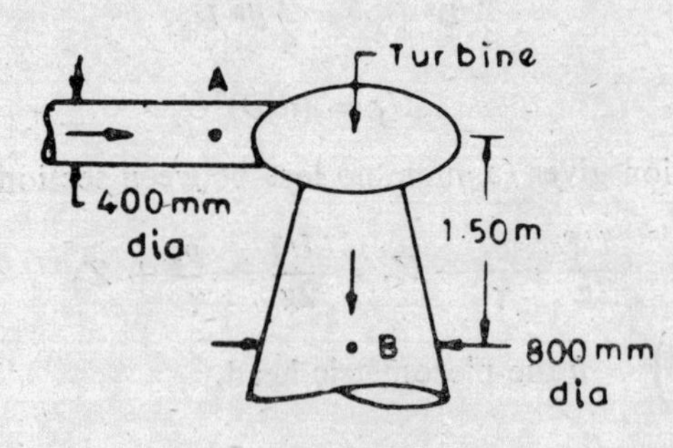

Fig. 3.11.

$$U_A = 0.40/(0.785 \times 0.40^2)$$
$$= 3.185 \text{ m/s}$$
$$U_B = 0.40/(0.785 \times 0.80^2)$$
$$= 0.796 \text{ m/s}$$

$$\frac{U_A^2}{2g} + \frac{p_A}{\gamma} + Z_A = \frac{U_B^2}{2g} + \frac{p_B}{\gamma} + Z_B + E_m$$

$$\therefore \quad \frac{3.185^2}{2 \times 9.806} + \frac{200}{9.787} + 1.5 = \frac{0.796^2}{2 \times 9.806} - \frac{35}{9.787} + 0 + E_m$$

$$0.517 + 20.435 + 1.500 = 0.0323 - 3.576 + E_m$$

$$E_m = 25.996 \text{ m}$$

$$\text{Energy} = \frac{0.40 \times 9787 \times 25.996}{1000} = 101.768 \text{ kW}$$

Actual energy output = 101.768 × 0.80 = 81.414 kW

3.13 If the velocity distribution in a pipe is given by $u/u_m = 1 - \left(\frac{r}{R}\right)^n$ determine expressions for average velocity and α.

$$Q = \int_0^R 2\pi r \left(1 - \left(\frac{r}{R}\right)^n\right) u_m dr$$

$$= 2\pi\, u_m \left[\frac{R^2}{2} - \frac{1}{R^n}\left(\frac{R^{n+2}}{n+2}\right)\right]_0^R = \pi R^2 u_m \left(\frac{n}{n+2}\right)$$

$$\therefore \quad \frac{Q}{\pi R^2} = U = u_m \left(\frac{n}{n+2}\right)$$

K.E. on the basis of velocity distribution

$$= \int_0^R 2\pi r dr \frac{\rho}{2} u_m^3 \left(1 - \left(\frac{r}{R}\right)^n\right)^3$$

$$= 2\pi \frac{\rho}{2} u_m^3 \left[\frac{R^2}{2} - \frac{3}{R^n}\frac{R^{n+2}}{(n+2)} + \frac{3}{R^{2n}}\frac{R^{2n+2}}{2(n+1)} - \frac{1}{R^{3n}}\frac{R^{3n+2}}{(3n+2)}\right]$$

$$= 2\pi \frac{\rho}{2} u_m^3 R^2 \times \frac{3n^3}{2(n+1)(n+2)(3n+2)}$$

$$= 3\pi\rho\, u_m^3 R^2 n^3/2(n+1)(n+2)(3n+2)$$

K.E. on the basis of mean velocity

$$= \pi R^2 \alpha \frac{\rho}{2} U^3$$

$$= \frac{\rho}{2} \pi R^2 \alpha\, u_m^3 n^3/(n+2)^3$$

Equating the two, one gets

$$\alpha = 3(n+2)^2/(n+1)(3n+2)$$

3.14 For the siphon shown in Fig. 3.12, determine the discharge and the maximum permissible value of h. Neglect frictional resistance.

Applying Bernoulli's equation between A and C,

$$0 + 0 + 6 = \frac{U_c^2}{2g} + 0 + 0$$

$$\therefore \quad U_c = \sqrt{2 \times 9.806 \times 6}$$

$$= 10.847 \text{ m/s}$$

$$Q = 0.785 \times 0.15^2 \times 10.847$$

$$= 0.194 \text{ m}^3\text{/s}$$

Actual discharge will be somewhat smaller due to frictional resistance

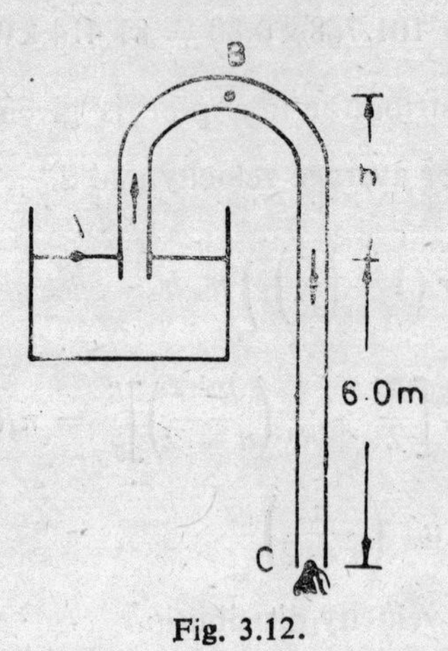

Fig. 3.12.

in the pipe. As h increases $\frac{p_B}{\gamma}$ will decrease. The pressure at B should be greater than the vapour pressure of water, otherwise due to release of dissolved air and water vapour, the continuity of water column will be broken and the siphon will cease to work For safe working, assume $p_B/\gamma = 2.0$ m absolute. Applying Bernoulli's equation between A and B (Take $p_A = 101.3$ kN/m²).

$$0 + 0 + \frac{101.3}{9.787} = h + 2.0 + \frac{10.847^2}{2 \times 9.806}$$

$$\therefore \quad h = 10.35-2.0-6.0$$

$$= 2.35 \text{ m}$$

3.15 A water jet of diameter D_0, velocity U_0 and inclined at an angle θ is discharged into the atmosphere as shown in Fig. 3.13. Assuming frictional resistance to be negligible and that the jet does not disintegrate, obtain

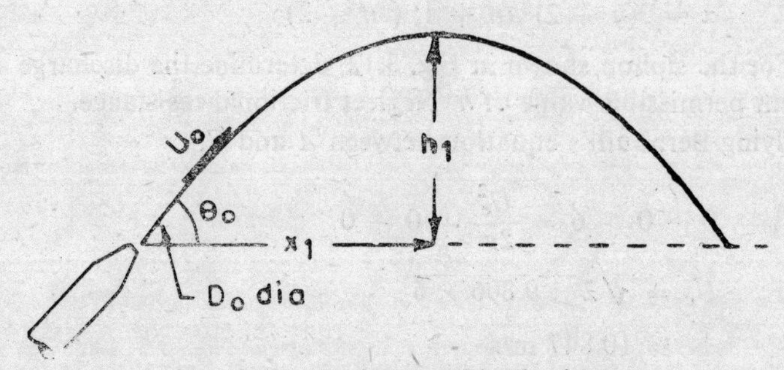

Fig. 3.13. Trajectory of inclined jet.

expressions for the maximum height to which it will rise and the horizontal distance it travels before it reaches the level of the nozzle.

Vertical component of jet velocity $= U_0 \sin\theta$

It will be zero when it reaches section 1. Applying Bernoulli's equation between 0 and 1

$$0 + 0 + \frac{U_0^2 \sin^2\theta}{2g} = 0 + h_1 + 0$$

$$\therefore \quad h_1 = \frac{U_0^2 \sin^2\theta}{2g}$$

The horizontal component of jet velocity, i.e. $U_0 \cos\theta$ remains unaltered. Continuity equation between sections 1 and 0 yields

$$\frac{\pi D_0^2}{4} U_0 = \frac{\pi D_1^2}{4} U_0 \cos\theta$$

$$\therefore \quad D_1 = D_0/\sqrt{\cos\theta}$$

The time required for water to travel from 0 to 1 is obtained in the following way

$$V = U_0 \sin\theta - gt$$

and at 1 $\quad V = 0 \quad \therefore \quad t = \dfrac{U_0 \sin\theta}{g}$

During this time the jet travels horizontal distance x_1

$$x_1 = U_0 \cos\theta \times \frac{U_0 \sin\theta}{g}$$

$$= U_0^2 \sin\theta \cos\theta/g$$

$$= U_0^2 \sin 2\theta/2g$$

Total horizontal distance travelled is $2x_1 = \dfrac{U_0^2 \sin 2\theta}{g}$.

It can be seen that $2x_1$ will be maximum when $2\theta = 90°$ i.e. when $\theta = 45°$.

3.16 When an ideal fluid with ambient velocity U_0 flows past a two dimensional circular cylinder, the velocity at the surface of the cylinder U is given by $U = 2U_0 \sin\theta$ (Fig. 3.14). Determine the pressure distribution around the cylinder, express it in dimensionless form and plot it.

Assuming the flow to be in plan so that Z remains same for all points, one can apply Bernoulli's equation between point 0 in the ambient flow and any point on the cylinder making an angle θ.

$$p_0 + \frac{\rho U_0^2}{2} = p + \frac{\rho U^2}{2}$$

$$\therefore \quad (p - p_0) = \Delta p = \frac{\rho U_0^2}{2}\left(1 - \frac{U^2}{U_0^2}\right)$$

or

$$\frac{\Delta p}{\rho U_0^2/2} = (1 - 4\sin^2\theta)$$

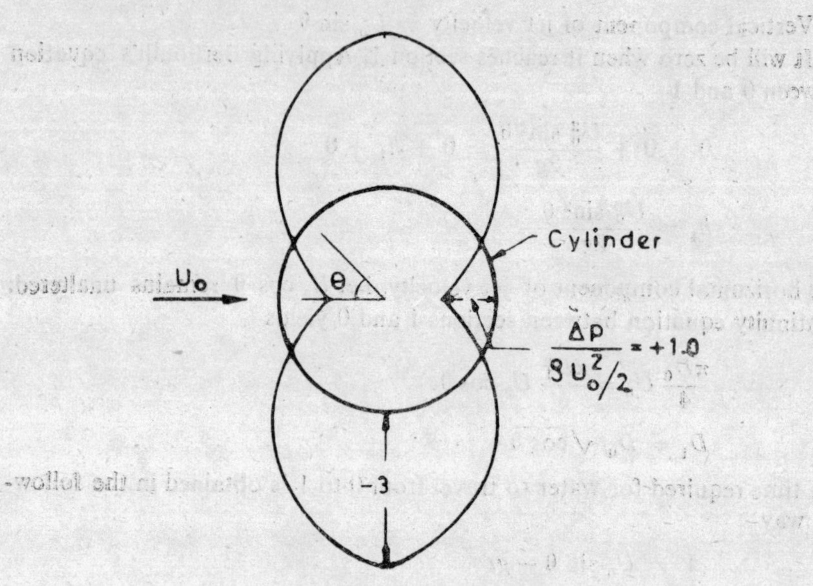

Fig. 3.14. **Pressure distribution around a 2–D cylinder.**

For various values of θ, dimensionless pressure distribution $\frac{\Delta p}{\rho U_0^2/2}$ is computed and plotted. Positive values are plotted inward while negative outward along radial direction. It can be seen that pressure distribution is symmetrical with respect to flow direction and also perpendicular to flow direction. Points where $U = 0$ and $\frac{\Delta p}{\rho U_0^2/2} = +1$ are known as stagnation points.

3.17 A boiler feed pump is supplied with water at 90°C and 75 kN/m² pressure absolute and is delivered at the same temperature and 10 000 kN/m² pressure absolute. Under steady state condition, heat of 400 J/N is lost by water while passing through the pump. Assuming inlet and outlet diameters of pump and their elevations to be same, determine the energy added by the pump if pump carries 0.30 m³/s discharge.

$$\frac{U_1^2}{2g} + \frac{p_1}{\gamma} + Z_1 + e_1 + H = \frac{U_2^2}{2g} + \frac{p_2}{\gamma} + Z_2 + e_2 - E_m$$

Since $U_1 = U_2$ and $Z_1 = Z_2$, $\left(\frac{U_1^2}{2g} + Z_1\right) = \frac{U_2^2}{2g} + Z_2$.

Further since the temperature of water remains the same, $e_1 = e_2$, and $H = -400$ J/N.

$$\therefore \quad E_m = \frac{p_2 - p_1}{\gamma} + 400$$

$$= \frac{(10{,}000 - 75) \times 10^3}{9787} + 400$$

$$= 1014.1 + 400 = 1414.4 \text{ J/N}$$

$\therefore$ Energy added by pump $= 1414.4 \times 0.3 \times 9787$

$$= 4{,}152{,}820 \text{ W} \quad \text{or} \quad 4152.82 \text{ kW}$$

3.18 Discuss the characteristics of flow obtained by superposition of radial inward flow on free vortex flow.

For radial inward flow, $V_r \times r =$ constant, A since $Q = 2\pi r V_r$. For free vortex flow $V_\theta . r =$ constant B.

Therefore, Resultant velocity

$$V = \sqrt{V_r^2 + V_\theta^2} = \left(\frac{A^2}{r^2} + \frac{B^2}{r^2}\right)^{1/2} = \frac{1}{r}\sqrt{A^2 + B^2}$$

Also angle between radius vector and stream line will be

$$\tan \alpha = \frac{V_\theta}{V_r} = \frac{B}{r}\frac{r}{A} = \frac{B}{A}$$

The streamlines will therefore form equiangular spirals as shown in Fig. 3.15. This type of flow is obtained in large scale tornadoes in which low pressure in the centre causes extensive damage. This is known as free spiral vortex.

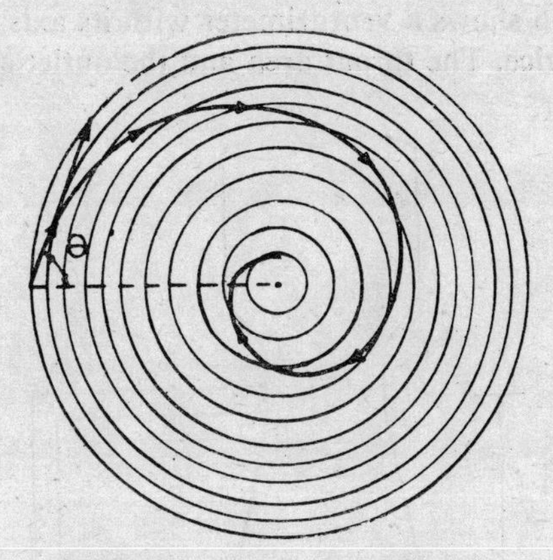

Fig. 3.15. Free spiral vortex.

3.19 A gas flows steadily in a pipe of constant area in such a way that the pressure and temperature at the upstream section are 703 kN/m² absolute and 510°C. At the downstream section, the pressure and temperature are 534 kN/m² and 305°C respectively. The mass densities at the two sections are 4.00 kg/m³ and 4.15 kg/m³ respectively. Take $C_v = 784.5 \cdot \dfrac{\text{Cal}}{\text{NK}}$ and determine the magnitude and transfer of heat assuming no shaft work, the change in velocity to be negligible and elevations of two sections to be same.

According to the energy equation:

$$\frac{U_1^2}{2g} + Z_1 + \frac{p_1}{g\rho_1} + e_1 + H = \frac{U_2^2}{2g} + Z_2 + \frac{p_2}{g\rho_2} + e_2 + E_m$$

As given in the problem $Z_1 = Z_2$, $U_1 \approx U_2$ and $E_m = 0$.
Hence

$$e_1 + \frac{p_1}{g\rho_1} + H = e_2 + \frac{p_2}{g\rho_2}$$

or

$$H = (e_2 - e_1) + \left(\frac{p_2}{g\rho_2} - \frac{p_1}{g\rho_1}\right)$$

However, $(e_2 - e_1) = C_v (T_2 - T_1)$ Cal where T_2 and T_1 are absolute temperatures.

$$T_1 = 273.16 + 510.00 = 783.16\text{K} \quad \text{and} \quad T_2 = 273.16 + 305$$
$$= 578.16 \text{ K}$$

$$H = 784.5\,(578.16 - 783.16) \times 4.187 + \left(\frac{534}{4.15} - \frac{703}{4.0}\right) \times \frac{10^3}{9.806}$$

$$= -673\,363.38 - 4800.66$$

$$= -678164.04 \text{ J/N} \quad \text{or} \quad -678.164 \text{ kJ/N}$$

3.20 Figure 3.16 shows a venturimeter with its axis vertical and arranged as a suction device. The throat area and the outlet area of the venturi are

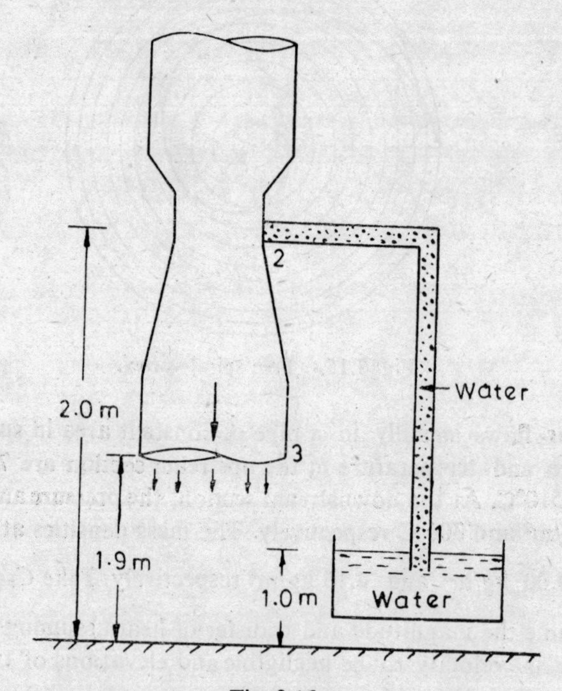

Fig. 3.16.

0.000 25 m^2 and 0.001 00 m^2 respectively. If the venturi discharges into the atmosphere, determine the minimum discharge in the venturi at which flow will occur up the suction pipe.

Apply Bernoulli's equation between sections 2 and 3

$$Z_2 + \frac{p_2}{\gamma} + \frac{U_2^2}{2g} = Z_3 + \frac{p_3}{\gamma} + \frac{U_3^2}{2g}$$

However according to continuity equation

$$0.00025\, U_2 = 0.001\, U_3 \quad \text{or} \quad U_2 = 4U_3$$

and since pressure at 3 is atmospheric, take $p_3 = 0$

$$2.0 + \frac{p_2}{\gamma} + \frac{16U_3^2}{2g} = \frac{U_3^2}{2g} + 1.9$$

$$\therefore \quad \frac{15\, U_3^2}{2g} = \left(-0.1 - \frac{p_2}{\gamma}\right)$$

Further, since water will rise to section 2

$$\frac{p_1}{\gamma} = \frac{p_2}{\gamma} + (Z_2 - Z_1) \quad \text{and} \quad p_1 = 0$$

$$\therefore \quad \frac{p_2}{\gamma} = -(Z_2 - Z_1) = -1.0$$

$$\therefore \quad \frac{15\, U_3^2}{2g} = +0.90 \quad \text{or} \quad U_2 = \sqrt{1.177} = 1.085 \text{ m/s}$$

$$\therefore \quad Q = 1.085 \times 0.001 = 0.001085 \text{ m}^3\text{/s} \quad \text{or} \quad 1.085 \text{ l/s}$$

3.21 Gaseous products of combustion enter a turbojet engine at a velocity and temperature of 25 m/s and 1000 K, and leave the turbine exit at 80 m/s and 700 K. If the mass flow rate through turbine is 40 kg/s, determine the power generated by turbine machine under steady flow conditions. Assume process to be adiabatic, gas to behave as perfect gas with $C_p = 122.37$ J/NK so that the enthalpy can be written as $C_p\,(T_2 - T_1)$.

$$\frac{U_1^2}{2g} + \frac{p_1}{\gamma} + Z_1 + e_1 + H = \frac{U_2^2}{2g} + \frac{p_2}{\gamma} + Z_2 + e_2 + E_m$$

$$Z_1 = Z_2, \text{ enthalpy change } \left(\frac{p_2}{\gamma} + e_2\right) - \left(\frac{p_1}{\gamma} + e_1\right) \text{ i.e.}$$

$h_2 - h_1 = C_p\,(T_2 - T_1)$, (Eq. 14.15). Further, $Z_1 = Z_2$ and for adiabatic process $H = 0$

$$\frac{25^2}{2g} + h_1 = \frac{80^2}{2g} + h_2 + E_m \quad \text{and} \quad h_2 - h_1 = C_p\,(T_2 - T_1)$$

$$\therefore \quad E_m = \frac{(-6400 + 625)}{9.806 \times 2} - 122.37\,(973.16 - 1273.16)$$

$$= -294.462 - 37078.11 = -37372.57 \text{ N/m}$$

Power output of turbine

$$P = \frac{37\,372.57 \times 40 \times 9.806}{1000} = 14\,659 \text{ kW}$$

PROBLEMS

3.1 In a converging pipe which is inclined upwards, the velocity U varies along its length as $U = 1.5 + 1.25\,S$ where S is measured along the pipe. If the pipe rises 0.5 m in 2 m length, determine the pressure gradients at the entrance and exit.

(– 4.318 kN/m² m and – 7.437 kN/m² m)

3.2 If a 300 mm diameter pipe carrying 0.212 m³ discharge of oil of relative density 0.8 has 200 kN/m² pressure at a section 10 m above the dam, determine total energy per unit mass of fluid.

(353.06 Nm/kg)

3.3 If velocity distribution in a pine is given by

$$u/u_m = \left(\frac{y}{R}\right)^{1/7}$$

where u is the velocity at a distance y from the boundary and u_m is the centre line velocity at $y = R$, R being the pipe radius, determine the of average velocity and α.

$\left(\frac{49}{60}\,u_m,\ 1.06\right)$

3.4 A pipe nozzle has the inlet diameter of 100 mm and outlet diameter of 50 mm, the difference in elevations of outlet and inlet being 0.25 m. If the inlet pressure is 15 kN/m² above the atmospheric pressure and the jet is discharged into atmosphere, determine the discharge.

(10.167 1/s)

3.5 The velocity distribution in laminar boundary layer over a flat plate is given by

$$\frac{u}{U_0} = \frac{3}{2}\left(\frac{y}{\delta}\right) - \frac{1}{2}\left(\frac{y}{\delta}\right)^3$$

where u is the velocity of distance y from the boundary and U_0 is the free stream velocity at $y = \delta$. Determine α. (1.677)

3.6 If the velocity distribution in an open channel with depth D follows a triangular profile i.e. zero velocity at the bed and velocity u_m at the water surface, what will be the value of α? (2.0)

3.7 If the velocity distribution in a pipe follows the law

$$u/u_m = \left(1 - \frac{r}{R}\right)^n,$$

obtain the expressions for mean velocity and α.

$$\left(2u_m/(n+1)(n+2),\ \frac{(n+1)^3(n+2)^3}{4(3n+1)(3n+2)}\right)$$

3.8 Velocity distribution in a 300 mm diameter pipe was measured with the help of pitot tube by dividing the pipe area into six equal areas by means of concentric circles. The velocities in these areas beginning at the centre of the pipe are (in m/s)
2.00, 1.91, 1.81, 1.68, 1.52, 1.43. Find the average velocity and α.
(1.725 m/s, 1.041)

3.9 A turbine is connected to a reservoir through a 300 mm diameter pipe. The water level in the reservoir is 40 m above the centre line of the turbine. Turbine discharges into atmosphere through a nozzle of 150 mm diameter attached to a 300 mm outlet pipe. If the water is discharged through the nozzle at velocity of 15 m/s, determine the energy developed by the turbine taking overall turbine efficiency as 80 per cent. (59.189 kW)

3.10 For a sudden expansion shown in Fig. 3.17

$Q = 0.283$ m³/s

$D_1 = 300$ mm, $D_2 = 600$ mm

$p_1 = 150$ kN/m² and head loss in expansion is given by

$p_2 = (U_1 - U_2)^2/2g$. Find p_2. (133.417 kN/m²)

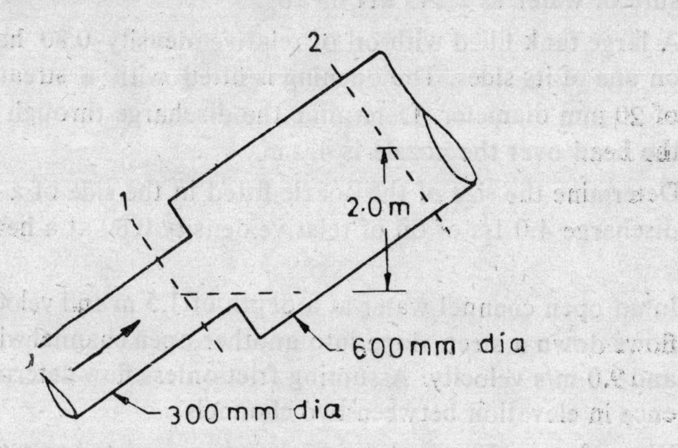

Fig. 3.17.

3.11 Water flows through a 90° bend cum diffuser at a discharge of 0.30 m³/s. The conditions at the two ends A and B are as follows :

Section A	Diameter 500 mm	Pressure 125 kN/m² $Z_A = 13$ m
Section B	300 mm	154 kN/m²; $Z_B = 10$ m

Determine the direction of flow and head loss.
(From B to A, 0.763 m)

3.12 Water at 20°C flows from sump to the pump at a velocity of 2.0 m/s through the suction pipe. Calculate the maximum height upto the centre line of the pump to which water can be lifted by the pump if the atmospheric pressure is 101.3 kN/m^2 and the vapour pressure is 0.348×10^3 N/m^2. Assume friction loss in the suction pipe to be 1.0 m. (9.110 m)

3.13 In the turbine in Ex. 3.10 if $p_A = 210$ kN/m^2, $p_B = -40$ kN/m^2 and the power developed is 100 kW, find the discharge passing through the turbine. (0.372 m^3/s)

3.14 Water is kept in a partially filled closed tank under pressure ***p***. A 50 mm hose taken out from the tank discharges water at 3.0 m/s velocity in the atmosphere 8 m above the water level in the tank. Assuming friction and other losses in the pipe to be 1.1 m, determine the pressure in the tank. (93.554 kN/m^2 gauge)

3.15 When the pressure at any section in a flowing liquid reduces to vapour pressure of the liquid at the liquid temperature, the liquid vapourises forming bubbles. These bubbles collapse in the region of high pressure. This phenomenon is known as cavitation. If a pipe of 200 mm diameter carrying 62.8 l/s of water at 20°C and 125 kN/m^2 gauge pressure is gradually contracted, determine the diameter of the constriction at which cavitation may start. Assume that the constriction takes place at 7.0 m elevation from the pipe. Take vapour pressure of water as 2.345 kN/m^2 abs. (67.12 mm)

3.16 A large tank filled with oil of relative density 0.80 has an opening on one of its sides. The opening is fitted with a streamlined nozzle of 20 mm diameter. Determine the discharge through the nozzle if the head over the nozzle is 4.0 m. (2.78 l/s)

3.17 Determine the size of the nozzle fitted in the side of a tank if it is to discharge 4.0 l/s of oil of relative density 0 80 at a head of 5.0 m. (22.68 mm)

3.18 In an open channel water at a depth of 1.5 m and velocity of 3.0 m/s, flows down a steep slope into another open channel with 0 5 m depth and 9.0 m/s velocity. Assuming frictionless flow determine the difference in elevation between two channels. (2.67 m)

3.19 Water from a 30 mm diameter nozzle connected to a fire hose must reach at a window 30 m above the ground level (Fig. 3.18). If the nozzle is held 2 m above the ground level determine the maximum distance x from the building where the fireman can stand and achieve his objective if the hose carries a discharge of 22.0 l/s. Also determine the angle θ of inclination of the nozzle.
(Hint: see Fig. 3.18; obtain an expression for x as a function of U_0 and t, and find value of x when $\frac{dx}{dt} = 0$). (65.116 m, 56.76°)

3.20 If a pump imparts 20 kW power to the water being pumped at the rate of 0.10 m^3/s, determine pressures at A and B (Fig. 3.19). (19.408 kN/m^2, 154.644 kN/m^2)

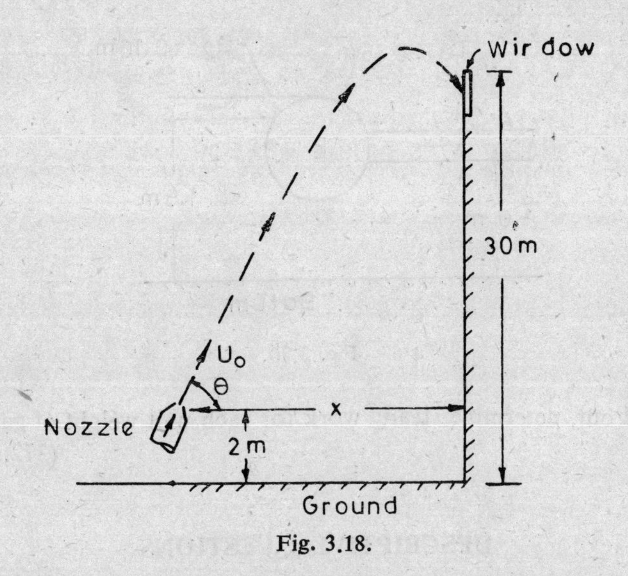

Fig. 3.18.

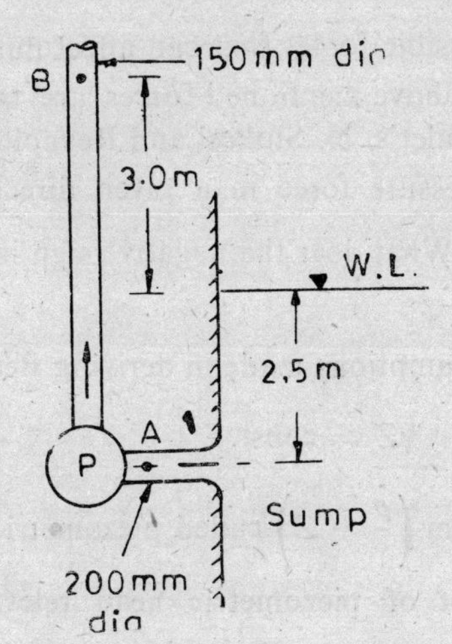

Fig. 3.19.

3.21 Water is pumped through the system shown in Fig. 3 20, at the rate of 0.10 m³/s. Neglect losses and calculate the power required by the pump. Take $p_1 = 20$ kN/m² and $p_2 = 195$ kN/m². (24.488 kW)

3.22 Steady flow through a compressor gave the following data:

$p_1 = 100$ kN/m², $p_2 = 700$ kN/m², $t_1 = 20$°C, $t_2 = 170$°C

$U_1 = 60.0$ m/s, $U_2 = 13.0$ m/s, $\rho_1 = 0.832$ kg/m³

$\rho_2 = 0.192$ kg/m³, $e_1 = 20\,276$ J/N, $e_2 = 31\,942$ J/N

If the outlet is 3.0 m above the inlet and 3800 J/N heat is transfer-

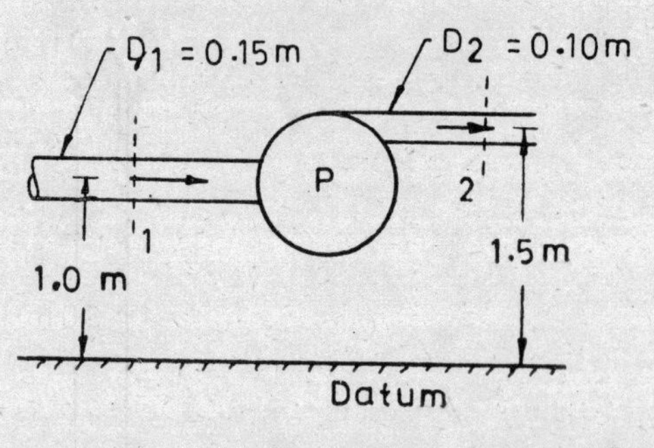

Fig. 3.20.

red out, determine steady work for each **unit weight** of gas.

(372.660 kJ/N)

DESCRIPTIVE QUESTIONS

3.1 List all the possible forces that can affect fluid motion.

3.2 Which of the above mentioned forces are taken into consideration in obtaining Euler's, N. Stokes' and Reynolds' equations of motion.

3.3 Show that pressure force in a given direction for unit volume of fluid is $-\dfrac{\partial p}{\partial x}$. What does the negative sign imply? What will (grad ϕ) mean?

3.4 List all the assumptions made in deriving Bernoulli's equation in the form $\dfrac{\rho V_s^2}{2} + p + \gamma Z = \text{const.}$

3.5 Why is the term $\left(\dfrac{p}{\gamma} + Z\right)$ called piezometric head?

3.6 Is the concept of piezometric head relevant for flow of gases? Explain.

3.7 Laminar flow takes place in a horizontal pipe at a constant discharge. How will total head vary from one stream line to another in the same section? Why?

3.8 Under what conditions can you apply Bernoulli's equation across different stream lines?

3.9 Four different types of piezometer's openings are shown below in Fig. 3.21. Which will measure the correct static pressure? In the other three cases state whether the measurement error will be + ve or – ve.

3.10 Bernoulli's equation in the form $\dfrac{U^2}{2g} + \dfrac{p}{\gamma} + Z = \text{constant}$, is derived under the assumption of incompressible fluid. Can one apply it for

(i) air flow through ventilation system (ii) motion of a car in atmosphere at 100 km/hr.

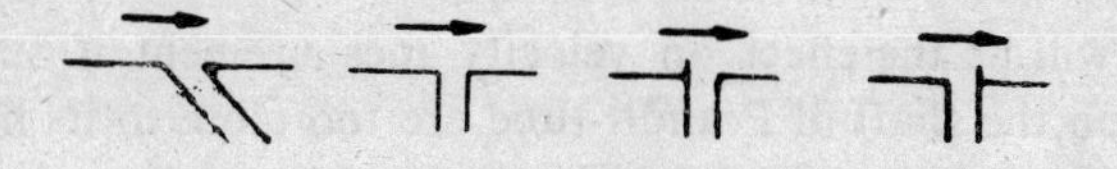

Fig. 3.21.

3.11 Why is pressure head a form of energy?

3.12 How is a venturimeter or orifice meter calibrated in a laboratory? Why is calibration required?

3.13 How is pitot tube calibrated?

3.14 In a flowing fluid system, indicate the conditions under which energy between two sections (i) will increase, (ii) will decrease.

3.15 Write approximate values of energy correction factors for 2-dimensional velocity distributions shown in Fig. 3.22.

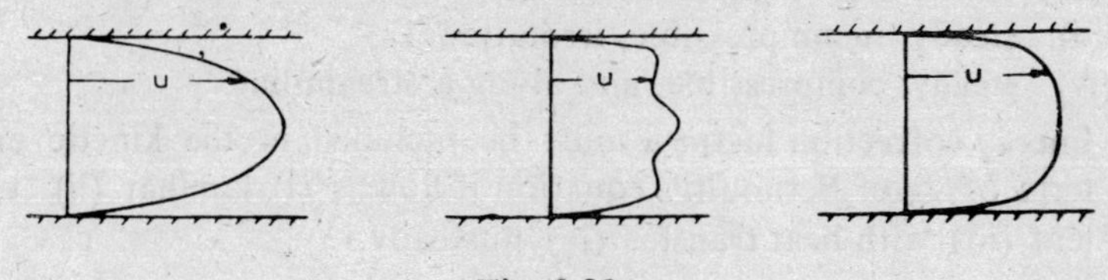

Fig. 3.22.

3.16 If a choice is given, which of the two devices, namely orifice meter and venturimeter will you use for flow measurement in a pipe? Why?

3.17 Can the flow occur from low pressure to high pressure region? Give an example.

3.18 Is the device shown in Fig. 3.23 a pump or a turbine? Explain.

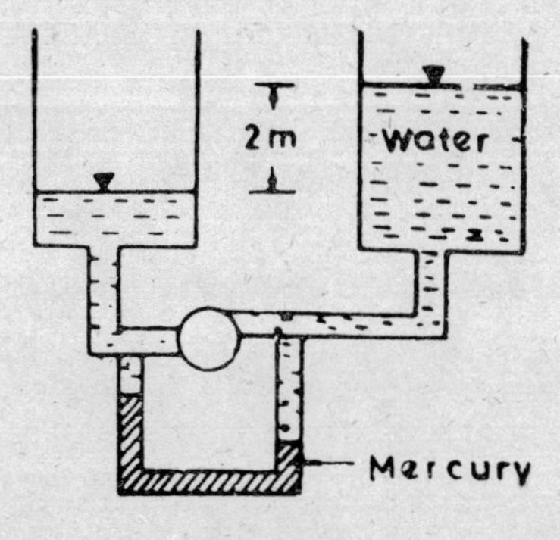

Fig. 3.23.

3.19 Consider a horizontal pipe in which a circular orifice is placed. Plot pressure distribution along the pipe wall and along the centre line for some distance upstream and some distance downstream of the orifice.

3.20 What will be the effect on velocity measurement if static pressure holes on the shaft of Prandtl-tube are too close to its head?

3.21 What are the undesirable effects of cavitation? How can the occurrence of cavitation be avoided?

3.22 Each term in the equation $p + \gamma Z + \frac{\rho U^2}{2} =$ constant, represents (i) force per unit volume (ii) force per unit weight (iii) energy per unit volume (iv) energy per unit weight.

3.23 Euler's equation of motion is a statement expressing (i) conservation of mass (ii) conservation of energy (iii) Newton's first law of motion (iv) Newton's second law of motion.

3.24 If in a flow field $p/\gamma + Z + U^2/2g =$ constant between any two points, flow must be

(i) steady, compressible, irrotational
(ii) unsteady, incompressible and irrotational
(iii) steady, incompressible, irrotational
(iv) steady, compressible, and along a stream line.

3.25 Energy correction factor α must be included in the kinetic energy term $U^2/2g$ of Bernoulli's equation if flow is (i) laminar (ii) turbulent (iii) with heat transfer (iv) unsteady.

3.26 Bernoulli's equation $P/\gamma + Z + U^2/2g =$ const. can be taken as nearly correct if the flow is (i) rapidly diverging (ii) rapidly converging (iii) of incompressible fluid (iv) steady.

3.27 Piezometric head is the sum of (i) elevation and K.E. heads (ii) elevation and pressure heads (iii) K.E. and pressure heads.

CHAPTER IV

Fluid Statics

4.1 FLUID PRESSURE

As defined earlier, the pressure p is given by $\lim_{\Delta A \to 0} \frac{\Delta F}{\Delta A}$, where ΔF is the force acting on area ΔA in the perpendicular direction to the plane of area. It is measured in N/m^2, pascal, kN/m^2 or bar.

One pascal $= 1$ N/m^2

1 bar $= 10^5$ N/m^2 or 100 kN/m^2

One bar is approximately equal to the atmospheric pressure at sea level (which is 101.325 kN/m^2). Pressure acts perpendicular to the surface. According to Pascal's law, pressure at a point in a static fluid is the same in all the directions.

For fluids that are incompressible and homogeneous (i.e. for which γ is constant)

$$p_2 - p_1 = \gamma h$$

or

$$\Delta p = \gamma h \tag{4.1}$$

where Δp is the difference of pressure between two points whose elevations differ by h. This is known as the *hydrostatic variation*. The term $\left(\frac{p}{\gamma} + Z\right)$, where Z is elevation above the datum, is known as the piezometric head. In static fluid $\left(\frac{p}{\gamma} + Z\right)$ is constant (Fig. 4.1). Various datums of pressure are shown in Fig. 4.2.

4.2 PRESSURE VARIATION IN STATIC COMPRESSIBLE FLUID

Under isothermal condition the pressure variation with elevation y above datum is given by

$$p = p_0 \left[\exp - (\gamma_0/p_0)y\right] \tag{4.2}$$

and for adiabatic condition,

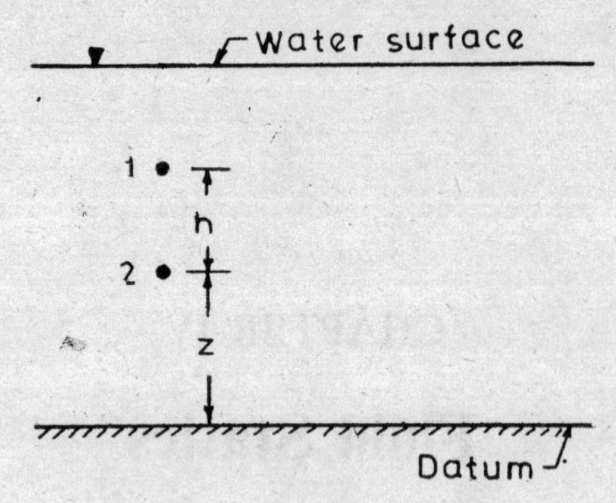

Fig. 4.1. Definition sketch.

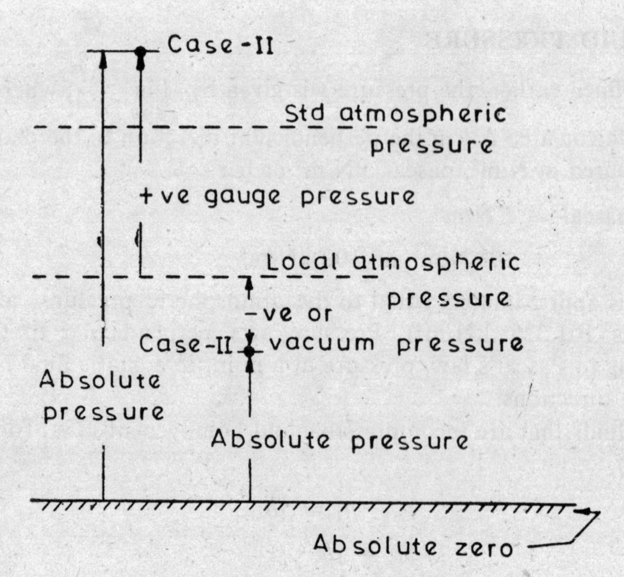

Fig. 4.2. Various pressure datums.

$$p = p_0 \left\{1 - \frac{k-1}{k} \frac{y}{(p_0/\gamma_0)}\right\}^{\frac{k}{k-1}} \tag{4.3}$$

where p_0 and γ_0 are the pressure and unit weight of fluid at $y = 0$. For the pressure variation existing in the atmosphere $k = 1.20$.

4.3 MANOMETERS

Manometer is a device used for determining pressure in a fluid by balancing it against a column of liquid in static equilibrium. It measures pressure at a point under condsideration with respect to the atmospheric pressure. By using U-tube type manometer, pressure difference between two points can be measured. Different types of manometers are shown in Fig. 4.3. Problems on manometers can be solved by following the steps given below:

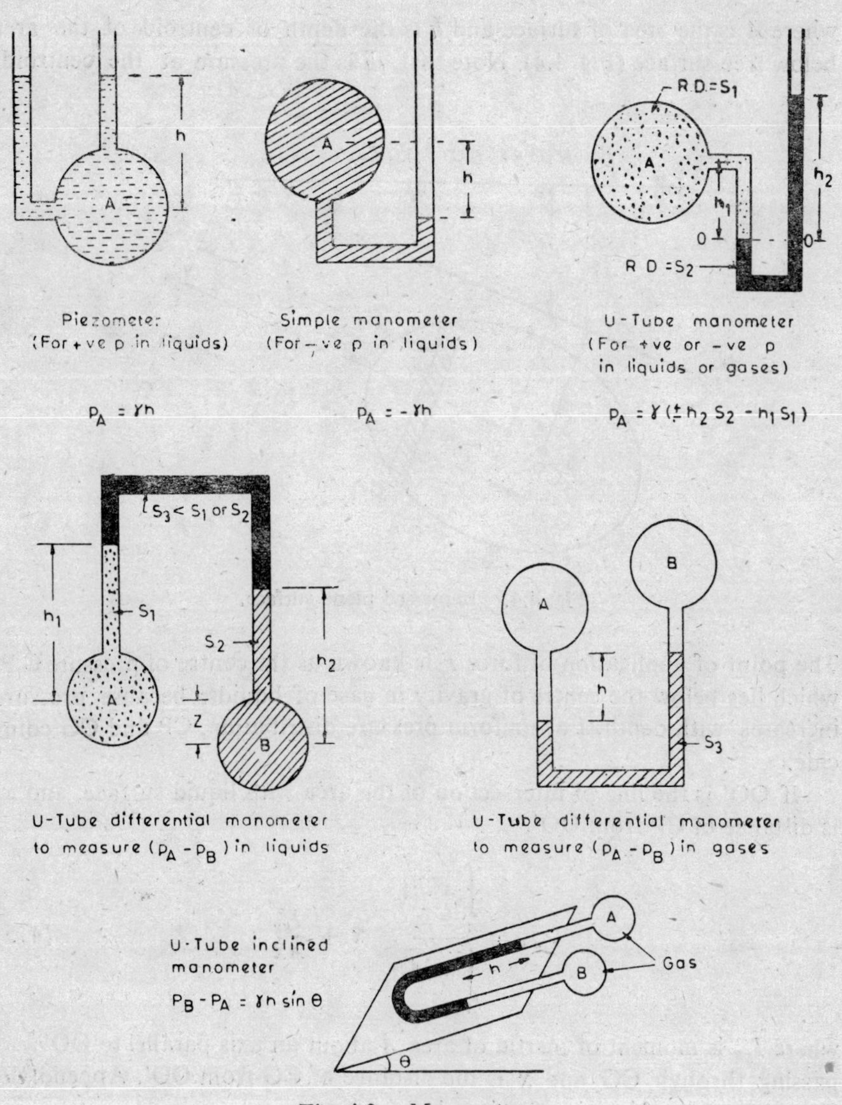

Fig. 4.3. Manometers.

—Start from one end and note the pressure there.

—Add change in the pressure from that end to the meniscus if there is lowering in elevation; subtract it if there is rise in elevation.

—Reach the other end of the gauge and equate the result with pressure there.

—Solve for the unknown.

4.4 FORCES ON IMMERSED PLANE SURFACES

Total force F on a plane surface immersed in a liquid is given by

$$F = \gamma A\bar{h} \tag{4.4}$$

where A is the area of surface and $\bar{h}$ is the depth of centroid of the area below free surface (Fig. 4.4). Note that $\gamma\bar{h}$ is the pressure at the centroid.

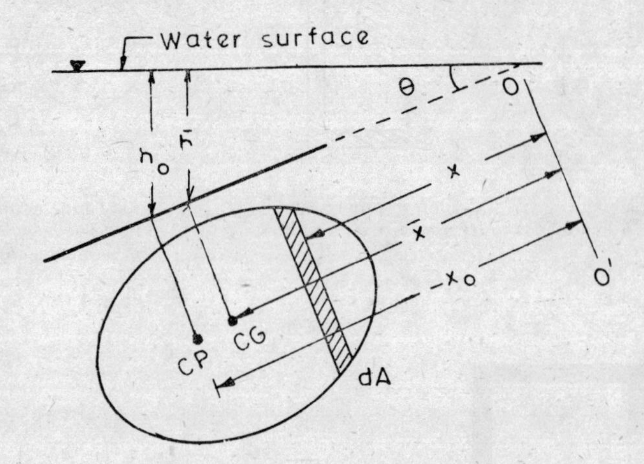

Fig. 4.4. Immersed plane surface.

The point of application of force F is known as the centre of pressure C.P. which lies below the centre of gravity in case of liquids, because pressure increases with depth. For uniform pressure distribution, CP and CG coincide.

If OO′ is the line of intersection of the area with liquid surface, and x_0 is distance of CP from OO′,

$$x_0 = \frac{\int_A x^2 dA}{\int_A x dA} = \bar{x} + \frac{I_{gg}}{A\bar{x}} \tag{4.5}$$

where I_{gg} is moment of inertia of area A about an axis parallel to OO′ and passing through CG and $\bar{x}$ is the distance of CG from OO′. Appendix *F* gives moments of inertia of some plane surfaces.

4.5 TOTAL FORCE ON IMMERSED CURVED SURFACES

Evaluation of total force R on a curved immersed surface can be made by determining the horizontal component R_x and the vertical component R_y of this force.

As regards the curved surface ABC, principles of statics yield the following relations:

$$\Sigma F_x = F_x - R_x = 0 \tag{4.6a}$$

$$\Sigma F_y = F_y + W - R_y = 0 \tag{4.6b}$$

Σ moments of all forces about axis passing through O = 0 (4.6c)

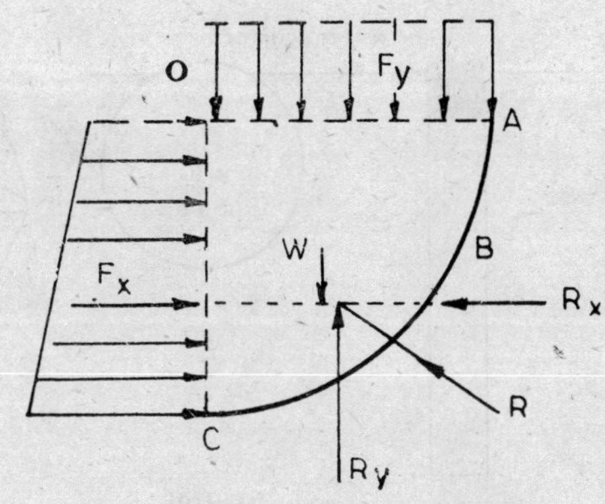

Fig. 4.5. Forces on a curved immersed surface.

Solution of these equations will give R_x, R_y and point of application of resultant force R. It can be seen from Eq. (4.6b) that R_y, the vertical component of the force on the curved surface is equal to weight of water above it. One can also consider a free body of fluid *AOC* and consider external force on it and write the condition of equilibrium. In that case R_x and R_y will be the components of the force exerted by the surface *ABC* on the free body. Such analysis leads to the same equations viz. Eqs. 4.6 a, b and c.

4.6 FLOATING BODIES

Basic principles of buoyancy and floation were first established by Archimedes (287–212 B.C.). These principles can be stated as follows. A body completely immersed in a fluid is acted upon by an upward buoyant force equal to the weight of fluid displaced by the body and it acts through the centre of gravity of displaced fluid. When the body is floating in a liquid, the weight of the body is equal to the buoyant force on the immersed part of the body. Therefore,

$$\text{Buoyant force } F_v = \gamma \forall \qquad (4.7)$$

where $\forall$ is the volume of immersed part of the body and γ is the specific weight of the liquid.

4.7 STABILITY OF FLOATING AND SUBMERGED BODIES

A body floating in a static liquid has vertical stability, i.e. a small vertical displacement results in an unbalanced force which tends to bring back the body to its original position. If a floating body is given a small angular displacement, a couple is set up which may tend to bring the body to its original position resulting in stable equilibrium, or can cause further angular displacement in the same direction resulting in an unstable equilibrium, or a couple may not be formed at all resulting in neutral equilibrium. A submerged body is rotationally stable if its centre of gravity is below the centre of buoyancy (Fig. 4.6).

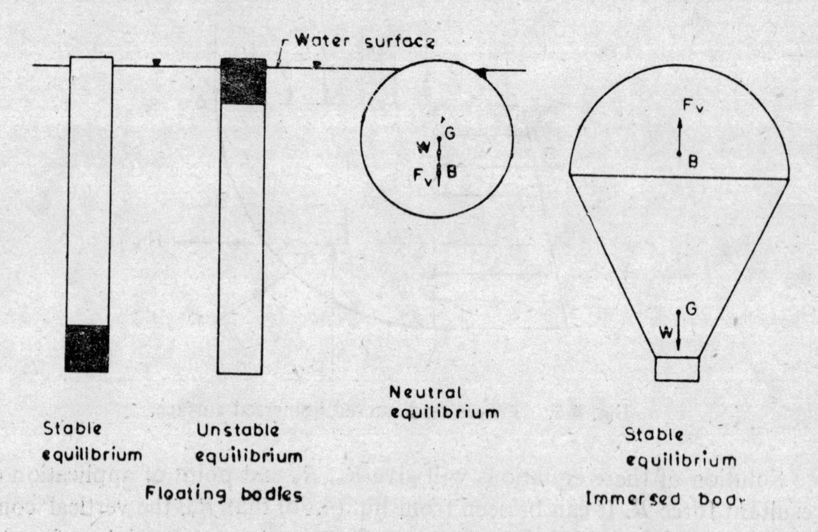

Fig. 4.6. Equilibrium of floating and immersed bodies.

The centre of curvature of a curve giving the locus of the centre of buoyancy obtained for small angular displacements of a floating body is known as the metacentre. The vertical distance between the centre of gravity G of the body and the metacentre M is known as the metacentric height. For a floating body to have rotational stability, M must lie above G. The distance MG is given by

$$MG = \frac{I}{\forall} - BG \tag{4.8}$$

where I is the moment of inertia of the area of water section $ABCD$ (Fig. 4.7) about the axis OO′ passing through O, and BG is the vertical distance between centre of gravity G and centre of bouyancy B. $\forall$ is the

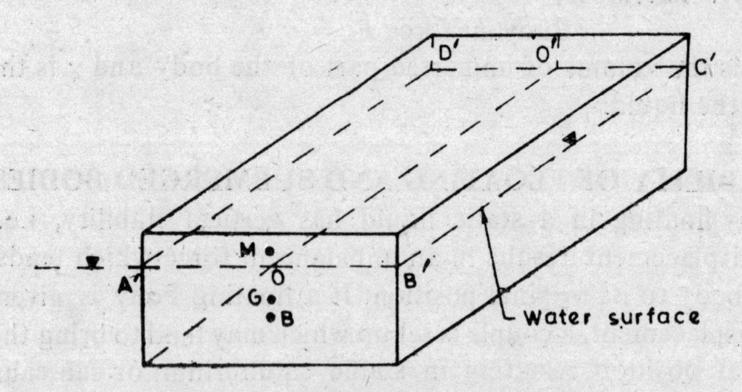

Fig. 4.7. Stability of floating body.

volume of immersed portion of the body. When a small angular displacement is given to the floating body it oscillates with a time period

$$T = 2\pi\sqrt{k^2/gMG}$$

where k is radius of gyration, of mass of floating body about axis oo′

4.8 FLUID MASS SUBJECTED TO UNIFORM ACCELERATION

If a liquid mass is subjected to a uniform horizontal acceleration a_x, its free surface will be inclined in the direction of motion at a slope of a_x/g. If a closed container filled with a fluid is given such an acceleration, the pressure gradient in the fluid in the direction of motion will be a_x/g.

If a liquid mass is given a vertical acceleration $\pm a_y$, the pressure difference between two points with elevation difference of h, is given by

$$(p_2 - p_1) = \gamma h \left(1 \pm \frac{a_y}{g}\right) \tag{4.9}$$

If a liquid mass with a free surface is rotated about a vertical axis (Fig. 4.8) at an angular velocity ω, the free surface is given by

$$Z = Z_0 + \frac{r^2\omega^2}{2g} \tag{4.10}$$

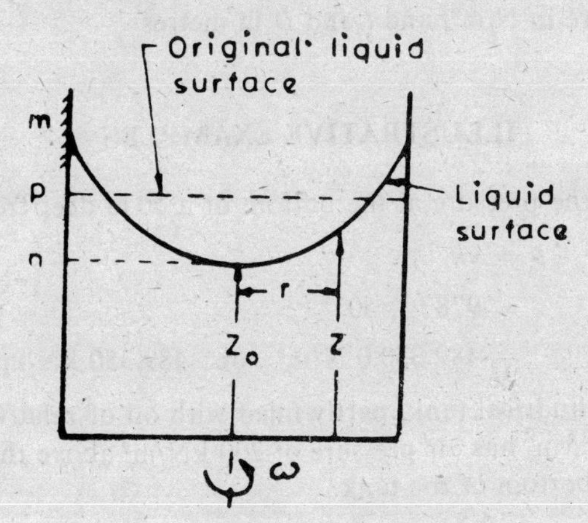

Fig. 4.8. Rotation at constant angular velocity.

This is the equation of paraboloid of revolution. Further $mn = 2\,np$. If a container is completely filled with a fluid and rotated about the vertical axis at an angular velocity ω, the pressure rise at a distance r from the axis of rotation is $\rho\omega^2 r^2/2$. If the closed cylinder is filled with gas and rotated about vertical axis, the variation of pressure along radial direction can be obtained on combining the dynamic equation $\partial p/\partial r = \rho r\omega^2$ with either

$$p/\rho = p_0/\rho_0 \quad \text{or} \quad p/\rho^k = p_0/\rho_0^k$$

as the case may be and integrating the resulting differential equation.

4.9 STRESSES IN PIPES DUE TO INTERNAL PRESSURE

When a pipe is subjected to a high inside pressure p, it will produce both longitudinal and circumferential stresses in it, S_L and S_c respectively. For a pipe of internal diameter D, thickness t and length l, they will be

$$2tL\, S_c = pDL$$

$$\text{or} \quad S_c = \frac{pD}{2t} \tag{4.10}$$

$$\text{and} \quad p\frac{\pi D^2}{4} = \pi\,(D + t)\, t\, S_L$$

$$\text{or} \quad S_L \approx \frac{pD}{4t} \tag{4.11}$$

Since $S_L = \frac{1}{2} S_c$, pipe is designed for circumferential stress. The thickness of the pipe of diameter D which will withstand the pressure p is related to allowable stress S by the equation

$$\frac{p}{S} = \frac{2\left(\dfrac{7}{8}\dfrac{t}{D} - \dfrac{0.0025}{D}\right)}{1 - 0.8\left(\dfrac{7}{8}\dfrac{t}{D} - \dfrac{0.0025}{D}\right)} \tag{4.12}$$

where S and p are in N/m², and t and D in metres.

ILLUSTRATIVE EXAMPLES

4.1 Determine the pressure at the bottom of a 50 m deep fresh water lake.

$$p = \gamma h$$
$$= 9787 \times 50$$
$$= 489\,350.0 \text{ N/m}^2 \quad \text{or} \quad 489.350 \text{ kN/m}^2$$

4.2 A closed cylindrical tank, partly filled with oil of relative density 0.80 to a depth of 1.5 m, has air pressure of 200 kN/m² above the oil. Find the pressure on the bottom of the tank.

$$\text{Since } p = 200 + (\text{R.D.})\,(\gamma)\,(h)$$
$$p = 200 + (0.80 \times 9.787 \times 1.5)$$
$$= 211.744 \text{ kN/m}^2$$

4.3 If atmosphere follows isothermal conditions, determine the density and pressure at an elevation of 2000 m above sea level if the mass density of air at atmospheric pressure of 100.325 kN/m² at sea level is 1.208 kg/m³.

For isothermal conditions p is given by

$$p = p_0 \exp\left[-(\gamma_0/p_0)\, y\right]$$

$$\text{Hence } p = 101.325 \exp\left(\frac{-(2000 \times 9.806 \times 1.208)}{101.325 \times 1000}\right)$$

$$= 80.200 \text{ kN/m}^2$$

But $\frac{p_0}{\rho_0} = \frac{p}{\rho}$ $\therefore$ $\rho = \frac{80.200 \times 1.208}{101.325}$

$$= 0.956 \text{ kg/m}^3$$

4.4 What will be the mass density of atmospheric air at the top of a mountain 8800 m above the sea level? Take $p_0 = 101.325$ kN/m², $\rho_0 = 1.208$ kg/m³ and $k = 1.20$.

For adiabatic condition p is given by

$$p = p_0 \left[1 - \frac{k-1}{k} \frac{y}{(p_0/\gamma_0)}\right]^{k/k-1}$$

$$p = 101.325 \left[1 - \frac{0.20}{1.20} \frac{8800}{\left(\frac{101.325 \times 1000}{9.806 \times 1.208}\right)}\right]^{\frac{1.20}{0.20}}$$

$$= 101.325 \times 0.8285^{6.0}$$

$$= 32.778 \text{ kN/m}^2$$

$$\rho = \rho_0 \left(\frac{p}{p_0}\right)^{1/k}$$

$$= 1.208 \left(\frac{32.778}{101.325}\right)^{0.8333}$$

$$= 0.472 \text{ kg/m}^3$$

4.5 Find the force required to lift a weight of 25.0 kN by means of a hydraulic press which has a ram of 200 mm and a plunger of 20 mm diameter. If the plunger has a stroke of 0.30 m and if it makes 20 strokes per minute, determine through what height is the weight lifted per minute? Also find the power required at the plunger.

The hydraulic press works on the principle of Pascal's law according to which the fluids transmit pressure equally in all directions (Fig. 4.9).

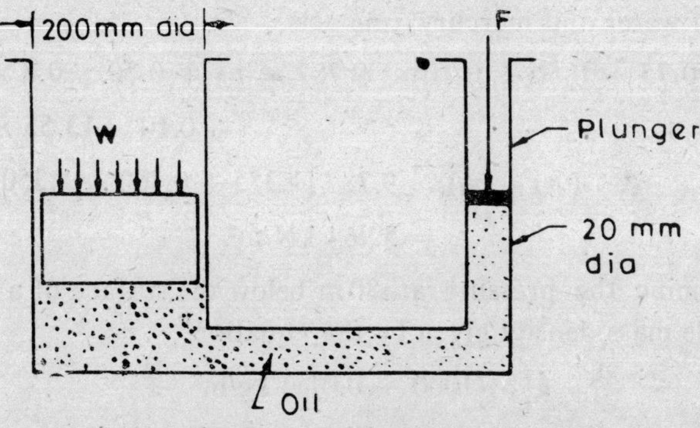

Fig. 4.9. Hydraulic press.

Pressure developed by ram $= W/A = \dfrac{25.00}{0.785 \times 0.20^2}$

$$= 796.178 \text{ kN/m}^2$$

which must be equal to Force/plunger area or $\left(\dfrac{W}{A} = \dfrac{F}{a}\right)$

$$\therefore \quad 796.178 = \frac{F}{0.785 \times 0.02^2} \quad \text{or} \quad F = 0.25 \text{ kN}$$

Thus a force of 0.25 kN is required to lift a weight of 25 kN.

Distance travelled by plunger in one minute $= 20 \times 0.30 = 6.0$ m

Distance through which the weight W will be moved $= \dfrac{6.0 \times 0.02^2}{0\ 20^2} = 0.06$ m

Power required $= \dfrac{0.06 \times 25.0}{60} = 0.0208$ kW

4.6 Determine the pressure difference $(p_A - p_B)$ in Fig. 4.10.

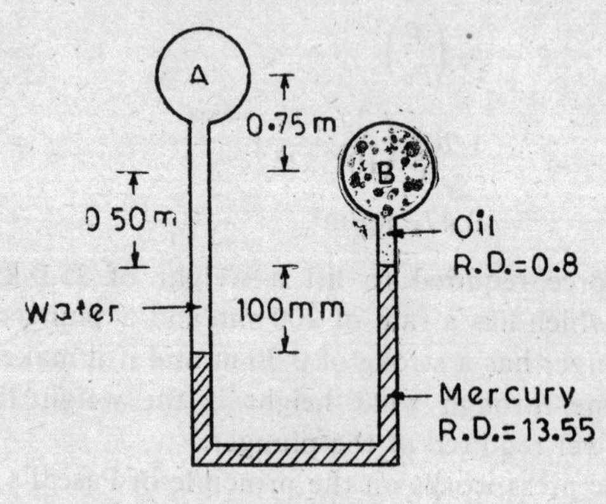

Fig. 4.10.

Equating the pressure in both limbs along the horizontal line at the interface of water and mercury, one gets,

$$p_A + (0.75 + 0.50 + 0.10) \times 9.787 = p_B + 0.50 \times 0.8 \times 9.787$$
$$+ 0.10 \times 13.55 \times 9.787$$

$$\therefore \quad (p_A - p_B) = 9.787\,[1.355 + 0.400 - 1.350]$$
$$= 3.964 \text{ kN/m}^2$$

4.7 Determine the pressure at 80 m below the surface of a liquid which has variable mass density given by the equation

$$\rho = (1000 + 0.01y) \text{ kg/m}^3$$

in which y is the distance measured below the free surface.

According to the condition for static equilibrium, if y is measured downwards from free surface

$$-\frac{\partial p}{\partial y} + \gamma = 0 \quad \text{or} \quad dp = \gamma dy$$

But $\qquad \gamma = \rho g = (1000 + 0.01\, y)\, g$

$$\therefore \quad dp = (1000 + 0.01\, y)\, g dy$$

Integration of above equation gives

$$p = \left(1000\, y + 0.01 \frac{y^2}{2}\right) g + C$$

where C is the constant of integration. Since gauge pressure p at the free surface (where $y = 0$) is zero, $C = 0$. Hence

$$p = \left(1000\, y + 0.01 \frac{y^2}{2}\right) g$$

$$\text{At} \quad y = 80, \; p = \left(1000 \times 80 + \frac{0.01 \times 80^2}{2}\right) \times 9.806$$

$$= 784\,793.792 \text{ N/m}^2 \quad \text{or} \quad 784.794 \text{ kN/m}^2$$

4.8 Determine $(p_A - p_B)$ and $(p_A - p_c)$ and express them in kN/m² (Fig. 4.11).

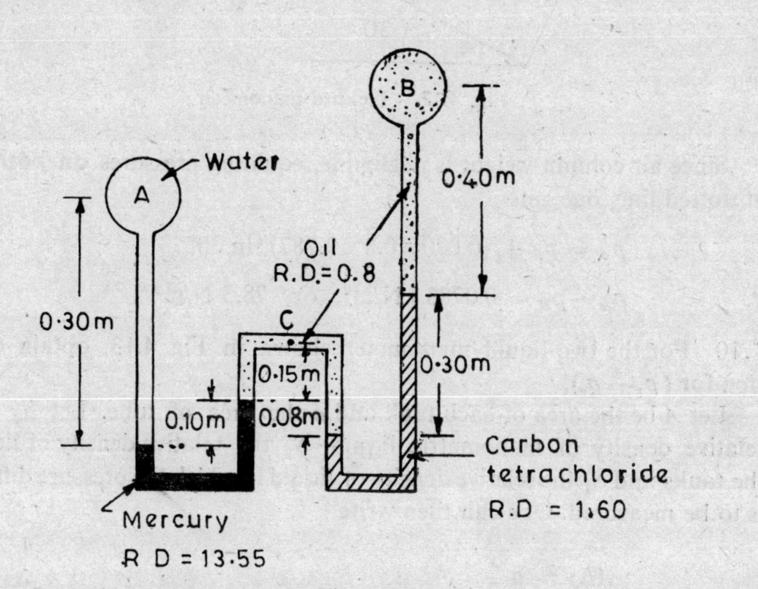

Fig. 4.11.

$$p_A + 0.30 \times 9.787 - 13.55 \times 0.10 \times 9.787$$
$$- 0.15 \times 0.80 \times 9.787 = p_c$$

$$(p_A - p_c) = 9.787\,(-0.30 + 1.355 + 0.120)$$
$$= 11.500 \text{ kN/m}^2$$

Similarly

$$p_A + 0.30 \times 9.787 - 13.55 \times 0.10 \times 9.787 + 0.08 \times 0.80 \times 9.787$$
$$- 0.30 \times 1.60 \times 9.787 - 0.40 \times 0.80 \times 9.787 = p_B$$

$$(p_A - p_B) = 9.787\,(-0.30 + 1.355 - 0.064 + 0.480 + 0.320)$$
$$= 9.787 \times 1.791$$
$$= 17.529 \text{ kN/m}^2$$

4.9 Determine $(p_A - p_B)$ in Fig. 4.12.

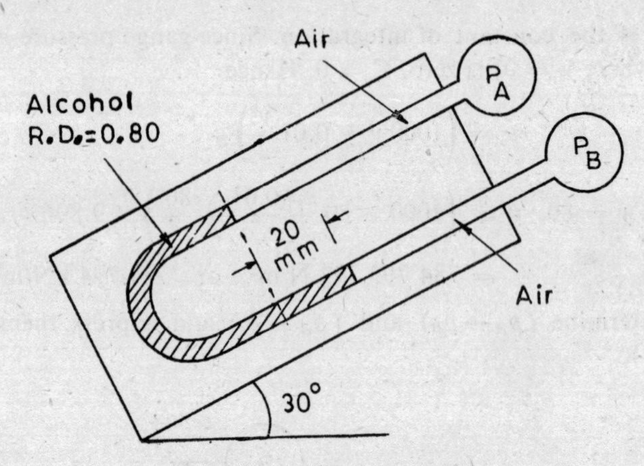

Fig. 4.12. Inclined manometer.

Since air column weight is negligible, equating pressures on both sides of dotted line, one gets

$$p_A = p_B + (0.020 \times 0.8 \times 9.787) \sin 30°$$

$$\therefore \quad p_A - p_B = 0.0783 \text{ kN/m}^2 \quad \text{or} \quad 78.3 \text{ N/m}^2$$

4.10 For the two-liquid manometer shown in Fig. 4.13, obtain expression for $(p_1 - p_2)$.

Let A be the area of each tank and a the area of tube. Let S_3 be the relative density of manometric liquid, S_2 the relative density of liquid in the tanks and S_1 the relative density of liquid in which the pressure difference is to be measured. One can then write

$$A\Delta y = a\,\frac{x}{2}$$

and

$$p_1 + \gamma S_1\,(x_1 + \Delta y) + \gamma S_2 \left(x_2 - \Delta y + \frac{x}{2}\right) - \gamma S_3 x$$
$$- \gamma S_2 \left(x_2 + \Delta y - \frac{x}{2}\right) - (x_1 - \Delta y) S_1 \gamma = p_2$$

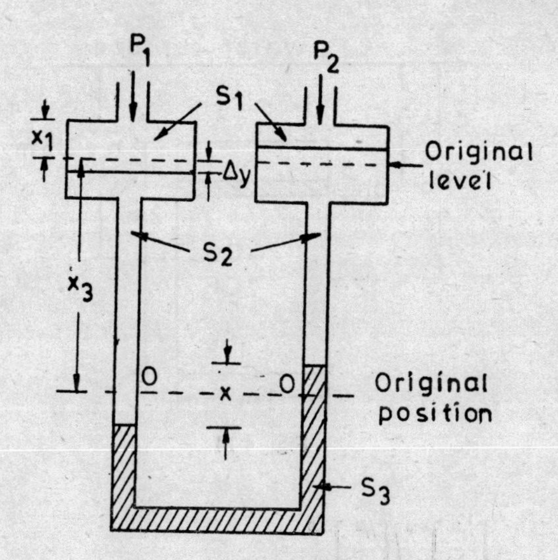

Fig. 4.13. Two Liquid manometer.

or
$$(p_1 - p_2) = xS_3\,\gamma + \gamma S_2\left(x_2 + \Delta y - \frac{x}{2} - x_2 + \Delta y - \frac{x}{2}\right)$$
$$+ \gamma S_1\,(x_1 - \Delta y - x_1 - \Delta y)$$
$$= \gamma(S_3 x - S_2 x + S_2 2\Delta y - 2\Delta y S_1)$$

Substituting $\Delta y = \frac{a}{A}\cdot\frac{x}{2}$ in the above equation, one gets

$$(p_1 - p_2)/\gamma = x\,\frac{a}{A}(S_2 - S_1) + x\,(S_3 - S_2)$$
$$= x\left[S_3 - S_2\left(1 - \frac{a}{A}\right) - S_1\frac{a}{A}\right]$$

In most of the cases a/A is much smaller than unity. Hence the above expression reduces to

$$(p_1 - p_2)/\gamma = x(S_3 - S_2)$$

4.11 Determine the total force and its point of application, for a 1 m wide and 2 m deep rectangular plane area immersed vertically with its top edge 1.5 m below the water surface (Fig. 4.14).

$$F = \gamma A\bar{h}$$
$$\bar{h} = 1.50 + 1.00 = 2.50 \text{ m}$$
$$F = 9.787 \times (1 \times 2) \times 2.5$$
$$= 48.935 \text{ kN}$$

$$h_0 = \int h^2 dA \Big/ \int h\,dA = \int_{1.5}^{3.5} h^2 dh \Big/ \int_{1.5}^{3.5} h\,dh$$

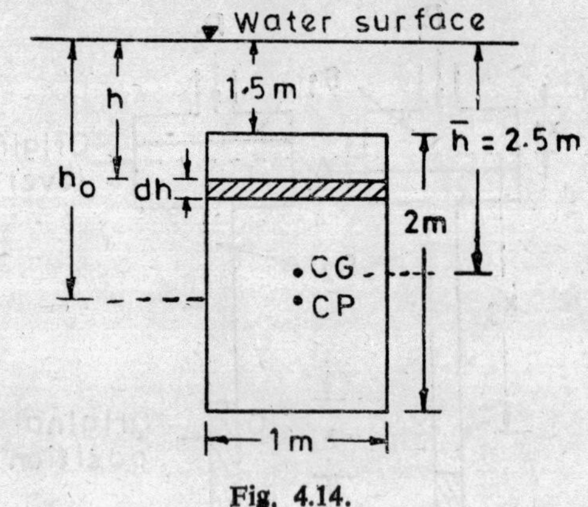

Fig. 4.14.

$$= \left[\frac{h^3}{3}\right]_{1.5}^{3.5} \Big/ \left[\frac{h^2}{2}\right]_{1.5}^{3.5}$$

$$h_0 = \tfrac{1}{3}\,(3.5^3 - 1.5^3)/\tfrac{1}{2}\,(3.5^2 - 1.5^2)$$

$$= \frac{39.5}{3} \cdot \frac{2}{(10)} = 2.633 \text{ m}$$

4.12 For a gate defined by $y = 2x^2$ as shown in Fig. 4.15, calculate the moment of hydrostatic force about the hinge.

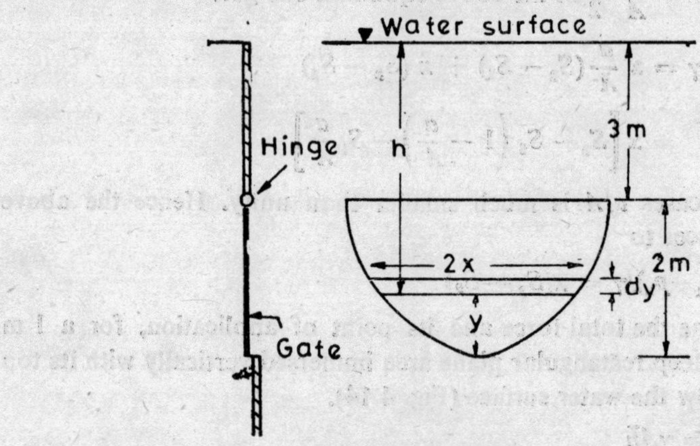

Fig. 4.15.

Since $y = 2x^2$, $x = \dfrac{y^{1/2}}{\sqrt{2}}$ and $2x = \sqrt{2y}$

$$\therefore \quad dA = 2x\,dy = \sqrt{2y}\,dy$$

$$dF = \gamma h\, dA$$

$$= \gamma(5 - y)\sqrt{2y}\,dy$$

$$\therefore \quad F = \int_0^2 \gamma \sqrt{2}\,(5-y)\, y^{1/2}\, dy$$

$$= 9.787 \times 1.414 \left[5 \times \frac{2}{3} y^{3/2} - \frac{2}{5} y^{5/2}\right]_0^2$$

$$= 99.170 \text{ kN}$$

If h_0 is the position of centre of pressure below the water surface,

$$h_0 = \frac{\int h^2 dA}{\int h dA} = \frac{\sqrt{2}\int_0^2 (5-y)^2 y^{1/2}\, dy}{\sqrt{2}\int_0^2 (5-y)\, y^{1/2}\, dy}$$

$$= \frac{\left[25 \times \frac{2}{3} y^{3/2} - 10 \times \frac{2}{5} y^{5/2} + \frac{2}{7} y^{7/2}\right]_0^2}{\left[5 \times \frac{2}{3} y^{3/2} - \frac{2}{5} y^{5/2}\right]_0^2}$$

$$= \frac{\left(25 \times \frac{2}{3} \times 2^{3/2} - 10 \times \frac{2}{5} \times 2^{5/2} + \frac{2}{7} \times 2^{7/2}\right)}{\left(5 \times \frac{2}{3} \times 2^{3/2} - \frac{2}{5} \times 2^{5/2}\right)}$$

$$= \frac{27.745}{7.165}$$

$$= 3.872 \text{ m below water surface}$$

Hence moment of F about the hinge is

$$M = 99.170 \times (3.872 - 3.000)$$

$$= 86.476 \text{ kN m}$$

4.13 Determine the total force and location of centre of pressure for a circular plate of 2 m diameter immersed vertically in water with its top edge 1.0 m below the water surface.

$$F = \gamma A \bar{h}$$

$$= 9.787 \times 0.785 \times 2^2 \times 2$$

$$= 61.462 \text{ kN}$$

Moment of inertia of circular plate about axis gg

$$I_{gg} = \frac{\pi D^4}{64} = \frac{3.142 \times 2^4}{64} = 0.785 \text{ m}^4$$

$$h_0 = \bar{h} + \frac{I_{gg}}{A\bar{h}} = 2.0 + \frac{0.785}{(0.785 \times 2^2) \times 2}$$

$$= 2.0 + 0.125$$

$$\therefore \quad h_0 = 2.125 \text{ m}$$

4.14 Determine the total force and position of centre of pressure for annular ring shown in Fig. 4.16.

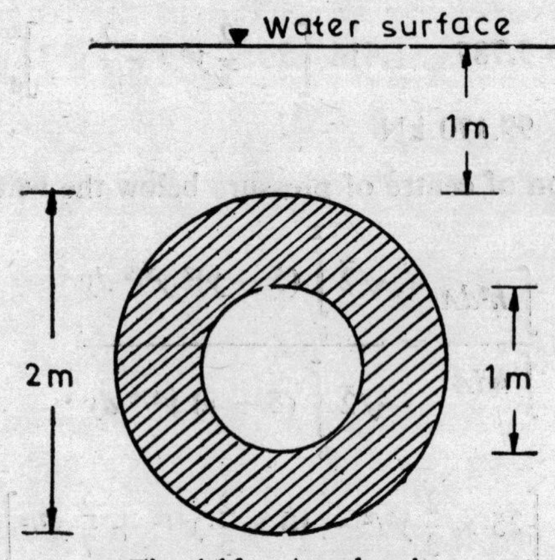

Fig. 4.16. Annular ring.

Here $\bar{h} = 2$ m and

$$A = \frac{\pi}{4}(2^2 - 1^2)$$

$$= 2.355 \text{ m}^2$$

$$\therefore \quad F = \gamma A\bar{h} = 9.787 \times 2.355 \times 2.0$$

$$= 46.097 \text{ kN}$$

h_{01} for 2 m dia plate is 2.125 m, and

$F_1 = 61.462$ kN (see Ex. 4.13)

In the same manner h_{02} for 1 m dia plate with its top edge 1.5 m below w.s. as well as F_2 the total force on it can be found as

$$F_2 = 9.787 \times 0.785 \times 1^2 \times 2$$

$$= 15.366 \text{ kN}$$

$$h_{02} = 2.0 + \frac{3.142 \times 1^4}{64 \times 3.142 \times 1^2 \times 2}$$

$$= 2.0313 \text{ m}$$

F on annular plate $= 61.462 - 15.366 = 46.096$ kN

If h_0 is the depth of centre of pressure of annular plane, taking moments of forces about w.s., one gets

$$h_0 \times 46.096 = 61.462 \times 2.125 - 15.366 \times 2.0313$$

$$\therefore \quad h_0 = (130.607 - 32.213)/46.096$$

$$= 2.135 \text{ m} \quad \text{below w.s.}$$

4.15 Determine the total force and position of centre of pressure for a trapezium of sides 2 m and 4 m and height 3 m, immersed in water at 30° inclination with the top edge 1 m below water surface (Fig. 4.17).

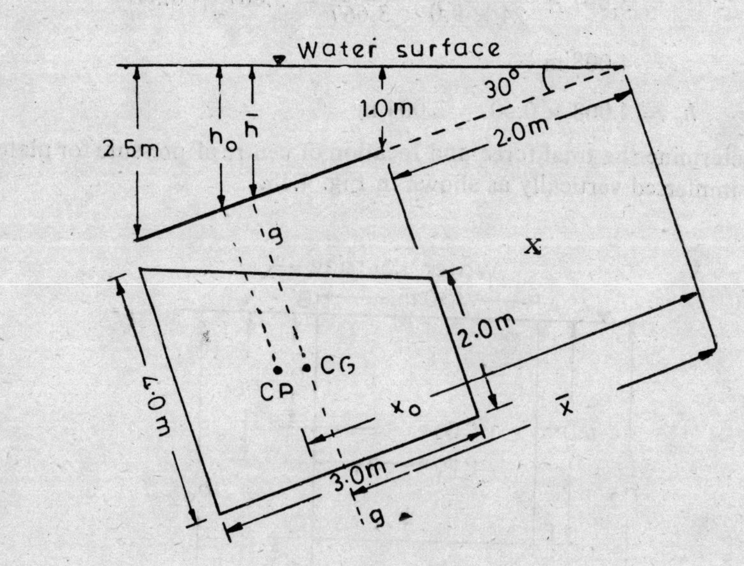

Fig. 4.17.

Centre of gravity of the area will be 1.333 m from 4 m edge.

$$\therefore \quad \bar{x} = 5.0 - 1.333$$
$$= 3.667 \text{ m}$$

and $\bar{h} = \bar{x} \sin 30°$
$$= 1.833 \text{ m}$$

$$A = \frac{2+4}{2} \times 3.0 = 9.00 \text{ m}^2$$

$$\therefore \quad F = \gamma A \bar{h} = 9.787 \times 9.00 \times 1.833$$
$$= 161.456 \text{ kN}$$

It can be shown that I_{gg} is given by

$$I_{gg} = \frac{h^3}{24}(3b - a)$$

where a = top side and

b = bottom side of trapezium

and $\bar{x} = 3.667$ m

$$\therefore \quad x_0 = \bar{x} + \frac{I_{gg}}{A\bar{x}}$$

$$= 3.667 + \frac{3^3(3 \times 4 - 2)}{24 \times 9 \times 3.667}$$

$$= 3.667 + \frac{3^3(10)}{24 \times 9.0 \times 3.667} = 3.667 + 0.341$$

$$= 4.008 \text{ m}$$

$$\therefore \qquad h_0 = 4.008 \times 0.50 = 2.004 \text{ m}$$

4.16 Determine the total force and location of centre of pressure for plate ABCDE immersed vertically as shown in Fig. 4.18.

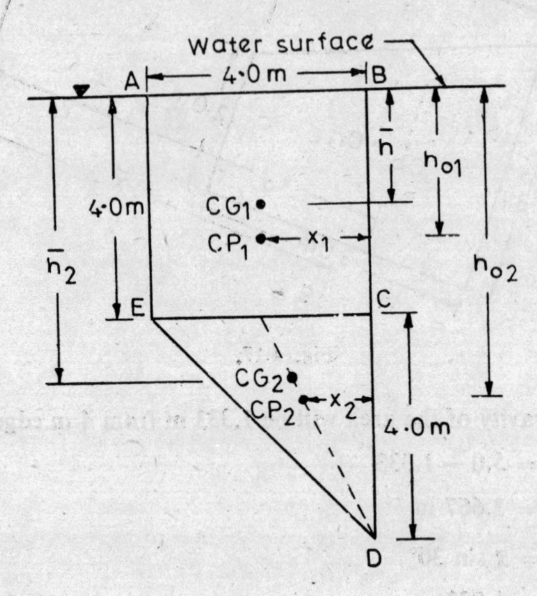

Fig. 4.18.

Consider the square area ABCE and triangular area CDE separately.
Force F_1 on square area

$$F_1 = 9.787 \times 16 \times 2$$
$$= 313.184 \text{ kN}$$

Force F_2 on triangular area is

$$F_2 = 9.787 \times \left(\frac{4 \times 4}{2}\right) \times 5.333 = 417.553 \text{ kN}$$

since C.G. of triangular area is $\left(4 + \frac{4}{3}\right)$ i.e. 5.333 m below w.s.

Therefore, total force

$$F = 313.184 + 417.553 = 730.737 \text{ kN}$$

F_1 will act at $h_{01} = 2.0 + \dfrac{4 \times 4^3}{12 \times (4 \times 4) \times 2}$

$= 2.0 + 0.667 = 2.667$ m

Similarly $\quad h_{02} = 5.333 + \dfrac{4 \times 4^3}{36} \times \dfrac{2}{4 \times 4 \times 5.333}$

$= 5.50$ m

The depth at which the resultant force will act can be determined by taking moments of forces F_1 and F_2 about water surface.

$$313.184 \times 2.667 + 417.553 \times 5.50 = 730.737 \times h_0$$

$\therefore \quad h_0 = 4.286$ m below the water surface

The horizontal location of centre of pressure can be obtained by taking the moments of F_1 and F_2 about BCD. The force F_1 acts 2 m from line BCD. The distance x_2 where force F_2 acts can be obtained from

$$\frac{2}{x_2} = \frac{4}{(2.50)} \quad \text{(from similarity of triangles)}$$

$\therefore \quad x_2 = \dfrac{5.0}{4} = 1.25$ m

$\therefore \quad 313.184 \times 2 + 417.553 \times 1.25 = 730.737 \times x_0$

$\therefore \quad x_0 = 1.573$ m

Hence coordinates of centre of pressure are 4.286 m below w.s. and 1.573 m from BCD.

4.17 A concrete wall of rectangular cross section is 7 m high and 3 m wide. It holds water to a depth of 5 m. Assuming the coefficient of friction μ between ground and concrete wall to be 0.60, check the stability of wall. Take specific weight for concrete as 20,000 N/m³ (Fig. 4.19).

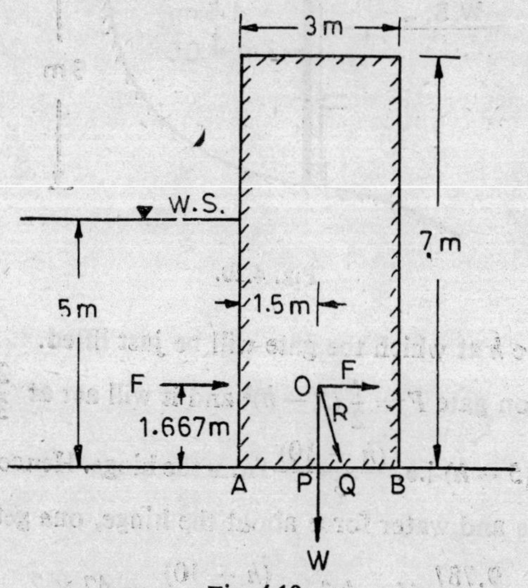

Fig. 4.19.

Consider width of wall in perpendicular direction as unity. Water force F will be

$$F = 9.787 \times (5 \times 1) \times 2.5$$
$$= 122.338 \text{ kN}$$

This will act at $\frac{10}{3}$ or 3.333 m below w.s.

The weight of wall W will be

$$W = 7 \times 3 \times 1 \times 20 = 420 \text{ kN}$$

It will act along the vertical line 1.5 m from upstream face. The resultant of F and W is R. Let it cut the base at Q.

$$\tan\theta = \frac{F}{W} = \frac{122.338}{420} = 0.291$$

$$\therefore \quad PQ = OP \tan\theta = 1.667 \times 0.291 = 0.486 \text{ m}$$

Since the resultant force intersects the base at Q within AB, the dam cannot overturn. Further since it falls within the middle 1/3rd of the base, there will be no tension on the upstream side. Further to avoid sliding, one must also ensure that μW is greater than F.

$W\mu = 420 \times 0.60 = 252$ kN. This being greater than 122.338, the wall is stable.

4.18 An L-shaped gate hinged at the corner is shown in Fig. 4.20. The coordinates of centre of gravity of the gate are given. If the gate weighs

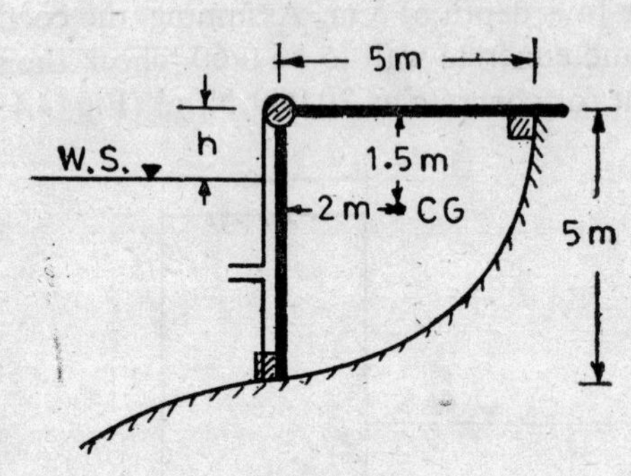

Fig. 4.20.

47 kN, determine h at which the gate will be just lifted.

Water force on gate $F = \frac{\gamma}{2}(5-h)^2$ and it will act at $\frac{2}{3}(5-h)$ below w.s. or $h + \frac{2}{3}(5-h)$ i.e. $\frac{(h+10)}{3}$ from the hinge. Hence taking moments of weight of gate and water force about the hinge, one gets

$$\frac{9.787}{2}(5-h)^2 \times \frac{(h+10)}{3} = 47 \times 2$$

or $$(5 - h)^2 (h + 10) = 57.63$$

Solving by trial procedure one gets $h = 2.885$ m.

4.19 determine the magnitudes and lines of action of horizontal and vertical components of the force acting on 2 m radius radial gate with 3 m width when its top edge is 4 m below the water surface. If the gate is weightless,

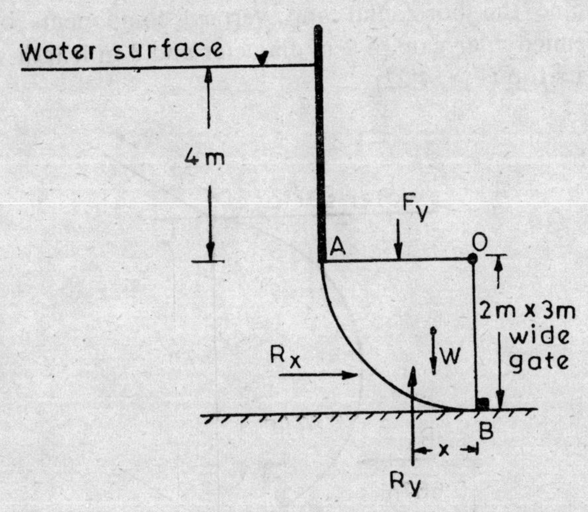

Fig. 4.21.

what will be the magnitude of F? What is the moment of force system about axis through O normal to the paper (Fig. 4.21)?

$$R_x = 9.787 \times (2 \times 3) \times 5$$

$$= 293.61 \text{ kN}$$

$$h_0 = 5 + \frac{3 \times 2^3}{12 \times (3 \times 2) \times 5}$$

$$= 5.067 \text{ m}$$

$\therefore$ R_x will act 5.067 m below water surface.

$$F_y = (2 \times 3) \times 9.787 \times 4$$

$= 234.889$ kN. It will act at 1 m from 0.

$$W = \frac{\pi R^2}{4} \times 3 \times 9.787 = \frac{3.142 \times 2^2 \times 3 \times 9.787}{4}$$

$= 184.505$ kN (W is weight of water in OAB)

W will act at $\frac{4}{3}\frac{R}{\pi}$ or $\frac{4 \times 2}{3 \times 3.142} = 0.849$ m from OB

$$\therefore \quad R_y = F_y + W = 234.889 + 184.505$$

$$= 419.394 \text{ kN}$$

Line of action of R_y can be found by taking moments of forces about OB

$$234.889 \times 1 + 184.505 \times 0.849 = 419.394\, x$$

$$\therefore \qquad x = 0.934 \text{ m from OB}$$

Since these forces are in equilibrium, $F = 0$ if the weght of gate is zero. Also moments of forces about the axis normal to the paper and through O will be zero.

4.20 Determine the horizontal and vertical components of the force acting on a semicircular gate of 2 m diameter and 3 m length when water stands up to its top (Fig. 4.22).

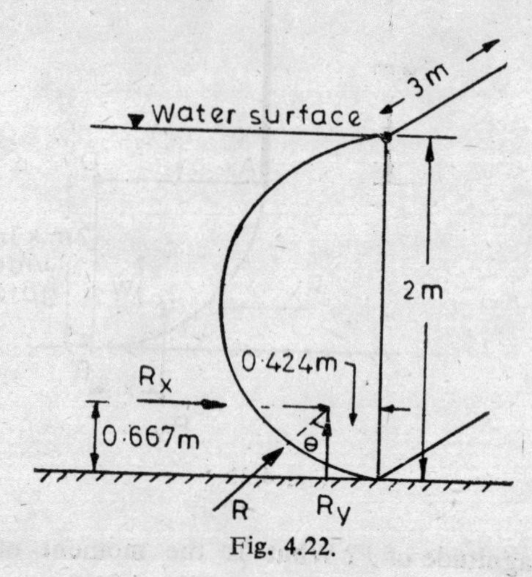

Fig. 4.22.

Horizontal force on curved surface

$$R_x = \text{force on its projected area on a vertical plane}$$
$$= 9.787 \times 3 \times 2 \times 1$$
$$= 58.722 \text{ kN}$$

It will act at $\frac{2}{3}$ or 0.667 m from the bottom.

The vertical force R_y is equal to weight of displaced water

$$R_y = 9.787 \times \frac{\pi \times 1^2}{2} \times 3$$
$$= 46.126 \text{ kN}$$

R_y will act through C.G. of displaced liquid. Hence its line of action will be $\frac{4R}{3\pi}$, i.e. 0.424 m from the vertical

$$R = \sqrt{58.722^2 + 46.126^2}$$
$$= 75.041 \text{ kN}$$

and $\qquad \tan\theta = \frac{R_x}{R_y} = \frac{58.722}{46.126} = 1.273 \quad \therefore \quad \theta = 51.85°$

4.21 Determine the magnitude and direction of the resultant force acting on the radial gate shown if its length is 4 m (Fig. 4.23).

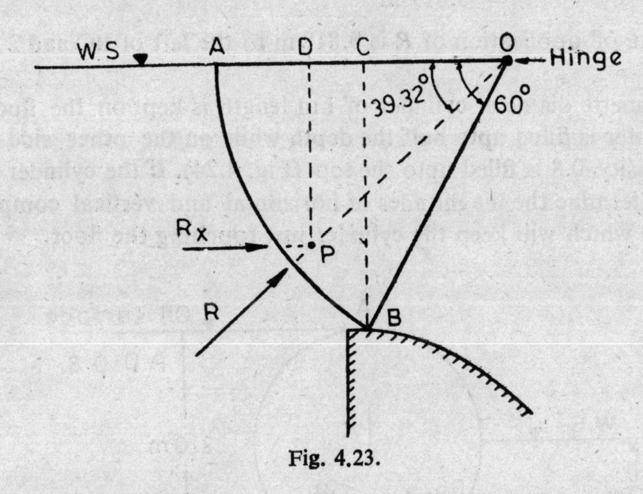

Fig. 4.23.

$$BC = 4 \sin 60°$$
$$= 3.464 \text{ m}$$
$$R_x = 9.787 \times (3.464 \times 4) \times \frac{3.464}{2}$$
$$= 234.881 \text{ kN}$$

It will act at $\frac{3.464}{3}$ or 1.155 m above B.

The vertical force, in its magnitude is equal to weight of water displaced, i.e. weight of volume equal to ABC × 4.

$$\text{Area ABC} = (\text{Area AOB} - \text{Area BCO})$$
$$= \frac{1}{2} R^2 \left(\frac{\pi}{3}\right) - \frac{3.464 \times 4 \cos 60}{2}$$
$$= 8.379 - 3.464$$
$$= 4.915 \text{ m}^2$$

$$\therefore \quad R_y = 4.915 \times 4 \times 9.787 = 192.412 \text{ kN}$$
$$R = \sqrt{234.881^2 + 192.412^2} = 303.630 \text{ kN}$$

If R makes an angle α with horizontal,

$\tan \alpha = \frac{R_y}{R_x} = 0.819$. Hence $\alpha = 39.32°$ and R must pass through O.

Since R_x acts at $(3.464 - 1.155) = 2.309$ m below water surface

$$OD = PD/\tan 39.32° = \frac{2.309}{0.819} = 2.819 \text{ m}$$

$$CD = OD - OC = 2.819 - 4 \cos 60°$$
$$= 0.819 \text{ m}$$

Hence point of application of R is 0.819 m to the left of BC and 2.309 m below w.s.

4.22 Two metre diameter cylinder of 1 m length is kept on the floor. On one side, water is filled upto half the depth while on the other side oil of relative density 0.8 is filled upto the top (Fig. 4.24). If the cylinder weighs 10.0 kN, determine the magnitudes of horizontal and vertical components of the force which will keep the cylinder just touching the floor.

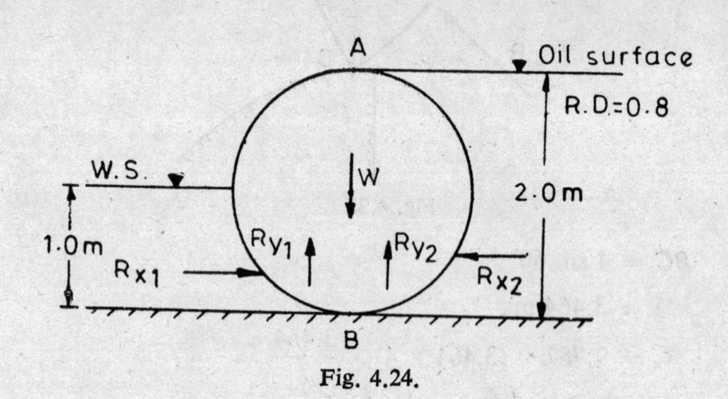

Fig. 4.24.

$$R_{x1} = 9.787 \times (1 \times 1) \times 0.5$$
$$= 4.894 \text{ kN}$$

This will act at 0.333 m from bottom.

$$R_{x2} = 9.787 \times 0.8 \times (2 \times 1) \times 1$$
$$= 15.659 \text{ kN. This will act at 0.667 m from bed.}$$

Hence (15.659 – 4.894) = 10.765 kN force acting towards right is required to hold the cylinder stationary. If it acts at a distance y, then taking moments about the bed, one gets

$$10.765 \times y + (4.894 \times 0.333) = 15.659 \times 0.667$$
$$\therefore \quad y = 0.819 \text{ m}$$

In the same way,

$R_{y1} = \dfrac{\pi \times 1^2}{4} \times 9.787 = 7.683$ kN and it will act along a vertical line 0.424 m to the left of AB

$R_{y2} = \dfrac{\pi \times 1^2}{2} \times 9.787 \times 0.8 = 12.300$ kN and it will act at $\dfrac{4 \times 1}{3\pi}$ or 0.424 m to the right of AB.

Since vertical forces must balance

$$\text{External force needed} = 7.683 + 12.300 - 10.00$$
$$= 9.983 \text{ kN.}$$

This external force is needed in vertically downward direction. To find out its line of action, take moments about the vertical line along which R_{y1} acts.

$$\therefore \quad (1.0 \times 0.424 + 9.983 \times x) = 12.300 \times 0.848$$

$$\therefore \quad x = 0.620 \text{ m}$$

4.23 A cylinder of radius R and length L holds the water behind it as shown in Fig. 4.25. Assuming the contact between the cylinder and wall to be smooth, determine the weight of the cylinder and horizontal force exerted against the wall. What will be the relative density of cylinder?

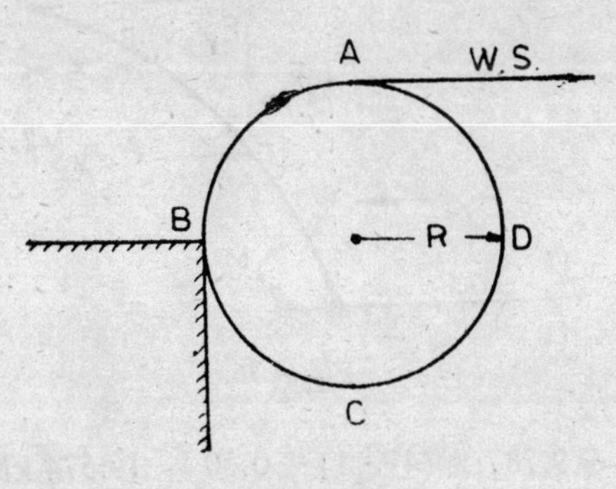

Fig. 4.25.

For equilibrium, the vertically upward force on the cylinder will be balanced by its weight.

$$R_v = R_{BCD} + R_{AD}$$

Since the vertical force on the curved surface is equal to weight of liquid vertically above it and extending up to actual or imaginary free surface,

$$R_v = \left(\frac{\pi R^2 L}{2} + 2R^2 L\right)\gamma - \left(R^2 L - \frac{\pi R^2 L}{4}\right)\gamma$$

$$= \frac{R^2 L\gamma}{4}(2\pi + 8.0 - 4.0 + \pi)$$

$$= \frac{R^2 L\gamma}{4}(3\pi + 4)$$

This must be equal to the weight of cylinder.

$$R_H = R_{ADC} - R_{BC}$$

$$= \gamma \times 2R \times L \times R - \gamma RL \times \frac{3R}{2}$$

$$= 2R^2 L\gamma - \frac{3}{2}R^2 L\gamma = \frac{1}{2}R^2 L\gamma$$

$$\text{Weight of cylinder } W = R_v = \frac{R^2 L\gamma}{4}(3\pi + 4)$$

Relative density of cylinder $= R^2L\gamma\,(3\pi + 4)/4\pi R^2 L\gamma$

$$= \frac{(3\pi + 4)}{4\pi} = \left(\frac{3}{4} + \frac{1}{\pi}\right)$$

4.24 A gate whose profile is given by $x = y^{1/2}$ holds water to a depth of 1.0 m behind it as shown in Fig. 4.26. If the gate width is 4.0 m, determine the moment M required to hold the gate in place.

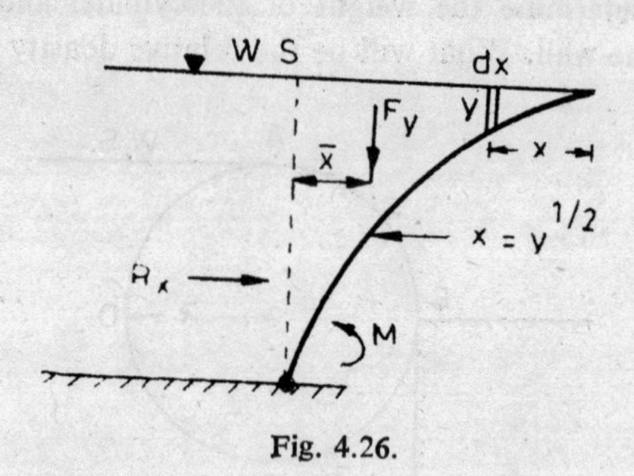

Fig. 4.26.

$$R_x = 9.787 \times (4 \times 1) \times 0.50 = 19.574 \text{ kN}$$

It will act at $\frac{1}{3}$ or 0.333 m from the bottom

$$F_y = \text{weight of liquid above the gate}$$

$$= 4\gamma \int_0^1 y dx \quad \text{but} \quad y = x^2$$

$$= 4\gamma \int_0^1 x^2 dx = \frac{4\gamma}{3}$$

$$= 9.787 \times \frac{4}{3} = 13.049 \text{ kN}$$

The vertical line along which F_y will act is obtained by taking moments of elementrary force $4\gamma y dx$ about y axis and equating it to $F_y\bar{x}$

$$\therefore \quad F_y\bar{x} = \int_0^1 4\gamma y\,(1 - x)\,dx = \int_0^1 4\gamma x^2\,(1 - x)\,dy$$

$$= 4\gamma\left(\frac{1}{3} - \frac{1}{4}\right) = \frac{\gamma}{3}$$

$$\therefore \quad \bar{x} = \frac{9.787}{3 \times 13.049} = 0.25 \text{ m}$$

Moment of F_x and F_y about Z axis passing through O is

$$M = 19.574 \times 0.333 + 13.044 \times 0.250 = 9.780 \text{ kN m}$$

Moment required to hold the gate will be 9.780 kN m

4.25 Obtain an expression for change in height of immersion of a hydrometer and relative density of the liquid (Fig. 4.27).

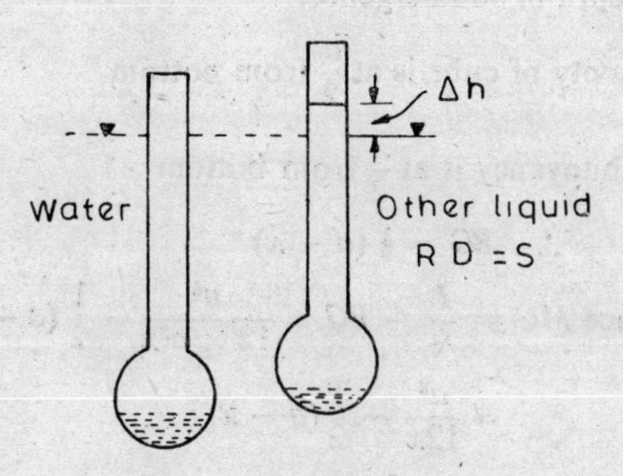

Fig. 4.27.

Let A be the cross-sectional area of the glass tube of the hydrometer and W be its weight in air. If $\forall$ is its volume then when it is immersed in water

$$W = \gamma \forall$$

If it is immersed in a liquid of relative density S which is greater than unity, the hydrometer will rise through height Δh.
In this case

$$W = (\forall - A\Delta h)\,\gamma S$$
$$= \gamma \forall$$

$$\therefore \quad A\Delta h S = \forall\,(S - 1) \quad \text{or} \quad \Delta h = \frac{\forall}{A}\left(1 - \frac{1}{S}\right)$$

4.26 A body weighs 29.361 kN in water and 26.425 kN in a liquid of relative density 1.60. Determine its volume and weight in air.

Let W be its weight in air and $\forall$ be its volume. Hence according to Archimedes principle,

$$\text{Weight in water} = W - \forall\gamma$$

and weight in liquid of relative density 1.6 $= W - \forall\gamma \times 1.6$

or

$$29.361 = W - 9.787\,\forall$$
$$26.425 = W - (9.787 \times 1.6)\,\forall$$

Solution of these two equations yields

$$\forall = 0.500 \text{ m}^3 \quad \text{and} \quad W = 34.255 \text{ kN}$$

4.27 A cube of side a and relative density S floats in water, determine the conditions for its stability if it is given an angular tilt.

According to Archimedes principle,

$$a^3 S\gamma = a^2 x\gamma$$

$$\text{or} \quad x = aS$$

where x is the depth of submergence.

Centre of gravity of cube is at $\frac{a}{2}$ from bottom

While centre of buoyancy is at $\frac{x}{2}$ from bottom

$$\therefore \quad BG = \tfrac{1}{2}(a - x)$$

$$\text{Since } MG = \frac{I}{\forall} - BG = \frac{a^4}{12 \times a^2 x} - \frac{1}{2}(a - x)$$

$$= \frac{a^2}{12x} - \frac{1}{2}(a - x)$$

Condition for stability is therefore:

It is stable if MG is + ve or $\frac{a^2}{12x} > \frac{1}{2}(a - x)$

It is unstable if MG is − ve or $\frac{a^2}{12x} < \frac{1}{2}(a - x)$

It is stable if $1 > 6\left(\frac{x}{a} - \frac{x^2}{a^2}\right)$

It is in neutral equilibrium if $1 = 6\left(\frac{x}{a} - \frac{x^2}{a^2}\right)$

It is unstable if $1 < 6\left(\frac{x}{a} - \frac{x^2}{a^2}\right)$

$$\text{But} \quad \frac{x}{a} = S$$

$$\text{Stable/unstable} \quad 1 \gtrless 6(S - S^2)$$

$$\text{or} \quad 0 \gtrless (-1 + 6S - 6S^2)$$

$$\text{or} \quad 6S^2 - 6S + 1 \gtrless 0$$

$$\text{or } S = \frac{6 \pm \sqrt{36 - 24}}{12}$$

$$= \frac{1}{2} \pm \frac{\sqrt{3}}{6}$$

$$= 0.50 \pm 0.289$$

$$= 0.789 \quad \text{or} \quad 0.211$$

Therefore, the relative density of cube should be greater than 0.789 or less than 0.211 for stability.

4.28 A wooden block of relative density 0.70 is 2 m × 2.0 m in plan and 1.0 m deep. Determine its metacentric height when it floats in water; (Fig. 4.28).

When it floats in water, the depth of immersion = 0.70 × 1.0 = 0.70 m

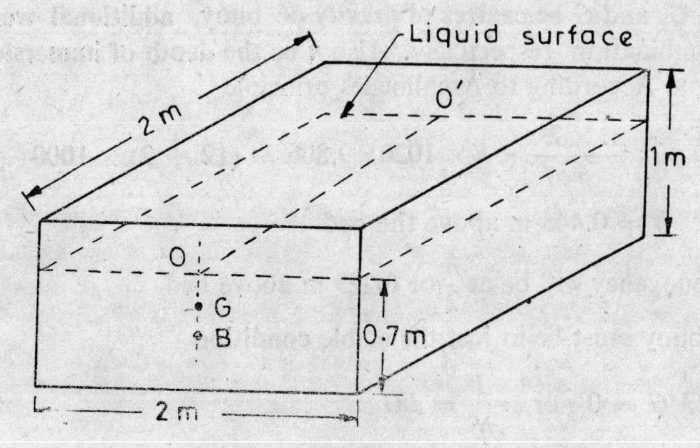

Fig. 4.28.

Distance of *CG* from bottom = 0.5 m

Distance of centre of buoyancy *B* from bottom = 0.35 m

$\therefore \; BG = 0.15$ m

$$MG = \frac{I}{\forall} - BG. \text{ Here } I = \frac{2 \times 2^3}{12} = \frac{4}{3} = 1.333 \text{ m}^4$$

$$\forall = 2 \times 2 \times 0.7 = 2.8 \text{ m}^3$$

$$= \frac{1.333}{2.80} - 0.15$$

$$= 0.326 \text{ m}$$

4·29 A cylindrical buoy 2.0 m in diameter and 1.5 m in height weighs 14 kN and floats in salt water of density 1020 kg/m³. Centre of gravity of buoy is 0.65 m above 0. If a load of 2 kN is placed on its top symmetrically, find the maximum height of the centre of gravity of this load above the bottom if the buoy is to remain in stable equilibrium (Fig. 4.29).

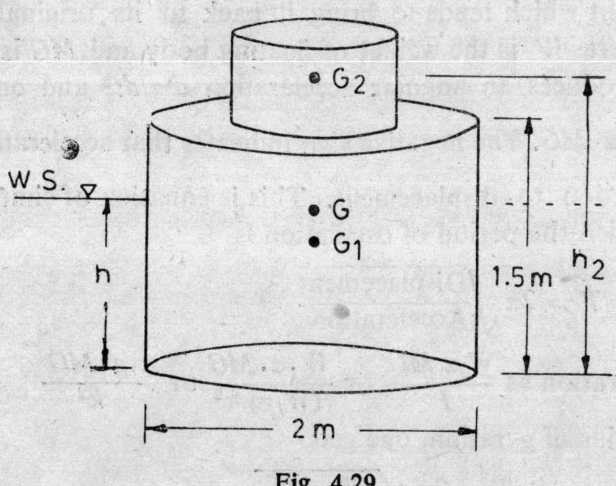

Fig. 4.29.

Let G_1, G_2 and G be centres of gravity of buoy, additional weight and of the combination respectively. Also h be the depth of immersion of the combination. According to Archimedes principle

$$\therefore \qquad \frac{\pi \times 2^2}{4} \times h \times 1020 \times 9.806 = (12 + 2) \times 1000$$

$$\therefore \qquad h = 0.445 \text{ m above the bed}$$

Centre of buoyancy will be at $\frac{h}{2}$ or 0.223 m above bed.

It the buoy must be in just the stable condition

$$MG = 0 \quad \text{or} \quad \frac{I}{\forall} = BG$$

Now $$I = \frac{\pi \times 2^4}{64} \quad \text{and} \quad \forall = \frac{\pi \times 2^2}{4} \times 0.445$$

$$\therefore \qquad BG = \frac{\pi \times 2^4}{64} \times \frac{4}{\pi \times 2^2 \times 0.445} = 0.562 \text{ m}$$

$$\therefore \qquad OG = OB + BG = 0.223 + 0.562 = 0.785 \text{ m}$$

where O is the centre of base. Taking moments of weights 2 kN and 12 kN about O, one can determine h_2.

$$2 \times h_2 + 14 \times 0.65 = 16 \times 0.785$$

$$\therefore \qquad h_2 = (16 \times 0.785 - 14 \times 0.65)/2$$

$$= 1.73 \text{ m}$$

4.30 A ship displaces 49 250 kN of sea water of density 1025 kg/m^3. The second moment of inertia of the water line section about axis through the centre is 12000 m^4 and the centre of buoyancy is 2 m below the centre of gravity. If the radius of gyration is 3.70 m, calculate the period of oscillation.

When a floating body is given a small angular displacement α, the righting moment which tends to bring it back to its original position is $W \cdot \alpha \cdot MG$ where W is the weight of floating body and MG is metacentric height. This produces an angular acceleration $d^2\alpha/dt^2$ and one can write $I\frac{d^2\alpha}{dt^2} = -W \cdot \alpha \cdot MG$. The negative sign indicates that acceleration is in the opposite direction to displacement. This is equation of simple harmonic motion for which the period of oscillation is

$$T = 2\pi\sqrt{\frac{\text{Displacement}}{\text{Acceleration}}}$$

Writing acceleration as $\frac{W.\alpha.MG}{I}$ or $\frac{W \cdot \alpha \cdot MG}{(W/g)\,k^2}$ or $\frac{\alpha \cdot g \cdot MG}{k^2}$

where k is radius of gyration, one gets

$$T = 2\pi\sqrt{k^2/g \cdot MG}$$

For the problem under consideration

$$MG = \frac{I}{\forall} - BG = \frac{12000}{(49250 \times 1000/1025 \times 9.806)} - 2.0$$

$$= 0.449 \text{ m}$$

$$\therefore \quad T = 2 \times 3.142\sqrt{3.7^2/9.806 \times 0.449}$$

$$= 11.081 \text{ s}$$

It can thus be seen that even though a large metacentric height increases stability, it also results in smaller period of oscillation which causes discomfort and excessive stress on the ship.

4.31 An open container partly filled with oil has 1 m free board. If the container is 10 m long, what is the maximum uniform horizontal acceleration that can be given to the container without spilling the oil?

If oil is not to spill, $\tan\theta = \frac{a_x}{g} = \frac{1}{10}$

$$\therefore \quad a_x = \frac{9.806}{10} = 0.981 \text{ m/s}^2$$

4.32 A tank shown in Fig. 4.30 contains oil of relative density 0.80. If it is given an acceleration of 5.0 m/s² along a 30° inclined plane in the upward direction, determine the slope of free surface and pressure at *b*.

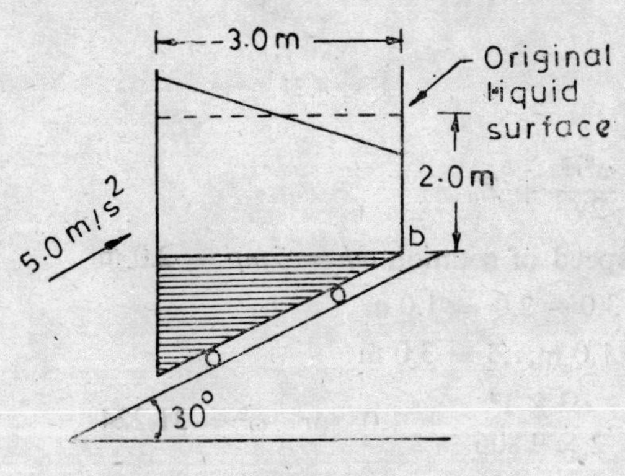

Fig. 4.30.

The acceleration a_s can be resolved into the horizontal and vertical components.

$$a_x = 5 \cos 30° = 4.33 \text{ m/s}^2$$

$$a_y = 5 \sin 30° = 2.50 \text{ m/s}^2$$

The water surface slope is given by $\tan\theta = \frac{a_x}{(a_y + g)} = \frac{4.33}{12.306}$

or $\tan\theta = 0.3519$ or $\theta = 19.38°$

Depth at $b = 2.0 - 1.5 \tan\theta = 2.00 - 0.528 = 1.472$ m

$$\frac{p_b}{\gamma} = \left(1 + \frac{a_y}{g}\right)h_b = \left(1.0 + \frac{2.500}{9.806}\right) \times 1.472$$

$$= 1.847 \text{ m}$$

$\therefore \quad p_b = 1.847 \times 9787 \times 0.80 = 14\,463.47$ N/m²

or $\quad 14\,463$ kN/m²

4.33 A 2.0 m diameter open cylindrical tank is filled to the depth of 2.0 m with water. The height of the tank is 3.0 m. Determine the speed of rotation about its vertical axis at which water will just spill out of the tank. What will be the depth at the centre and gauge pressure at bottom 0.50 m from the centre (Fig. 4.31)?

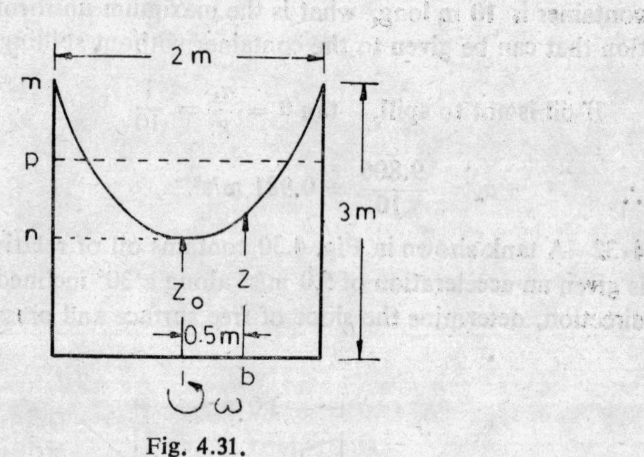

Fig. 4.31.

$$Z = \frac{\omega^2 r^2}{2g} + Z_0$$

At the desired speed of rotation $mh = 2mp = 2.0$ m

$\therefore \quad Z_0 = 3.0 - 2.0 = 1.0$ m

Also when $r = 1.0$ m, $Z = 3.0$ m

$$\therefore \quad 3 = \frac{\omega^2 \times 1^2}{2 \times 9.806} + 1.0 \quad \text{or} \quad \omega^2 = 31.224$$

or $\quad \omega = 6.263$ rad/s

$$\text{Since } \frac{2\pi N}{60} = w \quad \therefore \quad N = \frac{60 \times 6.263}{2 \times 3.142} = 59.8 \text{ rpm}$$

When $\quad r = 0.60$ m,

$$Z_b = \frac{6.263^2 \times 0.5^2}{2 \times 9.806} + 1.0 = 1.50 \text{ m}$$

$\therefore \quad p_b = \gamma Z_b = 9.787 \times 1.50 = 14.681$ kN/m²

4.34 An open cylindrical tank 0.50 m in diameter and 1 m in height is

completely filled with water and rotated about its axis at 240 rpm. Determine the radius upto which bottom will be exposed and the volume of water spilled out of the tank (Fig.4.32).

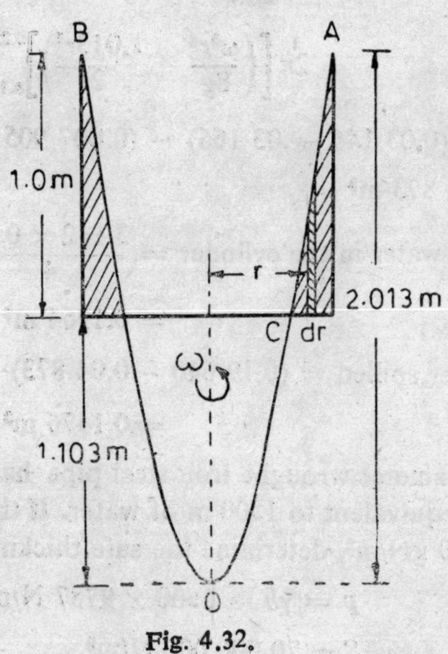

Fig. 4.32.

$$\text{Angular velocity } \omega = 2\pi N/60$$
$$= 2 \times 3.142 \times 240/60$$
$$= 25.136 \text{ rad/s}$$

$$Z = \frac{\omega^2 r^2}{2g} = 25.136^2 \times 0.25^2/2 \times 9.806$$
$$= 2.013 \text{ m}$$

Hence the imaginary paraboloid representing the w.s. is OAB whose height is 2.013 m. The water remaining in the tank assumes the shape shown by hatched lines. On imaginary paraboloid, point C is such that

$Z = (2.013 - 1.000) = 1.013$ m. Its radius r_c will be

$$\omega^2 r_c^2/2g = 1.013$$

or $$r_c = \sqrt{\frac{1.013 \times 2 \times 9.806}{(25.136)^2}} = 0.177 \text{ m}$$

Volume in elementary strip $= 2\pi r dr\,(Z - 1.013)$

$$= 2\pi \left(\frac{\omega^2 r^2}{2g} - 1.013\right) r dr$$

$$= 2\pi \left(\frac{\omega^2 r^3}{2g} - 1.013 r\right) dr$$

Volume of water remaining in the cylinder

$$= 2\pi \int_{0.177}^{0.25} \left(\frac{\omega^2 r^3}{2g} - 1.013r\right) dr$$

$$= 2\pi \left[\left(\frac{\omega^2 r^4}{8g} - \frac{1.013r^2}{2}\right)\right]_{0.177}^{0.25}$$

$$= 2\pi\ [(0.03\ 146 - .03\ 166) - (0.007\ 905 - 0.01\ 586)]$$
$$= 0.04\ 873\ \text{m}^3$$

Original volume of water in the cylinder $= \dfrac{3.142 \times 0.5^2}{4} \times 1.0$

$$= 0.1964\ \text{m}^3$$

$\therefore$ Volume of water spilled $= (0.19\ 640 - 0.04\ 873)$

$$= 0.1476\ \text{m}^3$$

4.35 A 300 mm diameter wrought iron steel pipe has to be designed to withstand pressure equivalent to 1500 m of water. If the allowable stress in wrought iron is 7000 kN/m², determine the safe thickness of the pipe.

$$p = \gamma h = 1500 \times 9787\ \text{N/m}^2$$
$$S = 70\ 000\ 000\ \text{N/m}^2$$
$$D = 0.30\ \text{m}$$

$$\frac{1500 \times 9787}{70\ 000\ 000} = \frac{2(2.917t - 0.008\ 33)}{(1 - 2.336t + 2.002)}$$

$$0.2097 - 4.889t + 0.00\ 419 = 5.834t - .01\ 666$$

$$\therefore \quad 10.723t = 0.2305$$

$$\therefore \quad t = 0.0215\ \text{m} \quad \text{or} \quad 21.50\ \text{mm}$$

PROBLEMS

4.1 Determine the pressure at the bottom of sea 1.0 km deep if relative density of sea water is 1030 kg/m³. Express it as gauge and absolute pressure if atmospheric pressure is 101.30 kN/m².
(10100.180 kN/m², 10201.480 kN/m²)

4.2 A tank contains 0.80 m of oil of relative density 0.860 below which there is 1.5 m of water and 0.20 m of mercury. Determine the pressure on the bottom of the tank, if relative density of mercury is 13.55. (47.934 kN/m²)

4.3 Determine the elevation at which the pressure at 20°C will be 75 kN/m² under isothermal conditions if the pressure and density at sea level at this temperature are 101.325 kN/m² and 1.208 kg/m³

respectively. Also determine the density of air at that elevation.
(2575.75 m, 0.894 kg/m^3)

4.4 If the pressure variation in the atmosphere follows adiabatic law, determine the pressure and density of air at 2575 m elevation above sea level if $p_0 = 101\ 325$ kN/m^2 and $\rho_0 = 1.208$ kg/m^3. Take $k = 1.40$ and 1.20. (73.962 kN/m^2, 0.965 kg/m^3; 74.402 kN/m^2, 0.935 kg/m^3)

4.5 A hydraulic press has a ram of 100 mm and a plunger of 12.5 mm. Determine the force required on the plunger to raise a weight of 10 000 N on the ram. If the plunger stroke is 0.40 m, through what distance will the weight be moved in 200 strokes? If the plunger makes 20 strokes per minute, determine the power required.
(156.2 N, 1.25 m, 0.02 083 kW)

4.6 A hydraulic press has 0 30 m diameter ram and 50 mm diameter plunger. Determine the force required to lift 25 kN weight. With a plunger stroke of 0.20 m and 20 strokes per minute frequency, calculate the time required to raise the load through 1.0 m and the power expended. (0.694 kN, 540.00 s, 0.0463 kW)

4.7 Combining the equations $\frac{p}{\rho} = RT$, $\frac{p}{\rho k} =$ constant, and

$$\frac{p}{p_0} = \left[1 - \frac{k-1}{k}\left(\frac{y}{p_0/\rho_0 g}\right)\right]^{k/(k-1)} \text{ show that}$$

$$T = T_0\left[1 - \frac{k-1}{k}\left(\frac{y}{p_0/\rho_0 g}\right)\right]$$

4.8 At the head of a mine, the pressure, density and temperature of atmosphere are 101.00 kN/m^2, 1.20 kg/m^3 and 20°C. Determine the value of k if the mine is 2000 m deep and the temperature increases at the rate of 0.009°C/m. (1.358)

4.9 If temperature varies with elevation y according to the relation $T = T_0(1 + my)$, show that

$$\frac{p}{p_0} = \left(\frac{1 + my}{1 + my_0}\right)^{-g/mRT_0}$$

where p_0 and T_0 are pressure and absolute temperature of reference elevation y_0.

(Hint: combine the equation for temperature variation with $\frac{p}{\rho} = RT$ and $dp = -\rho g dy$ and integrate)

4.10 Determine absolute pressure at A in Fig. 4.33 if atmospheric pressure is 101.2 kN/m^2. (97.481 kN/m^2)

4.11 Determine $p_B - p_A$ in Fig. 4.34 and express it in terms of metres of oil of relative density 0.80. (0.25 m).

4·12 If the mass density of a liquid with variable density is given by $\rho = 1000 + 0.008\ y^{3/2}$, determine the depth at which the pressure

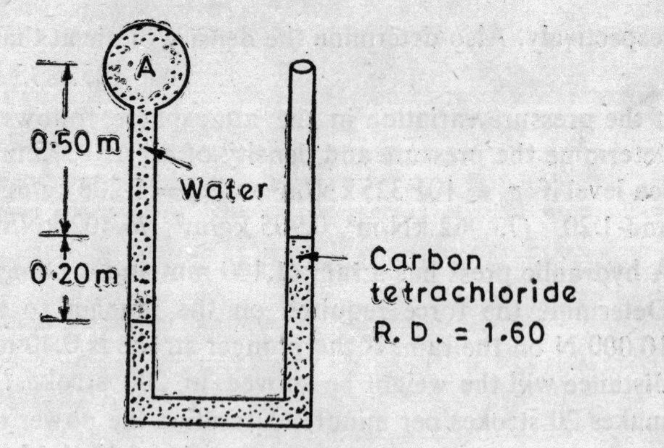

Fig. 4.33.

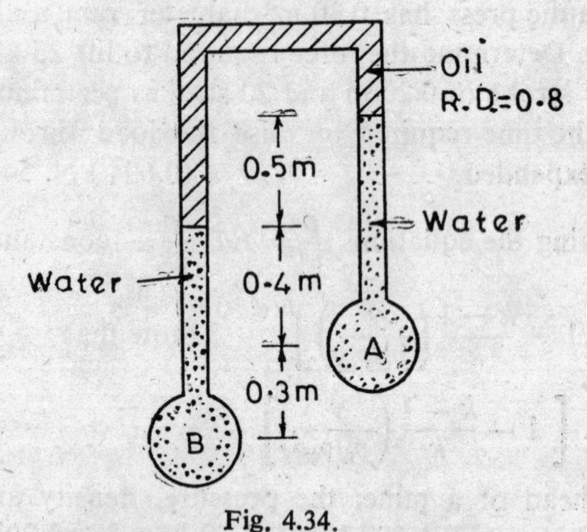

Fig. 4.34.

intensity will be 900 kN/m². (91.54 m)

4.13 Determine $(p_A - p_B)$ in Fig. 4.35 and express it in kN/m² (29.244 kN/m²)

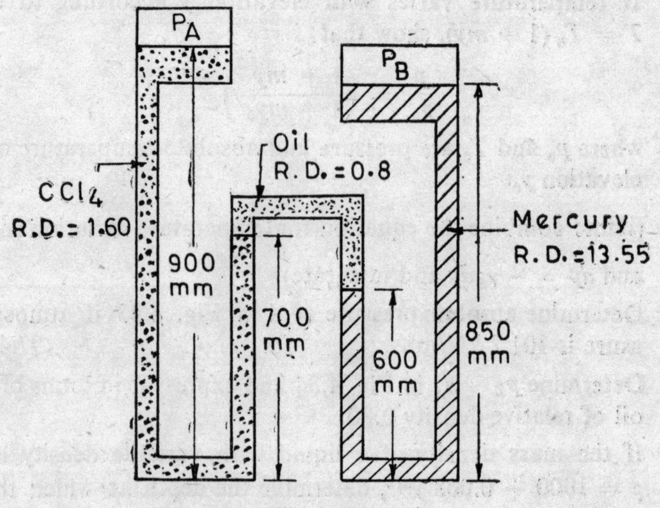

Fig. 4.35.

4.14 Determine the reading of the pressure gauge *B* in Fig. 4.36 for the situation shown. (27.404 kN/m²)

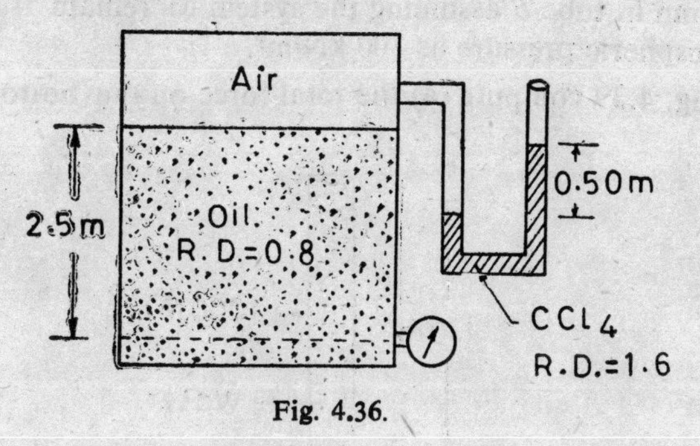

Fig. 4.36.

4.15 Determine gauge pressure at *A* in Fig 4.37. (−30.438 kN/m² gauge)

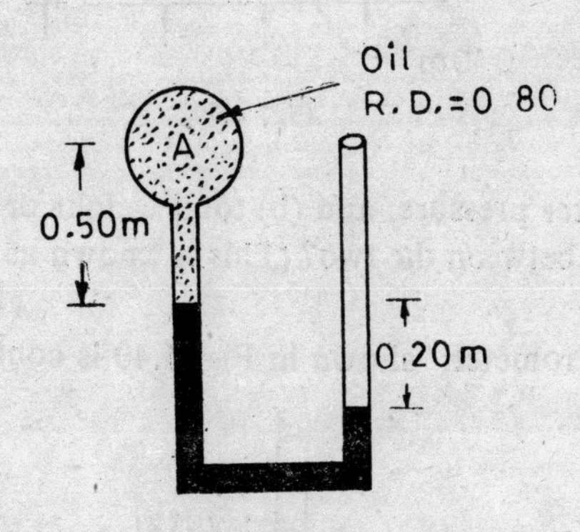

Fig. 4.37.

4.16 The container shown in Fig. 4.38 is initially filled with mercury. The

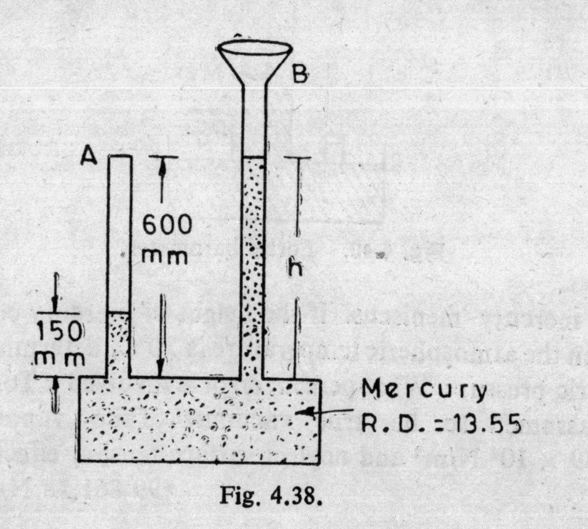

Fig. 4.38.

tube *A* is then closed and mercury added through the tube *B* until it is 150 mm high in tube *A*. Compute the height *h* of the mercury column in tube *B* assuming the system to remain isothermal. Take atmospheric pressure as 100 kN/m². (0.4013 m)

4.17 In Fig. 4.39 compute (a) the total force on the bottom of the tank

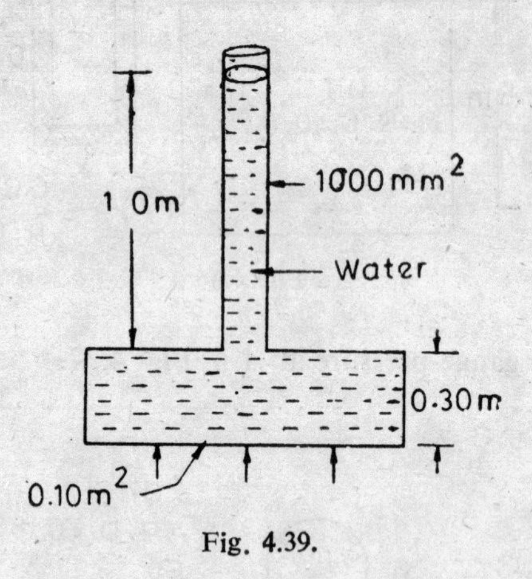

Fig. 4.39.

due to water pressure, and (b) total weight or water. Why is there difference between the two? (This is known as Pascal's paradox). (1272.21 N, 303.397 N)

4.18 Fortin barometer, shown in Fig. 4.40 is contaminated by water on

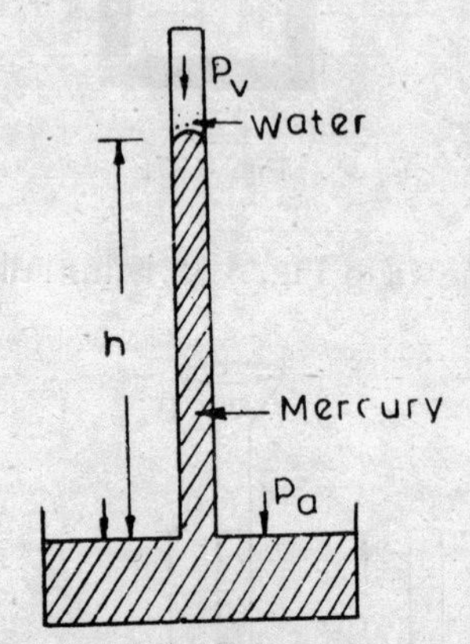

Fig. 4.40. Fortin barometer.

the mercury meniscus. If the height of mercury column is 735 mm when the atmospheric temperature is 20°C, determine the true barometric pressure. What percent error is involved if Torricellian vacuum is assumed to be true vacuum? Take vapour pressure as 0.239×10^4 N/m² and neglect surface tension effects. (99 861.18 N/m², 2.39 percent)

4.19 Determine the total force and location of centre of pressure in the following cases shown in Fig. 4.41.

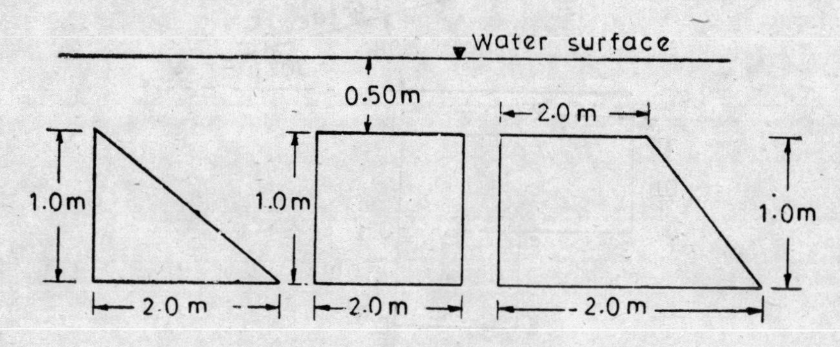

Fig. 4.41.

(11.418 kN, 1.214 m, 19. 574 kN, 1.0833 m, 30.992 kN, 1.1315 m)

4.20 An annular circular plate of 2 m external diameter and 1 m internal diameter is immersed vertically in water so that its lowest edge is 5 m below the water surface. Determine the total force and position of centre of pressure. (92.193 kN, 4.0781 m)

4.21 A right angled triangular plane plate with base B and height H is immersed vertically in water with the base coinciding with water surface. Show that the centre of pressure is located $\frac{H}{2}$ below the water surface and $B/4$ from the vertical side.

4.22 A circular plate of 4 m diameter is immersed in water with its plane making an angle of 30° with water surface and the top edge touching the water surface also. Determine the centre of pressure and the total force. (123.0 kN, 1.25 m)

4.23 Find total force and position of centre of pressure on a plane area shown in Fig. 4.42.

(41.595 kN, C.p. at 1.667 m depth and 0.794 m from AB)

4.24 An opening of 1 m depth and 3 m width is provided in the vertical side of a large tank. The water surface in the tank is 4 m above the top of opening. If the opening is closed by a plate which is held in place by 4 bolts placed at the corners, determine the force in each bolt.

(In each top bolt 31.816 kN, in each bottom bolt 34.246 kN)

4.25 A concrete dam of rectangular cross section is 10 m high and 3.5 m wide. If it is filled to a depth of 6 m, determime its stability. Take coefficient of friction between ground and concrete as 0.50 and unit weight of concrete as 20 kN/m³.

(Safe against sliding, safe against overturning, no tension any where at base)

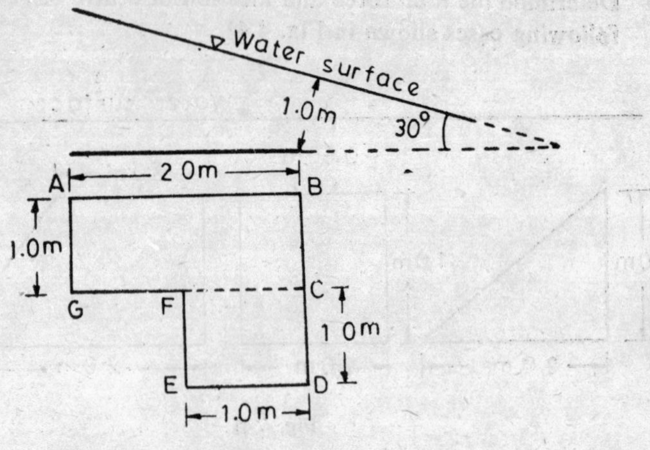

Fig. 4.42.

4.26 A concrete dam of trapezoidal cross section has its upstream face vertical. It is 3 m wide at the top, 8 m wide at the base and 16 m high. If it is filled to the top check its stability assuming unit weight of concrete as 23.5 kN/m³ and coefficient of friction as 0.60.

(Resultant cuts the base outside the middle third, but within the base. Hence it is safe for overturning but will cause tension. $F = 1252.736$ kN $> \mu W$. Hence it will be unsafe for sliding).

4.27 The flash board is held in place by two stops as shown in Fig. 4.43. Determine the distance y between them so that the flash board will tumble when water reaches the top. (1.0 m)

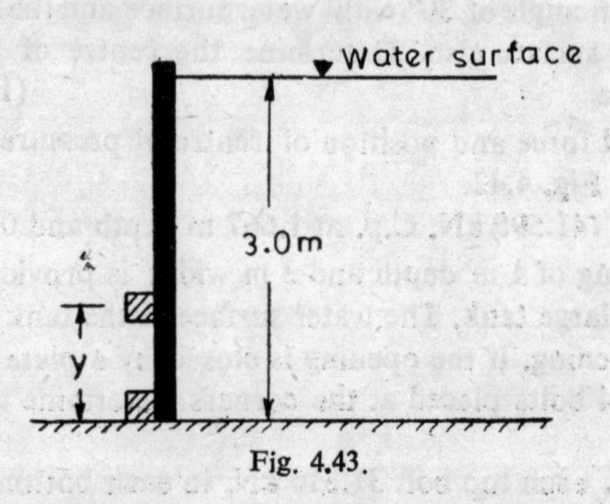

Fig. 4.43.

4.28 Compute the horizontal and vertical components of total force acting on the outer surface of a quadrant of a circular cylinder shown in Fig. 4.44. Also determine its point of application. The surface has 1 m in width in the perpendicular direction.

($R_x = 37.148$ kN, $R_y = 27.971$ kN, $y = 2.167$ m, $x = 1.381$ m)

4.29 A rhombus with diagonals a and b is immersed vertically in water so

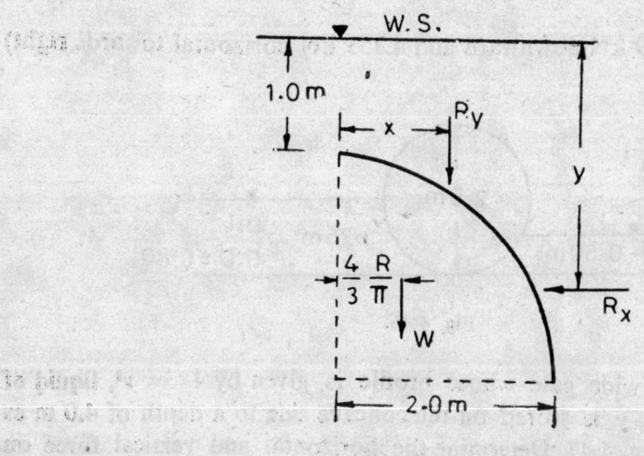

Fig. 4.44.

that the vertex is in the water surface and diagonal a is vertical. Show that the centre of pressure is at a depth of $\frac{7a}{12}$.

4.30 A 60° radial gate of 5 m radius and 3 m length stores water upto its top as shown in Fig. 4.45. Determine the components of the total force and its point of application.

(R_x = 367.013 kN, R_y = 66.561 kN, α = 10.28°)

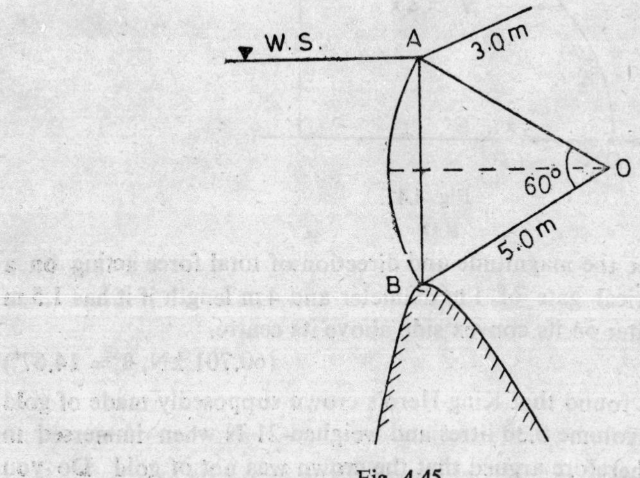

Fig. 4.45.

4.31 A 2.5 m diameter and 1.0 m long cylinder weighing 2.450 kN rests on the floor of a tank. On one side, oil of relative density 0.80 is poured to depth of 1.25 m while on the other side, water is poured to a depth of 0.60 m. Determine the horizontal and vertical components of the external force that must be applied to hold the cylinder in position.

(11.589 kN downward and 4.355 kN horizontal towards right)

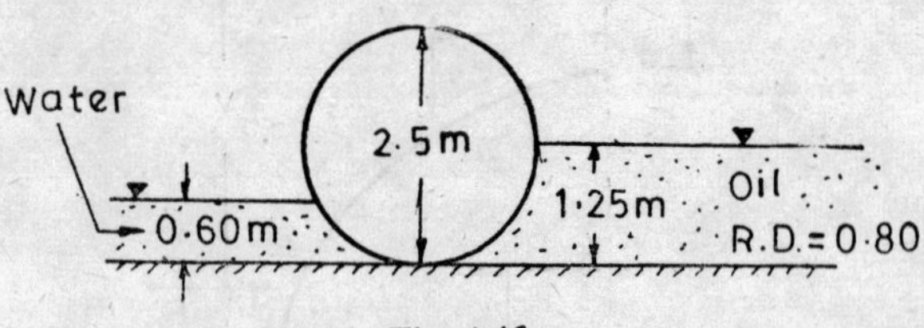

Fig. 4.46.

4.32 For a 5 m wide gate whose profile is given by $4x = y^2$, liquid of unit weight γ is stored on the concave side to a depth of 4.0 m as shown in Fig. 4.47. Determine the horizontal and vertical force on gate and their point of application.

(Horizontal: 40 γ acting 1.33 m above bed;

Vertical: $\frac{80}{3}\gamma$ acting 1.2 m from y axis)

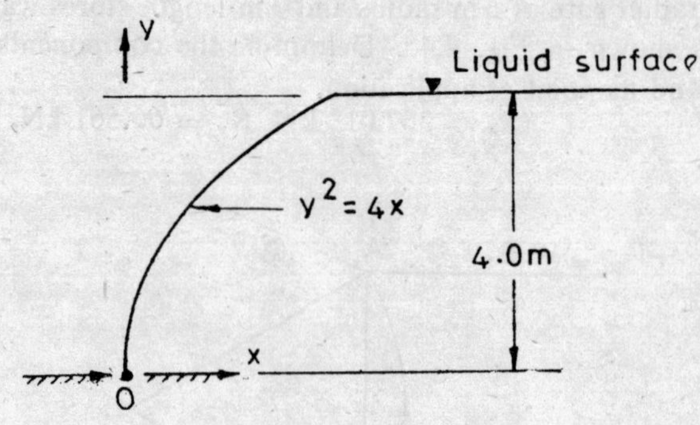

Fig. 4.47.

4.33 What will be the magnitude and direction of total force acting on a semi-cylindrical gate of 1 m diameter and 4 m length if it has 1.5 m depth of water on its convex side above its centre.

(60.701 kN, θ = 14.67°)

4.34 Archimedes found that King Hero's crown supposedly made of gold displaced a volume 0.30 litres and weighed 21 N when immersed in water. He therefore argued that the crown was not of gold Do you concur with his conclusion? Give reason. Take relative density of gold as 19.32. (yes)

4.35 A sphere of diameter 0.5 m and relative density 7.0 when completely submerged in a liquid causes a tension of 3.465 kN in a string from which it is suspended. What is the relative density of the liquid?

(1.59)

4.36 A hydrometer weighs 0.04 N and its stem is 5 mm in diameter. Compute the distance between the markings for liquids of relative densities 1.0 and 1.3. (48.06 mm)

4.37 If a solid conical buoy of height h and relative density S floats in water with its axis vertical and vertex upwards, show that $h(1-S)^{1/3}$ of its axis will be out of water.

4.38 A wooden block of 1 m × 1 m × 1 m dimensions and of relative density 0.70 floats in water. Determine the volume of concrete of relative density 2.5 that needs to be placed on it so that the block is just immersed in water. (0.12 m^3)

4.39 A conical buoy of 2 m radius and 4 m height floats in fresh water with its axis vertical and vertex downwards. Determine the range of the weight of the buoy if it is in stable equilibrium.
(83.97 kN and 164.00 kN)

4.40 What will be the maximum height h of a cylinder of diameter D and relative density S that can float in water in stable equilibrium with its axis vertical? ($h = D/\sqrt{8(1-S)S}$)

4.41 A solid cone of relative density 0.70 floats in fresh water with its axis vertical and vertex downwards. Determine the minimum value of R/h.
(0.3550)

4.42 Show that for a circular cylinder of radius R and height h floating in a liquid with its axis vertical, it will be under stable equilibrium if $R\sqrt{2} > h$ for any combination of mass densities of liquid and material provided the cylinder floats.

4.43 A barge 50 m in length, 12 m in width and 4 m in depth has a draft of 3.0 m when fully loaded. Under this condition the centre of gravity of the barge is 3.5 m above the bottom. What will be the metacentric height? (2.0 m).

4.44 A tank 3 m in length, 2 m wide and 2 m deep containing 1.5 m depth of water is given a constant horizontal acceleration of 3.0 m/s^2. Determine the w.s. slope. Will the water spill out? (0.3058, No)

4.45 In problem 4.44, how much water will spill out if the tank is given an acceleration of 4.0 m/s^2 in the horizontal direction? (0.672 m^3)

4.46 If a tank containing oil of relative density 0.8 to a depth of 2.0 m is given a vertically downward acceleration 4.90 m/s^2, determine the pressure at the bottom of the tank. (7.830 kN/m^2)

4.47 A tank shown in Fig. 4.48 is filled with oil of relative density 0.80 and given a uniform horizontal acceleration of 5.0 m/s^2. If tank is open to atmosphere at A, find pressures at B and C.
(−0.799 kN/m^2 and 22.846 kN/m^2 gauge)

4.48 An open circular cylinder of 1.0 m diameter and 2 m depth is completely filled with water and rotated about its axis at 45 rpm. Deter-

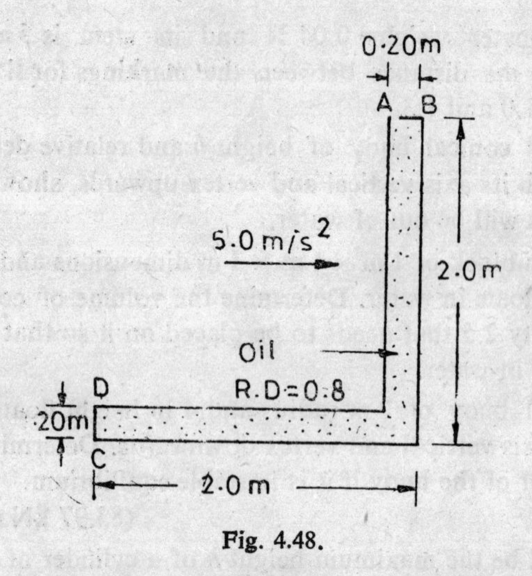

Fig. 4.48.

mine the depth at the axis and amount of water spilled.

(1.717 m, 0.111 m³)

4.49 A U tube whose centres of the limbs are 1.0 m apart is filled to a depth of 0.30 m with acetylene tetrabromide of relative density 2.96. The tube is then rotated about a vertical axis between the two limbs and 0.2 m away from one limb. At what speed should it be rotated so that the difference of liquid levels in the two limbs is 0.30 m.

(29.95 rpm)

4.50 A closed cylindrical tank is filled with oil of relative density 0.80, under a pressure of 200 kN/m² (Fig. 4.49). The tank is 2 m in

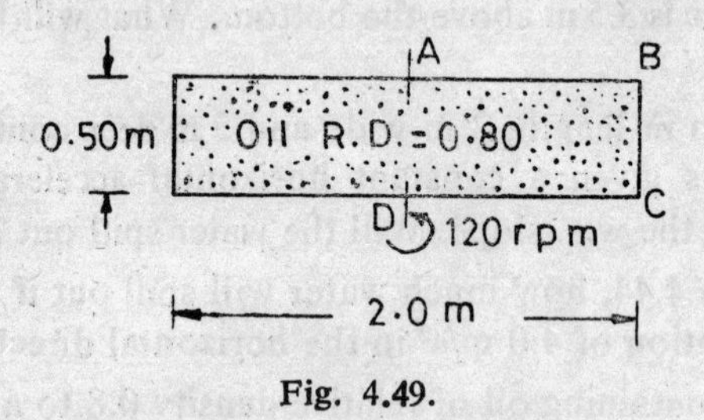

Fig. 4.49.

diameter and 0.5 m in height. If it is rotated at 120 rpm, determine pressures at A, B, C and D.

(200 kN/m², 263.060 kN/m², 203.915 kN/m², 266.975 kN/m²)

4.51 A closed cylinder containing gas at a constant temperature is rotated about its vertical axis at a constant angular velocity ω. Show that relative pressure variation with r is given by

$$p/p_0 = \exp(\omega^2 r^2/2C)$$

where $p/\rho = p_0/\rho_0 = C$ and p_0 and ρ_0 are initial pressure and density of gas.

4.52 A closed cylinder containing gas is rotated about its vertical axis at constant heat content. Show that the pressure variation in the radial direction is given by

$$p = \left[\frac{k-1}{k}\,\frac{\omega^2 r^2}{2C}\,1/k + p_0^{\frac{k-1}{k}}\right]^{\frac{k}{k-1}}$$

where $\dfrac{p}{\rho^k} = \dfrac{p_0}{\rho_0^{\,k}} = C.$

4.53 An open circular cylinder of radius R and height h is completely filled with water with its axis vertical and is rotated about its axis at an angular velocity ω. Determine ω such that the central portion of bottom of diameter R is exposed. ($\sqrt{8gh/3R^2}$)

4.54 What is the maximum pressure to which a 400 mm internal diameter steel pipe of 50 mm thickness can be subjected if the allowable stress in steel is 12×10^7 N/m^2? (26 980 kN/m^2)

DESCRIPTIVE QUESTIONS

4.1 What is hydrostatic pressure variation? Give two examples in which pressure distribution is hydrostatic and two in which it is nonhydrostatic.

4.2 What is Pascal's law? Prove it.

4.3 Is Pascal's law valid when a fluid mass is given an acceleration? Is it valid when there is relative motion between different fluid layers?

4.4 Is the concept of piezometric head relevant in case of gases? Explain.

4.5 Write Euler's equations of motion in cartesian coordinate system, and show that when $u = v = w = 0$, $(p_2 - p_1) = rh$ where h is the difference in elevation of two points.

4.6 What are the common liquids used in manometers? What conditions should it satisfy before you choose a manometric liquid?

4.7 Suggest manometric arrangement to measure a pressure difference of (i) 50 N/m^2, (ii) 20 kN/m^2 between two points along a pipe carrying water. How will the arrangement differ if the pipe carries air?

4.8 Suggest an arrangement to measure a gauge pressure of 100 kN/m^2 in a pipe carrying water.

4.9 What considerations govern the diameter of glass tube to be used in a manometer?

4.10 What will happen if two limbs of U tube manometer are of different diameter?

4.11 When will the centre of gravity and centre of pressure coincide in case of plane immersed surfaces ?

4.12 Why are different techniques adopted to calculate the total force and its point of application, in case of plane and curved immersed surfaces ?

4.13 What is Archimedes principle ? A body floats in between two fluids of specific weights γ_1 and γ_2. What will be the expression for buoyancy force ?

4.14 What is the significance of metacentric height ? For rotational stability is it enough that the floating body has as a large metacentric height as possible? Explain.

4.15 A water body is subjected to an acceleration in the vertically upward direction. At what acceleration will the pressure difference between two points separated by a vertical distance h, be zero?

4.16 Is the flow in a cylinder filled with water and rotating about vertical axis irrotational or rotational ? Can you explain the reason without any computations ?

4.17 Pascal's law is valid
(i) only when the fluid is at rest and is frictionless
(ii) when fluid is at rest, and when the frictionless fluid is in motion
(iii) only when fluid is at rest

4.18 Hydrostatic pressure variation is related to (i) constancy of $(p/\gamma + z)$ (ii) pressure variation in water only (iii) pressure variation in atmosphere.

4.19 A U-tube manometer measures difference in
(i) total energy (ii) pressure (iii) piezometric head

4.20 For a stable equilibrium of a submerged body
(i) B is above G (ii) G is above B (iii) B and G must coincide.
Here B and G are centres of buoyancy and gravity.

4.21 For an unstable equilibrium of a floating body
(i) M is above G (ii) M is below G (iii) M and G coincide.

4.22 If a cylindrical container is filled with water and is rotated about a vertical axis coinciding with the axis of cylinder, the pressure variation in the vertical will follow

(i) $p \sim$ depth (ii) $p \sim (\text{depth})^{1/2}$ (iii) $p \sim \dfrac{1}{\text{depth}}$

(iv) none of the above.

4.23 The centre of pressure will coincide with the centre of gravity if a plane surface is
(i) vertical (ii) horizontal (iii) immersed in a gas
(iv) none of the above.

CHAPTER V

Applications of Bernoulli's Equation

5.1 INTRODUCTION

In this chapter applications of Bernoulli's equation to closed-conduit and free-surface flows are illustrated. These include prediction of pressure distribution in steady two dimensional irrotational flows, cavitation, flow measuring devices such as orifices, mouthpieces, orifice plates, venturimeters, nozzles, bend meters, and weirs, spillways and gates. In all the cases frictional effects are neglected since the flow is rapidly converging and lengths involved are small.

5.2 PRESSURE DISTRIBUTION IN IRROTATIONAL FLOW

In the case of steady, two-dimensional irrotational flow with confining boundaries, there is a unique flow net and therefore a unique variation of velocity in its dimensionless form. Since changes in piezometric head (or presure if elevation remains constant) are related to velocity variation, the former can be determined from the latter once the flow net is obtained. With respect to Fig. 5.1,

$$n_0 U_0 = n_1 U_1 = n_2 U_2 \tag{5·1}$$

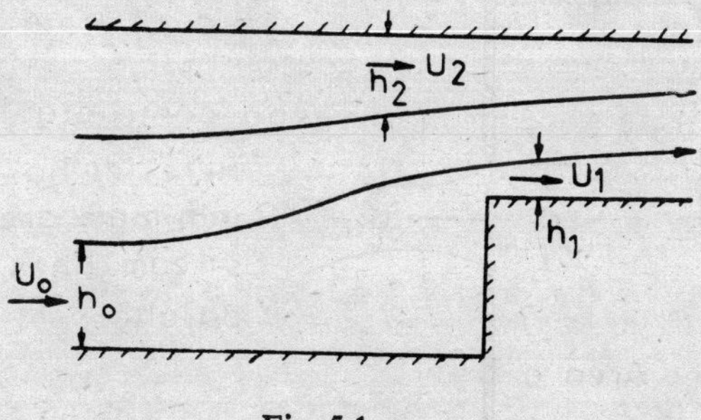

Fig. 5.1.

Hence application of Bernoulli's equation between sections 0 and 1 yields

$$\frac{(h_1 - h_0)}{U_0^2/2g} \quad \text{or} \quad \frac{(p_1 - p_0)}{\rho U_0^2/2} = 1 - \left(\frac{U_1}{U_0}\right)^2 \quad \text{or} \quad \left[1 - \left(\frac{n_0}{n_1}\right)^2\right] \tag{5.2}$$

Hence from the knowledge of relative spacing between consecutive streamlines, the variation of $(p_1 - p_0)/\rho U_0^2/2$ or $\Delta p/\rho U_0^2/2$ can be obtained. The parameter $U_0/\sqrt{2\,\Delta p/\rho}$ is known as Euler number and is the ratio of square root of inertial force per unit volume to pressure force per unit volume.

5.3 HYDRAULIC GRADE LINE AND TOTAL ENERGY LINE

The line joining the points at a height of $(p/\gamma + Z)$ at various sections along the flow is known as hydraulic grade line; any arbitrary datum can be used to find Z. The line joining points at the height of $\left(\frac{p}{\gamma} + Z + U^2/2g\right)$ at various sections is known as total energy line.

5.4 CAVITATION

When pressure at any point in a liquid becomes equal to the vapour pressure of the liquid, the liquid vapourises and develops vapour pockets or bubbles. These may break the continuity of flow as in case of siphons. Formation of bubbles, their transport to regions of high pressure, and subsequent collapse is known as cavitation. This is harmful and should be avoided.

$$\text{Cavitation number } \sigma = \frac{p - p_v}{\rho U_0^2/2}$$

where p is the absolute pressure at point under consideration, p_v is vapour pressure of liquid and U_0 is reference velocity. A minimum safe value of σ is usually prescribed.

5.5 FLOW THROUGH SMALL AND LARGE ORIFICES

For a small circular orifice of area a in the side of a tank, assume $d \ll H_1$ (Fig. 5.2). Then

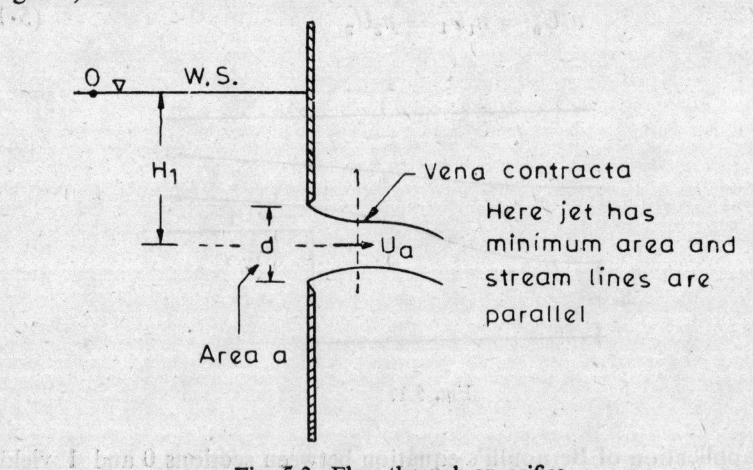

Fig. 5.2. Flow through an orifice.

$$C_c = \text{coeff. of contraction} = \frac{\text{Area of jet at vena contracta}}{\text{Area of opening}}$$

C_c is equal to 0.61 for $d \ll H_1$

Bernoulli's equation gives $U_t = \sqrt{2gH_1}$ and $U_a = C_v\sqrt{2gH}$

where U_t and U_a are the theoretical and actual velocities at vena contracta. The coefficient of velocity C_v is unity if there is no friction. Otherwise

$$C_v = 0.97 \quad \text{to} \quad 0.99.$$

$$Q = \text{area} \times \text{velocity}$$

$$\text{or} \quad Q = C_c a C_v \sqrt{2gH_1} = C_d\, a \sqrt{2gH_1} \tag{5.3}$$

where discharge coefficient

$$C_d = C_c C_v. \tag{5.4}$$

In the case of large orifice d is not very small as compared to H_1 or H_2 (Fig. 5.3). Hence

$$U_t \neq \sqrt{2g\left(\frac{H_1 + H_2}{2}\right)}$$

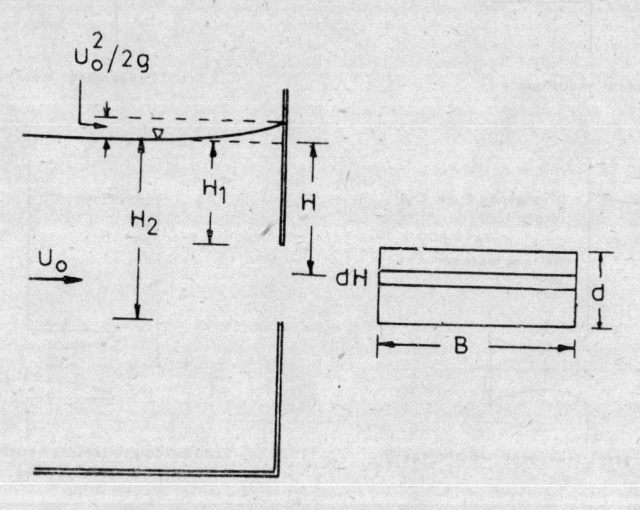

Fig. 5.3. Flow through a large orifice.

In this case $dQ = C_c B dH\sqrt{2gH}$ which on integration yields

$$Q = \frac{2}{3} C_c \sqrt{2g}\, B\, [H_2^{3/2} - H_1^{3/2}] \tag{5.5}$$

if velocity of approaching flow is neglected, or

$$Q = \frac{2}{3} C_c \sqrt{2g}\, B \left[\left(H_2 + \frac{U_0^2}{2g}\right)^{3/2} - \left(H_1 + \frac{U_0^2}{2g}\right)^{3/2}\right] \tag{5.6}$$

if velocity of approach is taken into account.

Values of C_c for large orifices are obtained from Table 5.1 by replacing b/B by $d/(2H_1 + d)$, if B is much greater than d.

5.6 FLOW THROUGH MOUTHPIECES

A mouthpiece is a relatively short pipe ($L \leqslant 2d$) which is fitted internally or externally to the orifice in the side of a tank. Details of these mouthpieces together with flow conditions and values of C_c and C_v are shown in Fig. 5.4. Mouthpieces are sometimes used as a flow measuring device. Calculations for mouthpieces are shown through illustrative examples.

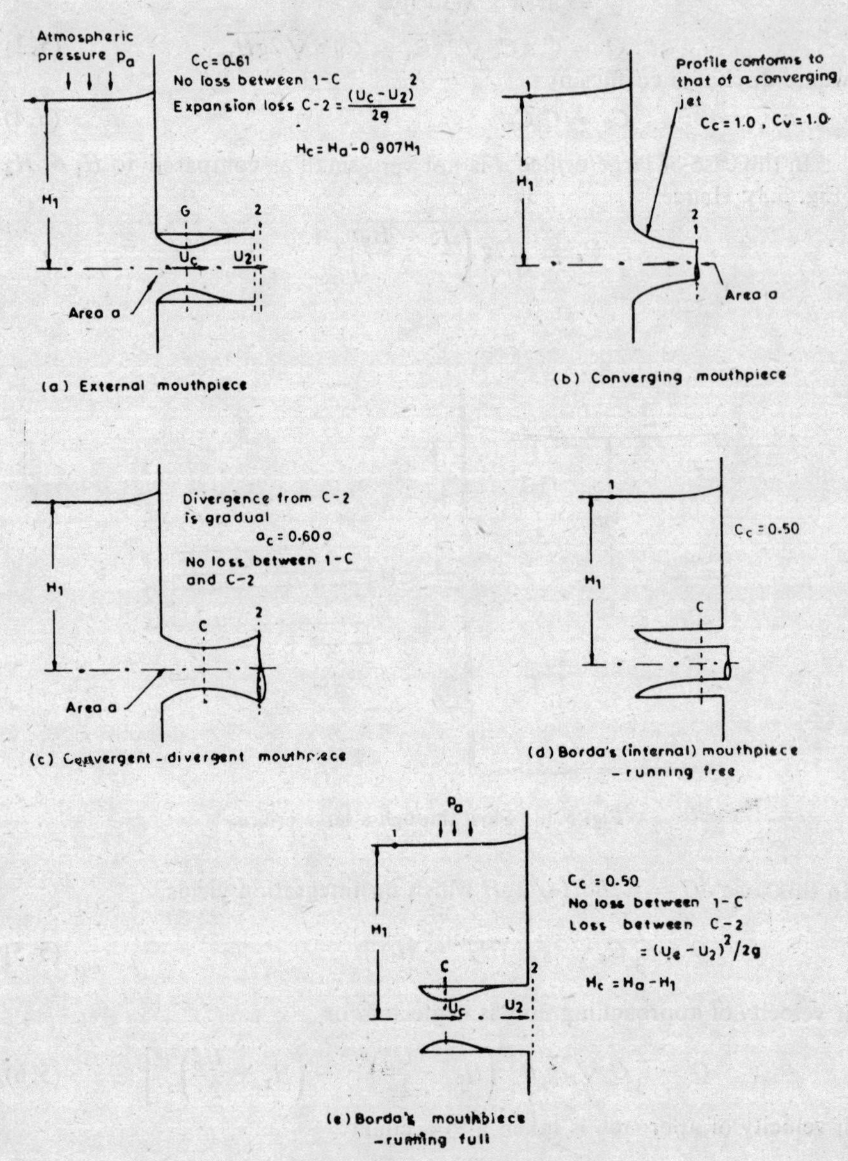

Fig. 5.4. Mouthpieces.

5.7 TWO AND THREE DIMENSIONAL ORIFICES AND FLOW MEASURING DEVICES IN PIPES

5.7.1 Two and three dimensional orifices

Two dimensional orifice is shown in Fig. 5.5. Application of Bernoulli's and continuity equations between sections 1 and 2 yields the discharge equation

$$q = C_d \, b \sqrt{2(p_1 - p_2)/\rho} \tag{5.7}$$

where

$$C_d = C_c / \sqrt{1 - C_c^2 (b/B)^2} \tag{5.8}$$

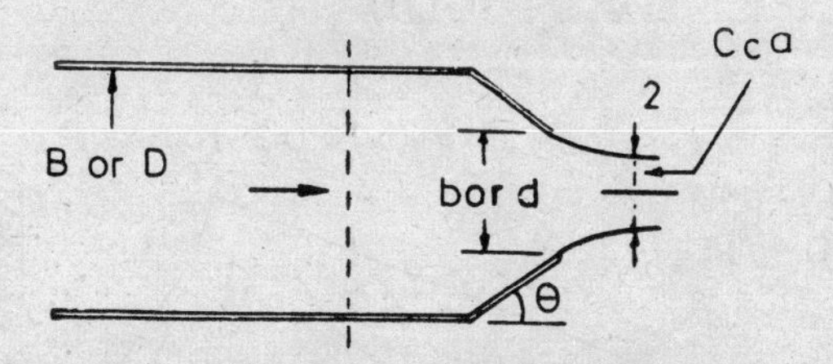

Fig. 5.5. Flow through conduit contraction.

No loss of energy between sections 1 and 2 has been assumed; hence $C_v = 1.0$. For circular orifice plate in a circular pipe, the corresponding equation is

$$Q = C_d \, a \sqrt{2(p_1 - p_2)/\rho} \tag{5.9}$$

where

$$C_d = C_c / \sqrt{1 - C_c^2 (d/D)^4} \tag{5.10}$$

In the above equations, q is the discharge per unit width in m^3/s m while Q is the discharge in m^3/s. C_c is a function of b/B or d/D and θ as shown in Table 5.1.

Table 5.1: Values of C_c for Conduit Contractions

b/B or d/D	θ			
	45°	90°	135°	180°
0	0.746	0.611	0.537	0.500
0.20	0.747	0.616	0.555	0.528
0.40	0.749	0.631	0.564	0.580
0.60	0.758	0.662	0.620	0.613
0.80	0.789	0.722	0.698	0.691
1.0	1.000	1.000	1.000	1.000

5.7.2. Thin plate Orifice meter

Orifice meter is a circular plate with a concentric circular opening fixed in a pipe. This is used for measurement of discharge by relating Q to difference of piezometric head $(h_1 - h_2)$ across the orifice plate (Fig. 5.6). The positions of pressure taps are also shown in the figure. The discharge equation is

$$Q = C_d\, a\sqrt{2g\,(h - h_2)} \quad \text{or} \quad C_d\, a\sqrt{2\Delta p/\rho} \tag{5.11}$$

$$\text{where} \quad C_d = \frac{C_c}{\sqrt{1 - C_c^2\left(\dfrac{d}{D}\right)^4}} \tag{5.12}$$

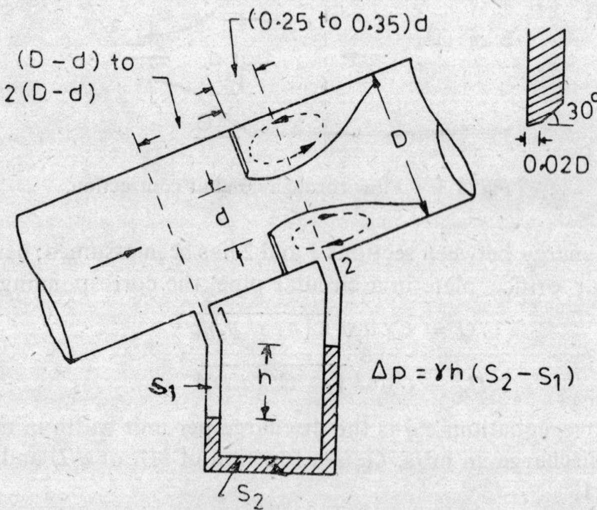

Fig. 5.6. Flow through orifice plate

For real fluids $C_d = f\left(\dfrac{U_2 d\rho}{\mu}, \dfrac{d}{D} \text{ and } M\right)$, where M is Mach number.

For values of $\dfrac{U_2 d\rho}{\mu}$ greater than 5×10^4 and M less than 0.40, the variation of C_d with d/D is as given in Table 5.2.

Table 5.2: Variation of C_d with d/D for orifice plates

$\dfrac{d}{D}$	0.20	0.30	0.40	0.50	0.60	0.70	0.80
C_d	0.58	0.59	0.60	0.63	0.67	0.72	0.78

5.7.3 Venturimeter

This is discussed in Chapter III. The discharge equation is

$$Q = \frac{C}{\sqrt{1-\left(\frac{d}{D}\right)^4}} \sqrt{2g\,(h_1 - h_2)} \qquad (5.13)$$

where the coefficient C is a function of $\frac{U_2 d_2}{u}$ as shown in Fig. 3.10.

5.7.4 Nozzle Meter or Flow nozzle

A nozzle meter is a short cylinder one end of which is flared to form a flange which can be clamped between pipe flanges. This end forms the curved entrance leading the flow into the throat (Fig. 5.7). Upstream pressure tap is located one pipe diameter from the inlet face. The downstream tap is located at $0.5D$ from inlet face. The discharge can be computed using Eq. (5.13) with C values ranging between 0.96 and 0.98.

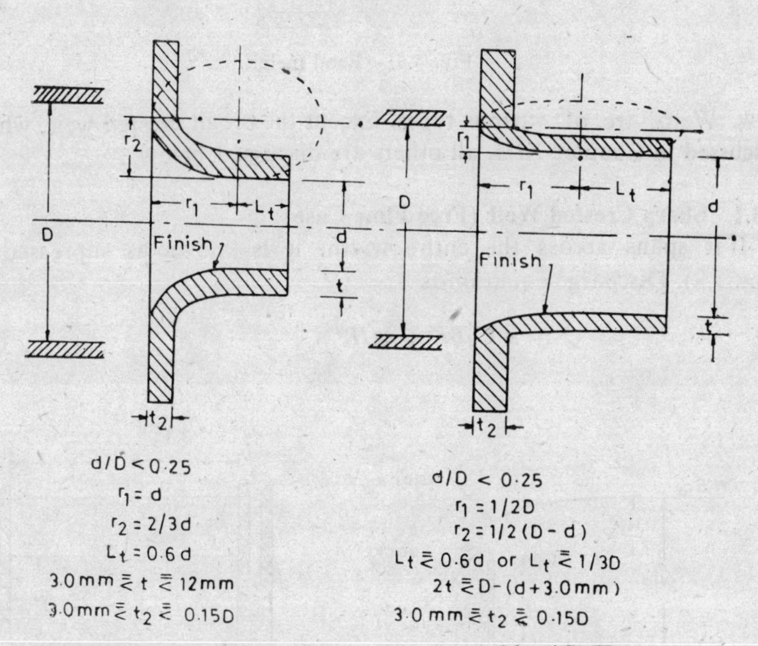

Fig. 5.7. Flow nozzles as recommended by ASME.

5.7.5 Bend Meter

A 90° pipe bend with moderate value of centre line bend radius R_0 to pipe diameter can be used as a flow measuring device if piezometric head difference across the bend $(h_1 - h_2)$ is measured. This difference is related to Q by the equation

$$Q = C_d A \sqrt{2g(h_1 - h_2)} \qquad (5.14)$$

where A is area of pipe, and C_d is equal to $\sqrt{R_0/2D}$ (Fig. 5.8).

5.8 WEIRS

Weir is a structure built across an open channel for measurement of

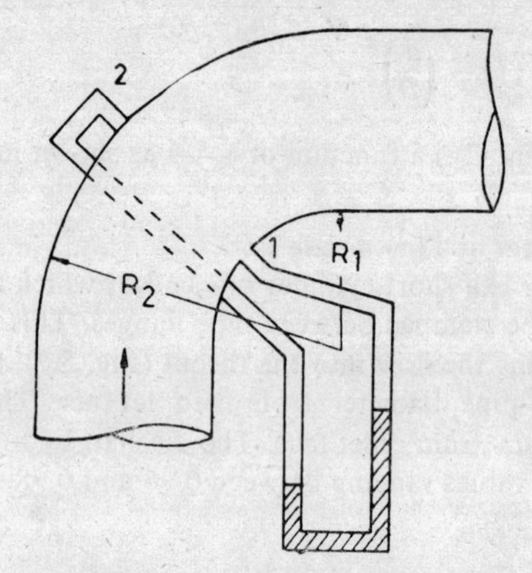

Fig. 5.8. Bend meter.

flow. Weirs are of various types. Except the broad crested weir, which is discussed in Chapter XIII, all others are discussed below.

5.8.1 Sharp Crested Weir (Free Flow Case)

If it spans across the entire width, it is known as supressed weir (Fig. 5.9). Discharge equation is

$$Q = \frac{2}{3} C_d BK \sqrt{2g} H_1^{3/2} \tag{5.15}$$

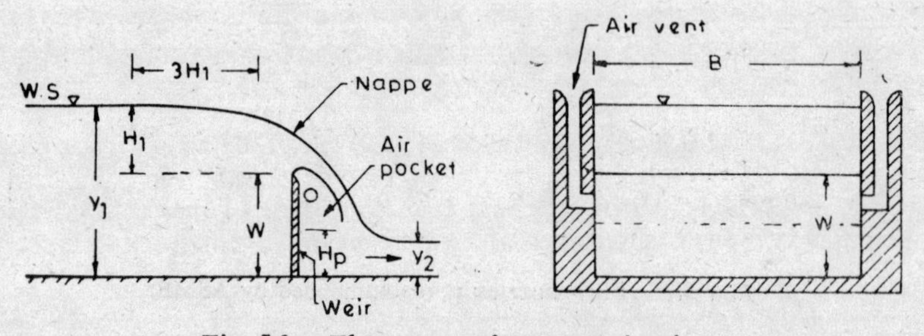

Fig. 5.9. Flow over a sharp crested weir.

where B is the channel width, H_1 is head over the weir ($= y_1 - W$) and discharge coefficient C_d is given by

$$C_d = 0.611 + 0.175 \, (H_1/W) \text{ for } \frac{H_1}{W} \leqslant 0.50 \tag{5.16}$$

where W is the height of weir. The coefficient K is a function of $R_e^{0.2} W_1^{0.60}$ as given in Table 5.3. If H_2 is less than zero, the weir acts as a free flow weir. Here H_2 is downstream depth of flow above the crest of the weir.

Table 5.3: Variation of K with $\mathrm{Re}^{0.2}\ W_1^{0.6}$ for sharp crested weirs and triangular notches

$\mathrm{Re}^{0.2}\ W_1^{0.6}$	30	60	100	400	600	10^3	and above
K	1.27	1.20	1.15	1.04	1.01	10	

Here $\mathrm{Re} = g^{1/2}H_1^{3/2}/\nu$ and $W_1 = gH_1^2\rho/\sigma$. The underside of the nappe has to be aerated in order to get a unique relationship between H_1 and Q. Air demand is given by the equation

$$q_{air} = \frac{0.10q}{(H_p/H_1)^{1.50}}$$

where q_a is air demand in m³/s m.

5.8.2 Contracted Sharp Crested Weir (Free Flow Case)

Here the channel width is greater than the length of weir and y_2 is less than W. Discharge equation proposed by Francis is

$$Q = \frac{2}{3} C_d \sqrt{2g}\,(B - 0.1nH_1)\left[\left(H_1 + \frac{U_0^2}{2g}\right)^{3/2} - \left(\frac{U_0^2}{2g}\right)^{3/2}\right] \qquad (5.17)$$

Here $C_d = 0.622$, n is the number of contractions and U_0 is the velocity in the approach channel. For symmetrically contracted weir $n = 2$, a symmetrically contracted weir with pier in between will have $n = 4$. Equation 5.17 is valid for B/H_1 greater than 3.0 and H_1/W less than one.

5.8.3 Triangular Notch or Weir

The discharge equation for triangular notch is (Fig. 5.10):

$$Q = \frac{8K}{15} C_d \sqrt{2g}\ \tan\frac{\theta}{2}\ H_1^{5/2} \qquad (5.18)$$

K depends on $\mathrm{Re}^{0.20}W_1^{0.60}$ as shown in Table 5.3. The coefficient C_d is a weak function of θ as shown below for $H_1/W \leqslant 0.40$ and $W/B \leqslant 0.20$.

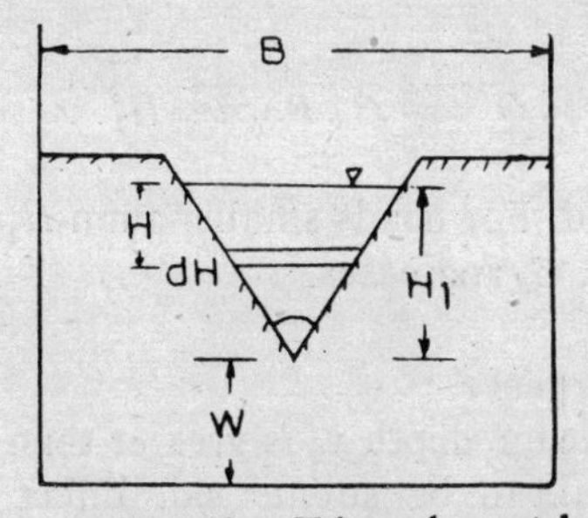

Fig. 5.10. Triangular notch.

Table 5.4: Variation of C_d with θ for triangular notches

θ	20°	40°	60°	80°	100°
C_d	0.595	0.581	0.577	0.577	0.580

5.8.4 Cipolletti Weir

It is a contracted trapezoidal weir with side slopes. $1V: \frac{1}{4} H$ (Fig. 5.11). At this side slope, the reduction in discharge of a rectangular contracted weir of length B due to two end contractions is compensated by additional discharge through two triangular portions. The discharge equation for cipolletti weir is Eq. 5.15 with K obtained from Table 5.3 and C_d from Eq. 5.16.

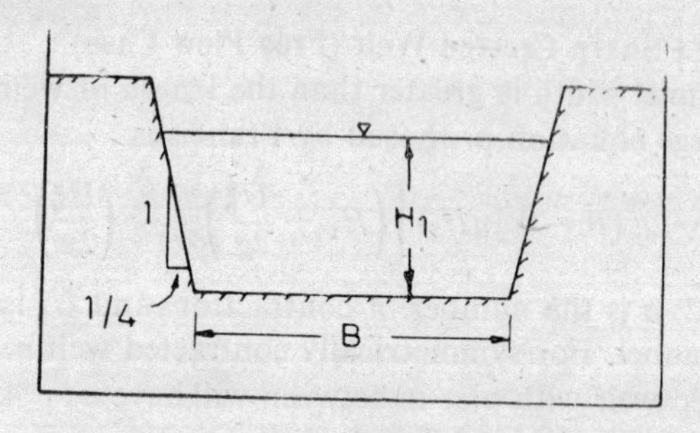

Fig. 5.11. Cipolletti weir.

5.8.5 Ogee Spillways

Spillways are provided on dams to allow excess flood flow, which cannot be stored, to flow in the downstreem channel. The profile of an ogee spillway conforms to the shape of the nappe of sharp crested weir of the same height as spillway and under the same head (Fig. 5.12). If H_1/W is less than 0.75, the discharge coefficient for the spillway can be taken as $C_d = 0.75$ in the equation

$$Q = \frac{2}{3} C_d B \sqrt{2g} H_1^{3/2} \tag{5.19}$$

where H_1 is design head. For heads smaller than H_1, C_d decreases while for heads greater than H_1, C_d increases.

5.8.6 Effect of submergence

When the downstream depth y_2 is greater than the weir height W, the flow over the weir is said to be submerged. Effect of submergence is to

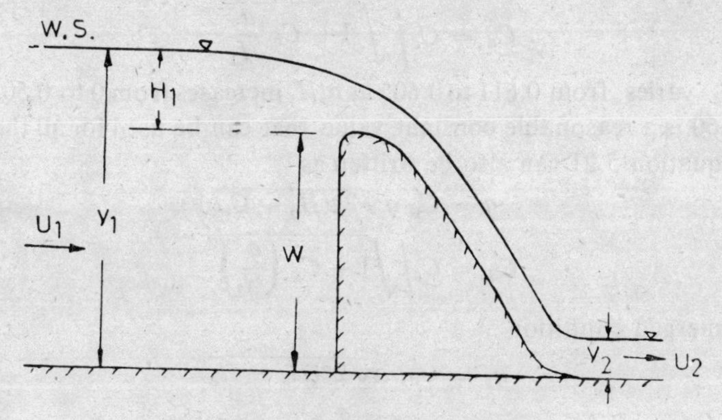

Fig. 5.12. Flow over a spillway.

reduce the discharge for the same H_1. This effect is described by the equation

$$Q_s/Q = \left\{1 - \left(\frac{H_2}{H_1}\right)^m\right\}^{0.385} \tag{5.20}$$

where Q_s is the discharge under submerged condition, Q is the discharge under free flow condition for same H_1, and m is the exponent of H_1 in the weir equation. Thus $m = 1.5$ for sharp crested weir, $m = 2.5$ for triangular notch etc. H_2 is the downstream depth above the weir crest. Equation 5.20 is applicable to rectangular suppressed and contracted weirs, triangular notches, Cipolletti weir, and parabolic and proportional weirs.

5.9 FLOW UNDER A SLUICE GATE

Flow under a sluice gate under free flow condition and submerged flow condition is shown in Fig. 5.13.

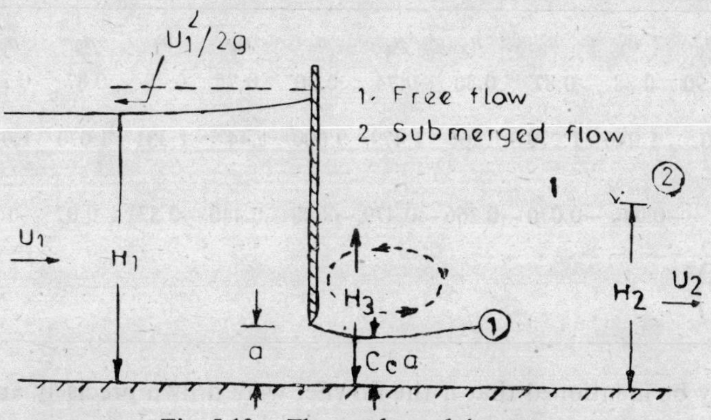

Fig. 5.13. Flow under a sluice gate.

For free flow case

$$q = C_d a \sqrt{2gH_1} \tag{5.21}$$

where $$C_d = C_c \Big/ \sqrt{1 + C_c \frac{a}{H_1}}$$

Here C_c varies from 0.611 to 0.605 as a/H_1 increases from 0 to 0.50. Hence $C_c = 0.60$ is a reasonable constant value that can be used for all the conditions. Equation 5.21 can also be written as

$$q = C_d' \, a\sqrt{2g(H_1 - C_c a)} \tag{5.22}$$

where $$C_d' = C_c \Big/ \sqrt{1 - C_c^2 \left(\frac{a}{H_1}\right)^2}$$

For submerged condition

$$q_s = C_{ds}' \, a\sqrt{2g(H_1 - H_3)} \tag{5.23}$$

where $C_{ds}' = C_d'$ with $C_c = 0.60$ for $\frac{a}{H_1}$ upto 0.70. The value of H_3 must be predicted for given H_1, H_2, a and C_c using momentum equation. This is given by

$$H_3/H_2 = \sqrt{1 - 2F_{r_2}^2 \left(\frac{H_2}{C_c a} - 1\right)} \tag{5.24}$$

where $F_{r2} = U_2/\sqrt{gH_2}$.

ILLUSTRATIVE EXAMPLES

5.1 For the flownet shown in Fig. 2.6 for the flow in a sharp two-dimensional bend, plot the nondimensional pressure distribution along the inner wall.

In Fig. 2.6, starting from the extreme left hand side, measure n, the average spacing between the boundary and stream line $\psi = 1$ at the centre of each mesh, and compute $\Delta p/\rho U_0^2/2$ as indicated below. Assuming flow to be uniform in the first mesh, n there is designated as n_0.

	n_0	n_1	n_2	n_3	n_4	n_5	n_6	n_7	n_8	n_9	n_{10}
cm	0.90	0.88	0.87	0.80	0.74	0.30	0.75	0.78	0.87	0.88	0.90
$\left(\frac{n_0}{n_i}\right)^2$	1.0	1.046	1.070	1.266	1 479	9.000	1.44	1.331	1.070	1.046	1.0
$\frac{\Delta p/U_0^2}{2} = 1 - \left(\frac{n_0}{n_i}\right)^2$	0	−0.046	−0.070	−0.266	−0.479	−8.00	−0.440	−0 331	−0.07	−0.046	0

It may be mentioned that if the flownet were drawn precisely and with fine mesh, at the sharp corner one would have got the spacing between boundary and next stream line as nearly zero giving an infinite velocity and hence $-\infty$ value of $\Delta p/\rho \frac{U_0^2}{2}$. This is indicated in Fig. 5.14.

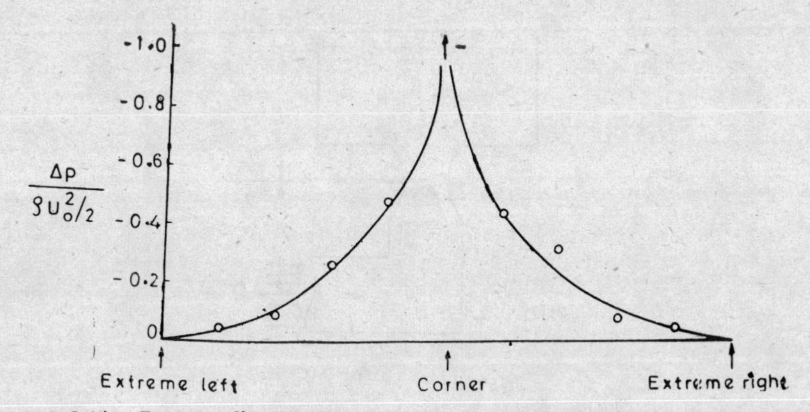

Fig. 5.14. **Pressure distribution on the inside wall of sharp corner in fig. 2.6.**

5.2 **A venturimeter is provided in a 200 mm diameter pipe for measurement of water discharge. For 100 kN/m² gauge pressure in the pipe, determine the diameter of throat of the venturimeter if it is to produce cavitation pressure there. The throat is 2.0 m higher than venturi inlet. Take atmospheric pressure as 101.3 kN/m² and vapour pressure as 2.39 kN/m² absolute. The pipe carries a discharge of 62.8 l/s.**

Hence $p_1 = 101.3 + 100.0 = 201.3$ kN/m² abs.

$$Z_1 = 0, \qquad d_1 = 0.200 \text{ m}$$

$$p_2 = 2.39 \text{ kN/m}^2, \quad Z_2 = 2.0, \quad d_2 = ?$$

$$U_1 = \frac{0.0628}{0.785 \times 0.2^2} = 2.0 \text{ m/s. Hence Bernoulli's equation yields}$$

$$\frac{p_1}{\gamma} + Z_1 + \frac{U_1^2}{2g} = \frac{p_2}{\gamma} + Z_2 + \frac{U_2^2}{2g}$$

$$\therefore \quad \frac{201.3}{9.787} + 0 + \frac{2.0^2}{2 \times 9.806} = \frac{2.39}{9.787} + 2.0 + \frac{U_2^2}{2g}$$

$$\therefore \quad \frac{U_2^2}{2g} = (20.568 + 0.204 - 0.244 - 2.0)$$

$$= 18.528 \qquad \therefore \quad U_2^2 = 363.371 \quad \text{or} \quad U_2 = 19.062 \text{ m/s}$$

$$\therefore \qquad a = 0.0628/19.062 = 0.003\,295 \text{ m}^2$$

$$\text{or} \qquad d = 0.06478 \text{ m} \quad \text{or} \quad 64\,78 \text{ mm}$$

5.3 **If $p_a = 100$ kN/m² and $p_v = 2.39$ kN/m², at what value of h will the cavitation just start in Fig. 5.15? Apply Bernoulli's equation between sections 0 and 1. Assume no frictional loss.**

$$\frac{100}{9.787} + 1.80 + 0 = \frac{2.39}{9.787} + 0 + \frac{U_1^2}{2 \times 9.806}$$

$$\therefore \quad U_1^2 = 11.7734 \times 2 \times 9.806$$

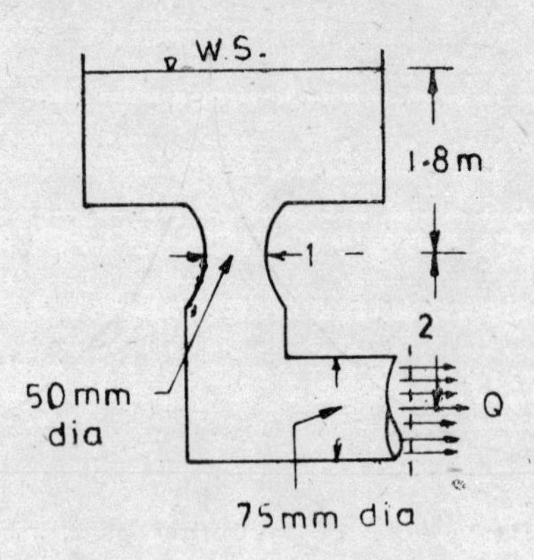

Fig. 5.15.

$$U_1^2 = 230.900 \quad \text{or} \quad U_1 = 15.195 \text{ m/s}$$

$$\therefore \quad Q = 15.195 \times 0.785 \times 0.05^2 = 0.029\ 82 \text{ m}^3\text{/s}$$

Also apply Bernoulli's equation between 0 and 2.

$$U_2 = 0.02982/0.785 \times 0.075^2 = 6.753 \text{ m/s}$$

$$\therefore \frac{100}{9.787} + (1.8 + h) + 0 = \frac{100}{9.787} + 0 + \frac{6.753^2}{2 \times 9.806}$$

$$\therefore \quad 1.8 + h = 2.325 \qquad h = 0.525 \text{ m}$$

5.4 A jet of water coming out from 50 mm diameter rounded nozzle attached to 100 mm diameter pipe is directed vertically downwards. If pressure in the 100 mm diameter pipe 0.20 m above the nozzle is 200 kN/m² gauge, determine the diameter of jet 5.0 m below the nozzle level (Fig. 5.16).

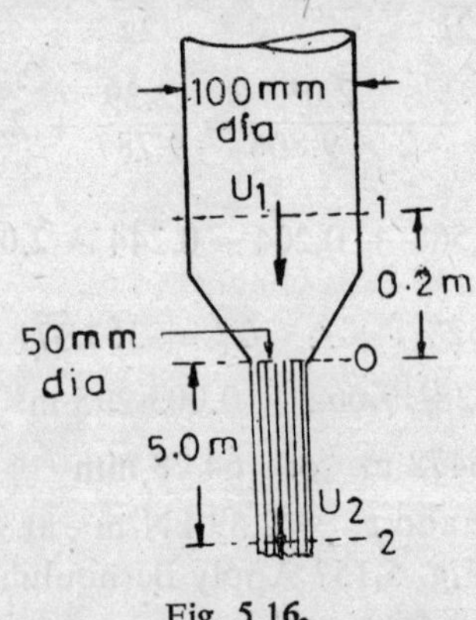

Fig. 5.16.

Continuity equation gives

$$0.785 \times 0.10^2 \times U_1 = 0.785 \times 0.05^2 \times U_0$$

$$U_0 = 4U_1$$

Apply Bernoulli's equation between sections 0 and 1, assuming no loss.

$$\therefore \quad \frac{200}{9.787} + 0.20 + \frac{U_1^2}{2 \times 9.806} = 0 + 0 + \frac{U_0^2}{2 \times 9.806}$$

Substitution of value of U_1 in terms of U_0 in the above equation yields

$$U_0^2 = 431.673 \quad \text{or} \quad U_0 = 20.776 \text{ m/s}$$

As the jet falls down, its velocity increases and hence the diameter decreases. If U_2 is the jet velocity 5.0 m below the nozzle

$$(U_2^2 - U_0^2) = 2 \times 9.806 \times 5.0$$

$$U_2^2 = 20.776^2 + 98.06$$

$$= 529.733 \qquad \therefore \quad U_2 = 23.016 \text{ m/s}$$

$$\therefore \quad d_2^2 \times 0.785 \times 23.016 = 0.05^2 \times 0.785 \times 20.776$$

$$\therefore \qquad d_2 = 0.0475 \text{ m} \quad \text{or} \quad 47.50 \text{ mm}$$

5.5 Oil of relative density 0.80 flows through a pipe line which changes in its size from 150 mm diameter at section A to 300 mm diameter at section B, section B being 4.5 m higher than section A. If gauge pressures at A and B are 200 kN/m² and 140 kN/m² respectively, determine the direction of flow and energy loss when the pipe carries discharge of 0.110 m³/s.

$$U_A = 0.110/0.785 \times 0.150^2 = 6.228 \text{ m/s}$$

$$U_B = 0.110/0.785 \times 0\,300^2 = 1.557 \text{ m/s}$$

Total energy at A, $E_A = \dfrac{p_A}{\gamma} + 0 + \dfrac{U_A^2}{2g}$

$$= \frac{200}{0.8 \times 9.787} + 0 + \frac{6.228^2}{2 \times 9.806}$$

$$= 25.544 + 1.978$$

$$= 27.522 \text{ m}$$

Total energy at B, $E_B = \dfrac{p_B}{\gamma} + 4.5 + \dfrac{U_B^2}{2 \times g}$

$$= \frac{140}{0.8 \times 9.787} + 4.5 + \frac{1.557^2}{2 \times 9.806}$$

$$= 17.881 + 4.500 + 0.124$$

$$= 22.505 \text{ m}$$

Since E_A is greater than E_B, flow will be from A to B.

Energy loss $= E_A - E_B = 27.522 - 22.505$

$$= 5.017 \text{ m of oil.}$$

5.6 A large tank resting on the floor is filled to a depth of 6.0 m. A sharp edged orifice of 10 mm diameter is located 2.5 m above the floor level.

For $C_c = 0.61$ and $C_v = 0.97$, determine the discharge and horizontal distance from the tank where the jet will strike the ground. Also determine C_d.

$$H = 6.0 - 2.5 = 3.5 \text{ m}$$

$$U_t = \sqrt{2gH} = \sqrt{2 \times 9.806 \times 3.5} = 8.285 \text{ m/s}$$

$$U_a = C_v U_t = 0.97 \times 8.285 = 8.0365 \text{ m/s}$$

Area of jet at vena contracta $= C_c a = 0.61 \times 0.785 \times 0.01^2$

$$a_c = 4.7885 \times 10^{-5} \text{ m}^2$$

$$\therefore \quad Q = U_a a_c = 8.0365 \times 4.7885 \times 10^{-5} = 3.848 \times 10^{-4} \text{ m}^3\text{/s}$$

$$= 0.3848 \text{ l/s.}$$

$$C_d = C_c C_v = 0.61 \times 0.97 = 0.5917$$

Vertical distance travelled by jet $y = 2.5$ m

If time required is t, $\quad 2.5 = \frac{1}{2} \times 9.806\, t^2$

$$\therefore \quad t = \sqrt{2 \times 2.5/9.806} = 0.714 \text{ s}$$

Horizontal distance travelled during this period, x will be

$$x = U_a \times t = 8.0365 \times 0714 = 5.739 \text{ m}$$

5.7 For the two orifices shown in Fig. 5.17, determine y_2 such that

$$x_2 = 3x_1/4.$$

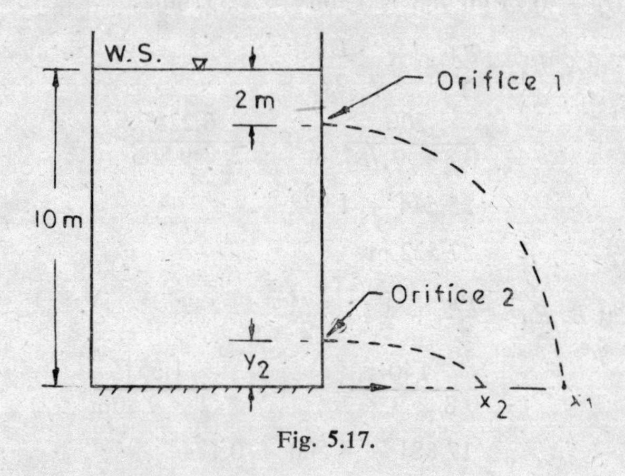

Fig. 5.17.

Consider the jet from 1st orifice.

$$y_1 = 10 - 2 = 8 \text{ m}$$

Since $\quad y_1 = \frac{1}{2} g t_1^2$

$$t_1 = \sqrt{2 \times 8/g} = 4/\sqrt{g} \text{ s}$$

and $\quad x_1 = \sqrt{2g \times 2} \times t_1 = 2\sqrt{g} \times \dfrac{4}{\sqrt{g}} = 8.0 \text{ m}$

$$\therefore \qquad x_2 = \frac{3x_1}{4} = \frac{3 \times 8}{4} = 6.0 \text{ m}$$

But $x_2 = U_2 t_2$ where $U_2 = \sqrt{2g(10 - y_2)}$ and $t_2 = \sqrt{\frac{2y_2}{g}}$

$$\text{or} \qquad 6 = \sqrt{2g(10 - y_2)} \times \sqrt{\frac{2y_2}{g}}$$

$$\therefore \qquad 9 = 10y_2 - y_2^2$$

$$\text{or} \qquad y_2^2 - 10y_2 + 9 = 0$$

$$\therefore \qquad (y_2 - 9)(y_2 - 1) = 0$$

$$\therefore \qquad y_2 = 9.0 \text{ m or } 1.0 \text{ m}$$

5.8 A large closed tank is partly filled with oil of relative density 0.80 to a depth of 3.0 m above 75 mm diameter orifice in its side. What should be the gauge pressure in the tank so that the jet has 0.788 kW power? Take $C_c = 0.60$ and $C_v = 0.92$.

Let p kN/m² be the gauge pressure in the tank. Application of Bernoulli's equation between oil surface and vena contracta of jet will give

$$\frac{p}{9.787 \times 0.8} + 3.0 + 0 = 0 + 0 + \frac{U_t^2}{2 \times 9.806}$$

where U_t is the theoretical velocity of jet.

$$\therefore \qquad U_t = \sqrt{2 \times 9.806\,(3 + 0.127\,72p)}$$

$$Q = C_c C_v a U_t$$

$$= 0.60 \times 0.92 \times 0.785 \times 0.075^2 \times U_t = 0.002\,437\,U_t$$

$$P = \text{power of jet} = \frac{1}{2}\,Q\rho\,U_a^2 = \frac{1}{2}\,Q\rho\,C_v^2 U_t^2$$

where U_a is the actual velocity of jet.

$$\therefore \qquad P = 0.50 \times 0.002437\,U_t \times 998 \times 0.80 \times 0.92^2 \times U_t^2$$

$$= 0.8234 U_t^3 \text{ W}$$

Substituting the value of U_t in the expression for P and equating the result with 788, one gets

$$788 = 0.8234[2 \times 9.806\,(3 + 0.127\,72p]^{3/2}$$

$$\text{or} \qquad 2 \times 9.806\,(3 + 0.127\,72p) = 97.135$$

$$p = 15.292 \text{ kN/m}^2$$

5.9 Determine the discharge flowing through a 0.30 m deep and 0.90 m wide orifice with 1.5 m head over its centre. The tank is 3 m deep and 2 m wide.

Here $b/B = \dfrac{0.30}{3.0} = 0.10$. Therefore, from Table 5.1, $C_c = 0.613$.

$$\therefore \quad H_1 = (1.50 - 0.15) = 1.35 \text{ m}$$

$$H_2 = (1.50 + 0.15) = 1.65 \text{ m}$$

Assuming velocity of approach to be zero,

$$Q = \frac{2}{3} \times 0.613 \times 0.90 \times \sqrt{2 \times 9.806}\ (1.65^{3/2} - 1.35^{3/2})$$

$$= 0.898 \text{ m}^3/\text{s}$$

$$\therefore \quad U_0 = 0.898/(2 \times 3) = 0.150 \text{ m/s}$$

$$\therefore \quad \left(H_1 + \frac{U_0^2}{2g}\right) = 1.35 + 0.00115 = 1.351\ 15 \text{ m}$$

$$\left(H_2 + \frac{U_0^2}{2g}\right) = 1.65 + 0.001\ 15 = 1.651\ 15 \text{ m}$$

Hence, if velocity head is taken into account,

$$Q = \frac{2}{3} \times 0.613 \times 0.90 \times \sqrt{2 \times 9.806}\ (1.651\ 15^{3/2} - 1.351\ 15^{3/2})$$

$$= 0.8982 \text{ m}^3/\text{s}.$$

The difference in the discharge calculated neglecting the velocity of approach and that considering the velocity of approach is small enough for the effect of velocity of approach to be neglected.

5.10 Obtain an expression for absolute pressure at the vena contracta for an external mouthpiece (Fig. 5.4 (a)).

Let a be the area of the mouthpiece. Hence area of jet at vena contracta $= C_c a$. According to continuity equation

$$C_c a\, U_c = U_2 a$$

$$\text{or} \quad U_c = U_2/C_c$$

As the jet contracts upto section C, there is little or no energy loss between sections 1 and C. As the jet expands between sections C and 2, expansion loss there will be given by the formula for sudden expansion in pipes (see Chapter VI), namely

$$(U_c - U_2)^2/2g \quad \text{i.e} \quad \frac{U_2^2}{2g}\left(\frac{1}{C_c} - 1\right)^2 \quad \text{or} \quad 0.409\,\frac{U_2^2}{2g} \text{ if } C_c = 0.61$$

is assumed. Bernoulli's equation can be applied between sections 0 and 2

$$\therefore \quad H_1 + 0 + 0 = 0 + 0 + \frac{U_2^2}{2g} + \frac{0.409\, U_2^2}{2g}$$

$$\therefore \quad U_2 = 0.842\sqrt{2gH_1}$$

$$\text{and} \quad U_c = U_2/C_c = \frac{0.842}{0.610}\sqrt{2gH_1} = 1.381\sqrt{2gH_1}$$

To determine absolute pressure at vena contracta, apply Bernoulli's equation between sections 0 and *C*. Let H_a be the water head corresponding to atmospheric pressure p_a.

$$H_a + H_1 + 0 = \frac{p_c}{\gamma} + \frac{U_c^2}{2g}$$

or Substituting the value of $U_c = 1.381\sqrt{2gH_1}$

$$H_1 + H_a = \frac{p_c}{\gamma} + 1.907\ H_1$$

$$\therefore \quad \frac{p_c}{\gamma} = (H_a - 0.907\ H_1)$$

5.11 What will be the discharge through a properly shaped converging external mouthpiece of 75 mm diameter under a head of 3.5 m?

In a convergent mouthpiece the diameter at the discharge end is 75 mm and it expands gradually in the upstream direction to conform with the jet profile (Fig. 5.4 (b)). Hence energy loss between sections 0 and 2 is zero.

$$Q = a\sqrt{2gH_1}$$

$$= 0.785 \times 0.075^2 \times \sqrt{2 \times 9.806 \times 3.5}$$

$$= 0.036\,58\ \text{m}^3/\text{s} \quad \text{or} \quad 36.58\ \text{l/s}$$

5.12 Determine the discharge through a 100 mm diameter internal mouthpiece under a head of 2.0 m when it flows freely. What will be the discharge and pressure at vena contracta if it runs full?

Free flow condition (Fig. 5.4 (d)):

$$2.0 + 0 + 0 = 0 + 0 + \frac{U_c^2}{2g}$$

$$\therefore \quad U_c = \sqrt{2 \times 2 \times 9.806} = 6.263\ \text{m/s}$$

Area at section *C* is $C_c a$ where $C_c = 0.50$

$$\therefore \quad a_c = 0.50 \times 0.785 \times 0.10^2 = 0.003\,925\ \text{m}^2$$

$$\therefore \quad Q = U_c a_c = 6.263 \times 0.003\,925 = 0.02458\ \text{m}^3/\text{s}$$

$$\text{or} \quad Q = 24.58\ \text{l/s}$$

Running full condition (Fig. 5.4 (e)):

Application of Bernoulli's equation between sections 0 and 2 yields

$$2.0 + 0 + 0 = 0 + 0 + \frac{U_2^2}{2g} + \frac{(U_c - U_2)^2}{2g}$$

Further $\quad U_c.C_c a = U_2 a \quad \text{or} \quad U_c = U_2/C_c = 2.0U_2$

$$\therefore \quad 2.0 = \frac{2U_2^2}{2g} \qquad \therefore \quad U_2 = \sqrt{2 \times 9.806} = 4.429\ \text{m/s}$$

$$Q = aU_2 = 0.785 \times 0.1^2 \times 4.429$$
$$= 0.03\,477 \text{ m}^3/\text{s} \quad \text{or} \quad 34.77 \text{ l/s}.$$

To determine pressure at vena contracta, apply Bernoulli's equation between sections 0 and C.

$$H_a + 2.0 + 0 = \frac{p_c}{\gamma} + \frac{U_c^2}{2g} \quad \text{and } U_c = 2U_2 = 2 \times 4.429 = 8.858 \text{ m/s}$$

$$\therefore \quad \frac{p_c}{\gamma} = H_a + 2.0 - \frac{8.858^2}{2 \times 9.806}$$
$$= (H_a - 2.0)$$

where H_a is head equivalent to the atmospheric pressure.

5.13 Discuss the theory of convergent-divergent mouthpiece and obtain an expression for ratio of areas at throat and outlet for maximum discharge condition.

The convergent portion of convergent-divergent mouthpiece is so shaped that it conforms to the shape of the jet issuing from an orifice upto the vena contracta (Fig. 5.4(c)). The divergent portion is such that its area increases very gradually. Hence it can be assumed that there is no energy loss between sections 0 and C, and also between C and 2. Bernoulli's equation then gives

$$H_1 + H_a = \frac{p_c}{\gamma} + \frac{U_c^2}{2g} = H_a + \frac{U_2^2}{2g}$$

For discharge to be maximum, p_c can be replaced by the vapour pressure p_v.

$$\therefore \quad U_c = \sqrt{2g\left(H_1 + H_a - \frac{p_v}{\gamma}\right)}$$

and $\quad U_2 = \sqrt{2gH_1}$

According to the continuity equation $aU_2 = a_cU_c$

$$\therefore \quad \frac{a_c}{a} = \frac{U_2}{U_c} = \sqrt{\frac{H_1}{\left(H_1 + H_a - \dfrac{p_v}{\gamma}\right)}}$$

5.14 For given H_1 and d_c, determine D such that Q is maximum (Fig. 5.18).

Apply Bernoulli's equation between 0 and 2.

$$\therefore \quad H_1 = \frac{U_2^2}{2g} + \frac{(U_c - U_2)^2}{2g}$$

and continuity equation gives

$$\frac{\pi}{4} d_c^2\, U_c = \frac{\pi}{4} D^2 U_2 = Q$$

$$\therefore \quad H_1 = \frac{1}{2g}\left[\left(\frac{4Q}{\pi D^2}\right)^2 + \left\{\left(\frac{4Q}{\pi d_c^2}\right) - \left(\frac{4Q}{\pi D^2}\right)\right\}^2\right]$$

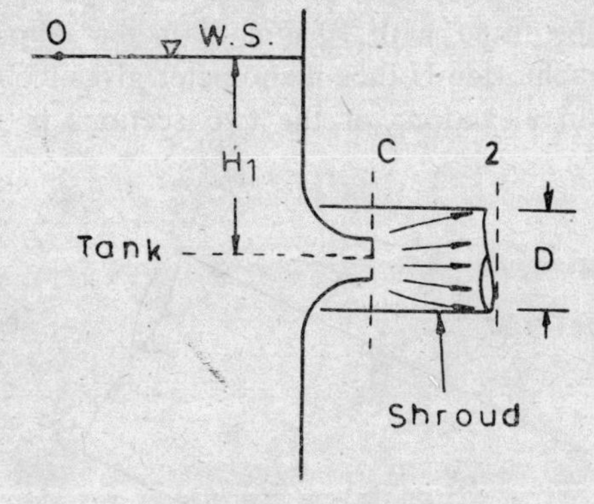

Fig. 5.18.

$$= \frac{8Q^2}{\pi^2 g}\left(\frac{1}{D^4} + \frac{1}{d_c^4} + \frac{1}{D^4} - \frac{2}{D^2 d_c^2}\right)$$

or
$$Q^2 = \frac{\pi^2 g H_1}{8} \frac{D^4 d_c^4}{(2d_c^4 - 2D^2 d_c^2 + D^4)}$$

For Q to be maximum $\frac{\partial Q}{\partial D} = 0$, which gives

$$D = \sqrt{2} d_c.$$

Substituting this value of D in the equation for Q, one gets the maximum value of Q as

$$Q_m^2 = \frac{\pi^2 g H_1}{8} \frac{4d_c^8}{(2d_c^4 - 4d_c^4 + 4d_c^4)} = \frac{\pi^2 g H_1 d_c^4}{4}$$

or
$$Q = \frac{\pi d_c^2}{2} \sqrt{gH_1} \quad \text{or} \quad \frac{\pi d_c^2}{2\sqrt{2}} \sqrt{2gH_1}$$

5.15 Determine the discharge of air flowing through 0.25 m deep closed two dimensional conduit at the end of which is an orifice of opening 100 mm with 45° angle of inclination. The pressure in the conduit is 150 kN/m². Take $\rho = 1.208$ kg/m².

From Table 5.1, for $\frac{b}{B} = \frac{100}{250}$ i.e. 0.40 and $\theta = 45°$

$$C_c = 0.749$$

$$\therefore \quad C_d = \frac{C_c}{\sqrt{1 - C_c^2\left(\frac{b}{B}\right)^2}} = \frac{0.749}{\sqrt{1 - 0.749^2(0.40)^2}} = 0.785$$

$$\therefore \quad q = C_d b\sqrt{2\Delta p/\rho} = 0.785 \times 0.100 \sqrt{\frac{2 \times 150}{1.208}} = 1.237 \text{ m}^3\text{/s m}$$

5.16 Determine the discharge of water through a 100 mm diameter inclined pipe fitted with 50 mm diameter shape-edged orifice, if carbon tetrachloride U tube manometer gives 0.50 m differential and difference in elevations of the two sections is 0.10 m as shown in Fig. 5.19.

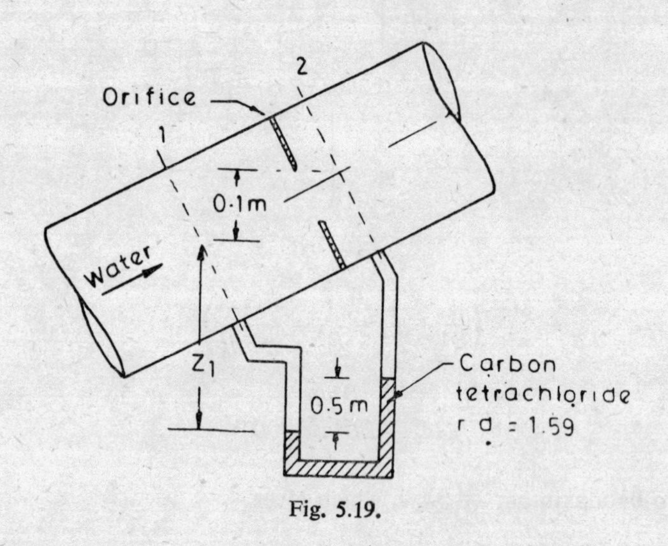

Fig. 5.19.

For $d/D = 0.50$, $C_e = 0.647$

$$\left(\frac{p_1}{\gamma} + Z_1\right) = \frac{p_2}{\gamma} + 0.1 + (Z_1 - 0.50) + 0.50 \times 1.59$$

or
$$\left(\frac{p_1}{\gamma} + Z_1\right) - \left(\frac{p_2}{\gamma} + Z_1 + 0.1\right) = \Delta h = 0.50(1.59 - 1)$$

or
$$\Delta h = 0.295 \text{ m}$$

where Δh is the difference in piezometric heads.

$$Q = \frac{C_c}{\sqrt{1 - C_c^2\left(\frac{d}{D}\right)^4}} \frac{\pi d^2}{4} \sqrt{2g\Delta h}$$

$$= \frac{0.647}{\sqrt{1 - 0.647^2(0.5)^4}} \times 0.785 \times (0.05)^2 \sqrt{2 \times 9.806 \times 0.295}$$

$$= 0.003\ 095 \text{ m}^3/\text{s} \quad \text{or} \quad 3.095 \text{ l/s.}$$

5.17 For the venturimeter of 150 mm × 75 mm dimensions, determine Δh in the mercury manometer if the pipe carries a discharge of 35.32 l/s of oil of relative density 0.80. Take $C = 0.97$.

$$Q = \frac{C \times a}{\sqrt{1 - \left(\frac{d}{D}\right)^4}} \sqrt{2g(h_1 - h_2)}$$

∴

where $(h_1 - h_2)$ is the difference in piezometric heads between sections 1 and 2.

$$\therefore \qquad (h_1 - h_2) = Q^2\left(1 - \left(\frac{d}{D}\right)^4\right) \Big/ C^2a^2 \times 2g$$

$$a = 0.785 \times 0.075^2 = 0.004\,416 \text{ m}^2$$

$$\therefore \qquad (h_1 - h_2) = \frac{0.035\,32^2\,(1 - 0.5^4)}{0.97^2 \times (.004\,416)^2 \times 2 \times 9.806} = 3.25 \text{ m}$$

but $\qquad (h_1 - h_2) = \Delta h\,(13.55 - 0.80)$

$$\therefore \qquad \Delta h = 3.25/12.75 = 0.255 \text{ m of mercury}$$

5.18 To straighten the flow in a 300 mm diameter pipe carrying water, a perforated plate having 25 sharp edged circular holes of 40 mm diameter is used. What will be the drop in pressure across the plate if the pipe carries 0.15 m³/s of water?

$$\frac{\text{Opening area}}{\text{Pipe area}} = \frac{25 \times 0.785 \times 0.04^2}{0.785 \times 0.3^2} = 0.444$$

$$\therefore \qquad \text{Effective } \frac{d}{D} = \sqrt{0.444} = 0.667$$

From Table 5.1 for $d/D = 0.667$, $\theta = 90°$

$$C_c = 0.680, \quad \text{Take} \quad C_v = 1.0$$

$$Q = 0.68 \times (25 \times 0.785 \times 0.04^2)\sqrt{2\Delta p/998} = 0.15$$

$$\therefore \qquad \Delta p = 24626.68 \text{ N/m}^2 \quad \text{or} \quad 24.627 \text{ kN/m}^2$$

Here it is assumed that the holes do not interfere with one another; hence C_c is read from Table 5.1. Such an assumption is not, however, fully justified.

5.19 Water discharge in a 100 mm diameter pipe ranges from 1.5 l/s to 20.0 l/s. Recommend the diameter of the orifice and manometric arrangement.

The discharge equation can be written as

$$(h_1 - h_2) = \frac{Q^2}{C_d^2 a^2 \times 2g}$$

Assume $d/D = 0.50$; hence from Table 5.1

$$\therefore \qquad C_d = \frac{C_c}{\sqrt{1 - C_c\,(d/D)^4}} = 0.63$$

and $\qquad a = 0.785 \times 0.05^2$

$$= 0.00196 \text{ m}^2$$

$$\therefore \qquad (h_1 - h_2)_{\min} = 1.5^2 \times 10^{-6}/(0.63^2 \times .00196^2 \times 2 \times 9.806)$$

$$= 0.0752 \text{ m} \quad \text{or} \quad 75.2 \text{ mm of water}$$

$$(h_1 - h_2)_{max} = 20^2 \times 10^{-6}/(0.63^2 \times 0.00196^2 \times 2 \times 9.806)$$
$$= 13.377 \text{ m of water.}$$

With a U-tube mercury manometer this will give 75.2/12.55 or 6.0 mm and 13.377/12.55 = 1.0659 m of Δh. The latter is too large. Hence d/D ratio should be increased. Let us take $d/D = 0.60$ for which $C_d = 0.67$ and

$$(h_1 - h_2)_{min} = 0.032 \text{ m} \quad \text{or} \quad 32 \text{ mm}$$

$$(h_1 - h_2)_{max} = 5.689 \text{ m.}$$

If mercury U-tube manometer is used for the entire discharge Δh for lowest flow will be 32/12.55 i.e. or 2.55 mm. This is too small. Hence it is recommended that low flows be measured using inverted U-tube manometer and high flows using U-tube manometer with mercury as the manometric liquid. The maximum Δh will be 5.689/12.55 or 0.453 m of mercury.

5.20 Many times, a 90° pipe bend with moderate value of R_0/D is used as a flow measuring device, by relating Q to pressure difference between outside and inside of the bend. Assume that the fluid flows at a constant velocity around the bend and obtain the discharge equation (Fig. 5.8).

According to Euler's equation

$$-\frac{\partial p}{\partial r} = -\rho \frac{U^2}{r}$$

or
$$dp = +\rho \frac{U^2}{r} dr$$

which on integration gives

$$(p_2 - p_1) = \rho U^2 \ln \frac{R_2}{R_1}$$

or
$$U = \sqrt{\frac{(p_2 - p_1)}{\rho \ln (R_2/R_1)}}$$

$\therefore$
$$Q = AU = A \sqrt{\frac{(p_2 - p_1)}{\rho \ln (R_2/R_1)}}$$

If this equation is compared with the standard form of discharge equation viz $Q = C_d \cdot A \dfrac{\sqrt{2(p_2 - p_1)}}{\rho}$, one gets

$$C_d = 1 \Big/ \sqrt{2 \ln \frac{R_2}{R_1}}$$

5.21 A rectangular conduit of 1.8 m width and 1.0 m depth has a 90° bend with inner and outer radii as 5.1 m and 6.9 m respectively. When it carries water, the pressure difference across the bend is equivalent to 0.225 m of water. Assuming the average velocity in the vertical to vary with r according to free vortex law, determine Q.

For free vortex flow $ur = C$

$$Q = \int_{5.1}^{6.9} 1 \times u dr = C \int_{5.1}^{6.9} \frac{dr}{r} = C \ln \frac{6.9}{5.1} = 0.3023\ C$$

$\therefore \quad 1 \times 1.8U = 0.3023\ C \quad$ or $\quad C = 5.9543\ U$

$\therefore \quad u = C/r = 5.9543\ U/r$

where U is average velocity over the cross-section.

According to Euler's equation of motion,

$$\frac{\partial p}{\partial r} = \frac{\rho u^2}{r} = \rho \times (5.9543)^2 \frac{U^2}{r^3}$$

$$\therefore \quad (p_2 - p_1) = \frac{\rho (5.9543)^2 U^2}{2} \left(\frac{1}{5.1^2} - \frac{1}{6.9^2} \right)$$

$$= \frac{\rho \times (5.9543)^2 \times 21.6 U^2}{2 \times 5.1^2 \times 6.9^2}$$

or
$$\frac{p_1 - p_1}{\gamma} = \Delta h = \frac{5.9543^2 \times 21.6}{2 \times 9.806 \times 5.1^2 \times 6.9^2} U^2 = 0.03153\ U^2$$

Substituting $\Delta h = 0.225$ one gets $U = 2.671$ m/s

$$\therefore \quad Q = 1 \times 1.8 \times 2.671$$

$$= 4.808 \text{ m}^3/\text{s}$$

5.22 Determine the discharge over 1.5 m high sharp crested weir fixed across 2.0 m wide rectangular channel when head over the weir is 0.05 m.

Here $\quad H_1 = 0.05$ m, $W = 1.5$ m $B = 2.0$ m

Assuming $\quad \nu = 1 \times 10^{-6}$ m²/s and $\sigma = 0.0735$ N/m

$$\text{Re} = g^{1/2} H_1^{3/2} / \nu = 9.806^{1/2} \times (0.05)^{3/2} / 1 \times 10^{-6} = 35010$$

$$W_1 = \rho g H_1^2 / \sigma = 998 \times 9.806 \times (0.05)^2 / 0.0735 = 332.87$$

$\therefore \quad \text{Re}^{0.2} W_1^{0.6} = 264.363$; for this value of $\text{Re}^{0.2} W_1^{0.6}$

$$K = 1.0895$$

For $\quad H_1/W = 0.05/1.5 = 0.0333$

$$C_d = 0.611 + 0.075 \times 0.0333 = 0.613$$

$$Q = \frac{2}{3} \times 0.613 \times 1.0895 \times \sqrt{2 \times 9.806} \times (0.05)^{3/2}$$

$$= 0.044\ 09 \text{ m}^3/\text{s} \quad \text{or} \quad 44.09 \text{ l/s}$$

5.23 In a 6.0 m wide rectangular channel with 1.2 m depth of flow a sharp crested weir of 2.5 m length and 0.60 m height is fixed symmetrically arcoss the channel width. If it flows free, determine the discharge.

Head over the weir $H_1 = (1.20 - 0.50) = 0.60$ m

Since weir is fixed symmetrically across the width, the number of contractions $n = 2$, and effective length

$$= (B - 0.1 \times 2 \times 0.60) = (2.50 - 0.12) = 2.38 \text{ m}$$

Neglecting the effect of velocity of approach, Q can be determined using Francis formula, namely

$$Q = \frac{2}{3} \times 0.622 \times \sqrt{2 \times 9.806} \times 2.38 \times 0.60^{3/2} = 2.035 \text{ m}^3/\text{s}$$

This value of Q can now be refined using velocity of approach. Velocity of approach $U_0 = \dfrac{2.035}{6.0 \times 1.2} = 0.282$ m/s and $\dfrac{U_0^2}{2g} = 0.00406$

$$\therefore \quad Q = \frac{2}{3} \times 0.622 \times \sqrt{2 \times 9.806} \times 2.38\ [(0.600 + 00406)^{3/2} - (0.00406)^{3/2}]$$

$$= 2.0510 \text{ m}^3/\text{s}$$

5.24 A sharp crested weir of 1.0 m height is fixed across 5.0 m wide channel. The depth of flow on the upstream and downstream sides of the weir are 1.50 m and 1.25 m respectively. Assume that the discharge over the weir under submerged condition is given by the sum of discharge through an orifice of channel width B and height $(H_2 - W)$ with head H_1, and discharge over a weir of same length B and height y_2 under a free flow condition with head $(H_1 - H_2)$. Assume appropriate value of C_d. Compare the answer with that of Prob. 5.30 (Fig. 5.20).

$$Q = Q_1 + Q_2$$

where Q_1 is flow over the weir and Q_2 is the discharge through the orifice.

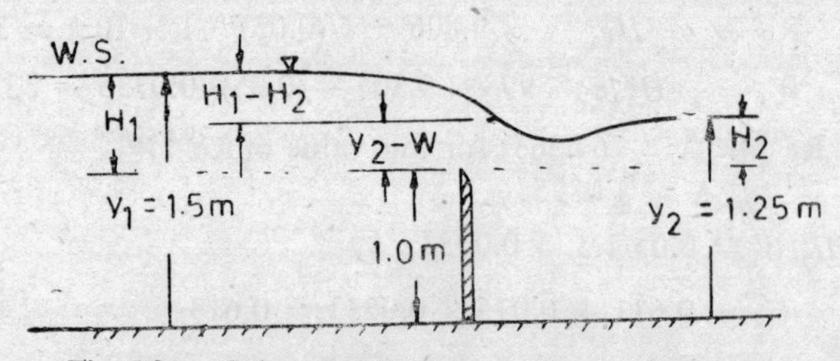

Fig. 5.20. Submerged flow over a sharp crested weir.

For weir $\dfrac{(H_1 - H_2)}{y_2} = \dfrac{0.25}{1.25} = 0.20$

$$\therefore \quad C_d = 0.611 + 0.075 \times 0.20 = 0.626$$

$$\therefore \quad Q_1 = \frac{2}{3} \times 0.626 \times 5.0 \times \sqrt{2 \times 9.806} \times (0.25)^{3/2}$$

$$= 1.155 \text{ m}^3/\text{s}$$

For large orifice for $\frac{b}{B_1} = \frac{0.25}{(0.25 + 2 \times 0.25)} = 0.333, C_c = C_d$ value can be obtained from Table 5.10 as $C_c = C_d = 0.629$.

Therefore, neglecting velocity of approach,

$$Q_2 = \frac{2}{3} \times 0.629 \times 5 \times \sqrt{2 \times 9.806}\,(0.50^{3/2} - 0.25^{3/2})$$

$$= 2.122 \text{ m}^3/\text{s}$$

$$\therefore \quad Q = Q_1 + Q_2 = 1.155 + 2.122 = 3.277 \text{ m}^3/\text{s}$$

On the other hand, if it is treated as submerged weir Q_s can be shown to be equal to 2.864 m³/s. It can therefore be seen that a simple assumption made as above will lead to an overestimation of discharge by

$$\frac{(3.277 - 2.864)}{2.864} \times 100 \quad \text{or} \quad 14.42 \text{ per cent}$$

5.25 Obtain the discharge equation for a triangular notch under free flow condition.

See Fig. 5.10. Consider flow through a small strip of height dh and width $2\,(H_1 - h)\tan\frac{\theta}{2}$, h m below the free surface. The discharge dQ is given by

$$dQ = C_d \times 2\,(H_1 - H)\tan\frac{\theta}{2}\,dH\,\sqrt{2gH}$$

Integrating the above expression from limits $H = 0$ to H_1, one gets

$$Q = 2\,C_d \tan\frac{\theta}{2}\sqrt{2g}\int_0^{H_1}(H_1 - H)\,H^{1/2}\,dH$$

$$= 2\,C_d \tan\frac{\theta}{2}\sqrt{2g}\left(\frac{2}{3}H_1 H^{3/2} - \frac{2}{5}H^{5/2}\right)_0^{H_1}$$

$$\text{or} \quad Q = \frac{8}{15}C_d\sqrt{2g}\tan\frac{\theta}{2}H_1^{5/2}$$

5.26 What will be the head required to carry a discharge of 2.75 m³/s through a 2.0 m wide gate at 0.30 m opening under free flow condition?

Assume $H_1 = 5.0$ m

$\therefore \quad a/H_1 = 0.30/5.0 = 0.06$. Taking $C_c = 0.60$, C_d value can be calculated

$$C_d = \frac{C_c}{\sqrt{1 + C_c\dfrac{a}{H_1}}} = 0.60/\sqrt{1 + 0.6 \times 0.06} = 0.59$$

$$\therefore \quad \text{Since } Q = C_d \times a \times B\sqrt{2gH_1}$$

$$2.75 = 0.59 \times 0.3 \times 2 \times \sqrt{2 \times 9.806\,H_1}$$

or $\quad H_1 = 3.077 \text{ m}$

With this value of H_1, new value of C_d is

$$C_d = 0.60/\sqrt{1 + 0.6 \times (0.3/3.077)} = 0.583$$

and $\quad 2.75 = 0.583 \times 0.3 \times 2.0 \times \sqrt{2 \times 9.806 \times H_1}$

$\therefore \quad H_1 = 3.151 \text{ m}$

No further trials are necessary and H_1 may be taken as 3.151 m.

5.27 In the case of a freely flowing sluice gate, the upstream depth is 5.0 m and gate opening is 1.5 m. Determine (i) the discharge per unit width, and (ii) water depth just upstream of the gate. What will be the discharge if water depth immediately downstream of the gate is 2.0 m? Compare this value with your estimation of discharge under submerged condition assuming the flow immediately downstream of the gate to be unaffected by submergence.

$$\frac{a}{H_1} = \frac{1.5}{5.0} = 0.30 \qquad \therefore \quad C_d = \frac{0.60}{\sqrt{1 + 0.3 \times 0.6}} = 0.552$$

(i) $\quad q = C_d\, a\sqrt{2gH_1} = 0.552 \times 1.5 \times \sqrt{2 \times 9.806 \times 5.0}$

$$= 8.199 \text{ m}^3/\text{s m}$$

(ii) $\quad U_1 = \dfrac{8.199}{5.0} = 1.640 \text{ m/s}$

$\therefore$ Depth just upstream of gate $= 5.0 + \dfrac{1.640^2}{2 \times 9.806}$

$$= 5.137 \text{ m}$$

For discharge under submerged condition

$$C'_{ds} = \frac{0.6}{\sqrt{1 - 0.6^2 \times 0.3^2}} = 0.610$$

$\therefore \quad q = 0.61 \times 1.5 \times \sqrt{2 \times 9.806 \times (5.0 - 2.0)}$

$$= 7.081 \text{ m}^2/\text{s m}$$

If one assumes that flow condition immediately downstream of the gate remains unaffected even after submergence, one can assume flow to take place only through $C_c.a$ depth, C_c remaining 0.60. Hence Bernoulli's equation gives

$$5.0 + \frac{U_1^2}{2 \times 9.806} = 2 + \frac{U_1^2}{2 \times 9.806}$$

and $\quad 5U_1 = 0.6 \times 1.5 \times U_2 \quad$ or $\quad U_2 = 5.556\, U_1$

$$\therefore \quad 5 + \frac{U_1^2}{2 \times 9.806} = 2.0 + \frac{5.556^2\, U_1^2}{2 \times 9.806}$$

which gives $U_1 = 1.404$ m/s

and $q = H_1U_1 = 5.0 \times 1.404 = 7.020$ m³/s m

5.28 When excess water on the upstream side of the dam is to be released during the flood, it is allowed to flow over part of the dam known as spillway (Fig. 5.12). The cross-sectional shape of the Ogee spillway conforms to the shape of nappe produced by a sharp crested weir for a given head. Determine the discharge over the spillway of 150 m length under a head of 1.5 m. What will be the depth of flow at the toe of the dam if dam is 50 m high?

For spillways $C_d = 0.75$ if $\frac{H_1}{W} < 0.75$

$$\therefore \quad Q = \frac{2}{3} C_d B \sqrt{2g}\, H_1^{3/2}$$

$$= \frac{2}{3} \times 0.75 \times 150 \times \sqrt{2 \times 9.806} \times 1.5^{3/3}$$

$$= 610.206 \text{ m}^3/\text{s}$$

Assuming no loss of energy between sections 1 and 2, one can write

$$H_1 + W + \frac{U_1^2}{2g} = H_2 + \frac{U_2^2}{2g}$$

and $(H_1 + W)U_1 = H_2U_2 = 610.206/150$ or 4.068

Substituting $H_1 = 1.50$ m, $W = 50$ m

$$51.5 + \frac{0.079^2}{2 \times 9.806} = H_2 + \frac{4.068^2}{2 \times 9.806 H_2^2}$$

or $$51.5003 = H_2 + \frac{0.8438}{H_2^2}$$

Solving by trial and error one gets $H_2 = 0.128$ m.

5.29 A sharp crested weir of 0.20 m height is fixed across a 0.50 m wide channel for the measurement of discharge. Two pipes of 10 mm diameter are provided, one on each side, for ventilating the nappe. At 0.15 m head over the weir, the pressure difference between outside and underside of the nappe is found to be 1.5 N/m². Determine the air demand for ventilating the nappe as a percentage of flow over the weir (Fig. 5.9).

When a nappe is formed, the air in the pocket underneath the nappe is continuously dissolved and entrained in the flow and pressure there is reduced. As a result the nappe is pu'led down and it affects the discharge-head relationship over the weir. Hence it is necessary to ventilate the weir.

$$C_d = 0.611 + 0.075 \times \frac{0.15}{0.20} = 0.667$$

$$\therefore \quad Q = \frac{2}{3} \times 0.667 \times 0.50 \times \sqrt{2 \times 9.806} \times (0.15)^{3/2}$$

$$= 0.0572 \text{ m}^3/\text{s}.$$

Assume that flow in the ventilating pipes is frictionless.

$\therefore$ Velocity of air in the ventilating pipe, $U = \sqrt{\dfrac{2\Delta p}{\rho}}$ or $U = \sqrt{\dfrac{2 \times 1.5}{1.208}}$

$$= 1.576 \text{ m/s}$$

$\therefore$ Discharge of air through two pipes $Q_a = 2 \times 0.785 \times 0.1^2 \times 1.576$

$\therefore$ $Q_a = 0.000\,247 \text{ m}^3\text{/s}$

$\therefore$ $\dfrac{Q_a}{Q} \times 100 = \dfrac{0.000\,247}{0.0572} \times 100 = 0.43$ per cent.

5.30 In many cases, such as in constant velocity sedimentation tanks, it is desirable to employ a weir form in which Q varies linearly with H. Such a weir is known as proportional weir or Sutro weir after the man who first designed it. Discuss the shape of such a weir and flow over it.

Sutro weir consists of a rectangular portion joined to a curved portion which has the profile equation

$$\frac{x}{b} = \left(1 - \frac{2}{\pi} \tan^{-1} \sqrt{(Z'/a)}\right)$$

This provides proportionality for all heads above line CD in Fig. 5.21.

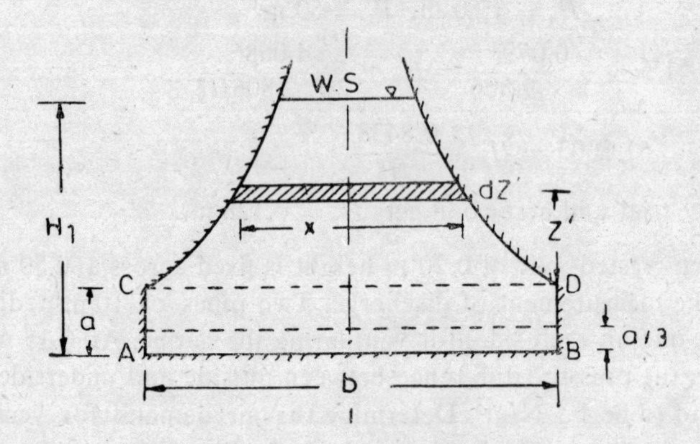

Fig, 5.21. Proportional weir.

The discharge equation for Sutro weir is

$$Q = C_d\, b(2ga)^{1/2} \left(H_1 - \frac{a}{3}\right)$$

The discharge coefficient C_d is mainly governed by the values of a and b. For practical ranges of a and b, these values are listed below in Table 5.5.

Table 5.5: C_d values of symmetrical Sutro weirs as a function of a and b.

a meters	b (meters)				
	0.150	0.230	0.300	0.330	0.460
0.006	0.608	0.613	0.617	0.618	0.619
0.015	0.606	0.611	0.615	0.617	0.618
0.030	0.603	0.608	0.612	0.613	0.614
0.046	0.601	0.606	0.610	0.611	0.612
0.061	0.599	0.604	0.608	0.609	0.610
0.076	0.598	0.602	0.606	0.608	0.609
0.091	0.597	0.602	0.606	0.607	0.608

This equation with the above C_d values should be used for $H_1 > 2a$ and $b/W > 1.0$ and B/b not less than 3.0.

PROBLEMS

5.1 Determine the pressure distribution on the outer boundary of sharp bend shown in Fig. 2.6.

5.2 In Fig. 5.22, determine the diameter of venturi throat which will give incipient cavitation condition there. Take $p_a = 100$ kN/m² and $p_v = 2.39$ kN/m². (183.9 mm)

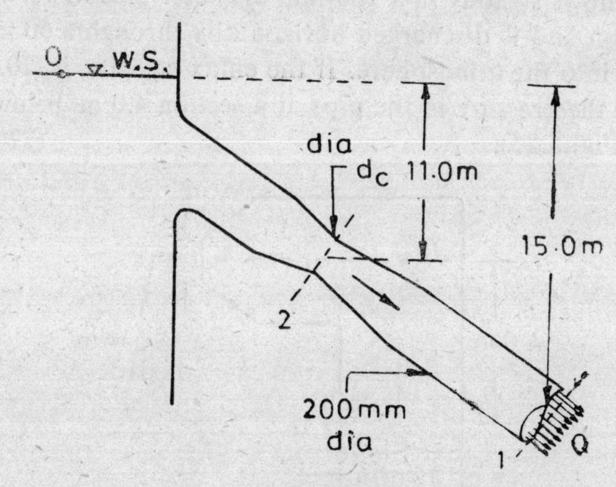

Fig. 5.22.

5.3 A reservoir is connected to 0.6 m diameter parallel plates kept 2 mm distance apart as shown in Fig. 5.23. Determine the discharge and gauge pressure at section 2. (0.0167 m^3/s, – 29.361 kN/m^2)

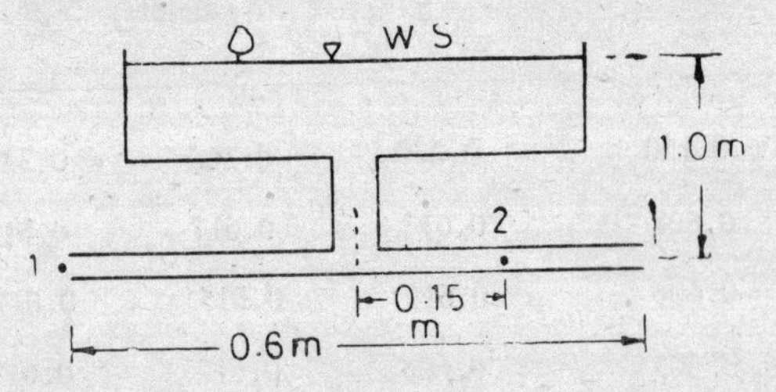

Fig. 5.23.

5.4 Determine the air discharge through open circuit wind tunnel shown in Fig. 5.24, whose cross section is 1 m × 1 m square. Assume $\rho = 1.208$ kg/m^3. (18.0 m^3/s)

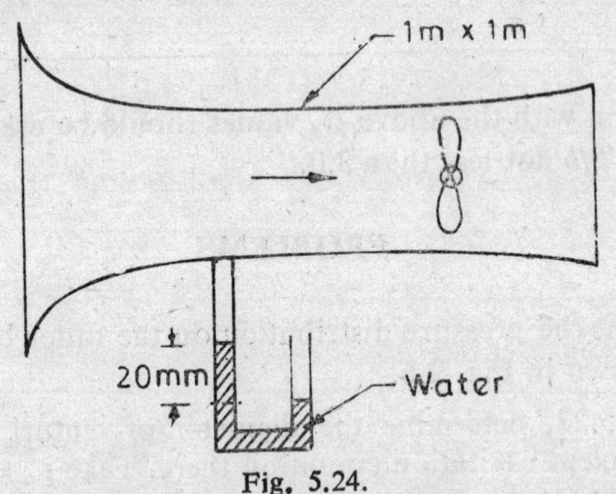

Fig. 5.24.

5.5 Water flows steadily in a 100 mm diameter pipe in vertically upward direction and is discharged horizontally through a 50 mm diameter nozzle into the atmosphere. If the efflux velocity is 20.0 m/s, what will be the pressure in the pipe at a section 4.0 m below the nozzle level (Fig. 5.25). (226.285 kN/m^2)

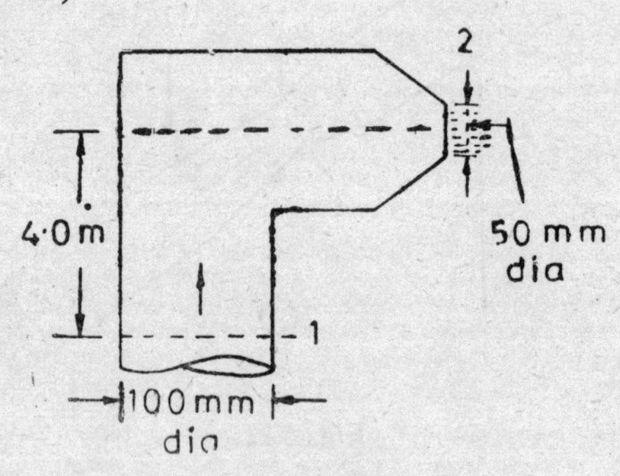

Fig. 5.25.

5.6 For a pipe bifurcation shown in Fig. 5.26 determine Q_0, Q_1 and Q_2, assuming no energy loss. (Note that pipes 1 and 2 discharge into atmosphere). (1.182 m³/s, 0.756 m³/s, 0.426 m³/s)

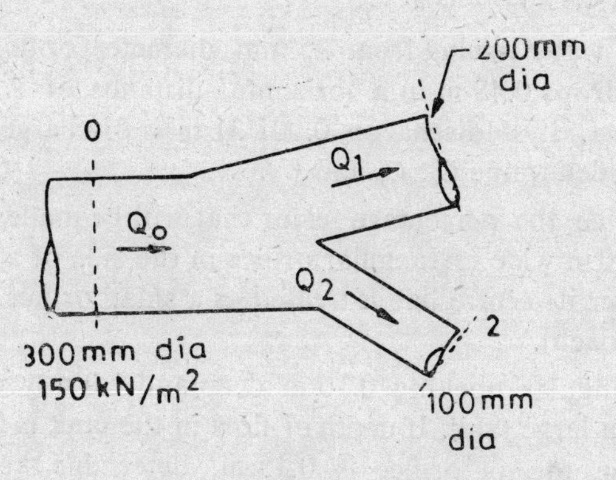

Fig. 5.26

5.7 For a pipe line system carrying water, the following data are given

$p_A = 300.00$ kN/m² $\qquad p_B = 310.326$ kN/m²

$Z_A = 10.00$ m $\qquad Z_B = 9.20$ m

$D_A = 200$ mm $\qquad D_B = 300$ mm

Energy loss between sections A and B is 0.40 m of water. Determine the discharge. (0.1256 m³/s)

5.8 Determine an expression for coefficient of velocity C_v for a sharp edged orifice located in the side of a vertical tank in terms of horizontal distance x, vertical distance y travelled by the jet and the head H over the orifice. ($C_v = \sqrt{x^2/4yH}$)

5.9 A jet of water, issuing from 5 mm diameter orifice working under a head of 2.0 m, was found to travel horizontal and vertical distances of 2.772 m and 1.0 m respectively. If C_c for the orifice is 0.61, determine the discharge. (0.073 48 l/s)

5.10 In Example 5.7, determine the location of the second orifice such that the two jets meet the ground at the same place. (8.0 m or 1.0 m)

5.11 Two small orifices are located in the side of a tank filled to a depth H_1. One orifice is located x m below the water surface, and other x m above the tank bottom. Show that both the jets will strike the ground at the same distance from the orifice. What is that distance? ($2\sqrt{x(H_1 - x)}$)

5.12 In order to determine experimentally the coefficients of contraction, velocity and discharge for a 100 mm diameter sharp orifice in the side of a tank, the following data were collected:

Diameter of jet at vena contracta as measured by calipers = 78.42 mm

$H = 3.60$ m $\quad Q = 0.0385$ m³/s

Obtain, C_c, C_v and C_d. (0.615, 0.949, 0.5837)

5.13 A jet of water issuing from 25 mm diameter orifice in the side of a tank drops 0.48 m in a horizontal distance of 1.39 m from vena contracta. If it discharges 0.001 31 m³/s discharge under a head of 1.07 m, determine C_c, C_v and C_d. (0.60, 0.97, 0.583)

5.14 Determine the percentage error that will be made if a 0.30 m deep and 0.70 m wide rectangular orifice in the side of a tank with 1.0 m head over its centre line is treated as a small orifice (Neglect velocity of approach). (9.57 percent)

5.15 A rectangular opening 0.10 m deep and 1.0 m wide is fitted in the side of a large tank. If depth of flow in the tank is 0.30 m and depth over the top of orifice is 0.15 m, determine the discharge. Take $C_c = 0.63$. (0.125 m³/s)

5.16 Determine the discharge through a 50 mm diameter external mouthpiece under a head of 3.0 m. Also determine the absolute pressure at vena contracta. Take $C_c = 0.62$ and $p_a = 101.30$ kN/m². (12.83 1/s, 7.691 m abs.)

5.17 A large tank is fitted with an external mouthpiece of 100 mm diameter in its side. If C_c for the vena contracta is 0.62 and entrance loss is given by $0.06\, U_c^2/2g$, determine the maximum permissible head and corresponding discharge at which lowest permissible pressure would prevail at the vena contracta. Take $p_v = 2.39 \times 10^3$ N/m³ and $p_a = 101.3$ kN/m². (12.63 m, 99.82 1/s)

5.18 What will be the diameter of a converging nozzle which discharges 38.86 l/s of water under a head of 7.0 m? (65 mm)

5.19 An internal mouthpiece with a diameter of 60 mm discharges under a head of 9.0 m. Determime the discharge and pressure at vena contracta. Take $H_a = 10.30$ m. (26.55 1/s, 1.30 m of water)

5.20 Determine the ratio of throat diameter to exit diameter of a convergent-divergent mouthpiece for the condition of maximum discharge if $H = 3.0$ m, $H_a = 10.3$ m and minimum allowable pressure at the throat is equivalent to 2.0 m water absolute. Assume no energy loss in the convergent section, and 20 per cent of head loss in sudden expansion for divergent section. If throat diameter is 50 mm, determine the discharge. (0.6735, 29.22 1/s)

5.21 Water is discharged at the rate of 53.0 l/s through 150 mm diameter pipe at the end of which is fixed an orifice of 75 mm diameter. What will be the gauge pressure within the pipe? (167.055 kN/m²)

5.22 Flow of air in a 0.50m × 0.50 m rectangular conduit is controlled by two 0.25 m × 0.50 m long plates hinged at the top and bottom of the conduit at its end as shown in Fig. 5.27. Obtain the co-ordinates of calibration curve (i.e. Q vs gate opening) for conduit pressure of 60 N/m². (Hint: for various values of b, determine θ_1. For known b/B and θ_1, obtain C_c and then Q using the equation

$$Q = \frac{C_c . b \times 0.30}{\sqrt{1 - C_c^2 \left(\frac{b}{0.50}\right)^2}} \sqrt{2 \Delta p/\rho}$$

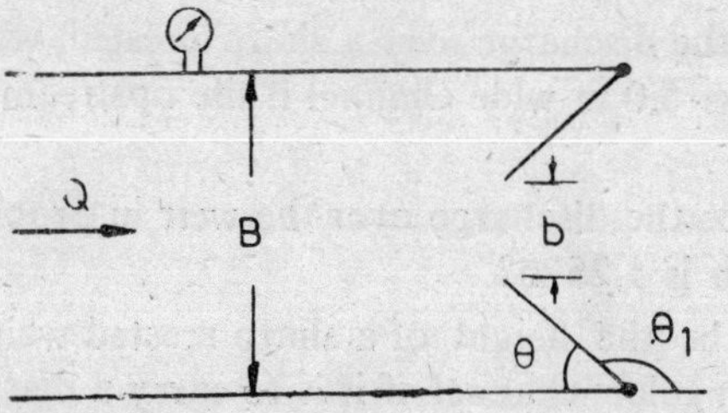

Fig. 5.27.

b	m	0.10	0.20	0.30	0.40
Q	m³/s	0.286	0.605	0.998	1.684

5.23 An orifice is fitted in a 150 mm diameter pipe. When the pipe carries 0.053 m³/s of water discharge, the pressure difference across the orifice is 167.242 kN/m². What is the diameter of the orifice? (75 mm)

5.24 A 200 mm diameter pipe carries 62.80 l/s of water at 20°C at a gauge pressure of 150 kN/m². As the pipe rises through 2.0 m it is constricted. Determine the minimum permissible diameter of constriction if vapour pressure of water is 2.39 kN/m² absolute and atmospheric pressure is 101.3 kN/m². Flow is in the upward direction. (61 mm)

5.25 What should be the diameter of throat of the venturimeter fixed in a vertical 0.30 m diameter pipe carrying 0.30 m³/s of air in the downward direction if the differential water manometer shows 30 mm head difference? Take $\rho = 1.208$ kg/m³ and $C = 0.97$. (0.1324 m)

5.56 For straightening the flow in a 0.50 m × 0.80 m duct, a series of 26 thin vertical strips of width 10 mm and clear distance 20 mm between them are fixed near the entrance of the duct. If the duct carries 0.10 m³/s of air, what will be the difference of pressure across the straightener? (10.108 N/m²)

5.27 Consider flow in a 90° rectangular bend of width b and depth h and inner and outer radii as R_1 and R_2. If average velocity in the vertical varies with r as $u = Cr$ show that

$$Q = \frac{hR_1R_2 \ln (R_2/R_1)}{\sqrt{R_2^2 - R_1^2}} \sqrt{\frac{2(p_2 - p_1)}{\rho}}$$

(Hint: first show that average velocity over the cross section is given by

$$U = \frac{c}{b} \ln (R_2/R_1).$$

5.28 Calculate the throat diameter of nozzle meter fixed in a 150 mm diameter pipe for measurement of kerosene (R.D. = 0.80) flowing through it when the U-tube mercury meanometer shows a differential of 0.102 m at a discharge of 17.66 l/s. Take $C = 0.97$.
(63.55 mm)

5.29 Determine the discharge over a sharp crested weir of height 1.0 m fixed across 5.0 m wide channel if the upstream depth is 1.5 m.
(3.3875 m^3/s)

5.30 What will be the discharge over the weir in problem 5.29 if downstream depth is 1.25 m? (2.8637 m^3/s)

5.31 What will be the height of a sharp crested weir fixed across 4.0 m wide rectangular channel if it is to carry a discharge of 1.50 m^3/s while maintaining an upstream depth of 1.20 m?
(Hint: Assume C_d and compute H_1, W and refine value of C_d).
(0.860 m)

5.32 If, in the measurement of discharge over the weir in a rectangular channel, two percent error is made in the measurement of head H_1, what will be the corresponding error in discharge? (3.0 percent)

5.33 A 10 m wide rectangular channel is provided with 1.5 m high sharp crested weir across its width with two intermediate piers of 1.0 m width. What discharge will be carried by the channel when the head over the weir is 0.80 m? (10.265 m^3/s)

5.34 A 3.5 m wide channel is to be provided a contracted sharp crested weir of 1.2 m height running free. If a maximum of 4.0 m^3/s water discharge is to be passed over the weir with upstream depth not exceeding 2.25 m, what should be the length of the weir? (2.19 m)

5.35 Obtain the discharge equation for Cipolletti weir (Fig. 5.14).

5.36 If the discharge over a 90° V notch is 1.50 m^3/s, determine the head causing the flow. Take $C_d = 0.62$. (1.01 m)

5.37 For a sluice gate fixed in 3.0 m wide channel, 4.0 m^3/s of water is carried under free flow condition when the upstream water depth is 2.5 m. Determine the gate opening. (0.330 m)

5.38 A sluice gate fixed in a wide rectangular channel discharges 5.0 m^3/s m water under submerged flow condition when depths on the upstream and downstream side of gate are 5.0 m and 2.5 m respectively. Determine the gate opening. (1.178 m)

5.39 What should be the length of Ogee spillway needed to pass the flood discharge of 1500 m^3/s over a dam of 60 m height if upstream water level can be raised to a maximum of 2.0 m above the crest of spillway. What would be the depth of flow at the toe of the dam?

(239.50 m, 0.180 m)

5.40 A sharp crested weir 0.30 m height is installed in a 100 m wide flume. With a head of 0.15 m over the head the two ventilating pipes were found to supply air at 0.40 percent of water discharge with pressure difference of 0.98 N/m² across the nappe. What is the diameer of ventilating pipes ? (15 mm)

DESCRIPTIVE QUESTIONS

5.1 Under what conditions can you neglect the energy loss between two sections for flow of a real fluid?

5.2 Why is pressure distribution in nondimensional form uniquely determined for steady two dimensional irrotational flow?

5·3 Why does fluid weight have no influence on the flow pattern if fluid is fully confined within the boundary?

5.4 Show that discharge coefficient for a discharge measuring device for pipes can be interpreted as Euler number.

5.5 Show that Euler number is a ratio of square root of inertial force per unit volume to pressure force per unit volume.

5.6 Under what conditions can cavitation occur? What are the undesirable effects of cavitation?

5·7 What are the characteristics of flow at vena contracta?

5.8 How will you decide whether a particular orifice is small or large?

5.9 Draw typical variation of C_d with d/D and $U_2 d \rho/u$ for an orifice plate fixed in a pipe.

5.10 Draw pressure variation (in the form of $\Delta p/(\rho U^2/2)$) along the centre line and along the wall, for orifice plate in a pipe. Show this starting from a few diameters upstream to few diameters downstream of orifice plate.

5.11 Is it permissible to fix a venturimeter in such a way that flow direction is reversed? Why?

5.12 What are the dimensionless parameters on which the bend meter coefficient C_d would depend?

5.13 Given a choice which of the two, namely venturimeter and orifice meter, will you choose for measurement of discharge in a pipe? Why?

5.14 Why is it necessary to have about 25D straight pipe length upstream and about 10D downstream of any device from measurement of flow in a pipe?

5·15 Show that discharge coefficient of a weir is proportional to Froude number.

5.16 What happens if sharp crested weir is not aerated?

5.17 Describe the flow condition on a spillway face when it is run (i) at design head, (ii) at head less than design head, and (iii) at head greater than design head.

5.18 Relation between C_d, C_c and C_v is

(i) $C_d = C_c/C_v$ (ii) $C_d = C_v/C_c$ (iii) $C_d = C_c C_v$

5.19 C_d is always (i) greater than C_c, (ii) less than C_c (iii) equal to C_c

5.20 Whether the orifice is large or small depends on
(i) absolute dimension of the orifice
(ii) depth over the orifice
(iii) ratio of orifice dimension to depth over it.

5.21 For a given Reynolds number, as d/D for an orifice increases, C_d will (i) increase (ii) remain constant (iii) decrease.

5.22 For a given discharge flowing over a sharp crested weir, the head required is (i) less (ii) more (iii) same, if the weir is submerged than when it flows free.

5.23 In the case of a free vortex
(i) outside energy has to be supplied to maintain the flow
(ii) $\frac{\partial}{\partial t}(MrV) = 0$
(iii) $\frac{\partial}{\partial t}(MV) = 0$
where M is mass, V is velocity and r is radial distance.

5.24 Match the discharge coefficient value with the following flow measuring devices:

(i) venturimeter	(a) 0.58 – 0.8
(ii) orifice plate	(b) 0.50
(iii) Borda's mouthpiece (running full)	(c) 0.93 – 0.98
(iv) Borda's mouthpiece (running free)	(d) 0.75

5.25 If head over the triangular notch is doubled, discharge will increase to (i) $2Q$ (ii) $2.828Q$ (iii) $5.657Q$ (iv) $4Q$

5.26 If all the losses are neglected, the pressure at the summit of a siphon is (i) independent of liquid density, (ii) is independent of Q in siphon (iii) depends on height of summit above reservoir level, velocity and mass density.

CHAPTER VI

Momentum Equation and Its Applications

6.1 LINEAR MOMENTUM EQUATION

Linear momentum is defined as

$$M_x = \rho.\delta A.u \cos \theta.u_x$$

In Fig. 6.1, δA is an elementary area, u is the velocity of flow through this area making an angle θ with normal and u_x is the component of u in the x direction. According to Newton's second law of motion, in a given direction, one can write

$$\Sigma F_x = \frac{d}{dt}(MU_x) = \begin{array}{l}\text{Rate of change of}\\ \text{momentum within}\\ \text{the control volume}\end{array} + \begin{array}{l}\text{Net rate of momentum}\\ \text{outflow through the}\\ \text{control volume in that}\\ \text{direction}\end{array} \quad (6.1)$$

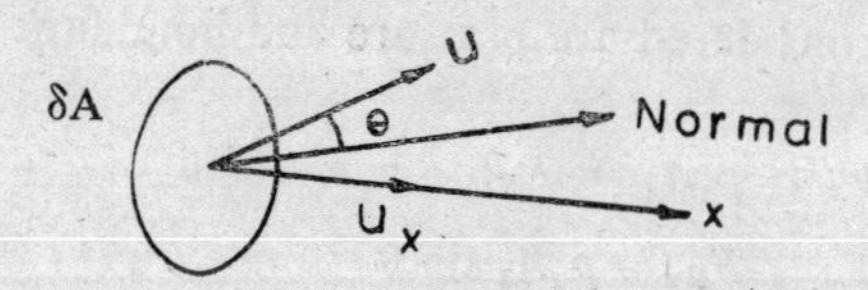

Fig. 6.1. Definition sketch.

Similar equations can be written in other directions y and z. This equation is applicable to an inertial system. This means that the control volume is fixed with the reference system or is moving at a constant speed in a straight line. For steady flows, the first term on the right hand side is zero, and the second term can be written as $\int_A \rho dA\, u \cos \theta\, u_x$ where A is the surface of control volume. While applying the momentum equation, the control volume should be so chosen that to its left and right the flow is uniform and all the nonuniformity is contained within the control volume. Further, all the fluid should enter through one face and leave through the other, the remaining surfaces of the control volume acting as boundaries.

Forces acting on the fluid in the control volume are the components of body force (i.e. fluid weight), surface forces namely friction and pressure force, and forces exerted by the boundary on the fluid in control volume. When other forces are relatively large, the frictional force is neglected.

Momentum equation can be derived from Euler's equation of motion as shown in Example 6.1.

Consider the forces on a bend-cum-reducer placed in the horizontal plane (Fig. 6.2). Here fluid weight can be neglected. Assume U_1 and U_2 to be uniform over areas A_1 and A_2. Hence momentum equation gives

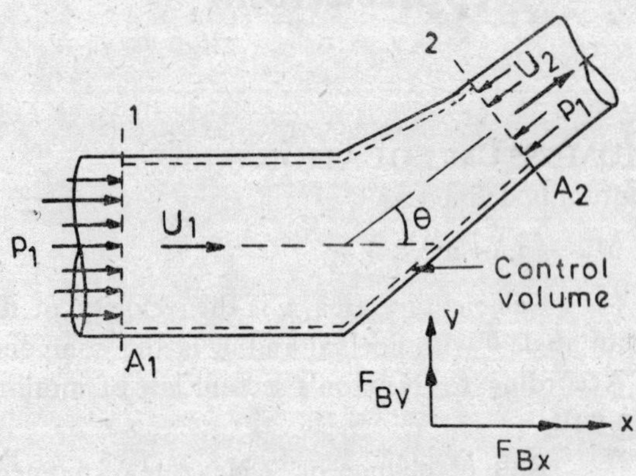

Fig. 6.2. Forces on bend cum reducer

$$\Sigma F_x = Q\rho\,(U_2 \cos\theta - U_1) \tag{6.2}$$

$$\Sigma F_y = Q\rho\,(U_2 \sin\theta - 0) \tag{6.3}$$

External forces to be considered are pressure and boundary forces F_{Bx} and F_{By} (assumed +ve).

$$\Sigma F_x = p_1 A_1 + F_{Bx} - p_2 A_2 \cos\theta \tag{6.4}$$

$$\Sigma F_y = F_{By} - p_2 A_2 \sin\theta \tag{6.5}$$

Solution of the above equations for known $A_1, A_2, U_1, U_2, \rho, \theta$ and p_1 and p_2 will give F_{Bx} and F_{By}.

6.2 MOMENTUM CORRECTION FACTOR β

If velocities U_1 and U_2 at sections 1 and 2 in Fig. 6.2 are not constant over the cross section, the momentum flux through any section must be calculated taking into account actual velocity distribution. In this regard momentum correction factor β is defined as

$$\beta = \frac{\text{Momentum flux through an area computed using the actual velocity distribution}}{\text{Momentum flux through the area computed using the average velocity}}$$

$$= \int_A \rho u dA \, u_x / AU\rho \, U_x = \frac{1}{A} \int_A \left(\frac{u}{U}\right)^2 dA \qquad (6.6)$$

where U and U_x are the average velocity and its component in x direction, while u and u_x are the local velocity and its component in x direction respectively. Normally β is assumed equal to unity for turbulent flows. However, such an assumption can give erroneous results in case of laminar flow.

6.3 APPLICATIONS

Linear momentum equation finds many applications such as determination of forces exerted on stationary and moving plates, curved stationary and moving vanes, forces acting on bodies in air and structures in open channels, boundary layer growth, propellers, wind mills, rockets, hydraulic jump in open channels and shock waves in compressible fluids. Corresponding equation in rotational flows (section 6.4) finds applications in sprinklers, pumps and turbines. These applications are shown through illustrative examples.

6.4 ANGULAR MOMENTUM

Moment of momentum about a given axis is known as the angular momentum. For a mass δm rotating about z axis in xy plane,

$$\text{Angular momentum} = (\delta m \, U_y x - \delta m \, U_x \, y)$$

See Fig. 6.3. Further rate of change of angular momentum is equal to the external torque τ_z applied.

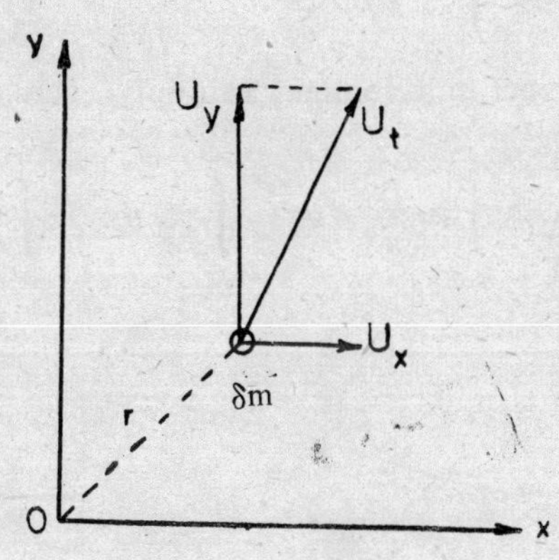

Fig. 6.3. **Definition sketch.**

$$\therefore \quad \tau_z = \frac{d}{dt} (\delta m U_y x - \delta m U_x y) \qquad (6.7)$$

which can be simplified to

$$\tau_z = Q\rho \, (U_{t2} \, r_2 - U_{t1} \, r_1) \qquad (6.8)$$

for steady flow where U_{t2} and r_2 are the tangential velocity and radius at exit section 2 while U_{t1} and r_1 are the corresponding quantities at section 1.

6.5 APPLICATIONS OF ENERGY AND MOMENTUM EQUATIONS

Application of energy equation alone or momentum equation alone (along with continuity equation) may not give the complete information about the forces and flow field in a given problem. Hence in many instances these two equations are used together with the continuity equation. Analysis of Borda's internal mouthpiece, flow in a sudden expansion and hydraulic jump in open channels are a few such examples.

ILLUSTRATIVE EXAMPLES

6.1 Obtain linear momentum equation from Euler's equation of motion.

In obtaining energy equation, each term in Euler's equation along a stream line is multiplied by a small distance δs along the streamline and is then integrated. Since each term in Euler's equation represents force, the integration terms give work done. One can multiply each term of Euler's equation by a small volume $d\forall$ and integrate it over the control volume $\forall$ which contains the nonuniformity of flow. This yields linear momentum equation.

Consider one dimensional flow along x direction of an ideal fluid (Fig. 6.4). x axis need not be horizontal. Hence according to Euler's equation,

$$\rho\left(\frac{\partial u}{\partial t} + u\frac{\partial u}{\partial x}\right) = X_1 - \frac{\partial p}{\partial x}$$

where X_1 is body force per unit volume. Multiply each term by $d\forall$ and integrate. Hence

$$\int_{\forall} \rho \frac{\partial u}{\partial t}\, d\forall + \int_{\forall} \rho u \frac{\partial u}{\partial x}\, d\forall = \int_{\forall} X_1\, d\forall - \int_{\forall} \frac{\partial p}{\partial x}\, d\forall$$

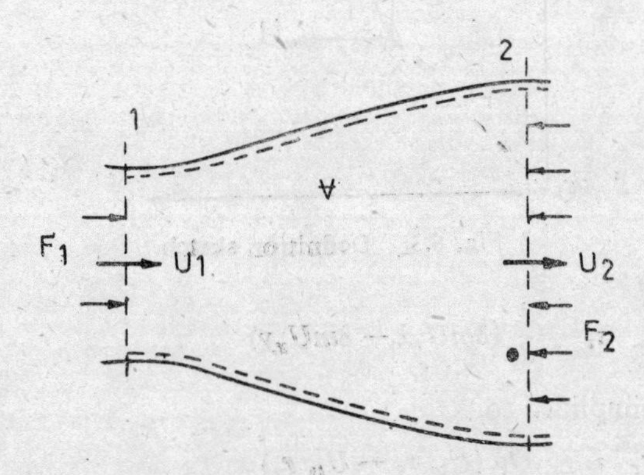

Fig. 6.4. Definition sketch.

One can write $d\forall = dAdx$. Hence

$$\underset{\substack{\forall \\ 1}}{\frac{\partial}{\partial t}\int}(\rho d\forall)\,u + \underset{\substack{A\;x \\ 2}}{\int\int}\rho u\frac{\partial u}{\partial x}\,dxdA = \underset{\substack{\forall \\ 3}}{\int}X_1\,d\forall - \underset{\substack{A\;x \\ 4}}{\int\int}\frac{\partial p}{\partial x}\,dxdA$$

In the first term $\rho d\forall = dM$ the mass of fluid in $d\forall$; henee $\rho d\forall u$ is the momentum of mass dM in x direction. So the 1st term represents the rate of change of momentum of mass within the control volume.

The second term can be written as

$$\int_A\int_X \rho u\frac{\partial u}{\partial x}\,dxdA = \int_A \rho\,(udA)\,du = \int_A \rho\,dQ\,(u_2 - u_1)$$

Here u_1 and u_2 are the velocities at sections 1 and 2 over area dA. If U_1 and U_2 are constant velocities over the sections 1 and 2, the above term reduces to $\rho Q\,(U_2 - U_1)$. If there is variation of u over the area A and if U_1 and U_2 are average velocities, this term can be written as $\rho Q\,(\beta_2 U_2 - \beta_1 U_1)$. This represents net rate of momentum outflow through control volume in x direction.

The third term $\int_\forall X_1 d\forall$ = component of fluid weight in x direction, say X.

The fourth term can be interpreted as follows

$$-\int_\forall \frac{\partial p}{\partial x}\,d\forall = -\int_A\int_x \frac{\partial p}{\partial x}\,dxdA = \int (p_1 - p_2)\,dA$$

p_1 and p_2 acting over the area dA may vary over A. Hence,

$$\int_A (p_1 dA - p_2 dA) \text{ can be written as } (F_1 - F_2).$$

Hence momentum equation becomes

$$\frac{\partial}{\partial t}\text{(Momentum of mass within control volume)} + \rho Q\,(\beta_2 U_2 - \beta_1 U_1)$$

$$= X + (F_1 - F_2) = \Sigma F_x$$

When one deals with real fluids ΣF_x also includes boundary friction. In all the cases it also includes other external forces exerted by boundary on the fluid.

6.2 A circular jet of a fluid of mass density ρ and diameter d, strikes a stationary plate held perpendicular to flow with velocity U, as shown in Fig. 6.5. Determine the force F_{Bx} exerted by the plate on the fluid.

Assume there is no friction. The pressure everywhere is atmospheric. Hence, the only force in x direction to be considered is F_{Bx}. Assume it to be acting in the direction shown.

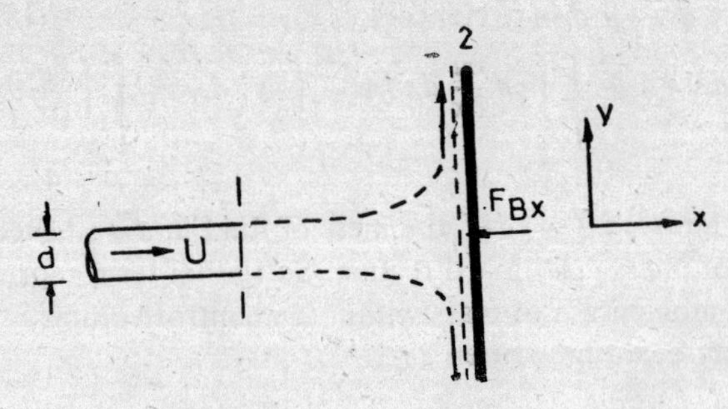

Fig. 6.5 A jet striking a stationary plate normally.

$$\Sigma F_x = -F_{Bx} = \frac{\pi d^2}{4} U\rho (0 - U)$$

$$-F_{Bx} = -\frac{\pi d^2}{4} \rho U^2$$

i.e. $F_{Bx} = \frac{\pi d^2}{4} \rho U^2$ and will act in negative x direction. Force exerted by the fluid on the plate will be in +ve x direction and of same magnitude, viz. $\frac{\pi d^2}{4} \rho U^2$. It may be mentioned that since there is no friction and pressure at sections 1 and 2 is atmospheric, velocity along the plate at section 2 is also U; further there is no force acting on the plate in y direction.

6.3 A 40 mm diameter jet of water strikes normally against a plate held in position and causes a force of 900 N on the plate. Determine the jet velocity and the discharge.

$$Q = \frac{\pi d^2 U}{4} = 0.785 \times 0.040^2 \, U$$

$$900 = 0.785 \times 0.40^2 \times U \times 998 \times U$$

$$U^2 = 717.997 \quad \text{or} \quad U = 26.795 \text{ m/s}$$

$$Q = 0.785 \times 0.040^2 \times 26.795$$

$$= 0.0337 \text{ m}^3/\text{s}$$

6.4 A jet of water issues from 20 mm diameter fire hose at the end of which a 5.0 mm diameter nozzle is fixed. If pressure at section 1 is 200 kN/m² gauge, determine force exerted by nozzle on the flow (Fig. 6.6).

Firstly apply Bernoulli's equation between sections 1 and 2 to determine Q. Since $p_2 = 0$

$$\therefore \quad \frac{200}{9.787} + \frac{U_1^2}{2 \times 9.806} = 0 + \frac{U_2^2}{2 \times 9.806}$$

and continuity equation gives

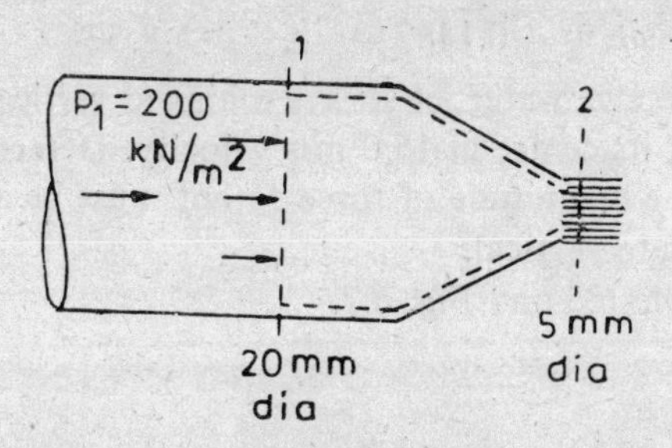

Fig. 6.6.

$$0.785 \times 0.02^2\ U_1 = 0.785 \times 0.005^2 U_2$$

or $$16U_1 = U_2$$

or $$20.435 = \frac{255U_1^2}{2 \times 9.806}$$

$\therefore$ $$U_1 = 1.254 \text{ m/s and } U_2 = 16 \times 1.254 = 20.059 \text{ m/s}$$

$$Q = 0.785 \times 0.02^2 \times 1.254 = 3.937 \times 10^{-4} \text{ m}^3\text{/s}$$

$\therefore$ Momentum equation gives

$$(200 \times 10^3 \times 0.785 \times 0.02^2) + F_{Bx}$$

$$= 3.937 \times 10^{-4}\ (20.059 - 1.254) \times 998$$

$$62.8 + F_{Bx} = 7.391$$

$$F_{Bx} = 62.800 - 7.391 = -\ 55.409 \text{ N}$$

6.5 A thin 0.40 m long plate weighing 10 N is hinged at its top edge and allowed to rotate about the horizontal axis. A jet of air of 25 mm diameter strikes the plate normally at its centre at a velocity of 50.0 m/s. Determine the angle that the plate will make with the vertical, under equilibrium condition (Fig. 6.7). Take $\rho = 1.208$ kg/m³.

Under equilibrium condition, the clockwise moment of weight W about the axis passing through O must be balanced by anticlockwise moment of force exerted by the fluid on the plate.

Component of fluid force on the plate in the direction perpendicular to plate $= Q\rho U \cos\theta$

$\therefore$ Moment of fluid force in anticlockwise direction

$$= Q\rho U \cos\theta \times OB; \text{ but } OB = \frac{l}{2\cos\theta}$$

$$\text{Moment of } W \text{ about } O = W\frac{l}{2}\sin\theta$$

$$\frac{l}{2\cos\theta} \times Q\rho U \cos\theta = W\frac{l}{2}\sin\theta$$

$$\sin\theta = \frac{Q\rho U}{W} = \frac{0.785 \times 0.25^2 \times 1.208 \times 50^2}{10}$$

$$\sin\theta = 0.1482 \qquad \therefore \quad \theta = 8.52°$$

6.6 A 40 mm diameter water jet strikes a hinged vertical plate of 800 N weight, normally at its centre at 15.0 m/s velocity. Determine (i) the angle of deflection, (ii) the magnitude of force F that must be applied at its lower edge to keep the plate vertical.

Refer to Example 6.5 and Fig. 6.7.

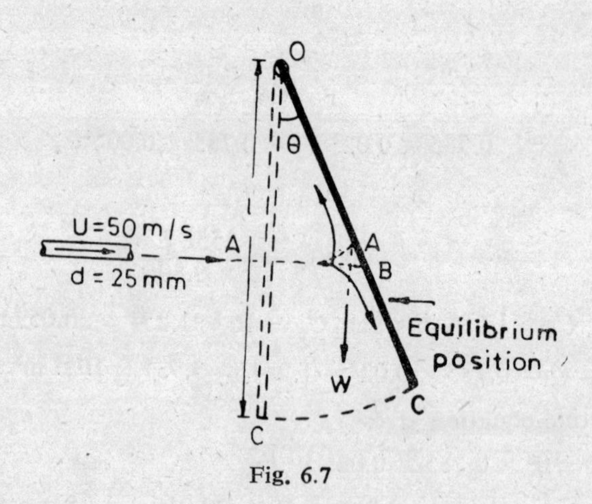

Fig. 6.7

$$\sin\theta = \frac{Q\rho U}{W} = \frac{0.785 \times 0.040^2 \times 998 \times 15^2}{800}$$

$$= 0.3525$$

$$\therefore \qquad \theta = \sin^{-1}(0.3525) = 20.64°$$

If a force F is applied at C towards left so that the plate remains vertical, the sum of moments of fluid force and force F about O must be zero; or

$$Q\rho U \times \frac{l}{2} = Fl$$

$$\therefore \qquad F = Q\rho U/2$$

$$= \frac{0.785 \times 0.040^2 \times 998 \times 15^2}{2}$$

$$= 141.017 \text{ N}$$

6.7 If a single vertical plate, on which the jet of area A strikes normally at a velocity U, moves in the same direction at a velocity u, determine the work done on the plate and the efficiency of the jet (Fig. 6.8).

The problem can be made a steady state one by applying velocity u to the left to the jet as well as plate.

$\therefore$ Fluid striking the plate $= A(U-u)\rho$

Force on the plate $F = A\rho(U-u)^2$

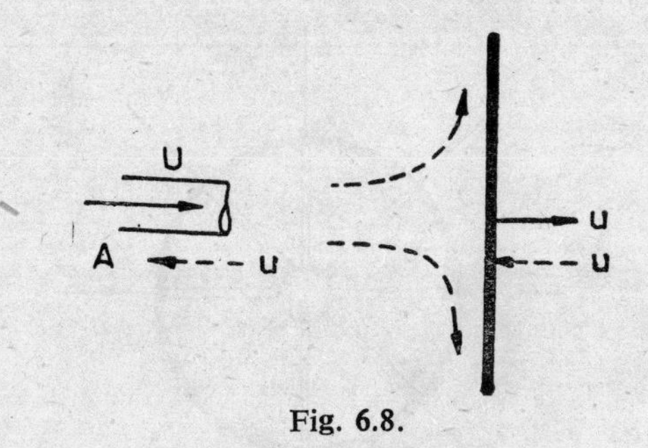

Fig. 6.8.

Work done per second = Force × velocity

$$= A\rho(U-u)^2 \times u.$$

The efficiency of the jet, η, can be defined as the work done per second by the jet divided by kinetic energy of the jet. Hence

$$\eta = \frac{A\rho(U-u)^2u}{A\rho U^3/2} = \frac{2u}{U}\left(1-\frac{u}{U}\right)^2$$

η will be maximum when $d\eta/d\left(\frac{u}{U}\right) = 0$

$$\text{or} \quad -\frac{2u}{U} + \left(1-\frac{u}{U}\right) = 0 \quad \text{i.e.} \quad \frac{u}{U} = \frac{1}{3}$$

$$\eta_{max} = \frac{8}{27} \quad \text{or} \quad 29.63 \text{ percent}$$

6.8 A 25 mm diameter jet of water discharging at a velocity of 30.0 m/s strikes normally on a plate also moving in the same direction (at velocity smaller than that of jet). If the work done per second is 1.5 kW, determine the velocity of the plate.

$$A\rho(U-u)^2u = 1500$$

$$\therefore \quad (30.0-u)^2u = \frac{1500}{0.785 \times 0.025^2 \times 998} = 3063.452$$

Solve this equation by assuming different values of u

$$\therefore \quad u = 4.839 \text{ m/s}.$$

6.9 A jet of cross-sectional area A and velocity U strikes a series of flat blades of an undershot wheel moving at a velocity of u in the same direction (Fig. 6.9). Determine the work done per second on the plate and the efficiency of the jet.

Here it can be assumed that there will always be one plate or blade in front of the jet. Hence

Discharge striking the plate = AU

Force on the plate = $A\rho U(U-u)$

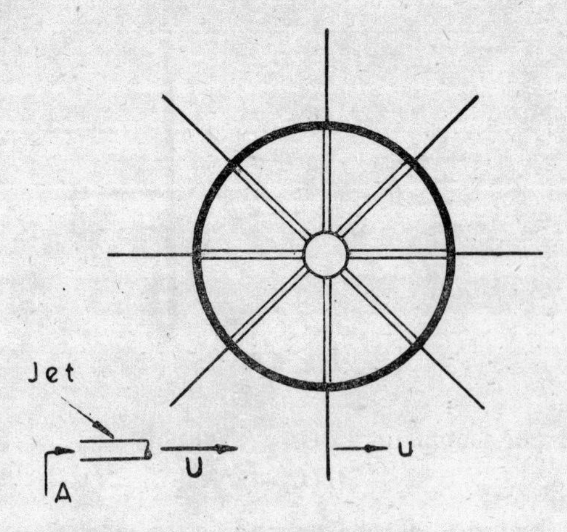

Fig. 6.9

Work done per second $= A\rho U(U-u)u$

$$\eta = \frac{A\rho U(U-u)u}{\rho \dfrac{A}{2} U^3} = 2\left(1-\frac{u}{U}\right)\frac{u}{U}$$

$$\eta \text{ will be maximum when } \frac{d\eta}{d\left(\dfrac{u}{U}\right)} = 0 \text{ i.e. when } \frac{u}{U} = \tfrac{1}{2}$$

$$\eta_{max} = 2 \times 0.5 \times 0.5$$
$$= 0.50 \text{ or } 50 \text{ percent}$$

6.10 Determine the resultant force (and its direction) on the vane shown in Fig. 6.10 if a water jet of 50 mm dia and 20.0 m/s velocity strikes the vane tangentially and is deflected without friction.

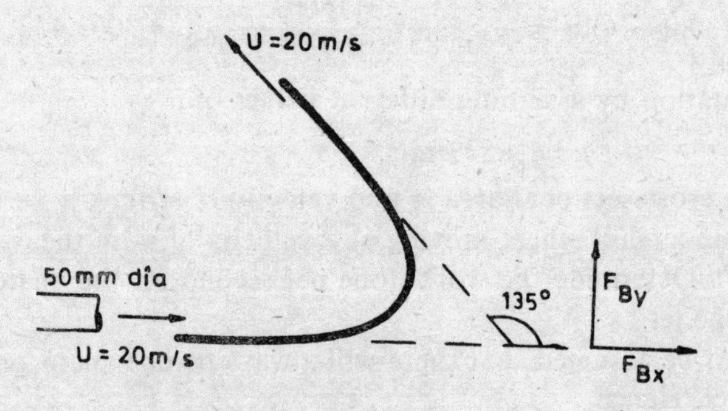

Fig. 6.10.

$$Q = 0.785 \times 0.050^2 \times 20$$
$$= 0.039\ 25 \text{ m}^3/\text{s}$$

Since pressure is atmospheric everywhere, the only external forces to be considered are F_{Bx} and F_{By} the components of force exerted by vane on the fluid. Further, since there is no friction, using Bernoulli's equation it can be shown that U remains constant along the vane.

$$F_{Bx} = Q\rho\,(-U \cos 45° - U)$$
$$= 0.039\,25 \times 998\,(-20 \times 0.7071 - 20)$$
$$= -1337.393 \text{ N} \qquad \therefore \quad F_{fx} = 1337.393 \text{ N}$$
$$F_{By} = Q\rho\,(U \sin 45° - 0)$$
$$= 0.03925 \times 998 \times 20 \times 0.7071$$
$$= 553.963 \text{ N}$$
$$\therefore \quad F_{fy} = -553.963 \text{ N}$$

where F_{fx} and F_{fy} are components of force exerted by the fluid on the vane.

$$F_f = \sqrt{1337.393^2 + 553.963^2}$$
$$= 1447.583 \text{ N}$$
$$\tan\theta = -\frac{553.963}{1337.393} = -0.4142$$
$$\therefore \quad \theta = -22.5° \text{ with } x\text{-axis}$$

6.11 Obtain general expressions for components of force exerted by a single moving curved vane when a jet strikes and leaves the vane tangentially (Fig. 6.11).

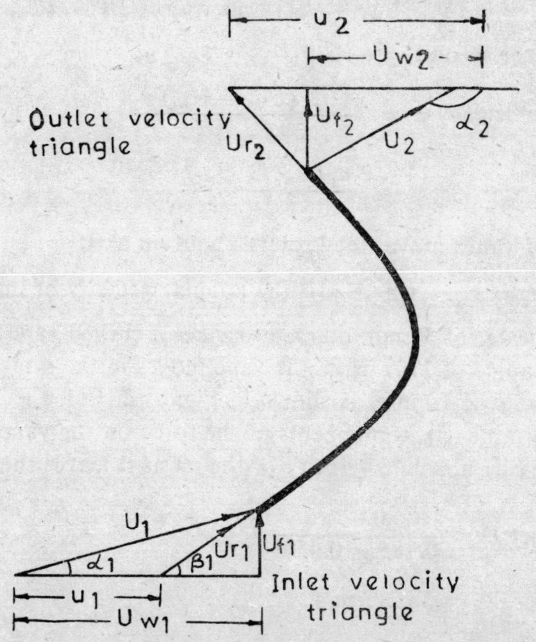

Fig. 6.11. Velocity triangles for a moving vane.

Let U_1 and U_2 : Inlet and outlet absolute velocities of jet

U_{r1} and U_{r2}: Relative velocities of the jet with respect to vane, at inlet and outlet

u_1 and u_2: Vane velocities at inlet and outlet

U_{w1} and U_{w2}: Horizontal components of absolute velocities of jet at inlet and outlet

U_{f1} and U_{f2}: Vertical components of absolute velocities of jet at inlet and outlet

α_1: Angle that the jet makes with direction of motion of vane at the inlet

α_2: Angle between absolute velocity of jet at outlet and direction of motion of vane

β_1: Angle between U_{r1} and u_1; it is known as the vane angle at inlet

β_2: Angle between U_{r2} and u_2; it is known as the vane angle at outlet.

Assuming no friction, Bernoulli's equation will indicate that $U_{r1} = U_{r2}$. Further for a single vane $u_1 = u_2$.

Discharge striking the single vane $= AU_{r1}$

where A is cross-sectional area of jet.

$$F_{Bx} = \rho AU_{r1}\,(U_{w2} - U_{w1}) \text{ and } F_{By} = \rho AU_{r1}\,(U_{f2} - U_{f1})$$

$$\text{Work done per second} = \rho AU_{r1}\,(U_{w2} - U_{w1})\,u_1$$

$$\text{Vane efficiency } \eta = \frac{\rho AU_{r1}\,(U_{w2} - U_{w1})\,u_1}{\frac{1}{2}\,\rho AU_{r1}U_1^2}$$

$$= \frac{2\,u_1\,(U_{w2} - U_{w1})}{U_1^2}$$

When a series of vanes move, they rotate about an axis.

$\therefore\ u_1 = r_1\omega,\ u_2 = r_2\omega$ where ω is the angular velocity of the vanes.

6.12 A jet of water of 50 mm diameter strikes a curved vane and is deflected through an angle of 170° while the vane moves in the same direction as the jet at a velocity of 10 m/s, as shown in Fig. 6.12. For a jet discharge of 0.075 m³/s, determine (i) components of the force on the vane, (ii) power developed, and (iii) absolute velocity of the jet as it leaves the vane.

$$\text{Velocity of jet } U = \frac{0.075}{0.785 \times 0.050^2}$$

$$= 38.217 \text{ m/s}$$

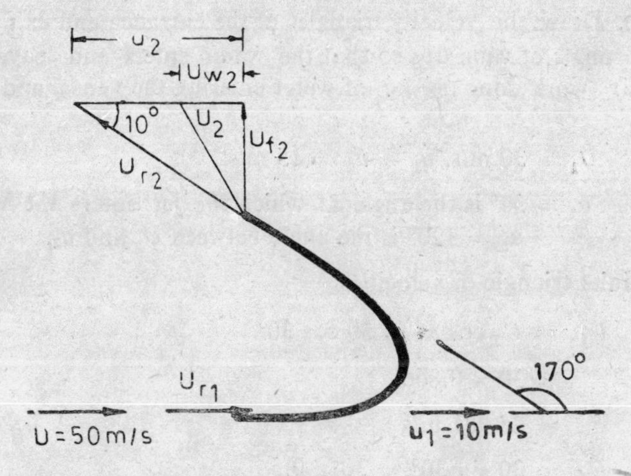

Fig. 6.12.

Discharge striking the vane

$$= 0.785 \times 0.050^2 (U - u)$$
$$= 0.785 \times 0.050^2 (38.217 - 10.0)$$
$$= 0.0554 \text{ m}^3/\text{s}$$

U_{r_1} Relative velocity of jet w.r.t. vane $= (38.217 - 10.000)$

$$= 28.217 \text{ m/s} = U_{r2}$$

$$U_{f_2} = 28.217 \sin 10°$$
$$= 4.900 \text{ m/s}$$

$U_{w2} = U_{r_2} \cos 10° - u_2$ (note here $u_2 = u_1 = 10$ m/s and $U_{r_2} = U_{r1}$)

$$= (27.788 - 10)$$
$$= 17.788 \text{ m/s}$$

$$\therefore \quad U_2 = \sqrt{U_{w_2}^2 + U_{f_2}^2} = \sqrt{17.788^2 + 4.900^2} = 18.451 \text{ m/s}$$

$$F_{Bx} = 0.0554 \times 998 (-17.788 - 28.217)$$
$$= -2543.580 \text{ N}$$

$$\therefore \quad F_{fx} = 2543.580 \text{ N}$$

$$F_{By} = 0.0554 \times 998(4.90 - 0)$$
$$= 270.917 \text{ N} \qquad \therefore \quad F_{fy} = -270.917 \text{ N}$$

$$\text{Power developed} = \frac{2543.58 \times 10}{1000} = 25.436 \text{ kW}$$

Absolute velocity of jet as it leaves vane is $U_2 = 18.451$ m/s

6.13 A jet of water having a velocity of 30.0 m/s impinges on a series of vanes with a velocity of 15.0 m/s. The jet makes an angle of 30° to the direction of motion of vanes when entering and leaves at an angle of 120°,

(Fig. 6.11). Draw the velocity triangles at the entrance and exit and determine (i) angle of vane tips so that the water enters and leaves without shock, (ii) work done per kg of water entering the vanes, and (iii) the efficiency.

$$U_1 = 30 \text{ m/s}, \; u_1 = u_2 = 15 \text{ m/s}$$

$\alpha_1 = 30°$ is the angle at which the jet enters the vane. Also $\alpha_2 = 120°$ is the angle between U_2 and u_2.

From the inlet triangle of velocities

$$U_{w1} = U_1 \cos \alpha_1 = 30 \cos 30°$$

$$= 30 \times 0.866$$

$$= 25.981 \text{ m/s}$$

$$U_{f1} = 30 \sin 30° = 30 \times 0.5$$

$$= 15 \text{ m/s}$$

$$\therefore \quad \tan \beta_1 = \frac{15}{(25.981 - 15)} = \frac{15}{10.981} = 1.366$$

$$\therefore \quad \beta_1 = 53.793°$$

Also $U_{r1} = \dfrac{U_{f1}}{\sin 53.793°} = 18.59$ m/s

Consider the outlet triangle $U_{r2} = U_{r1} = 18.59$ m/s

$$\frac{18.59}{\sin 120°} = \frac{15}{\sin (60 - \beta_2)}$$

$$\therefore \quad \sin (60 - \beta_2) = \frac{15 \times 0.866}{18.59} = 0.699$$

$$\therefore \quad 60 - \beta_2 = 44.33$$

$$\therefore \quad \beta_2 = 15.67°$$

Work done per kg of water $= (U_{w1} + U_{w2})u$

However $U_{w2} = U_{r2} \cos 15.67° - u$

$$= 18.59 \times 0.9628 - 15$$

$$= 17.899 - 15 = 2.899 \text{ m/s}$$

$$\therefore \quad \text{Work done} = (25.981 + 2.899) \times 15$$

$$= 28.880 \times 15 = 433.2 \text{ Nm/kg}$$

$$\text{Efficiency} = \frac{\text{work done per kg of water}}{\text{energy of jet per kg of water}}$$

$$= \frac{433.2}{\frac{1}{2} \times 30^2} = 0.963 \quad \text{or} \quad 96.3 \text{ per cent}$$

6.14 A vane with a turning angle θ can move on a cart in the same direction as the jet of area A and velocity U which strikes on it tangentially. Assume the cart to be stationary and as the jet strikes it, the cart moves at a velocity u. Find an expression for $u(t)$ if the cart weighs W N. Neglect friction (Fig. 6.13).

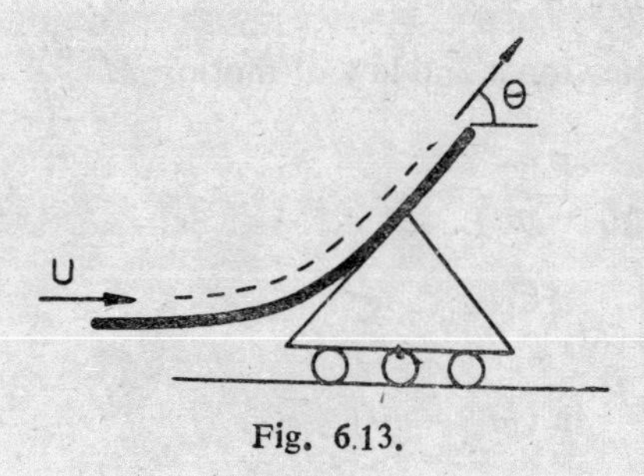

Fig. 6.13.

Relative velocity of jet $= (U - u)$

Force exerted by the jet on the vane in x direction will be

$$F_{fx} = A\rho(U - u)\,[(U - u)\cos\theta]$$

$$= A\rho(U - u)^2\,(1 - \cos\theta)$$

Because of this unbalanced force, cart alongwith the vane will accelerate.

According to Newton's second law,

$$M\frac{du}{dt} = A\rho\,(U - u)^2\,(1 - \cos\theta)$$

where $M = W/g$

$$\int\frac{du}{(U - u)^2} = \int\frac{A\rho g}{W}(1 - \cos\theta)\,dt$$

Integration of the above yields

$$\frac{1}{(U - u)} = \frac{A\rho g}{W}(1 - \cos\theta)\,t + C$$ where C is the constant of integration. At $t = 0$, $u = 0$ $\therefore$ $C = \dfrac{1}{U}$

$$\therefore \quad \frac{1}{U - u} - \frac{1}{U} = \frac{A\rho g}{W}(1 - \cos\theta)\,t = Bt$$

where $B = \dfrac{A\rho g}{W}(1 - \cos\theta)$. Hence $\dfrac{u}{U(U - u)} = Bt$

or $$\frac{u}{U} = \frac{BUt}{1 + BUt}$$

It can be seen that as $t \to \alpha$, $u/U \to 1$ or $u = U$.

6.15 A vehicle moves at a velocity U_0 along level ground under the action of a constant force F. At time $t = 0$ mass begins leaving the vehicle vertically at a constant rate of m per unit time, through a hole in the bottom while the vehicle continues to move under the action of force, F. Determine the variation of velocity of the vehicle as a function of time if its initial mass is M_0.

Mass at any time $M = (M_0 - mt)$

According to Newton's 2nd law of motion $M \dfrac{dU}{dt} = F$

$$\therefore \quad \frac{dU}{dt} = \frac{F}{(M_0 - mt)}$$

$$\therefore \quad \int dU = \int \frac{F dt}{(M_0 - mt)} + C$$

$$\therefore \quad U = -\frac{F}{m} \ln (M_0 - mt) + C$$

The constant of integration can be evaluated from the condition that

$$\text{at } t = 0, U = U_0$$

$$\therefore \quad C = U_0 + \frac{F}{m} \ln M_0$$

$$\therefore \quad U = U_0 + \frac{F}{m} \ln \frac{M_0}{(M_0 - mt)}$$

6.16 For the sluice gate shown in Fig. 6.14, determine the force required to hold the gate in place. Assume gate width to be 2.0 m.

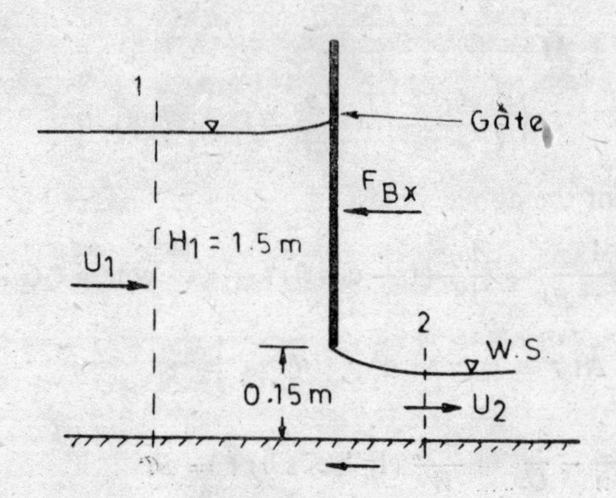

Fig. 6.14.

Assume $C_c = 0.60$

$$C_d = \frac{0.60}{\sqrt{1 + 0.6 \dfrac{0.15}{1.50}}} = 0.583$$

$$q = C_d a \sqrt{2gh}$$

$$= 0.583 \times 0.15 \times \sqrt{2 \times 9.806 \times 1.5}$$

$$= 0.474 \text{ m}^3/\text{s m}$$

$$U_1 = \frac{0.474}{1.50} = 0.316 \text{ m/s}$$

$$H_2 = 0.150 \times 0.60 = 0.090 \text{ m} \quad \text{and} \quad U_2 = \frac{0.474}{0.09} = 5.267 \text{ m/s}$$

Apply momentum equation between sections 1 and 2. The boundary frictional force F_f is assumed to be very small and neglected.

$$\therefore \quad \left(\frac{1}{2}\gamma H_1^2 B - \frac{1}{2}\gamma H_2^2 B - F_{BX}\right) = Q\rho(U_2 - U_1)$$

$$\therefore \quad \frac{1}{2} \times 2 \times 9787\,(1.5^2 - 0.09^2) - F_{BX} = 0.474 \times 2 \times 998\,(5.267 - 0.316)$$

$$\therefore \quad F_{Bx} = 17\,255.94 \text{ N} \quad \text{or} \quad 17.256 \text{ kN}$$

6.17 Obtain the relationships for thrust and efficiency for a ship propelled by a reaction jet (Fig. 6.15).

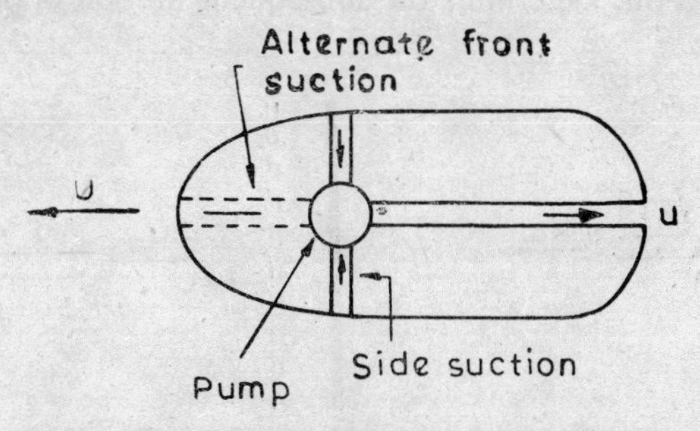

Fig. 6.15. Ship propelled by a reaction jet.

Here the water is taken in, either from the sides or from the front and discharged at a high velocity from the rear. Let U be the forward velocity of the ship and u be the absolute velocity of the jet coming out from rear of ship.

$\therefore$ Relative velocity of jet with respect to ship $U_r = u + U$

Jet discharge $= A\rho U_r$ kg/s where A is area of the jet.

Thrust on the ship $= A\rho U_r u$

Work done per second $= A\rho_r u U_r u$

K.E. supplied per second $= \frac{1}{2} A\rho U_r U_r^2$

$$\eta = \frac{A\rho U_r\, u\, U}{\frac{1}{2} A\rho U_r\, U_r^2} = \frac{2uU}{U_r^2} \quad \text{or} \quad \frac{2uU}{(U+u)^2}$$

η will be maximum when $\frac{d\eta}{du} = 0$ for given U

$$\frac{d\eta}{du} = 2U - \frac{4uU}{(u + U)} = 0$$

$$u = U$$

$$\eta_{max} = \frac{2U^2}{4U^2} = 0.50 \quad \text{or} \quad 50 \text{ per cent}$$

If the suction is in the front, K.E. of the jet as it enters is $\frac{U^2}{2}$ per unit mass and when it leaves it is $\frac{U_r^2}{2}$.

$$\therefore \qquad \eta = \frac{A\rho U_r uU}{\frac{A\rho U_r}{2}(U^2 - U_r^2)} = \frac{2uU}{(U + U_r)\,u} = \left(\frac{2u}{U + U_r}\right)$$

6.18 A plate of mass M is suspended and is subjected to action from two equal horizontal jets in opposite directions. The jet area is A and velocity U_0. At time $t = 0$ the plate is suddenly set into motion with velocity U_0 in positive x direction. Determine the subsequent motion of the plate, i.e. $x(t)$ (Fig. 6.16)

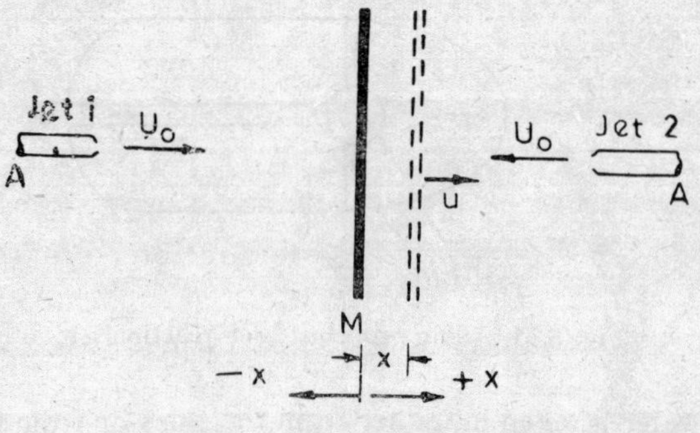

Fig. 6.16.

Let the plate be at x at any time t and let its velocity be u.

$\therefore$ Force on the plate due to jet 1, $F_1 = A\rho\,(U_0 - u)^2$

Force on the plate due to jet 2, $F_2 = A\rho\,(U_0 + u)^2$

$\therefore$ Resultant force on the plate $F = A\rho\,[(U_0 - u)^2 - (U_0 + u)^2]$

$$= -\,4A\rho U_0 u$$

Plate acceleration $du/dt = \frac{\text{Force}}{\text{mass}} = -\frac{4A\rho U_0 u}{M}$

Integration of this differential equation yields

$$\int \frac{du}{u} = -\frac{4A\rho U_0}{M}\int dt$$

$$\therefore \qquad \ln u = -\frac{4A\rho U_0}{M} t + C$$

With the boundary condition, at $t = 0, u = U_0 \quad \therefore \quad c = \ln U_0$

$$\therefore \qquad \ln u = -\frac{4A\rho U_0}{M} t + \ln U_0$$

or
$$u = U_0 \exp\left[-\left(\frac{4A\rho U_0}{M}\right) t\right]$$

However
$$u = \frac{dx}{dt}$$

$$\therefore \qquad dx = U_0 \exp\left[-\left(\frac{4A\rho U_0}{M}\right) t\right] dt$$

$$\therefore \qquad x = -\frac{M}{4A\rho} \exp\left[-\left(\frac{4A\rho U_0}{M}\right) t\right] + C_1$$

The constant C_1 is determined from the condition that at $t = 0, x = 0$

$$\therefore \qquad C_1 = \frac{M}{4A\rho}$$

$$x = \frac{M}{4A\rho}\left[1 - \exp\left(-\frac{4A\rho U_0}{M} t\right)\right]$$

6.19 In a wide rectangular open channel, the veloctiy distribution in the vertical follows the law

$$\frac{u}{u_m} = \left(\frac{y}{D}\right)^n$$

where u_m is the surface velocity, D is the depth of flow and u is the velocity at a distance y above the bed (Fig. 6.17). Determine the average velocity and momentum correction factor. What are their values when $n = \frac{1}{7}$?

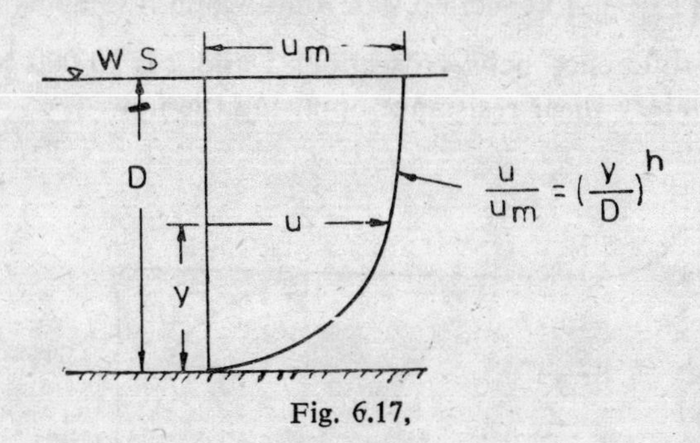

Fig. 6.17,

$$q = \int_0^D u\,dy$$

$$\therefore \qquad q = \frac{u_m}{D^n}\int_0^D y^n\, dy$$

$$= \frac{u_m}{D^n}\frac{D^{n+1}}{n+1} \quad \text{or} \quad \frac{u_m D}{n+1}$$

$\therefore$ Average velocity $U = \dfrac{q}{D} = \dfrac{u_m}{n+1}$

Also $\quad \beta U^2 D\rho = \int_0^D \rho u^2 dy = \dfrac{\rho u_m^2}{D^{2n}} \int_0^D y^{2n} dy$

$$= \frac{\rho u_m^2}{D^{2n}} \frac{D^{2n+1}}{2n+1}$$

Substituting value of $U = \dfrac{u_m}{n+1}$ in the above equation and solving for β, one gets

$$\beta = \frac{(n+1)^2}{(2n+1)}$$

when $n = \dfrac{1}{7}$

$$U = \frac{u_m}{1+\frac{1}{7}} \quad \text{or} \quad \frac{7\,u_m}{8}$$

$$\beta = \frac{(8/7)^2}{(9/7)} = \frac{64}{63} \quad \text{or} \quad 1.0159$$

6.20 Oil of relative density 0.80 flows through a 75 mm diameter pipe having bell mouthed entrance, at an average velocity of 3.0 m/s. Because of the frictional resistance and subsequent formation of boundary layer, the uniform velocity distribution at section 1 (Fig. 6.18) changes to the distribution $\dfrac{u}{u_m} = \left(1 - \dfrac{r^2}{R^2}\right)$ at section 2 beyond which it remains unchanged. If the pressure difference between sections 1 and 2 is 10 000 N/m², calculate the total boundary shear resistance opposing the flow.

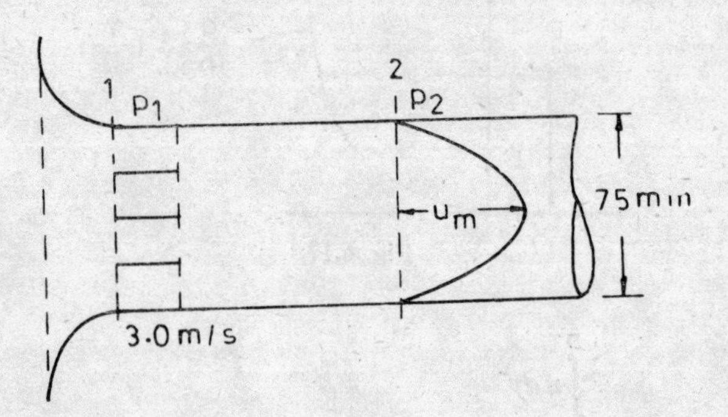

Fig. 6.18.

First obtain u_m using continuity equation

$$\frac{\pi}{4} \times 0.075^2 \times 3.0 = \int_0^R 2\pi r u_m \left(1 - \frac{r^2}{R^2}\right) dr$$

$$\therefore \quad u = 6.0 \text{ m/s}$$

Momentum correction factor at section 1 is unity, or $\beta_1 = 1$. Momentum correction factor at section 2 can be calculated as follows

$$\beta_2 = \frac{\int_0^R \rho 2\pi r \, dr \, u_m^2 \left(1 - \frac{r^2}{R^2}\right)^2}{\frac{\pi}{4} R^2 \rho \, U_2^2} \quad \text{with } U_1 = U_2 = 3.0 \text{ m/s and } u_m = 6.0 \text{ m/s}$$

This gives $\beta_2 = 1.333$. Hence momentum equation can be written as

$$p_1 A_1 - p_2 A_2 - F_s = \rho Q \,(U_2\beta_2 - U_1\beta_1)$$

$$\text{or} \quad F_s = (0.785 \times 0.075^2 \times 10\,000) - 998 \times 0.8 \times 0.785 \times .075^2 \times 3^2 (1.333 - 1.0)$$

$$= 44.156 - 10.560$$

$$= 31.596 \text{ N}$$

6.21 Experiments conducted in a free air stream gave velocity distributions on the upstream and downstream of two dimensional cylinder as shown in Fig. 6.19. Apply momentum equation and determine the profile drag of the cylinder per unit length. Assume $\rho = 1.208$ kg/m^3 and pressures at sections 1 and 2 to be equal. Velocities at downstream end are measured at 100 mm interval. What will be its C_D value?

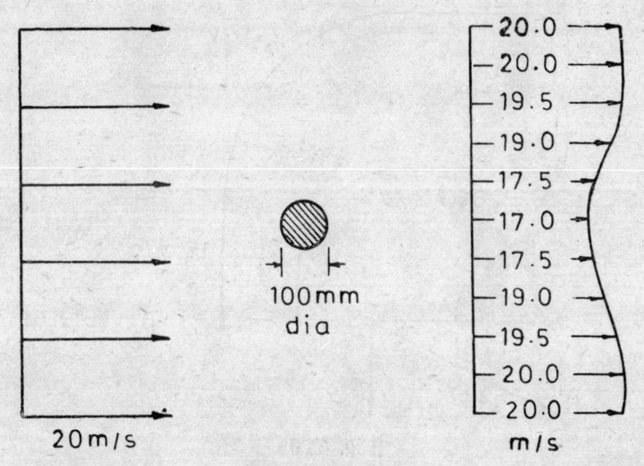

Fig. 6.19.

If a strip of height dy is taken and u_1 and u_2 are the average velocities at upstream and downstream sections, then $\rho u_2 dy$ is the mass rate of flow through this strip at d.s. end and its velocity has changed by $(u_2 - u_1)$.

$$\therefore \quad \text{Loss of momentum} = \rho u_2 (u_2 - u_1)\, dy$$

$\therefore$ Total loss of momentum

$$= 2 \times 0.10 \times 1.208\,[(16.25^2 - 16.25 \times 20) + (18.25^2 - 18.25 \times 20) + (19.25^2 - 19.25 \times 20) + (19.75 - 19.75 \times 20)]$$

$$= 0.2416\,(-60.938 - 31.938 - 14.438 - 4.938)$$

$$= -27.120 \text{ N}$$

$\therefore$ $F_{Bx} = -27.120$ N is the force exerted by the cylinder on the fluid.

Fluid will exert a drag of 27.120 N in the direction of flow. If drag coefficient C_D is defined as

$$C_D = \frac{\text{Drag force}}{(\text{Projected area perpendicular to flow})} \times \frac{\rho U^2}{2}$$

$$= \frac{27.120 \times 2}{1 \times 0.1 \times 1.208 \times 20^2}$$

$$= 1.122$$

6.22 Water flows through a 90° reducer-bend. The pressure at the inlet is 206 kN/m² (gauge) where the cross-sectional area is 0.01 m². At the exit section, the area is 0.0025 m² and the velocity is 15 m/s. The pressure at the exit is atmospheric. Determine the force required to hold the bend in place (Fig. 6.20).

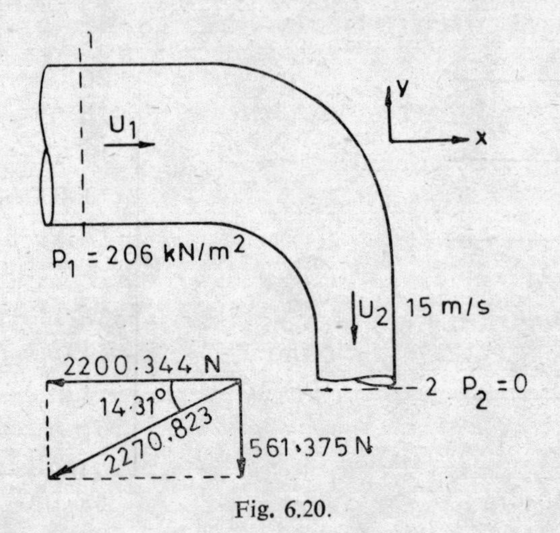

Fig. 6.20.

$$Q = U_2 A_2 = 15 \times 0.0025 = 0.0375 \text{ m}^3/\text{s}$$

$$U_1 = \frac{0.0375}{0.010} = 3.75 \text{ m/s}$$

Assume the bend is horizontal and in xy plane. Momentum equation in x direction gives

$$Q\rho\,(0 - U_1) = F_{Bx} + p_1 A_1$$

$$\therefore \quad -998 \times 0.0375 \times 3.75 = F_{Bx} + 206\,000 \times 0.01$$

$$\therefore \quad F_{Bx} = -\ 140.344 - 2060$$

$$= -\ 2200.344 \text{ N}$$

Similarly momentum equation in y direction yields

$$Q\rho\,(-15 - 0) = p_2 A_2 + F_{By} \quad \text{but} \quad p_2 = 0$$

$$\therefore \quad -0.0375 \times 998 \times 15 = F_{By}$$

$$\text{or} \quad F_{By} = -\ 561.375 \text{ N}$$

$$F_B = \sqrt{(-2200.344)^2 + (-561.375)^2}$$

$$= 2270.827 \text{ N}$$

$$\tan\theta = \frac{561.375}{2200.344} = 0.2551 \quad \therefore\ \theta = 14.31^\circ \text{ in the third quadrant,}$$

6.23 In Example 6.22, assume that the bend does not discharge into the atmosphere but 0.0025 m² area conduit continues. Further assume that the head loss in the reducer-bend is

$$h_L = K_b \frac{U^2}{2g}$$

where $K_b = 1$ and U is the average velocity $\dfrac{U_1 + U_2}{2}$. Determine the force exerted by the flow on the elbow.

$$U = \frac{15.000 + 3.750}{2} = 9.375 \text{ m/s}$$

$$h_L = 9.375^2/2 \times 9.806 = 4.481 \text{ m}$$

Apply Bernoulli's equation between sections 1 and 2 and find p_2.

$$\frac{p_1}{\gamma} + \frac{U_1^2}{2g} = \frac{p_2}{\gamma} + \frac{U_2^2}{2g} + h_L$$

$$\therefore \frac{p_2}{9787} = \frac{206000}{9787} + \frac{3.75^2}{2 \times 9.806} - \frac{15^2}{2 \times 9.806} - 4.481$$

$$= 21.048 + 0.717 - 11.473 - 4.481$$

$$= 5.801 \text{ m}$$

$$\therefore \quad p_2 = 5.801 \times 9787 = 56\,774.387 \text{ N/m}^2$$

Momentum equation in x direction gives

$$0.0375 \times 998\,(0 - 3.75) = 0.01 \times 206\,000 + F_{Bx}$$

$$\therefore \quad F_{Bx} = -\ 2200.344 \text{ N} \qquad F_{fx} = 2200.344 \text{ N}$$

Momentum equation in y direction gives

$$0.0375 \times 998\,(-15 - 0) = 56\,774.387 \times 0.0025 + F_{By}$$

$$F_{By} = -561.375 - 141.936$$

$$= -703.311 \text{ N} \quad \therefore \quad F_{fy} = 703.311 \text{ N}$$

$$F_f = \sqrt{2200.344^2 + 703.311^2}$$

$$= 2310.013 \text{ N}$$

$$\tan\theta = 703.311/2200.344 = 0.3195$$

$$\theta = 17.72°, \text{ in the first quadrant.}$$

6.24 Use continuity, momentum and energy equations and show that head loss in a sudden expansion in a pipe is given by $h_L = \dfrac{(U_1 - U_2)^2}{2g}$. What assumptions are involved in the derivation?

Control volume chosen is shown in Fig. 6.21.

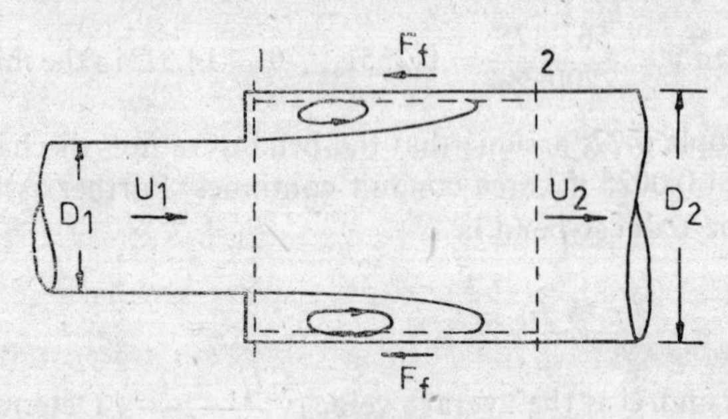

Fig. 6.21.

Continuity equation: $Q = A_1U_1 = A_2U_2$ (1)

Apply momentum equation between sections 1 and 2 assuming

(i) Flow is steady

(ii) Velocity distribution at sections 1 and 2 is uniform; hence $\beta_1 = \beta_2$

(iii) Pressure at annular ring at section 1 is p_1; this is verified by experiments

(iv) Boundary friction F_f is negligible.

(v) Pipe is horizontal and hence component of gravity in the direction of flow is zero.

$$\therefore \quad p_1A_1 + p_1\,(A_2 - A_1) - p_2A_2 = Q\rho\,(U_2 - U_1)$$

or $A_2\,(p_1 - p_2) = Q\rho\,(U_2 - U_1)$ (2)

Combining equations (1) and (2), one gets

$$\frac{(p_2 - p_1)}{\rho} = U_1U_2 - U_2^2 \qquad (3)$$

Energy equation gives

$$\frac{p_1}{\gamma}+\frac{U_1^2}{2g}=\frac{p_2}{\gamma}+\frac{U_2^2}{2g}+h_L \tag{4}$$

Eliminate $\frac{(p_2-p_1)}{\rho}$ from Equation (3) and (4) to get

$$h_L=(U_1-U_2)^2/2g$$

6.25 A rocket travels at 600 m/s and discharges its exhaust gases at 700 m/s velocity relative to the rocket. The mass rate of exhaust gases is 4.5 kg/s and the exit area is 0.07 m². Assuming absolute pressure at the exit to be 9.6 N/cm² and outside pressure to be 8.25 N/cm², determine the thrust on the rocket (Fig. 6.22).

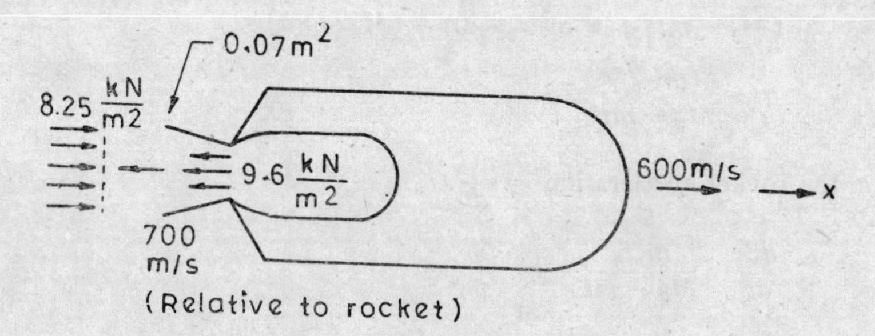

Fig. 6.22.

The velocity 600 m/s of the rocket does not come in the calculations since exit velocity of gas given is relative to rocket.

4.5 (– 700 – 0) = change in momentum per sec.

External forces causing this change of momentum of the gas

$$= -F_{Bx}+0.07(82\,500-96\,000).$$

Here F_{Bx} is force exerted by rocket on the gas. Equating these two, one gets

$$F_{Bx}=2205\text{ N} \quad \text{(acting to the left)}$$

Force exerted on the rocket will be 2205 N acting to the right.

6.26 A small rocket which has initial weight of 2000 N is to be launched vertically. The rocket consumes 4.0 kg/s of fuel and exhausts it at atmospheric pressure with a velocity of 1000 m/s relative to the rocket. Determine the initial acceleration of the rocket and its velocity after 10 seconds. Assume air resistance to be negligible (Fig. 6.23).

Mass of rocket at any time

$$M=M_0-mt$$

where M_0 initial mass of rocket and fuel, and m is the rate at which fuel is burnt. Since gases are exhausted at outside pressure

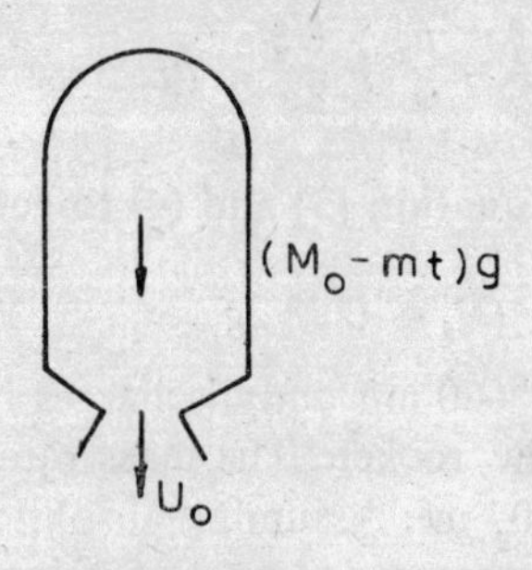

Fig. 6.23.

$$(M_0 - mt)\,a = m(U_0 - 0) - (M_0 - mt)\,g$$

$$\therefore \qquad a = \frac{mU_0}{M_0 - mt} - g$$

Here a, the rocket acceleration $= \dfrac{dU}{dt}$

$$\therefore \qquad \frac{dU}{dt} = \frac{mU_0}{M_0 - mt} - g$$

or
$$\int dU = \int \frac{mU_0}{(M_0 - mt)}\,dt - \int g\,dt$$

$$\therefore \qquad U = -\,gt - U_0 \ln (M_0 - mt) + C$$

The constant of integration C can be evaluated from the condition that at $t = 0$, $U = 0$.

$$\therefore \qquad C = U_0 \ln M_0$$

$$\therefore \qquad U = -\,gt + U_0 \ln \left(\frac{M_0}{M_0 - mt}\right)$$

For data given in the problem at $t = 0$

$$a = \frac{4.0 \times 1000}{2000/9.806} - 9.806$$

$$= 9.806 \text{ m/s}^2$$

Also at $\quad t = 10$ s

$$U = -\,(9.806 \times 10) + 1000 \ln \frac{2000/9.86}{\left(\dfrac{2000}{9.806} - 40\right)}$$

$$= 120.245 \text{ m/s}$$

6.27 Apply continuity, momentum and energy equations to a propeller in a fluid stream as shown in Fig. 6.24 and obtain expressions for power input and theoretical efficiency.

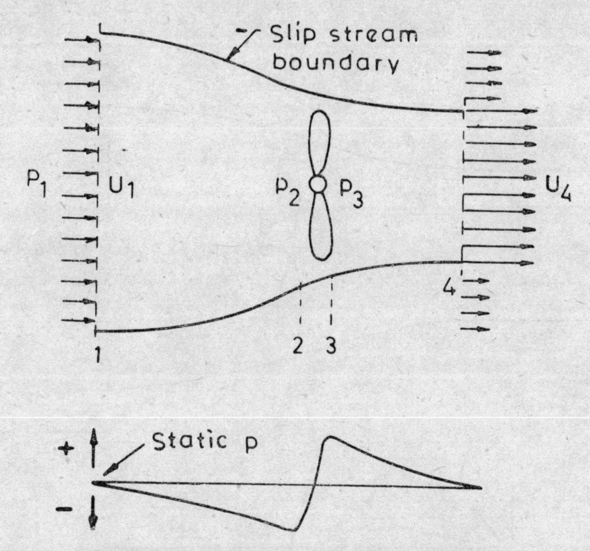

Fig. 6.24. Propeller in a fluid stream.

The propeller changes the momentum of the fluid within which it is submerged and thus develops a thrust that is used for propulsion. A propeller with its slip-stream boundary and velocity distributions at the two ends is shown in Fig. 6.24. Slipstream is nothing but the body of air affected by the propeller. Propeller is assumed to be stationary and fluid is assumed to be frictionless and incompressible.

The flow is accelerated as it approaches the propeller. In passing through the propeller, the fluid has its pressure increased. The flow accelerates from 3 to 4 and its pressure is reduced. Applying momentum equation between sections 1 and 4

$$\rho Q\,(U_4 - U_1) = F_{Bx} = A(p_3 - p_2)$$

where F_{Bx} is the force exerted by the propeller on the flow. Pressures at section 1 and 4 are the same. Here A is the area swept over by the propeller blades. Since $Q = AU$

$$\rho U(U_4 - U_1) = p_3 - p_2$$

Further, Bernoulli's equation between sections 1 and 2, and between 3 and 4 yields,

$$p_1 + \frac{\rho U_1^2}{2} = p_2 + \tfrac{1}{2}\rho U^2$$

$$p_3 + \tfrac{1}{2}\rho U^2 = p_4 + \tfrac{1}{2}\rho U_4^2$$

However $p_1 = p_4$

$$\therefore \qquad (p_3 - p_2) = \tfrac{1}{2}\rho\,(U_4^2 - U_1^2)$$

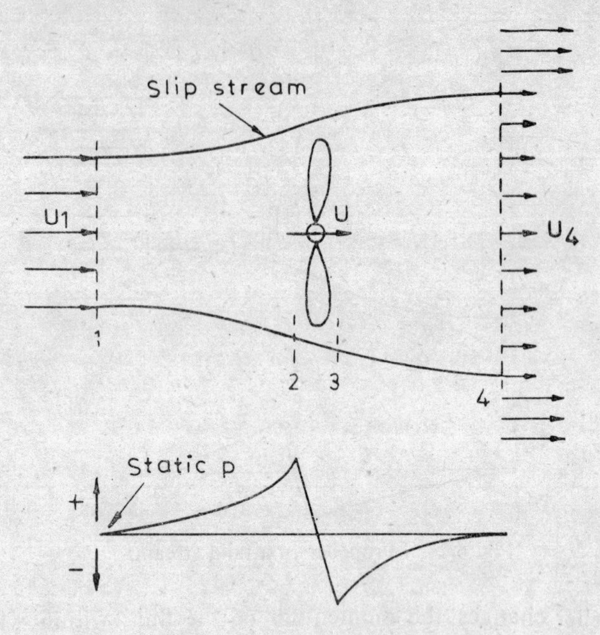

Fig. 6.25. Flow past a windmill.

Equating the values of $(p_3 - p_2)$ from the momentum and energy equations, one gets

$$\rho\, U(U_4 - U_1) = \tfrac{1}{2}\,\rho\,(U_4 + U_1)\,(U_4 - U_1)$$

$$\therefore \qquad U = \left(\frac{U_4 + U_1}{2}\right)$$

Work done by the propeller per second $= F_{Bx} U_1$

$$= \rho Q(U_4 - U_1)U_1$$

Power input = work done by propeller per second + kinetic energy remaining in slipstream

$$= \rho Q(U_4 - U_1)U_1 + \frac{\rho Q}{2}\,(U_4 - U_1)^2$$

$$= \frac{Q\rho}{2}\,(U_4^2 - U_1^2)$$

Theoretical efficiency = power output/power input

$$= \frac{\rho Q\,(U_4 - U_1)U_1}{\dfrac{Q\rho}{2}\,(U_4^2 - U_1^2)}$$

$$= \frac{2U_1}{U_4 + U_1} = \frac{U_1}{U}$$

6.28 Discuss the principle of windmills and obtain the condition for maximum power and maximum efficiency.

Windmill extracts power from the wind. Windmill output is used for pumping water, grinding coarse grains, sawing wood or for its conversion into electricity. As the conventional sources of energy are depleting fast, windmills can be a potential alternate source of energy in developing countries.

As seen in Ex. 6.27, there is a static pressure increase across a propeller. However, in windmills there is a decrease in static pressure across the windmill rotor since energy is extracted. With reference to Fig. 6.25,

Rate of work done by fluid = Rate of decrease of kinetic energy

$$\dot{m}\, U(U_1 - U_4) = \frac{\dot{m}}{2}(U_1^2 - U_4^2)$$

where $\dot{m}$ is the mass flow rate. This gives $U = \frac{1}{2}(U_1 + U_4)$

Also $\dot{m} = \rho A U = \rho A \left(\frac{U_1 + U_4}{2}\right)$

Power P which is the rate of work done by the fluid

$$= \dot{m}\left(\frac{U_1^2 - U_4^2}{2}\right) = A\rho\left(\frac{U_1 + U_4}{2}\right)\left(\frac{U_1^2 - U_4^2}{2}\right)$$

$$= \frac{A\rho}{4}(U_1 + U_4)(U_1^2 - U_4^2)$$

For P to be maximum $\frac{\partial P}{\partial U_4} = 0$ for given wind velocity U_1,

$$\frac{\partial P}{\partial U_4} = \frac{A\rho}{4}[(U_1 + U_4)(-2U_4) + (U_1^2 - U_4^2)] = 0$$

$$U_1^2 - 2U_1U_4 - 3U_4^2 = 0$$

or

$$(U_1 - 3U_4)(U_1 + U_4) = 0$$

$\therefore$

$$U_1 = 3U_4 \quad \text{or} \quad U_4 = U_1/3$$

Substituting this value of U_4 in the expression for power, one gets

$$\text{Maximum power } P_m = \frac{A\rho}{4}\frac{4}{3}U_1 \times \frac{8}{9}U_1^2 = \frac{8}{27}A\rho U_1^3$$

Power available in wind $= \frac{1}{2}A\rho U_1^3$

$$\text{Maximum efficiency} = \frac{\text{Maximum power}}{\text{Power available}} = \frac{8}{27}\frac{A\rho U_1^3 \times 2}{A\rho U_1^3}$$

$$= \frac{16}{27} \quad \text{or} \quad 0.593, \text{ i.e. } 59.3\%$$

Actual efficiency will be somewhat lower due to friction and flow loss across the blades. In actual windmills, the rotor consists of 2 to 3 propeller like blades with blades covering a small fraction of area swept by blades. In India 2 to 4 m diameter blades are used; however in foreign countries blades as large as 37.5 m in diameter are used, generating 100 kW power at 8.00 m/s wind speed.

6.29 The sprinkler shown in Fig. 6.26 has nozzles of 5 mm diameter and carries a total discharge of 0.20 l/s. Determine the angular speed of rotation of the sprinkler and the torque required to hold the sprinkler stationary. Neglect friction.

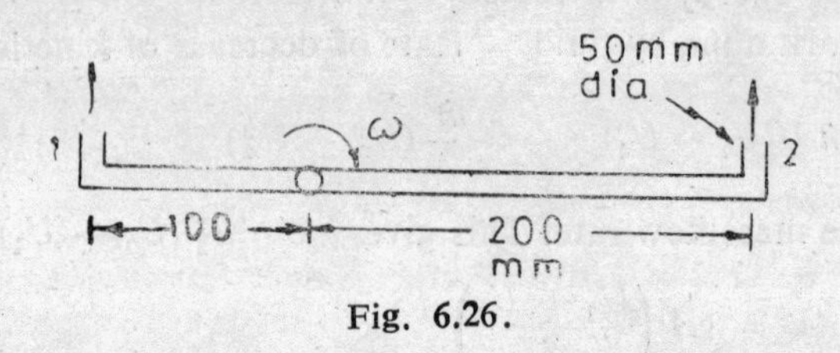

Fig. 6.26.

$$\text{Nozzle area} = 0.785 \times 0.005^2$$
$$= 1.9625 \times 10^{-5} \text{ m}^2$$

$$\text{Flow through each nozzle} = \frac{0.20}{2} = 0.1 \text{ l/s} \quad \text{or} \quad 10^{-4} \text{ m}^3/\text{s}$$

$$\therefore \quad \text{Relative velocity of flow through nozzle} = \frac{10^{-4}}{1.9625 \times 10^{-5}}$$
$$= 5.096 \text{ m/s}$$

The sprinkler will rotate in clockwise direction at an angular speed of rotation ω. Since there is no friction and no external torque is applied to the system, and since the initial moment of momentum of fluid entering the system is zero, the moment of momentum of the fluid leaving the system must be zero.

$$\text{or} \quad U_{t1}\, Q\rho r_1 - U_{t2}\, Q\rho r_2 = 0$$

where U_{t1} and U_{t2} are the absolute velocities of the jets at nozzles 1 and 2.

$$U_{t1} = (5.096 + 0.1\,\omega) \quad U_{t2} = (5.096 - 0.2\,\omega)$$

$$\therefore \quad 10^{-4} \times 998 \times 0.10\,(5.096 + 0.1\,\omega)$$
$$= 10^{-4} \times 998 \times 0.20\,(5.096 - 0.2\,\omega)$$

$$\therefore \quad \omega = 10.192 \text{ rad/s}$$

$$\text{or} \quad \text{since } \frac{2\pi N}{60} = \omega,\ N = \frac{60 \times 10.192}{2 \times 3.142}$$
$$= 97.31 \text{ rpm}$$

Torque required to hold the sprinkler stationary

$$= Q\rho U_r r_2 - Q\rho U_r r_1$$
$$= 10^{-4} \times 998 \times 5.096\,(0.20 - 0.10) = 0.5086 \times 10^{-1} \text{ N m}$$

6.30 A lawn sprinkler with equal arm lengths r discharges water at relative velocity U_r through nozzles of area A at angle θ to the tangent as shown in Fig. 6.27. Find the steady state angular velocity of sprinkler assuming no friction at the pivot.

Fig. 6.27.

Tangential component of absolute velocity of jet issuing from the nozzle will be $= (U_r \cos\theta - r\omega)$.

This contributes to moment of momentum about z axis through pivot. Initial moment of momentum of water coming out through vertical pipe is zero.

Final moment of momentum $= AU_r\rho\,(U_r \cos\theta - r\omega)\,r$. Since no external torque is applied, this must be zero.

$$\therefore \quad \omega = \frac{U_r \cos\theta}{r}$$

Torque required to hold the sprinkler in place

$$T = 2AU_r\rho\,(U_r \cos\theta)\,r$$

6.31 This problem illustrates the principle of jet pump. In a jet pump shown in Fig. 6.28, a water jet of diameter 10 mm and velocity 30 m/s is directed along the axis of the pipe of diameter 25 mm. The jet entrains a secondary stream of velocity 15 m/s. Considering one dimensional flow and neglecting the skin friction, determine (1) average velocity of flow at section 2 (ii) pressure difference between sections 1 and 2 and (iii) rate of mechanical energy dissipation in watts.

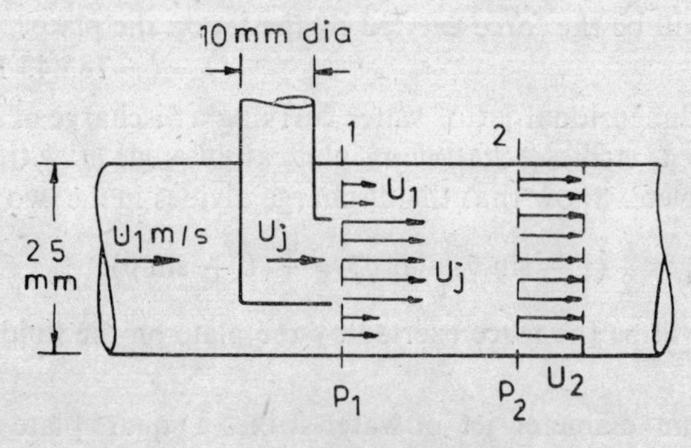

Fig. 6.28. Jet-pump.

Let U_j be the average velocity of jet, U_1 velocity of flow in pipe at section 1 and U_2 the average velocity at section 2. Assume pressure to be constant over the section 1. Momentum equation gives

$$A_2 (p_1 - p_2) = \rho A_2 U_2^2 - \rho A_j U_j^2 - \rho A_1 U_1^2$$

Continuity equation gives

$$A_2 U_2 = A_1 U_1 + A_j U_j$$

and
$$A_1 = A_2 - A_j$$

$$A_j = 0.785 \times 0.01^2 = 7.850 \times 10^{-5} \text{ m}^2$$

$$A_2 = 0.785 \times 0.025^2 = 49.063 \times 10^{-5} \text{ m}^2$$

$$A_1 = 41.213 \times 10^{-5} \text{ m}^2$$

$$\therefore \quad \frac{41.213 \times 10^{-5} \times 15 + 7.85 \times 10^{-5} \times 30}{49.063 \times 10^{-5}} = U_2$$

$$\therefore \quad U_2 = 17.40 \text{ m/s}$$

$$\therefore \quad 49.063 \times 10^{-5} (p_1 - p_2) = 998 [49.063 \times 10^{-5} \times 17.40^2 - 7.85 \times 10^{-5} \times 30^2 - 41.213 \times 10^{-5} \times 15^2]$$

$$= 998 \times 10^{-5} \times (-1483.61)$$

$$\therefore \quad (p_1 - p_2) = -30\,178.40 \text{ N/m}^2$$

Rate of energy dissipation = Rate of inflow of energy—Rate of outflow of energy

$$= (41.213 \times 10^{-5}) \times 15 \times 998 \times \left[\frac{p_1}{\rho} + \frac{U_1^2}{2}\right] + 7.85 \times 10^{-5} \times 998$$

$$\times 30 \left(\frac{p_1}{\rho} + \frac{U_j^2}{2}\right) - 49.063 \times 10^{-5} \times 998 \times 17.40 \left[\frac{p_2}{\rho} + \frac{U_2^2}{2}\right]$$

$$= 204.23 \text{ N m/s} \quad \text{or} \quad \text{W}$$

PROBLEMS

6.1 Determine the force exerted by a fixed plate on a water jet of 100 mm diameter striking the plate normally at a velocity of 10.0 m/s. What will be the force exerted by the jet on the plate?

(– 783.43 N, 783.43 N)

6.2 A two dimensional jet of water carrying a discharge of q m³/s m with velocity U strikes a stationary plate at an angle θ with the normal to the plate. Show that the discharge divides in the two directions as

$$q_1 = \frac{q}{2}(1 - \sin\theta) \text{ and } q_2 = \frac{q}{2}(1 + \sin\theta)$$

What will be the force exerted by the plate on the fluid?

$(- q\rho U \cos\theta)$

6.3 A 20 mm diameter jet of water strikes a square plate at its centre, the plate being hinged at its top edge about a horizontal axis and hanging vertically. If the plate deflects through an angle of 30°, determine the velocity of the jet. Take weight of the plate as 58 N. Mention any assumptions made.

(13.605 m/s. Effect of weight of water striking on the plate is neglected)

6.4 A vertically upward moving jet of water issuing from a 50 mm diameter nozzle at 8.5 m/s velocity strikes a fixed horizontal plate kept 1.5 m above the nozzle. Determine the force exerted on the plate. (Hint: Find the velocity of jet at 1.5 m above nozzle) (112.283 N)

6.5 A 50 mm diameter jet of water discharging at 18.0 m/s velocity strikes normally on a plate moving in the same direction at 5.0 m/s velocity. Determine the force exerted by the jet on the plate, work done on the plate and efficiency of the jet.

(331 N, 1.655 kW, 29 per cent)

6.6 A 25 mm diameter jet of water strikes normally against a plate moving in the same direction at 6.0 m/s velocity. If the work done by the jet on the plate is 2000 W, determine the velocity of the jet.

(32.092 m/s)

6.7 A jet of water 75 mm in diameter and moving with a velocity of 12.0 m/s strikes a series of flat plates normally. If the plates are moving in the same direction as the jet with a velocity of 9.0 m/s, determine the force on the plate, work done per second and the efficiency. (158.645 N, 1.428 kW, 37.50 percent)

6.8 A jet of water of 75 mm diameter and 40.0 m/s velocity strikes a stationary curved vane and deflects the jet through 30° as shown in Fig. 6.29. Determine the force exerted by the vane on the jet.

(3649.799 N, 105° with +ve x axis)

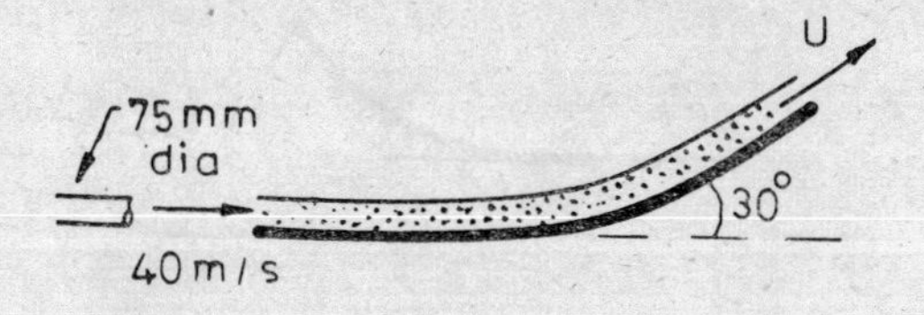

Fig. 6.29.

6.9 A circular jet of 50 mm diameter and 40.0 m/s velocity enters a fixed curved vane at an angle of 30° with the x axis and leaves the vane at 20° with negative x axis as shown in Fig. 6.30. Determine the components of force exerted by the vane on the fluid.

(– 5658.538 N, – 495.128 N)

6.10 A jet of water with a velocity of 60.0 m/s is deflected by a vane moving at 25.0 m/s in a direction of 30° by a vane moving at 25.0

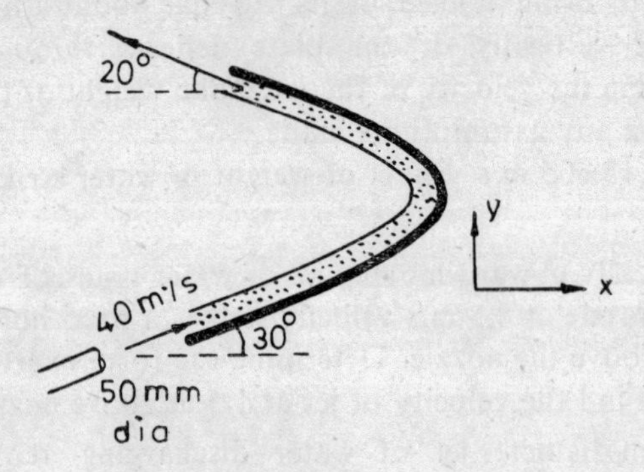

Fig. 6.30.

m/s in a direction at 30° to the direction of the jet. The water leaves the blade normal to the motion of the vane. Determine the vane angles for no shock at the entry or the exit. Take $U_{r2} = 0.85\ U_{r1}$.
($\beta_1 = 48.055°$, $\beta_2 = 43.182°$)

6.11 A 75 mm diameter jet of water impinges on a series of hemispherical cups and is deflected through 180°. If the velocity of the jet is 40.0 m/s and that of the cup 10.0 m/s in the same direction, determine the work done per second by the water striking the cups.
(105.763 kW)

6.12 A vane with a turning angle of 60° moves with a constant velocity of 6.0 m/s as shown in Fig. 6.31. Water flowing out from a nozzle 60 mm diameter strikes the vane tangentially at 30 m/s velocity. Determine the force on the vane. (1624.425 N – 60° with x axis)

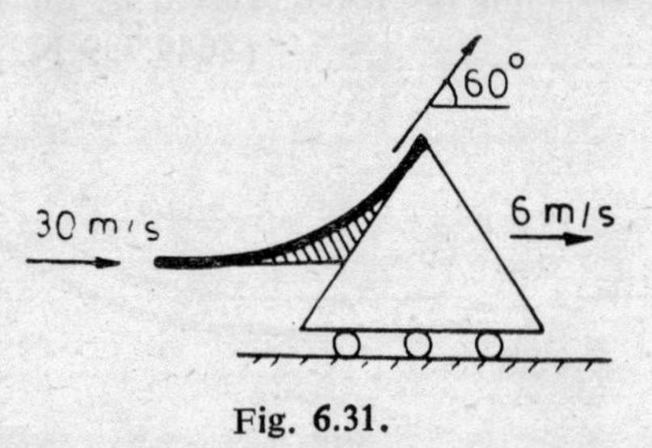

Fig. 6.31.

6.13 In Example 6.14 obtain expression for $x(t)$.

$$x = \frac{1}{BU}\,(BUt - \ln(BUt + 1)$$

6.14 An irrigation sprinkler unit mounted on a cart discharges water at a velocity of 30 m/s at an angle of 45° with the horizontal. If the 50 mm nozzle is 2.0 m above the ground, calculate the moment that tends to overturn the cart (Fig. 6.32). (2492.482 Nm anticlockwise)

6.15 A plastic container weighing 100 N (empty) has 0.1 m² cross-sectional

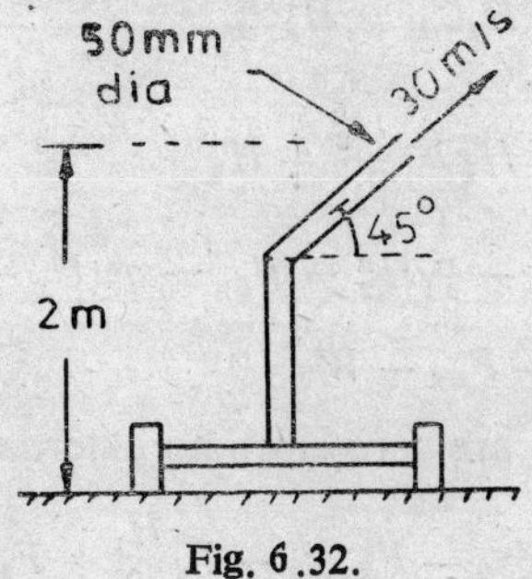

Fig. 6.32.

area and two equal sized openings on opposite sides as shown in Fig. 6.33. The container is kept on a weighing balance and water flows into it from the top through a jet of 0.01 m² area and 8.0 m/s velocity. If under steady flow condition the depth of water in the tank is 1.0 m, determine the scale reading. (1580.095 N)

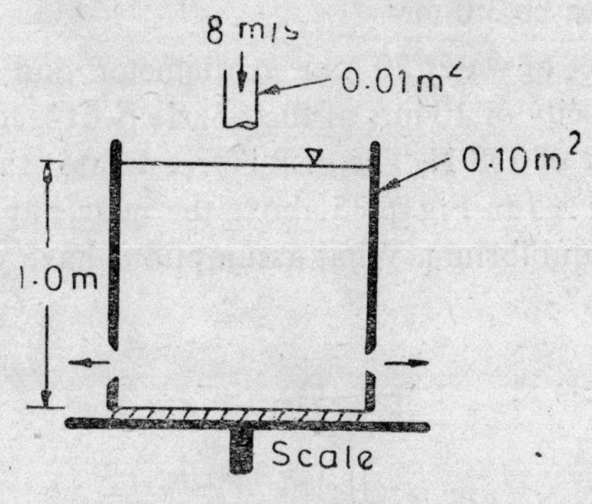

Fig. 6.33.

6.16 Write continuity, momentum and energy equations for flow through a gate shown in Fig. 6.34, if H_1 and H_2 are known. What assumptions are made in writing these equations?

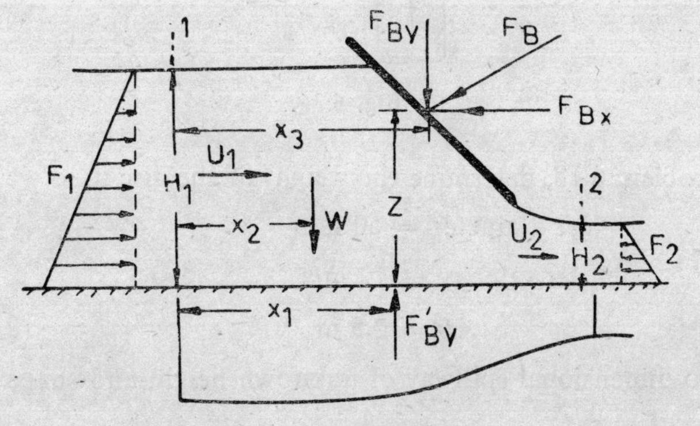

Fig. 6.34.

$$\Big[q = H_1 U_1 = H_2 U_2$$

$$\frac{U_1^2}{2g} + H_1 = \frac{U_2^2}{2g} + H_2$$

$$\tfrac{1}{2}\gamma H_1^2 - \tfrac{1}{2}\gamma H_2^2 - F_{Bx} = q\rho(U_2 - U_1)$$

$$F'_{By} - F_{By} - W = 0$$

Moments about A are zero since no external torque is applied.

$$F'_{By} \cdot x_1 - Wx_2 - F_{By} x_3 - F_1 \frac{H_1}{3} + F_2 \frac{H_2}{3} + F_{Bx} z$$

$$= -\frac{q\rho U_2 H_2}{2} + \frac{q\rho U_1 H_1}{2} \Big]$$

6.17 Assuming $C_c = 0.60$, determine the force required to hold the gate in position if upstream depth is 8.0 m, the opening of vertical lift gate is 0.30 m and the gate works under free flow condition. Assume gate width to be 3.0 m. (858.336 kN)

6.18 A circular jet of water 50 mm in diameter and having a vertically upward velocity of 10 m/s at the nozzle is deflected by a horizontal circular disk of 150 N. The disk is free to move vertically. Determine the distance AB in Fig. 6.35 above the nozzle at which the plate will be held in equilibrium. What assumptions have you made in solving the problem? (2.11 m)

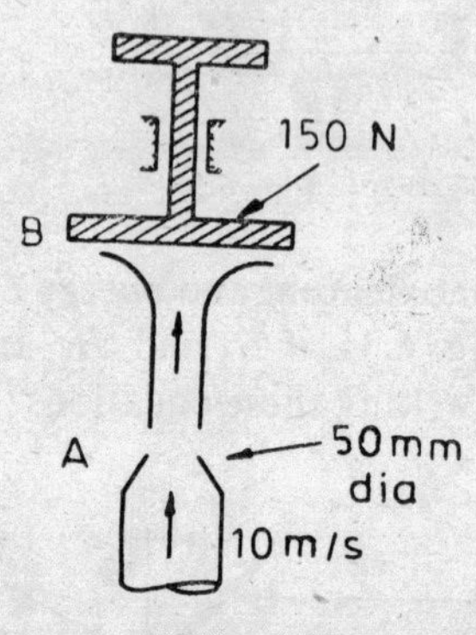

Fig. 6.35.

6.19 In problem 6.18, determine the weight of the disc if

Jet diameter = 50 mm

Jet velocity = 12 m/s

AB = 2.5 m (229.031 N)

6.20 A two dimensional spillway of unknown height and shape has 1.5 m and 0.50 m as the depths upstream and downstream as shown in Fig. 6.36. Calculate the magnitude and direction of the horizontal force on the structure. (– 8213.838 N)

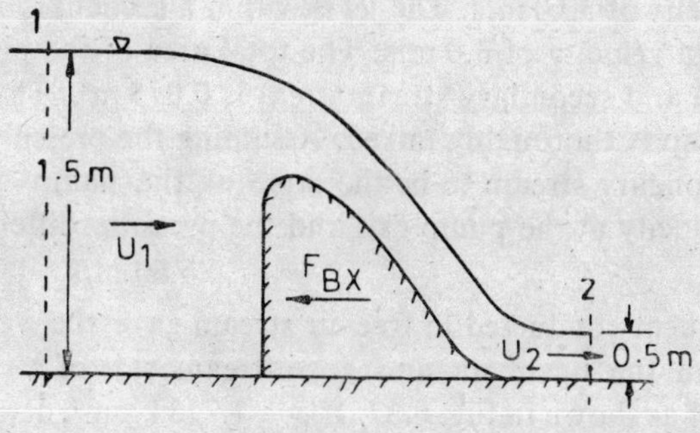

Fig. 6.36.

6.21 A boat driven by a reaction jet discharges water at a velocity of 15.0 m/s relative to the boat when the jet opening is 0.15 m diameter. If the velocity of the boat is 18.0 km/hr, determine the resistance to the motion of the boat, power of the jet and its efficiency. Assume the inlet openings to be on sides.

(2649.69N, 13.248 kW, 44.44 percent)

6.22 Obtain expressions for average velocity and momentum correction factor β for pipe flow in which the velocity distribution is given by

$$\frac{u}{u_m} = 1 - \left(\frac{r}{R}\right)^n$$

$$\left(\frac{nu_m}{n+2}, \frac{(n+2)^2}{(n+1)(n+2)}\right)$$

6.23 The velocity distribution in a two dimensional laminar boundary layer on a flat plate is given by

$$\frac{u}{U} = \left(\frac{3}{2}\eta - \frac{1}{2}\eta^2\right) \text{ where } \eta = \frac{y}{\delta}$$

Here U_0 is the free stream velocity at $y = \delta$. Determine average velocity U and β. $\left(\frac{7}{12} U_0,\ 1.249\right)$

6.24 Flow in a wide rectangular open channel follows the velocity distribution $\frac{u}{u_m} = \left(\frac{2}{3} + \frac{1}{3}\frac{y}{D}\right)$ where u_m is the surface velocity and D the depth of flow. Obtain the mean velocity and momentum correction coefficient β. $\left(\frac{5}{6} u_m,\ 1.0133\right)$

6.25 The velocity distribution for laminar flow between parallel plates kept distance B apart is given by $u = \frac{-1}{2\mu}\frac{\partial p}{\partial x}(By - y^2)$. Obtain expression for U and find value of β. $\left(-\frac{\partial p}{\partial x}\frac{B^2}{12\mu}, 1.2\right)$

6.26 A jet pump as shown in Fig. 6.25 has a jet area of 0.010 m² and a jet velocity of 30.0 m/s. The jet is within a secondary stream of water having a velocity of 3.0 m/s. The total area of the pipe (i.e. sum of jet area and secondary stream area) is 0.075 m². The water leaving the pump is thoroughly mixed. Assuming the pressures of the jet and the secondary stream to be the same at the pump inlet, determine the velocity at the pump exit and the pressure difference $(p_1 - p_2)$.
(6.60 m/s, – 102.235 kN/m²)

6.27 Experiment conducted in free air stream gave the velocity distributions on the upstream and downstream side of a two dimensional aerofoil as shown in Fig. 6.37. If $\rho = 1.226$ kg/m³, determine the resistance offered by aerofoil to flow per unit width. (0.537 N)

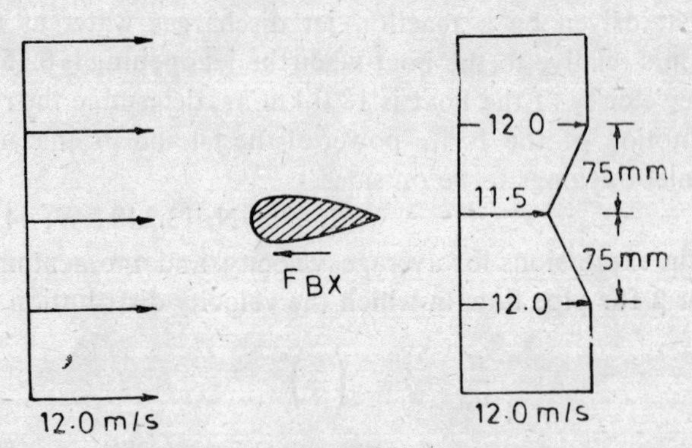

Fig. 6.37.

6.28 A 300 mm diameter 150° bend discharges 0.350 m³/s of water in the atmosphere (Fig. 6.38). If the pressure at section 1 is 150 kN/m² (gauge), determine the force required to hold the bend in place. Assume the bend to be in horizontal plane.
(10 863.886 N, 4.57° with – ve x axis)

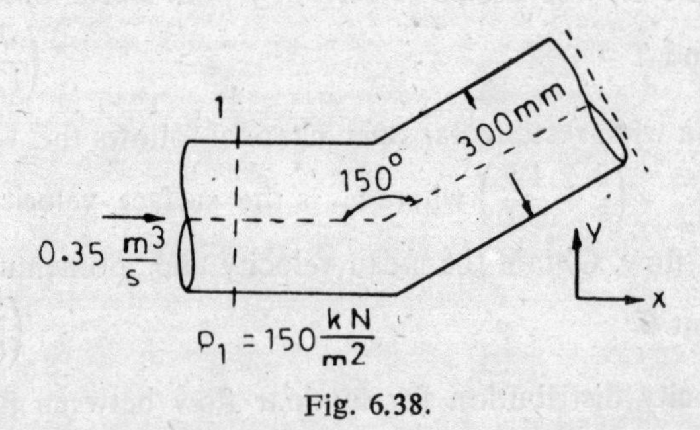

Fig. 6.38.

2.29 A 120° bend-cum reducer has 300 mm diameter at inlet and 200 mm diameter at the other end as shown in Fig. 6.39. When the bend-cum reducer carries 0.30 m^3/s of water, pressure at section 1 is 210 kN/m^2. Assume no energy loss in the bend and determine the components of force exerted by the bend on the flow. Assume the weight of bend plus water in it to be 1500 N. (– 12 018.749 N, 3627.879 N)

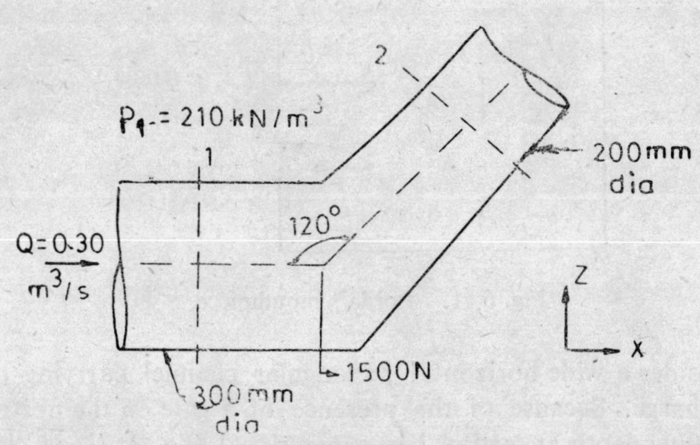

Fig. 6.39.

6.30 A 300 mm diameter pipe is reduced to 50 mm diameter by means of a nozzle. If the pipe carries 65 l/s of water which is then discharged into the atmosphere, determine the force exerted by water on the pipe assembly. Neglect frictional loss. (36.540 kN).

6.31 For the pipe junction shown in Fig. 6.40, determine the components of force exerted by the junction on the flow of water.

$D_1 = 200$ mm $\quad D_2 = 150$ mm $\quad D_3 = 300$ mm

(10199 N, – 3893 N)

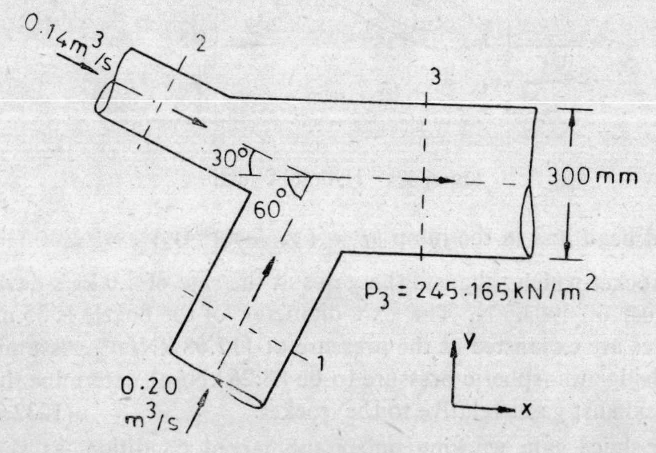

Fig. 6.40.

6.32 Use continuity, momentum and energy equations to show that contraction coefficient for Borda's mouthpiece in Fig. 6.41 is 0.50.

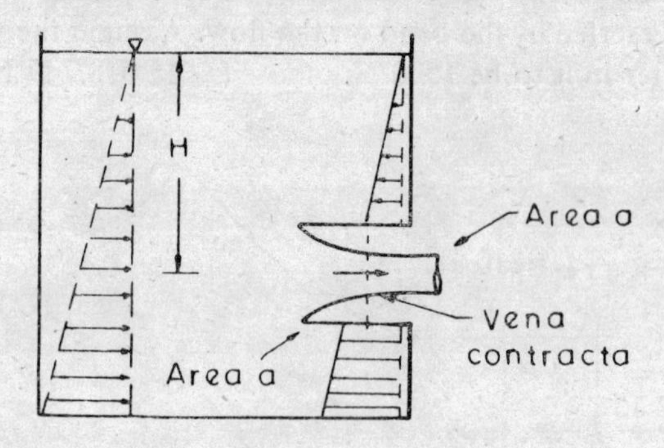

Fig. 6.41. Borda's mouthpiece.

6.33 Consider a wide horizontal rectangular channel carrying q m³/s m discharge. Because of the presence of a gate on the upstream side the flow depth at section 1 is supercritical (i.e. $U_1/\sqrt{gy_1}$ is greater than 1.0). If downstream depth is adequate a hydraulic pump forms (Fig. 6.42). After the hydraulic jump the flow at section 2 is subcritical. Make suitable assumptions and show that

$$\frac{y_2}{y_1} = \frac{1}{2}\sqrt{1 + 8\frac{U_1^2}{gy_1}} - 1$$

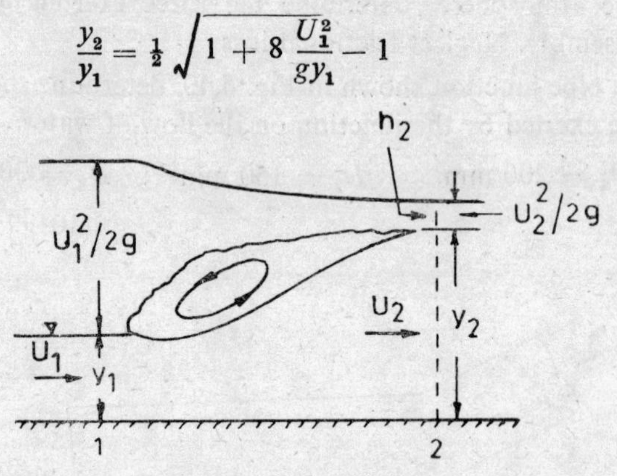

Fig. 6.42. Hydraulic jump.

and head loss in the jump $h_L = (y_2 - y_1)^3/4y_1y_2$.

6.34 A rocket which exhausts the gases at the rate of 4.0 kg/s develops a thrust of 4903.3 N. The exit diameter of the nozzle is 35 mm and gases are exhausted at the pressure of 117.68 kN/m². Assuming the outside atmospheric pressure to be 88.26 kN/m², determine the speed of exhaust gases relative to the rocket. (1232.71 m/s)

6.35 For sluice gate working under submerged condition, as shown in Fig. 6.43, prove that the depth H_3 is given by the relation

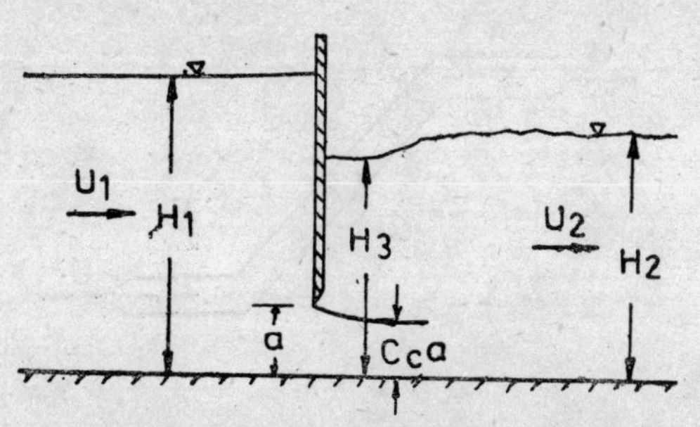

Fig. 6.43. Submerged flow in a sluice gate.

$$\frac{H_3}{H_2} = \sqrt{1 - 2F_{r2}^2\left(\frac{H_2}{aC_c} - 1\right)}$$

where $F_{r2} = U_2/\sqrt{gH_2}$.

6.36 A rocket while travelling discharges its exhaust gases at 1500 m/s velocity. The mass rate of exhaust gases is 6.0 kg/s and exit area is 0.08 m². If the absolute pressure at exit is 10 N/cm² and outside is 8.00 N/cm², determine the speed of the rocket when the thrust on it is 2500 N. (816.667 m/s)

6.37 An experimental solid propellant rocket has a mass of 9 kg, 6 kg of which is the fuel. The rocket is directed vertically upwards from rest, and it burns fuel at a constant rate of 0.20 kg/s, exhausting gases at a velocity of 450 m/s relative to the rocket. Assume the pressure at the exit is atmospheric and air resistance is negligible. Calculate the rocket velocity after 20 seconds and the distance travelled during this period. (244.88 m/s, 9441.48 m)

6.38 A lawn sprinkler shown in Fig. 6.44 discharges 0.314 l/s through both the nozzles. Assuming no friction, determine its speed of rotation. (381.92 rpm)

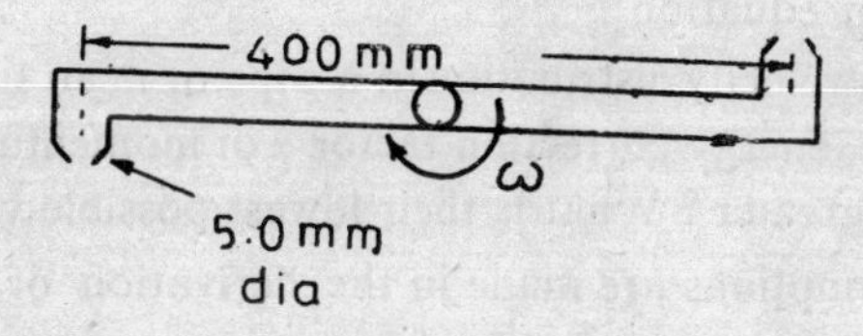

Fig. 6.44

6.39 In Example 6.29, determine the speed of rotation and torque required to hold the sprinkler in place if $r = 0.25$ m, diameter of nozzle $= 6.0$ mm, $V_r = 6.0$ m/s and $\theta = 30°$. (198.45 rpm, 0.44 Nm)

6.40 Determine the torque about point O required to prevent the rotation of sprinkler shown in Fig. 6.45. ($2A\rho U^2 x$)

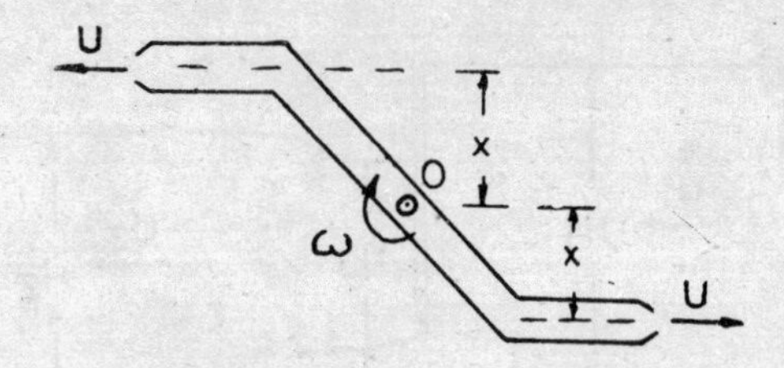

Fig. 6.45

6.41 An airplane travelling at 360 km/hr through still air ($\rho = 1.20$ kg/m^3) discharges 1000 m^3/s air through its two 2 m diameter propellers. Determine its theoretical efficiency, thrust of each propeller, pressure difference across the propeller and theoretical power of each.
(62.8 percent, 71 083.2 N, 22 637.96 N/m^2, 11 318.981 kW)

6.42 A propeller must produce a thrust of 9090 N to drive a plane at 280 km/hr speed. Determine the ideal diameter of the propeller if it operates at 90 percent efficiency. Assume ρ for air to be 1.20 kg/m^3.
(2.19 m)

6.43 An experimental windmill at Sandusky (Ohio, U.S.A.) consists of two blades of 37.5 m diameter. What will be the maximum power developed at a wind speed of 8.0 m/s ? Take $\rho = 1.208$ kg/m^3. If the actual experiment produced only 100 kW, what is the reason for the discrepancy ? (202.3 kW)

DESCRIPTIVE QUESTIONS

6.1 Explain the terms momentum flux, control volume and inertial frame.

6.2 Momentum and energy equations are both derived from Euler's equations of motion. What is the essential difference in their derivation ?

6.3 Define momentum correction factor β. What is its significance in momentum equation ?

6.4 For a given velocity distribution in a pipe or in an open channel, which of the two, energy correction factor α or momentum correction factor β, will be greater ? What is their lowest possible value ?

6.5 What assumptions are made in the derivation of linear momentum Eq. 6.1 ?

6.6 How is the control volume chosen while applying the momentum equation ?

6.7 In the application of momentum equation to bends, etc. why is the air pressure force on the enclosing boundary not considered ?

6.8 If the motion of devices shown in Fig. 6.46 is restricted to x direction alone, state whether each will move in $+x$ or $-x$ direction. Assume pressure at entrance and exit to be essentially same.

Fig. 6.46.

6.9 Which of the following conditions yields zero value of instantaneous time rate of change of linear momentum within the control volume ?

(i) steady flow

(ii) steady flow, and $\frac{\partial}{\partial x}(\rho u)+\frac{\partial}{\partial y}(\rho v)+\frac{\partial}{\partial z}(\rho w)=0$

(iii) steady flow and fluid with uniform and steady mass density

(iv) unsteady flow and fluid of constant mass density.

6.10 Which of the following assumptions are made in the analysis of impingment of a jet on a moving plate ?

(i) friction between the jet and the plate is neglected

(ii) the flow from the nozzle is steady

(iii) the jet leaves without any velocity

(iv) plate moves at a constant velocity

(v) momentum of the jet is unchanged.

6.11 What assumptions are made in the derivation of energy loss equation for sudden expansion in a pipe ? Will the equation be applicable to laminar flow ? Why ?

6.12 Briefly discuss the capabilities of energy equation and momentum equation, and the type of data needed for their applications.

6.13 A control volume is (i) an isolated system (ii) a closed system (iii) a region fixed in space.

6.14 Which of the following statements are true, when a steady jet strikes a fixed plate held perpendicular to flow

(i) absolute speed does not change along the plate

(ii) pressure force needs to be taken into account

(iii) frictional force between the jet and the plate is negligible

(iv) as the jet strikes the plate, cross-sectional area is unaltered.

6.15 For a given velocity distribution in a pipe

(i) $\alpha=\beta$ (ii) α is greater than β (iii) β is greater than α.

6.16 State whether the following statements are true or false

(i) Momentum equation can give the actual pressure or velocity variation in the control volume.

(ii) It will give final velocity of stream or pressure force.

(iii) If the boundary is not long enough, frictional force can be neglected.

(iv) Energy loss can be estimated by using momentum equation alone.

CHAPTER VII

Dimensional Analysis and Similitude

7.1 DIMENSIONAL ANALYSIS

Dimensional analysis is a method by which one obtains certain information about a given phenomenon on the assumption that the phenomenon is governed by a dimensionally homogeneous equation amongst the dependent and the independent variables. After the dimensional analysis is carried out one gets a complete set of dimensionless parameters.

If the dimensions of the terms on the left hand side and right hand side of a given equation are identical, the equation is said to be dimensionally homogeneous. All dimensionally homogeneous equations can be expressed in the form of equations involving only dimensionless groups. If the equation is dimensionally homogeneous, it can be used in any consistent system of units e.g. CGS, SI or FPS, and the numerical constants appearing in the equation have no dimensions; as such they remain unchanged.

Dimensional analysis is used in checking the correctness of a given equation, changing the equation from one system of units to the other, simplifying theoretical analysis, reducing the number of variables with which one has to deal with in studying a problem, analysis of experimental data and in the analysis of models.

7.2 BUCKINGHAM'S π THEOREM

If there are n variables which govern a certain phenomenon and if these variables involve m primary units, then the phenomenon can be described by $(n - m)$ independent dimensionless parameters. Primary units to be used in fluid mechanics problems are either M, L, T and θ or F, L, T and θ.

Buckingham's π theorem stated in the above manner will have a few exceptions as shown in Example 7.5. Hence it is alternately stated as follows. The number of dimensionless parameters in a complete set will be equal to the total number of variables minus the rank of their dimensional matrix.

7.3 VARIABLES IN FLUID MECHANICS

The variables with which one deals in fluid mechanics can be classified into variables related to geometry, fluid and flow. Some of these are listed in Table 7.1 along with their dimensions.

Table 7.1 Dimensions of physical quantities in fluid mechanics

Quantity	Notation	Mass system	Force system
		Geometric variables	
Area	a, A	L^2	L^2
Angle	α, θ	$M^0L^0T^0$	$M^0L^0T^0$
Length	l	L	L
Volume	$\forall$	L^3	L^3
		Variables related to fluid	
Dynamic viscosity	μ	$ML^{-1}T^{-1}$	$FL^{-2}T$
Kinematic viscosity	ν	L^2T^{-1}	L^2T^{-1}
Mass density	ρ	ML^{-3}	$FL^{-4}T^2$
Specific weight	γ	$ML^{-2}T^{-2}$	FL^{-3}
Surface tension	σ	MT^{-2}	FL^{-1}
Volume modulus of elasticity	E	$ML^{-1}T^{-2}$	FL^{-2}
		Variables related to flow	
Acceleration	a	LT^{-2}	LT^{-2}
Angular acceleration	—	T^{-2}	T^{-2}
Angular momentum	—	ML^2T^{-1}	FLT
Angular velocity	ω	T^{-1}	T^{-1}
Force	F	MLT^{-2}	F
Frequency	f	T^{-1}	T^{-1}
Moment	—	ML^2T^{-2}	FL
Momentum	M	MLT^{-1}	FT
Power	P	ML^2T^{-3}	FLT^{-1}
Pressure or stress	p	$ML^{-1}T^{-2}$	FL^{-2}
Rotation		T^{-1}	T^{-1}
Strain	—	$M^0L^0T^0$	$M^0L^0T^0$
Torque	T	ML^2T^{-2}	FL
Velocity	U, V	LT^{-1}	LT^{-1}
Work or energy	W	ML^2T^{-2}	FL

7.4 PROCEDURES OF DIMENSIONAL ANALYSIS

Dimensional analysis can be performed by any one of the following three methods. It may be mentioned that even though the three methods mentioned below look different outwardly, basically they are the same.

7.4.1 Matrix Method

After the n variables to be considered are known, write them in one line and form a dimensional matrix by writing dimensions of each in terms of M, L, T or F, L, T in column form. Then determine the rank of matrix; r; $(n - r)$ will be the number of dimensionless parameters. Write homogeneous linear algebraic equations whose coefficients are the numbers in the rows of the dimensional matrix and solve them assigning arbitrary values to $(n - r)$ variables. This is illustrated in Example 7.8.

7.4.2 Raleigh's method

If $x_1, x_2, x_3, \ldots x_n$ are the variables and x_1 is the dependent variable, assume the relationship for x_1 as

$$x_1 = C\, x_2^a\, x_3^b\, x_4^c \ldots \tag{7.1}$$

Write the dimensions of each quantity in the above equation. Equating the exponents of M, L, T on both sides of the equation, one gets three linear equations among $a, b, c, d \ldots$. Express all others in terms of any three, say a, b, c, substitute their values in Eq. 7.1, and simplify.

7.4.3 Method of repeating variables

After the variables governing the phenomenon are identified, choose repeating variables equal in number to the number of primary units (or rank of dimensional matrix) used in describing the variables. Do not choose the dependent variable as one of the repeating variables. The repeating variables should contain all the primary units among themselves and they should not combine among themselves to form a dimensionless group. For most of the problems in fluid mechanics, a characteristic geometric variable such as length or area, one fluid property such as mass density, and one flow characteristic such as velocity are a good choice. Combine each non-repeating variable with repeating variables to form dimensionless groups.

7.5 SIMILITUDE

Model studies are usually conducted to find solutions to numerous complicated problems in hydraulic engineering and fluid mechanics. In order that results obtained in the model studies correctly represent the behaviour of the prototype (actual structure), the following three similarities must be ensured between the model and the prototype.

7.5.1. Geometric similarity

For geometric similarity to exist between the model and the prototype, the ratios of corresponding lengths in the model and in the prototype must be same and the included angles between two corresponding sides must be the same. Models which are not geometrically similar are known as geometrically distorted models.

7.5.2 Kinematic similarity

Kinematic similarity is similarity of motion. If at the corresponding (or homologous) points in the model and in the prototype, the velocity or acceleration ratios are same and the velocity or acceleration vectors point in the same direction, the two flows are said to be kinematically similar.

7.5.3 Dynamic similarity

Dynamic similarity is the similarity of forces. The flows in the model and in the prototype are dynamically similar if at all the corresponding points, identical types of forces are parallel and bear the same ratio. In dynamic similarity, the force polygons of the two flows can be superimposed by change in force scale.

7.6 IMPORTANT DIMENSIONLESS PARAMETERS

Dimensional analysis will yield various dimensionless parameters. Some will be length ratios of mass densities, specific weights and velocities etc. The force ratios obtained in dimensional analysis acquire great significance. Their structure and significance are given below:

Reynolds number $$Re = \frac{\text{Inertial force/volume}}{\text{Viscous force/volume}} = \frac{Ul\rho}{\mu}$$

Froude number $$Fr = \sqrt{\frac{\text{Inertial force/volume}}{\text{Gravity force/volume}}} = \frac{U}{\sqrt{gi}}$$

$$= \frac{\text{flow velocity}}{\text{Celerity of a small gravity wave}}$$

Euler number $$E = \sqrt{\frac{\text{Inertial force/volume}}{\text{Pressure force/volume}}} = \frac{U}{\sqrt{\Delta p/\rho}}$$

Drag coefficient $$C_D = \frac{\text{External force/volume}}{\text{Inertial force/volume}} = \frac{F}{L^2\rho U^2/2}$$

Weber number $$W = \frac{\text{Inertial force/volume}}{\text{Surface tension force/volume}} = \frac{\rho U^2 l}{\sigma}$$

$$= \left(\frac{\text{Flow velocity}}{\text{Velocity of a small ripple on liquid surface}}\right)^2$$

Mach number $$M = \sqrt{\frac{\text{Inertial force/volume}}{\text{Compressibilty force/volume}}} = \frac{U}{\sqrt{E/\rho}}$$

= (flow velocity/velocity of pressure wave in fluid)

7.7 MODEL SCALES

Models are designed and model results converted into prototype results by identifying the most important fluid force (other than inertia) which influences the fluid motion and then ensuring the equality of corresponding dimensionless number, e.g. *Re*, *Fr* or *M* in the model and prototype. Accordingly models can be classified as models following Reynolds law, Froude or Mach law. The corresponding scale ratios can be then obtained. In the discussion of models, scale ratio is defined as the value of a particular quantity in model to that in the prototype. Thus velocity scale $U_r = U_m/U_p$ where U_m and U_p are velocities in the model and prototype respectively. Table 7.2 lists various scale ratios for models governed by equality of *Re*, *Fr* and *M*.

Table 7.2 Scale Ratios

Scale ratio	Reynolds law	Froude law	Mach law
Velocity U_r	$\mu_r/l_r\rho_r$	$l_r^{1/2}\, g_r^{1/2}$	$E_r^{1/2}/\rho_r^{1/2}$
Acceleration a_r	$\mu_r^2/\rho_r^3\, l_r^3$	g_r	$E_r^{3/2}/\rho_r l_r$
Time t_r	$l_r^2\rho_r/\mu_r$	$t_r^{1/2}/g_r^{1/2}$	$l_r\rho_r^{1/2}/E_r^{1/2}$
Discharge Q_r	$l_r\mu_r/\rho_r$	$l_r^{5/2}\, g_r^{1/2}$	$l_r^2E^{1/2}/\rho_r^{1/3}$
Force F_r	μ_r/ρ_r	$l_r^3\rho_r g_r$	$l_r^2\, E_r^{3/2}$
Pressure p_r	$\mu_r^2/\rho_r l_r^2$	$l_r\rho_r g_r$	$E_r^{3/2}$
Power P_r	$\mu_r^3/\rho_r^2 l_r$	$l_r^{7/2}\rho_r g_r^{3/2}$	$l_r^2E_r/\rho_r^{1/2}$

If there is no free surface or if there is no likelihood of wave formation (e.g. motion of a torpedo) or if the flow is completely immersed or enclosed, gravitational effects are unimportant and hence equality of *Fr* is not necessary. Similarly for flows with free surface, if length dimension related to flow, e.g. depth, is greater than 10 mm, surface tension force can be neglected and hence equality of *W* is not necessary. In free surface models, characteristic Reynolds number UD/ν should be greater then 1500. If Mach number is less than 0.40, compressibility effects are unimportant. These considerations decide whether a particular phenomenon is to be modelled with equality of *Re*, *Fr* or *M* number.

ILLUSTRATIVE EXAMPLES

7.1 This example illustrates the use of dimenional analysis in the conversion of dimensionally nonhomogeneous equation from one system of units to other. When water is taken out from a reservoir through a tunnel of diameter D at a velocity U, vortices can form on the water surface if the depth H is not adequate The minimum depth H_c required for vortex-free flow is given by Gordon's equation

$$H_c = 0.40\, UD^{1/2}$$

in fps system of units. Change it to SI system of units.

If the equation is dimensionally homogeneous, the constant will have no dimensions. Dimensions of 0.40 are

$$[0.40] = \frac{[H_c]}{[U]\,[D^{1/2}]} = \frac{L\,T}{L.L^{1/2}} = \frac{T}{L^{1/2}} \quad \text{or} \quad \frac{s}{ft^{1/2}}$$

for converting this equation into SI units, one will use s and m as the corresponding units. Since 1 ft = 0.305 m,

$$1\ \text{ft}^{1/2} = 0.305^{1/2} = 0.552.$$

Hence in SI units the equation will be

$$H_c = \frac{0.40}{0.552}\, UD^{1/2} = 0.725\, UD^{1/2}$$

7.2 According to Hazen William's formula the average velocity U for tnrbulent flow in a pipe is given by

$$U = 1.32\, C_H R^{0.63}\, S^{1/2}$$

where U is in fps, R is the hydraulic radius in ft and S is the slope of energy line. The coefficient C_H is a function of pipe material. Obtain the corresponding equation in SI units

$$[1.32] = [U]/[R^{0.63}]\,[S^{1/2}]$$

$$= \frac{L}{T}\frac{1}{L^{0.63}} = L^{0.37}/T$$

Hence the equation is not dimensionally homogeneous. Since 1 ft = 0.305 m, the value of 1.32 in the equation in SI units will be

$$1.32 \times (0.305)^{0.37} = 0.851$$

∴ In SI units Hazen William's equation takes the form

$$U = 0.851\, R^{0.63}\, S^{1/2}$$

7.3 Check whether the following equations are dimensionally homogeneous.

(i) $\Delta p = \dfrac{32\mu U l}{D^2}$

(ii) $U = \dfrac{\text{Const.}}{n} \, g^{1/2} \, R^{2/3} \, S^{1/2}$

where $[n] = L^{1/6}$.

Writing the dimensions of quantities

$$\text{(i)} \quad 32 = \frac{[\Delta p]\ [D^2]}{[\mu]\,[U]\,[l]} = \frac{M}{LT^2}\,\frac{LT}{M}.\frac{L^2}{L}\,\frac{T}{L} = M^0L^0T^0.$$

Hence the equation is dimensionally homogeneous

$$\text{(ii) [Const.]} = \frac{[U]\ [n]}{[g^{1/2}]\ [R^{2/3}]\ [S^{1/2}]}$$

$$= \frac{L}{T}\,\frac{L^{1/6}}{1}\,\frac{T}{L^{1/2}}\,\frac{1}{L^{2/3}}\,\frac{L}{L}$$

$$= M^0L^0T^0$$

Hence this equation also is dimensionally homogeneous.

7.4 Form dimensionless parameters from the following variables

(i) $\dfrac{\partial p}{\partial x}$, μ, Q and D

Since a certain combination of these variables will yield a dimensionless parameter, assume

$$\frac{\partial p}{\partial x} = \text{const. } \mu^a Q^b D^c$$

$$\frac{M}{T^2\,L^2} = \left(\frac{M}{LT}\right)^a \left(\frac{L^3}{T}\right)^b (L)^c$$

$$M \rightarrow \quad 1 = a$$

$$L \rightarrow -2 = -a + 3b + c$$

$$T \rightarrow -2 = -a - b$$

$$\therefore \quad a = 1,\ b = 2 - a = 1$$

$$\text{and} \quad c = a - 3b - 2 = 1 - 3 - 2 = -4$$

$$\therefore \quad \pi = \frac{(\partial p/\partial x)\ D^4}{\mu Q}$$

(ii) U, D, ρ and μ

One can assume $U^a D^b \rho^c \mu^d$ is dimensionless.

$$\therefore \quad M^0L^0T^0 = \left(\frac{L}{T}\right)^a (L)^b \left(\frac{M}{L^3}\right)^c \left(\frac{M}{LT}\right)^d$$

$$M \rightarrow \quad 0 = c + d \qquad \therefore \quad c = -d$$

$$L \rightarrow \quad 0 = a + b - 3c - d$$

$$T \rightarrow \quad 0 = -a - d \quad \therefore \quad a = -d$$

Hence from 2nd equation one gets $b = 3c + d - a$

$$= -3d + d + d$$

$$= -d$$

$\therefore$ The dimensionless parameters is $U^{-d} D^{-d} \rho^{-d} \mu^{d}$

or $\left(\frac{UD\rho}{\mu}\right)^{-d}$ where d can have any value.

When $d = -1$, one gets Reynolds number.

7.5 Obtain expression for capillary rise in a small diameter tube partly immersed in a liquid.

From physical observation it is known that the liquid rises upto a certain height and remains stationary. The rise h will depend on tube diameter D, and coefficient of surface tension σ. Since the liquid is stationary, μ does not influence the phenomenon. Further since the problem is of statics, ρ as an index of inertia or opposition to motion will not come in the picture. Instead specific weight γ should be included since gravity force is important.

$$h = f(\gamma, \sigma, D)$$

Since it is a statics problem, F, L, T can be used as primary units.

$$[h] = L, [\gamma] = F/L^3, [\sigma] = F/L, [D] = L$$

Hence only two primary units F, L are required to describe, four variables.

$$\therefore \quad (n - m) = 4 - 2 = 2.$$

There will thus be two dimensionless parameters. Using γ and D as repeating variables

$$\pi_1 = hD^a. \text{ Naturally } \pi_1 = \frac{h}{D}$$

$$\pi_2 = \sigma D^a \gamma^b$$

$$\therefore \quad F^0 L^0 = \left(\frac{F}{L}\right) (L)^a \left(\frac{F}{L^3}\right)^b$$

$$\therefore \quad 0 = 1 + b \quad \text{and} \quad 0 = -1 + a - 3b$$

$$\text{or} \quad b = -1, \; a = -2 \quad \therefore \quad \pi_2 = \frac{\sigma}{\gamma D^2}$$

$$\therefore \quad \frac{h}{D} = f(\sigma/\gamma D^2)$$

It may be seen that if M, L, T were chosen as primary units, $(n - m) = 4 - 3 = 1$; on the other hand, as shown above, one gets two dimensionless parameters. This would look like an exception to Buckingham's π theorem. However, it is not an exception because in true sense only two primary units, namely F and L or $\frac{M}{T^2}$ and L are necessary to describe four variables.

7.6 If period of oscillation t of a mass m suspended from a spring depends on the amplitudes of the oscillation, and the stiffness of the spring K, obtain the form of equation for t.

The stiffness of the spring or spring constant K is defined as restitutive force required for unit displacement.

Hence dimensions of K are F/L or M/T^2

$$\therefore \quad t = f(s, m, K)$$

Using Raleigh's method assume the form of equation as

$$t = \text{const. } m^a K^b s^c$$

$$\therefore \quad T = [M]^a \, [M/T^2]^b \, [L]^c$$

$$\left.\begin{aligned} M &\rightarrow a + b = 0 \\ L &\rightarrow c = 0 \\ T &\rightarrow -2b = 1 \end{aligned}\right\} \quad \therefore \quad \begin{aligned} b &= -1/2 \\ a &= 1/2 \\ c &= 0 \end{aligned}$$

$$\therefore \quad t = \text{const.} \sqrt{\frac{m}{K}}$$

Experimentally the constant can be determined and it is found to be 2π. Hence

$$t = 2\pi \sqrt{m/K}$$

7.7 Assume that bacterium propels itself by rotating its tail (known as flagellum) in the form of a spiral (Fig. 7.1). If the thrust F developed by the bacterium depends on the rotational speed N of flagellum, mass density ρ and dynamic viscosity μ of the fluid and the diameter of D coiled flagellum, perform the dimensional analysis

$$F = f(N, D, \rho, \mu)$$

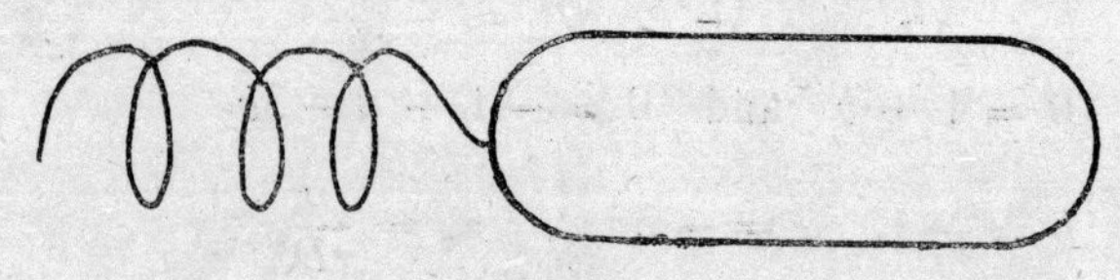

Fig. 7.1.

Since M, L, T are the three primary units involved, we will choose three repeating variables. Choose the repeating variables as D (geometric characteristics), ρ (fluid characteristics) and N (flow characteristics).

$\therefore$ $(n - m) = (5 - 3) = 2$ dimensionless parameters will be obtained, say π_1 and π_2 by combining N, D, ρ with F, and N, D, ρ with μ respectively.

$$\pi_1 = F N^a D^b \rho^c$$

$$M^0L^0T^0 = \left(\frac{ML}{T^2}\right)\left(\frac{1}{T}\right)^a (L)^b \left(\frac{M}{L^3}\right)^c$$

$$\left.\begin{aligned} M &\rightarrow 1 + c = 0 & \text{or} \quad c &= -1 \\ L &\rightarrow 1 + b - 3c = 0 & b &= -4 \\ T &\rightarrow -2 - a \quad = 0 & a &= -2 \end{aligned}\right\} \therefore \pi_1 = \frac{F}{D^4N^2\rho}$$

$$\pi_2 = \mu N^a D^b \rho^c$$

$$\therefore \quad M^0L^0T^0 = \left(\frac{M}{LT}\right)\left(\frac{1}{T}\right)^a (L)^b \left(\frac{M}{L^3}\right)^c$$

$$\left.\begin{aligned} M &\rightarrow 0 = 1 + c \quad \therefore & c &= -1 \\ L &\rightarrow 0 = -1 + b - 3c & b &= -2 \\ T &\rightarrow 0 = -1 - a & a &= -1 \end{aligned}\right\} \therefore \pi_2 = \frac{\mu}{ND^2\rho} \quad \text{or} \quad \frac{ND^2\rho}{\mu}$$

$$\therefore \quad \frac{F}{D^4N^2\rho} = f\left(\frac{ND^2\rho}{\mu}\right)$$

7.8 The drag force F on a body of dimension D depends on viscosity μ, mass density ρ, volume modulus of elasticity of fluid E, and its velocity U. Perform dimensional analysis using matrix method.

$$F = f(U, D, \rho, \mu, E)$$

or

$$f(F, U, D, \rho, \mu, E) = 0$$

The dimensional matrix for these variables is

	1	2	3	4	5	6
	F	μ	E	ρ	U	D
	a	b	c	d	e	f
M	1	1	1	1	0	0
L	1	−1	−1	−3	1	1
T	−2	−1	−2	0	−1	0

The first step is to compute the rank of the matrix. It can be seen that atleast one 3rd order determinant is non zero e g. determinant on the extreme left. It there is at least one 3rd order determinant non-zero, the rank of the matrix is 3. In determining the rank the columns can be rearranged since the way in which the variables were written was arbitrary. It can be seen that it is preferable to write nonrepeating variables first and repeating variables at the end.

∴ Number of dimensionless parameters $= 6 - 3 = 3$

Let a, b, c, d, e, f be the exponents of F, μ, E, ρ, U, D. These are written on the top in the dimensional matrix.

Since $\pi = F^a \mu^b E^c \rho^d U^e D^f$

$$M^0L^0T^0 = \left(\frac{ML}{T^2}\right)^a \left(\frac{M}{LT}\right)^b \left(\frac{M}{LT^2}\right)^c \left(\frac{M}{L^3}\right)^d \left(\frac{L}{T}\right)^e (L)^f$$

which gives
$$a + b + c + d = 0$$
$$a - b - c - 3d + e + f + 0$$
$$-2a - b - 2c - e = 0$$

It can be seen that three homgeneous linear algebraic equations can be written easily by inspection of dimensional matrix. This is so because the coefficients in each equation are a row of numbers in the dimensional matrix. Since there are six unknowns and three equations, any values can be assigned to 3 unknowns a, b, c, and above equations will give corresponding values of d, e, f. This has to be done 3 times. Thus assume as follows

(i) If $a = 1, b = c = 0$ ∴ $a + d = 0$ and $d = -1$

$$a - 3d + e + f = 0$$
$$-2a - e = 0 \quad \therefore \quad e = -2$$

∴ $\pi_1 = F/\rho U^2 D^2$ and $f = -3 - 1 + 2 = -2$

(ii) If $a = c = 0$ and $b = 1$ $d = -1$

$$-b - 3d + e + f = 0$$

∴ $\pi_2 = \dfrac{\mu}{UD\rho}$ $\quad -b - e = 0 \quad \therefore \quad e = -1$

(iii) If $a = b = 0$ and $c = 1$ it can be similarly seen that

$d = 1$ and $e = -2$, $f = 0$

∴ $\pi_3 = E/U^2\rho$

Thus $$\frac{F}{D^2\rho U^2} = f\left(\frac{UD\rho}{\mu}, \frac{U^2}{E/\rho}\right)$$

or $$C_D = f(Re, M)$$

The solution can be neatly arranged in matrix form as follows

	F	μ	E	ρ	U	D
	a	b	c	d	e	f
π_1	1	0	0	-1	-2	-2
π_2	0	1	0	-1	-1	-1
π_3	0	0	1	-1	-2	0

7.9 Solve Example 7.8 using Releigh's method.

Since $F = f(U, D, \rho, \mu, E)$, one can write the functional relationship, as

$$F = \text{const. } U^a D^b \rho^c \mu^d E^e$$

Using M, L, T, as primary units, one can write

$$\frac{ML}{T^2} = \left(\frac{L}{T}\right)^a (L)^b \left(\frac{M}{L^3}\right)^c \left(\frac{M}{LT}\right)^d \left(\frac{M}{LT^2}\right)^e$$

or $\quad M \to 1 = c + d + e$

$$L \to 1 = a + b - 3c - d - e$$

$$T \to -2 = -a - d - 2e$$

These three equations can be solved for a, b, c in terms of d and e as

$$a = 2 - d - 2e$$

$$b = 2 - d$$

$$c = 1 - d - e$$

$$\therefore \quad F = \text{const. } U^{(2-d-2e)} D^{(2-d)} \rho^{(1-d-e)} \mu^d E^e$$

$$= \text{const. } U^2 D^2 \rho \left(\frac{\mu}{UD\rho}\right)^d \left(\frac{E}{\rho U^2}\right)^e$$

If one strictly restricts to the above analysis, one can write

$$\frac{F}{\dfrac{D^2 \rho U^2}{2}} = \text{const. } \left(\frac{\mu}{UD\rho}\right)^d \left(\frac{E}{\rho U^2}\right)^e$$

which restricts the form of relation between C_D, Fe and M. Hence one uses the freedom to generalise the result as

$$C_D = f(Re, M)$$

7.10 Consider the functional relationship between drag force F, diameter of a sphere D, its velocity U, and fluid properties ρ and μ, namely

$$F = f(D, U, \rho, \mu)$$

Eliminate from this equation each of the dimensions M, L and T in turn and form the dimensionless parameters.

Only F, μ and ρ contain M. Hence eliminate M by first dividing F and μ by ρ. Hence

$$\frac{F}{\rho} = f\left((U, D, \frac{\mu}{\rho}, \rho\right)$$

However, ρ cannot now occur singly on the right hand side of above equation, since it is the only term in the new equation involving M. Hence the above equation must be written as

$$\begin{array}{ccccc} F/\rho & = & f(U, & D, & \mu/\rho) \\ \left(\frac{ML}{T^2}\frac{L^3}{M}\right) & & \frac{L}{T}, & L, & \left(\frac{M}{LT}\frac{L^3}{M}\right) \end{array}$$

Now T can be eliminated by using U. Thus

$$\frac{F}{\rho U^2} = f\left(U, D, \frac{\mu}{\rho U}\right)$$

Again U cannot occur singly in the above equation since it is the only term inovlving T. Hence above equation becomes

$$\begin{array}{cc} \frac{F}{\rho U^2} = f\left(D\ \frac{\mu}{\rho U}\right) \\ L^2 \qquad\quad L\ \ L \end{array}$$

Now eliminate L by using D. Thus

$$\frac{F}{\rho U^2 D^2} = f\left(\frac{\mu}{\rho UD}\right)$$

Thus we have successfully called upon the repeating variables ρ, U, D to form dimensionless terms.

7.11 If viscosity of μ of a gas depends on its mass density ρ, mean velocity of molecules $\bar{u}$, diameter of molecule d and mean free path λ, determine expression for μ

$$\mu = f(d, \bar{u}, \rho, \lambda)$$

Three primary units M, L, T are required to express the five variables. Hence number of dimensionless parameters, is equal to $(5 - 3)$ i.e. two. Choose d, ρ and $\bar{u}$ as the repeating variables. Hence

$$\pi_1 = \mu d^a \rho^b \bar{u}^c$$

or $$M^0L^0T^0 = \left(\frac{M}{LT}\right)(L)^a\left(\frac{M}{L^3}\right)^b\left(\frac{L}{T}\right)^c$$

It can be seen that $a = -1 \quad b = -1 \quad c = -1$

$$\therefore \qquad \pi_1 = \frac{\mu}{\bar{u}d\rho}; \quad \text{similarly} \quad \pi_2 = \frac{\lambda}{d}$$

Hence $\dfrac{\mu}{\bar{u}d\rho} = f\left(\dfrac{\lambda}{d}\right)$.

Further information cannot be obtained from dimensional analysis. The nature of function f can be determined only by analysis of experimental data. However intuitively one can write

$$\frac{\mu}{\bar{u}d\rho} = \text{const.}\left(\frac{\lambda}{d}\right)^p$$

Maxwell found that μ is independent of density of gas. This result is obtained if it is assumed that $\rho \sim \dfrac{1}{\lambda}$ which is true and $p = 1$.

$$\therefore \qquad \mu = \text{const. } \bar{u}d\rho\left(\frac{\lambda}{d}\right)^p$$

$$= \text{const. } \bar{u}d\,\frac{k}{\lambda}\cdot\frac{\lambda}{d}$$

$$= \text{const. } k.\bar{u}$$

Since increase in temperature increases $\bar{u}$, μ increases with increase in temperature; here k is constant.

7.12 Lord Raleigh was interested in the vibration of a spherical drop of diameter D which is formed when liquid issues from a circular orifice. When the drop is slightly deformed from its spherical shape and left free, on account of surface tension σ it vibrates about its position of equilibrium with frequency f. If

$$f = \phi(\sigma, \rho, D, g)$$

perform dimensional analysis for f.

There are five variables and three primary units. Hence number of dimensionless parameters will be $(5 - 3) = 2$. Choose ρ, D and g as the repeating variables.

$$\therefore \qquad \pi_1 = f\,\rho^a D^b g^c$$

$$\therefore \qquad M^0L^0T^0 = \left(\frac{1}{T}\right)\left(\frac{M}{L^3}\right)^a (L)^b \left(\frac{L}{T^2}\right)^c$$

It is obvious that $a = 0$. Also

$$b + c = 0 \quad \text{and} \quad 0 = -1 - 2c$$

$$\therefore \qquad c = -\tfrac{1}{2} \text{ and } b = +\tfrac{1}{2} \quad \therefore \quad \pi_1 = f\sqrt{\frac{D}{g}}$$

$$\pi_2 = \sigma\,\rho^a D^b g^c$$

$$\therefore \qquad M^0L^0T^0 = \left(\frac{M}{T^2}\right)\left(\frac{M}{L^3}\right)^a (L)^b \left(\frac{L}{T^2}\right)^c$$

$$\therefore \quad 1 + a = 0 \quad \text{or} \quad a = -1$$

$$-3a + b + c = 0 \qquad b = 3a - c = -3 + 1 = -2$$

$$-2 - 2c = 0 \quad \text{or} \quad c = -1$$

$$\therefore \quad \pi_2 = \frac{\sigma}{\rho D^2 g} \quad \text{or} \quad \frac{\sigma}{\gamma D^2}$$

$$\therefore \quad f\sqrt{\frac{D}{g}} = \phi\left(\frac{\sigma}{D^2 g}\right)$$

7.13 The following example illustrates the advantages of working with components of vector quantities such as velocity or displacement instead of the vector itself.

A jet of liquid is discharged with initial velocity V at an angle θ with the horizontal. Determine R the range i.e., the horizontal distance it can travel.

Since gravity affects the flow, one can write

$$R = f(V, \theta, g).$$

Using Raleigh's method one can write R = const. $V^a \theta^b g^c$.

Since θ is dimensionless,

$$L = \left(\frac{L}{T}\right)^a \left(\frac{L}{T^2}\right)^b$$

$$\therefore \quad a + b = 1 \quad \therefore \quad -b = 1 \quad \text{or} \quad b = -1$$

$$-a - 2b = 0 \qquad a = 2$$

$$\therefore \quad R = \text{const.}\ \frac{V^2}{g}$$

One can only suspect that the constant c will depend on θ, but nothing more can be known. However, if we consider u and v as the components of V and take them as variables in place of V and θ,

$$R = C\,(u)^a (v)^b (g)^c$$

where C is a constant. Now to express them in primary units we take L_x and L_y as the lengths in x and y directions

$$\therefore \quad L_x = \left(\frac{L_x}{T}\right)^a \left(\frac{L_y}{T}\right)^b \left(\frac{L_y}{T^2}\right)^c$$

Solving this, one gets $a = 1$, $b = 1$ and $c = -1$

$$\therefore \quad R = C.\frac{uv}{g}$$

$$= C\frac{u}{V} \cdot \frac{v}{V}\frac{V^2}{g}$$

$$= C\cos\theta . \sin\theta\ \frac{U^2}{g}$$

$$= C\sin 2\theta\ \frac{U^2}{2g}$$

7.14 If pressure loss Δp for laminar flow in a pipe depends on fluid discharge Q, pipe length l, diameter D and the viscosity μ, perform dimensional analysis

$$\Delta p = f(Q, l, D, \mu).$$

Since there are five variables and three primary units, two dimensionless parameters will be formed. Choose Q, D and μ as the repeating variables.

$$\therefore \quad \pi_1 = \Delta p\, Q^a \mu^b D^c$$

$$\therefore \quad M^0L^0T^0 = \left(\frac{M}{LT^2}\right)\left(\frac{L^3}{T}\right)^a\left(\frac{M}{LT}\right)^b (L)^c$$

$$\therefore \quad 0 = 1 + b \qquad \therefore\ b = -1 \qquad a = -1 \text{ and}$$

$$0 = -1 + 3a - b + c \qquad c = 1 - 3a + b = 1 + 3 - 1 = 3$$

$$0 = -2 - a - b$$

$$\therefore\ \pi_1 = \frac{\Delta p D^3}{Q\mu}$$

π_2 is obviously $1/D$. Hence

$$\frac{\Delta p D^3}{Q\mu} = f\left(\frac{l}{D}\right)$$

No further information is obtained from dimensional analysis. However, if one realises that as l increases Δp will increase proportionally, then $\Delta p \sim 1$. Hence $f\left(\frac{1}{D}\right) = C\frac{l}{D}$, where C is constant.

$$\therefore \quad \frac{\Delta p D^3}{Q\mu} = C\frac{l}{D} \quad \text{or} \quad \Delta p = C\frac{Q\mu l}{D^4}$$

7.15 The pressure change across a diffuser of circular cross-section depends on the discharge Q, inlet and outlet diameters D_1 and D_2, diffuser length l, and mass density ρ and viscosity μ of the fluid. Perform dimensional analysis.

The functional relation for Δp can be written as

$$\Delta p = f(Q, \rho, \mu, D_1, D_2, l)$$

with seven variables and three primary units, there will be four dimensionless parameters. Choose Q, D_1 and ρ as the repeating variables. Hence the four dimensions parameters can be expressed as

$$\pi_1 = \Delta p Q^a D_1^b \rho^c, \qquad \pi_2 = \mu Q^a D_1^b \rho^c$$

$$\pi_3 = D_2 Q^a D_1^b \rho^c \qquad \pi_4 = l\, Q^a D_1^b \rho^c$$

It is obvious that π_3 and π_4 will be

$$\pi_3 = D_2/D_1 \quad \text{and} \quad \pi_4 = l/D_1$$

For π_1

$$M^0L^0T^0 = \left(\frac{M}{LT^2}\right)\left(\frac{L^3}{T}\right)^a (L)^b \left(\frac{M}{L^3}\right)^c$$

$$\therefore \quad \left.\begin{aligned} 0 &= 1 + c \\ 0 &= -1 + 3a + b - 3c \\ 0 &= -2 - a \end{aligned}\right\} \quad \begin{aligned} \therefore \quad c &= -1 \\ a = -2 \text{ and } b &= 3c - 3a + 1 \\ &= -3 + 6 + 1 \\ &= 4 \end{aligned}$$

$$\therefore \qquad \pi_1 = \frac{\Delta p D_1^4}{Q^2 \rho}$$

In the same way it can be shown that $\pi_2 = \dfrac{\mu D_1}{\rho Q}$.

$$\therefore \qquad \frac{\Delta p D_1^4}{Q^2 \rho} = f\left(\frac{Q\rho}{D_1\mu}, \frac{D_2}{D_1}, \frac{l}{D_1}\right)$$

7.16 The discharge Q over a sharp crested weir of height W spanning across a rectangular channel of width B depends on the head H over the weir, gravitational acceleration g and the fluid properties ρ, μ and σ. Perform dimensional analysis for Q.

$$Q = f(B, W, H, g, \rho, \mu, \sigma)$$

Since discharge per unit width, q, is a more logical parameter one can write

$$\frac{Q}{B} = q = f(B, W, H, g, \rho, \mu, \sigma)$$

Choosing H, g and ρ as the repeating variables, one will get $(8 - 3) = 5$ dimensionless parameters which will be

$$\pi_1 = \left(\frac{Q}{B}\right) H^a \rho^b g^c, \quad \pi_2 = \frac{B}{H}, \quad \pi_3 = \frac{W}{H}$$

$$\pi_4 = \mu H^a \rho^b g^c \qquad \pi_5 = \sigma H^a \rho^b g^c$$

$$\therefore \quad \text{For } \pi_1 \quad M^0 L^0 T^0 = \left(\frac{L^2}{T}\right) (L)^a \left(\frac{M}{L^3}\right)^b \left(\frac{L}{T^2}\right)^c$$

$$\therefore \quad b = 0, \quad 0 = 2 + a + c \quad \text{and} \quad 0 = -1 - 2c$$

$$\therefore \quad c = -\frac{1}{2}, \quad a = -2 + \frac{1}{2} = -\frac{3}{2}$$

$$\therefore \qquad \pi_1 = \frac{Q}{BH\sqrt{gH}}$$

In the same way it can be shown that

$$\pi_4 = \frac{\mu}{H\sqrt{gH}} \quad \text{and} \quad \pi_5 = \frac{\sigma}{\rho g H^2}$$

$$\therefore \qquad \frac{Q}{BH\sqrt{gH}} = f\left(\frac{\rho H\sqrt{gH}}{\mu}, \frac{\rho g H^2}{\sigma}, \frac{B}{H}, \frac{W}{H}\right)$$

7.17 The power P required by the pump is a function of discharge Q, head

H, g, viscosity μ and mass density ρ of the fluid, speed of rotation N and impeller diameter D. Obtain the relevant dimensionless parameters.

Since H comes in the picture as the potential energy, it is convenient to consider (gH) as one variable, which represents potential energy per unit mass of fluid.

$$P = f(Q, gH, \mu, \rho, N, D)$$

With seven variables and three primary units, there will be $(7 - 3) = 4$ dimensionless parameters. Choose ρ, N, D as the repeating variables.

$$\therefore \quad \pi_1 = P\rho^a N^b D^c \quad \therefore \quad M^0 L^0 T^0 = \left(\frac{ML^2}{T^3}\right)\left(\frac{M}{L^3}\right)^a \left(\frac{1}{T}\right)^b (L)^c$$

$$\therefore \quad \left.\begin{aligned} 0 &= 1 + a \\ 0 &= 2 - 3a + c \\ 0 &= -3 - b \end{aligned}\right\} \quad a = -1, b = -3, c = -5$$

$$\therefore \quad \pi_1 = P/\rho N^3 D^5$$

In the same manner $\quad \pi_2 = Q\rho^a N^b D^c = \dfrac{Q}{ND^3}$

$$\pi_3 = \frac{gH}{N^2D^2}, \quad \pi_4 = ND^2\rho/\mu$$

$$\therefore \quad \frac{P}{\rho N^3 D^5} = f\left(\frac{Q}{ND^3}, \frac{gH}{N^2D^2}, \frac{ND^2\rho}{\mu}\right)$$

7.18 Hydrometer of mass m and cross-sectional area A floats in a liquid of mass density ρ. If it is gently pressed down in the liquid through the amplitude x, determine its period of oscillation t.

$$t = f(m, A, \rho, g, x)$$

Since the phenomenon is affected by gravity, g is included in the above functional relation. Using Raleigh's method, one can write

$$t = C\, m^a \rho^b A^c g^d x^e$$

or

$$M^0 L^0 T^1 = (M)^a \left(\frac{M}{L^3}\right)^b - (L^2)^c \left(\frac{L}{T^2}\right)^d (L)^e$$

$$\therefore \quad \left.\begin{aligned} a + b &= 0 \\ -3b + 2c + d + e &= 0 \\ -2d &= 1 \end{aligned}\right\} \quad \begin{aligned} \therefore\ b &= -a \\ d &= -\frac{1}{2} \\ c &= -\frac{3}{2}a + \frac{1}{4} - \frac{e}{2} \end{aligned}$$

$$\therefore \quad t = C\, m^a \rho^{-a} A^{(-\frac{3}{2}a + \frac{1}{4} - e/2)} g^{-1/2} x^e$$

$$\therefore \quad \frac{tg^{1/2}}{A^{1/4}} = C\left(\frac{m}{\rho A^{3/2}}\right)^a \left(\frac{x}{\sqrt{A}}\right)^e$$

Here C is numerical constant. Or one can write a more general expression as

$$\frac{tg^{1/2}}{A^{1/4}} = f\,(m/gA^{3/2} \;\; x/\sqrt{A})$$

7.19 In general, one can postulate that the velocity of propagation of surface waves in liquids depends on g, liquid depth h, wave length λ, mass density ρ of liquid and its surface tension σ. Perform dimensional analysis.

$$U = f\,(g, H, \lambda, \rho, \sigma)$$

There are 6 variables and three primary units. Hence three dimensionless groups will be formed. Choosing g, λ and ρ as the repeating variables the three dimensionless groups can be written as

$$\pi_1 = Ug^a\lambda^b\rho^c$$

$$\pi_2 = hg^a\lambda^b\rho^c$$

$$\pi_3 = \sigma g^a\lambda^b\rho^c$$

In the manner as described earlier, values of a, b and c can be determined for all the three relations. The resulting dimensionless parameters are

$$\pi_1 = U/\sqrt{g\lambda}, \qquad \pi_2 = h/\lambda \quad \text{and} \quad \pi_3 = \sigma/\lambda^2 g\rho$$

Therefore one can write $\quad \dfrac{U}{\sqrt{g\lambda}} = f\left(\dfrac{h}{\lambda}, \; \sigma/\lambda^2\, g\rho\right).$

7.20 From Euler's equations of motion, obtain the conditions for similarity of flows of ideal fluids.

When the differential equations or empirical equations governing the flow phenomenon are known, expressing them in dimensionless form gives conditions for similarity. This is shown below with reference to Euler's equation. Consider Euler's equation in x direction.

$$\frac{\partial u}{\partial t} + u\frac{\partial u}{\partial x} + v\frac{\partial u}{\partial y} + w\frac{\partial u}{\partial z} = X - \frac{1}{\rho}\frac{\partial p}{\partial x} \tag{1}$$

Let U_p, D_p, ρ_p and p_{op} be the characteristic velocity, length dimension, mass density of fluid and ambient pressure in the prototype. The following non-dimensional variables are formed

$$\bar{u} = u/U_p, \; \bar{v} = v/U_p, \; \bar{w} = w/U_p, \; \bar{x} = x/D_p, \; \bar{y} = y/D_p,$$

$$\bar{z} = z/D_p, \; \bar{p} = p/p_{op}, \; \bar{t} = tU_p/D_p, \; \bar{\rho} = \rho/\rho_p$$

$$\therefore \quad \frac{\partial}{\partial t} = \frac{\partial}{\partial \bar{t}} \cdot \frac{\partial \bar{t}}{\partial t} = \frac{U_p}{D_p}\frac{\partial}{\partial \bar{t}}; \quad \frac{\partial}{\partial x} = \frac{\partial}{\partial \bar{x}}\frac{\partial \bar{x}}{\partial x} = \frac{1}{D_p}\frac{\partial}{\partial \bar{x}}$$

Similarly $\quad \dfrac{\partial}{\partial y} = \dfrac{\partial}{\partial \bar{y}} \cdot \dfrac{\partial \bar{y}}{\partial y} = \dfrac{1}{D_p}\dfrac{\partial}{\partial \bar{y}}, \; \dfrac{\partial}{\partial z} = \dfrac{\partial}{\partial \bar{z}} \cdot \dfrac{\partial \bar{z}}{\partial z} = \dfrac{1}{D_p}\dfrac{\partial}{\partial \bar{z}}$

Substitution of these variables gives the Euler's equation (1) in the form

$$\frac{U_p^2}{D_p}\left(\frac{\partial\bar{u}}{\partial\bar{t}}+\bar{u}\frac{\partial\bar{u}}{\partial\bar{x}}+\bar{v}\frac{\partial\bar{u}}{\partial\bar{y}}+\bar{w}\frac{\partial\bar{u}}{\partial\bar{z}}\right)=X_p-\frac{p_{op}}{\rho_p D_p}\left(\frac{1}{\rho}\frac{\partial\bar{p}}{\partial\bar{x}}\right)$$

or dividing all terms by U_p^2/D_p, one gets

$$\frac{\partial\bar{u}}{\partial\bar{t}}+\bar{u}\frac{\partial\bar{u}}{\partial\bar{x}}+\bar{v}\frac{\partial\bar{u}}{\partial\bar{y}}+\bar{w}\frac{\partial\bar{u}}{\partial\bar{z}}=\frac{X_p D_p}{U_p^2}-\frac{p_{op}/\rho_p}{U_p^2}\frac{1}{\bar{\rho}}\frac{\partial\bar{p}}{\partial\bar{x}} \tag{2}$$

In a similar manner, Euler's equation can be non-dimensionalised for geometrically similar model using U_m, D_m, ρ_m and p_{om} as the characteristic quantities. One will then get

$$\frac{\partial\bar{u}}{\partial\bar{t}}+\bar{u}\frac{\partial\bar{u}}{\partial\bar{x}}+\bar{v}\frac{\partial\bar{u}}{\partial\bar{y}}+\bar{w}\frac{\partial\bar{u}}{\partial\bar{z}}=\frac{X_m D_m}{U_m^2}-\frac{p_{om}/\rho_m}{U_m^2}\frac{1}{\bar{\rho}}\frac{\partial\bar{p}}{\partial\bar{x}} \tag{3}$$

If the two flows are dynamically similar, Eqs. 2 and 3 must be identical. This is true if

$$U_p^2/X_p D_p = U_m^2/X_m D_m$$

i.e. Froude number in the model and in the prototype are equal. Also

$$U_p^2/(p_{op}/\rho_p) = U_m^2/(p_{om}/\rho_m)$$

Since $$\frac{U_p^2\rho_p}{p_{op}}=\frac{kU_p^2}{kp_{op}/\rho_p}=k\left(\frac{U_p}{C_p}\right)^2=kM^2$$

where k is the adiabatic constant C_p is velocity of pressure wave and M is Mach number, equality of Mach number is also required. However if M is very small, this condition for similarity is waived and hence equality of Froude number is considered adequate for dynamic similarity. Thus equality of Froude number ensures equality of drag coefficient or discharge coefficients which are forms of Euler number.

7.21 For models governed by gravity force, obtain the scaling ratios for time, discharge, force and power.

Since the gravity force is predominant, Froude number should be kept the same in model and prototype.

$$\therefore \quad \frac{U_m}{\sqrt{g_m l_m}}=\frac{U_p}{\sqrt{g_p l_p}}$$

where the subcripts m and p refer to quantities related to the model and the prototype.

Hence $$U_r=\frac{U_m}{U_p}=\frac{g_m^{1/2}}{g_p^{1/2}}\cdot\frac{l_m^{1/2}}{l_p^{1/2}}=g_r^{1/2}\,l_r^{1/2}$$

where subscript r refers to the scale ratio.

$$t_r=\frac{l_r}{U_r}=l_r/g_r^{1/2}\,l_r^{1/2}=l_r^{1/2}\,/g_r^{1/2}$$

Since $Q = AU,\ Q_r = A_r U_r = l_r^2\, U_r = l_r^2\, g_r^{1/2}\, l_r^{1/2}$

$\therefore\quad Q_r = l_r^{5/2}\, g_r^{1/2}$

Also since F_r = mass × acceleration

= mass density × volume × acceleration

$F_r = \rho_r l_r^3 g_r$

Power $P = FU$

Hence $P_r = F_r U_r$

$= \rho_r l_r^{7/2}\, g_r^{3/2}$

7.22 A geometrically similar model of a spillway discharges 0.10 m³/s discharge per metre width at a head of 0.14 m. If the scale ratio of model is 1:10, determine the prototype head and discharge per metre length of spillway.

According to Froude's law $Q_r = l_r^{5/2}$ since $g_r = 1.0$.

But $Q_r = \dfrac{Q_m}{Q_p} = \dfrac{q_m l_m}{q_p l_p} = q_r l_r$

$\therefore\quad q_r = \dfrac{Q_r}{l_r} = l_r^{5/2}/l_r$

$= l_r^{3/2} = 1/10^{1.5} = \dfrac{1}{31.623}$

$\therefore\quad q_p = q_m \times 31.623$

$= 0.10 \times 31.623 = 3.162$ m³/s m

Also $\dfrac{h_m}{h_p} = l_r \quad \therefore \quad h_p = 0.14 \times 10 = 1.4$ m

7.23 A model of a torpedo is to be tested in a laboratory in a towing tank. If the prototype velocity is to be 8.0 m/s at 15°C and if maximum speed that can be achieved in the towing tank is 24 m/s, determine the model scale that can be chosen.

Since torpedo will be sufficiently deep under water, there are no surface waves. Hence the model will be governed by equality of Reynolds number.

$$\frac{U_m l_m}{\nu_m} = \frac{U_p l_p}{\nu_p}.$$

Assuming $\nu_m = \nu_p$, $U_r = 1/l_r$. However $U_r = \dfrac{24}{8} = 3$

$\therefore\quad l_r = \frac{1}{3}$

7.24 A flow meter when tested in a laboratory gives a pressure drop of 100 kN/m² for a discharge of 0.10 m³/s in a 150 mm diameter pipe. If a geometrically similar model is tested in 600 mm dia pipe at identical conditions of fluid, determine the corresponding discharge and pressure drop in the model.

$$l_r = \frac{600}{150} = 4.0$$

For dynamic similarity $\dfrac{U_r l_r}{\nu_r} = 1$ and $\nu_r = 1$

$$\therefore \qquad U_r = \frac{1}{l_r} = 0.25$$

$$Q_r = l_r^2\, U_r = l_r^2 \frac{1}{l_r} = l_r = 4.0$$

$$\therefore \qquad \frac{Q_m}{0.10} = 4.0 \quad \therefore \quad Q_m = 0.4 \text{ m}^3/\text{s}$$

Also for complete similarity

$$\Delta p_r = \rho_r U_r^2 \quad \text{and} \quad \rho_r = 1$$

$$\therefore \qquad \Delta p_r = (0.25)^2 \quad \therefore \quad \Delta p_m = (0.25)^2 \times 100$$

$$= 6.25 \text{ kN/m}^2$$

7.25 A model of an aeroplane is to be tested in a wind tunnel operating under a pressure of 15 atmospheres. If the prototype is to fly at 500 km/hr under atmospheric condition, what would be the corresponding speed of the model if it is built to a scale of 1 : 12? If the model experiences a force of 600 N, what would be the corresponding force on prototype? Also find the power required by the prototype.

Model will be governed by equality of Reynolds number. Hence

$$U_r = \frac{\mu_r}{\rho_r l_r}$$

In compressed air it is assumed that the dynamic viscosity remains unaltered. Hence $\mu_r = 1.0$. Further, for constant temperature, equation of state gives $p_m/\rho_m = p_p/\rho_p$ or $\rho_r = p_r = 15/1 = 15$.

$$\therefore \qquad U_r = 1/\rho_r l_r = 12/15 = 0.80$$

$$\therefore \qquad U_m = 500 \times 0.80 = 400 \text{ km/hr}$$

Also $\qquad F_r = \rho_r l_r^2 U_r^2$

$$= 15 \times \frac{1}{144} \times 0.80^2 = 0.0667$$

$$\therefore \qquad F_m/F_p = 0.0667 \quad \text{or} \quad F_p = 600/0.0667 = 8995.50 \text{ N}$$

since power = force × velocity

$$= 8995.50 \times \frac{500 \times 1000}{3600}$$

$$= 1249\ 375.31 \text{ N} \quad \text{or} \quad 1249.375 \text{ kN}$$

7.26 A centrifugal pump was tested in the laboratory by constructing a 1 : 8 scale geometrically similar model. The model consumed 5 kW power working at 5 m head and 450 rpm speed. If the prototype is to work under 80 m head, determine its power requirement, speed, and the discharge ratio.

For rotating machines the following parameters will have the same value in the model and in the prototype

$$\frac{gH}{N^2D^2}, \frac{P}{\rho N^3D^5} \frac{Q}{ND^3}$$

Hence $\frac{g_rH_r}{N_r^2 D_r^2} = 1$ and $g_r = 1\ 0$. Therefore, $N_r = \frac{H_r^{1/2}}{D_r} = \left(\frac{5}{80}\right)^{1/2} \left(\frac{8}{1}\right)$

$$= 2.0$$

$$\therefore \quad \frac{N_m}{N_p} = 2 \quad \text{or} \quad N_p = \frac{450}{2} = 225 \text{ rpm}$$

In the same way $\quad P_r = N_r^3 D_r^5$

$$= 2^3 \left(\frac{1}{8}\right)^5 = \frac{1}{4096}$$

$$\therefore \quad \frac{P_m}{P_p} = \frac{1}{4096} \quad \text{or} \quad P_p = 4096 \times 5 = 20\ 480 \text{ kW.}$$

Further $\quad Q_r = N_r D_r^3$

$$= 2 \times \left(\frac{1}{8}\right)^3 = \frac{1}{256}$$

7.27 What should be the roughness of model of a stream built to a scale of 1 : 200, if the roughness of the prototype stream in terms of Manning's n is 0.03 ?

Manning's equation can be written for the model and prototype from which one gets

$$\frac{U_m}{U_p} = \frac{n_p}{n_m} \left(\frac{R_m}{R_p}\right)^{2/3} \left(\frac{S_m}{S_p}\right)^{1/2}$$

or $\quad U_r = R_r^{2/3}\ S_r^{1/2}/n_r$

However, for undistorted model, the scales for vertical and horizontal distances being same $S_r = 1.0$. If R is assumed to be equal to depth h, $R_r = h_r$. Further since equality of Froude number is to be maintained, $U_r = h_r^{1/2}$. Hence one gets

$$U_r = h_r^{1/2} = \frac{h_r^{2/3}}{n_r} \quad \text{or} \quad n_r = h_r^{1/6}$$

$$\therefore \quad n_m = n_p/(200)^{1/6}$$

$$= \frac{0.03}{200^{1/6}} = 0.012$$

7.28 A model of a river carrying a discharge of 3500 m³/s has a depth of 2.25 m and width 1500 m. From the point of view of availability of space the horizontal scale of 1 : 400 is chosen. Assuming slope scale to be unity, determine the depth and discharge scales.

$$\text{Velocity in prototype} = U_p = \frac{3500}{2.25 \times 1500} = 1.037 \text{ m/s}$$

If a geometrically similar model is designed, the depth in the model will be $h_m = \dfrac{2.25}{400} = 0.005\,625$ m or 0.5625 cm. This depth is too small. For testing the model at lower discharges, the depth will be still smaller. Further the flow in the model may become laminar, while it is turbulent in the prototype.

$$\text{Since} \quad U_r = \sqrt{h_r} \qquad U_m = U_p/\sqrt{400}$$
$$= 0.0519 \text{ m/s}$$

$$\text{and} \qquad Re_m = \frac{0.0519 \times 0.005625}{10^{-6}} = 291.3$$

This is smaller than 1500 and hence flow in model is laminar.

Reynolds number in the model can be increased by either increasing velocity or by increasing the depth. For same h_m if U_m is to be increased S_m must be increased. However it is stipulated that $S_p = S_m$. Hence h_m must be increased using $h_r > l_r$.

$$\text{Since} \quad \frac{U_m}{U_p} = \sqrt{h_r} \qquad U_m = \sqrt{h_r}\, U_p$$

$$\text{and} \qquad \frac{h_m}{h_p} = h_r \qquad h_m = h_r h_p$$

$$\therefore \qquad \frac{U_m h_m}{\nu_m} = \frac{U_p h_p}{\nu_m} h_r^{3/2} \text{ which must be greater than 1500.}$$

Taking ν_m for water to be 1×10^{-6} m²/s and substituting values of U_p and h_p, one gets

$$\frac{1.037 \times 2.25}{10^{-6}} h_r^{3/2} \geqslant 1500$$

$$\therefore \quad h_r \geqslant \frac{1}{134}$$

$$\text{Assume} \qquad h_r = \frac{1}{125}$$

$$Q_r = U_r l_r h_r = l_r h_r^{3/2}$$

$$= \frac{1}{400 \times 125^{3/2}} = \frac{1}{509\,017}$$

$$\therefore \qquad Q_m = \frac{3500}{509\,017} \times 10^3 = 6.88 \text{ l/s}$$

7.29 In a tidal model horizontal scale ratio l_r is 1/6000. If 12 hour tide is to be reproduced in the model in one minute, what should be the depth scale ?

$$U_r = \sqrt{h_r}, \quad \text{and} \quad t_r = \frac{l_r}{U_r} = l_r/h_r^{1/2}$$

But
$$t_r = \frac{1}{12 \times 60} = 1/720$$

$$\therefore \quad \frac{1}{720} = \frac{1}{6000} \cdot \frac{1}{h_r^{1/2}} \quad \text{or} \quad h_r^{1/2} = \frac{720}{6000}$$

$$\therefore \quad h_r = \frac{1}{69.44}$$

7.30 A 1 : 25 scale model of a ship with surface area of 4.0 m² and length 6.0 m when tested in the laboratory at 2.0 m/s gave a total drag force of 40 N. Calculate the drag force and power of the prototype when it is cruising at the corresponding speed. Assume 20°C as the temperature of water and calculate skin friction by the formula $C_f = 0.074/Re_L^{1/5}$.

This example illustrates the analysis of ship-model data, where model is governed by Froude's law but Reynolds number effect cannot be neglected. As regards frictional resistance refer to Chapter IX.

First calculate frictional resistance in the model.

$$(Re_L)_m = \frac{2.0 \times 6.0}{1 \times 10^{-6}} = 1.2 \times 10^7$$

$$\therefore \quad C_f = \frac{0.074}{(1.2 \times 10^7)^{1/5}} = 0.00284$$

$\therefore$ Frictional force in model $F_{fm} = \dfrac{C_f A \rho U^2}{2}$

or
$$F_{fm} = \frac{0.00284 \times 4.0 \times 998 \times 2^2}{2}$$

$$= 22.675 \text{ N}$$

Since the total resistance is equal to wave resistance F_w plus frictional resistance

$$40 = F_{wm} + 22.675$$

$$\therefore \quad F_{wm} = 17.325 \text{ N}.$$

Since wave resistance is dependent on Froude number, prototype wave resistance can be determined by using force scale ratio for Froude's law (see Table 7.2).

$$F_r = l_r^3 \rho_r g_r$$

$$\therefore \quad \frac{F_{wm}}{F_{wp}} = \left(\frac{1}{25}\right)^3 \times 1 \times 1 \qquad \therefore \quad F_{wp} = 17.325 \times 25^3$$

$$= 270\ 703.125 \text{ N}$$

The prototype frictional resistance is now calculated using the formula for C_f.

$$(Re_L)_p = \frac{U_p L_p}{\nu_p}, \quad \frac{U_m}{U_p} = l_r^{1/2} = \frac{1}{5}$$

$$\therefore \quad U_p = 5 \times 2 = 10 \text{ m/s} \qquad l_p = 25 \times l_m = 25 \times 6 = 150 \text{ m}$$

$$\therefore \quad (Re_L)_p = \frac{10 \times 150}{10^{-6}} = 1.5 \times 10^9$$

At this Reynolds number C_f is given by the formula (See chapter IX)

$$C_f = \frac{0.455}{(\log Re_L)^{2.58}} = 0.001494$$

$$\therefore \quad F_{fp} = \frac{C_f A_p \rho_p U_p^2}{2} = \frac{0.001494 \times (4 \times 25^2) \times 998 \times 10^2}{2}$$

$$= 186\ 378.161 \text{ N}$$

Total resistance for prototype $F_p = F_{fp} + F_{wp}$

$$= 186\ 378.161 + 270,\ 703.125$$

$$= 457,\ 081.286 \text{ N} \quad \text{or} \quad 457.081 \text{ kN}$$

Power $= F_p U_p$

$$= 457.081 \times 10 = 4570.81 \text{ kW}$$

7.31 An aircraft flies at 1800 km/hr in air at 80 kN/m² pressure and – 23.16°C temperature. A 1:10 scale model of the aircraft is to be tested in a compressed air wind tunnel at prototype speed. When experiments are done at the same temperature, namely – 23.16°C and viscosity of air is independent of pressure, what will be the prototype drag if drag in the model is 15 N? What pressure should be maintained in the tunnel? Take $E = 1.10 \times 10^5$ N/m².

In general, dynamic similarity between the model and prototype of an aircraft will require equality of *Re* and *M* numbers.

Velocity of sound in air $C = \sqrt{kgRT}$

Hence C will remain the same in the model and prototype because T is same. Further aircraft speed is maintained the same in model and in prototype. Hence M will be the same in both. For equality of Reynolds number

$$\frac{U_m \rho_m l_m}{\mu_m} = \frac{U_p \rho_p l_p}{\mu_p}$$

$U_m = U_p$, and $\mu_m = \mu_p$ since $T_m = T_p$ and pressure rise does not affect μ significantly. Hence equality of Re number can be attained if $\rho_m l_m = \rho_p l_p$

$$\therefore \quad \frac{\rho_p}{\rho_m} = \frac{l_m}{l_p} = \frac{1}{10} \quad \therefore \quad \rho_m = 10\rho_p$$

but at constant temperature $\frac{p}{\rho}$ = const. or $p \sim \rho$. Hence to achieve $10\rho_p$ density there has to be ten fold increase in pressure in the wind tunnel. Hence $p_m = 80 \times 10 = 800$ kN/m². For equality of Reynold number

$$F_r = \mu_r^2/\rho = \frac{1}{10} \quad \therefore \quad F_p = 10F_m = 150 \text{ N}$$

7.32 A river carries a discharge of 16 000 m³/s at 8.0 m depth and 0.0025 slope when its width is 400 m. About 15 km reach of the river is to be reproduced in the laboratory where 30 m space and a maximum of 0.2 m³/s discharge are available. Determine horizontal and vertical scales, model slope and roughness scale.

$$\text{Length scale } l_r = 30/15 \times 1000 = \frac{1}{500}$$

$$\text{Discharge scale } Q_r = 0.20/16000 = 1/80000$$

Use of undistorted model will give model depth of 800/500 = 1.60 cm which is too small. Hence model must be distorted.

$$Q_r = l_r h_r^{3/2}, \text{ hence } \frac{1}{80000} = \frac{1}{500} \cdot h_r^{3/2}$$

$$\therefore \quad h_r^{3/2} = \frac{500}{80000} = \frac{1}{1600}$$

$$\text{or} \quad h_r = \frac{1}{29.48}, \text{ say } \frac{1}{30}$$

This will give model depth $h_m = 800/30 \simeq 27$ cm

In order to reduce the distortion one can modify the vertical scale such that about 10 cm depth is available in model i.e. $h_r = 10/600 = 1/60$

It must be ensured that R_e in model is greater than 1500.

$$U_p = \frac{16000}{400 \times 8} = 5.0 \text{ m/s}$$

$$\therefore \quad U_m/U_p = \sqrt{h_r} = \frac{1}{7.746}$$

$$U_m = \frac{5.0}{7.746} = 0.645 \text{ m/s}$$

$$h_m = \frac{8.00}{60} = 0.133 \text{ m}$$

$$\therefore \quad Re_m = \frac{0.645 \times 0.133}{10^{-6}} = 85\,785$$

which is adequate (being greater than 1500)

$$\therefore \quad S_r = \frac{h_r}{l_r} = \frac{S_m}{S_p} \quad \therefore \quad S_m = 0.0025 \times \frac{500}{60}$$

$$= 0.02082$$

Also $\quad U_r = h_r^{1/2} = \frac{1}{n_r} h_r^{2/3} \left(\frac{h_r}{l_r}\right)^{1/2}$

$$\therefore \quad n_r = h_r^{2/3} \, l_r^{1/2} = \frac{500^{1/2}}{60^{2/3}} = \frac{22.36}{15.347}$$

$$= 1.456$$

$$\therefore \quad l_r = 1:500, \; h_r = 1:60, \; S_m = 0.02083, \; n_r = 1.456$$

PROBLEMS

7.1 The critical velocity U_c in fps at which sediment of size d mm and relative density (ρ_s/ρ) just moves on the bed of an alluvial stream is given by the equation

$$U_c = 0.50 \, d^{4/9} \left(\frac{\rho_s}{\rho} - 1\right)^{1/2}$$

Check whether the equation is dimensionally homogeneous. If not, obtain the value of numerical constant when U_c is expressed in m/s and d in m. (No, 3.283)

7.2 If the perimeter P of an alluvial channel is related to channel forming discharge Q by the equation

$$P = 2.667 \sqrt{Q}$$

where P is in ft and Q is in ft³/s. Modify the formula for its use in SI units $(P = 4.83\sqrt{Q})$

7.3 Form dimensionless parameter from the following variables:

(i) F, ρ, U, l (ii) U, γ, ρ, l

(iii) $\frac{\partial u}{\partial y}, y, \rho, \mu$ (iv) p, γ, l

(v) F, l, μ, U

(vi) ω, ν, t and Γ, where ω is angular velocity and Γ circulation

(vii) f, l, U where f is frequency (viii) $l, t, \nu,$

(ix) $\Delta p, \rho, U, C$, where C is velocity of pressure wave

(x) F, ρ, μ (xi) σ, g, ρ, μ

(xii) $\delta, U, \mu, \frac{\partial p}{\partial x}$, where δ is boundary layer thickness

$$\left(F/\rho l^2U^2,\ U^2\rho/\gamma l,\ \frac{\partial u}{\partial y}\,y^2\rho/\mu,\ p/\gamma l,\ F/\mu Ul,\ \omega\nu t/\Gamma\right.$$

$$\left. fl/U,\ \nu t/l^2,\ \Delta p/\rho CU,\ F\rho/\mu^2,\ \sigma^3\rho/g\mu^4,\ \frac{\delta^2}{\mu U}\,\frac{\partial p}{\partial x}\right)$$

7.4 Below are given some dimensionally homogeneous equations. Convert them into equations among dimensionally homogeneous parameters and verify Buckingham's π theorem.

(i) $\Delta p = 32uU\,l/D^2$

(ii) $h_f = flU^2/2gD$, where f is dimensionless friction factor

(iii) $U = \dfrac{C}{d^{1/6}}\,h^{2/3}\,S^{1/2}$, where d and h are length parameters and S is slope, C is a constant

(iv) $\tau_0 = C\sqrt{U^3\rho\mu/x}$

7.5 If velocity of propagation U of a surface wave in deep liquids depends on its wave length λ, mass density ρ and gravitational acceleration g, obtain expression for U using Raleigh's method. $(U/\sqrt[5]{g\lambda} = C)$

7.6 The critical depth y_c in a triangular channel depends on discharge Q, g, and central angle θ, obtain expression for y_c. $y_c/\sqrt[5]{Q^2/g} = f(\theta)$

7.7 If velocity of pressure wave C in a fluid depends on mass density ρ and viscosity μ of the fluid and the pressure p, obtain expression for C using Raleigh's method. $C = \text{const.}\sqrt{p/\rho}$

7.8 The velocity of propagation U of surface waves in shallow liquids depends on ρ, g, water depth h and wave length λ. Obtain expression for U. $U/\sqrt{gh} = f(h/\lambda)$

7.9 Assume mean free path λ of molecules of a gas to be dependent on n, the number of molecules per unit volume of gas, and the diameter d of the molecule. Obtain expression for λ. $\lambda/d = f(nd^3)$

7.10 An elastic pendulum is formed by attaching a box of volume V with fluid of mass density ρ to a weightless spring of spring constant K (spring constant K is the force required for unit displacement of spring). Find the period of oscillation of the pendulum t, if it additionally depends on g. $tg^{1/2}/V^{1/6} = f(K/\rho\, g\forall^{2/3})$

7.11 Solve problem 7.10 using mass, volume, time and length as primary units. $t = C\sqrt{V\rho/K}$

7.12 If the depth after the hydraulic jump, h_2, in an open channel depends on the initial flow depth h_1, discharge per unit width q and g, perform the dimensional analysis. $h_2/h_1 = f(q/h_1\sqrt{gh_1})$

7.13 The minimum average velocity U_c in an open channel at which uniform bed sediment of size d just starts moving, is known as critical velocity and it depends on mass density ρ of the liquid, flow depth

h, and $\Delta\gamma_s$ the difference in specific weights of sediment and liquid. Obtain functional relationship for U_c.

$$U_c \Big/ \sqrt{\frac{\Delta\gamma_s d}{\rho}} = f(h/d)$$

7.14 Show that the frictional torque T of a disc of diameter D rotating at an angular velocity ω in a liquid of mass density ρ and viscosity μ is given by $T/\rho\omega^2D^3 = f(\omega D^2\rho/\mu)$.

7.15 The volume $\forall$ of a drop of liquid that forms at the end of a tube before it falls under the action of gravity, depends on tube diameter D, σ, and γ. Obtain expression for $\forall$. $\forall/D^3 = f(\sigma/\gamma D^2)$

7.16 If the drag force F on geometrically similar bodies depends on the characteristic size D of the body, ρ, μ and flow velocity U, perform dimensional analysis in such a manner that the resulting solution gives direct solution for F as well as U.

$$F/\rho\mu^2 = f(UD\rho/\mu)$$

7.17 When the drag force F on the body depends on ρ, D, μ and U, one gets (see Ex. 7.10), $F/D^2\rho U^2 = f\left(\frac{UD\rho}{\mu}\right)$. Determine the form of this relationship for F (i) for small values of Reynolds number when F is proportional to U, and (ii) for large values of Reynolds number when F is almost independent of Re since major drag is due to separation. $F = C\,D\mu U$, $F = C\,\rho D^2U^2$

7.18 If the resistance F offered to the motion of a ship depends on its velocity U, characteristic length l, ρ, μ and g perform dimensional analysis for F. $F/\rho l^2U^2 = f(Ul\rho/\mu, U/\sqrt{gl})$

7.19 If the drag force F on geometrically similar aeroplanes is a function U, l, ρ, μ and E, determine the form of functional relationship for F.

$$F/l^2\rho U^2 = f(Ul\rho/\mu, U/\sqrt{E/\rho})$$

7.20 The efficiency η of geometrically similar fans depends on mass density ρ and viscosity μ of the air, angular velocity ω, diameter of blades D and the discharge Q. Perform dimensional analysis for η.

$$\eta = f(Q/\omega D^3, \rho\omega D^2/\mu)$$

7.21 The shear stress τ_0 on the bed of an alluvial channel at which the sediment of uniform size d forming the bed just starts moving depends on the difference in specific weights of sediment and water, $\Delta\gamma_s$, and μ and ρ for water. Perform dimensional analysis in such a way that an explicit solution for τ_0 is obtained (Hint: τ_0 should not be the repeating variable).

$$\tau_0/\Delta\gamma_s d = f\left(\frac{\Delta\gamma_s d^3\rho}{\mu^2}\right)$$

7.22 The pressure difference Δp across an orifice plate of diameter d fixed in a pipe of diameter D depends on Q, ρ and μ. Perform dimensional analysis taking Δp as the dependent variable.

$$\Delta p d^4/\rho Q^2 = f(Q\rho/\mu d, D/d)$$

7.23 In problem 7.22 perform dimensional analysis in such a manner that an explicit solution for Q is obtained.

$$Q^2\rho/\Delta p d^4 = f(D/d, \Delta p.\rho d^2/\mu^2)$$

7.24 If the stagnation pressure p_s in a compressible fluid depends on the ambient velocity U_0, ambient pressure p_0 mass density of the ambient fluid ρ_0 and adiabatic constant k, obtain expression for $(p_s - p_0)$ using dimensional analysis.

$$(p_s - p_0)/\rho_0 U_0^2 = f(U/\sqrt{p_0/\rho_0}, k)$$

7.25 In a rarified gas, the drag force on geometrically similar bodies depends on its characteristic length l, mass density ρ, volume modulus of elasticity E, velocity U and mean free path λ of molecules. Perform dimensional analysis for F. $F/\rho l^2 U^2 = f(U/\sqrt{E/\rho}, \lambda/l)$

7.26 Small droplets of liquid are formed when liquid jet breaks up in spray and fuel injection processes. The resulting droplet diameter d depends on liquid density ρ, μ, σ, jet diameter D and jet velocity U. Perform dimensional analysis for d.

$$d/D = f\left(\frac{\rho U^2 D}{\sigma}, \frac{UD\rho}{\mu}\right)$$

7.27 For a cone bearing, the torque T required to rotate the shaft at an angular velocity ω depends on the clearance d between shaft and the bearing, half angle θ of a cone, height of cone H and μ. Obtain expression for T using dimensional analysis. $T/\omega\mu H^3 = f(\theta, d/H)$

7.28 If the boundary layer thickness δ in flows with pressure gradient depends on the distance x, U, ρ, μ, and $\frac{\partial p}{\partial x}$, perform dimensional analysis for δ.

$$\delta/x = f\left(Ux\rho/\mu, \left(\frac{\partial p}{\partial x}\right) x/\rho U^2\right)$$

7.29 A "Viscous drag" pump delivers high pressure fluid by utilising viscosity of the fluid to draw it through small clearance passage. Assuming that the rise in pressure Δp depends on ρ, μ, rotor diameter D, clearance space b and rotational speed N of the shaft, obtain expression for Δp through dimensional analysis.

$$\Delta p/\rho D^2 N^2 = f(ND^2\rho/\mu, b/D)$$

7.30 For turbulent flow, the velocity u, at a distance y from the boundary depends on wall shear τ_0, ρ, μ and surface roughness K_s. Determine the functional form of velocity distribution law.

$$u/\sqrt{\tau_0/\rho} = f\left(\frac{\rho y\sqrt{\tau_0/\rho}}{\nu}, \frac{y}{K_s}\right)$$

7.31 If power P required to transport a fluid through length l of pipe of diameter D and surface roughness k_s depends on Q, ρ and μ. perform dimensional analysis.

$$PD^4/\rho Q^3 = \frac{l}{D} f\left(\frac{Q\rho}{D\mu}, k_s/D\right)$$

7.32 The thickness of the boundary layer δ formed on a flat plate of surface roughness k_s depends on the free stream vetocity U, distance x from the leading edge, ρ and μ. Obtain expression for δ. Also obtain expression for local shear τ_0 at that place.

$$\delta/x \quad \text{or} \quad \tau_0/\rho U^2 = f(Ux\rho/\mu, k_s/x)$$

7.33 If darg force F depends on U, D, ρ, g, σ, μ and E perform dimensional analysis for F using matrix method. Use U, D and ρ as repeating variables.

$$F/\rho U^2 D^2 = f\left(UD\rho/\mu,\ U/\sqrt{gD},\ \frac{\rho U^2 D}{\sigma},\ U/\sqrt{E/\rho}\right)$$

7.34 Fall velocity ω of a sediment particle of size d depends on mass density ρ and viscosity μ of the fluid, the difference in specific weights of sediment and fluid $(\gamma_s - \gamma)$, and the diameter D of the cylindrical container in which it falls. Perform dimensional analysis for ω.

$$\rho\omega^2/(\gamma_s - \gamma)\, d = f(D/d,\ d^3(\gamma_s - \gamma)\, \rho/\mu^2)$$

7.35 The pressure loss Δp along the bend in a pipe of diameter D depends on velocity of flow U, pipe roughness k_s, bend radius r, bend angle θ, and fluid properties ρ and μ. Obtain the complete set of dimensionless parameters.

$$\Delta p/\rho U^2 = f(UD\rho/\mu,\ k_s/D,\ r/D,\ \theta)$$

7.36 For a triangular notch the discharge Q flow over it depends on head H, central angle θ, ρ, μ, g and σ. Obtain general expression for Q.

$$Q/H^2\sqrt{gH} = f(\theta,\ H\sqrt{gH}\,\rho/\mu,\ \rho H^2 g/\sigma)$$

7.37 The time t required to lower the depth of water in a tank from H_1 to H_2 depends on cross-sectional area A of the tank, area of orifice a and g. Perform dimensional analysis.

$$t\sqrt{g/H_1} = f(\sqrt{a}/H_1,\ \sqrt{A}/H_2,\ H_2/H_1) \quad = f\left(\frac{a}{A},\ \frac{H_2}{H_1}\right)$$

7.38 Solve Example 7.17 using matrix method.

7.39 Navier Stokes' equation in x direction for laminer flow of an incompressible fluid is

$$\frac{\partial u}{\partial t} + u\frac{\partial u}{\partial x} + v\frac{\partial u}{\partial y} + w\frac{\partial u}{\partial z} = X - \frac{1}{\rho}\frac{\partial p}{\partial x} + \frac{\mu}{\rho}\left(\frac{\partial^2 u}{\partial x^2} + \frac{\partial^2 u}{\partial y^2} + \frac{\partial^2 u}{\partial z^2}\right)$$

Using the method employed in Example 7.20, obtain conditions for two laminar flows to be similar.

7.40 Show that, if a model is to be designed to maintain equality of Re and Fr in the model and prototype, and if the same liquid is to be used in both, the model has to be as large as the prototype.

7.41 What should be the ratio of kinematic viscosities of the liquid in the model and in the prototype if the model and prototype have equality of Re and Fr numbers?

7.42 Obtain the scale ratios for velocity, time, discharge, acceleration, pressure, force and power when a model is governed by equality of

Weber number.

$U_r = \sigma_r^{1/2} / \rho_r^{1/2} \, l_r^{1/2}$, $t_r = l_r^{3/2} \, \rho_r^{1/2} / \sigma_r^{1/2}$, $Q_r = l_r^{3/2} \, \sigma_r^{1/2} / \rho_r^{1/2}$

$a_r = \sigma_r / \rho_r l_r^2$, $p_r = \sigma_r / l_r$, $F_r = \sigma_r l_r$, $P_r = l_r^{1/2} \, \sigma_r^{3/2} / \rho_r^{1/2}$

7.43 A geometrically similar model of scale 1: 10 is built to study wave motion on a beach. Determine the scale ratios for velocity, time, acceleration and force if mass densities in the prototype and model are 1030 kg/m³ and 1000 kg/m³ respectively.
(0.316, 0.316, 0.1, 0.000 909)

7.44 A venturimeter of diameter 0.75 m fixed in a pipe of 1.0 m diameter carries water at 1.57 m³/s. If the performance of this venturimeter is to be tested on 1:5 scale model using air as a fluid, determine the model discharge, take $\nu_p = 1 \times 10^{-6}$ m²/s and $\nu_m = 1.5 \times 10^{-5}$ m²/s.
(4.71 m³/s)

7.45 The drag on a sonar transducer is to be predicted from wind tunnel model tests. The prototype transducer of 300 mm diameter is to be towed at 3.0 m/s in water. If the model is of 150 mm diameter, determine its speed in air. If the drag in the model is 25 N, what will be the corresponding drag in the prototype?
Take $\rho_p = 1000$ kg/m³, $\rho_m = 1.21$ kg/m³, $\mu_p = 10^{-3}$ kg/m s and $\mu_m = 1.85 \times 10^{-5}$ kg/m s. (91.737 m/s, 88.37 N)

7.46 A tidal model is built with horizontal scale of 1: 5000 and vertical scale of 1:100. What will be the tidal period in the model if in nature it is 13 hrs? (1.56 minutes)

7.47 A 20 m high dam has a 60 m spillway operating at 4.0 m head at which it carries a discharge of 1000 m³/s. A geometrically similar model of this spillway is to be constructed. If the discharge available in the laboratoy is 0.30 m²/s, determine the height, length, and head over the spillway. ($l_r = 25.0$, 0.80 m, 2.4 m, 0.16 m)

7.48 A fan is to be used in a region where mass density of air is 0.96 kg/m³ and is designed to deliver 5.7 m³/s of air against a static resistance of 50 mm of water column, with an efficiency of 65 per cent. Calculate power input, pressure rise and power output at test site where $\rho = 1.20$ kg/m³. (5.364 kW, 62.5 mm, 3.486 kW)

7.49 A 500 mm diameter fan operating at 12 revolutions per second requires 3.5 kW energy to deliver 3.25 m³/s air at a static load of 250 N/m². Estimate the performance of a geometrically similar fan whose diameter is 1000 mm and operates at 6.0 revolutions per second. (13.0 m³/s, 500 N/m², 14.0 kW)

7.50 An axial flow water pump has to deliver 12 m³/s discharge at a head of 20 m of water. Calculate the air flow delivery rate and pressure rise for 1:3 scale model using air as a fluid in the model. Take $\rho_a = 1.208$ kg/m³. Model and prototype are run at the same speed.
(0.44 m³/s, 26.325 N/m²)

7.51 An open channel model is constructed with horizontal scale of 1:100 and vertical scale of 1:25. What will be the bed slope in the model if the slope in the prototype is 1 in 1000? (1 in 250)

7.52 A tidal period of an estuary is known to be 15.2 hrs. What will be the corresponding period in a geometrically similar model of scale 1:1000? (0.481 hrs)

7.53 A centrifugal pump is required to run at 600 rpm and deliver 1.2 m^3/s of oil of mass density 960 kg/m^3 and viscosity 0.115 N s/m^2. A 1:4 scale model of the pump is to be tested at 1200 rpm using air as a fluid for which ρ and μ are 1.2 kg/m^3 and 1.8×10^{-5} N s/m^2. Determine discharge through the model, ratios of Reynolds numbers in the model and prototype, and the power required for the prototype if model requires 10 W. (0.0375 m^3/s, 1.0, 1024 kW)

7.54 Obtain the scale ratios for velocity, time, acceleration and force, for the models governed by equality of M number.

$$\left(\sqrt{E_r/\rho_r},\ l_r\rho_r^{1/2}/E_r^{1/2},\ E_r/\rho_r l_r,\ l_r^3 E_r\right)$$

7.55 If an undistorted open channel model has to satisfy equality of Froude number and also Manning's equation, show that $n_r = l_r^{1/6}$, where n_r is scale for Manning's roughness.

7.56 A 1:50 scale model of a ship 200 m long and with 5000 m^2 surface area is tested for prototype velocity of 30 km/hr. The model gave a total resistance of 6.238 N. Determine the power required to overcome resistance to motion. Use $C_f = 0.074/\text{Re}_L^{1/5}$ and $C_f = 0.455/(\log_{10} \text{Re}_L)^{2.58}$ for the model and prototype respectively where frictional resistance is given by $\left(\text{area} \times C_f \times \frac{\rho U^2}{2}\right)$.

(3691.886 kW)

DESCRIPTIVE QUESTIONS

7.1 What is the basic premise in the method of dimensional analysis?

7.2 What are dimensionally homogeneous and nonhomogeneous equations? Give two examples each from fluid mechanics/hydraulic engineering.

7.3 What is the complete set of dimensionless parameters obtained from given variables?

7.4 What are the usages of dimensional analysis? Discuss each with an example.

7.5 Is the choice of primary units to be used (M, L, T or F, L, T) dependent on the nature of the problem under consideration? Explain with examples.

7.6 State the conditions that must be satisfied by the repeating variables

7.7 If several variables fulfill the conditions for repeating variables, what determines their final choice?

7.8 List the variables which govern the rate of erosion of soil from soil surface due to impact of raindrops.

7.9 List the variables that govern the depth of scour around a bridge pier in an alluvial river.

7.10 How are kinematic and dynamic similarity related?

7.11 State whether Re, Fr or M law will govern the models of

(i) Laminar flow through soil mass
(ii) Hydraulic jump
(iii) Torpedo
(iv) Passenger ship
(v) Bullet
(vi) Aeroplane travelllng at 1500 km/hr
(vii) Harbours
(viii) Venturimeter carrying water
(ix) Spillway of a dam
(x) Submarine

7.12 Form dimensionless parameter from

(i) Re and Euler number which does not involve velocity
(ii) Froude number and Reynolds number which does not involve velocity.

7.13 What are the types of distortions in the model that you know? Give one example of each.

7.14 Explain why river models have to be distorted.

7.15 When a particular phenomenon is governed by both, Re and Fr normally the model is designed using Fr number equality and Re number effect is accounted for. Discuss with the help of ship model.

7.16 How can equality of Re and *M* number be satisfied simultaneously?

7.17 Give both the kinematic and dynamic significance of Re, *W* and *M*.

7.18 Which will be the correct dimensionless parameter formed from

δ, μ, U and $\partial p/\partial x$?

(i) $\dfrac{\partial p}{\partial x}\dfrac{\mu U}{\delta}$ (ii) $\dfrac{uU\delta}{\partial p/\partial x}$ (iii) $\dfrac{\partial p}{\partial x}\dfrac{\delta U}{\mu}$ (iv) $\dfrac{\partial p}{\partial x}\dfrac{\delta^2}{\mu U}$

7.19 A 1:50 scale model of a ship offers 10 N as the wave force. The corresponding force in prototype will be (i) 500 N (ii) 2500 0 N (iii) 125 0000 N (iv) None of these.

7.20 Model of an aircraft travelling at supersonic speed is to be tested with equality in Re and *M*. Then

(i) Model must be tested in ordinary open circuit wind tunnel.

(ii) Model must be tested in compressed air wind tunnel at any suitable temperature.
(iii) Model must be tested in compressed air tunnel at the prototype temperature.
(iv) None of these.

7.21 Which of the following statements are true?
(i) Number of repeating variables will always equal the number of primary units used
(ii) The repeating variables can combine among themselves to form a dimensionless group
(iii) Number of dimensionless groups = number of variables minus rank of dimensional matrix
(iv) If dependent variable is chosen as the repeating variable, one gets implicit relationship for the dependent variable
(v) Under no condition dependent variable should be chosen as the repeating variables

7.22 Gravity affects the flow phenomenon if
(i) Flow takes place through inclined conduits flowing full
(ii) There is free surface and wave formation can take place
(iii) a body moves deep under water but there is a free surface
(iv) two fluids of different mass densities flow one over the other.

7.23 For most of the problems in fluid mechanics, the repeating variables which will give significant dimensionless parameters are
(i) U, μ, ρ (ii) μ, g, ρ (iii) U, L, ρ (iv) L, ρ, μ.

7.24 The dimensionless parameter $\mu U/EL$ is a ratio of (i) inertial force to viscous force (ii) compressibility force to inertial force (iii) viscous force to compressibility force.

7.25 Vortex formation in a sump for a pump is governed by equality of
(i) Reynolds number (ii) Euler number (iii) Froude number (iv) Mach number.

7.26 If there are N variables and m primary units are required to describe them, the maximum number of dimensionless groups formed will be
(i) $(N-m)$ (ii) $N!/(N-m-1)!$ (iii) $N!/(m+1)!\,(N-m-l)!$ (iv) $N!/(m+1)!$

7.27 Dimensionless number formed from F_D, ρ and μ is
(i) $F_D/\rho\mu^2$ (ii) $F_D/\rho\mu$ (iii) $\dfrac{F\mu}{\rho}$ (iv) $F_D\rho/\mu^2$

7.28 The dependent variable should not be chosen as the repeating variable because
(i) formation of dimensionless parameters becomes difficult
(ii) wrong dimensionless parameters are obtained
(iii) solution of problem becomes implicit
(iv) none of the above.

7.29 For the study of drag on a sphere, C_D and Re are appropriate dimensionless parameters if

(i) information about drag force is sought

(ii) information about sphere diameter is sought

(iii) information about velocity causing a given drag force is sought.

7.30 If performance of a long conical diffuser in a pipe line carrying water is to be studied on a model, the following conditions must be satisfied

(i) geometric similarity (ii) geometric similarity and equality of Re, M (iii) geometric similarity and equality of M (iv) geometric similarity and equality of Re.

CHAPTER VIII

Laminar Flow

8.1 INTRODUCTION

In the preceding chapters, primarily the flow of an ideal fluid has been discussed. In the case of Newtonian fluids, the flows can be classified as laminar or turbulent depending on characteristic Reynolds number $\frac{Ul\rho}{\mu}$, where l is the characteristic length. Laminar flows are discussed in this chapter. The following are the characteristics of laminar flow:

1. 'No slip' at the boundary, i.e. because of viscosity, velocity of fluid at $y = 0$ is zero if boundary is stationary or is equal to velocity of the boundary if it is in motion.
2. Because of viscosity there is shear between fluid layers which is given by $\tau = \mu \frac{du}{dy}$ for flow in x direction.
3. The flow is rotational.
4. There is continuous dissipation of energy due to viscous shear and energy must be supplied externally to maintain the flow.
5. There is no mixing between different fluid layers except by molecular motion, which is very small.
6. Flow remains laminar as long as $Ul\rho/\mu$ is less than what is known as the critical value of Reynolds number.
7. Energy loss is proportional to first power of velocity and first power of viscosity.

Laminar flow occurs in capilliary tubes, blood veins, in the case of flow past tiny bodies, in lubrication bearings, underground flow, etc. The characteristics 1, 3 and 4 are true for turbulent flow also.

8.2 NAVIER-STOKES' EQUATIONS

Consider laminar flow of incompressible Newtonian fluids when body force, pressure force and forces due to viscosity affect the motion. Navier-Stokes' equation are the dynamic equations governing the flow under such co ndition. In Cartesian co-ordinate system these are

$$\left.\begin{aligned}
\rho\left(\frac{\partial u}{\partial t}+u\frac{\partial u}{\partial x}+v\frac{\partial u}{\partial y}+w\frac{\partial u}{\partial z}\right)&=X-\frac{\partial p}{\partial x}+\mu\left(\frac{\partial^2 u}{\partial x^2}+\frac{\partial^2 u}{\partial y^2}+\frac{\partial^2 u}{\partial x^2}\right)\\
\rho\left(\frac{\partial v}{\partial t}+u\frac{\partial v}{\partial x}+v\frac{\partial v}{\partial y}+w\frac{\partial v}{\partial z}\right)&=Y-\frac{\partial p}{\partial y}+\mu\left(\frac{\partial^2 v}{\partial x^2}+\frac{\partial^2 v}{\partial y^2}+\frac{\partial^2 v}{\partial z^2}\right)\\
\rho\left(\frac{\partial w}{\partial t}+u\frac{\partial w}{\partial x}+v\frac{\partial w}{\partial y}+w\frac{\partial w}{\partial z}\right)&=Z-\frac{\partial p}{\partial z}+\mu\left(\frac{\partial^2 w}{\partial x^2}+\frac{\partial^2 w}{\partial y^2}+\frac{\partial^2 w}{\partial z^2}\right)
\end{aligned}\right\} \tag{8.1}$$

Continuity equation is

$$\frac{\partial u}{\partial x}+\frac{\partial v}{\partial y}+\frac{\partial w}{\partial z}=0$$

These are 2nd order non-linear partial differential equations in x, y, z and t with u, v, w and p as unknowns. These can be solved analytically only in case of some simplified cases, when appropriate boundary conditions are specified.

The corresponding equations in cylindrical polar co-ordinate system (see Fig. 8.1 for notations) are

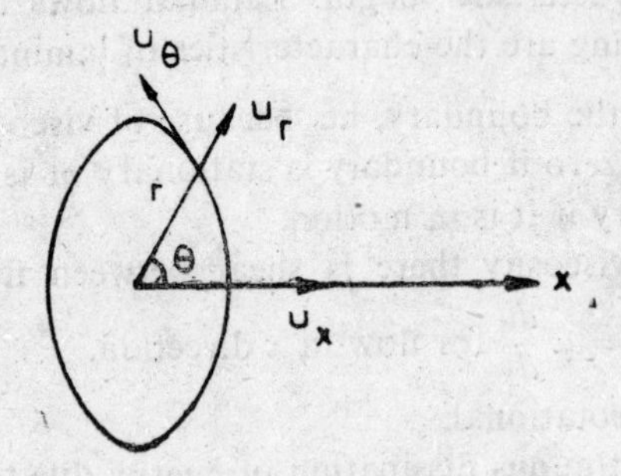

Fig. 8.1. Definition sketch.

$$\left.\begin{aligned}
&\rho\left(\frac{\partial u_r}{\partial t}+u_r\frac{\partial u_r}{\partial r}+\frac{u_\theta}{r}\frac{\partial u_r}{\partial \theta}-\frac{u_\theta^2}{r}+u_x\frac{\partial u_r}{\partial x}\right)=X_r-\frac{\partial p}{\partial r}\\
&\quad+\mu\left(\frac{\partial^2 u_r}{\partial r^2}+\frac{1}{r}\frac{\partial u_r}{\partial r}-\frac{u_r}{r^2}+\frac{1}{r^2}\frac{\partial^2 u_r}{\partial \theta^2}-\frac{2}{r^2}\frac{\partial u_\theta}{\partial \theta}+\frac{\partial^2 u_r}{\partial x^2}\right)\\
&\rho\left(\frac{\partial u_\theta}{\partial t}+u_r\frac{\partial u_\theta}{\partial r}+\frac{u_\theta}{r}\frac{\partial u_\theta}{\partial \theta}+\frac{u_\theta u_r}{r}+u_x\frac{\partial u_\theta}{\partial x}\right)=X_\theta-\frac{\partial p}{r\partial \theta}\\
&\quad+\mu\left(\frac{\partial^2 u_\theta}{\partial r^2}+\frac{1}{r}\frac{\partial u_\theta}{\partial r}-\frac{u_\theta}{r^2}+\frac{1}{r^2}\frac{\partial^2 u_\theta}{\partial \theta^2}+\frac{2}{r^2}\frac{\partial u_r}{\partial \theta}+\frac{\partial^2 u_\theta}{\partial x^2}\right)\\
&\rho\left(\frac{\partial u_x}{\partial t}+u_r\frac{\partial u_x}{\partial r}+\frac{u_\theta}{r}\frac{\partial u_x}{\partial \theta}+u_x\frac{\partial u_x}{\partial x}\right)=X_x-\frac{\partial p}{\partial x}\\
&\quad+\mu\left(\frac{\partial^2 u_x}{\partial r^2}+\frac{1}{r}\frac{\partial u_x}{\partial r}+\frac{1}{r^2}\frac{\partial^2 u_x}{\partial \theta^2}+\frac{\partial^2 u_x}{\partial x^2}\right)
\end{aligned}\right\} \tag{8.2}$$

Continuity equation

$$\frac{\partial u_r}{\partial r} + \frac{u_r}{r} + \frac{1}{r}\frac{\partial u_\theta}{\partial \theta} + \frac{\partial u_x}{\partial x} = 0$$

Here X_r, X_θ and X_x are the components of body force per unit. Volume in r, θ and x directions respectively, and u_r, u_θ and u_x velocities in these directions.

For steady two dimensional flows which are uniform, one gets from Eq. 8.1

$$-\frac{\partial p}{\partial x} + \mu \frac{\partial^2 u}{\partial y^2} = 0$$

or

$$\frac{\partial p}{\partial x} = \frac{\partial \tau}{\partial y} \tag{8.3}$$

Corresponding equation in axisymmetric flow is

$$\frac{\partial p}{\partial x} = \frac{2}{r}\tau = \frac{2\mu}{r}\frac{\partial u}{\partial r} \tag{8.4}$$

8.3 HAGEN-POSEUILLE'S EQUATION FOR LAMINAR FLOW IN PIPES

The velocity distribution for steady laminar flow in a circular pipe is given by (Fig. 8.2)

$$u = \left(-\frac{1}{4\mu}\frac{\partial p}{\partial x}\right)(R^2 - r^2) \tag{8.5}$$

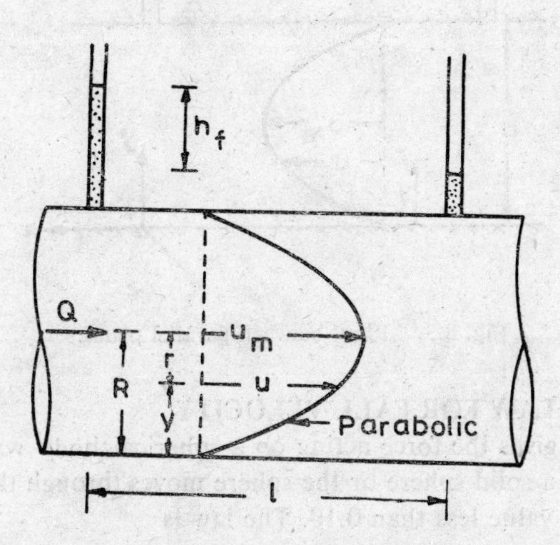

Fig. 8.2. Laminar flow in a pipe.

which on integration and simplification gives

$$U = -\frac{R^2}{8\mu}\frac{\partial p}{\partial x} \tag{8.6}$$

where U is the average velocity of flow. It can be seen that

$$u_m = 2U \tag{8.7}$$

where u_m is the maximum velocity which occurs at the centre of the pipe. Also Eq. 8.6 gives

$$\Delta p = \gamma h_f = \frac{32\mu Ul}{D^2} = \frac{128Q\mu l}{\pi D^4} \tag{8.8}$$

which is known as Hagen-Poiseull's equation. It is valid for $UD\rho/\mu$ less than 2100. For flow between parallel plates, corresponding equations are

$$u = -\frac{1}{2\mu}\frac{\partial p}{\partial x}(By - y^2) \tag{8.9}$$

$$U = -\frac{\partial p}{\partial x}\frac{B^2}{12\mu} \tag{8.10}$$

and $$(p_1 - p_2) = \Delta p = \gamma h_f = \frac{12\mu Ul}{B^2} = \frac{12\mu ql}{B^3} \tag{8.11}$$

where B is the spacing between the parallel plates and q is discharge per unit width, i.e. $q = UB$ (Fig. 8.3).

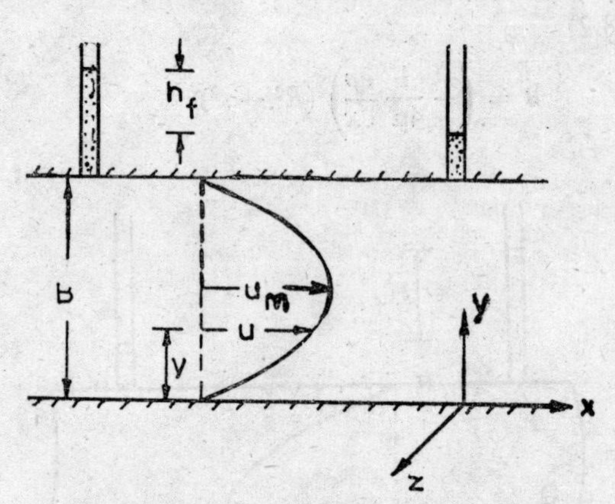

Fig. 8.3. Flow between parallel plates.

8.4 STOKES' LAW FOR FALL VELOCITY

Stokes' law gives the force acting on a spherical body when either the fluid flows past a solid sphere or the sphere moves through the fluid at velocity U at $Ud\rho/u$ value less than 0.10. The law is

$$F_D = 3\pi d\mu U \tag{2.12}$$

where F_D is the force exerted by the fluid on the sphere and d is the diameter of sphere. The motion of sphere will be resisted by F_D. Assumptions in the derivation of Stokes' law are

1. The body is spherical in shape
2. No slip condition prevails
3. The fluid is infinite in extent and quiescent
4. Inertial force is small and viscous force predominates. Hence $Ud\rho/\mu$ is less than 0.10
5. Only a single sphere is considered.

If the particle is falling under the action of gravity, at a constant velocity ω known as the terminal fall velocity

$$F_D = \pi d^3(\gamma_s - \gamma)/6$$

or

$$\frac{\pi d^3}{6}(\gamma_s - \gamma) = 3\pi d\mu\omega$$

or

$$\omega = \frac{d^2}{18\mu}(\gamma_s - \gamma) \tag{8.13}$$

This is also known as Stokes' law. If drag coefficient C_D is defined as

$$C_D = F_D \Big/ \frac{\pi d^2}{4}\frac{\rho\omega^2}{2},$$

one gets, in Stokes' range,

$$C_D = \frac{24}{\text{Re}} \tag{8.14}$$

For $0.1 \leqslant \text{Re} \leqslant 2.0$, Oseen has given the equation for C_D as

$$\left.\begin{aligned} C_D &= \frac{24}{R_e}\left(1 + \frac{3}{16}R_e\right) \\ \text{or} \qquad C_D &= \frac{24}{R_e} + 4.5 \end{aligned}\right\} \tag{8.15}$$

For variation of C_D with Re for larger values of Re, see Chapter XII. In many problems a group of particles fall under the action of gravity. Their fall velocity ω_c is given by

$$\omega_c = c^n\omega \tag{8.16}$$

where ω_c is the fall velocity of particle of size d at concentration c defined as $c = \dfrac{\text{volume of liquid}}{\text{volume of suspension}}$. The exponent n as given by Scholl as $n = 3.65$ for c greater than 0.60. Maude and Whitmore recommend,

$$n = 4.4\ \text{Re}^{-0.08} \qquad \text{if} \quad 1 < \text{Re} < 2500$$

$$n = 2.35 \qquad \text{if} \quad \text{Re} > 2500$$

The theory of fall velocity of sediment particles is useful in the design of grit chambers and settling tanks in water supply and sewage treatment works and in the design of air-clarifiers in air pollution control.

8.5 DARCY'S LAW

According to Darcy the superficial velocity U of flow through soil mass ($U = Q/A$ where A is total area of pores and solid particles) when flow is laminar is given by (Fig. 8.4)

$$U = Ki = -K\frac{\partial h}{\partial L} \tag{8.17}$$

where i is hydraulic gradient. K is known as the coefficient of permeability and is given by

$$K = \frac{d^2 \gamma e^2}{18\mu(1-e)} \text{ as long as } \frac{Ud\rho}{\mu} < 1.0$$

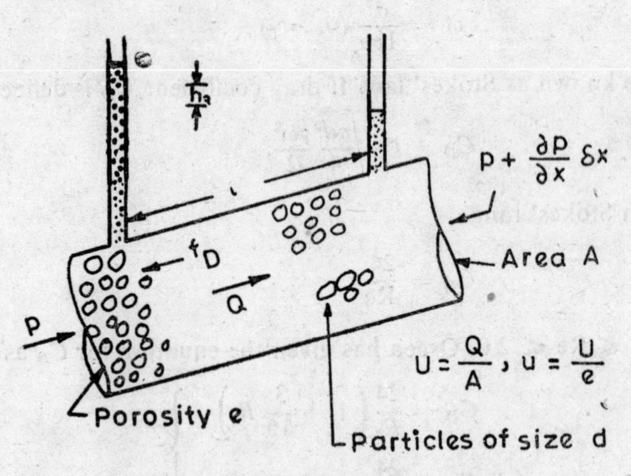

Fig. 8.4. Flow through porous material.

Here e is the porosity defined as volume of pores divided by the bulk volume. The negative sign in Eq. 8.17 implies that flow is in the direction of decreasing piezometric head. This expression is based on drag model for flow through porous material. If one considers a simplified capillary tube model, in which pores are assumed to form straight capillaries, one gets

$$K = \frac{d^2 \gamma e}{k\mu}$$

where k is a numerical constant depending on shape of pores. Darcy's law is useful in the study of flow through filters, earth dams and ground water flow studies. If $Ud\rho/\mu$ is greater than unity, Darcy's law needs modifications.

8.6 FLUIDIZATION

When a fluid is passed upwards at increasing velocities through a column of granular material of size d, a stage is reached when all the particles are entrained in the flow and the bed is in a fluid state. This is known as fluidization and the corresponding superficial velocity U_f as fluidization velocity (U_f = discharge/gross area). At this stage pressure gradient equals submerged weight of particles in a unit volume.

$$U_f = \frac{(\gamma_s - \gamma)\, d^2 e^2}{18\mu} \tag{8.18}$$

where e is the porosity of material. This process is often used in chemical engineering and in back washing of gravity flow filters in water purification studies. Equation 8.18 is valid within Stokes' range.

8.7 VISCOMETERS

Viscometer is an instrument used for determination of viscosity of liquids. It is based on laminar flow in an apparatus where fall velocity, discharge along with head loss, or torque is measured and viscosity determined. These are shown in Fig. 8.5 along with corresponding equations.

8.8 LUBRICATION MECHANICS

Whenever there is rubbing or sliding between two metallic surfaces due to relative linear or rotational motion between the two, friction and corresponding loss of energy, mechanical wearing and generation of heat take place. This is avoided by providing different types of bearings. The hydrodynamical analysis of bearings is based on the theory of laminar flow. The common types of bearings are slipper or sliding bearing, journal bearing, collar bearing, conical bearing, etc. Their analysis is explained through illustrative examples and problems.

8.9 TRANSITION FROM LAMINAR TO TURBULENT FLOW

Laminar flow takes place only at small values of Reynolds number. For pipe flow $UD\rho/\mu$ must be less than 2100; corresponding value of Reynolds number for open channels, defined as $UR\rho/\mu$, is 500, Here R is the hydraulic radius which is the ratio of area to the wetted perimeter. The Reynolds number at which flow ceases to be laminar is known as the critical Reynolds number. Characteristic of laminar flow and its transition to turbulent flow in the case of pipes were studied by Osborne Reynolds through his well-known experiments.

Laminar flow changes into turbulent flow through repeated breakdown of external disturbances imposed on the flow in the form of velocity or pressure fluctuations. These are caused by faulty inlet conditions, vibrations of pump and pipe, etc. Local instability of flow occurs when the parameter χ (chi) defined as

$$\chi = \rho y^2 \frac{du}{dy} \Big/ \mu$$

exceeds a value of 500.

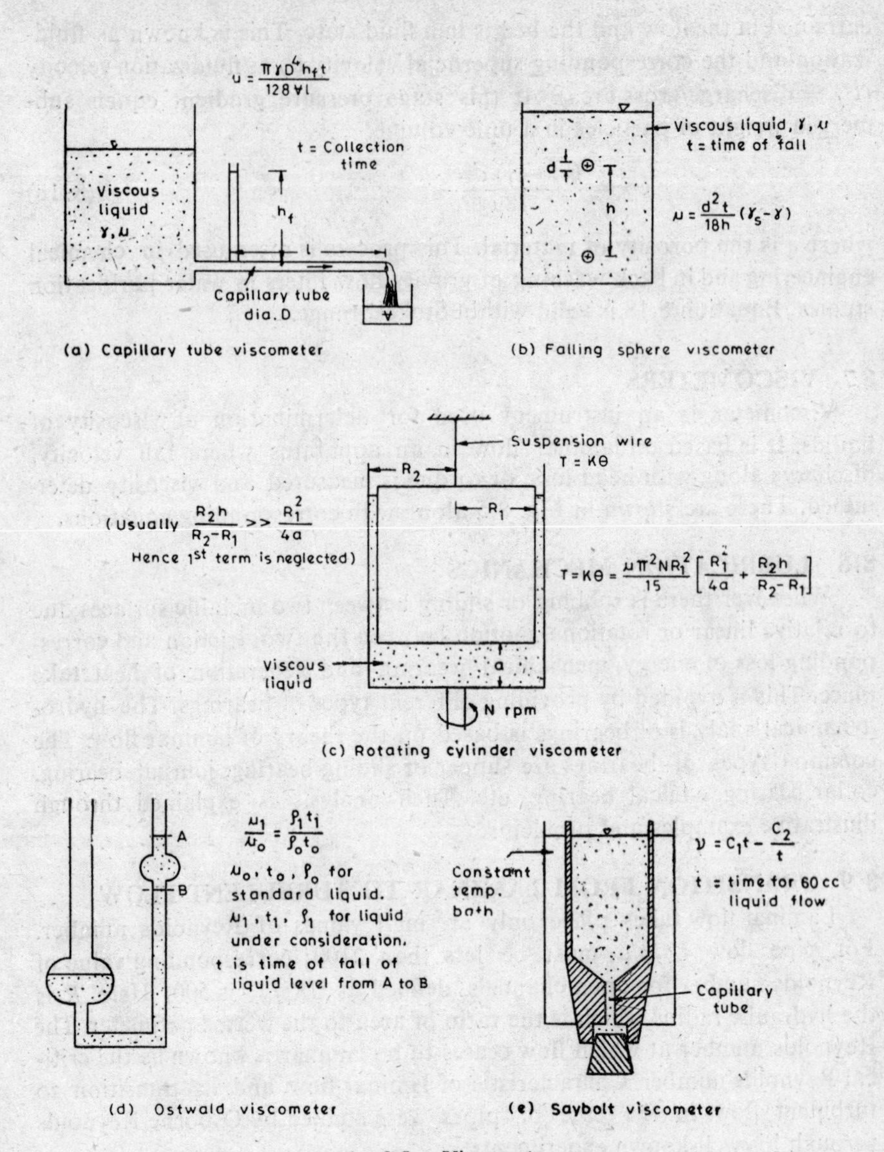

Fig. 8.5. Viscometers.

ILLUSTRATIVE EXAMPLES

8.1 Consider steady uniform laminar flow of an incompressible fluid between two parallel plates which are kept at distance B apart. Assume plates to be sufficiently wide in z direction. (Fig. 8.3). Reduce the N.S. equations to appropriate differential equation and obtain expression for velocity distribution and average velocity. Assume plates to be horizontal.

Because the flow is steady $\frac{\partial}{\partial t} = 0$; since flow is uniform $\frac{\partial}{\partial x} = 0$

$\left(\text{however, } \frac{\partial p}{\partial x} \text{ is not zero as explained below}\right)$; further $\frac{\partial}{\partial z} = 0$ since flow is two dimensional. Also it is obvious that $v = w = 0$. Therefore continuity equation gives

$$\frac{\partial u}{\partial x} = 0$$

which on integration yields $u = f(y)$ since it cannot be a function of x, z and t. Further, substitution of above conditions in Eqs. 8.1 gives

$$0 = -\frac{\partial p}{\partial x} + \mu \frac{\partial^2 u}{\partial^2 y}$$

$$0 = -\frac{\partial p}{\partial y}$$

$$0 = -\frac{\partial p}{\partial z}$$

The last two equations indicate that $p = p(x)$ i.e. it depends on x alone. This can mean $\frac{\partial p}{\partial x}$ to be dependent on x. However if $\frac{\partial p}{\partial x}$ is a function of x, solution of $0 = -\frac{\partial p}{\partial x} + \mu \frac{\partial^2 u}{\partial y^2}$ would yield $u = f(x, y)$ which is incompatible with solution of continuity equation which yielded $u = f(y)$. Hence one concludes that $\frac{\partial p}{\partial x}$ is constant or pressure varies linearly in x-direction.

Integration of the equation

$$\frac{\partial^2 u}{\partial y^2} = \frac{1}{\mu} \frac{\partial p}{\partial x}$$

yields

$$\frac{\partial u}{\partial y} = \frac{y}{\mu} \frac{\partial p}{\partial x} + c,$$

and

$$u = \frac{y^2}{2\mu} - \frac{\partial p}{\partial x} + c_1 y + c_2$$

where c_1 and c_2 are constants of integration. The boundary conditions are $u = 0$ when $y = 0$ and also $u = 0$ when $y = B$. The first condition gives $c_2 = 0$.

The second gives

$$c_2 = -\frac{B}{2\mu} \frac{\partial p}{\partial x}.$$

Hence

$$u = -\frac{1}{2\mu} \frac{\partial p}{\partial x} (By - y^2)$$

This gives the velocity distribution law. The discharge per unit width will be

$$q = \int_0^B u dy = -\frac{1}{2\mu} \frac{\partial p}{\partial x} \left[\frac{B^2 y}{2} - \frac{y^3}{3} \right]_0^B$$

or
$$q = -\frac{B^3}{12\mu}\frac{\partial p}{\partial x}$$

Since average velocity $U = \frac{q}{B}$,

$$U = \frac{-B^2}{12\mu}\frac{\partial p}{\partial x}$$

But
$$-\frac{\partial p}{\partial x} = (p_1 - p_2)/l \quad \therefore \quad p_1 - p_2 = \frac{12\mu U l}{B^2}$$

8.2 When a flat plate kept in an infinite viscous fluid is suddenly set in motion in its own plane at a velocity U, N.S. equations reduce to the differential equation (see Prob. 8.1)

$$\rho\frac{\partial u}{\partial t} = \mu\frac{\partial^2 u}{\partial y^2}.$$

Show that $\frac{\partial \tau}{\partial t} = \nu\frac{\partial^2 \tau}{\partial y^2}$, where τ is shear in the x direction on a plane perpendicular to y axis.

$$\frac{\partial u}{\partial t} = \left(\frac{\mu}{\rho}\right)\frac{\partial^2 u}{\partial y^2}$$

Differentiate this equation w.r.t. y to get

$$\frac{\partial^2 u}{\partial y \partial t} = \left(\frac{\mu}{\rho}\right)\frac{\partial^3 u}{\partial y^3};$$

$$\therefore \quad \frac{\partial}{\partial t}\left(\frac{\partial u}{\partial y}\right) = \frac{1}{\rho}\frac{\partial^2}{\partial y^2}\left(\mu\frac{\partial u}{\partial y}\right) \quad \text{if } \mu \text{ is constant.}$$

Writing $\frac{\partial u}{\partial y} = \frac{\tau}{\mu}$ on left hand side of the above equation one obtains

$$\frac{1}{\mu}\frac{\partial \tau}{\partial t} = \frac{1}{\rho}\frac{\partial^2 \tau}{\partial y^2} \quad \text{or} \quad \frac{\partial \tau}{\partial t} = \nu\frac{\partial^2 \tau}{\partial y^2}$$

8.3 For steady laminar flow between parallel plates, $B = 2.0$ mm, $\mu = 5 \times 10^{-2}$ kg/m s and $\frac{\partial p}{\partial x} = -10$ kN/m³. Determine q, maximum shear and maximum velocity.

Since
$$u = \frac{1}{2\mu}\frac{\partial p}{\partial x}(By - y^2)$$

$$\frac{\partial u}{\partial y} = -\frac{1}{2\mu}\frac{\partial p}{\partial x}(B - 2y)$$

$$\therefore \quad \tau = \mu\frac{\partial u}{\partial y} = \frac{1}{2}\frac{\partial p}{\partial x}(B - 2y)$$

Maximum shear τ will occur at $y = 0$.

$$\therefore \quad \tau = -\frac{B}{2}\frac{\partial p}{\partial x} = \frac{2 \times 10^{-3}}{2} \times 10{,}000 = 10 \text{ N/m}^2$$

Average velocity
$$U = -\frac{B^2}{12\mu}\frac{\partial p}{\partial x}$$

$$= \frac{0.002^2}{12 \times 5 \times 10^{-2}} \times 10000 = 0.0667 \text{ m/s}$$

$$q = UB = 0.0667 \times .002$$

$$= 1.333 \times 10^{-4} \text{ m}^3\text{/s m} \quad \text{or} \quad 0.1333 \text{ l/s m}$$

Maximum velocity will occur at $y = B/2$,

$$u_m = -\frac{1}{8\mu}\frac{\partial p}{\partial x}B^2 = \frac{10{,}000 \times 0.002^2}{8 \times 5 \times 10^{-2}} = 0.10 \text{ m/s}$$

8.4 For steady laminar flow in a circular pipe, obtain expression for average velocity and power expended.

Consider a cylindrical element of fluid of radius r and length δx. The forces acting on it are due to shear and pressure as shown in Fig. 8.6. For steady uniform flow, the summation of the forces on the element in the direction of flow must be zero since there is no acceleration. Hence

$$\pi r^2 p - \left(p + \frac{\partial p}{\partial x}\delta x\right)\pi r^2 - 2\pi r \delta x \tau = 0$$

or

$$-\frac{r}{2}\frac{\partial p}{\partial x} = \tau$$

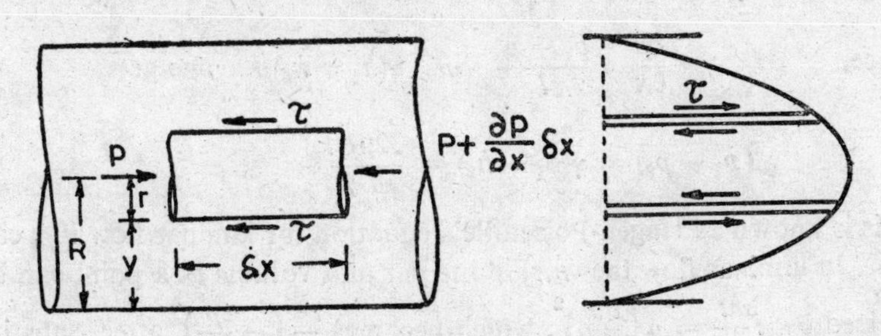

Fig. 8.6 Balance of forces in pipe flow

However $\tau = \mu \dfrac{\partial u}{\partial y}$ and since $y = R - r$, $dy = -dr$

$$\tau = -\mu \frac{\partial u}{\partial r};$$

substitution of this in the above equation yields

$$\frac{\partial u}{\partial r} = \frac{r}{2\mu}\frac{\partial p}{\partial x}$$

Integration of this equation yields

$$u = \frac{r^2}{4\mu}\frac{\partial p}{\partial x} + c$$

where c is the constant of integration, which can be determined from the boundary condition $u = 0$.

When $r = R$. Hence $\quad c = -\dfrac{R^2}{4\mu}\dfrac{\partial p}{\partial x}$

and $$u = -\frac{1}{4\mu}\frac{\partial p}{\partial x}(R^2 - r^2)$$

This gives the distribution of velocity in the vertical. The maximum velocity u_m will occur at $r = 0$ and it will be

$$u_m = -\frac{R^2}{4\mu}\frac{\partial p}{\partial x}$$

$$Q = \int_0^R 2\pi r dr\, u$$

$$= -\frac{1}{4\mu}\left(\frac{\partial p}{\partial x}\right) 2\pi \left(R^2\frac{r^2}{2} - \frac{r^4}{4}\right)_0^R$$

or $$Q = -\frac{\pi R^4}{8\mu}\frac{\partial p}{\partial x}$$

$$\therefore \qquad \frac{Q}{\pi R^2} = U = -\frac{R^2}{8\mu}\frac{\partial p}{\partial x} = -\frac{D^2}{32\mu}\frac{\partial p}{\partial x}$$

Since, $\quad -\dfrac{\partial p}{\partial x} = \dfrac{p_1 - p_2}{l}$ or $\gamma(h_1 - h_2)/l$, one gets

$$(p_1 - p_2) = \gamma(h_1 - h_2) = \frac{32\mu U l}{D^2}$$

This is known as Hagen-Poiseuille's equation for laminar flow in a circular pipe. In laminar flow the work done per unit volume at a point can be expressed by $\tau\dfrac{du}{dy} = \mu\left(\dfrac{du}{dy}\right)^2$, which becomes $\dfrac{r^2}{4u}\left(-\dfrac{\partial p}{\partial x}\right)^2$ after substitution of expression for $\left(\dfrac{\partial u}{\partial y}\right)$.

$$\therefore \quad \text{Work done per second} = \int_0^R 2\pi r dr\,\frac{r^2}{4u}\left(-\frac{\partial p}{\partial x}\right)^2 = \frac{\pi R^4}{8\mu}\left(\frac{\partial p}{\partial x}\right)^2$$

But $$Q = \frac{\pi D^4(-\partial p/\partial x)}{128\mu} = \frac{\pi R^2}{8\mu}\left(-\frac{\partial p}{\partial x}\right)$$

Hence $$P = Q\left(-\frac{\partial p}{\partial x}\right).$$

8.5 Oil of mass density 800 kg/m³ and dynamic viscosity 0.02 poise flows through 50 mm diameter pipe of length 500 m at the rate of 0.19 l/s. Determine (i) Reynolds number of flow, (ii) centreline velocity, (iii) pressure gradient, (iv) loss of pressure in 500 m length, (v) wall shear stress and (vi) power required to maintain the flow.

$$U = \frac{Q}{\pi D^2/4} = \frac{0.19 \times 10^{-3}}{0.785 \times 0.05^2} = 0.0968 \text{ m/s}$$

$$\therefore \qquad \text{Re} = \frac{UD\rho}{\mu} = (0.0968 \times 0.05 \times 800)/0.002$$

$$= 1936.3$$

Hence the flow is laminar

$$u_m = 2U = 2 \times 0.0968 = 0.1936 \text{ m/s}$$

$$(p_1 - p_2) = 32\mu Ul/D^2$$

$$= \frac{32 \times 0.002 \times 0.0968 \times 500}{0.05^2} = 1239.04 \text{ N/m}^2$$

$$\therefore \qquad \frac{\partial p}{\partial x} = \frac{p_1 - p_2}{l} = \frac{1239.04}{500} = 2.478 \text{ N/m}^3$$

$$\tau_0 = (p_1 - p_2)\frac{D}{4l}$$

$$= (1239.04) \times 0.05/(4 \times 500)$$

$$= 0.03098 \text{ N/m}^2$$

Power required $= Q(p_1 - p_2)$

$$= 0.19 \times 10^{-3} \times 1239.04$$

$$= 0.2354 \text{ W}$$

8.6 Obtain expression for Darcy-Weisbach friction factor f for laminar flow in a pipe.

Darcy-Weisbach friction factor f is defined by the relation $f = 4\tau_0 \Big/ \frac{\rho U^2}{2}$ where τ_0 is the wall shear and U is the average velocity of flow (see Chapter X). For laminar flow in a pipe,

$$u = \frac{-1}{4\mu}\frac{\partial p}{\partial x}(R^2 - r^2) \text{ and } U = -\frac{R^2}{8\mu}\frac{\partial p}{\partial x}$$

as shown in Example 8.3. Hence

$$u = \frac{2U}{R^2}(R^2 - r^2)$$

$$\therefore \qquad \tau_0 = \mu\left(-\frac{\partial u}{\partial r}\right)_{r=R} = \frac{2U}{R^2}\mu \times 2R = \frac{4\mu U}{R} = \frac{8\mu U}{D}$$

$$\therefore \qquad f = \frac{4 \times 8\mu U}{D}\frac{2}{\rho U^2} = \frac{64\mu}{UD\rho} \text{ or } \frac{64}{\text{Re}}$$

8.7 Kerosene oil flows upwards through inclined parallel plates at the rate of 2.0 l/s per metre width. The distance between the parallel plates is 10 mm and the inclination of plates is 20° with the horizontal. Determine the difference of pressure between two sections 10 m apart. Take $\rho = 800$ kg/m³ and $\mu = 2 \times 10^{-3}$ kg/m s.

$$U = \frac{q}{B} = \frac{2 \times 10^{-3}}{10 \times 10^{-3}} = 0.20 \text{ m/s}$$

$$\therefore \quad (h_1 - h_2) = \frac{12\mu U l}{\gamma B^2} = \frac{12 \times 2 \times 10^{-3} \times 0.20 \times 10}{9.806 \times 800 \times 0.01^2}$$

$$= 0.0612 \text{ m}$$

But $$h_1 - h_2 = \left(\frac{p_1}{\gamma} + z_1\right) - \left(\frac{p_2}{\gamma} + z_2\right)$$

and $$z_2 - z_1 = 10 \sin 20° = 10 \times 0.3420 = 3.42 \text{ m}$$

$$\therefore \quad 0.0612 = \left(\frac{p_1}{\gamma} - \frac{p_2}{\gamma}\right) - 3.42$$

$$\therefore \quad (p_1 - p_2) = 3.4812 \times 9.806 \times 800$$

$$= 27\ 309.32 \text{ N/m}^2$$

8.8 Determine the optimum diameter of the pipe required to carry 100 l/s of crude oil ($\rho = 950$ kg/m³, $\mu = 8 \times 10^{-2}$ kg/m s) and still maintaining laminar flow. Also determine the power required for its transport over one kilometre.

$$\text{Re} = \frac{UD\rho}{\mu} = \frac{4Q\rho}{\pi D\mu}$$

$$\therefore \quad 2100 = (4 \times 0.10 \times 950)/(3.142 \times D \times 8 \times 10^{-2})$$

$$\therefore \quad D = 0.720 \text{ m}$$

$$\Delta p = \frac{128 Q\mu l}{\pi D^4} = \frac{128 \times 0.10 \times 8 \times 10^{-2} \times 1000}{3.142 \times 0.72^4}$$

$$= 1212.73 \text{ N/m}^2$$

$$= Q \times \Delta p = 0.1 \times 1212.73 \text{ W}$$

$$= 121.273 \text{ W} \quad \text{or} \quad 0.213 \text{ kW.}$$

8.9 Consider steady uniform flow between two parallel plates kept distance B apart under the action of pressure gradient. If the top plate moves at uniform velocity $\bar{U}$ in x direction, obtain generalised velocity distribution law. This is known as generalised Couette flow. Also determine the pressure gradient required for zero flow.

In Example 8.1 it is shown that flows between parallel plates is governed by the differential equation

$$\frac{\partial^2 u}{\partial y^2} = \frac{1}{\mu}\frac{\partial p}{\partial x}$$

Integration of this equation twice w.r.t. y yields

$$\frac{\partial u}{\partial y} = \frac{y}{\mu}\frac{\partial p}{\partial p} + c_1$$

$$u = \frac{y^2}{2\mu}\frac{\partial p}{\partial x} + c_1 y + c_2$$

where the constants of integration c_1 and c_2 can be obtained from the boundary conditions $u = 0$ at $y = 0$ and $u = \bar{U}$ at $y = B$

$$\therefore \qquad c_2 = 0 \text{ and } c_1 = \left(\frac{\bar{U}}{B} - \frac{B}{2\mu}\frac{\partial p}{\partial x}\right)$$

Hence velocity distribution law becomes

$$u = \frac{y^2}{2\mu}\frac{\partial p}{\partial x} - \frac{By}{2\mu}\frac{\partial p}{\partial x} + \bar{U}\frac{y}{B}$$

or

$$\frac{u}{\bar{U}} = \frac{y}{B} - \frac{1}{2\mu\bar{U}}\frac{\partial p}{\partial x}(By - y^2)$$

$$= \frac{y}{B} + \left(-\frac{B^2}{2\mu\bar{U}}\frac{\partial p}{\partial x}\right)\left(\frac{y}{B} - \frac{y^2}{B^2}\right)$$

This velocity distribution can be plotted as $\frac{u}{\bar{U}}$ vs $\frac{y}{B}$ for various values of $P = \left(-\frac{B^2}{2\mu\bar{U}}\frac{\partial p}{\partial x}\right)$ as shown in Fig. 8.7. When the parameter P is zero i.e. there is no pressure gradient, velocity distribution follows linear flow $u/\bar{U} = y/B$. This flow in known as simple Couette flow.

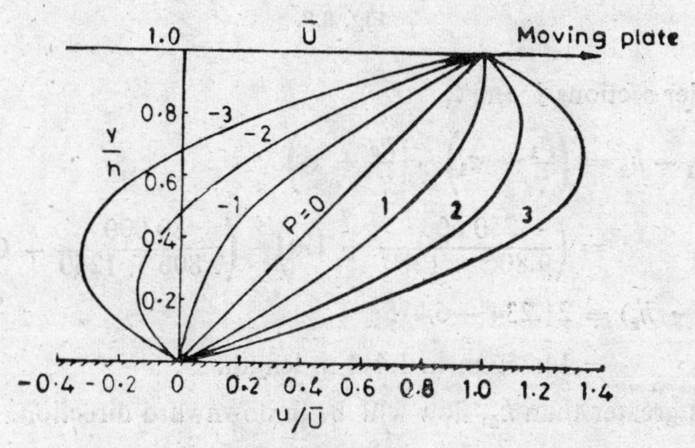

Fig. 8.7. Velocity distribution in Couette flow.

The equation $\frac{u}{\bar{U}} = \frac{y}{B} + P\left(\frac{y}{B} - \frac{y^2}{B^2}\right)$ can be integrated over the depth B to obtain the expression for discharge per unit width B as

$$\frac{q}{\bar{U}} = B\left(\frac{1}{2} + \frac{P}{6}\right)$$

Hence for $q = 0,\ p = -3$ or $\dfrac{\partial p}{\partial x} = \dfrac{6u\bar{U}}{B^2}$

8.10 Laminar flow takes place between parallel plates 10 mm apart. The plates are inclined at 45° with the horizontal. For oil of viscosity 0.90 kg/m s and mass density of 1260 kg/m³, the pressures at two points 1.0 m vertically apart are 80 kN/m² and 250 kN/m² when the upper plate moves at 2.00 m/s velocity relative to the lower plate but in opposite direction to flow (Fig. 8.8). Determine (i) velocity distribution (ii) maximum velocity and (iii) shear stress on the top plate.

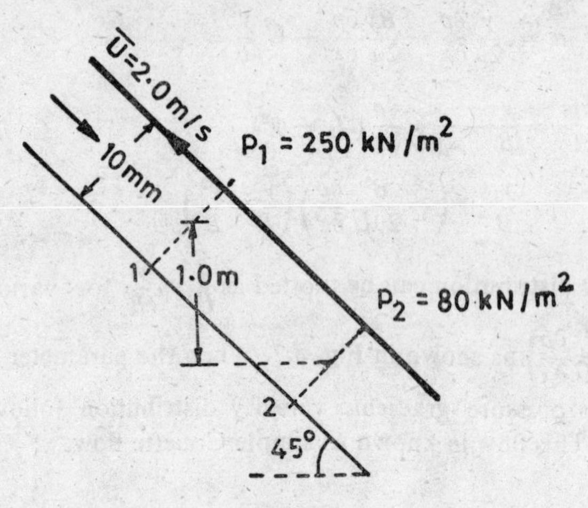

Fig. 8.8.

Consider sections 1 and 2

$$h_1 - h_2 = \left(\frac{p_1}{\gamma} + z_1\right) - \left(\frac{p_2}{\gamma} + z_2\right)$$

$$= \left(\frac{250\,000}{9.806 \times 1260} + 1.0\right) - \left(\frac{80\,000}{9.806 \times 1260} + 0\right)$$

or $\quad (h_1 - h_2) = 21.234 - 6.475$

$$= 14.759 \text{ m in } 1.414 \text{ m length.}$$

Since h_1 is greater than h_2, flow will be in downward direction.

$$\frac{\partial h}{\partial x} = -\frac{14.759}{1.414} = -10.438$$

and $\quad \dfrac{\partial p}{\partial x} = \gamma \dfrac{\partial h}{\partial x} = -10.438 \times 1260 \times 9.806$

$$= -128.967.33 \text{ N/m}^2 \quad \text{or} \quad -128.97 \text{ kN/m}^2$$

Using results from Example 8.9

$$\frac{u}{\bar{U}} = \frac{y}{B} + \left(\frac{-B^2}{2\mu\bar{U}}\frac{\partial p}{\partial x}\right)\left(\frac{y}{B} - \frac{y^2}{B^2}\right)$$

But $\bar{U} = -2.0$ m/s, $B = 0.01$ m, $\mu = 0.90$ kg/m s and $\frac{\partial p}{\partial x} = -128\ 967.33$ N/m². Hence above equation yields

$$u = 516.486y - 71648.6y^2$$

To find the value of y at which u is maximum, set $\frac{du}{dy} = 0$

$$\frac{du}{dy} = 516.486 - 143297.2y = 0 \quad \text{or} \quad y = 3.604 \times 10^{-3}\ \text{m}$$

$$\therefore \quad u_m = (516.486 \times 0.003\ 604) - (71\ 648.2 \times 0.003\ 604^4)$$

$$= 1.8614 - 0.9306 = 0.9308\ \text{m/s}$$

$$\tau_0 = u\left(\frac{du}{dy}\right)_{y=0\cdot01} = 0.90\ (516.486 - 143\ 297.2 \times 0.01)$$

$$= -824.837\ \text{N/m}^2$$

8.11 For steady laminar flow, determine an expression for pressure loss across a conical contraction (Fig. 8.9).

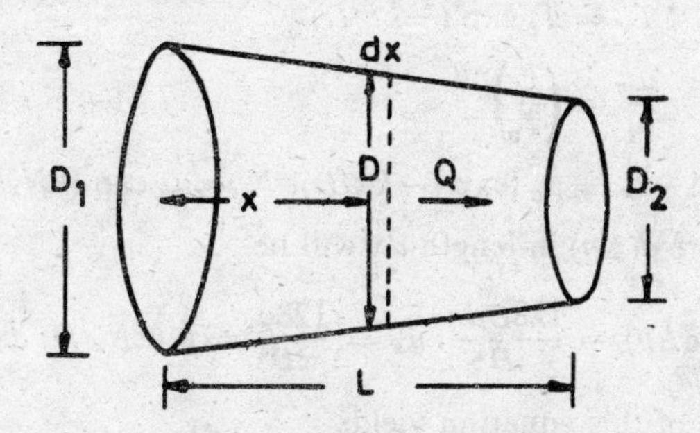

Fig. 8.9. Conical contraction.

From geometry it can be seen that

$$D = D_2 + (D_1 - D_2)\frac{(l - x)}{l}$$

Since for laminar flow in circular pipes the loss of pressure is given by Hagen-Poiseuille's equation, loss of pressure in length dx will be given by

$$d(\Delta p) = \frac{128\mu Q}{\pi D^4}\,dx$$

where D is given by the above equation. Further,

$$dD = \frac{-(D_1 - D_2)}{l} dx \quad \text{or} \quad dx = \frac{-l\, dD}{(D - D_2)}$$

Substitution of the values of dx in the expression for $d(\Delta p)$ yields

$$d(\Delta p) = \frac{-128\mu Q}{\pi D^4} \frac{l}{(D_1 - D_2)} dD$$

Integration of this equation gives

$$\Delta p = \frac{128uQ}{3\pi} \frac{l}{(D_1 - D_2)} \left(\frac{1}{D^3}\right)_{D_1}^{D_2}$$

or

$$\Delta p = \frac{128\mu Q l}{3\pi (D_1 - D_2)} \left(\frac{1}{D_2^3} - \frac{1}{D_1^3}\right)$$

8.12 When heated oil is pumped from the oil well to its destination through an uninsulated pipe line, the temperature of oil T_x at any distance x decreases along the pipe length due to heat transfer to surrounding environment. As a first approximation assume that the temperature variation can be expressed as $T_x/T_0 = \exp(-kx/D)$ where T_0 is temperature at $x = 0$ and k is a constant. Further dynamic viscosity of oil can be assumed to vary with temperature as $\mu \sim T^{-N}$ where N is constant. Obtain an expression for pressure loss in length x if p_0, T_0, μ_0 represent the pressure, temperature (in K) and viscosity at inlet for discharge Q flowing in pipe of diameter D.

$$T_x = T_0 \exp(-kx/D)$$

and

$$\frac{\mu_x}{\mu_0} = \left(\frac{T_x}{T_0}\right)^{-N}$$

$$\therefore \quad \mu_x = \mu_0 [\exp(-kx/D)]^{-N} = \mu_0 \exp(kN_x/D)$$

Loss of pressure $d(\Delta p)$ in length dx will be

$$d(\Delta p) = \frac{128 Q dx}{\pi D^4} \cdot \mu_x = \frac{128\mu_0}{\pi D^4} \exp(kN_x/D)\, dx$$

Integration of this equation yields

$$\Delta p = (p_0 - p_x) = \frac{128 Q\, \mu_0}{\pi D^4} \frac{D}{kN} [\exp(kNx/D]_0^x$$

$$= \frac{128 Q \mu_0}{\pi D^3\, kN} [\exp(kN_x/D) - 1]$$

8.13 To determine the viscosity of benzene, an experiment was conducted using capillary tube viscometer (Fig. 8.5a). The data obtained were

Capillary tube diameter $D = 2.0$ mm

Length of tube $l = 0.50$ m

Head loss $h_f = 0.20$ m

Volume collected $\forall = 189$ cc

Time of collection $t = 100$ s

Mass density of benzene $\rho = 860$ kg/m³

Determine the dynamic viscosity of benzene.

$$Q = \frac{189}{100} = 1.89 \text{ cc} = 1.89 \times 10^{-6} \text{ m}^3\text{/s}$$

$$\therefore \qquad \mu = \pi\gamma h_f D^4/128Ql$$

$$\mu = \frac{3.142 \times 860 \times 9.806 \times 0.20 \times (2 \times 10^{-3})^4}{128 \times 1.89 \times 10^{-6} \times 0.5}$$

$$= 7.010 \times 10^{-4} \text{ kg/m s}$$

8.14 A sphere of diameter 1.0 mm falls through 335 mm in 100 s in a viscous liquid. If the relative densities of the sphere and the liquid are 7.0 and 0.96 respectively, determine the dynamic viscosity of the liquid.

Assume that Stokes' law is valid. Hence

$$\omega = \frac{d^2}{18\mu}(\gamma_s - \gamma) \quad \text{and} \quad \omega = \frac{335}{100} = 3.35 \text{ mm/s}$$

$$\therefore \qquad \mu = \frac{d^2}{18\omega}(\gamma_s - \gamma)$$

$$= \frac{(1 \times 10^{-3})^2 \times 998 \times 9.806(7.00 - 0.96)}{18 \times 3.35 \times 10^{-3}}$$

$$= 0.9803 \text{ kg/m s}$$

Check the value of Re.

$$\text{Re} = \frac{3.35 \times 10^{-3} \times 1 \times 10^{-3} \times 998 \times 0.96}{0.9803} = 0.00327$$

Hence the assumption about applicability of Stokes' law is valid.

8.15 A 20 mm diameter lead sphere of relative density 11.38 has a terminal fall velocity of 0.30 m/s in an oil of relative density 0.90. Determine the viscosity of the oil.

Assume Stokes' law to be valid. Hence

$$\mu = \frac{d^2}{18\omega}(\gamma_s - \gamma) = \frac{(0.02)^2}{18 \times 0.3}(11.38 - 0.90) \times 998 \times 9.806$$

$$= 7.597 \text{ kg/m s}$$

$$\therefore \qquad \text{Re} = \frac{0.02 \times 0.30 \times 0.9 \times 998}{7.597} = 0.709$$

Since this is greater than 0.10, Stokes' law is not applicable. Hence use Oseen's equation,

$$C_D = \frac{24}{\text{Re}}\left(1 + \frac{3}{16}\text{Re}\right)$$

$$C_D = \frac{4d}{3}\frac{(\gamma_s - \gamma)}{\rho\omega^2}$$

$$= \frac{4 \times 0.02 \times 10.48 \times 998 \times 9.806}{3 \times 0.90 \times 998 \times 0.3^2} = 33.833$$

Substituting this value in Oseen's equation, one gets

$$33.833 = \frac{24}{\text{Re}} + 4.5$$

or $$\text{Re} = \frac{\omega d\rho}{\mu} = \frac{24}{(33.833 - 4.5)}$$

$$= 0.818$$

$$\therefore \quad \mu = \frac{0.30 \times 0.02 \times 998 \times 0.9}{0.818} = 6.588 \text{ kg/m s}$$

8.16 Determine the fall velocity of 0.06 mm sand particles at 20°C in water when their concentration is 0.20 by volume.

Fall velocity of single particle can be determined from Stokes' law.

$$\omega = \frac{d^2}{18\mu}(\gamma_s - \gamma) = \frac{0.06^2 \times 10^{-6} \times 9787 \times 1.65}{18 \times 10^{-3}}$$

$$= 3.230 \times 10^{-3} \text{ m} \quad \text{or} \quad 3.23 \text{ mm/s}$$

$$\text{Re} = \frac{\omega d\rho}{\mu} = \frac{3.23 \times 10^{-3} \times 0.06 \times 10^{-3} \times 998}{1 \times 10^{-3}}$$

$$= 0.193$$

Hence it can be assumed that Stokes' equation is valid. For using Scholl's equation, one must determine c

$$c = \frac{1}{1.0 + 0.20} = 0.833$$

Since c is greater than 0.60, $\quad n = 3.65$

$$\therefore \quad \frac{\omega_c}{\omega} = c^n = 0.833^{3.65}$$

$$\therefore \quad \omega_c = 1.658 \text{ mm/s}$$

To use Moude and Whitmore's method, compute Re

$$\text{Re} = 0.193 \quad \therefore \quad n = 4.4(0.193)^{-0.08} = 5.019$$

$$\therefore \quad \omega_c = 3.230 \times 10^{-3} \times (0.833)^{5.019}$$

$$= 1.291 \times 10^{-3} \text{ m/s}$$

$$\therefore \quad \text{Re} = \frac{1.291 \times 10^{-3} \times 0.06 \times 10^{-3} \times 998}{1 \times 10^{-3}} = 0.0771$$

$$\therefore \quad n = 4.4(0.0771)^{-0.08} = 5.40$$

$$\therefore \quad \omega_c = 3.23 \times 10^{-3}(0.833)^{5.40} = 1.204 \times 10^{-3} \text{ m/s}$$

$\therefore$ Now Re value $= 1.204 \times 10^{-3} \times 0.06 \times 10^{-3} \times 998/1 \times 10^{-3}$

$$= 0.0719$$

$$\therefore \quad n = 4.4(0.0719)^{-0.08} = 5.431$$

$$\therefore \quad \omega_c = \omega(0.833)^{5.431} = 1.197 \times 10^{-3} \text{ m/s}$$

Since the two successive values of ω_c namely 1.204×10^{-3} and 1.197×10^{-3} are sufficiently close, no further refinement is necessary. Note that as an approximation, value of n in Moude and White's equation has been obtained from $n = 4.4(\text{Re})^{-0.08}$ even though Re < 1.0.

8.17 Design a circular basin as a settling tank for removal of flocculating material from water to be supplied to the city at the rate of 10,000 m^3/day for drinking purposes. Flocculating particles are produced by coagulation (by addition of alum, etc.) and experimentally it is found that 20 m/day (known as overflow rate) velocity will effect satisfactory removal at a depth of 4.0 m. Determine the size of settling tank and inflow arrangement.

Assume a single tank is provided. Since $Q = UA$

$$A = \frac{10000}{20} = 500 \text{ m}^2 = \frac{\pi D^2}{4}$$

$$\text{Diameter of tank} = \left(\frac{500 \times 4}{3.142}\right)^{1/2} = 25.23 \text{ m, say 25 m}$$

The liquid should enter at the centre near the surface without much turbulence. Hence inlet arrangement is through a riser as shown in Fig. 8.10.

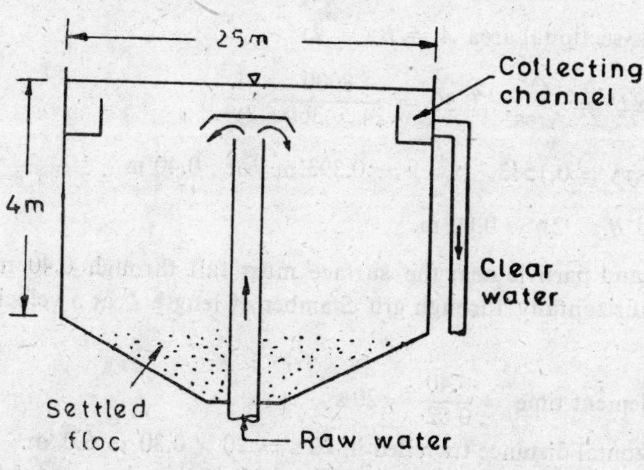

Fig. 8.10. Settling tank.

Also water will be collected through a circular channel along the periphery. Of course some arrangement in the form of a rotating arm and scraper is to be made to remove settled material.

The concept of overflow rate can be explained with the help of rectangular settling basin. Consider a discharge Q entering in such a basin of width B, length L and depth h. The flow will be detained in the basin for time t

$$t = \frac{LBh}{Q}$$

If ω is fall velocity of particle to be removed, it must fall through distance h in this time. Hence

$$t = \frac{h}{\omega}$$

Equating the two expressions for t one gets

$$\frac{LBh}{Q} = \frac{h}{\omega} \quad \text{or} \quad \omega = \frac{Q}{LB}$$

The term discharge/surface area of basin is known as the overflow rate and it corresponds to settling velocity of particle that is 100 percent removed.

8.18 In the treatment of waste water from cities. a channel type grit-chamber or settling tank is provided to remove particles of 0.02 mm diameter and relative density 2.65. Their fall velocity is found to be 0.20 m/s. Design the channel dimensions if flow through velocity of 0.30 m/s is to be maintained, for a design discharge of 8000 m³ per day.

Assume channel width = 2 × depth

$\therefore$ Cross-sectional area $A = By = 2y^2$

$$\text{Since } U = \frac{Q}{\text{Area}}, \quad 0.30 = \frac{8000}{24 \times 3600} \frac{1}{2y^2}$$

$$\therefore \quad y^2 = 0.1543 \quad \therefore \quad y = 0.393 \text{ m} \quad \text{or} \quad 0.40 \text{ m}$$

and $B = 2y = 0.80$ m.

The sand particle near the surface must fall through 0.40 m while it travels horizontally through grit chamber of length L at a velocity of 0.30 m/s.

$$\therefore \quad \text{Settlement time} = \frac{0.40}{0.02} = 20 \text{ s}$$

Horizontal distance travelled in 20 s = 20 × 0.30 = 6.0 m.

$\therefore$ The grit chamber should be 0.80 m wide, 0.4 m deep and 6.0 m long.

It must be mentioned that the above analysis is much simplified. In the actual case, the fall velocity of particles will be affected by their concentration in flow as indicated by Eq. 8.16 and also by turbulence in the grit-chamber or settling basin. Hence actual fall velocity is very difficult to determine. One can arbitrarily reduce the fall velocity of a single particle by a factor say 2.0, to take into account the effect of turbulence and concentration and design the chamber.

8.19 Briefly discuss the working of settling chamber for removal of fine particles from air.

In an air pollution control system also, principle of sedimentation is used to remove fine suspended particles. Normallv, enlarged area settling chambers are provided in which average velocity is reduced to less than 0.30 m/s (Fig. 8.11).

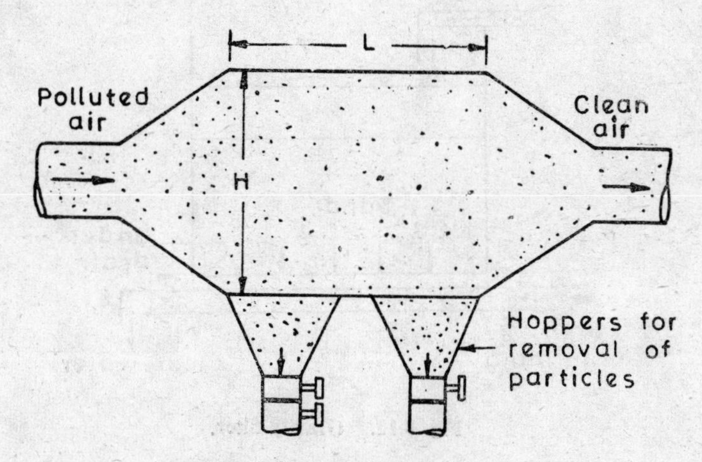

Fig. 8.11. Settling chamber.

If H is height of chamber

$$\frac{H}{\omega} = \text{time } t \text{ required for settling}$$

$$\text{If } U = 0.30 = \frac{L}{t}$$

$$\therefore \quad \frac{L}{U} = \frac{H}{\omega}. \text{ If fall of particle is in Stokes' range,}$$

$$\frac{UH}{L} = \omega = \frac{d^2}{18\mu}(\gamma_s - \gamma_a)$$

where γ_s and γ_a are specific weights of particle and air. Neglect γ_a since it is too small.

$$\therefore \quad d = \left(\frac{18\mu UH}{L\gamma_s}\right)^{1/2}$$

Because the flow in the settle chamber is not quiet and particles are falling in high concentrated flow, the fall velocity is normally reduced by a factor of two (arbitrarily). Hence above equation becomes

$$d = \left(\frac{18 \times 2\mu UH}{L\gamma_s}\right)^{1/2}$$

These chambers are found to remove particles upto 0.05 to 0.02 mm in diameter.

8.20 What will be the water discharge through 0.60 m thick filter of cross-sectional area 15 m² if it is made of 0.20 mm diameter uniform sand? Assume head loss through the filter as 0.20 m, water viscosity 1×10^{-3} kg/m s and porosity of filter 0.50 (Fig. 8.12).

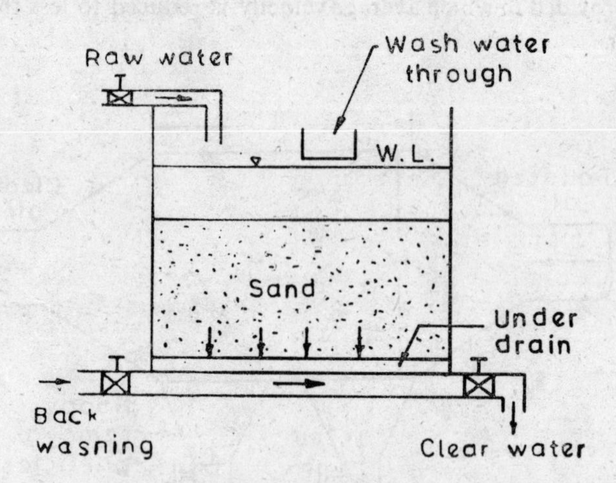

Fig. 8.12. Gravity filter.

Since coefficient of permeability K is given

$$K = \frac{d^2\gamma e^2}{18\mu\,(1-e)}$$

$$K = \frac{(0.2 \times 10^{-3})^2 \times 9787 \times 0.5^2}{18 \times 10^{-3} \times 0.50} = 0.0109 \text{ m/s}$$

$\therefore$ $$U = -Ki$$

$$= 0.0109 \times \frac{0.20}{0.60} = 0.00363 \text{ m/s}$$

Check value of $$\text{Re} = \frac{0.00363 \times 0.0002 \times 998}{10^{-3}} = 0.725$$

Since it is less than unity, flow is laminar.

$\therefore$ $$Q = AU = 15 \times 0.000363 = 0.05445 \text{ m}^3/\text{s}$$

When filter gate clogged due to arresting of matter, it is to be washed. The arrangement for backwashing is shown in Fig. 8.12.

8.21 Two fixed water bodies with constant depths h_0 and h_1 are separated by a permeable medium of permeability K. Obtain expression for discharge per unit width (Fig. 8.13).

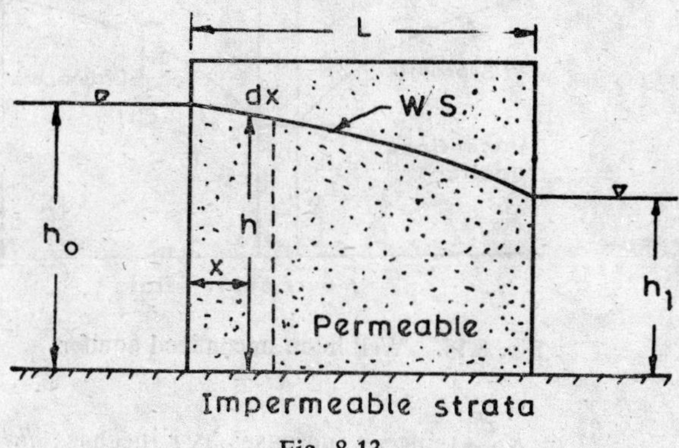

Fig. 8.13.

According to Darcy's law

$$U = -Ki = -K\frac{dh}{dx}$$

For unit width, area = $(h \times 1)$

$$\therefore \qquad q = hU = -Kh\frac{dh}{dx}$$

or
$$hdh = -\frac{q}{K}dx$$

Integration of this equation yields

$$\frac{h^2}{2} = -\frac{q}{K}x + c$$

With the boundary condition $h = h_0$ at $x = 0$. Hence $c = \frac{h_0^2}{2}$ and

$$\frac{h^2}{2} = \frac{-qx}{K} + \frac{h_0^2}{2} \quad \text{or} \quad q = \frac{K}{2x}(h_0^2 - h^2)$$

When
$$x = l, h = h_1$$

$$\therefore \qquad q = \frac{K}{2l}(h_0^2 - h_1^2)$$

8.22 Obtain an expression for flow into a circular well of radius R_w in an unconfined aquifer.

An unconfined aquifer is one in which water table serves as the upper surface of zone of saturation (Fig. 8.14).

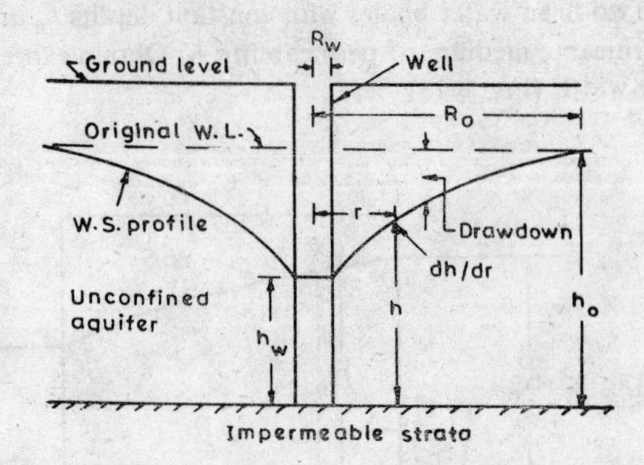

Fig. 8.14. Well in an unconfined aquifer.

$$U = K\frac{dh}{dr} \text{ (since } h \text{ increases as } r \text{ inceases } dh/dr \text{ is positive)}$$

Area of porous material at distance $r = 2\pi rh$

$$\therefore \quad Q = 2\pi rhK\frac{dh}{dr}$$

$$\therefore \quad hdh = \frac{Q}{2\pi r}\frac{dr}{K}$$

Integration of this equation yields

$$\frac{h^2}{2} = \frac{Q}{2\pi K}\ln r + c_1$$

or

$$h^2 = \frac{Q}{\pi K}\ln r + C$$

When $r = R_w, h = h_w$

$$\therefore \quad C = h_w^2 - \frac{Q}{\pi K}\ln R_w, \text{ and}$$

$$Q = K\pi\frac{(h^2 - h_w^2)}{\ln(r/R_w)}$$

If one substitutes $h = h_0$ when $r = R_0$

$$Q = K\pi\frac{(h_0^2 - h_w^2)}{\ln(R_0/R_w)}$$

The radius R_0 is known as the radius of influence, since h is affected upto this radius. Suggested values of R_0 range from 150 m to 300 m.

8.23 In a 300 mm diameter well in a confined aquifer of 15.0 m thickness and 30.0 m/day permeability, 10 m drawdown is observed with 150 m radius of influence. Determine the well discharge.

See problem 8.23 and Fig. 8.18. The discharge equation for a well in confined aquifer is

$$Q = 2\pi Kb \frac{(h_0 - h_w)}{\ln (R_0/R_w)}$$

Here $K = \dfrac{30.0}{24 \times 3600} = 3.472 \times 10^{-4}$ m/s, $h_0 - h_w = 10$ m

and $R_0/R_w = \dfrac{150}{0.150} = 1000$

$$\therefore \quad Q = \frac{2 \times 3.142 \times 0.000\,347\,2 \times 15 \times 10}{\ln (1000)} = 0.04\,738 \text{ m}^3/\text{s}$$

8.24 **Show that the general expression for incipient fluidization velocity is**

$$U_f^2 = \frac{4\,(\gamma_S - \gamma)\, g\, d\, e^3}{3\gamma C_D}$$

where C_D is the drag coefficient of sediment particle, with actual velocity in pores.

As mentioned in Section 8.5, when incipient fluidization velocity is reached, pressure gradient is equal to submerged weight of particles in a unit volume.

$$\frac{\partial p}{\partial z} = \gamma \frac{\partial h}{\partial z} = (1 - e)\,(\gamma_S - \gamma) \tag{1}$$

Under such a condition, the gradient $\dfrac{\partial h}{\partial z}$ will be related to the velocity also.

$$\text{Drag on individual particle} = f_D = \frac{\pi d^2}{4} C_D \frac{\rho u^2}{2}$$

where u is velocity in pores. The number of particles in unit volume, N is given by

$$N = (1 - e) / \frac{\pi d^3}{6}$$

$$\text{Total resisting force} = f_D N = \frac{\pi d^2}{4} C_D \frac{\rho u^2}{2} \times \frac{(1 - e)}{\pi d^3/6}$$

$$= \frac{3 C_D \rho u^2 (1 - e)}{4d}$$

Further, change in piezometric head will cause a force per unit volume in the direction of flow which will be balanced by the resisting force

$$\therefore \quad e\gamma \frac{\partial h}{\partial z} = \frac{3 C_D \rho u^2 (1 - e)}{4d} = \frac{3 C_D \rho U_f^2 (1 - e)}{4d\, e^2} \tag{2}$$

The term e with $\dfrac{\partial h}{\partial z}$ occurs because e is the pore space in unit volume; on the right hand side $u = U_f/e$, where U_f is fluidization velocity. Equating the two values of $\frac{\partial h}{\partial x}$ from Equations 1 and 2, one gets

$$\frac{3C_D\rho U_f^2(1-e)}{4e^3\rho g d} = \frac{(1-e)(\gamma_s-\gamma)}{\rho g}$$

$$\therefore \quad U_f^2 = \frac{4(\gamma s-\gamma)\, e^3\, d}{3C_D\rho} \quad \text{or} \quad \frac{4}{3}\frac{(\gamma s-\gamma)\, e^3\, dg}{C_D\gamma}$$

This relation can be simplified for two cases.

(i) When particles are small and flow around them is laminar,

$$C_D = \frac{24\mu e}{U_f d\rho}, \text{ substitution of which gives}$$

$$U_f^2 = \frac{(\gamma s-\gamma)\, d^2 e^2}{18\mu}$$

(ii) For reasonably large particles $C_D = 0.40$

$$\therefore \quad U_f^2 = 3.333\left(\frac{\gamma s-\gamma}{\gamma}\right) g\, e^3 d$$

8.26 Obtain an expression for torque on the inner cylinder of the rotating cylinder viscometer (Fig. 8.5 c).

When outer cylinder rotates at N rpm, velocity of fluid close to outer cylinder will be $2\pi R_2 N/60$, while that at the surface of inner cylinder (which will remain stationary under equilibrium condition) will be zero.

$$\therefore \quad \frac{du}{dy} = \frac{2\pi R_2 N}{60\,(R_2 - R_1)} \quad \text{and} \quad \tau_0 = \mu\frac{du}{dy} = \frac{2\pi R_2 N\mu}{60\,(R_2-R_1)}$$

Torque on the inner cylindrical surface = area $\times\, \tau_0 \times R_1$

$$= (2\pi R_1 h)\left(\frac{2\pi R_2 N\mu}{60\,(R_2-R_1)}\right) R_1$$

$$T_1 = \frac{\pi^2 R_1^2 R_2 h\, N\mu}{15\,(R_2-R_1)}$$

If a is small, there will be some contribution of torque from the bottom.

Shear at any distance r on the bottom $= \frac{2\pi r N\mu}{60a}$

Area on which this acts $= 2\pi r dr$

$\therefore$ Torque on small area $2\pi r dr$, $dT_2 = \frac{2\pi r N\mu}{60a}\, 2\pi r dr\, r$

or $$dT_2 = \frac{4\pi^2 N\mu}{60a}\, r^3 dr$$

Integration gives

$$T_2 = \frac{\pi^2 N\mu\, R^4}{60a}$$

$$\therefore \quad (T_1 + T_2) = T = \frac{\mu\pi^2 N R_1^2}{15}\left(\frac{R_1^2}{4a} + \frac{R_2 h}{R_2 - R_1}\right)$$

Normally a is kept large so that contribution of T_2 in T is very small and can be neglected. T is independently obtained from deflection of cylinder through an angle θ. $T = K\theta$ where K is coefficient of torsional resistance of the wire from which inner cylinder is suspended.

8.26 For a rotating cylinder viscometer

Inner diameter = 100 mm

Outer diameter = 108 mm

Depth of immersion of inner cylinder = 200 mm

Clearance at bottom = 4.0 mm

N = 100 rpm, Total torque = 0.393 Nm

Determine the viscosity of liquid.

$R_1 = 0.050$ m $\quad R_2 = 0.054$ m, $\quad h = 0.20$ m, $\quad a = 0.004$ m

$N = 100$ rpm $\quad T = 0.393$ Nm

Considering contribution from bottom to the torque also,

$$T = \frac{\mu\pi^2 N R_1^2}{15}\left(\frac{R_1^2}{4a} + \frac{R_2 h}{(R_2 - R_1)}\right)$$

$$= \frac{\mu \times 3.142^2 \times 100 \times 0.05^2}{15}\left(\frac{0.05^2}{4 \times 0.004} + \frac{0.054 \times 0.20}{0.004}\right)$$

$$\therefore \quad 0.393 = 0.1645\,\mu\,(0.1562 + 2.700)$$

$$= 0.4699\,\mu$$

$$\therefore \quad \mu = 0.836 \text{ kg/m s}$$

If torque on the bottom is neglected, the term 0.1562 in the bracket above will disappear.

$$\therefore \quad 0.393 = 0.1645 \times 2.70\,\mu$$

or

$$\mu = 0.885 \text{ kg/m s.}$$

8.27 Determine the power absorbed in the collar bearing shown in Fig. 8.15. Express it in kW.

At a distance r from the centre

$$\tau = \frac{2\pi r N}{60 b}\,\mu$$

$$\therefore \quad dF = 2\pi r dr \times \frac{2\pi r N}{60 b}\,\mu$$

$$dP = dF \times u$$

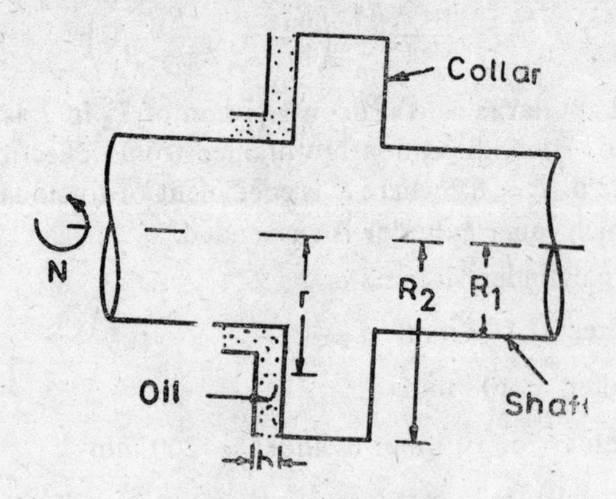

Fig. 8.15. Collar bearing.

$$dP = 2\pi r dr \times \frac{2\pi r N\mu}{60b} \times \frac{2\pi r N}{60}$$

$$= \frac{8\pi^3 N^2 \mu}{3000b} r^3 dr$$

$$\int dP = P = \frac{8\pi^3 N^2 \mu}{3600b} \left(\frac{r^4}{4}\right)_{R_1}^{R_2}$$

$$= \frac{\pi^3 N^2 \mu (R_2^4 - R_1^4)}{1800b} \text{ W} \quad \text{or} \quad \frac{\pi^3 N^2 \mu (R_2^4 - R_1^4)}{1800\,000b} \text{ kW}$$

8.28 A round wire of radius R_1 and length L is coated continuously by drawing it at a uniform velocity U through a pipe of radius R_2 containing the coating liquid. If liquid is maintained at a uniform pressure everywhere, obtain the equation for power required to draw it (Fig. 8.16).

For axisymmetric flow, one can use N.S. equation in cylindrical polar coordinate system, viz. Eq. 8.2 with the following conditions

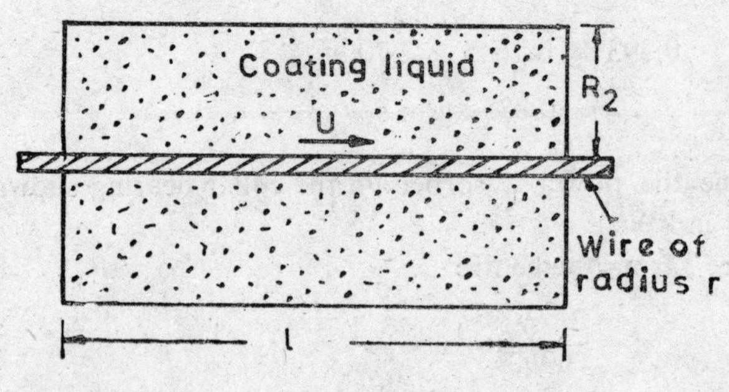

Fig. 8.16.

$$\frac{\partial}{\partial t} = 0, \frac{\partial}{\partial \theta} = 0, \frac{\partial}{\partial x} = 0$$

$$u_r = u_\theta = 0, \quad X_x = X_r = X_\theta = 0, p = \text{const.}$$

Hence Eq. 8.2 reduces to

$$\mu \left(\frac{\partial^2 u_x}{\partial r^2} + \frac{1}{r} \frac{\partial u_x}{\partial r} \right) = 0$$

or

$$\left(\frac{\partial^2 u_x}{\partial r^2} + \frac{1}{r} \frac{\partial u_x}{\partial r} \right) = 0 \text{ if } \mu \neq 0$$

By substitution of $\eta = \dfrac{\partial u_x}{\partial r}$, this differential equation becomes

$$\frac{\partial \eta}{\partial r} + \frac{1}{r} \eta = 0 \quad \text{or} \quad \frac{1}{r} \frac{\partial}{\partial r} (\eta r) = 0$$

i.e.

$$\frac{1}{r} \frac{\partial}{\partial r} \left(r \frac{\partial u_x}{\partial r} \right) = 0$$

Integration of this equation gives

$$\frac{\partial u_x}{\partial r} = \frac{c_1}{r} \quad \text{and} \quad u_x = c_1 \ln r + c_2$$

with the boundary conditions, at $r = R_1 \quad u_x = U$

and at $\quad r = R_2, \quad u_x = 0$

$$\therefore \quad \left. \begin{aligned} U &= c_1 \ln R_1 + c_2 \\ 0 &= c_1 \ln R_2 + c_2 \end{aligned} \right\} \quad \therefore \quad c_1 = \frac{U}{\ln (R_1/R_2)}$$

$$\text{and} \quad c_2 = -c_1 \ln R_2 = \frac{-U \ln R_2}{\ln (R_1/R_2)}$$

$$\therefore \quad u_x = \frac{U}{\ln (R_1/R_2)} \ln r + \frac{U}{\ln (R_1/R_2)} \ln \left(\frac{1}{R_2} \right)$$

$$u_x = \frac{U}{\ln (R_1/R_2)} \ln \left(\frac{r}{R_2} \right)$$

$$\therefore \quad \frac{\partial u_x}{\partial r} = \frac{U}{\ln (R_1/R_2)} \frac{R_2}{r} \frac{1}{R_2}$$

$$\text{Shear on the wire} = \tau_0 = \mu \left(\frac{\partial u_x}{\partial r} \right)_{r=R_1}$$

$$\tau_0 = \frac{U}{\ln (R_1/R_2)} \frac{1}{R_1}$$

Work done per second, $P = 2\pi R_1 l \times \tau_0 \times U$

$$\therefore \quad P = 2\pi R_1 l \times \frac{\mu U^2}{\ln (R_1/R_2)} \frac{1}{R_1}$$

$$P = 2\pi l \mu U^2 / \ln (R_1/R_2)$$

8.29 A continuous belt dipped in a chemical bath travels vertically upwards at a velocity V thereby carrying a liquid film of thickness h up against gravity. Assuming the flow to be fully developed laminar flow with zero pressure gradient, obtain an expression for discharge per unit width of belt, q, in terms of V, h, ρ, g and μ (Fig. 8.17).

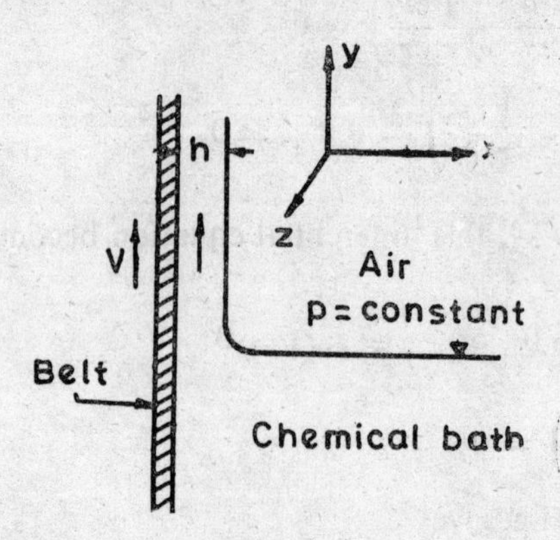

Fig. 8.17.

For N.S. equations in Cartesian system (Eq. 8.1) one can write

$$\frac{\partial}{\partial t} = 0, \ \frac{\partial}{\partial y} = 0, \ \frac{\partial}{\partial z} = 0$$

$u = 0$, $w = 0$, p = constant everywhere and $X = 0$, $Y = -\gamma$

and $\quad Z = 0$

Hence from N.S. equations, one gets

$$0 = -\rho g + \mu \frac{\partial^2 v}{\partial x^2}$$

$$\therefore \quad \frac{\partial^2 v}{\partial x^2} = \frac{\rho g}{\mu}, \text{ integration of which yields}$$

$$\frac{\partial v}{\partial x} = \frac{\rho g}{\mu} x + c_1$$

and

$$v = \frac{\rho g x^2}{2\mu} + c_1 x + c_2$$

The boundary conditions are at $x = 0$, $v = V$ $\therefore$ $c_2 = V$

Also at $x = h$, $\frac{\partial v}{\partial x} = 0$ $\therefore$ $c_1 = \frac{\rho g}{\mu} h$

$$\therefore \quad v = \frac{\rho g}{2\mu} x^2 - \frac{\rho g}{\mu} hx + V$$

$$q = \int_0^h v dx = \left(\frac{\rho g}{6\mu} x^3 - \frac{\rho g}{2\mu} hx^2 + Vx \right)_0^h$$

$$\therefore \quad q = \left(\frac{g}{6\nu} h^3 - \frac{g}{2\nu} h^2 + Vh\right)$$

or

$$q = \left(Vh - \frac{g}{3\nu} h^3\right)$$

8.30 For flow in a circular pipe having laminar flow, determine the value of y at which the stability parameter chi, $\chi = \frac{\rho y^2 du/dy}{\mu}$ is maximum. Express the corresponding χ_m value in terms of Reynolds number.

For flow in a circular pipe

$$u = -\frac{1}{4\mu}\frac{\partial p}{\partial x}(R^2 - r^2) \quad \text{and} \quad U = -\frac{R^2}{8\mu}\frac{\partial p}{\partial x}$$

$$\therefore \quad u = \frac{2U}{R^2}(R^2 - r^2)$$

$$\therefore \quad \frac{du}{dr} = \frac{-4Ur}{R^2}. \text{ Further } \quad y = R - r$$

$$\therefore \quad dy = -dr$$

$$\therefore \quad \frac{du}{dy} = \frac{4U}{R^2}(R - y)$$

$$\therefore \quad \chi = \frac{\rho y^2 du/dy}{\mu} = \frac{4U\rho}{R^2\mu}(R - y)y^2$$

$$\therefore \quad \frac{\partial \chi}{\partial y} = 0 = \frac{4U\rho}{R^2\mu}(2Ry - 3y^2)$$

$$\therefore \quad \chi \text{ will be maximum at } y = \frac{2R}{3}$$

$$\therefore \quad \chi_m = \frac{4R^2\rho}{9\mu} \times \frac{4U}{R^2}\frac{R}{3} = \frac{UR\rho}{\mu} \times \frac{16}{27} \quad \text{or} \quad \frac{8}{27}\frac{UD\rho}{\mu}$$

since flow becomes unstable when χ reaches a value of 500 or more,

$$500 = \frac{8}{27}\text{Re}$$

$$\therefore \quad \text{Re} = \frac{500 \times 27}{8} = 1687.5$$

It may be remembered that flow becomes unstable at this value of Reynolds number; onset of turbulence will occur at somewhat higher Re of 2100 as reported by many investigators.

8.31 Determine the discharge at which flow of kerosene at 20°C will cease to be laminar in 100 mm diameter conduit.

From Appendix A, $\rho = 800$ kg/m³ and $\mu = 2 \times 10^{-3}$ kg/m s for kerosene at 20°C. Since the flow changes from laminar to turbulent at Reynolds number of 2100

$$2100 = \frac{UD\rho}{\mu} = \frac{U \times 0.10 \times 800}{2 \times 10^{-3}}$$

$$\therefore \quad U = \frac{2100 \times 2.0 \times 10^{-3}}{0.1 \times 800} = 0.0525 \text{ m/s}$$

$$\therefore \quad Q = \frac{\pi D^2}{4} U = 0.785 \times 0.100^2 \times 0.0525$$

$$= 4.121 \times 10^{-4} \text{ m}^3\text{/s or } 0.4121 \text{ l/s}$$

PROBLEMS

8.1 An infinitely long plate is kept parallel to x axis in an infinite viscous fluid and it is suddenly set in motion in its own plane at a velocity U in x direction. Show that N.S. equations can be reduced to the differential equation $\rho \frac{\partial u}{\partial t} = \mu \frac{\partial^2 u}{\partial y^2}$. What are the boundary conditions? (For $t < 0$, $u = 0$ for all y; for $t > 0$, $u = U$ at $y = 0$ and $u = 0$ at $y \to \infty$)

8.2 What sized pipe should be installed to carry 7.0 l/s of fuel oil with a permissible head loss of 2.0 m per 100 m length of pipe if relative density and kinematic viscosity of oil are 0.90 and 2.04 St respectively? (0.1312 m)

8.3 For laminar flow through 75 mm diameter pipe the wall shear stress is 50 N/m² at a discharge of 7.5 l/s of oil of mass density 900 kg/m³. Determine the viscosity of oil and check the Reynolds number of flow. (0.2762 kg/m s, 415)

8.4 Obtain an expression for Darcy Weisbach friction factor f for laminar flow between parallel plates, in terms of Reynolds number $UB\rho/\mu$. ($f = 48/\text{Re}$)

8.5 30 l/s of castor oil are to be transported over 1000 m at a permissible energy gradient of 0.02. If oil temperature is 20°C, determine the diameter of the pipe. Also determine the power required and Reynolds number of flow. Take $\mu = 0.98$ kg/m s and $\rho = 960$ kg/m³. (0.2824 m, 5.648 kW, 132.62)

8.6 Obtain the expression for q and average velocity $\bar{U}$ of flow for generalised Couette flow discussed in Example 8.9.

$$\left(q = \frac{\bar{U}B}{2} - \frac{B^3}{12\mu}\frac{\partial p}{\partial x},\ U = \frac{\bar{U}}{2} - \frac{B^2}{12u}\frac{\partial p}{\partial x}\right)$$

8.7 A masonry tank with wall thickness of 1.0 m develops a 1.0 mm deep and 15 mm wide crack across the wall near the bottom of the tank. If the water in the tank is 2.0 m deep, estimate the leakage from the reservoir in one hour. Assume water temperature to be 20°C. (88.074 l)

8.8 For steady laminar flow in a circular pipe, Navier Stokes' equations in cylindrical polar co-ordinate system (Eq. 8.2) reduce to

$$0 = \frac{\partial p}{\partial x} + \mu \left(\frac{\partial^2 u_x}{\partial y^2} + \frac{1}{r} \frac{\partial u_x}{\partial r} \right).$$

Integrate this equation with appropriate boundary conditions to get velocity distribution law $u_x = u = -\frac{1}{4\mu} \frac{\partial p}{\partial x} (R^2 - r^2)$ (See Example 8.28)

8.9 If the velocity distribution for fully developed laminar flow in a pipe is given by $u = -\frac{1}{4\mu} \frac{\partial p}{\partial x} (R^2 - r^2)$, determine the distance from the pipe axis at which the velocity is equal to the average velocity of flow. ($r/R = 0.707$)

8.10 Determine α and β values for laminar flow in a pipe. (2, 1.333)

8.11 A discharge of 10 l/s per meter width, of oil of relavite density 0.80 and dynamic viscosity 0.98 N s/m^2 flows between parallel plates with a spacing of 40 mm. What should be the inclination of the plates with the horizontal so that the flow takes place at constant pressure? ($\theta = 13.57°$ sloping in the direction of flow)

8.12 Steady laminar flow takes place between two horizontal plates kept distance B apart. The lower plate moves in its own plane at velocity $-U_1$ while the top plate moves in the direction of flow at velocity $+U_2$. Obtain the expression for velocity distribution in terms of U_1, U_2, $\frac{\partial p}{\partial x}$, μ and B.

$$\left(u = U_2 \frac{y}{B} - U_1 \left(1 - \frac{y}{B} \right) - \frac{\partial p}{\partial x} \frac{1}{2\mu} (By - y^2) \right)$$

8.13 Considering only friction loss, determine the difference in liquid levels between two tanks 245 m apart connected by a 150 mm diameter pipe carrying 31.0 l/s of oil of kinematic viscosity 0.00028 m^2/s. (17.451 m)

8.14 For steady laminar flow in a circular pipe, obtain an expression for boundary shear stress τ_0 in terms of Q, μ and D. ($\tau_0 = 32\ \mu Q/\pi D^3$)

8·15 Glycerine flows at the rate of 7.85 l/s through 1.0 m high vertical conical contraction of inlet and outlet diameters 200 mm and 100 mm respectively. If the mass density and kinematic viscosity of glycerine are 1260 kg/m^3 and 6.63×10^{-4} m^2/s respectively, determine the pressure difference across the contraction. What assumptions are made in the calculation. (13.727 kN/m^2, $\alpha_1 = \alpha_2 = 1$)

8.16 How much time will be required to collect 50 cc of kerosene in a capillary tube viscometer of capillary diameter 2.0 mm, length 0.50 m under a head of 0.30 m, if ρ and μ for kerosene are 800 kg/m³ and 2.0×16^3 kg/m s respectively. (54.1 s)

8.17 What will be the terminal fall velocity of 0.02 mm mist droplet in air at 20°C? Take $\rho = 1.208$ kg/m³ and $\mu = 1.85 \times 10^{-5}$ kg/m s. (11.74 mm/s)

8.18 Determine the velocity at which a 50 mm diameter air bubble will rise in an oil of mass density 960 kg/m³ and dynamic viscosity of 0.98 kg/m s. (Neglect changes in bubble size as it rises) (0.0134 m/s)

8.19 Calculate the largest diameter of sand particle (Relative density 2.65) which will settle in water at 20°C still obeying Stokes' law. (0.0482 mm)

8.20 A spherical sand particle of relative density 2.65 falls in water at 20°C, at a terminal fall velocity of 15 mm/s. Determine its diameter. (0.160 mm)

8.21 Drinking water is to be supplied at the rate of 10 000 m³/day. Determine the size of the filter, if 0.60 m deep filter of 0.20 mm sand is to be provided and permissible head loss is 0.40 m. Take porosity as 0.40 and water temperature as 20°C. (29.922 m²)

8.22 Determine the maximum apparent velocity at which an upward flow filter made of sand of uniform diameter 0.10 mm can be operated. Take $e = 0.50$ and assume $\mu = 10^{-3}$ kg/m s. Assume the fall of particle to be in Stokes' range. (2.243×10^{-3} m/s)

8.23 A confined aquifer, also known as artesian aquifer, occurs where ground water is confined under pressure greater than the atmospheric pressure by the overlying relatively impermeable strata (Fig. 8.18). Show that the discharge in a radial flow into a well of R_w in a confined aquifer of depth b is given by

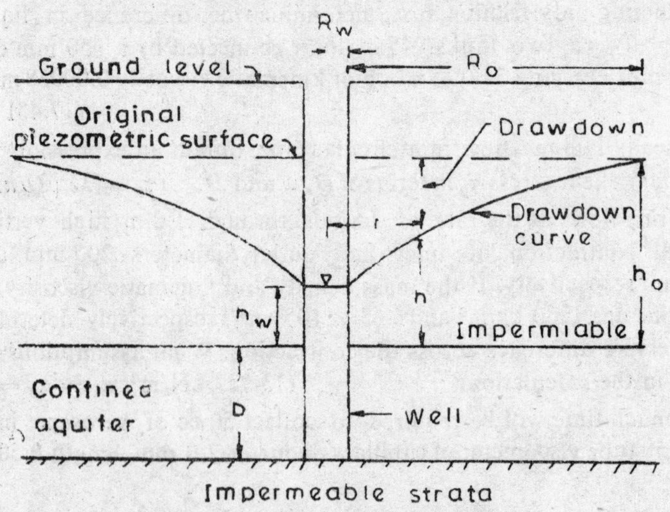

Fig. 8.18. Well in a confined aquifer.

$$Q = 2\,Kb\,\frac{(h_0 - h_w)}{\ln\,(R_0/R_w)}.$$

8.24 A 300 mm diameter well in an unconfined stratum shows a drawdown of 3.6 m and 2.2 m at 17 m and 45 m from the well respectively. Determine the permeability of the stratum if the well gives 0.10 m^3/s discharge under equilibrium condition. Assume water table to be 30 m above the impermeable stratum 300 m away from the well. What will be the depth of water in the well ?

(34.56 m/day, 17.182 m)

8.25 For a rotating cylinder viscometer (Fig. 8.5C). $R_1 = 30$ mm

$$R_2 = 32 \text{ mm},\ h = 75 \text{ mm and } \mu = 0.075 \text{ poise.}$$

Determine the torque on the inner cylinder if the outer cylinder rotates at 180 rpm. Neglect the torque contribution from the bottom.

(9.956×10^{-4} Nm)

1.26 Oil of dynamic viscosity 0.230 Ns/m^2 is filled in space between two concentric cylinders of radii 120 mm and 125 mm and immersion depth of 0.30 m. It the torque produced on the inner cylinder is 0.98 Nm, determine the speed of rotation of outer cylinder. Neglect the contribuion to torque from the bottom of inner cylinder.

(59.95 rpm)

8.27 A parallel plate viscometer consists of two circular plates, the lower one being stationary and upper one rotating at an angular velocity ω (Fig. 8.19). Obtain an expression for torque on the lower plate.

$$\left(T = \frac{\pi\omega\mu R^4}{4h}\right)$$

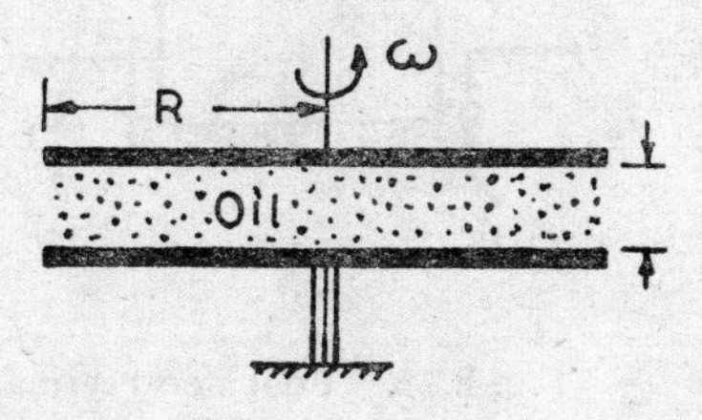

Fig. 8.19

8.28 Show that the power absorbed by the journal bearing show in Fig. 8.20 is

$$P = \left(\frac{8\pi^3\,L\mu N^2 R^3}{3600\,000b}\right)\text{kW}$$

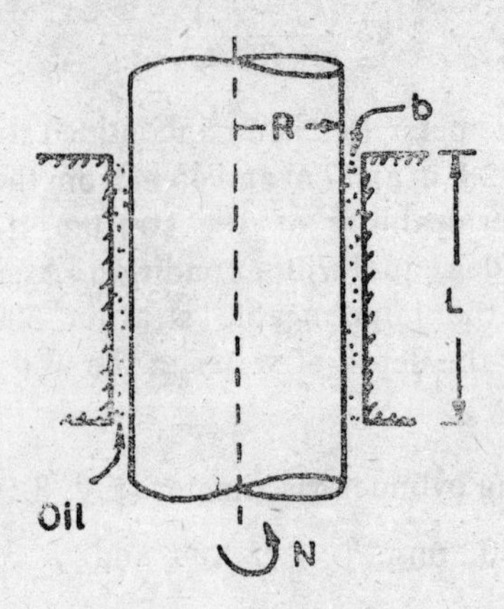

Fig. 8.20. Journal bearing.

8.29 For the foot step bearing shown in Fig. 8.21, determine the power required to overcome viscous resistance.

$$\left(\frac{4\pi^3 N^2 R^2 \omega}{2000b}\ \text{kW}\right)$$

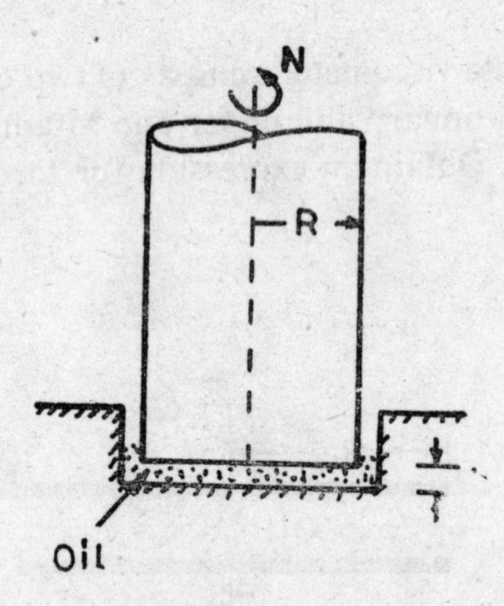

Fig. 8.21. Foot step bearing.

8.30 Obtain an expression for the torque required to rotate the shaft at an angular velocity ω in the cone bearing shown in Fig. 8.22.

$$\left(T = \frac{\pi\omega\mu}{2b}\ H^4 \tan^3 \sec \theta\right)$$

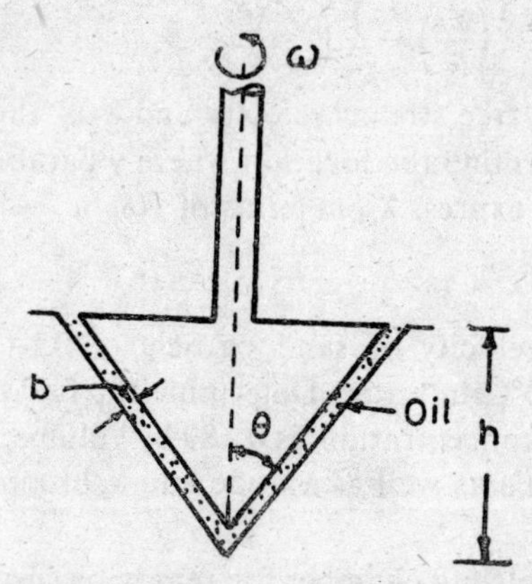

Fig. 8.22. Cone bearing.

8.31 Determine the torque required to rotate a vertical shaft of 100 mm diameter at 750 rpm if the lower end of the shaft rests in a foot-bearing. The clearance of 0.50 mm between the shaft and the bearing surface is filled with oil of dynamic viscosity 1.5 poise.

(0.2314 N m)

8.32 A flat belt is moved parallel to a fixed horizontal surface, at a velocity of 3.0 m/s as shown in Fig. 8.23. Calculate (i) the pressure

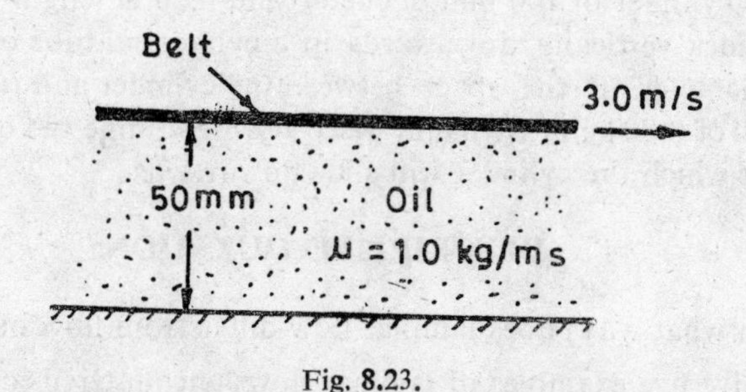

Fig. 8.23.

gradient required to cause zero shear stress at the surface and (ii) the discharge for zero shear on the belt.

($2400\ N/m^3$, $0.10\ m^3/s\ m$)

8.33 For laminar flow between parallel plates, determine the location where stability parameter $\chi = \rho y^2 \frac{du}{dy} \Big/ \mu$ will be maximum and express this maximum value in terms of Reynolds number $UB\rho/\mu$.

$\left(\frac{B}{3}, \frac{2Re}{9}\right)$

8.34 The velocity distribution in laminar boundary layer over a flat plate with zero pressure gradient is given by

$$\frac{u}{U} = \frac{3}{2}\left(\frac{y}{\delta}\right) - \frac{1}{2}\left(\frac{y}{\delta}\right)^3$$

where U is free stream velocity and δ is the boundary layer thickness. Determine the location where χ parameter will attain maximum value and express χ_m in terms of $U\delta\rho/\mu$.

$\left(0.707\delta,\ 0.375\ \frac{U\delta\rho}{\mu}\right)$

8.35 The fall velocity of sand particle of 0.14 mm is found to be 15.0 mm/s at 13°C in water. Determine the fall velocity of the particles when its concentration is 0.189 by volume. Take $\mu = 0.0012$ kg/m s. Use Scholl's as well as Maude and Whitmore's equations.

(8.0 mm/s, 6.81 mm/s)

8.36 Design a settling chamber for removing suspended particulate matter in air with 100 percent efficiency, with the data:

Horizontal velocity = 0.30 m/s

Air temperature = 17°C

Particle relative density = 2.0

Chamber length = 7.5 m

Height = 1.5 m

What size of particles will it remove ? (0.041 mm)

8.37 A cylinder of 100 mm diameter and 0.20 m length weighing 10.0 N slides vertically downwards in a cylindrical tube of 102 mm inside diameter. If the space between the cylinder and tube is filled with oil of 0.80 kg/m s dynamic viscosity, determine the constant velocity at which the cylinder will slide downwards. (0.199 m/s)

DESCRIPTIVE QUESTIONS

8.1 In what ways does laminar flow differ from flow of an ideal fluid?

8.2 Give five examples of laminar flows encountered in every day life.

8.3 Why is no analytical solution of complete N.S. equations yet available ?

8.4 In what region of pipe flow across its diameter will the greatest heat be generated in laminar flow ?

8.5 Show that Reynolds number is the ratio of inertial force per unit volume to viscous force per unit volume.

8.6 Is turbulence always undesirable ? If your answer is yes, give reasons. If your answer is negative give examples where turbulence is advantageous.

8.7 Under what conditions, need the body forces be included in the dynamic analysis using N.S. equations ?

8.8 Below are given three flow phenomena and three critical values of Reynolds numbers. Match them correctly and mention what velocity and length are used in each case while forming Reynolds number. Re values are 1.0, 0.10, 2100.

Flow	Correct values of Re	U	l
Pipe flow	—	—	—
Flow past sphere	—	—	—
Flow through porous material	—	—	—

8.9 In what ways does laminar flow differ from turbulent flow ?

8.10 Which of the following statements are true for laminar flow ?
(i) laminar flow is rotational
(ii) loss of head is proportional to square of velocity
(iii) loss of head is proportional to first power of viscosity
(iv) Velocity is constant over the cross-section
(v) Other quantities remaining same, increase in diameter will increase critical Re value
(vi) For modelling subcritical laminar flow ($Fr < 1.0$) in open channels, equality of Reynolds number is not essential.

8.11 Why does expression for Δp given by Hagen-Poiseuille's equation not involve ρ?

8.12 Shear stress at the boundary for laminar flow in a pipe is
(i) directly proportional to D^2 for given Q and μ
(ii) inversely proportional to D^3 for given Q and μ
(iii) directly proportional to μ^2 for given Q and D
(iv) is proportional to μ for given Q and D
(v) is proportional to Q for given Q and μ.

8.13 Friction factor f for laminar flow in a pipe is given by
(i) $f = 64\,Re$, (ii) $f = 1.328/Re^{1/2}$,
(iii) $f = \dfrac{16}{Re}$ (iv) $f = \dfrac{64}{Re}$.

8.14 Briefly explain the difference between onset of instability in laminar flow and the transition.

8.15 Briefly describe Reynolds experiment and mention its significance.

8.16 Distinguish between lower critical and upper critical velocities.

8.17 Should there be an upper limit for rotational speed of rotating cylinder viscometer? Why?

8.18 Velocity distribution for laminar flow between parallel plates is (i) constant (ii) parabolic (iii) logarithmic (iv) linear (v) none of the above.

8.19 Is it reasonable to use fall velocity of a single particle in the analysis of settling tanks ? If not, what other factors need to be taken into account?

8.20 Critical velocity in a pipe.
(i) increases as D increases
(ii) is independent of D
(iii) increases as viscosity increases
(iv) is independent of fluid density
(v) decreases as diameter increases.

8.21 What are the relative merits of drag model and capillary tube model for Darcy's law ?

8.22 What is the percentage increase in discharge of an unconfined well if its diameter is doubled, other factors remaining the same ?

8.23 Energy correction factor α for laminar flow in a circular pipe is
(i) 0.5 (ii) 1.0 (iii) 1.5 (iv) 2.0

1.24 Momentum correction factor β for laminar flow in a circular pipe is
(i) 0.5 (ii) 1.0 (iii) 1.33 (iv) 1.50

8.25 For the same discharge of a given liquid carried by a circular pipe having laminar flow, reduction of diameter to half will increase energy gradient by a factor of
(i) 2 (ii) 4 (iii) 8 (iv) 16

8.26 For laminar flow in a circular pipe, the velocity distribution will be given by
(i) $\frac{u}{u_m} = \left(1 - \frac{r^2}{R^2}\right)$ (ii) $\frac{u}{u_m} = \left(1 - \frac{r}{R}\right)$
(iii) $\frac{u}{u_m} = \left(1 - \frac{r^3}{R^3}\right)$ (iv) none of the above.

8.27 Instability of laminar flow will occur where
(i) u is maximum (ii) $\frac{du}{dy}$ is maximum
(iii) $y\rho \frac{du}{dy}/\mu = 500$ (iv) $y\rho \frac{du}{dy}/\mu = 500.$

CHAPTER IX

Boundary Layer Theory

9.1 INTRODUCTION

Consider steady flow of a real fluid of low viscosity and reasonably high Reynolds number past a flat plate kept parallel to the flow. Because of no slip condition, fluid sticks to the wall and there is a region close to the boundary where fluid velocity is retarded. This layer adjacent to the boundary where fluid is retarded is known as the boundary layer (Fig. 9.1). Boundary

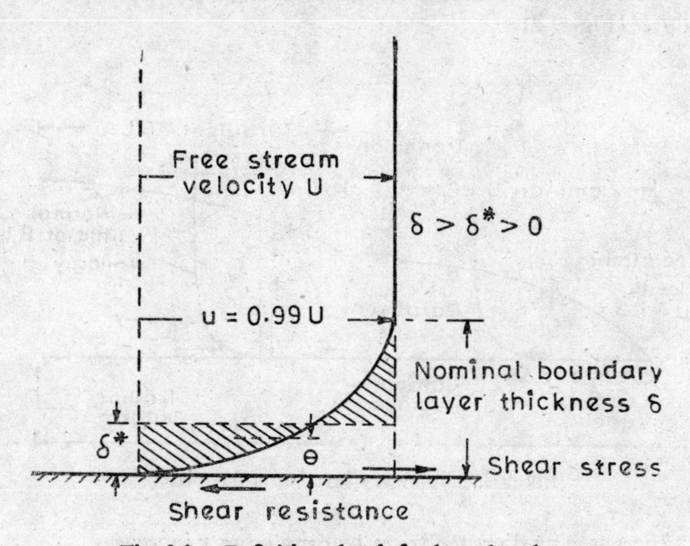

Fig. 9.1. Definition sketch for boundary layer.

layer is formed whenever there is relative motion between the boundary and the fluid. Since $\tau_0 = \mu \left(\frac{du}{dy}\right)_{y=0}$, the fluid exerts a shear stress on the boundary and boundary exerts an equal and opposite force on fluid known as the shear resistance.

9.2 BOUNDARY LAYER DEFINITIONS AND CHARACTERISTICS

Nominal boundary layer thickness δ: It is the distance adjacent to wall such that at $y = \delta$, $u = 0\ 99U$.

Displacement thickness δ^*: Displacement thickness indicates the distance by which external streamlines are shifted due to boundary layer formation. It is defined by the equation

$$\delta^* U = \int_0^\infty (U - u)\, dy \approx \int_d^\delta (U - u)\, dy \tag{9.1}$$

Momentum thickness θ: Momentum thickness indicates the distance through which momentum flux flowing in the absence of boundary layer is equal to deficiency of momentum flux due to boundary layer formation. It is given by the equation

$$\rho U^2 \theta = \int_0^\infty \rho u(U - u)\, dy \approx \int_0^\delta \rho u\,(U - u)\, dy \tag{9.2}$$

Main characteristics of boundary layers can be explained by considering the boundary layer formed on a flat plate kept parallel to flow of fluid of velocity U (Fig. 9.2).

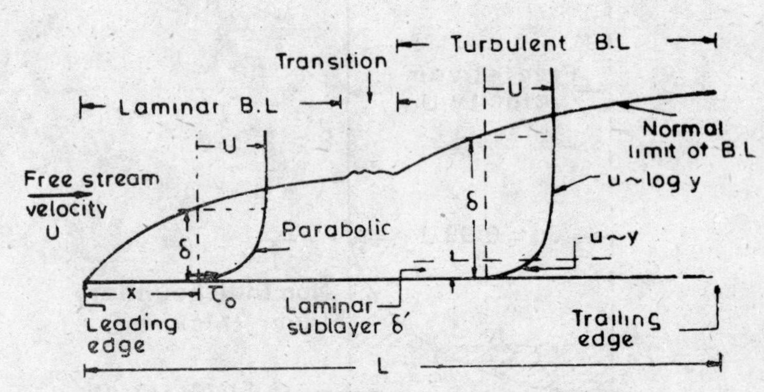

Fig. 9.2. **Boundary layer on a flat plate.**

1. δ increases as distance from leading edge x increases.
2. δ decreases as U increases.
3. δ increases as kinematic viscosity increases.
4. $\tau_0 \approx \mu \left(\dfrac{U}{\delta}\right)$; hence τ_0 decreases as x increases. However, when boundary layer becomes turbulent (see below), it shows a sudden increase and then decreases with increasing x.

5. If U increases in the downstream direction, i.e. $\partial p/\partial x$ is $-$ve, boundary layer growth is reduced.
6. If U decreases in the downstream direction, i.e. $\partial p/\partial x$ is $+$ve, flow near the boundary is further retarded, b.L. growth is faster and boundary layer is susceptible to separation.
7. All characteristics of the boundary layer on flat plate, such as variation of δ, τ_0 or force F, are governed by inertial and viscous forces; hence they are functions of either Ux/ν or UL/ν.
8. When Ux/ν is less than 5×10^5, boundary layer is laminar and velocity distribution in boundary layer is parabolic. When $Ux/\nu > 5 \times 10^5$ the boundary layer on that portion is turbulent.
9. Critical value or Ux/ν at which boundary layer changes from laminar to turbulent depends on turbulence in ambient flow, surface roughness, pressure gradient, plate curvature and temperature difference between fluid and boundary.
10. Velocity distribution in laminar boundary layer follows parabolic law while that in turbulent boundary layer follows logarithmic low or power law.
11. In turbulent boundary layer a thin layer known as laminar sublayer δ' exists near the boundary, if it is smooth.

9.3 LAMINAR BOUNDARY LAYER

Local drag coefficient c_f is defined as

$$c_f = \tau_0/\rho \frac{U^2}{2} \tag{9.3}$$

where τ_0 is the local shear stress at any distance x from the leading edge, while average drag coefficient C_f is defined as

$$\frac{F/Bl}{\rho U^2/2} = C_f \tag{9.4}$$

where F is the force on the plate of length L and width B. For laminar boundary layer δ, c_f and C_f are given by

$$\delta/x = 5.0/\sqrt{\text{Re}_x} \tag{9.5}$$

$$c_f = 0.664/\sqrt{\text{Re}_x} \tag{9.6}$$

$$C_f = 1.328/\sqrt{\text{Re}_L} \tag{9.7}$$

where $\text{Re}_x = Ux\rho/\mu$ and $\text{Re}_L = UL\rho/\mu$. Equations (9.5), (9.6) and (9.7) are valid if ReL is less than 5×10^5. Velocity distribution in laminar boundary layer can be expressed as $u/U = f(y/\delta) = f(\eta)$ which should satisfy the following boundary conditions:

$$\left.\begin{array}{l}\eta = 0,\ u = 0,\ \text{i.e. } f = 0\\ \eta = 1,\ u = U,\ \text{i.e. } f = 1\end{array}\right\}. \text{ These are essential.}$$

$$\eta = 1,\ \partial f/\partial \eta = \frac{\partial^2 f}{\partial \eta^2} = 0. \quad \text{These are desirable.}$$

Here $\eta = y/\delta$.

Typical distribution for velocity in laminar boundary layer is

$$\frac{u}{U} = \frac{3}{2}\left(\frac{y}{6}\right) - \frac{1}{2}\left(\frac{y}{\delta}\right)^3 = \frac{3}{2}\eta - \frac{1}{2}\eta^3$$

9.4 TURBULENT BOUNDARY LAYER

If boundary layer is turbulent from the beginning (this can be done by providing roughness near leading edge) and if Re_L is between 5×10^5 and 2×10^7, δ, c_f and C_f are given by the equations

$$\delta/x = 0.377/Re_x^{1/5} \tag{9.8}$$

$$c_f = 0.059/Re_x^{1/5} \tag{9.9}$$

$$C_f = 0.074/Re_L^{1/5} \tag{9.10}$$

If part of the plate is covered with laminar boundary layer and $Re_L < 2 \times 10^7$, C_f is given by

$$C_f = \frac{0.074}{Re_L^{1/5}} - \frac{1700}{Re_L} \tag{9.11}$$

For higher value of Re_L (5×10^5 to 10^9), δ/x, c_f and C_f are given by

$$\delta/x = 0.22/Re_x^{1/6} \tag{9.12}$$

$$c_f = 0.370/(\log Re_x)^{2.50} \tag{9.13}$$

$$C_f = \frac{0.455}{(\log Re_L)^{2.58}} - \frac{1700}{Re_L} \tag{9.14}$$

It may be mentioned that at high value of Re_L, the corrective term 1700/ReL in Eqs. (9.11) and (9.14) becomes negligible. Hence Eq. (9.14) reduces to

$$C_f = 0.455/(\log Re_L)^{2.58} \tag{9.14a}$$

Figure 9.3 show these equations for C_f plotted as C_f vs Re_L. Velocity distribution in turbulent boundary layer is given by

$$\frac{u}{U} = \left(\frac{y}{\delta}\right)^{1/n} \tag{9.15}$$

where n changes from 5 to 10 as Reynolds number $U\delta/\nu$ increases. It can also be expressed by the law

$$\frac{u - U}{u_*} = f\left(\frac{y}{\delta}\right). \tag{9.16}$$

Atmospheric boundary layer is invariably turbulent. Its analysis is

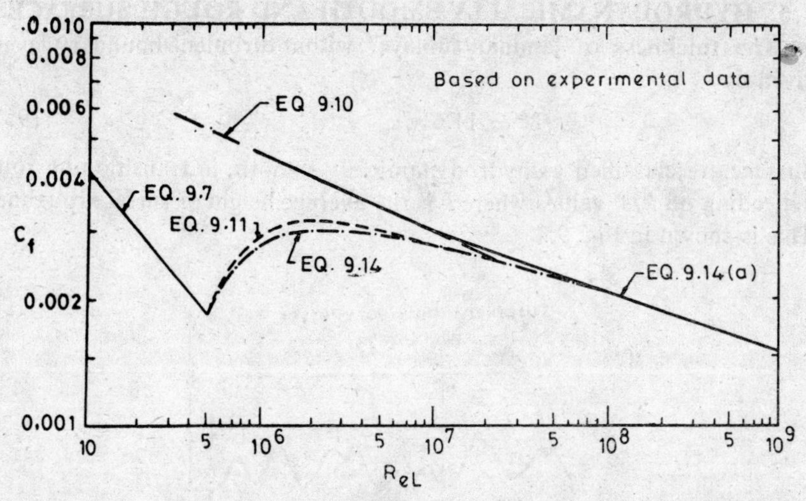

Fig. 9.3. Variation of C_f with Re_L for boundary layer on a flat plate.

complicated because of wide variation in surface roughness and some uncertainty in knowledge of ambient velocity U, also known as geostrophic wind velocity. The velocity distribution in atmospheric turbulent boundary layer is given by

$$\frac{u}{u_*} = \frac{2.3}{K} \log \frac{(y-d)}{k} \tag{9.17}$$

where $u_* = \sqrt{\tau_0/\rho}$ called shear velocity, K is Karman's universal constant $(= 0.40)$; d is known as zero plane displacement length and k characteristic roughness height such that at $(y - d) = k$, $u = 0$. For crops of height h, d and k are given by the relations

$$d = 0.63\ \text{h} \quad \text{and} \quad k = 0.15\ \text{h}$$

Velocity distribution in turbulent atmospheric boundary layer can also be expressed by Eq. (9.15). Table 9.1 shows average values of k, δ and n for different surfaces.

Table 9.1 Values of k, δ and n for atmospheric boundary layers

Description	Open sea	Flat open country	Wood land forest	Urban areas
k (mm)	0.25–3.5	35	350	4000
δ (m)	—	300	425	560
$1/n$	0.10–0.13	0.16	0.281	0.40

9.5 HYDRODYNAMICALLY SMOOTH AND ROUGH SURFACES

The thickness of laminar sublayer within turbulent boundary layer is given by

$$\delta' = 11.6\ \nu/u_* \tag{9.18}$$

Surface are classified as hydrodynamically smooth, in transition or rough depending on k/δ' values where k is the average height of surface roughness. This is shown in Fig. 9.4.

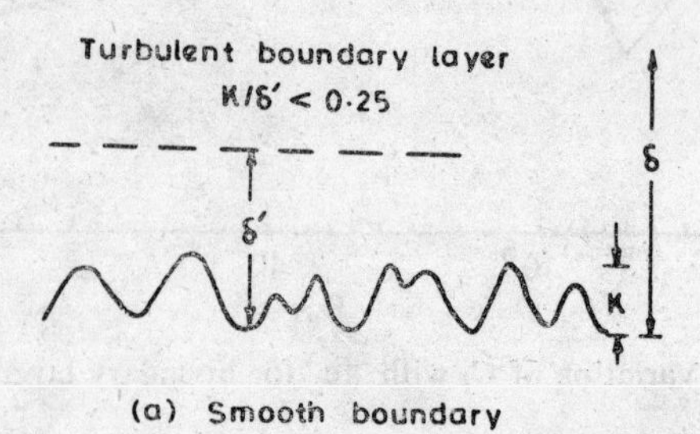

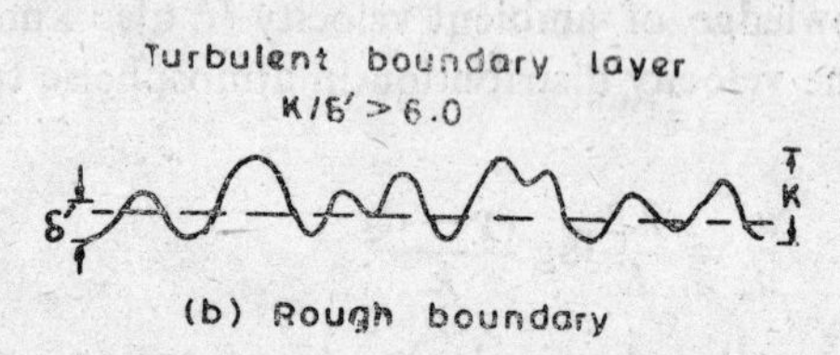

Fig. 9.4. Smooth and rough boundaries.

If k/δ' is less than 0.25, the surface is hydrodynamically smooth and c_f or C_f are functions of Reynolds number. If k/δ' is greater than 6.0, the laminar sublayer is completely pierced by roughness and the roughness is exposed to turbulent flow. In such a case boundary is hydrodynamically rough. For rough boundary c_f and C_f are functions of relative roughness (i.e. roughness height to characteristic flow length). For $6.0 > k/\delta' > 0.25$, boundary is in transition regime.

9.6 APPLICATION OF MOMENTUM EQUATION

Application of momentum equation to steady flow of an incompressible fluid in case of laminar boundary layer on a flat plate, yields (Fig. 9.5)

$$\frac{\tau_0}{\rho} = U^2 \frac{d\theta}{dx} \tag{9.19}$$

Corresponding equation for flow with pressure gradient is

$$U^2 \frac{d\theta}{dx} + (2\theta + \delta^*)\, U \frac{dU}{dx} = \frac{\tau_0}{\rho} \tag{9.20}$$

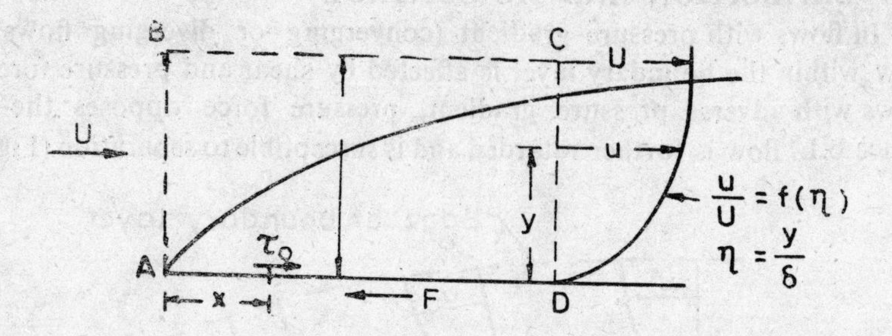

Fig. 9.5. Definition sketch.

These equations are true for turbulent boundary layer only as an approximation since changes in momentum flux due to turbulent fluctuations are not accounted for. The equation 9.19 or 9.20 can be solved if one assumes velocity distribution law within boundary layer and if variation of U with x is known.

9.7 ESTABLISHMENT OF FLOW IN PIPES

When laminar flow enters a conduit from a reservoir (Fig. 9.6) a boundary layer forms in the pipe and at distance L_e it reaches the centreline. Beyond L_e, flow is fully developed and velocity distribution is given by Hagen-Poiseuille's law

$$\frac{L_e}{D} = 0.07 \text{ Re} \tag{9.21}$$

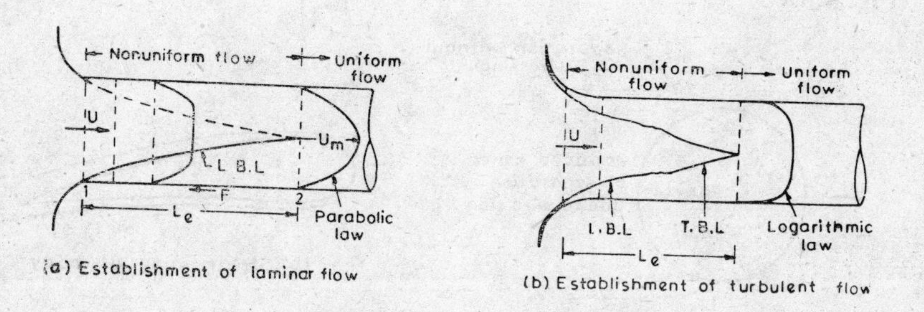

Fig. 9.6 Establishment of flow in pipes.

If Re is large, boundary layer changes from laminar to turbulent before it reaches the centre. Here

$$L_e/D \approx 50 \tag{9.22}$$

In the length of establishment, pressure loss is partly due to flow acceleration along centreline and partly due to frictional resistance. It is only beyond L_e that Hagen-Poiseuille's equation for laminar flow or Darcy-Weisbach equation for turbulent flow is applicable.

9.8 SEPARATION AND ITS CONTROL

In flows with pressure gradient (converging or diverging flows), the flow within the boundary layer is affected by shear and pressure forces. In flows with adverse pressure gradient, pressure force opposes the flow; hence b.L. flow is further ratarded and is susceptible to separation (Fig. 9.7).

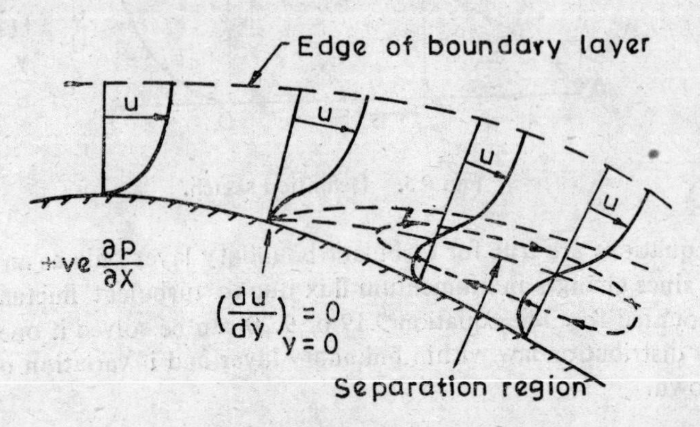

Fig. 9.7. Boundary layer separation.

Separation will take place when $\left(\frac{du}{dy}\right)_{y=0} = 0$. Separation occurs in diffusers, open channel transitions, on blades of turbines, pumps, fans, on aerofoils, etc. Separation can be retarded or avoided by the following methods (i) tripping the b.L. from laminar to turbulent by provision of surface roughness, (ii) streamlining the body shape, (iii) sucking the retarded fluid and (iv) injecting high velocity fluid in boundary layer. These are shown in Fig. 9.8.

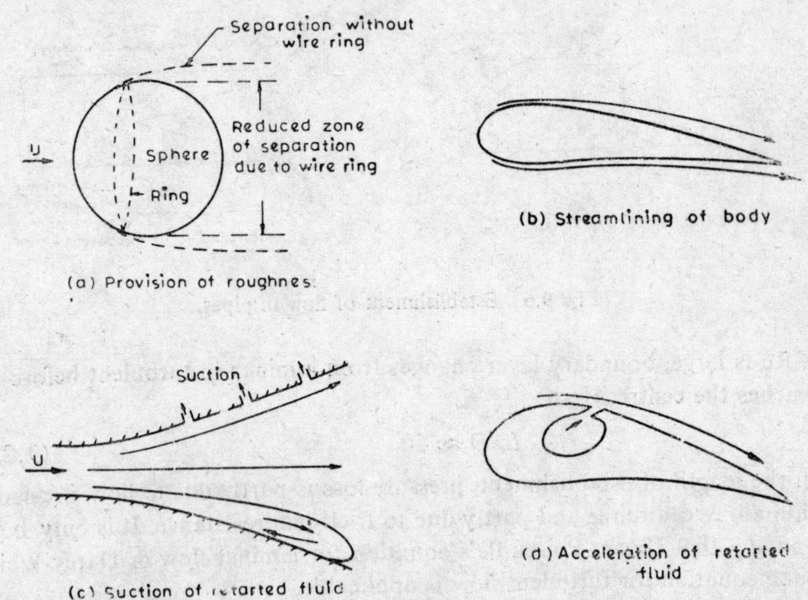

Fig. 9.8. Methods of boundary layer control.

ILLUSTRATIVE EXAMPLES

9.1 Below are given the wind tunnel measurements of velocity at different elevations from the boundary:

y mm	1	2	5	7	10	15	20	25	50	75	100	150	200
u m/s	6.10	6.80	7.98	8.40	8.95	9.58	10.05	10.25	10.40	10.30	10.60	10.20	10.42

What is the free stream velocity? Determine δ the thickness of nominal boundary layer; also determine the exponent n in the equation $u/U = (y/\delta)^{1/n}$. Is the boundary layer laminar or turbulent?

The velocity distribution is plotted on natural scale in Fig. 9.9. It can

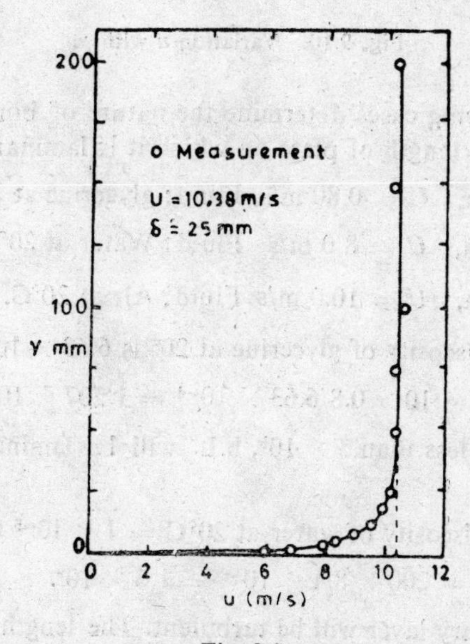

Fig. 9.9.

be seen that away from the boundary, the velocity is constant. The average of last five points will yield 10.38 m/s which can be taken as free-stream velocity U. According to definition of boundary layer δ

$$y = \delta,\ u = 0.99U = 0.99 \times 10.38 \text{ or } 10.28 \text{ m/s}.$$

From Fig. 9.10 it can be seen that this happens at $y = 25$ mm.

$$\therefore \qquad U = 10.28 \text{ m/s},\ \delta = 25 \text{ mm}.$$

In order to determine n, one must plot u/U vs y/δ on log-log scale and take the slope of the straight line. This is done in Fig. 9.10. where only values of u for y upto 50 mm have been plotted. Value of n comes out to be 6. Hence b.L. is turbulent.

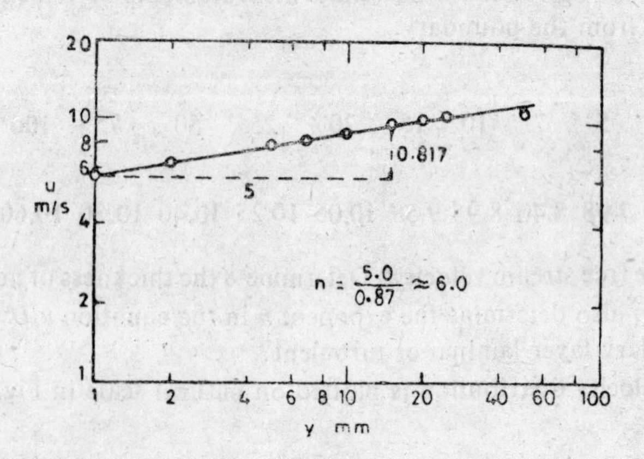

Fig. 9.10. Variation u with y.

9.2 In the following cases determine the nature of boundary layer and if it is turbulent, the length of plate on which it is laminar.

(i) $L =$ 10.0 m, $U = 0.80$ m/s, Fluid: glycerine at 20°C

(ii) $L =$ 200.0 m, $U =$ 8 0 m/s Fluid: Water at 20°C

(iii) $L =$ 2.0 m, $U = 10.0$ m/s Fluid: Air at 20°C.

(i) Kinematic viscosity of glycerine at 20° is 6.63×10^{-4} m²/s

$$\therefore \quad UL/\nu = 10 \times 0.8/6.63 \times 10^{-4} = 1.207 \times 10^4$$

Since this is less than 5×10^5, b.L. will be laminar over the whole length.

(ii) Kinematic viscosity of water at 20°C $= 1 \times 10^{-6}$ m²/s

$$\therefore \quad UL/\nu = 200 \times 8/1 \times 10^{-6} = 1.6 \times 10^9.$$

Here boundary layer will be turbulent. The length x of the plate on which boundary layer will be laminar will be

$$Ux/\nu = 5 \times 10^5 \quad \text{or} \quad x = \frac{5 \times 10^5 \times 10^{-6}}{8.0} = 0.0625 \text{ m}$$

which is very small and can be neglected for all practical purposes.

(iii) Kinematic viscosity of air at 20°C $= 1.53 \times 10^{-5}$ m²/s

$$\therefore \quad UL/\nu = 2 \times 10/1.53 \times 10^{-5} = 1.307 \times 10^6$$

This Reynolds number being slightly greater than 5×10^6, the front portion of plate will have laminar boundary layer and the rear portion, turbulent

$$Ux/\nu = 5 \times 10^5 \quad \therefore \quad x = \frac{5 \times 10^5 \times 1.53 \times 10^{-5}}{10}$$

or $\quad x = 0.765$ m is the length on which boundary layer will be laminar.

9.3 Air flows at 30.0 m/s on a wide plate kept parallel to the air stream. Assuming the laminar boundary to follow the velocity distribution $u/U = y/\delta$, determine the rate of flow across section *bc*, if section *cd* is located 0.25 m downstream of the leading edge of the plate, where boundary layer thickness is 0.15 mm (see Fig. 9.11).

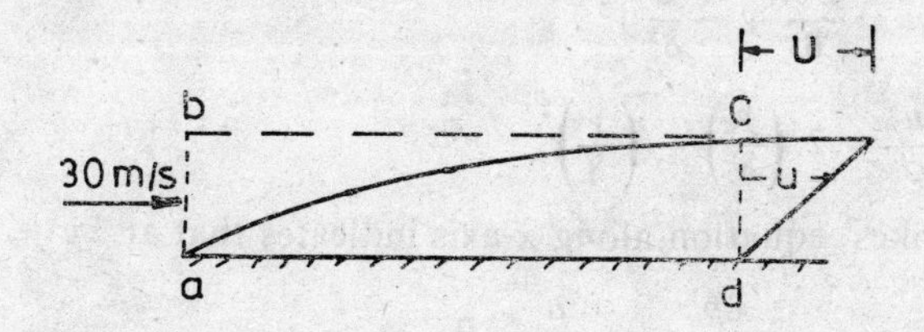

Fig. 9.11.

$$cd = 0.15 \text{ mm} = ab$$

At section *ab*, velocity is constant equal to 30.0 m/s.

$\therefore$ Rate of flow through $ab = 0.15 \times 10^{-3} \times 30$

$$= 4.5 \times 10^{-3} \text{ m}^3/\text{s m}$$

Rate of flow through $cd = 0.15 \times 10^{-3} \times$ average velocity

But since at section *cd* velocity varies linearly and velocity at $c = 0.99U \approx 30.0$ m/s, average velocity will be 15.0 m/s

Rate of flow through $cd = 0.15 \times 10^{-3} \times 15$

$$= 2.25 \times 10^{-3} \text{ m}^3/\text{s m}$$

$\therefore$ Rate of outward flow through *bc*

= Rate of inflow through *ab*—Rate of outflow through *cd*

$= (4.50 \times 10^{-3} - 2.25 \times 10^{-3})$

$= 2.25 \times 10^{-3}$ m³/s m.

9.4 If velocity distribution in laminar boundary layer over a flat plate is assumed to be given by second order polynomial $u = a + by + cy^2$, determine its form using the necessary boundary conditions.

The first b.c. it must satisfy is $u = 0$ at $y = 0$

$\therefore \quad a = 0$; hence $u = by + cy^2$.

The second condition that must be satisfied is at $y = \delta$, $u = U$. Hence

$$U = b\delta + c\delta^2 \qquad (1)$$

Also at $\quad y = \delta, \dfrac{du}{dy} = 0$

$$\therefore \quad \left(\frac{du}{dy}\right)_{y=\delta} = b + 2c\delta = 0 \qquad (2)$$

Solution of equations 1 and 2 yield $b = \dfrac{2U}{\delta}$ and $c = -\,U/\delta^2$. Hence velocity distribution law will be

$$u = \frac{2U}{\delta}\,y - \frac{U}{\delta^2}\,y^2$$

or $$\frac{u}{U} = 2\left(\frac{y}{\delta}\right) - \left(\frac{y}{\delta}\right)^2$$

9.5 N. Stokes' equation along x-axis indicates that at $y = 0$,

$$-\frac{\partial p}{\partial x} + \mu\frac{\partial^2 u}{\partial y^2} = 0$$

for steady two dimension flow with pressure gradient. In the case of two dimensional boundary layer, the pressure gradient $-\dfrac{\partial p}{\partial x} = \rho U\dfrac{\partial U}{\partial x}$ where U is the free stream velocity. Assume that velocity distribution in boundary layer on a flat plate with pressure gradient is given by

$$\frac{u}{U} = f(\eta) = a\eta + b\eta^2 + c\eta^3 + d\eta^4 \quad \text{where} = \frac{y}{\delta}$$

The boundary conditions it must satisfy are

$$f(0) = 0,\ f(1) = 1,\ f'(1) = 0$$

and $\quad \mu\left(\dfrac{\partial^2 u}{\partial y^2}\right)_{y=0} = -\,\rho U\dfrac{\partial U}{\partial x}$. Obtain the velocity distribution.

Substitution of these boundary conditions in the expression for $\dfrac{u}{U}$ gives the following equations:

$$\left.\begin{aligned} a + b + c + d &= 1 \\ a + 2b + 3c + 4d &= 0 \\ 2b + 6c + 12d &= 0 \\ \text{and} \quad 2b = -\frac{\delta^2}{\nu}\frac{\partial U}{\partial x} &= -\lambda \end{aligned}\right\}$$

Here $\lambda = \dfrac{\delta^2}{\nu}\dfrac{\partial U}{\partial x}$. Solution of these equations yields

$$a = 2 + \frac{\lambda}{6},\ b = -\frac{\lambda}{2},\ c = -2 + \frac{\lambda}{2},\ d = 1 - \frac{\lambda}{6}.$$

$$\therefore \quad \frac{u}{U} = \left(2 + \frac{\lambda}{6}\right)\eta + \left(-\frac{\lambda}{2}\right)\eta^2 + \left(-2 + \frac{\lambda}{2}\right)\eta^3 + \left(1 - \frac{\lambda}{6}\right)\eta^4$$

$$= (2\eta - 2\eta^3 + \eta^4) + \frac{\lambda}{6}(\eta - 3\eta^2 + 3\eta^3 - \eta^4)$$

$$\therefore \quad \frac{u}{U} = (2\eta - 2\eta^3 + \eta^4) + \frac{\lambda}{6}\eta(1 - \eta)^3$$

This velocity distribution law was used by Karman and Pohlhausen to study development of laminar boundary layers in flows with pressure gradient. It can be seen that when $\frac{\partial p}{\partial x} = 0$, $\lambda = 0$ and velocity distribution reduces to $\frac{u}{U} = (2\eta - 2\eta^3 + \eta^4)$ for flat plate with zero pressure gradient.

9.6 Assume that, at the edge of laminar boundary layer, the inertial force per unit volume is of the same order of magnitude as the viscous force per unit volume and obtain expressions for δ and τ_0 along the length of flat plate.

$$\text{Inertial force/volume} = \rho u \frac{\partial u}{\partial x} \quad \text{or} \quad \frac{\rho U^2}{x}$$

$$\text{Viscous force/volume} = \mu \frac{\partial^2 u}{\partial y^2} \quad \text{or} \quad \frac{\mu U}{\delta^2}$$

or $$\frac{\rho U^2}{x} = k\frac{\mu U}{\delta^2}$$

or $$\frac{\delta^2}{x^2} = \frac{k\mu}{\rho U x} \quad \text{and} \quad \frac{\delta}{x} = \frac{k_1}{\sqrt{\text{Re}_x}}$$

where k and k_1 are constants. Blasius found the value of $k_1 = 5.0$.

$$\text{Similarly} \quad \tau_0 = \mu\left(\frac{\partial u}{\partial y}\right)_{y=0} \approx \frac{\mu U}{\delta}$$

Substitution of the value of δ from above equation yields

$$\tau_0 = \frac{\mu U}{k_1 x}\sqrt{\frac{U x \rho}{\mu}}$$

Thus local drag coefficient c_f is given by

$$c_f = \frac{\tau_0}{\rho U^2/2} = \frac{\mu U}{k_1 x}\frac{2}{\rho U^2}\sqrt{\frac{U x \rho}{\mu}}$$

$$= \text{const}/\sqrt{\text{Re}_x}$$

Blasius found the constant to be 0.664.

9.7 A flat plate of 2.0 m width and 4.0 m length is kept parallel to air flowing at 5.0 m/s velocity at 15°C. Determine the length of plate over which the boundary layer is laminar, shear at the location where boundary layer ceases to be laminar, and total force on both sides on that portion of plate where the boundary layer is laminar. Take

$$\rho = 1.208 \text{ kg/m}^3 \text{ and } \quad \nu = 1.47 \times 10^{-5} \text{ m}^2/\text{s}.$$

$$\frac{UL}{\nu} = \frac{5 \times 4}{1.47 \times 10^{-5}} = 1.361 \times 10^6$$

Hence on the front portion, boundary layer is laminar and on the rear, it is turbulent.

$$\frac{Ux}{\nu} = 5 \times 10^5 \quad \therefore \quad x = \frac{5 \times 10^5 \times 1.47 \times 10^{-5}}{5},$$

$$= 1.47 \text{ m}$$

Hence boundary layer is laminar on 1.47 m length of the plate.

$$\delta = \frac{5 \times 1.47}{\sqrt{5 \times 10^5}} = 0.01039 \text{ m} \quad \text{or} \quad 1.039 \text{ cm.}$$

$$c_f = \frac{0.664}{\sqrt{5 \times 10^5}} = 0.000939$$

$$\therefore \qquad \tau_0 = 0.000939 \times \frac{1.208 \times 5^2}{2}$$

$$= 0.01418 \text{ N/m}^2$$

Force on 1.47 m length (on both sides) will be

$$F = 2 \times (2 \times 1.47)\, C_f\, \frac{\rho U^2}{2}$$

and
$$C_f = \frac{1.328}{\sqrt{5 \times 10^5}} = 0.001878$$

$$\therefore \qquad F = 2 \times (2 \times 1.47) \times 0.001878 \times 1.208 \times \frac{5^2}{2}$$

$$= 0.1667 \text{ N}$$

9.8 A wind tunnel has a cross-section at its inlet of 1.0 m by 1.0 m and a length of 10 m. Wind at uniform velocity of 15 m/s enters the tunnel at 20°C. Determine the cross-sectional dimensions at the end of the test section which will yield zero pressure gradient along its length. Assume velocity distribution in turbulent boundary layer to follow the law $\frac{u}{U} = \left(\frac{y}{\delta}\right)^{1/5}$.

Because of the growth of boundary layer, the centre line or free stream velocity increases and hence there will be −ve pressure gradient in the tunnel. To make the free stream velocity at both the sections the same (so that $\frac{\partial p}{\partial x} \approx 0$), each side needs to be displaced backwards by δ^*, the displacement thickness. Hence tunnel dimension at exit will be $(1 + 2\delta^*)$ (Fig. 9.12).

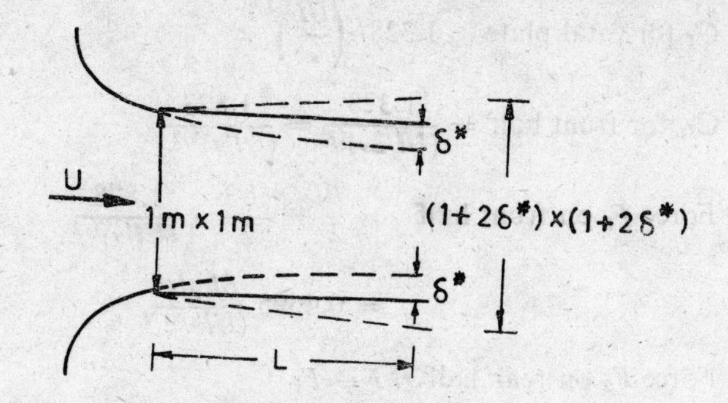

Fig. 9.12

$$\frac{UL}{\nu} = \frac{15 \times 10}{(1.53 \times 10^{-5})} = 0.9804 \times 10^7 \quad \text{or} \quad 9.804 \times 10^6$$

Assuming the boundary layer to be turbulent from the beginning (this can be achieved by providing roughness at the entrance)

$$\frac{\delta}{x} = \frac{0.377}{(Re_x)^{1/5}} = \frac{0.377}{(9.804 \times 10^6)^{1/5}} = 0.01507 \text{ m}$$

or $\delta = 0.01507 \times 10 = 0.1507$ m

Since the velocity distribution in the turbulent boundary layer is given by $u/U = (y/\delta)^{1/5}$

$$\therefore \quad \delta^* = \delta \int_0^1 (1 - \eta^{1/5})\, d\eta \quad \text{where} \quad \eta = \frac{y}{\delta}$$

$$= \delta \left[\eta - \frac{5}{6} \eta^{6/5} \right]_0^1$$

$$= \delta \left(1 - \frac{5}{6} \right) = \frac{\delta}{6}$$

$$\therefore \quad \delta^* = \frac{0\ 1507}{6} = 0.0251 \text{ m}$$

$\therefore \quad 1 + 2\delta^* = 1 + 2 \times 0.0251 = 1.0502$ m

$\therefore$ Cross-section of tunnel at the exit will be 1.0502 m $\times$ 1.0502 m.

9.9 Find the ratio of friction drag on the front half and rear half of the flat plate kept at zero incidence in a stream of uniform velocity, if the boundary layer is laminar over the whole plate.

Reynolds number for whole plate $= \frac{Ul}{\nu}$

For front half of the plate Re_L will be $\frac{Ul}{2\nu}$

C_f for total plate $= 1.328/\left(\dfrac{Ul}{\nu}\right)^{1/2}$

$$\therefore \quad C_{f1} \text{ for front half} = \frac{1.328}{(Ul/2\nu)^{1/2}} = \frac{1.878}{(Ul/\nu)^{1/2}}$$

$$\text{Force } F_1 \text{ on front half} = \frac{Bl}{2} \times \frac{\rho U^2}{2} \times \frac{1.878}{(Ul/\nu)^{1/2}}$$

$$= 0.4695 \frac{Bl\,\rho U^2}{(Ul/\nu)^{1/2}}$$

$$\therefore \quad \text{Force } F_2 \text{ on rear half} = F - F_1$$

$$= (0.6640 - 0.4695)\frac{Bl\,\rho U^2}{(Ul/\nu)^{1/2}}$$

$$= 0.1945 \frac{Bl\,\rho U^2}{(Ul/\nu)^{1/2}}$$

$$\therefore \quad \frac{F_1}{F_2} = \frac{0.4695}{0.1945} = 2.414.$$

9.10 A submarine can be assumed to have cylindrical shape with rounded nose. Assuming its length to be 55 m and diameter 6.0 m, determine the total power required to overcome boundary friction if it cruises at 8.0 m/s velocity in sea water at 20°C ($\rho = 1030$ kg/m³, $\nu = 1 \times 10^{-6}$ m²/s).

$$\frac{UL}{\nu} = \frac{6 \times 55}{10^{-6}} = 4.40 \times 10^8$$

The length over which boundary layer will be laminar is given by $\dfrac{Ux}{\nu} = 5 \times 10^5$.

$$\therefore \quad x = \frac{5 \times 10^5 \times 10^{-6}}{8}$$

$$= 0.0625 \text{ m}$$

This being very small, contribution to total drag from laminar boundary layer is negligible; hence C_f is given by

$$C_f = \frac{0.455}{(\log_{10} Re_L)^{2.58}} = \frac{0.455}{(\log_{10} 4.4 \times 10^8)^{2.58}} = 0.00\,1743$$

$$\text{Area} = \pi Dl = 3.142 \times 6.0 \times 55 = 1036.86 \text{ m}^2$$

$$\therefore \quad F = 1036.86 \times 0.001\,743 \times \frac{1030 \times 8^2}{2}$$

$$= 59\,566.86 \text{ N}$$

$$P = \frac{FU}{1000} \text{ kW} = \frac{59\,566.86 \times 8}{1000}$$

$$P = 476.535 \text{ kW}$$

9.11 If laminar boundary layer prevails on a flat plate, the average drag coefficient C_{fl} is given by $C_{ft} = \frac{1.328}{\text{Re}_L^{1/2}}$.

If the boundary layer over the entire plate is turbulent, the average drag coefficient C_{ft} is given by $C_{ft} = \frac{0.074}{\text{Re}_L^{1/5}}$.

Obtain an expression for C_f when part of the plate is covered by laminar boundary layer. Assume the transition to occur at Reynolds number of 5×10^5.

Let L be the total length of plate, B its width and l that portion of L on which boundary layer is laminar.

$\therefore$ Drag F_l on length $l = \frac{1.328}{\text{Re}_l^{1/2}} \times Bl \times \frac{\rho U^2}{2}$

where $\text{Re}_l = 5 \times 10^5$.

The drag F_{tL} on length L when boundary layer is turbulent is

$$F_{tL} = \frac{0.074}{\text{Re}_L^{1/5}} \times BL \times \rho \frac{U^2}{2}$$

Drag on length l when boundary layer is turbulent will be

$$F_{tl} = \frac{.074}{\text{Re}_l^{1/5}} \times Bl\ \frac{\rho U^2}{2}$$

$\therefore$ Drag on length $(L - l)$ will be

$$F_{t(L-l)} = 0.074B\frac{\rho U^2}{2}\left[\frac{L}{\text{Re}_L^{1/5}} - \frac{l}{\text{Re}_l^{1/5}}\right]$$

Hence total drag on the plate when both laminar and turbulent boundary layers are present on the plate is

$$F = F_l + F_{t(L-l)} = \frac{B\rho U^2}{2}\left[\frac{1.328l}{\text{Re}_l^{1/2}} + \frac{0.074L}{\text{Re}_L^{1/5}} - \frac{0.074l}{\text{Re}_l^{1/5}}\right]$$

However since U and ν are constant, one can write

$$\frac{l}{L} = \frac{\text{Re}_l}{\text{Re}_L} \quad \text{or} \quad l = \frac{L\text{Re}}{\text{Re}_L}.$$

Substitution of this value of l in the above equation

$$F = \frac{B\rho U^2}{2}\left[\frac{1.328L}{\text{Re}_l^{1/2}}\frac{\text{Re}_l}{\text{Re}_L} + \frac{0.074L}{\text{Re}_L^{1/5}} - \frac{0.074L}{\text{Re}_l^{1/5}}\frac{\text{Re}_l}{\text{Re}_L}\right]$$

$$= \frac{LB\rho U^2}{2}\left[\frac{0.074}{\text{Re}_L^{1/5}} - \left(\frac{0.074\,\text{Re}_l^{4/5}}{\text{Re}_L} - \frac{1.328\,R_{el}^{1/2}}{\text{Re}_L}\right)\right]$$

Noting that $\frac{F}{BL\rho U^2/2} = C_f$ and substituting $\text{Re}_l = 5 \times 10^5$, the above equation yields

$$C_f = \left(\frac{0.074}{\text{Re}_L} - \frac{1742}{\text{Re}_L}\right)$$

Compare this equation with Eq. 9.11.

9.12 A smooth rectangular plate of 1.0 m width and 20 m length, when towed through water at 20°C lengthwise, experiences drag of 1440 N on both the sides. Determine (i) average drag coefficient, (ii) velocity of the plate, and (iii) boundary layer thickness at the edge of the plate.

$$F = 2 \times 1 \times 20 \times C_f \frac{\rho U^2}{2} = 1440$$

$$\therefore \quad C_f U^2 = 0.07214$$

Since U is not known, Re_l cannot be computed. Hence assume any reasonable value of C_f between 0.005 and 0.001 (Fig. 9.3). Let us assume $C_f =$ 0.0015.

$$\therefore \quad 0.005\, U^2 = 0.7214$$

and $\quad U^2 = 48.093 \quad$ or $\quad U = 6.935$ m/s

$$\therefore \quad \frac{UL}{\nu} = \frac{6.935 \times 20}{1 \times 10^{-6}} = 1.387 \times 10^8$$

For such high Re_L value, C_f can be computed by the equation

$$C_f = \frac{0.455}{(\log 1.387 \times 10^8)^{2.58}} = 0.002034$$

$\therefore \quad 0.002034 \times U^2 = 0.07214$ which gives $U = 5.955$ m/s.

Again Re_l can be computed and procedure repeated to get more and more refined value of U. After two more trials, one gets $U = 5.889$ m/s and $C_f = 0.00208$

$\therefore \quad \text{Re}_l = \dfrac{20 \times 5.889}{1 \times 10^{-6}} = 1.178 \times 10^8$. For Re_L values in this range,

δ is given by

$$\frac{\delta}{x} = \frac{0.22}{\text{Re}_L^{1/6}} = \frac{0.22}{(1.178 \times 10^8)^{1/6}} = 0.009\,93$$

$$\therefore \quad \delta = 0.009\,93 \times 20 = 0.1986 \text{ m.}$$

9.13 Boeing 727 has a wing of chord length 6.0 m. For standard air, determine the location on the wing where boundary layer will change from laminar to turbulent when it travels at 260 km/hr. What will be the location of this transition if it travels at the same speed at 20,000 m altitude.

$$260 \text{ km/hr} = \frac{260 \times 1000}{3600} = 72.22 \text{ m/s}$$

For standard air condition, $\rho = 1.225$ kg/m³ and $\nu = 1.461 \times 10^{-5}$ m²/s. Hence

$$\frac{Ux}{\nu} = 5 \times 10^5 \quad \text{or} \quad x = \frac{5 \times 10^5 \times 1.461 \times 10^{-5}}{72.22} = 0.101 \text{ m}$$

At 20,000 m altitude (see Appendix E)

$$\nu = 1.599 \times 10^{-4} \text{ m}^2/\text{s}$$

Since

$$Ux/\nu = 5 \times 10^5,\ x = \frac{5 \times 10^5 \times 1.599 \times 10^{-4}}{72.22} = 1.107 \text{ m}$$

Hence it can be seen that substantial portion of the wing will have laminar boundary layer at an altitude of of 20 000 m.

9.14 By application of momentum equation, obtain relationship between local shear stress and momentum thickness variation along x for laminar boundary layer on a flat plate. Also obtain expressions for $\frac{\delta}{x}$ c_f and C_f.

Consider the control volume as shown in Fig. 9.5.

Assume that (i) pressure is constant, (ii) fluid is incompressible, and (iii) flow is steady. Hence the only external force acting on the control volume is $-F_D$ the force exerted by the boundary. Momentum equation gives,

$$\therefore \quad -F_D = \begin{pmatrix}\text{Rate of momentum flux}\\ \text{going out through } CD\end{pmatrix} + \begin{pmatrix}\text{Rate of momentum flux}\\ \text{going out through } BC\end{pmatrix} - \begin{pmatrix}\text{Rate of momentum flux}\\ \text{coming in through } AB\end{pmatrix}$$

$$-F_D = \int_0^h \rho u^2 dy + \int_0^h \rho(U-u)\,U dy - \int_0^h \rho U^2 dy$$

$$= \int_0^h \rho\,(u-U)\,dy$$

But magnitude of $F_D = \int_0^x \tau_0\,dx$

$$\therefore \quad \int_0^x \tau_0\,dx = \int_0^h \rho\,u\,(U-u)\,dy$$

Differentiating both sides w.r.t. x one gets

$$\frac{d}{dx}\left(\int_0^x \tau_0 dx\right) \text{ i.e. } \tau_0 = \rho\frac{d}{dx}\int_0^h u\,(U-u)\,dy$$

But $\int_0^h u\,(U-u)\,dy = U^2\theta$ where θ is the momentum thickness of boundary layer.

$$\therefore \quad \frac{\tau_0}{\rho} = U^2\frac{d\theta}{dx}$$

Expressions of ρ, c_f and C_f can be obtained if one assumes velocity distribution law, in the form

$$\frac{u}{U} = f(\eta) \text{ where } \eta = \frac{y}{\delta}.$$

$$\therefore \qquad \theta = \int_0^\delta \frac{u}{U}\left(1 - \frac{u}{U}\right) dy = \delta \int_0^1 f(1 - f)\, d\eta$$

$\int_0^1 f(1-f)\, d\eta$ will have a constant value α_1 which depends on the function f. Hence $\theta = \alpha_1 \delta$.

Similarly

$$\tau_0 = \mu \left(\frac{\partial u}{\partial y}\right)_{y=0} = \frac{\mu U}{\delta} \left(\frac{\partial f}{\partial \eta}\right)_{\eta=0}$$

Let $\left(\frac{\partial f}{\partial \eta}\right)_{\eta=0} = \beta_1 \quad \therefore \quad \tau_0 = \frac{\mu U}{\delta} \beta_1$

β_1 will also have a constant value for given function $f(\eta)$. Substitution of these values of θ and τ_0 in the momentum equation $\frac{\tau_0}{\rho} = U^2 \frac{d\theta}{dx}$, gives

$$\frac{\mu U \beta_1}{\rho \delta} = U^2 \alpha_1 \frac{d\delta}{dx}.$$

Integration of this equation gives

$\frac{\delta^2}{2} = \frac{u \beta_1 x}{U \rho \alpha_1} + C$, where C is the constant of integration which can be evaluated from the condition $\delta = 0$ at $x = 0$. Hence $C = 0$.

$$\therefore \qquad \frac{\delta}{x} = \sqrt{\frac{2B_1}{\alpha_1}} \frac{1}{\text{Re}_x^{1/2}}$$

Again $\tau_0 = \frac{\mu U \beta_1}{\delta}$. Substitution of the above value of δ in this equation and subsequent simplification yields

$$\frac{\tau_0}{\rho U^2/2} = c_f = \sqrt{2\alpha_1 \beta_1} \frac{1}{\text{Re}_x^{1/2}}$$

Again $F_D = \int_0^L \tau_0\, dx$. Substitution of the value of τ_0 from the above equation and subsequent simplification gives

$$\frac{F}{BL\rho \frac{U^2}{2}} = C_f = 2\sqrt{2\alpha_1 \beta_1} \frac{1}{\text{Re}_L^{1/2}}$$

9.15 If velocity distribution in the boundary layer is given by $\frac{u}{U} = 2\eta - \eta^2$, obtain expressions for $\frac{\delta}{x}$, c_f, C.

$$\alpha_1 = \int_0^1 f(1-f)d\eta = \int_0^1 (2\eta - \eta^2)(1 - 2\eta + \eta^2)d\eta$$

$$= \int_0^1 (2\eta - 4\eta^2 + 2\eta^3 - \eta^2 + 2\eta^2 - \eta^4)d\eta$$

$$= \int_0^1 (2\eta - 5\eta^2 + 4\eta^3 - \eta^4)\, d\eta$$

$$= \left(\frac{2}{2} - \frac{5}{3} + \frac{4}{4} - \frac{1}{5}\right) = \frac{2}{15}$$

$$\beta_1 = \left(\frac{df}{d\eta}\right)_{n-0} = 2$$

$$\therefore \quad \frac{\delta}{x} = \sqrt{\frac{2\beta_1}{\alpha_1}}\,\frac{1}{Re_x^{1/2}} = \frac{(30)^{1/2}}{Re_x^{1/2}}$$

$$\therefore \quad \frac{\mu}{x} = \frac{5.477}{Re_x^{1/2}}$$

$$\sqrt{2\alpha_1\beta_1} = \sqrt{2 \times \frac{2}{15} \times 2} = 0.730$$

$$\therefore \quad c_f = \frac{0.73}{R_{ex}^{1/5}}$$

and $$C_f = \frac{2\sqrt{2\alpha_1\beta_1}}{Re_L^{1/2}} = \frac{2 \times 0.73}{Re_L^{1/2}} = \frac{1.46}{Re_L^{1/2}}$$

9.16 If velocity distribution is laminar boundary layer is given by $\frac{u}{U} = \frac{3}{2}\eta - \frac{1}{2}\eta^3$, obtain expressions for θ/δ, $\frac{\delta^*}{\theta}$, c_f and $\frac{\delta}{x}$

$$\alpha_1 = \int_0^1 f(1-f)\, d\eta$$

$$= \int_0^1 \left(\frac{3}{2}\eta - \frac{1}{2}\eta^3\right)\left(1 - \frac{3}{2}\eta + \frac{1}{2}\eta^3\right) d\eta$$

$$= \int_0^1 \left(\frac{3}{2}\eta - \frac{9}{4}\eta^2 + \frac{3}{4}\eta^4 - \frac{1}{2}\eta^3 + \frac{3}{4}\eta^4 - \frac{1}{4}\eta^6\right) d\eta$$

$$= \frac{3}{4} - \frac{3}{4} + \frac{3}{20} - \frac{1}{8} + \frac{3}{20} - \frac{1}{28}$$

$$= \frac{39}{280}$$

If $$\alpha_2 = \int_0^1 (1 - f)\, d\eta$$

$$= \int_0^1 \left(1 - \frac{3}{2}\eta + \frac{1}{2}\eta^3 \right) d\eta$$

$$= \left[\eta - \frac{3}{4}\eta^2 + \frac{1}{8}\eta^4 \right]_0^1 = \left(1 - \frac{3}{4} + \frac{1}{8} \right)$$

$$= 3/8$$

$$\beta_1 = \left(\frac{d\partial}{d\eta}\right)_{\eta=0} = \left(\frac{3}{2} - \frac{3}{2}\eta^2\right)_{\eta=0} = 3/2$$

$$\therefore \quad \frac{\theta}{\delta} = \alpha_1 = 0.139, \quad \frac{\delta^*}{\theta} = \frac{\alpha_2}{\alpha_1} = 2.815,$$

$$c_f = \frac{\sqrt{2\alpha_1\beta_1}}{\sqrt{\text{Re}_x}} = \frac{0.646}{\sqrt{\text{Re}_x}},$$

$$\frac{\delta}{x} = \sqrt{\frac{2\beta_1}{\alpha_1}} \frac{1}{\sqrt{\text{Re}_x}} = 4.64/\sqrt{\text{Re}_x}$$

Since $$C_f = \frac{2\sqrt{2\alpha_1\beta_1}}{\sqrt{\text{Re}_l}} = \frac{1.292}{\sqrt{\text{Re}_l}}$$

9.17 For a passenger ship of 300 m length and 12.0 m draft, 11000 kW power is required when it cruises at 40 km/hr. If $\rho = 1030$ kg/m³ and $\nu = 1 \times 10^{-6}$ m²/s, determine the combined form and wave resistance of the ship.

$$U = 40 \times 1000/3600 = 11.11 \text{ m/s}$$

$$\therefore \quad \text{Re}_l = 11.11 \times 300/10^{-6} = 3.333 \times 10^9$$

At this Reynolds number, the boundary layer will be turbulent on almost the whole length. Hence

$$C_f = 0.455/(\log 3.333 \times 10^9)^{2.58}$$

$$= 0.001\ 358$$

$$F_{\text{friction}} = 2 \times 300 \times 12 \times 0.001\ 358 \times \frac{1030 \times 11.11^2}{2}$$

$$= 621{,}649.79 \text{ N} \quad \text{or} \quad 621.650 \text{ kN}$$

Since total power required $P = FU$

Total force $F = \dfrac{11\,000}{11.11} = 990.01$ kN

Since

$$F = F_{\text{friction}} + (F_{\text{form}} + F_{\text{wave}})$$

$$(F_{\text{form}} + F_{\text{wave}}) = 990.01 - 621.65 = 368.36 \text{ kN}$$

9.18 If the laminar boundary layer on a flat plate has the velocity distribution

$$f = \frac{u}{U} = \frac{3}{2}\eta - \frac{1}{2}\eta^3$$

where $\eta = y/\delta$, determine maximum value of parameter χ (Chi) in terms of $U\delta/\nu$ and Ux/ν.

Since

$$\frac{u}{U} = \frac{3}{2}\eta - \frac{1}{2}\eta^3$$

$$\frac{du}{dy} = U\frac{du}{d\eta}\frac{d\eta}{dy} = \frac{U}{\delta}\left(\frac{3}{2} - \frac{3}{2}\eta^2\right); \text{ also } y = \delta\eta$$

$$\therefore \qquad \chi = \frac{\rho y^2\, du/dy}{\mu} = \frac{\rho U\delta}{\mu}\left(\frac{3}{2}\eta^2 - \frac{3}{2}\eta^4\right)$$

For χ to be maximum

$$\frac{d\chi}{dy} = \frac{d\chi}{d\eta}\frac{1}{\delta} = \frac{\rho U\delta}{\mu}(3\eta - 6\eta^3) = 0$$

or $\qquad \eta^2 = 0.5 \quad \text{and} \quad \eta = 0.707$

$$\therefore \qquad \chi_{\max} = \frac{\rho U\delta}{\mu}\left(\frac{3}{2} \times 0.5 - \frac{3}{2} \times 0.5^2\right) = \frac{3}{8}\frac{U\delta\rho}{\mu}$$

Since flow becomes unstable at $\chi_{\max} = 500$.

Therefore, instability will occur in laminar boundary layer with the above velocity distribution when

$$\frac{U\delta\rho}{\mu} = \frac{500 \times 8}{3} = 1330.$$

However, for this velocity distribution (see Ex. 9.16)

$$\frac{\delta}{x} = 4.64/\sqrt{\text{Re}_x}$$

$$\therefore \qquad \delta = \frac{4.64\nu^{1/2}x^{1/2}}{U^{1/2}}$$

$$\therefore \qquad \frac{U\delta}{\nu} = 4.64\,\frac{U^{1/2}x^{1/2}}{\nu^{1/2}} = 4.64\left(\frac{Ux}{\nu}\right)^{1/2}$$

$$\therefore \quad \frac{Ux}{\nu} = \left(\frac{U\delta}{\nu}\right)^2 \frac{1}{4.64^2} = \frac{1330^2}{4.64^2} = 82\ 161$$

$\therefore \quad \dfrac{Ux}{\nu}$, when flow becomes unstable, is 82 161.

9.19 By a simple analysis show that the establishment length for laminar flow in a pipe in dimensionless form is proportional to pipe Reynolds number, i.e.

$$\frac{L_e}{D} = \text{const. Re}$$

What assumptions have you made in the derivation?

See Fig. 9.6. At the entrance to the pipe the free stream velocity is U. When the laminar boundary layer reaches the centre line, centre line velocity $u_m = 2U$. As a result, the pressure decreases in downstream direction and hence boundary layer development is under −ve pressure gradient and boundary layer growth is suppressed. For simplicity, neglect this effect and assume that the development of boundary layer takes place at constant velocity U. Hence

$$\text{At} \quad x = L_e \quad \delta = \frac{D}{2}$$

$$\text{and} \quad \frac{\delta}{x} = \frac{5}{(Ux/\nu)^{1/2}}$$

Substitution of these values yields

$$\frac{D}{2L_e} = \frac{5}{(UL_e/\nu)^{1/2}} = \frac{5}{(UD/\nu)^{1/2}} \left(\frac{L_e}{D}\right)^{1/2}$$

$$\therefore \quad \left(\frac{UD}{\nu}\right)^{1/2} \frac{1}{10} = \left(\frac{L_e}{D}\right)^{1/2} \quad \text{or} \quad \frac{L_e}{D} = 0.01\ \text{Re}$$

As mentioned earlier experimentally the constant is found to be 0.07.

The assumptions made in the derivation are

(i) Actually free stream velocity increases in x direction but is assumed to be constant equal to U. Therefore the effect of pressure gradient on boundary layer development is neglected.

(ii) Curved boundary effect is negligible.

9.20 If turbulent boundary layer on a flat plate has the velocity distribution given by $\dfrac{u}{u_*} = 8.74 \left(\dfrac{u_* y}{\nu}\right)^{1/7}$, obtain the relationship for variation of δ/x.

Since at $y = \delta$, $u = U$

$$\frac{U}{u_*} = 8.74 \left(\frac{u_*\delta}{\nu}\right)^{1/7}$$

Dividing the first expression by the second, one gets $\frac{u}{U} = \left(\frac{y}{\delta}\right)^{1/7}$ and as seen in Prob. 9.21 for such a velocity distribution law

$$\theta/\delta = \frac{n}{(n+2)(n+1)} = \frac{7}{72}$$

$$\therefore \qquad \frac{d\theta}{dx} = \frac{7}{72}\frac{d\delta}{dx}$$

Also from the equation $U/u_* = 8.74\left(\frac{u_*\delta}{\nu}\right)^{1/7}$, one gets

$$u_* = \frac{U^{7/8}}{8.74^{7/8}}\left(\frac{\nu}{\delta}\right)^{1/8} = \frac{U^{7/8}}{6.6653}\left(\frac{\nu}{\delta}\right)^{1/8}$$

and

$$\frac{\tau_0}{\rho} = u_*^2 = \frac{U^{7/4}}{44.426}\left(\frac{\nu}{\delta}\right)^{1/4}$$

Momentum equation (see Ex. 9.14) gives the relation

$$\frac{\tau_0}{\rho} = U^2\frac{d\theta}{dx}$$

On substitution of values of $\frac{\tau_0}{\rho}$ and $\frac{d\theta}{dx}$ from above, one gets

$$\frac{U^{7/4}}{44.426}\left(\frac{\nu}{\delta}\right)^{1/4} = U^2\frac{7}{72}\frac{d\delta}{dx}$$

$$\therefore \qquad \delta^{1/4}d\delta = 0.2315\left(\frac{\nu}{U}\right)^{1/4}dx$$

or

$$\frac{4}{5}\delta^{5/4} = 0.2315\left(\frac{\nu}{U}\right)^{1/4}x$$

since the constant of integration is zero. This gives

$$\left(\frac{\delta}{x}\right)^{5/4} = \frac{0.2315\times 5}{4}\frac{1}{\text{Re}_x^{1/4}}$$

or

$$\frac{\delta}{x} = \frac{0.371}{\text{Re}_x^{1/5}}$$

9.21 Show that in the length of establishment for laminar flow in a circular horizontal pipe, the frictional force of the pipe wall is given by

$$F = \pi R^2\left(p_1 - p_2 - \frac{1}{3}\rho U^2\right)$$

where U is the average velocity of flow in the pipe of radius R, and p_1 and p_2 are the pressures at sections 1 and 2 respectively when the pipe is horizontal.

As seen earlier, the velocity distribution at section 2 is fully developed and is given by

$$\frac{u}{U} = 2\left(1 - \left(\frac{r}{R}\right)^2\right)$$

The momentum correction factor β for this velocity distribution is 1.33. Hence one can apply momentum equation between sections 1 and 2 in Fig. 9.6(a); the forces to be considered are pressure and friction. Hence

$$(\pi R^2 \rho U^2 \beta - \pi R^2 \rho U) = \pi R^2 (p_1 - p_2) - F$$

$$\therefore \quad F = \pi R^2 (p_1 - p_2 - 0.33 \rho U^2)$$

9.22 Determine the ratio of δ^*/θ for laminar flow in a circular pipe.

Here the velocity distribution is given by

$$u = -\frac{R^2}{4\mu}\frac{\partial p}{\partial x}\left(1 - \frac{r^2}{R^2}\right) = u_m\left(1 - \frac{r^2}{R^2}\right)$$

If δ^* is the displacement thickness, by extending the definition of b.L. on plate to axisymmetric flow, one can write

$$u_m\left\{\pi R^2 - \pi (R - \delta^*)^2\right\} = \int_0^R 2\pi r dr\, (u_m - u)$$

Left hand side $\approx 2\pi R \delta^* u_m$ if δ^{*2} is neglected.

$$\text{R.H.S.} = 2\pi \int_0^R \left(u_m - u_m + u_m \frac{r}{R^2}\right) r dr$$

$$= \frac{2\pi u_m}{R^2}\frac{R^4}{4} = \frac{\pi u_m R^2}{2}$$

$$\therefore \quad 2\pi R \delta^* u_m = \frac{\pi u_m R^2}{2} \quad \text{or} \quad \delta^* = \frac{R}{4}$$

In the same way, the momentum thickness θ can be defined for this flow as

$$2\pi R \theta u_m^2 \rho = \int_0^R 2\pi r dr\, \rho u\, (u_m - u)$$

$$\therefore \quad R\theta = \int_0^R \frac{u}{u_m}\left(1 - \frac{u}{u_m}\right) dr$$

$$= \int_0^R \left(1 - \frac{r^2}{R^2}\right)\frac{r^2}{R^2}\, r dr$$

$$= \left(\frac{R^2}{4} - \frac{R^2}{6}\right) = \frac{R^2}{12} \quad \therefore \quad \theta = \frac{R}{12}$$

$$\therefore \quad \delta^*/\theta = \frac{12}{4} = 3,$$

9.23 Determine whether the pipe will act as hydrodynamically smooth, in transition or rough in the following cases:

(i) $D = 300$ mm, $l = 50$ m, drop in pressure $= 4.2$ kN/m^2

$k = 0.02$ mm, $\rho = 998$ kg/m^3, $\nu = 0.95 \times 10^{-6}$ m^2/s

(ii) $\tau_0 = 638.78$ N/m^2, $\rho = 998$ kg/m^3, $\nu = 10^{-6}$ m^2/s and

$k = 2.0$ mm for rivetted steel pipe.

In the first case,

$$\tau_0 = (p_1 - p_2)\frac{D}{4l} = \frac{4.2 \times 10^3 \times 0.30}{4 \times 50} = 6.3 \text{ N/m}^2$$

$$u_* = \sqrt{\tau_0/\rho} = \sqrt{6.3/998} = 0.07945 \text{ m/s}$$

$$\delta' = 11.6\nu/u_* = (11.6 \times 0.95 \times 10^{-6})/0.07945$$

$$= 138.704 \times 0.95 \times 10^{-6} \text{ m} \quad \text{or} \quad 0.1318 \text{ mm}$$

$$\therefore \quad k/\delta' = 0.02/0.1318 = 0.151$$

Since $k/\delta' < 0.25$, the boundary will act as hydrodynamically smooth boundary.

In the second case,

$$u_* = \sqrt{638.78/998} = 0.80 \text{ m/s}$$

$$\delta' = 11.6\nu/u_* = \frac{11.6 \times 10^{-6}}{0.80} = 14.5 \times 10^{-6} \text{ m} \quad \text{or} \quad 0.0145 \text{ mm}$$

$$\therefore \quad k/\delta' = 2.0/0.145 = 137.93.$$

Since this value is greater than 6.0, the boundary will act as rough boundary.

9.24 In the case of wind flowing over an agricultural field, velocity at 4.0 m above the ground was found to be 12.0 m/s. If the crop height is 1.5 m, determine the boundary shear and velocity at 20 m above the ground.

$$h = 1.5 \text{ m}; \quad d = 0.63 \quad h = 0.63 \times 1.5 = 0.945 \text{ m}$$

$$k = 0.15 \ h = 0.15 \times 1.5 = 0.225 \text{ m}$$

Also Karman constant $K = 0.40$.

Since $\dfrac{u}{u_*} = \dfrac{2.3}{K} \log \dfrac{y-d}{k}$

$$\frac{12}{u_*} = \frac{2.3}{0.40} \log \frac{(4 - 0.945)}{0.225}$$

$$\therefore \quad u_* = 1.842 \text{ m/s}$$

$$\therefore \quad \tau_0 = \rho u_*^2 = 1.208 \times 1.842^2$$

$$= 4.0998 \text{ N/m}^2$$

Also $\dfrac{u}{1.842} = \dfrac{2.3}{0.40} \log \left(\dfrac{20 - 0.945}{0.225}\right)$

$$\therefore \quad u = 20.419 \text{ m/s.}$$

PROBLEMS

9.1 Velocity measurements in the vertical were made in an open circuit wind tunnel of cross section 0.30 m × 0.30 m at a certain section along its length. These are recorded below

y mm	2	4	6	8	10	12	16	20	500	100	150
u m/s	4.01	4.61	4.98	5.29	5.53	5.74	5.95	6.0	6.11	5.90	6.06

Plot y vs u on natural scale. Determine free stream velocity U and δ. Also plot u/U vs y/δ on log-log scale and determine n in the equation $u/U = (y/\delta)^{1/n}$. (6.0 m/s, 15 mm, 5.0)

9.2 Classify the boundary layers in the following cases, indicating the length over which the boundary layer is laminar, if it is turbulent.

(i) $U = 10.0$ m/s, $L = 8.0$ m, fluid: air at 20°C

(ii) $U = 3.0$ m/s, $L = 1.50$ m, fluid: glycerine at 20°C

[(i) turbulent, 0.765 m, (ii) laminar]

9.3 For the velocity distribution given in Problem 9.1, obtain δ^* and θ values by numerical integration. (3.03 mm, 1.82 mm)

9.4 If velocity distribution within the laminar boundary layer on a flat plate is assumed to be of the form $u = a \sin(by) + c$ where a, b, c are constants, determine the velocity distribution law.

$$\left(u = U \sin\left(\frac{\pi}{2}\frac{y}{\delta}\right)\right)$$

9.5 If velocity distribution law in laminar boundary layer over a flat plate is assumed as

(i) $u = ay + by^3$, and

(ii) $u = ay + by^3 + cy^4$

apply the appropriate boundary conditions and determine the velocity distribution laws.

$$\text{(i)} \quad \frac{u}{U} = \frac{3}{2}\left(\frac{y}{\delta}\right) - \frac{1}{2}\left(\frac{y}{\delta}\right)^3$$

$$\text{(ii)} \quad \frac{u}{U} = 2\left(\frac{y}{\delta}\right) - 2\left(\frac{y}{\delta}\right)^3 + \left(\frac{y}{\delta}\right)^4$$

9.6 If $u/U = [1 - \exp(5y/\delta)]$ represents the velocity distribution law within the boundary layer, check which boundary conditions does it satisfy exactly and which boundary conditions it satisfies approximately and to what accuracy.

9.7 Compare the boundary layer thickness on a flat plate 0.67 m from the leading edge (i) in air at 20°C and (ii) in water at 20°C, if in both cases fluid velocity is 1.25 m/s. (13.55 mm, 15.11 mm)

9.8 Air at 20°C ($\rho = 1.208$ kg/m^3 ; $\mu = 1.85 \times 10^{-5}$ kg/m s) flows over a 2.0 m wide plate at 10.0 m/s velocity. Determine (i) τ_0 and δ at a place where the boundary layer ceases to be laminar (ii) Force on one side of the plate in the laminar region and (iii) boundary layer thickness at 10.0 m from leading edge.

(0.055 41 N/m^2, 5.414 mm; 0.1736 N; 0.1634 m)

9.9 Air flows over a 1.0 m wide and 0.60 m long flat plate at a velocity of 8.0 m/s. If the kinematic velocity of air is 1.44×10^{-5} m^2/s, determine (i) δ at the end of the plate (ii) shear stress at 0.30 m from the leading edge of the plate, and (iii) total drag force on both sides of the plate. Take $\rho = 1.208$ kg/m^3.

(5.2 mm, 0.0622 N/m^2, 0.107 N)

9.10 Find the ratio of friction drag on the front half and rear half of the flat plate kept parallel to a flowing stream when the boundary layer on the whole plate is turbulent, C_f being given by

$$C_f = 0.074/\mathrm{Re}_L{}^{1/5} \qquad (1.349)$$

9.11 For laminar boundary layer on a flat plate held parallel to a stream of uniform velocity, determine the location of the section where drag up to that section is twice the drag on the remaining portion.

(0.444 L)

9.12 A two metre wide and 5.0 m long plate when towed through water at 20°C experiences a drag of 30.38 N on both the sides. Determine the velocity of the plate and its length over which the boundary layer is laminar. (1.0 m/s, 0.50 m)

9.13 A small boat of 20 m length and 1.5 m draft is propelled in water at 10.0 m/s velocity. Determine the shear stress at 10 m downstream of the leading edge and thickness of laminar sublayer there. Also determine the power required for propulsion.

(86.363 N/m^2, 0.0394 mm, 57.934 kW)

9.14 If velocity distribution in laminar boundary layer is given by $\frac{u}{U} = \eta$, where $\eta = \frac{y}{\delta}$, obtain values/expressions for θ/δ, δ/x, δ^*/θ.

$$\left(0.333, \frac{3.64}{\sqrt{\mathrm{Re}_x}}, 2.998\right)$$

9.15 Assume velocity distribution in laminar boundary layer over a flat plate to follow the law

$$\frac{u}{U} = \sin\left(\frac{\pi}{2}\frac{y}{\delta}\right).$$ Obtain expressions for $\frac{\delta^*}{\theta}$ and δ/x.

$$\left(\frac{2(\pi - 2)}{(4 - \pi)}, 4.80/\sqrt{\mathrm{Re}_x}\right)$$

9.16 Laminar boundary layer over a flat plate has the velocity distribution given by

$$\frac{u}{U} = 2\left(\frac{y}{\delta}\right) - \left(\frac{y}{\delta}\right)^2$$

Determine values of θ/δ, δ^*/θ and expression for $\dfrac{\delta}{x}$.

$$\left(0.133,\ 1.373,\ \frac{5.477}{\sqrt{Re_x}}\right)$$

9.17 Water flows over a 0.30 m long and 0.1 m wide flat plate at 15.0 m/s parallel to it. Calculate (i) drag force on that portion of the plate over which the boundary layer is laminar and (ii) total drag force on both sides of the plate. Assume $\rho = 998$ kg/m³ and $\nu = 1 \times 10^{-6}$ m²/s. (1.392 N, 20.735 N)

9.18 A rectangular plate of sides a and b is towed through water once in the direction of a at the velocity U_a and then in the direction of b at the velocity U_b. If in both the cases the boundary layer is laminar, determine the ratios of U_a and U_b which will give equal force on the plate. $U_a/U_b = (a/b)^{1/3}$

9.19 One meter wide and 25 m long plate is towed through water at 20°C parallel to its plane, at such a velocity that it experiences a total drag of 1800 N on both of its sides. Determine C_f, U and δ. (0.002 08, 5.89 m/s, 0.20 m)

9.20 A 300 m long passenger ship has 12 m draft. Assuming the ship's surface to act as a flat plate, determine the total friction drag when the ship travels at 45 km/hr. Take $\rho = 1000$ kg/m³ and $\nu = 1 \times 10^{-6}$ m²/s.
Also determine the power required to overcome this resistance. (751.681 kN, 9396.01 kW)

9.21 If velocity distribution in turbulent boundary layer over a flat plate is given by $u/U = (y/\delta)^{1/n}$, show that

$$\delta^*/\delta = \frac{1}{n+1} \quad \text{and} \quad \theta/\delta = \frac{h}{(n+2)(n+1)}$$

9.22 Determine the length of the pipe required to establish flow of 230 l/s of glycerine at 20°C in 150 mm diameter pipe. What will be this length if the same discharge of water is carried through the pipe at the same temperature? (3.093 m, 7.50 m)

9.23 Two velocity distribution laws are given for the laminar boundary layer, namely (i) $\dfrac{u}{U} = \dfrac{3}{2}\eta - \dfrac{1}{2}\eta^3$ and (ii) $\dfrac{u}{U} = -2\eta + \dfrac{1}{2}\eta^3$.

State whether the boundary layer is separated or not. (Note: check the sign of (du/dy) at $y = 0$.) (No, yes)

9.24 For laminar flow between parallel plates obtain the value of δ^*/θ. Use the velocity distribution law

$$\frac{u}{u_m} = 4\left(\frac{y}{B} - \frac{y^2}{B^2}\right). \tag{2.5}$$

9.25 Determine whether the pipe will act as hydrodynamically smooth, in transition or rough in the following cases:

(i) $D = 100$ mm, $Q = 3.925$ l/s, $\rho = 1260$ kg/m³, $\nu = 6.65 \times 10^{-4}$ m²/s and $k = 1.0$ mm

(ii) $D = 300$ mm, $\frac{\partial p}{\partial x} = 84$ N/m³, $\rho = 998$ kg/m³, $\nu = 1 \times 10^{-6}$ m²/s and $k = 0.20$ mm.

[(i) laminar and hence roughness does not affect flow, (ii) in transition]

9.26 In a flat open country, wind velocity at 10 m above the ground is 30.0 m/s. Estimate geostrophic wind velocity U and velocity at 150 m above the ground. (51.7 m/s, 46.27 m/s)

9.27 Over a field with standing crop 15.0 m/s, wind velocity was observed at 5.0 m above the ground level. If shear at the ground is 5.0 N/m² and ρ for air is 1.20 kg/m³, estimate the crop height. (1.437 m)

DESCRIPTIVE QUESTIONS

9.1 Give four examples in every day life where formation of boundary layer is important.

9.2 What boundary conditions must be satisfied by the velocity distribution in laminar boundary layer over a flat plate?

9.3 Which of the following velocity distributions satisfy the essential boundary conditions for laminar boundary layer on a flat plate?

(i) $\frac{u}{U} = \eta$ (ii) $\frac{u}{U} = \cos\left(\frac{\pi}{2}\eta\right)$ (iii) $\frac{u}{U} = 2\eta - \eta^2$

(iv) $\frac{u}{U} = \sin\left(\frac{\pi}{2}\eta\right)$ (v) $\frac{u}{U} = 2\eta - \eta^3$ (vi) $\frac{n}{U} = \eta^2$

(vii) $\frac{u}{U} = \frac{3}{2}\eta - \frac{1}{2}\eta^3$ here $\eta = y/\delta$.

9.4 What is the physical significance of displacement thickness of boundary layer?

9.5 What will be the ratio of shear stresses at two sections on a flat plate, where $\frac{x_1}{x_2} = 3$ if the boundary layer on the whole plate is laminar?

9.6 Will the laminar boundary layer on a flat plate held at zero incidence always turn into turbulent boundary layer at $Re_x = 5\times10^5$? Explain.

9.7 Why is it necessary to control the growth of boundary layer on most of the bodies? What methods are used for such a control?

9.8 Sketch and compare the boundary layer development on a flat plate in the following two cases: (i) when the plate is non-porous and (ii) when the plate is porous and fluid is sucked through the pores at a constant rate.

9.9 Is the flow within the boundary layer rotational or irrotational?

9.10 It is stated that the pressure distribution within the boundary layer is determined by the outside flow which can be treated as inviscid. Explain.

9.11 If velocity distribution within the turbulent boundary layer follows the law, $u/U = (y/\delta)^{1/n}$, where $n = 7$, why is the shear τ_0 not equal to infinity at $y = 0$?

9.12 Which one of the following statements is true? For boundary layer formation on a rough plate the sequence of flow from the leading edge to the trailing edge is

(i) laminar, transition, smooth, transition from smooth to rough, and rough.
(ii) laminar, transition, rough, transition from rough to smooth, and smooth.
(iii) laminar, turbulent, smooth.
(iv) laminar, turbulent, rough.

9.13 Other conditions remaining the same, which of the following velocity distributions is more susceptible to separation under favourable conditions?

(i) $u/U = (y/\delta)^{1/6}$, (ii) $u/U = (y/\delta)^{1/10}$.

9.14 For boundary layer development on a flat plate, draw C_f vs Re_L diagram for Re_L values ranging from 10^3 to 10^9, showing on it various equations and limits of their applicability.

9.15 Give five examples in every day life where separation takes place. Draw rough sketches to indicate the flow pattern in each case. In each case mention how separation can be minimised or eliminated.

9.16 Give one illustration in which separation is beneficial.

9.17 An aeroplane is in flight at sea level and also at a much higher altitude. For the same speed of the aeroplane, will the boundary layer thickness at the edge of the wing be larger at sea level or at higher altitude?

9.18 The laminar boundary layer on a flat plate increases in its thickness as (i) x, (ii) $x^{4/5}$ (iii) $x^{1/7}$ (iv) $x^{1/2}$ (v) none of the above.

9.19 Boundary layer thickness on a flat plate

(i) increases faster when boundary layer is turbulent
(ii) increases faster when boundary layer is laminar
(iii) increases at the same rate whether boundary layer is laminar or turbulent

9.20 Separation takes place when

(i) $\left(\frac{du}{dy}\right)_{y=0}$ is +ve (ii) $\left(\frac{du}{dy}\right)_{y=\delta} = 0$ (iii) $\left(\frac{\partial u}{\partial y}\right)_{y=0} = 0$

(iv) $\left(\frac{\partial u}{\partial y}\right)_{y=0}$ is −ve.

CHAPTER X

Turbulent Flow

10.1 CHARACTERISTICS OF TURBULENT FLOW

As mentioned earlier, for a real fluid there is a fixed value of characteristic Reynolds number which, when exceeded, gives rise to turbulent flow. Turbulent flow occurs more often in nature and in engineering applications than laminar flow. Characteristics of turbulent flow can be summarised as follows;

1. 'No Slip' condition is satisfied at the boundary.
2. Turbulence is generated due to instability of flow in regions of high shear, i.e. near the boundary or at the interface of two moving layers. The former is called 'wall turbulence' while the latter is called 'free turbulence'.
3. Local velocity components, pressure, force or any other quantity associated with flow (such as local concentration of sediment) show random fluctuations. Hence the instantaneous value is the sum of a time-averaged value and a fluctuating component (Fig. 10.1).

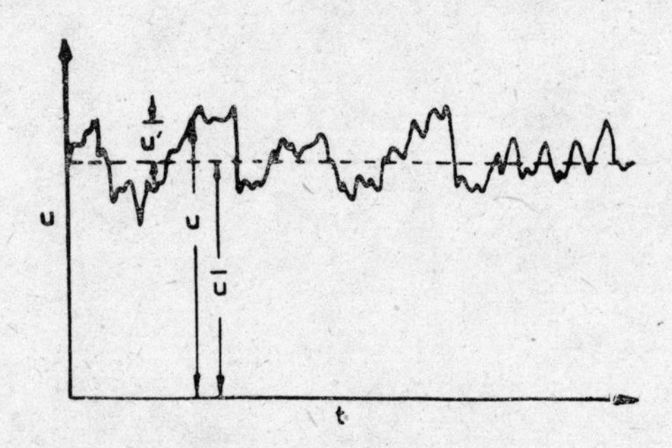

Fig. 10.1. Definition sketch.

$$\left.\begin{aligned} u &= \bar{u} + u' \\ v &= \bar{v} + v' \\ w &= \bar{w} + w' \\ p &= \bar{p} + p' \\ F &= \bar{F} + F' \end{aligned}\right\} \tag{10.1}$$

where the quantities with bar indicate time averaged value, e.g.

$$u = \frac{1}{T}\int_0^T u dt \tag{10.2}$$

Here T is the sampling time and quantity with prime indicates the fluctuating part which can be +ve, −ve or zero. Evidently $\overline{u'} = \overline{v'} = \overline{w'} = 0$. These fluctuations in u, v, w and p are zero at $y = 0$.

4. Vigorousness of turbulence at a given point is measured by turbulence intensity. For mean flow in x direction, it is defined as

$$\mathrm{I} = \sqrt{\frac{1}{3}(\overline{u'^2} + \overline{v'^2} + \overline{w'^2}} \Big/ \bar{u} \tag{10.3}$$

5. In pipe flow $\overline{u'^2} > \overline{v'^2} > \overline{w'^2}$.
6. Turbulent flow is characterised by the presence of circulatory fluid masses known as eddies. At any time, eddies of a wide spectrum of sizes are present in the flow. The largest eddy will be of the size of the flow itself, e.g. pipe diameter or depth of flow in channel. The smallest eddy will be extremely small.
7. Presence of eddies in the flow makes it capable of efficient transport of momentum, mass or engery across the flow.
8. Large eddies capture energy from the mean flow; they break into smaller and smaller eddies until eddy Reynolds number, (size of eddy × characteristic turbulence velocity/kinematic viscocity), becomes very small. Ultimately the small eddies die out and their energy is dissipated through viscous shear. As a result, turbulence causes much greater energy loss than in laminar flow. Head loss is proportional to U^n, where n varies between 1.75 and 2.
9. Presence of turbulent fluctuations in velocities causes additional normal and tangential stresses at any point. These constitute what is known as Reynolds stress tensor which has nine components: 3 normal stresses and 6 tangential stresses. These are given as

$$\begin{vmatrix} -\rho\overline{u'^2} & -\rho\overline{u'v'} & -\rho\overline{u'w'} \\ -\rho\overline{v'u'} & -\rho\overline{v'^2} & -\rho\overline{v'w'} \\ -\rho\overline{w'u'} & -\rho\overline{w'v'} & -\rho\overline{w'^2} \end{vmatrix}$$

The normal stresses are $-\rho\overline{u'^2}$, $-\rho\overline{v'^2}$, $-\rho\overline{w'^2}$, the remaining being tangential stresses. Naturally $\rho\overline{u'v'} = \rho\overline{v'u'}$ etc.

10.2 TURBULENT SHEAR

For flow in a pipe or wide open channel, Boussinesq expressed the turbulent shear stress $\bar{\tau}$ as

$$\bar{\tau} = \eta \frac{d\bar{u}}{dy} = \rho\epsilon\frac{d\bar{u}}{dy} \tag{10.4}$$

where η is eddy dynamic viscosity and ϵ is eddy kinematic viscosity. Normally η and ϵ are much greater than μ and ν respectively and are functions of flow characteristics. From dimensional considerations, one can write that $\epsilon \sim$ (eddy size) $\times$ characteristic turbulence velocity, e.g. $\sqrt{\overline{v'^2}}$.

According to Prandtl's mixing length hypothesis

$$\bar{\tau} = -\rho\overline{u'v'} = \rho l^2\left(\frac{d\bar{u}}{dy}\right)^2 \tag{10.5}$$

where l is the mixing length. The variation of l/R with y/R for pipe flow is given by

$$\frac{l}{R} = K\frac{y}{R}\left(1 - 1.1\frac{y}{R} + 0.6\frac{y^2}{R^2} - 0.15\frac{y^3}{R^3}\right)$$

which reduces to

$$l = Ky \tag{10.6}$$

where K is known as Karman constant and has a value of 0.40. If one makes the assumption that for small values of $y, \tau = \tau_0$ the boundary shear stress, one can integrate Eq. 10.5 to get the velocity distribution laws for turbulent flow near hydrodynamically smooth and rough boundaries. Figure 10.2 shows variation of $\overline{u'^2}$, $\overline{v'}$ and $\overline{u'v'^2}$ for flow in conduits, while Fig. 10.3 shows variation of l/R and ϵ.

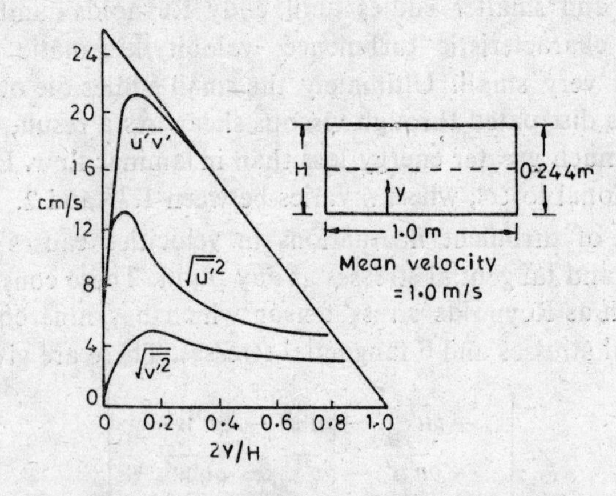

Fig. 10.2. Variation of $\sqrt{\overline{u'^2}}$, $\sqrt{\overline{v'^2}}$ and $\overline{u'v'}$ for conduit flow.

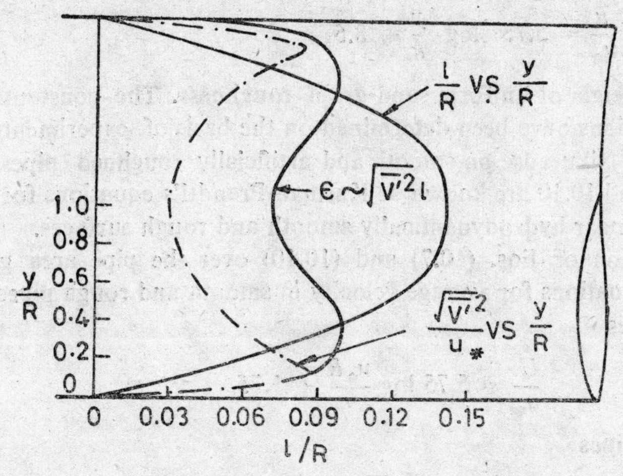

Fig.10.3. Variation l/R, $\frac{\sqrt{\overline{v'^2}}}{u_*}$ and ϵ with y/R.

10.3 VELOCITY DISTRIBUTION IN TURBULENT FLOW

Integration of Eq. 10.5 using Eq. 10.6 and a boundary condition, gives the following equations for velocity distribution.

(a) Near hydrodynamically smooth surface

$$\frac{u}{u_*} = 5.75 \log \frac{u_* y}{\nu} + 5.5 \quad \text{for } \frac{u_* y}{\nu} > 70 \qquad (10.7)$$

$$\frac{u}{u_*} = 11.5 \log \frac{u_* y}{\nu} - 3.05 \quad \text{for } 70 > \frac{u_* y}{\nu} > 5.0 \qquad (10.8)$$

$$\frac{u}{u_*} = \frac{u_* y}{\nu} \quad \text{for } \frac{u_* y}{\nu} < 5.0 \qquad (10.9)$$

as in Fig. 10.4.

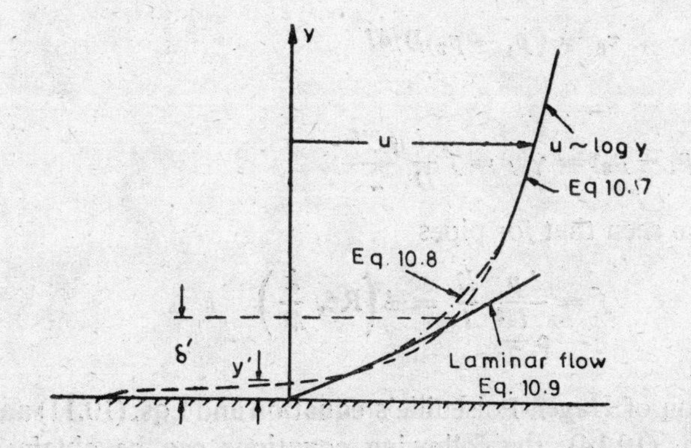

Fig. 10.4. Velocity distribution near hydrodynamically smooth surface.

(b) Near hydrodynamically rough surface

$$\frac{u}{u_*} = 5.75 \log \frac{y}{k} + 8.5 \tag{10.10}$$

where k is height of uniform sand grain roughness. The constants in the above equations have been determined on the basis of experimental data, collected by Nikuradse on smooth and artificially roughned pipes. Equations 10.7 and 10.10 are known as Karman-Prandtl's equations for velocity distribution near hydrodynamically smooth and rough surfaces.

Integration of Eqs. (10.7) and (10.10) over the pipe area gives the following equations for average velocity in smooth and rough pipes.
Smooth pipes

$$\frac{U}{u_*} = 5.75 \log \frac{u_* R}{\nu} + 1.75 \tag{10.11}$$

Rough pipes

$$\frac{U}{u_*} = 5.75 \log \frac{R}{k} + 4.75 \tag{10.12}$$

From Eqs. (10.7) and (10.11), or (10.10) and (10.12) one gets

$$\frac{u - U}{u_*} = 5.75 \log \frac{y}{R} + 3.75 \tag{10.13}$$

Thus the dimensionless velocity defect $\frac{u - U}{u_*}$ depends on the relative distance y/R alone and not on the nature of pipe roughness.

10.4 FRICTION FACTOR VARIATION IN SMOOTH AND ROUGH PIPES

Since Darcy-Weisbach friction factor is defined as

$$f = \frac{4\tau_0}{\rho U^2/2} \tag{10.14}$$

and

$$\tau_0 = (p_1 - p_2)D/4l \tag{10.15}$$

one gets

$$(p_1 - p_2) = \gamma h_f = f \frac{l}{D} \frac{\rho U^2}{2} \tag{10.16}$$

It has also been seen that for pipes

$$f = \frac{\Delta p}{\rho \frac{U^2}{2}} \cdot \frac{D}{l} = \phi\left(\text{Re}, \frac{k}{D}\right) \tag{10.17}$$

By manipulation of Hagen-Poiseuille's equation and Eqs. (10.11) and (10.12) along with Eq. (10.14), the following equations can be obtained for the variation of f.

Laminar flow :

$$f = \frac{64}{\text{Re}} \quad \text{for Re} < 2100.$$

Turbulent flow:
Blasius (Empirical) equation

$$f = \frac{0.316}{\text{Re}^{1/4}} \tag{10.18}$$

for hydrodynamically smooth pipes if $10^5 > \text{Re} > 4000$

Smooth pipes:

$$\left.\begin{aligned} \frac{1}{\sqrt{f}} &= 2 \log \text{Re}\sqrt{f} - 0.80 \quad &(10.19)\text{ (a)} \\ f &= 0.0032 + \frac{0.221}{\text{Re}^{0.237}} \quad &(10.19)\text{ (b)} \end{aligned}\right\} 4 \times 10^7 > \text{Re} > 4000$$

Rough pipes:

$$\frac{1}{\sqrt{f}} = 2 \log \frac{R}{k} + 1.74 \tag{10.20}$$

In fact Eqs. 10.19(a) and 10.20 can be written in a unified manner as

$$\left(\frac{1}{\sqrt{f}} - 2 \log \frac{R}{k}\right) = 2 \log \frac{\text{Re}\sqrt{f}}{R/k} - 0.80 \tag{10.21}$$

$$= 2 \log 65.6 \frac{k}{\delta'} - 0.8$$

When all data of Nikuradse are plotted with $\frac{1}{\sqrt{f}} - 2 \log \frac{R}{k}$ as ordinate and $\frac{\text{Re}\sqrt{f}}{R/k}$ as abscissa, they fall together. Figure 10.5 shows this variation. Limit of smooth pipe equations is

$$\frac{\text{Re}\sqrt{f}}{R/k} \leqslant 17 \quad \text{or} \quad \frac{k}{\delta'} < 0.25$$

and limit for rough pipe equations is

$$\frac{\text{Re}\sqrt{f}}{R/k} \geqslant 400 \quad \text{or} \quad k/\delta' \geqslant 6.0.$$

It may be noted that these limiting values of k/δ' or $\text{Re}\sqrt{f}/R/k$ are different for commercial pipes.

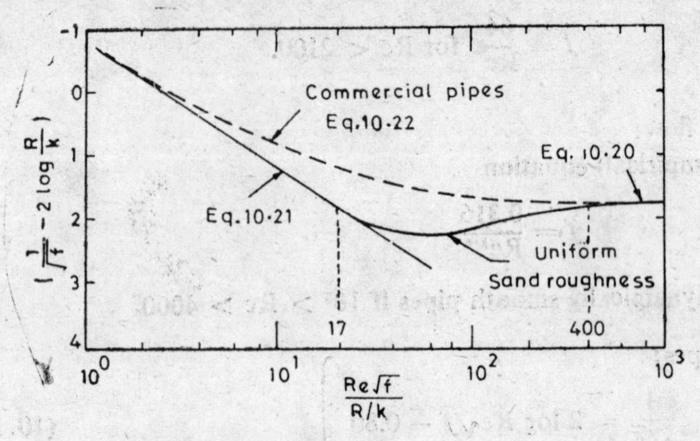

Fig. 10.5. Transition function.

10.5 FRICTION FACTORS FOR COMMERCIAL PIPES

The surface roughness of commercial pipes is not as uniform as the roughness of artificially roughened pipes of Nikuradse. Hence to standardise the roughness of commercial pipes, one defines equivalent sand grain roughness such that when a pipe of the same diameter as the given commercial pipe is coated with uniform sand grains of this diameter, it gives the same limiting value of f (i.e. at high Reynolds number when viscous effects are the equinegligible) as the commercial pipe. Table 10.1 lists the average values of k, valent sand grain roughness, for different pipe materials.

Table 10.1. Average values of k for different materials

Pipe material	k (*mm*)
Glass, drawn brass tubing, plastic, perspex, fibreglass etc.	smooth (0.0025)
Wrought iron, steel	0.045
Asphalted cast iron	0.120
Galvanised iron	0.150
Cast iron	0.260
Water mains	1.20
Concrete	0.30 to 3.0
New precast concrete pipes	0.90 to 9.0

Figure 10.6 shows variation for f with Re and $k/2R$ for commercial pipes. This is commonly known as Moody's diagram. Figure 10.5 also shows

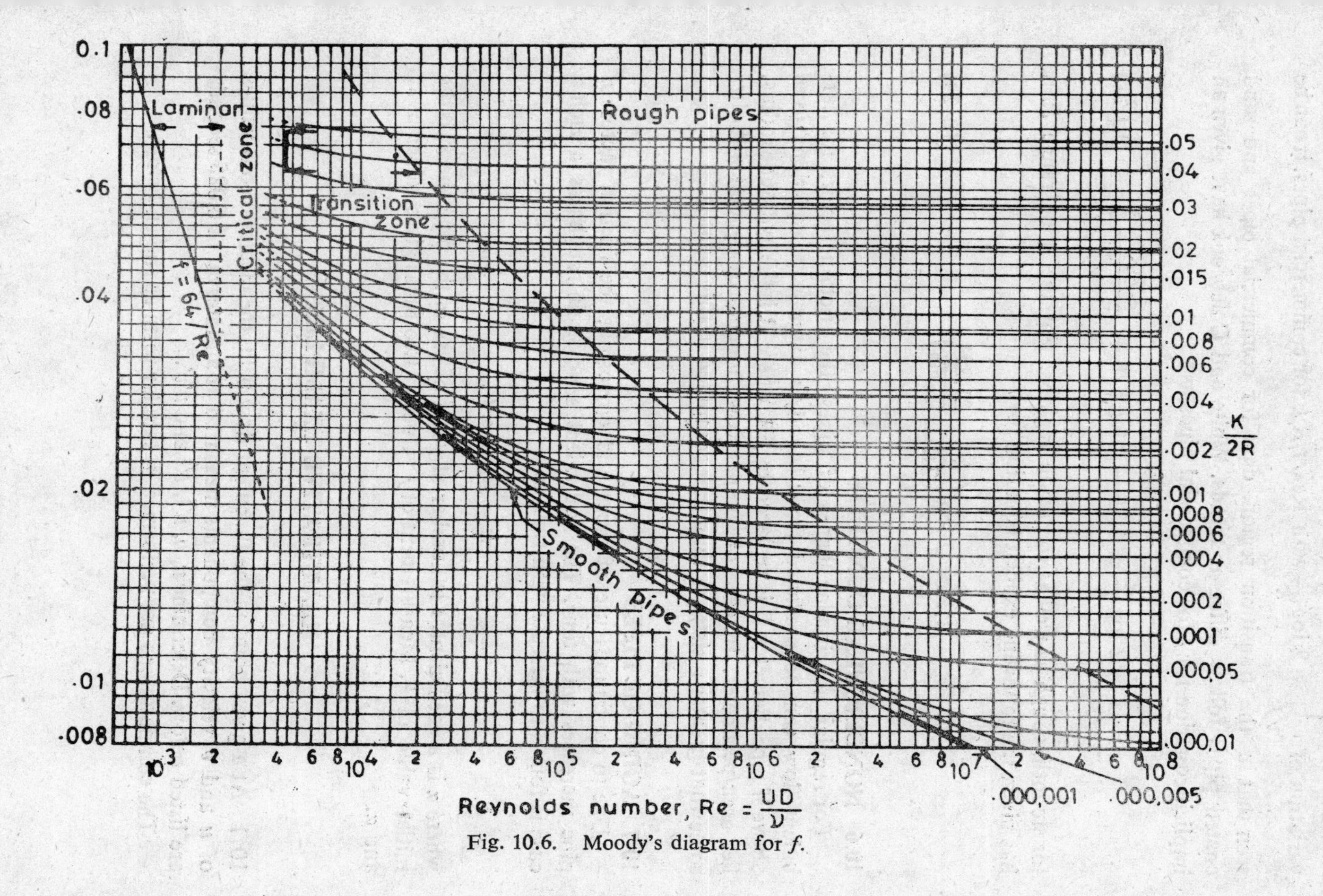

Fig. 10.6. Moody's diagram for f.

variation of $\frac{1}{\sqrt{f}} - 2\log\frac{R}{k}$ with $\text{Re}\sqrt{f}R/k$ for commercial pipes. It can be seen that in the transition region, data for commercial pipes and sand coated pipes follow different trends. White and Colebrook have given an implicit equation for f for commercial pipes as:

$$\frac{1}{\sqrt{f}} - 2\log\frac{R}{k} = 1.74 - 2\log\left(1 + \frac{18.7\,R/k}{\text{Re}\sqrt{f}}\right) \quad (10.22)$$

for smooth, rough as well as transition regions. For the same range Jain has given the following explicit equation for f

$$\frac{1}{\sqrt{f}} = 1.14 - 2\log\left(\frac{k}{2R} + \frac{21.25}{\text{Re}^{0.90}}\right) \quad (10.23)$$

10.6 NONCIRCULAR CONDUITS

For regular noncircular shapes of pipes, the friction factors for turbulent flow can be predicted from Moody's diagram or equations given above if pipe diameter D is replaced by 4R where R is the hydraulic radius, i.e. area/perimeter. In case of laminar flow, noncircular conduit will offer somewhat greater resistance than that computed in this manner.

10.7 AGING OF PIPES

Due to continuous use of pipes and corrosion, the friction factor of pipe increases with time. The equivalent sand grain roughness in such a case is found to increase with time as

$$k = k_0 + \alpha\, t$$

where α is a coefficient, t is number of years of use of the pipe and k_0 is the initial value of k. Measurement of f values at two times gives values of k_0 and α.

ILLUSTRATIVE EXAMPLES

10.1 At a given point in the turbulent flow field, the instantaneous values of u and v velocity components measured at an interval of 0.025 seconds are listed below. Determine $\bar{u}$, $\bar{v}$, $\overline{u'^2}$, $\overline{v'^2}$ and $\overline{u'v'}$.

The computations are performed in tabular form as shown below.

n	u m/s	$U' = u - \bar{u}$ m/s	u'^2 m^2/s^2	v m/s	$v' = v - \bar{v}$ m/s	v'^2 m^2/s^2	$u'v'$ m^2/s^2
1	7.93	2.523	6.366	0.180	−0.046	0.0021	−0.0053
2	3.69	−1.717	2.948	0.740	0.514	0.264	−0.8825
3	4.21	−1.197	1.433	−2.050	−3.762	14.153	4.5031
4	4.30	−1.107	1.225	0.550	0.324	0.105	−0.3587
5	5.37	−0.037	0.001	−0.900	−1 426	1.268	0.0417
6	2 62	−2.787	7.767	−0.080	−0.306	0.094	0.8528
7	4.85	−0.557	0.310	1.490	1.264	1.598	−0.7040
8	4.97	−0.437	0.191	0.160	−0.066	0.004	0.0228
9	8.54	3.133	9.816	0.520	0.294	0.086	0.9211
10	7.50	2.093	4.381	0 520	0.294	0.036	0.6153
11	7.62	2.213	4.897	−1.230	−1.456	2.120	−3.2260
12	2.87	−2.537	6.436	0.980	0.754	0.569	−1.9129
13	8.75	3.343	11.176	2.230	2.004	4.016	6.6993
14	4.15	−1.257	1.580	1.380	1.154	1.332	−1.4505
15	6.43	1.023	1.047	−0.570	−0.796	0.634	−0.8143
16	5.18	−0.227	0.052	−1.280	−1.506	2.268	0.3419
17	3.96	−1.447	2.094	0.760	0.534	0.285	−0.7727
18	4.65	−0.757	0.573	0.480	0.254	0.065	−0.1923
19	7.56	2.153	4.635	0.250	0.024	0.001	0.0517
20	2.99	−2.417	5.842	+0.380	0.154	0.024	−0.3722

$\Sigma u = 108.14, \quad \Sigma u'^2 = 72.77, \quad \Sigma v = 4.51 \quad \Sigma v'^2 = 28.974 \quad \Sigma u'v' = -9.2959$

$$\bar{u} = \frac{108.14}{20}, \quad \overline{u'^2} = \frac{72.77}{20}, \quad \bar{v} = \frac{4.51}{20}, \quad \overline{v'^2} = 1.448 \text{ m}^2/\text{s}^2 \quad \overline{u'v'} = -0.465 \text{ m}^2/\text{s}^2$$

$$= 5.407 \text{ m/s} \qquad = 3.639 \qquad = 0.226 \text{ m/s}$$

Hence $\bar{u} = 5.407$ m/s, $\bar{v} = 0.226$ m/s, $\overline{u'^2} = 3.639$ m²/s², $\overline{v'^2} = 1.448$ m²/s² and $\overline{u'v'} = -0.465$ m²/s². It may be mentioned that if long time record were available, v would be zero.

10.2 Explain how one can calculate the mixing length distribution in turbulent flow in a pipe from velocity distribution and mean flow data.

Shear is linearly distribued along the depth. Hence

$$\tau = \tau_0 \left(1 - \frac{y}{R}\right).$$

Also the bed shear τ_0 is related to pressure difference $(p_1 - p_2)$ in pipe of diameter D and length l by the relation $(p_1 - p_2)\frac{D}{4l} = \tau_0$ and τ is related to velocity gradient and mixing length by the relation $\tau = \rho l^2 \left(\frac{du}{dy}\right)^2$. Hence one gets

$$\tau = \tau_0 \left(1 - \frac{y}{R}\right) = \rho l^2 \left(\frac{du}{dy}\right)^2$$

$$\therefore \quad u_*^2 \left(1 - \frac{y}{R}\right) = l^2 \left(\frac{du}{dy}\right)^2 \quad \text{or} \quad \frac{l}{R} = \frac{u_* \left(1 - \frac{y}{R}\right)^{1/2}}{R\left(\frac{du}{dy}\right)}$$

τ_0 or u_* can be determined from the known values of $(p_1 - p_2)$, D and l. Hence in addition, if the velocity distribution data are available, $\frac{du}{dy}$ at various y values can be determined. Thus $\frac{l}{R}$ can be computed and plotted against $\frac{y}{R}$.

10.3 Two open wagons A and B are moving on parallel tracks in the same direction at velocities of 20.0 m/s and 30 m/s respectively. If gravel is transported from B to A at the rate of 1000 kg per second, determine the tangential force acting on A (Fig. 10.7).

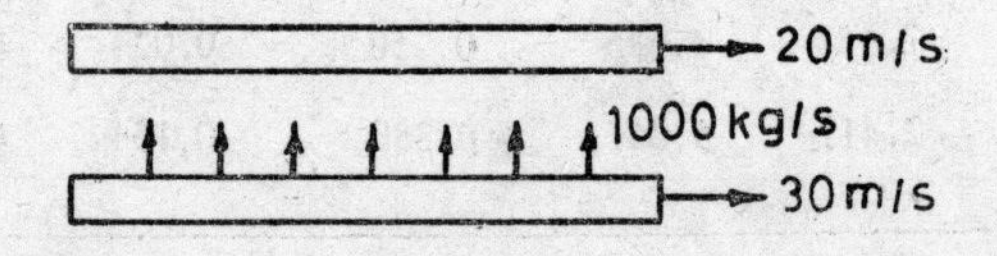

Fig. 10.7.

Mass transferred from B to A per unit time = 1000 kg/s

Relative velocity of B w.r.t. A = 10 m/s

$\therefore$ Force acting on A = 1000 (10 − 0) = + 10,000 N

This force will be in the forward direction.

10.4 For flow in open channels assume turbulent shear to be constant $\tau = \tau_0$ and the mixing length variation with y given by

$$l = 0.40\, y \quad \text{for } y \leqslant 0.2D$$

$$l = 0.08D \quad \text{for } y \geqslant 0.2D$$

where D is the depth of flow. Obtain the velocity distributlon law which will satisfy the boundary condition $u = U$ at $y = D$.

$$\tau = \tau_0 = \rho l^2 \left(\frac{du}{dy}\right)^2$$

Hence for $y \leqslant 0.2D$ $\qquad u_* = 0.40y \dfrac{du}{dy}$

which on integration gives $u = 2.5\, u_* \ln y + C_1$ (1)

For $y \geqslant 0.2D$ $\qquad u* = 0.08D \left(\dfrac{du}{dy}\right)$

which on integration gives $\quad u = \dfrac{u_*}{0.08D}\, y + C_2$ (2)

Equation 2 must satisfy the b.c. $u = U$ at $y = D$; hence

$$C_2 = U - 12.5\, u_*$$

and $\qquad u = \dfrac{u_*}{0.08} \dfrac{y}{D} + U - 12.5\, u_*$

or $\qquad u = U - 12.5\, u_* \left(1 - \dfrac{y}{D}\right)$ (3)

Further Eqs. 1 and 3 must give the same velocity at $y = 0.20D$.

$\therefore \qquad 2.5\, u_* \ln 0.2D + C_1 = U - 12.5\, u_* \times 0.80$

$\therefore \qquad C_1 = U - 10u_* - 2.5\, u_* \ln 0.2D$

Substitution of this value of C_1 in Eq. 1 yields

$$u = 2.5u* \ln y + U - 10u_* - 2.5u_* \ln 0.2D$$

$$\frac{u - U}{u_*} = \left(2.5 \ln \frac{5y}{D} - 10\right)$$

10.5 Assume that the flow near a smooth boundary can be divided into laminar sublayer of thickness δ' in which the flow is laminar and turbulent flow above it. Assuming $\tau = \tau_0$ for both the regions, show that the velocity distribution in the turbulent region is given by

$$\frac{u}{u_*} = 2.5 \ln \frac{u_* y}{\nu} + B$$

where $B = 11.6 - 2.5 \ln 11.6$.

In laminar sublayer, $\tau = \tau_0 = \mu \dfrac{du}{dy}$

$$\frac{\tau_0}{\rho} = \nu \frac{du}{dy} \quad \text{or} \quad \frac{u_*^2}{\nu} = \frac{du}{dy}$$

Integration of this equation yields $\frac{u}{u_*} = \frac{u_* y}{\nu} + c_1$

The constant of integration $c_1 = 0$ since at $y = 0$, $u = 0$.

Hence $$\frac{u}{u_*} = \frac{u_* y}{\nu} \tag{1}$$

gives velocity distribution law in the laminar sublayer of thickness

$$\delta' = \frac{11.6\nu}{u_*}.$$

Velocity at the edge of laminar sublayer will be

$$\frac{u_{\delta'}}{u_*} = \frac{u_*}{\nu} \times \frac{11.6\nu}{u_*} = 11.6 \quad \text{or} \quad u_{\delta'} = 11.6\, u_*.$$

In the turbulent region

$$\tau = \tau_0 = \rho l^2 \left(\frac{du}{dy}\right)^2$$

and $l = Ky$. This gives

$$\frac{u}{u_*} = \frac{1}{K} \ln y + c_1$$

The constant c_1 in this equation can be evaluated from the condition that at $y = \delta'$, $u = 11.6\, u_*$

$$\therefore \quad \frac{u_{\delta'}}{u_*} = \frac{1}{K} \ln \frac{11.6\nu}{u_*} + c_1$$

or $$c_1 = \frac{u_{\delta'}}{u_*} - \frac{1}{K} \ln \frac{11.6\nu}{u_*}$$

$$\therefore \quad \frac{u}{u_*} = \frac{1}{K} \ln y + \frac{u_{\delta'}}{u_*} - \frac{1}{K} \ln \frac{11.6\nu}{u_*}$$

$$= \frac{1}{K} \ln \left(\frac{u_* y}{11.6\nu}\right) + \frac{u_{\delta'}}{u_*}$$

but $$\frac{u_{\delta'}}{u_*} = 11.6 \quad \therefore \quad \frac{u}{u_*} = \frac{1}{K} \ln \left(\frac{u_* y}{\nu}\right) + \left(11.6 - \frac{1}{K} \ln 11.6\right)$$

Substituting $K = 0.4$, one gets the velocity distribution law as

$$\frac{u}{u_*} = 2.5 \ln \left(\frac{u_* y}{\nu}\right) + B$$

where $B = (11.6 - 2.5 \ln 11.6) = 5.47$.

10.6 When 50 mm diameter smooth pipe carries water at 20°C, the flow Reynolds number is 10^4. Determine u_* and total frictional resistance in 200 m length of the pipe.

$$\text{Re} = \frac{UD}{\nu} = 10^4 \quad \therefore \quad U = \frac{10^4 \times 10^{-6}}{0.05} = 0.20 \text{ m/s}$$

For smooth pipe and Re $<$ 10^5

$$= \frac{0.316}{\text{Re}^{1/4}} = 0.0316$$

$$\therefore \qquad \tau_0 = \frac{f\rho}{4}\frac{U^2}{2} = \frac{0.0316}{4} \times \frac{998 \times 0.2^2}{2}$$

$$= 0.1577 \text{ N/m}^2$$

$\therefore$ Frictional resistance $F = \pi D l \tau_0$

$$= 3.142 \times 0.050 \times 200 \times 0.1577 = 49.55 \text{ N}$$

Also $\qquad u_* = \sqrt{\tau_0/\rho} = \sqrt{0.1577/998} = 0.01257$ m/s

10.7 Compare the mixing abilities of two domestic mixers; the first one has 100 mm diameter blades rotating at 1000 rpm; the second one has 150 mm diameter blades rotating at 1500 rpm.

The mixing ability will be proportional to their eddy kinematic viscosities which in turn will be proportional to the product of their diameter and rotational speed. Hence

$$\frac{\text{Mixing ability of first mixer}}{\text{Mixing ability of second mixer}} = \frac{\epsilon_1}{\epsilon_2} = \frac{l_1\sqrt{\overline{v_1'^2}}}{l_2\sqrt{\overline{v_2'^2}}} = \frac{100 \times 1000}{150 \times 1500}$$

$$= \frac{1}{2.25} \quad \text{or} \quad 0.444$$

10.8 According to Karman's similarity hypothesis, the mixing length is given by

$$l = K\frac{du/dy}{d^2u/dy^2}$$

Assuming $\tau = \tau_0$ for small values of y, obtain the velocity distribution equation for turbulent flow in a pipe. The two boundary conditions to be satisfied are:

(i) du/dy becomes very large as $y \to 0$

(ii) at $y = R$, $u = U$

$$\tau = \tau_0 = \rho l^2 \left(\frac{du}{dy}\right)^2$$

$$= \rho K^2 (du/dy)^4 \bigg/ \left(\frac{d^2u}{dy^2}\right)^2$$

$$\therefore \qquad u_* = -K(du/dy)^2/d^2u/dy^2$$

(Note: The $-$ve sign is given in the above expression since d^2u/dy^2 will be $-$ve)

Let $\dfrac{du}{dy} = f \quad \therefore \quad \dfrac{d^2u}{dy^2} = \dfrac{df}{dy}$ and $\dfrac{d}{dy}\left(\dfrac{1}{f}\right) = -\left(\dfrac{f'}{f^2}\right)$

$$\therefore \qquad \frac{K}{u_*} = -\frac{d^2u/dy^2}{(du/dy)^2} = \frac{d}{dy}\left(\frac{1}{du/dy}\right)$$

Integration of this equation yields

$$\frac{1}{du/dy} = \frac{Ky}{u_*} + C$$

Substitution of the first boundary condition, viz. as $y \to 0$, $\dfrac{du}{dy} \to$ large gives $C = 0$.

$$\therefore \qquad \frac{du}{dy} = \frac{u_*}{Ky}$$

which on integration gives

$$\frac{u}{u_*} = \frac{1}{K} \ln y + C_1$$

But at $y = R$, $u = u_m$

$$\therefore \qquad C_1 = \frac{u_m}{u_*} - \frac{1}{K} \ln R$$

and

$$\frac{u}{u_*} = \frac{u_m}{u_*} + \frac{1}{K} \ln \frac{y}{R}$$

$$\therefore \qquad \frac{u - u_m}{u_*} = \frac{2.302}{K} \log \frac{y}{R}$$

With $K = 0.40$, this gives

$$\frac{u - u_m}{u_*} = 5.75 \log \frac{y}{R}$$

10.9 From equations for velocity distribution in smooth and rough pipes, obtain the equation

$$\frac{u_m}{U} = 1.326\sqrt{f} + 1.0$$

One can start with the equation

$$\frac{u - U}{u_*} = 5.75 \log \frac{y}{R} + 3.75$$

At $y = R$, $u = u_m$. Hence $\dfrac{u_m - U}{u_*} = 3.75$

or

$$\frac{u_m}{u_*} = 3.75 + \frac{U}{u_*}.$$

However, since $\tau_0 = \frac{f}{4} \rho \frac{U^2}{2}$, one gets $u_*^2 = \frac{f}{8} U^2$

or $$u_* = U\sqrt{f/8}$$

Substitution of this value of u_* in the above equation and subsequent simplification yields

$$\frac{u_m}{U} = 1.326\sqrt{f} + 1.0$$

10.10 Water at 20°C flows in a 250 mm diameter cast iron pipe with the centreline velocity of 3.0 m/s. Determine the average velocity of flow.

From Table 10.1, k for cast iron is 0.26 mm. Also at 20°C

$$\nu = 10^{-6} \text{ m}^2/\text{s}$$

$k/D = 0.26/250 = 0.00104$. Also knowing centreline velocity of 3.0 m/s, Re can be of the order of $\frac{3 \times 0.25}{10^{-6}}$ or 7.5×10^5. Hence using Moody's diagram, one gets $f = 0.023$.

Since

$$\frac{u_m}{U} = 1.326\sqrt{f} + 1 = 1.326\sqrt{0.023} + 1 = 1.2011$$

$$\therefore \quad U = 3.0/1.2011 = 2.498 \text{ m/s}$$

$$\therefore \quad \text{Re} = \frac{2.498 \times 0.25}{10^{-6}} = 6.244 \times 10^5$$

for which at $k/D = 0.00104$, $f = 0.024$

$$\frac{u_m}{U} = 1.326\sqrt{0.024} + 1.0 = 1.2054$$

$$\therefore \quad U = 3.0/1.2054 = 2.489 \text{ m/s}$$

Since two successive values of U are sufficiently close, no further refinement is warranted.

10.11 150 mm diameter pipe with $k = 0.01$ mm carries water at 20°C over a length of 100 m with a pressure loss of 26.613 kN/m². State whether the pipe will act as smooth or rough pipe. Also determine maximum velocity, average velocity and discharge.

$$(p_1 - p_2)\frac{D}{4l} = \tau_0 \quad \therefore \quad \tau_0 = \frac{26613 \times 0.150}{4 \times 100}$$

$$= 9.98 \text{ N/m}^2$$

$$\therefore \quad u_* = \sqrt{\tau_0/\rho} = \sqrt{9.98/998} = 0.1 \text{ m/s}$$

Laminar sublayer thickness δ' is given as

$$\delta' = \frac{11.6\nu}{u_*} = \frac{11.6 \times 10^{-6}}{0.10} \text{ m} = 0.116 \text{ mm}$$

$$\therefore \qquad k/\delta' = 0.01/0.116 = 0.0862$$

which is less than 0.25; hence the pipe will act as hydrodynamically smooth pipe.

$$\therefore \qquad \frac{u}{u_*} = 5.75 \log \frac{u_* y}{\nu} + 5.50$$

Maximum velocity u_m will occur when $y = 0.075$ m

$$u_m = 0.10 \left(5.75 \log \frac{0.10 \times 0.075}{10^{-6}} + 5.50\right) = 2.778 \text{ m/s}$$

$$\frac{U}{u_*} = 5.75 \log \frac{u_* R}{\nu} + 1.75 = 24.0316$$

$\therefore$ Average velocity $U = 2.403$ m/s

$$Q = \frac{\pi}{4} D^2 U = 0.785 \times 0.15^2 \times 2.403 = 0.0424\ 43 \text{ m}^3\text{/s}$$

or $\qquad Q = 42.443\ l/\text{s}$

10.12 A smooth pipe carries 0.30 m³/s of water discharge with a head loss of 3.0 m per 100 m length of the pipe. If the water temperature is 20°C, determine the diameter of the pipe.

$$\text{Re} = \frac{QD\rho}{(0.785D^2)\mu}$$

Assuming $\dfrac{\mu}{\rho} = 10^{-6}$ m²/s

$$\text{Re} = \frac{0.30}{0.785D \times 10^{-6}} = \frac{3.8217 \times 10^5}{D} \tag{1}$$

Also $\qquad h_f = \dfrac{8flQ^2}{\pi^2 g D^5}$

Substituting $h_f = 3.0$ m, $l = 100$ m, $Q = 0.30$ m³/s, one gets

$$D^5 = 0.2481 f \tag{2}$$

Assume $f = 0.02$

$\therefore \quad D^5 = 0.2481 \times 0.02 = 0.004\ 962$ and $D = 0.346$ m

$$\therefore \qquad \text{Re} = \frac{3.8217 \times 10^5}{0.346} = 1.105 \times 10^6$$

Since the pipe acts as smooth, f is given by

$$f = 0.0032 + \frac{0.221}{\text{Re}^{0.237}}$$

$$= 0.0032 + 0.221/(1.105 \times 10^6)^{0.237}$$

$$= 0.0114$$

$\therefore$ Use of Eq. 2 gives $D^5 = 0.2481 \times 0.0114$

and $D = 0.309$ m

Reynolds number value can be further improved

$$\text{Re} = \frac{3.8217 \times 10^5}{0.309} = 1.236 \times 10^6$$

and $f = 0.0032 + 0.221/(1.236 \times 10^6)^{0.237}$

$$= 0.0112$$

Since the two successive values of f are sufficiently close, one can take the correct pipe diameter as 0.309 m and use the next standard diameter.

10.13 In a 500 mm diameter rough pipe carrying water at 20°C, the velocity and velocity gradient at 50 mm from the wall are 4.42 m/s and 5.523×10^{-3} s^{-1} respectively. Determine the water discharge, roughness height, friction factor, wall shear and the pressure gradient.

For rough pipe

$$\frac{u}{u_*} = 5.75 \log \frac{y}{k} + 8.5$$

and $$\frac{du}{dy} = \frac{2.5\, u_* K}{y}$$

$$\therefore \quad \frac{4.42}{u_*} = 5.75 \log_{10} \frac{0.050}{k} + 8.50$$

and $$5.523 \times 10^{-3} = \frac{2.5k\, u_*}{0.05}$$

Substitution of value of u_* from 2nd equation in the first and its solution gives $k = 0.50$ mm.

$$u_* = \frac{0.05 \times 5.523 \times 10^{-3}}{2.5 \times 0.0005} = 0.221 \text{ m/s}$$

$$\therefore \quad \tau_0 = \rho u_*^2 = 998 \times 0.221^2 = 48.743 \text{ N/m}^2$$

The equation for average velocity in a rough pipe is

$$\frac{U}{u_*} = 5.75 \log \frac{R}{k} + 4.75$$

$$\therefore \quad \frac{U}{u_*} = 5.75 \log \frac{250}{0.5} + 4.75$$

$$\frac{U}{u_*} = 20.269$$

$$\therefore \quad U = 20.269 \times 0.221 = 4.479 \text{ m/s}$$

$\therefore \qquad Q = 0.785 \times 0.50^2 \times 4.479 = 0.879\ \text{m}^3/\text{s}$

$$\frac{1}{\sqrt{f}} = 2 \log \frac{R}{k} + 1.74 = 7.138$$

$\therefore \qquad f = 0.0196$

Since

$$(p_1 - p_2)\frac{D}{4l} = \tau_0, \qquad \frac{(p_1 - p_2)}{l} = \frac{4 \times 48.743}{0.50} = 389.944\ \text{N/m}^2\ \text{m}$$

10.14 A smooth pipe of 200 mm diameter carries oil of mass density 850 kg/m³ and kinematic viscosity 3.0×10^{-6} m²/s at a discharge of 31.40 *l*/s. What will be the wall shear, thickness of laminar sublayer, velocity at the edge of laminar sublayer, and maximum velocity?

$$U = 0.0314/0.785 \times 0.200^2 = 1.0\ \text{m/s}$$

$$\text{Re} = \frac{1.0 \times 0.200}{3 \times 10^{-6}} = 6.667 \times 10^4$$

$$U/u_* = 5.75 \log \frac{u_* R}{\nu} + 1.75$$

Substituting $U = 1.0$ m/s, $R = 0.100$ m and $\nu = 3 \times 10^{-6}$ m²/s, one gets

$$\frac{1}{u_*} = 5.75 \log u_* + 27.757$$

This can be solved by trial and error to get $u_* = 0.0494$ m/s.

$\therefore \qquad \tau_0 = \rho u_*^2 = 850 \times 0.0494^2 = 2.0743\ \text{N/m}^2$

$$\delta' = \frac{11.6\nu}{u_*} = \frac{11.6 \times 3 \times 10^{-6}}{0.0494} = 704.453 \times 10^{-6}\ \text{m or } 0.704\ \text{mm}$$

To determine velocity at $y = \delta'$, use the equation

$$\frac{u}{u_*} = 11.5 \log \frac{u_* y}{\nu} - 3.05$$

$$= 11.5 \log \frac{(0.0494 \times 704.453 \times 10^{-6})}{3 \times 10^{-6}} - 3.05$$

$$= 9.191$$

$\therefore \qquad u = 9.191 \times 0.0494 = 0.454\ \text{m/s}$

$$\frac{u_m}{u_*} = 5.75 \log \frac{0.0494 \times 0.10}{3 \times 10^{-6}} + 5.5 = 23.995$$

$\therefore \qquad u_m = 23.995 \times 0.0494 = 1.185\ \text{m/s}$

10.15 A 300 mm diameter pipe carries oil of mass density 950 kg/m³ and dynamic viscosity 1.0 N s/m² at a discharge of 141.3 *l*/s over a distance of

1.0 km. If the viscosity of oil changes by a factor of ten due to increase in temperature, compare the costs of pumping the oil if the same quantity of oil is to be pumped.

$$U = 0.1413/0.785 \times 0.30^2 = 2.0 \text{ m/s}$$

In the initial condition $\mathrm{Re}_1 = \dfrac{UD\rho}{\mu_1} = \dfrac{2.0 \times 0.30 \times 950}{1.0} = 570$

Hence the flow is laminar and

$$h_{f_1} = \frac{32\mu U l}{\gamma D^2} = \frac{32 \times 1.0 \times 2.0 \times 1000}{950 \times 9.806 \times 0.30^2} = 76.304 \text{ m}$$

$$P_1 = \frac{\gamma Q h_{f_1}}{1000} = \frac{0.1413 \times 950 \times 9.806 \times 76.304}{1000}$$

$$= 100.440 \text{ kW}$$

When the temperature is increased, viscosity will decrease.

$$\therefore \qquad \mu_2 = \frac{1}{10} = 0.10 \text{ N s/m}^2$$

$$\mathrm{Re}_2 = \frac{2 \times 0.30 \times 950}{0.10} = 5700$$

Flow will be turbulent,

$$f = \frac{0.316}{\mathrm{Re}_2^{0.25}} = \frac{0.316}{(5700)^{0.25}} = 0.036\,37$$

$$h_{f_2} = \frac{0.036\,37 \times 1000 \times 2^2}{2 \times 9.806 \times 0.30} = 24.716 \text{ m}$$

$$P_2 = (0.1413 \times 950 \times 9.806 \times 24.716)/1000$$

$$= 32.534 \text{ kW}$$

$$\frac{P_2}{P_1} = \frac{32.524}{100.440} = 0.3239$$

10.16 Determine the size of steel pipe required to carry water at 30 l/s if the permissible energy gradient is 0.05. Will the boundary act as smooth or in transition?

$$h_f/l = \frac{8fQ^2}{\pi^2 g D^5}$$

$$\therefore \qquad D^5 = \frac{8 \times 0.03^2 f}{3.142^2 \times 9.806 \times 0.05}$$

or $$D^5 = 0.001\,488 f \qquad (1)$$

$$\text{Re} = \frac{4Q}{\pi D \nu} \quad \text{or} \quad \text{Re} = \frac{4 \times 0.03}{3.142 \times D \times 1 \times 10^{-6}}$$

$$\text{Re} = 38192/D \tag{2}$$

Further for steel pipe $k = 0.045$ mm.

Assume $f = 0.025 \quad \therefore \quad D^5 = 0.001\,488 \times 0.025$

or $D = 0.130$ m

$$\therefore \qquad \text{Re} = \frac{38192}{0.13} = 2.938 \times 10^5 \quad \text{and} \quad \frac{k}{D} = \frac{0.045}{130} = 0.000\,346$$

From Moody's diagram, for these values of Re and k/D, $f = 0.0175$.

$$\therefore \qquad D^5 = 0.001\,488 \times 0.0175 = 0.000\,026\,04$$

$$\text{or} \qquad D = 0.121 \text{ m}, \ \text{Re} = \frac{381\,92}{0.121} = 3.156 \times 10^5$$

$$\text{and} \qquad \frac{k}{D} = \frac{0.045}{121} = 0.000\,372$$

Again Moody's diagram gives $f = 0.018$ for these values of Re and k/D.

$$\therefore \qquad D^5 = (0.001\,488 \times 0.018) = 0.000\,0268$$

$$D = 0.122 \text{ m}$$

Since this va lue differs very little from earlier value of 0.121 m, no further improvement is necessary. Hence $D = 0.122$ m.

$$\tau_0 = \frac{(p_1 - p_2)}{l} \frac{D}{4} = 0.05 \times 9787 \times \frac{0.122}{4} = 14.925 \text{ N/m}^2$$

$$u_* = \sqrt{14.925/998} = 0.122 \text{ m/s}$$

$$\delta' = \frac{11.6 \times 10^{-6}}{0.122} = 0.000\,0951 \text{ m} \quad \text{or} \quad 0.0951 \text{ mm}$$

$$k/\delta' = \frac{0.045}{0.0951} = 0.473$$

Since k/δ' lies between 0.25 and 6.0, the boundary will be in transition.

10.17 Tests on a 500 mm diameter commercial pipe indicated that head loss in 100 m length of pipe at different discharges of water is as given below.

Q l/s	40	200	400	800
h_f m	0.01	0.210	0.820	3.27

(i) Determine the equivalent sand grain roughness of the pipe.

(ii) What is the maximum water discharge at which this pipe will act as smooth pipe ?

(iii) What is the minimum discharge at which this pipe will act as rough pipe ?

The values of U and f corresponding to the above discharges are computed and listed below.

Q l/s	40	200	400	800
U m/s	0.204	1.019	2.038	4.076
f	0.0235	0.0198	0.0194	0.0193

It can be seen that f value, at higher discharges, is nearly constant at 0.0193. Hence friction factor equation for rough pipes can be used, namely

$$\frac{1}{\sqrt{f}} = 2 \log \frac{R}{k} + 1.74$$

or
$$1/\sqrt{0.0193} = 2 \log \frac{250}{k} + 1.75$$

or
$$k = 0.933 \text{ mm}$$

Also when pipe just ceases to act as smooth pipe

$$\text{Re}\sqrt{f}/R/k = 17 \quad \therefore \quad \text{Re}\sqrt{f} = \frac{17 \times 250}{0.933} = 9108.6$$

Also
$$\frac{1}{\sqrt{f}} = 2 \log R_e\sqrt{f} - 0.8$$

$$= 7.1189 \qquad \therefore \; f = 0.0197$$

and
$$\text{Re} = 9108.6/\sqrt{0.0197} = 64600$$

$$\therefore \quad \frac{U \times 0.500}{10^{-6}} = 64600 \quad \text{or} \quad U = \frac{64600 \times 10^{-6}}{0.50}$$

$$= 1.292 \text{ m/s}$$

$$\therefore \quad Q = 0.785 \times 0.50^2 \times 1.292 = 0.253 \text{ m}^3\text{/s}$$

When the pipe acts just as rough pipe

$$f = 0.0193 \quad \text{and} \quad k/\delta' = 6.0$$

But
$$\delta' = \frac{11.6\nu}{u_*} \qquad \therefore \; u_* = \frac{11.6 \times 10^{-6} \times 6}{0.933 \times 10^{-3}}$$

$$= 0.0746 \text{ m/s}$$

$$\tau_0 = \frac{f}{4}\frac{\rho U^2}{2} \quad \text{and} \quad \tau_0 = \rho u_*^2$$

$$\therefore \quad \frac{u_*^2}{U^2} = \frac{f}{8} \text{ and } U = u_* \sqrt{\frac{8}{f}} = 0.0746 \sqrt{\frac{8}{0.0193}}$$

$$\therefore \quad U = 1.519 \text{ m/s}$$

$$\therefore \quad Q = 0.785 \times 0.50^2 \times 1.519 = 0.298 \text{ m}^3/\text{s}$$

10.18 One meter diameter pipe is to carry a water discharge of 1.0 m³/s at the minimum loss of energy. What will be the permissible height of surface roughness?

$$U = 1.0/0.785 \times 1.0^2 = 1.274 \text{ m/s}$$

$$\text{Re} = 1.274 \times 1.0/10^{-6} = 1.274 \times 10^6$$

For minimum loss of energy pipe should act as smooth and

$$\text{Re}\sqrt{f}/R/k = 17$$

For smooth pipe

$$f = 0.0032 + \frac{0.221}{(1.274 \times 10^6)^{0.237}} = 0.0111$$

and
$$\frac{\text{Re}\sqrt{f}}{R/k} = 17 = \frac{1.274 \times 10^6 \times \sqrt{0.0111}}{R/k}$$

$$\therefore \quad R/k = \frac{1.274 \times 10^6 \times \sqrt{0.0111}}{17} = 7895.54$$

$$\therefore \quad k = \frac{0.50}{7895.54} \times 10^3 \text{ mm} = 0.0633 \text{ mm}$$

10.19 Water enters the brass tubes of a heater at a velocity of 1.5 m/s and at a temperature of 40°C. Water leaves the tubes at a temperature of 80°C. The temperature of the wall of the tube is 98°C. If the brass tubes are of 15 mm internal diameter and 5.0 m length, determine the head loss in each tube.

The bulk or average temperature of water can be taken as the average of inlet and outlet temperatures i.e. (40 + 80)/2 or 60°C.

At this temperature

$$\nu = 3.8 \times 10^{-7} \text{ m}^2/\text{s}$$

$$\text{Re} = \frac{1.5 \times 0.15 \times 10^{-3}}{3.8 \times 10^{-7}} = 5.921 \times 10^4$$

For smooth pipe at this Reynolds number, f can be calculated using Blasius equation

$$f = 0.316/(59210)^{1/4} = 0.0202$$

The dynamic viscosities of the water at the bulk temperature (60°C) and wall temperature (98°C) are $\mu = 4.8 \times 10^{-4}$ and $\mu_w = 2.9 \times 10^{-4}$ kg/m s respectively. The corrected friction factor f_c is given by

$$f_c = f/\psi$$

where
$$\psi = \left(\frac{\mu}{\mu_w}\right)^{0.14} \quad \text{if Re} < 2100$$

and $$\psi = \left(\frac{\mu}{\mu_w}\right)^{0.25} \quad \text{if} \quad \text{Re} > 2100$$

Hence

$$f_c = \frac{0.0202}{\left(\frac{4.8 \times 10^{-4}}{2.9 \times 10^{-4}}\right)^{0.25}} = \frac{0.0202}{1.073} = 0.0188$$

$$h_f = \frac{flU^2}{2gD} = \frac{0.0188 \times 5 \times 1.5^2}{2 \times 9.806 \times 0.015} = 0.719 \text{ m}$$

10.20 For a hydrodynamically smooth pipe, obtain an expression for ϵ/ν in terms of $\frac{y}{R}$ and Reynolds number of flow. Calculate the maximum value of ϵ/ν at Reynolds numbers of 10^5, 10^6 and 10^7.

$$\tau = \tau_0\left(1 - \frac{y}{R}\right) = \rho\epsilon\frac{du}{dy}$$

But for turbulent flow

$$\frac{du}{dy} = \frac{u_*}{Ky}$$

where K is Karman constant equal to 0.40.

$$\therefore \qquad \tau_0\left(1 - \frac{y}{R}\right) = \rho\epsilon\frac{u_*}{Ky}$$

or dividing both sides by ν and simplifying one gets

$$\frac{\epsilon}{y} = \frac{u_* y}{\nu} K\left(1 - \frac{y}{R}\right) = K\frac{u_* R}{\nu}\left(\frac{y}{R} - \frac{y^2}{R^2}\right)$$

However

$$\frac{u_* R}{\nu} = \frac{2UR}{\nu}\frac{1}{2}\frac{u_*}{U} = \text{Re}\frac{1}{2}\sqrt{\frac{f}{8}} = \text{Re}\sqrt{\frac{f}{32}}$$

$$\therefore \qquad \frac{\epsilon}{\nu} = K\text{Re}\sqrt{\frac{f}{32}}\left(\frac{y}{R} - \frac{y^2}{R^2}\right)$$

For given K, Re (and hence f), $\frac{\epsilon}{\nu}$ will be maximum when $\frac{y}{R} = \frac{1}{2}$.

$$\therefore \qquad \left(\frac{\epsilon}{\nu}\right)_{max} = \frac{K\text{Re}}{4}\sqrt{\frac{f}{32}}$$

For $K = 0.40$ and Re $= 10^5$, 10^6 and 10^7, following values of f from the equation $f = 0.0032 + \frac{0.221}{\text{Re}^{0.237}}$ and $(\epsilon/\nu)_{max}$ can be calculated

Re	f	$(\epsilon/\nu)_{max}$
10^5	0.0176	234.52
10^6	0.0116	1903.94
10^7	0.00805	15850.94

Thus it can be seen that the eddy kinematic viscosity is much larger than the kinematic viscosity of fluid, and it increases with increase in Reynolds number of flow.

10.21 Determine the pressure loss in a square conduit of 100 mm size carrying water at the rate of 0.025 m³/s at 20°C if the pipe length is 20 m. Assume the surface roughness to be 0.50 mm.

$$\text{Area} = 0.1 \times 0.1 = 0.01 \text{ m}^2, \text{ Perimeter} = 0.40 \text{ m}$$

$$U = 0.025/0.01 = 2.5 \text{ m/s}$$

$$\text{Hydraulic radius } R = A/P = \frac{0.01}{0.40} = 0.025 \text{ m}$$

$$\therefore \quad \text{Re} = \frac{4UR}{\nu} = \frac{4 \times 2.5 \times 0.025}{10^{-6}} = 2.5 \times 10^5$$

$$k/4R = \frac{0.50 \times 10^{-3}}{4 \times 0.025} = 0.005$$

$\therefore$ From Moody's diagram $f = 0.03$

$$\therefore \quad h_f = \frac{flU^2}{2g \times 4R} = \frac{0.03 \times 20 \times 2.5^2}{8 \times 9.806 \times 0.025} = 1.912 \text{ m}$$

10.22 A certain 100 mm diameter pipe which acts as rough pipe gave 0.026 and 0.030 as its friction factor values after five and ten years of operation. Estimate the friction factor after 15 years of service.

First determine surface roughness values K_5 and K_{10} after 5 and 10 years' of service.

$$\frac{1}{\sqrt{0.026}} = 2 \log \frac{50}{k_5} + 1.74$$

$$6.202 = 5.138 - 2 \log k_5 \quad \therefore \quad k_5 = 0.294 \text{ mm}$$

Similarly $$\frac{1}{\sqrt{0.030}} = 2 \log \frac{50}{k_{10}} + 1.74$$

$$5.7735 = 5.1380 - 2 \log k_{10} \qquad k_{10} = 0.481 \text{ mm}$$

Since $$k = k_0 + \alpha t$$

$$\left.\begin{aligned} k_5 &= k_0 + 5\alpha \\ k_{10} &= k_0 + 10\alpha \end{aligned}\right\} \quad \therefore \quad \frac{0.481 - 0.294}{5} = \alpha = 0.0374$$

and $k_0 = 0.107$ mm

$$\therefore \quad k_{15} = 0.107 + (0.0374 \times 15)$$

$$= 0.668 \text{ mm}$$

$$\therefore \quad \frac{1}{\sqrt{f_{15}}} = 2 \log \frac{50}{0.668} + 1.74 = 5.4884$$

$$\therefore \quad f_{15} = 0.0332$$

10.23 **If the velocity distribution for turbulent flow in a pipe is given by**

$$u/u_* = 8.74 \left(\frac{u_* y}{\nu}\right)^{1/7}, \text{ obtain the expression for } f.$$

$$Q = \int_0^R 2\pi r u dr$$

$$= \int_0^R 2\pi r u_* \times 8.74 \left(\frac{u_*(R-r)}{\nu}\right)^{1/7} dr \quad \text{since} \quad y = R - r$$

$$= \frac{2\pi u_*^{8/7}}{\nu^{1/7}} \times 8.74 \int_0^R (R-r)^{1/7} r dr$$

Let $\quad (R - r) = t \quad \therefore \quad -dr = dt \quad \text{and} \quad r = (R - t)$

$$Q = \frac{2\pi \times 8.74\, u_*^{8/7}}{\nu^{1/7}} \int_R^0 -(R-t) t^{1/7} dt$$

$$= 2\pi \times 8.74 \times \frac{46}{120} \frac{R^{15/7} u_*^{8/7}}{\nu^{1/7}}$$

$$\frac{Q}{\pi R^2 u_*} = \frac{U}{u_*} = \frac{2 \times 8.74 \times 46}{120} \left(\frac{u_* R}{\nu}\right)^{1/7}$$

$$= 7.138 \left(\frac{u_* R}{\nu}\right)^{1/7}$$

or $\quad \dfrac{U}{u_*} = 7.138 \left(\dfrac{UR}{\nu}\right)^{1/7} \left(\dfrac{u_*}{U}\right)^{1/7}$

This gives $\quad \dfrac{U}{u_*} = 5.583 \left(\dfrac{UR}{\nu}\right)^{1/8}$

However $\quad \tau_0 = \dfrac{f}{4} \dfrac{U^2}{2} \quad \therefore \quad 8 \left(\dfrac{u_*}{U}\right)^2 = f$

$$\therefore \quad \frac{U}{u_*} = \sqrt{\frac{8}{f}}$$

$$\therefore \qquad \sqrt{\frac{8}{f}} = \frac{5.583}{2^{1/8}}\left(\frac{UD}{\nu}\right)^{1/8}$$

$$\therefore \qquad f = 0.305/\text{Re}^{1/4}$$

compare this equation with Blasius equation for f.

PROBLEMS

10.1 Below are given the instantaneous values of velocity in x direction measured at 3 m above the ground in the atmospheric boundary layer. These are given in m/s and were recorded at a time interval of one second.

13.10, 13.15, 13.05, 13.20, 12.95, 13.16, 13.02, 12.85,
12.85, 13.00, 13.09. 12.97, 13.12, 13.00, 12.93, 13.11,
13.09, 12.81, 13.26, 13.00.

Determine $\bar{u}$, $\overline{u'^2}$ and $\sqrt{\overline{u'^2}}/\bar{u}$. (130.355 m/s, 0.0137m²/s², 0.0090)

10.2 Two wagons A and B are moving on parallel tracks at velocities of 80 km/hr and 50 km/hr in the same direction. If goods are transferred from B to A at the rate of 1500 kg/s, determine the tangential force on wagon A. (– 12 499.833 N)

10.3 For steady two dimensional flow between parallel plates, assume the shear stress to vary linearly with y and eddy kinematic viscosity to be constant equal to E. Determine the velocity distribution law.
$$\left[\frac{u}{u^*} = \frac{u_* B}{E}\left(\frac{y}{B} - \frac{y^2}{B^2}\right)\right]$$

10.4 For a special type of turbulent flow, assume that the mixing length distribution is given by $l = Ky^2$. Further assume that the turbulent shear $\tau \approx \tau_0$ and determine the form of velocity distribution law using the boundary condition $u = U$ as $y \to \infty$. Does it satisfy the boundary condition at the wall? $\left(u = U\dfrac{u_*}{ky}, \text{ no}\right)$

10.5 Compare the eddy kinematic viscosities in two pipes of diameter 300 mm and 600 mm respectively carrying equal discharge of the same fluid. (Ratio = 2.0)

10.6 Measurements in a fully developed turbulent flow in a pipe indicate that the velocity midway between the pipe wall and pipe centre is 0.90 times the centreline velocity. Determine the average velocity U in times of u_m. What is the value of k/D if pipe acts as rough pipe? (U/u_m = 0.7833, k/D = 0.0156)

10.7 If $\dfrac{u - U}{u_*} = 5.75 \log \dfrac{y}{R} + 3.75$ is valid for turbulent flow in pipes, show that $\dfrac{u_m - u}{U} = -2.03\sqrt{f} \log\left(\dfrac{y}{R}\right)$.

10.8 A smooth pipe of 150 mm diameter gives the centreline velocity of 3.0 m/s when carrying water at 20°C. Determine τ_0, loss of pressure in 50 m length, discharge and friction factor.
(11.64 N/m², 15 520 N/m², 0.04642 m³/s, 0.0136)

10.9 A 300 mm diameter mild steel pipe carries 0.15 m³/s of water discharge at 20°C and requires 3.824 kW power for its transport over 100 m. Determine friction factor, surface roughness height k, shear, and whether the pipe acts as smooth or rough pipe. Also determine velocity at 75 mm from the pipe wall.
(0.017, 1.2 mm, 9.561 N/m², rough, 1.843 m/s)

10.10 Crude oil of relative density 0.85 and kinematic viscosity 5×10^{-6} m²/s is to be pumped through a 300 mm diameter cast iron pipe ($k = 0.30$ mm). If the pipe can withstand a pressure of 1100 kN/m², how far should the pumping stations be if the pipe has to carry 6000 m³/day of the oil? (less than 34 km)

10.11 Determine the diameter of a smooth pipe carrying oil of mass density 850 kg/m³ and dynamic viscosity 3×10^{-3} kg/m s at the rate of 98.125 l/s if permissible energy loss is 0.01093 m per metre length. (250 mm)

10.12 A new galvanised iron pipe of 300 mm diameter carries air at 20°C with a pressure gradient of 0.644 N/m²/m. Determine the nature of the pipe, friction factor and air discharge. Take $k = 0.120$ mm.
(Smooth, 0.0187, 0.292 m³/s)

10.13 A what location will the local velocity for turbulent flow in a pipe be equal to the average velocity of flow? Will the result depend on whether the pipe acts as hydrodynamically smooth or rough?
$\left(\frac{y}{R} = 0.2228, \text{ no}\right)$

10.14 A pipe of 150 mm diameter carries water at 20°C at the rate of 100 l/s. Assuming the average height of surface roughness to be 1.0 mm determine f, τ_0 and u_m. (0.0332, 132.667 N/m², 7.03m/s)

10.15 A steel pipe of 0.60 m diameter has to carry oil of mass density 850 kg/m³ and dynamic viscosity 2×10^{-2} N s/m² at the rate of 650 l/s over 50 km length. Determine the power input required if the efficiency of the pump is 85 per cent. (2722.15 kW)

10.16 For flow of water at 20°C, upto what velocity of flow will a steel pipe of 100 mm diameter act as smooth? What will be its friction factor at that velocity? (1.464 m/s, 0.0166)

10.17 Determine the reduction in discharge carried by 1000 m long smooth pipe of 150 mm diameter due to reduction in water temperature from 40°C to 20°C, assuming power input to remain same, if it carries 40 l/s at 40°C temperature, Assume ν values as 1×10^{-6} m²/s and 6×10^{-7} m²/s at 20°C and 40°C respectively. (1.355 l/s)

10.18 At a given water discharge, a certain galvanised iron pipe acts as hydrodynamically rough and has a friction factor of 0.02. What will be the ratio of costs of pumping water at the same discharge through 10 km length of this pipe and a galvanised iron pipe three fourth its diameter, if the latter also acts as rough?

$\left(\text{Hint: first show that cost} \sim \frac{f}{D^5}\right)$ (0.2207)

10.19 In a hydrodynamically rough pipe of 100 mm diameter, the ratio of velocities at 10 mm and 30 mm from the pipe wall is 0.838. Determine the average height of the wall roughness, shear stress at the wall and mean velocity of flow if velocity at 30 mm is 1.90 m/s.
(1.025 mm, 12.519 N/m², 1.619 m/s)

10.20 A pipe of 1.0 diameter and surface roughness 2.0 mm carries natural gas at 20°C. The observed centreline velocity is 15.0 m/s. Determine the average velocity of flow, bottom shear and velocity at 200 mm from the wall. Assume natural gas characteristics to be the same as that of air. Hence $\rho = 1.208$ kg/m², $\mu = 1.85 \times 10^{-5}$ N s/m².
(12.464 m/s, 0.549 N/m², 13.48 m/s)

10.21 At 0.30 m and 1.20 m from a wall the velocities of air measured were 5.8 m/s and 6.1 m/s respectively. If $\rho = 1.20$ kg/m³ and $\mu = 1.50 \times 10^{-5}$ kg/m s, what is the largest surface roughness value k which will make this surface still smooth? (0.46 mm)

10.22 A water drainage pipe is to be installed so that 2.0 m³/day water discharge flows through it solely by gravity. If the pipe is made of concrete with $k = 0.30$ mm and of diameter 1.20 m, determine the slope of the pipe required to obtain the desired flow. (0.094 23)

10.23 Determine the pressure drop in 100 m length of a rectangular conduit 750 mm × 500 mm made from mild steel if it carries air at a rate of 5.0 m³/s at 20°C. (250.42 N/m²)

10.24 The pipe wall roughness height k for new cast iron pipe is 0.26 mm. After five years of service it becomes 0.40 mm. Assuming the pipe which is 150 mm to act as rough pipe, determine its friction factor after 15 years in service and conmpare it with initial friction factor value. (0.0225, 0.0295)

10.25 If velocity distribution for turbulent flow in a pipe is given by $\frac{u}{u_*} = A\left(\frac{u_* y}{\nu}\right)^n$, where A and n are constants, show that

$$f \sim (\text{Re})^{-2n/n+1}.$$

10.26 In a 300 mm diameter rough concrete pipe, the pressure gradient required to pass the same discharge of water was 1.3 times its original value after a lapse of five years. Estimate the energy required after 15 years for passing the same discharge. (1.727)

DESCRIPTIVE QUESTIONS

10.1 Explain how it is possible to maintain the flow in a pipe laminar even at a Reynolds number as high as 20 000.

10.2 Show that $\overline{u'} = 0$.

10.3 Why is energy loss in turbulent flow much larger than that in laminar flow?

10.4 Where is turbulence first generated? How does it get 'mixed' with the flow?

10.5 Explain the difference between wall turbulence and free turbulent shear flow. Give two examples of each.

10.6 Draw velocity distribution in a pipe at Reynolds numbers of 500, 500 and 500 000. Comment on their shapes.

10.7 Draw a neat sketch of velocity distribution near a hydrodynamically smooth surface indicating clearly the velocity distribution in transition region. Give equations for velocity distribution in each region.

10.8 List the similarities and dissimilarities between μ and η.

10.9 Briefly describe Nikuradse's experiments on pipe flow and the results obtained therefrom.

10.10 Draw velocity distribution in a pipe for the same average velocity (i) when it acts as smooth, and (ii) when it acts as a rough pipe.

10.11 Can you visualise the shape of f vs R_e curve for a smooth pipe as Re becomes very large? Give reasons.

10.12 What is the physical significance of the parameter $\mathrm{Re}\sqrt{f}R/k$?

10.13 For the same cross-sectional area, discharge, fluid properties and boundary characteristics, which one of the circular and square pipe will offer greater resistance? Why?

10.14 What is the effect of Reynolds number on

(i) Smoothness or roughness of boundary,

(ii) Ratio of maximum to mean velocity,

(iii) n in $h_f = rQ^n$ for pipe flow,

(iv) $\frac{\epsilon}{\nu}$ value at a given relative location in pipe?

10.15 What factors are responsible for 'aging' of the pipe?

10.16 Two dimensional air jet issues from a tank and is discharged into a calm atmosphere. Sketch, how the jet will diffuse, velocity distribution at different sections and turbulence intensity distribution across the jet.

10.17 If two pipe flows are to be dynamically similar, what conditions must be satisfied between the model and prototype?

10.18 It can be seen from Fig. 10.4 that the values of $\frac{\text{Re}\sqrt{f}}{R/k}$ at which artificially roughened and commercial pipes cease to behave as smooth and begin to act as rough are different. What is the reason for such a behaviour?

10.19 A sustained turbulence is obtained when

(i) Reynolds number is large
(ii) Reynolds number is low
(iii) Mean shear is present
(iv) Velocity is constant over the cross section.

10.20 According to Prandtl's mixing length theory, turbulent shear is given by

(i) $\rho l \left(\frac{du}{dy}\right)$ (ii) $\rho l \left(\frac{du}{dy}\right)^2$ (iii) $\rho l^2 \left(\frac{du}{dy}\right)^2$ (iv) $\rho l^2 \left(\frac{du}{dy}\right)$

10.21 The turbulence quantities $\overline{u'v'}$, $\overline{u'^2}$ and $\overline{v'^2}$ for pipe flow are maximum

(i) at the wall (ii) at the centre
(iii) slightly away from the wall.

10.22 Velocity distribution in laminar sub-layer is given by

(i) $u/u_* = 11.5 \log_e \frac{u_* y}{\nu} - 3.05$

(ii) $u/u_* = 5.75 \log_{10} \frac{u_* y}{\nu} + 5.5$

(iii) $\frac{u}{u_*} = \frac{u_* y}{\nu}$ (iv) none of the above.

10.23 For hydrodynamically smooth pipes f depends on

(i) Re and R/k (ii) R/k (iii) Re only.

10.24 $\text{Re}\sqrt{f}/R/k$ is proportional to

(i) δ'/k (ii) δ'/R (iii) k/δ' (iv) k/R

10.25 Equivalent roughness k for commercial pipes changes with time t as

(i) $k = k_0 e^{-\alpha t}$ (ii) $k = k_0 + \alpha t$ (iii) $k = k_0 - \alpha t$

CHAPTER XI

Problems in Pipe Flow

11.1 INTRODUCTION

In applications of principles of pipe flow in engineering, one has to deal with various forms of transitions in the pipe line system where flow area, flow direction and cross-sectional shape are changed. Such transitions incur additional energy loss which must be taken into account in calculations of power requirement. This loss becomes more important when the pipe length is small than when it is large. Similarly, in many problems, especially in water supply projects, one has to deal with pipe network. Computations involved in such situations are discussed in this chapter.

11.2 SUDDEN EXPANSIONS AND DIFFUSERS

It has been shown in Chapter VI that application of energy, momentum and continuity equations gives the following expression for head loss in a sudden expansion

$$h_L = (U_1 - U_2)^2/2g = \left(1 - \frac{A_1}{A_2}\right)^2 \frac{U_1^2}{2g} \qquad (11.1)$$

The hydraulic grade line and variation of total energy in a sudden expansion are shown in Fig. 11.1.

When a pipe discharges into a large reservoir

$$A_2 = \infty \quad \therefore \quad A_1/A_2 = 0 \quad \therefore \quad h_L = \frac{U_1^2}{2g} \qquad (11.2)$$

When a pipe discharges into a reservoir, energy loss takes place because of conversion of kinetic energy into heat energy due to viscous shear. This loss is known as exit loss.

In the case of conical diffusers, the increase in cross-sectional area from A_1 to A_2 takes place gradually. The diffuser length depends on the ratio A_2/A_1 and the diffuser angle θ. Head loss h_L in a diffuser can thus be expressed as

$$h_L = K(U_1 - U_2)^2/2g$$

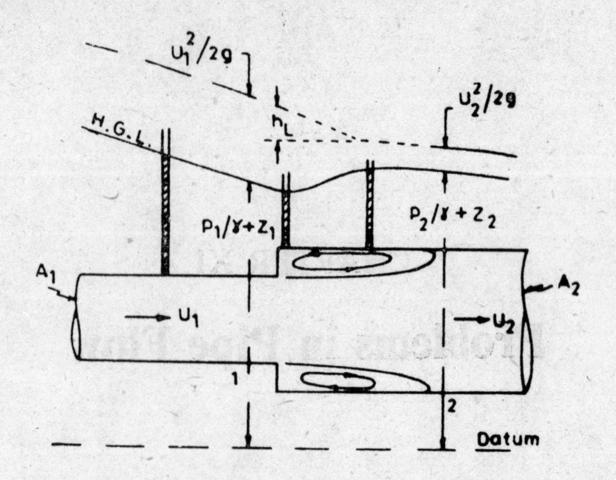

Fig. 11.1. Flow in a sudden expansion.

where $K = f\left(\frac{A_2}{A_1}, \theta, \text{Re}_1\ k/D_1\right)$.

Here Re_1 is Reynolds number in the first pipe and $\frac{k}{D_1}$ is the relative roughness in the same pipe. These two parameters affect K values only when diffusers or contractions are long and boundary friction is quite a significant portion of the total resistance. Typical variation of K with A_2/A_1 and θ is shown in Fig. 11.2. Two types of efficiencies can be defined for diffusers which convert velocity head into pressure head.

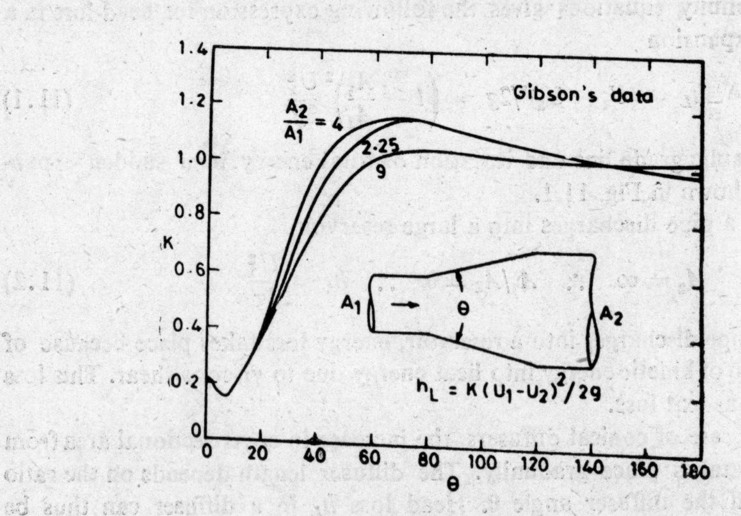

Fig. 11.2. Variation of K with θ and A_2/A_1 for diffusers.

Energy efficiency of diffuser,

$$\eta_e = \left(\frac{\alpha_1 U_1^2}{2g} - h_L\right) \Big/ \frac{\alpha_1 U_1^2}{2g} \tag{11.3}$$

Pressure efficiency of a diffuser,

$$\eta_p = (p_2 - p_1) \Big/ \left(\frac{\alpha_1 U_1^2}{2g} - \frac{\alpha_2 U_2^2}{2g}\right) \tag{11.4}$$

where α_1 and α_2 are the kinetic energy correction factors at sections 1 and 2 respectively.

11.3 CONTRACTIONS AND CONTRACTING TRANSITIONS

Energy loss in sudden contraction is partly due to contraction of flow between sections 1 and 0 and partly due to expansion of flow from section 0 to 2 (Fig. 11.3). Or

$$h_L = K_1 \frac{U_0^2}{2g} + \frac{(U_0 - U_2)^2}{2g}$$

which can be simplified to

$$h_L = \left[\frac{K_1}{C_c^2} + \left(\frac{1}{C_c} - 1\right)^2\right] \frac{U_2^2}{2g} = K \frac{U_2^2}{2g} \tag{11.5}$$

Values of coefficient of contraction $C_c = \dfrac{A_0}{A_2}$ and K_1 depend on A_2/A_1 ratio.

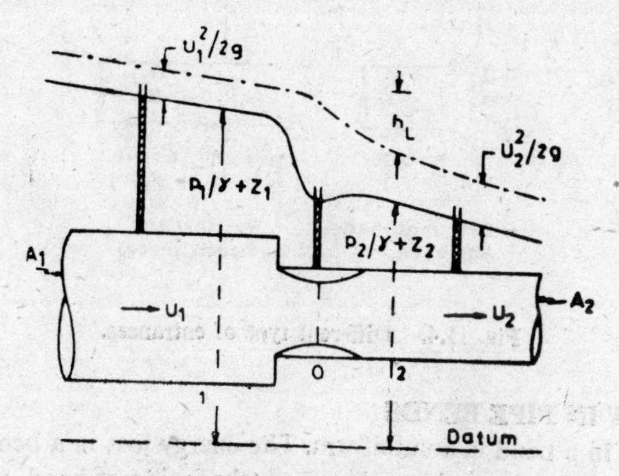

Fig. 11.3. Flow in a sudden contraction.

The coefficient of contraction C_c is related to A_2/A_1 by the empirical relation

$$C_c = 0.62 + 0.38\,(A_2/A_1)^3.$$

Table 11.1 gives values of K as a function of A_2/A_1.

Table 11.1 Values of contraction coefficient K as related to A_2/A_1

A_2/A_1	0	0.10	0.20	0.30	0.40	0.50	0.60	0.70	0.80	0.90	1.0
K	0.50	0.46	0.41	0.36	0.30	0.24	0.18	0.12	0.06	0.02	0

If the contraction is gradual or conical in shape, K will naturally be smaller than the value listed in Table 11.1 for the same A_2/A_1 ratio. Head loss at the entrance from reservoir to the pipe depends on its shape as can be seen from Fig. 11.4. Avoidance of separation reduces the loss coefficient. The entrance shape is usually chosen such that excessively low pressure does not occur anywhere in the entrance reach.

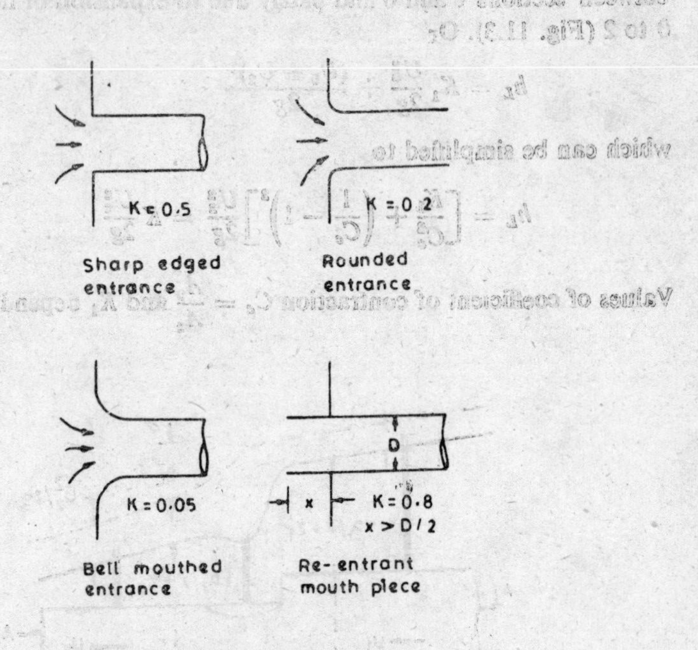

Fig. 11.4. Different type of entrances.

11.4 FLOW IN PIPE BENDS

The flow in a bend is nonuniform. The energy loss in a bend is due to pronounced separation taking place near the inside of bend, generation of secondary circulation, formation of weak zone of stagnation near outer side of bend and increased level of turbulence and its dissipation in the downstream portion of bend. Flow conditions in a bend alongwith variation of peizometric head and total energy along the bend are shown in Fig. 11.5. Figure 11.6 shows the variation of K with R_b/D and θ at Re value of 10^6.

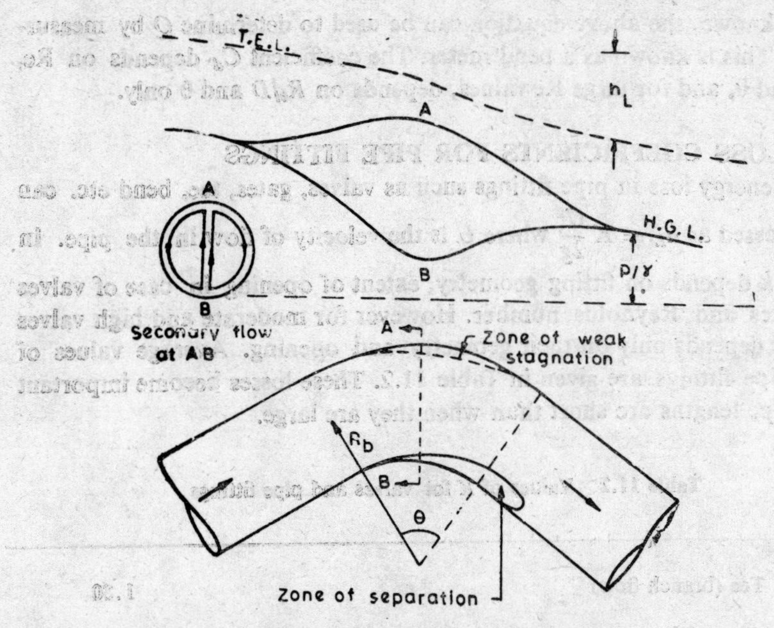

Fig. 11.5. Flow in a pipe bend.

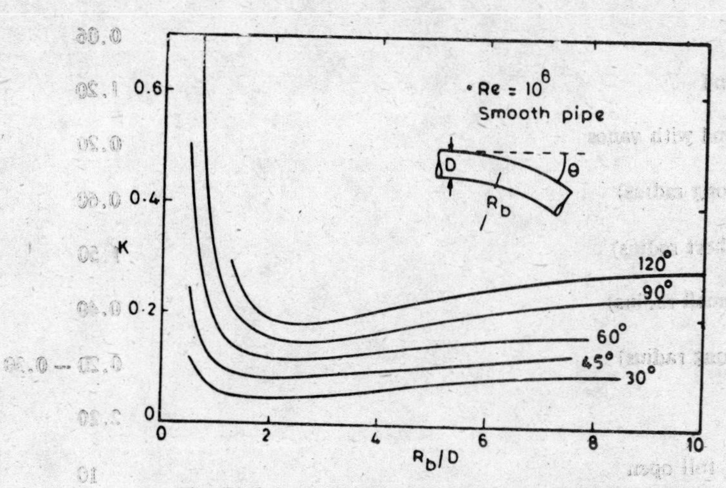

Fig. 11.6. Variation of K with R_b/D and θ for bends.

Here $K = h_L \Big/ \frac{U^2}{2g}$. In general K is a function of R_b/D, θ, k/D and Reynolds number. Energy loss in a bend can be reduced by providing vanes.

Figure 11.5 shows that there is difference of pressure Δp between the outside and the inside of the bend. This can be related to the discharge Q flowing through the bend by the equation

$$Q = C_d A \sqrt{2\Delta p/\rho}$$

If C_d is known, the above equation can be used to determine Q by measuring Δp. This is known as a bend meter. The coefficient C_d depends on Re, R_b/D and θ, and for large Re values, depends on R_b/D and θ only.

11.5 LOSS COEFFICIENTS FOR PIPE FITTINGS

The energy loss in pipe fittings such as valves, gates, tee, bend etc. can be expressed as $h_L = K\frac{U^2}{2g}$ where U is the velocity of flow in the pipe. In general k depends on fitting geometry, extent of opening in case of valves and gates and Reynolds number. However for moderate and high valves of Re, k depends only on their geometry and opening. Average values of k for pipe fittings are given in Table 11.2. These losses become important when pipe lengths are short than when they are large.

Table 11.2 Values of K for valves and pipe fittings

Standard Tee (branch flow)	1.80
Standard Tee (line flow)	0.40
Couplings	0.06
Sharp 90° bend	1.20
Sharp 90° bend with vanes	0.20
90° Elbow (long radius)	0.60
90° Elbow (short radius)	1.50
45° Elbow (small radius)	0.40
45° Elbow (long radius)	0.20 – 0.30
180° Bend	2.20
Globe valve, full open	10
Gate valve full, $\frac{3}{4}$, $\frac{1}{2}$, $\frac{1}{4}$ open	0.20, 1.15, 5.6, 24.0
Angle valve full, $\frac{3}{4}$, $\frac{1}{2}$, $\frac{1}{4}$ open	2.5
Switch valve full, $\frac{3}{4}$, $\frac{1}{2}$, $\frac{1}{4}$ open	2.4
Foot valve full, $\frac{3}{4}$, $\frac{1}{2}$, $\frac{1}{4}$ open	10

(These are typical average values. Actual values will depend on the make of the fitting).

11.6 EQUIVALENT LENGTH OF PIPE

If $h_L = K\,U^2/2g$ gives energy loss due to any nonuniformity in a pipe flow, e.g. due to bend, valve, etc., one can determine equivalent length l_e of pipe of diameter D and friction factor f which will give the same frictional loss as h_L for the same discharge. Hence

$$K\frac{U^2}{2g} = \frac{fl_e}{D}\frac{U^2}{2g}$$

or

$$l_e = KD/f$$

Equivalent length of l_2 of diameter D_2 and friction factor f_2, in terms of length l_1 of pipe of diameter D_1 and friction factor f_2 can be computed from the formula

$$\frac{f_1 l_1}{D_1^5} = \frac{f_2 l_2}{D_2^5}$$

11.7 SIPHONS

Siphons are used to carry water from higher elevation 1 (Fig. 11.7) to a lower elevation 3, over an intermediate higher elevation 2. Bernoulli's equation yields

$$H = \frac{U^2}{2g}\left(1 + K + \frac{fl_{1,3}}{D}\right)$$

where D is the diameter of pipe and K is minor loss coefficient and $l_{1,3}$ is the pipe length between the sections 1 and 3.

The absolute pressure at 2 is given by

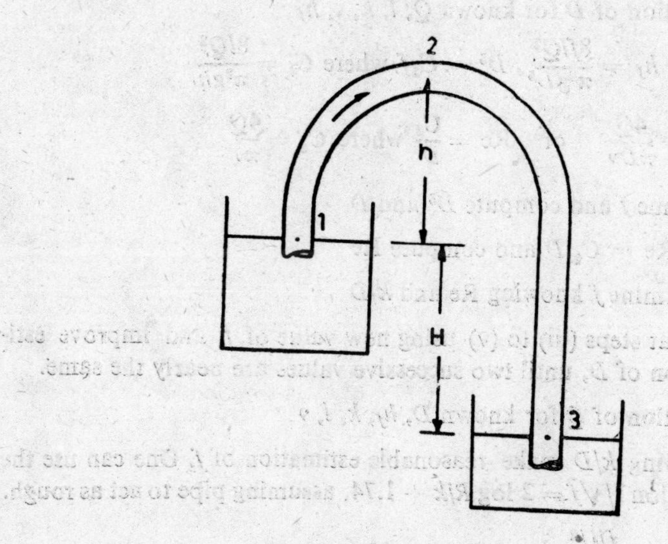

Fig. 11.7. Siphon.

$$\frac{p_2}{\gamma} = \frac{p_1}{\gamma} - h - \frac{V^2}{2g}\left(1 + K' + \frac{fl_{1,2}}{D}\right)$$

where K' is minor loss coefficient between 1 and 2, and $l_{1,2}$ is the pipe length between these sections. The pressure at 2 should not fall below 2 m of water absolute. If f is known, then solution for H or the discharge Q is straightforward. But if f is not given, solution for Q is one of successive approximations.

11.8 PIPE FLOW PROBLEMS

In problems related to pipe flow, one has to deal with six variables, namely ν, l, D, k, Q and h_f out of which five are known and the sixth is to be determined. In reality ν, k and l are always known. Hence there are three types of problems in which h_f D or Q has to be determined. The equations known are

$$h_f = \frac{flU^2}{2gD} = \frac{8flQ^2}{\pi^2 gD^5}$$

and Moody's diagram or corresponding equations for f in terms of Re and k/D.

(a) Determination of h_f for known Q, D, l, k, ν

(i) Determine $U = 4Q/\pi D^2$, and $\text{Re} = \dfrac{UD\rho}{\mu}$ and k/D.

(ii) Determine f using Moody's diagram or Eq. 10.22 or 10.23.

(iii) $h_f = flU^2/2gD$

(b) Determination of D for known Q, l, k, ν, h_f

(i) Since $h_f = \dfrac{8flQ^2}{\pi^2 gD^5}$, $D^5 = C_1 f$ where $C_1 = \dfrac{8lQ^2}{\pi^2 gh_f}$

(ii) $Re = \dfrac{4Q}{\pi D\nu}$ or $\text{Re} = \dfrac{C_2}{D}$ where $C_2 = \dfrac{4Q}{\pi\nu}$

(iii) Assume f and compute D^5 and D

(iv) Use $\text{Re} = C_2/D$ and compute Re

(v) Determine f knowing Re and k/D

(vi) Repeat steps (iii) to (v) using new value of f and improve estimation of D, until two successive values are nearly the same.

(c) Determination of Q for known D, h_f, k, l, ν

(i) Knowing k/D make reasonable estimation of f. One can use the equation $1/\sqrt{f} = 2 \log R/k + 1.74$, assuming pipe to act as rough.

(ii) Use $h_f = \dfrac{flU^2}{2gD}$ and determine U.

(iii) Determine $\text{Re} = \frac{UD\rho}{\mu}$ and refine the value of f.

(iv) Repeat steps (ii) and (iii) and refine values of f and U until no further refinement is possible. Compute

$$Q = \frac{\pi D^2}{4} U.$$

Direct solution of (b) and (c) type of problems can also be obtained using the following empirical equation

$$\left(\frac{Q}{k\nu}\right)\left(\frac{k}{D}\right)^{8/3} = 2.30\left(\frac{k^3 gS}{\nu^2}\right)^{0.517}$$

where $S = h_f/l$.

11.9 PARALLEL PIPES

When two or more pipes are connected in parallel as shown in Fig. 11.8, the following conditions govern the flow:

$$Q = Q_1 + Q_2 + Q_3 = \sum_1^n Q_i$$

$$h_{f_1} = h_{f_2} = h_{f_3} = h_{f_i}$$

$$Z_A - Z_B = h_{f_i} = \frac{f_i\, l_i\, U_i^2}{2gD_i}$$

where Z_A and Z_B are piezometric heads at A and B. Two types of problems arise in case of parallel pipes.

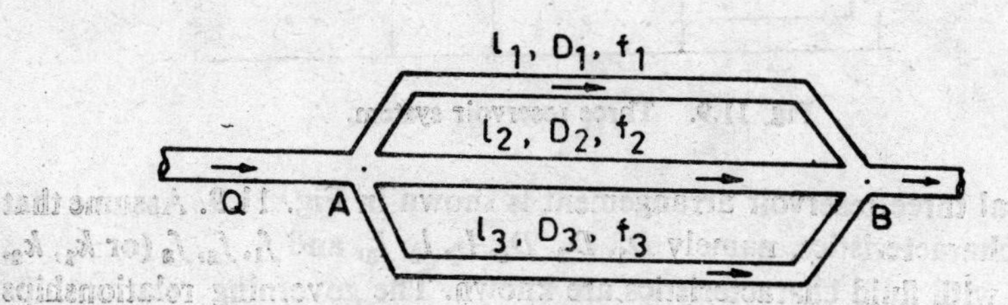

Fig. 11.8. Parallel pipes.

(a) To determine Q for known h_f and pipe characteristics, i.e. l_i, D_i and f_i.

(i) First determine $Q_i = \sqrt{\pi^2 g D_i^5 h_f / 8 f_i\, l_i}$

(ii) $Q = \sum_1^n Q_i$

If f_i is not known, one must assume f_i, calculate Q_i as discussed above and then recalculate values of f_i using Moody's diagram.

(b) For given Q and pipe characteristics, determination of Q_i and h_f

(i) Assume discharge Q'_1 in pipe 1 and determine U'_1.

(ii) $h'_f = \dfrac{f_1 l_1 U'^2_1}{gD_1}$

(iii) For known h'_f calculate U'_2 and U'_3, and hence Q'_2 and Q'_3.

(iv) $Q' = Q'_1 + Q'_2 + Q'_3$.

(v) The required distribution of Q is

$$Q_1 = \frac{Q}{Q'} Q'_1,\ Q_2 = \frac{Q}{Q'} Q'_2,\ Q_3 = \frac{Q}{Q'} Q'_3.$$

(vi) Compute h_f for known Q_1.

11.10 BRANCHING PIPES

Typical problems of branching pipes are those of interconnected reservoirs, i.e., three or four reservoir problems.

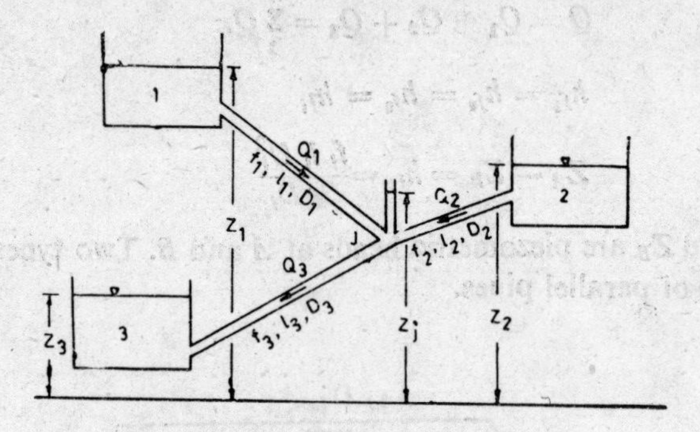

Fig. 11.9. Three reservoir system.

Typical three reservoir arrangement is shown in Fig. 11.9. Assume that the pipe characteristics, namely D_1, D_2, D_3, l_1, l_2, l_3, and f_1, f_2, f_3 (or k_1, k_2, k_3) along with fluid characteristics are known. The governing relationships are

(i) Moody's diagram or corresponding equations for f

(ii) $h_f = flU^2/2gD$

(iii) At the junction $\sum_1^n Q_i = 0$, i.e. the flow coming into the junction must be equal to flow going out of the junction.

Three types of problems can be visualised, in respect of Fig. 11.9.

(a) To find Q_2, Q_3 and Z_3 when Z_1 and Z_2 and Q_1 are known.

(i) Determine h_{f_1} and piezometric head of junction Z_j.

(ii) Knowing Z_j and Z_2, determine $h_{f_2} = (Z_j - Z_2)$ or $(Z_2 - Z_j)$ and Q_2 along with its direction.

(iii) Determine Q_3, h_{f_3} and hence Z_3.

(b) To determine Z_2, Q_1, and Q_2 when Z_1, Z_3 and Q_2 are known.

(i) $Z_1 - Z_3 = h_{f_1} + h_{f_3}$ is known. Also $(Q_1 - Q_3)$ or $(Q_3 - Q_1)$ i.e. Q_2 is known.

(ii) Assume Z_j and determine $(Z_1 - Z_j) = h_{f_1}$ and $Z_j - Z_3 = h_{f_3}$.

(iii) Compute Q_1 and Q_3. Compare Q_2 with computed value by plotting Z_j against $Q_2 - |Q_1 - Q_3|$ for various assumed values of Z_j.

(iv) That value of Z_j is correct which gives $Q_2 - |Q_1 - Q_3| = 0$.

(v) For this Z_j, determine Q_1 and Q_3.

(c) To determine Q_1, Q_2, Q_3 given Z_1, Z_2 and Z_3.

(i) Assume $Z_j = Z_2$ and find Q_1 and Q_3. If $Q_1 > Q_3$, $Z_j > Z_2$, otherwise $Z_j < Z_2$.

(ii) Assume Z_j and determine h_{f_1}, h_{f_2}, h_{f_3} such that $h_{f_1} = Z_1 - Z_j$, $h_{f_2} = |Z_j - Z_2|$, $h_{f_3} = |Z_j - Z_3|$.

(iii) Compute Q_1, Q_2, Q_3.

(iv) Check if these values satisfy the continuity equation at the junction. If not, assume another value of Z_j and recompute Q_1, Q_2, Q_3.

(v) By plotting Z_j vs $Q_2 - |Q_1 - Q_3|$ correct value of Z_j can be interpolated and Q_1, Q_2 and Q_3 computed.

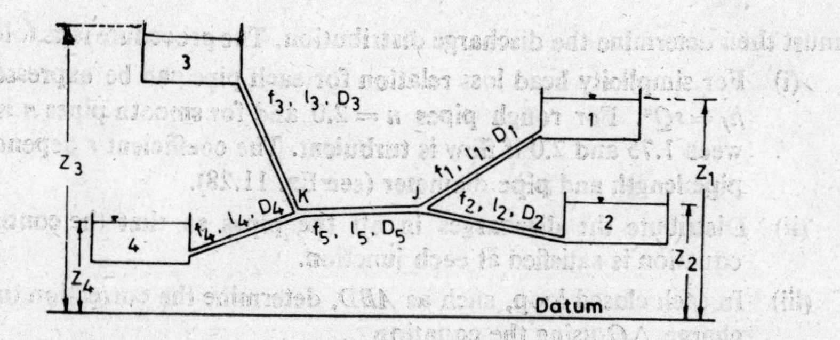

Fig. 11.10.

In the case of four-reservoirs problem, as before one can assume pipe characteristics and fluid characteristics to be known. With respect to Fig. 11.10, assume Z_1, Z_2, Z_3 and Z_4 to be known and one wants to determine Q_1, Q_2, Q_3 and Q_4. This is achieved in the following way.

1. Assume Z_j. Knowing Z_1 and Z_2, determine Q_1 and Q_2.
2. Determine $(Q_1 + Q_2)$ or $(Q_1 - Q_2)$ which flows in pipe 5.

3. Determine h_{f_5} and hence Z_{jk}.
4. Knowing Z_4 and Z_5 determine h_{f_4} and h_{f_5} and hence Q_4 and Q_5.
5. Check if continuity equation is satisfied at junction k.
6. If continuity equation is not satisfied at junction k, assume a new value of Z_j and repeat steps (i) through (vi), till continuity equation is satisfied.

11.11 PIPE NETWORKS

In pipe networks such as *ABCD* (Fig. 11.11), discharges are given at the nodes *A, B, C, D* and pipe characteristics such as *f, l, D* are known. One

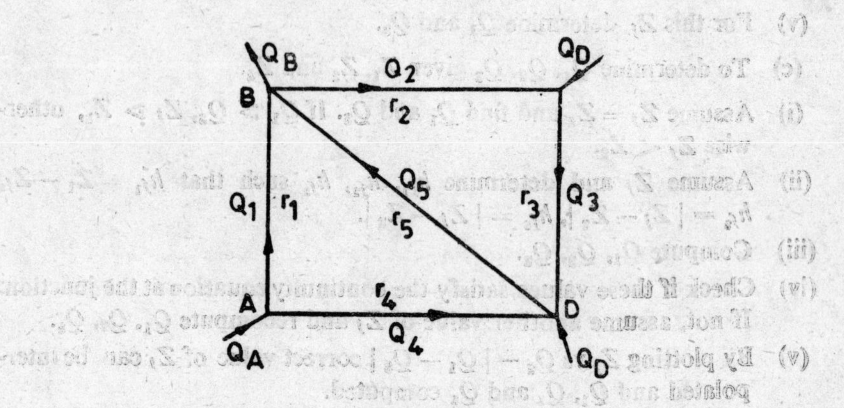

Fig. 11.11.

must then determine the discharge distribution. The procedure is as follows:

(i) For simplicity head loss relation for each pipe can be expressed as $h_f = rQ^n$. For rough pipes $n = 2.0$ and for smooth pipes n is between 1.75 and 2.0 if flow is turbulent. The coefficient r depends on pipe length and pipe diameter (see Ex. 11.28).

(ii) Distribute the discharges in all the pipes so that the continuity equation is satisfied at each junction.

(iii) In each closed loop, such as *ABD*, determine the correction in discharge ΔQ using the equation

$$\Delta Q = \frac{-\Sigma rQ^n}{\Sigma nrQ^{n-1}}$$

where head loss in each pipe is given by $h_f = r_i Q_i^n$. Head loss is considered +ve when clockwise and −ve when anticlockwise. ΔQ is clockwise when +ve and anticlockwise when −ve.

(iv) Correct discharges in all pipes taking into consideration corrections in all the loops and repeat step 3.

(v) Stop when ΔQ in each loop is negligibly small.

ILLUSTRATIVE EXAMPLES

11.1 Determine the discharge in a pipe of 200 mm diameter which suddenly expands to 400 mm diameter and in which the hydraulic grade line rises by 10 mm in the expansion.

Apply Bernoulli's equation between section just before expansion and section just after expansion. Since $U = \frac{4Q}{4D^2}$,

$$\left(\frac{p_1}{\gamma} + Z_1\right) + \frac{8Q^2}{\pi^2 g D_1^4} = \left(\frac{p_2}{\gamma} + Z_2\right) + \frac{8Q^2}{\pi^2 g D_2^4} + \frac{1}{2g}\left(\frac{4Q}{\pi D_1^2} - \frac{4Q}{\pi D_2^2}\right)^2$$

The last term on right hand side represents the energy loss in the sudden expansion.

$$\therefore \quad \left(\frac{p_2}{\gamma} + Z_2\right) - \left(\frac{p_1}{\gamma} + Z_1\right) = 0.01 = \frac{8Q^2}{\pi^2 g}\left\{\frac{1}{D_1^4} - \frac{1}{D_2^4}\right\} - \frac{8Q^2}{\pi^2 g}\left\{\frac{1}{D_1^2} - \frac{1}{D_2^2}\right\}^2$$

$$\therefore \quad 0.010 = \frac{8Q^2}{\pi^2 g}\{(625 - 39.0625) - (25 - 6.25)^2\}$$

$$= \frac{Q^2}{12.1} \times 234.375$$

$$\therefore \quad Q^2 = (12.1 \times 0.010)/234.375 = 0.000\,5163$$

$$Q = 0.022\,722 \text{ m}^3/\text{s} \quad \text{or} \quad 22.722 \ l/\text{s}$$

11.2 A horizontal sudden expansion in a pipe expends 4.446 kW of power when the pipe diameters are 200 mm and 400 mm before and after the expansion. Determine the discharge and pressure difference $(p_2 - p_1)$ in the expansion.

$$\text{Expansion loss } h_L = \frac{(U_1 - U_2)^2}{2g} = \frac{Q^2}{2g}\left(\frac{1}{A_1} - \frac{1}{A_2}\right)^2$$

or

$$h_L = \frac{Q^2}{2 \times 9.806}\left[\frac{1}{0.785 \times 0.2^2} - \frac{1}{0.785 \times 0.4^2}\right]^2$$

$$= 29.089\,Q^2$$

$$\text{Power} \quad P = Q\gamma h_L/1000$$

$$\therefore \quad 4.446 = Q^3 \times 9787 \times 29.089/1000$$

$$\therefore \quad Q^3 = 0.01562 \quad \text{and} \quad Q = 0.250 \text{ m}^3/\text{s}$$

Since the pipe is horizontal

$$\frac{p_1}{\gamma} + \frac{U_1^2}{2g} = \frac{p_2}{\gamma} + \frac{U_2^2}{2g} + \frac{(U_1 - U_2)^2}{2g}$$

$$\therefore \quad (p_2 - p_1) = \left(\frac{U_1^2 - U_2^2}{2g} - \frac{(U_1 - U_2)^2}{2g}\right) \times 9787$$

$$U_1 = \frac{0.250}{0.785 \times 0.2^2} = 7.961 \text{ m/s}, \; U_2 = \frac{0.25}{0.785 \times 0.4^2} = 1.990 \text{ m/s}.$$

$$\therefore \quad (p_2 - p_1) = (3.030 - 1.818) \times 9787$$

$$= 11\,864.61 \text{ N/m}^2 \quad \text{or} \quad 11.865 \text{ kN/m}^2$$

11.3 A 150 mm diameter pipe line is contracted to 100 mm diameter. If the pipe line carries a discharge 40 *l*/s of water, determine the pressure difference $(p_1 - p_2)$ across the contraction if the two sections are having an elevation difference 0.30 m, section 2 being higher in elevation.

In Fig. 11.3,

$$U_1 = 0.04/0.785 \times 0.15^2 = 2.265 \text{ m/s}$$

$$U_2 = 0.04/0.785 \times 0.10^2 = 5.096 \text{ m/s}$$

From Table 11.1, $K = 0.27$ for $A_2/A_1 = (0.1/0.15)^2 = 0.444$

$$\text{Contraction loss } h_L = 0.27 \frac{U_2^2}{2g} = 0.358 \text{ m}$$

$$\therefore \quad \frac{p_1}{\gamma} + Z_1 + \frac{U_1^2}{2g} = \frac{p_2}{\gamma} + Z_2 + \frac{U_2^2}{2g} + h_L$$

$$\therefore \quad (p_1 - p_2)/\gamma = (Z_2 - Z_1) + \frac{1}{2g}(U_2^2 - U_1^2) + h_L$$

$$= 0.30 + \frac{1}{2 \times 9.806}(5.096^2 - 2.265^2) + 0.358$$

$$= 0.300 + 1.063 + 0.358$$

$$= 1.721 \text{ m}$$

$$\text{or} \quad (p_1 - p_2) = 1.721 \times 9.787 = 16.843 \text{ kN/m}^2$$

11.4 In a pipe of diameter D carrying a discharge of Q_0 at its upstream end, a discharge of q per unit length is withdrawn uniformly along its length. Obtain an expression for head loss h_f in length L (Fig. 11.12).

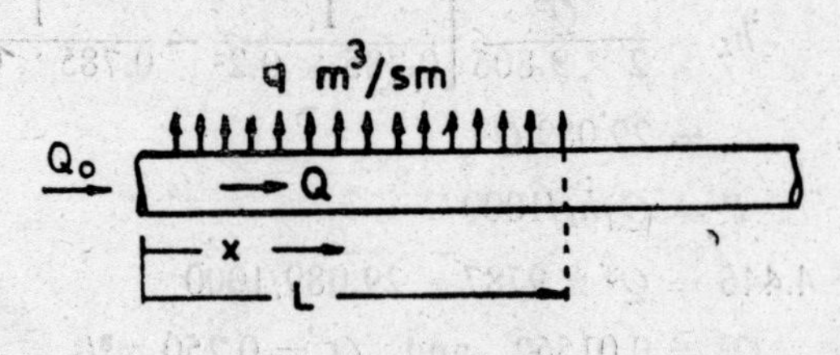

Fig. 11.12.

Discharge is withdrawn at the rate of q per unit length. Hence withdrawal in distance $x = qx$.

$\therefore$ Discharge in pipe at any distance x, $Q = (Q_0 - qx)$

$$dh_f = \frac{8fQ^2dx}{\pi^2 gD^5} = \frac{8f(Q_0 - qx)^2\,dx}{\pi^2 gD^5}$$

$$\therefore \qquad h_f = \frac{-8f}{3\pi^2 g D^5 q}\left[(Q_0 - qx)^3\right]_{Q_0}^{(Q_0-Lq)}$$

$$= \frac{8f}{3\pi^2 g D^5 q}[Q_0^3 - (Q_0 - qL)^3]$$

11.5 A 300 mm diameter pipe connected to a reservoir has the following nonuniformities in its length of 300 m.

Rounded entrance ($K = 0.2$)

Sudden expansion to 400 mm diameter pipe

Sudden contraction to 300 mm diameter pipe (take $K = 0.215$)

Sharp bend (with vanes) $K = 0.20$

A gate valve fully open $K = 10$

Assume pipe friction coefficient to be 0.012 and estimate the discharge under a dead of 40 m, if pipe discharges into atmosphere.

Express all the losses in terms of velocity in 300 mm diameter pipe.

$$\text{Entrance loss} = 0.2\frac{U^2}{2g}$$

$$\text{Expansion loss} = \left(1 - \frac{A_1}{A_2}\right)^2 \frac{U^2}{2g} = 0.563\frac{U^2}{2g}$$

$$\text{Contraction loss} = 0.215\frac{U^2}{2g}$$

$$\text{Bend loss} = 0.2\frac{U^2}{2g}$$

$$\text{Gate value loss} = 10\frac{U^2}{2g}$$

$$\text{Exit loss} = \frac{U^2}{2g}$$

$$\text{Friction loss} = \frac{flU^2}{2gD} = \frac{0.10 \times 300}{6 \times 0.30}\frac{U^2}{2g} = 12\frac{U^2}{2g}$$

$$\text{Total loss} = (0.20 + 0.563 + 0.215 + 0.20 + 10 + 1.0 + 12)\frac{U^2}{2g}$$

$$= 24.178\frac{U^2}{2g}$$

$$\therefore \qquad 40 = 24.178\frac{U^2}{2g} \quad \therefore \quad U^2 = 32.446$$

or $\qquad U = 5.696$ m/s

$$\therefore \qquad Q = \frac{\pi D^2}{4}U = 0.785 \times 0.30^2 \times 5.696 = 0.4024 \text{ m}^3/\text{s}$$

In this particular problem, the minor loss coefficient (i.e.

$$0.2 + 0.5638 + 0.215 + 0.2 + 10.0 + 1.0) \text{ i.e. } 12.178$$

is nearly the same as friction loss coefficient 12. Hence minor losses must be taken into account. If the pipe were 3000 m long, their ratio would have been $\frac{12.178}{120}$ or 10%. Thus in longer pipes minor losses are some times neglected. However this cannot be done in pipes of small length.

11.6 Consider a horizontal pipe line connecting two reservoirs as shown in Fig. 11.13. The difference of water level in two tanks is 50 m. Considering the friction and nonuniformities listed below, determine the discharge in m³/s and draw the total energy line and hydraulic grade line. Take $f = 0.015$.

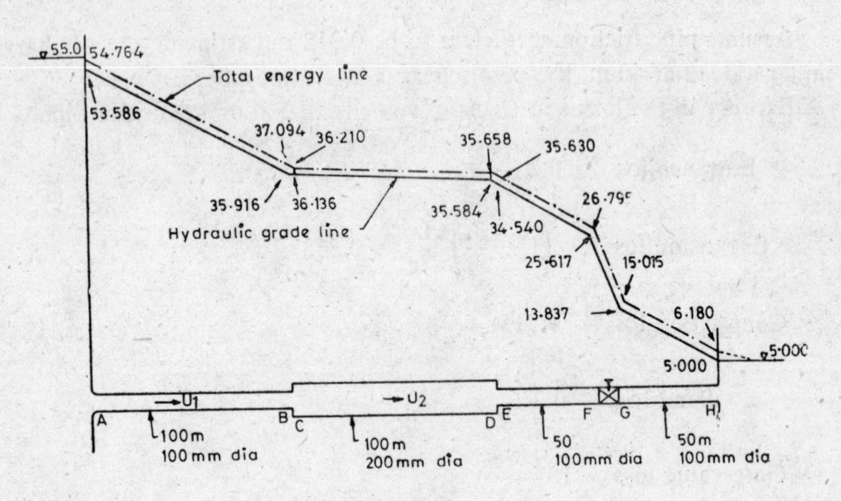

Fig. 11.13.

(i) Rounded entrance at A; $h_L = 0.20 \frac{U_1^2}{2g}$

(ii) Friction in 100 m length AB, $h_f = \frac{0.015 \times 100}{0.10} \frac{U_1^2}{2g} = 15 \frac{U_1^2}{2g}$

(iii) Sudden expansion from 100 m to 200 mm diameter

$$h_L = (1 - 0.25) \frac{U_1^2}{2g} = 0.75 \frac{U_1^2}{2g}$$

(iv) Friction in 100 m length CD of 200 mm diameter pipe

$$h_f = \frac{0.015 \times 100}{0.200} \left(\frac{U_2^2}{2g}\right) = 0.4688 \frac{U_1^2}{2g} \text{ since } U_1 = 4U_2$$

(v) Sudden contraction at DE, $h_L = 0.385 \frac{U_2^2}{2g} = 0.024 \frac{U_1^2}{2g}$

(vi) Friction in 50 m length of 100 mm diameter pipe

$$h_f = \frac{0.015 \times 50}{0.100} \frac{U_1^2}{2g} = 7.5 \frac{U_1^2}{2g}$$

(vii) Head loss in globe valve at F, $h_L = 10 \frac{U_1^2}{2g}$

(viii) Friction in 50 m length of 100 mm diameter pipe GH

$$h_f = 7.5 \frac{U_1^2}{2g}$$

(ix) Exit loss at H, $h_L = \frac{U_1^2}{2g}$

$$\therefore \quad 50 = (0.2 + 15.0 + 0.75 + 0.4688 + 0.024 + 7.5 + 10.0 + 7.5 + 1.0) \frac{U_1^2}{2g}$$

$$\therefore \quad 50 = 42.4428 \frac{U_1^2}{2g} \quad \text{or} \quad U_1^2 = 23.104 \quad \text{and} \quad U_1 = 4.807 \text{ m/s}$$

$$Q = 0.785 \times 0.10^2 \times 4.807 = 0.037\ 73 \text{ m}^3\text{/s}$$

Also $\frac{U_1^2}{2g} = 1.178$ and $\frac{U_2^2}{2g} = 0.0736$

(i) Energy loss at entrance $= 0.20 \times 1.178 = 0.236$ m

$$\therefore \quad \frac{p_A}{\gamma} = 55 - 1.178 - 0.236 = 53.586 \text{ m}$$

Total energy at $A = 53.586 + 1.178 = 54.764$ m

(ii) h_f in length $AB = 15 \times 1.178 = 17.670$ m

$$\therefore \quad \frac{p_B}{\gamma} = \frac{p_A}{\gamma} - 17.670 = 53.586 - 17.670 = 35.916 \text{ m}$$

Total energy at $B = 35.916 + 1.178 = 37.094$ m

(iii) Loss in sudden expansion, $h_L = 0.75 \times 1.178 = 0.884$ m

$$\therefore \quad 37.094 = \frac{p_c}{\gamma} + 0.0736 + 0.884$$

$$\therefore \quad \frac{p_c}{\gamma} = 36.136 \text{ m}$$

Total energy at $C = 36.136 + 0.0736 = 36.21$ m

(iv) Frictional loss in CD, $h_f = 0.4688 \times 1.178 = 0.552$ m

$$\therefore \quad \frac{p_D}{\gamma} = 36.136 - 0.552 = 35.584 \text{ m}$$

Total energy at $D = 35.584 + 0.0736 = 35.658$ m

(v) Loss in contraction, $h_L = 0.024 \times 1.178 = 0.0283$ m

$\therefore$ Total energy at $E = 35.658 - 0.0283 = 35.630$ m

$$\frac{p_E}{\gamma} = 35.630 - 1.178 = 34.452 \text{ m}$$

(vi) Frictional loss in EF, $h_f = 7.5 \times 1.178 = 8.835$ m

$$\therefore \quad \frac{p_F}{\gamma} = 34.452 - 8.835 = 25.617 \text{ m}$$

Total energy at $F = 25.617 + 1.178 = 26.795$ m

(vii) Head loss in globe valve $= 10 \times 1.178 = 11.780$

$$\therefore \quad \frac{p_G}{\gamma} = 25.617 - 11.780 = 13.837 \text{ m}$$

Total energy at $G = 13.837 + 1.178 = 15.015$ m

(viii) Frictional loss in GH, $h_f = 7.5 \times 1.178 = 8.835$ m

Total energy at $H = 15.015 - 8.835 = 6.180$ m

$$\frac{p_H}{\gamma} = 6.180 - 1.178 = 5.0 \text{ m}$$

In Fig. 11.13 the total energy line and hydraulic grade line are plotted.

11.7 A 300 mm diameter cast iron pipe is 30 m long and is connected in series to 250 mm diameter cast iron pipe of 90 m length. If the pipe carries a water discharge of 0.14 m³/s at 20°C, find the equivalent length of pipe in terms of 250 mm dia pipe. Neglect contraction loss.

$$U_1 = \frac{0.14}{0.785 \times 0.30^2} = 1.982 \text{ m/s}$$

$$Re_1 = (1.982 \times 0.30)/10^{-6} = 5.94 \times 10^5$$

For cast iron $k = 0.26$ mm; hence $\dfrac{k}{D} = \dfrac{0.26}{300} = 0.000867$

Moody's diagram gives $f_1 = 0.0195$ for $\dfrac{k}{D} = .000867$ and

$Re_1 = 5.94 \times 10^5$.

$$U_2 = \frac{0.14}{0.785 \times 0.25^2} = 2.854 \text{ m/s}$$

$Re_2 = (2.854 \times 0.25)/10^{-6} = 7.135 \times 10^5$ and

$$\frac{k}{D} = \frac{0.26}{250} = 0.00104$$

Hence from Moody's diagram $f_2 = 0.0190$

$$h_f = h_{f_1} + h_{f_2} = \frac{0.0195 \times 30 \times 1.982^2}{2 \times 9.806 \times 0.30} + \frac{0.0190 \times 90 \times 2.854^2}{2 \times 9.806 \times 0.25}$$

$$= 0.391 + 2.841$$

$$= 3.232 \text{ m}$$

Total length of 250 mm dia, $l_e = \dfrac{3.232 \times 2 \times 9.806 \times 0.25}{0.0190 \times 2.854^2} = 102.39$ m

One can alternatively determine equivalent length for 300 mm diameter pipe and add it to 250 mm diameter pipe length.

11.8 A pipe system consists of the following

50 m of 300 mm diameter pipe of $f = 0.025$

90° bend with $K = 0.50$

300 mm valve $K = 1.0$

Sudden expansion to 400 mm diameter pipe

100 m of 400 mm diameter pipe of $f = 0.02$

400 mm valve with $K = 1.5$

Convert this pipe system to an equivalent 300 mm diameter pipe of $f = 0.02$ if velocity in 300 mm diameter pipe is 3.0 m/s.

(i) Pipe of 300 mm diameter — 50 m

(ii) Equivalent length of bend $= \dfrac{0.5 \times 300}{0.025}$ $= 6.0$ m

(iii) Equivalent length for valve $= \dfrac{KD}{f} = \dfrac{1.0 \times 0.300}{0.025}$ $= 12.0$ m

(iv) Velocity in 400 mm diameter pipe will be

$$3.0 \times \left(\frac{0.30}{0.40}\right)^2 = 1.6875 \text{ m/s}$$

$\therefore$ Head loss in expansion $h_L = \dfrac{(3.0 - 1.6875)^2}{2 \times 9.806}$ $= 0.0878$ m

(v) Hence equivalent length for expansion

$$= \frac{0.0878 \times 2 \times 9.806 \times 0.3}{0.025 \times 3^2} = 2.297 \text{ m}$$

(vi) Equivalent length for 100 m of 400 mm diameter pipe of $f = 0.02$,

$$= \frac{0.02 \times 100}{0.025}\left(\frac{0.3}{0.4}\right)^5 = 18.980 \text{ m}$$

Total equivalent length $= 89.365$ m

11.9 A 300 mm diameter pipe with friction factor of 0.018 has a 90° elbow with loss coefficient of 1.9 and 200 mm diameter pipe of 50 m length with friction factor of 0.021. Determine their equivalent lengths in terms of 300 mm diameter pipe.

$$\text{Since} \quad \frac{KU^2}{2g} = f\frac{l_e}{D}\frac{U^2}{2g}$$

$$l_e = \frac{KD}{f}$$

$$\therefore \quad \text{Equivalent length for elbow} = \frac{1.9 \times 0.300}{0.018} = 31.667 \text{ m}$$

Also since

$$\frac{8f_1l_1Q^2}{\pi^2gD_1^5} = \frac{8f_2l_2Q^2}{\pi^2gD_2^5}, \; l_2 = \frac{l_1f_1}{f_2}\left(\frac{D_2}{D_1}\right)^5$$

$$l_2 = \frac{50 \times 0.021}{0.018}\left(\frac{300}{200}\right)^5 = 442.969 \text{ m}$$

11.10 A bend meter in 100 mm diameter pipe has a discharge coefficient of 0.90. What discharge of water through this meter is indicated when the differential manometer across the bend shows 250 mm of mercury (r.d = 13.55)?

$$\Delta p = 0.250 \times (13.55 - 1.00) \times 9787 \text{ N/m}^2$$

$$\therefore \quad Q = 0.9 \times 0.785 \times 0.10^2 \sqrt{\frac{2 \times 0.25 \times 12.55 \times 9787}{998}}$$

$$= 0.055\,42 \text{ m}^3\text{/s}$$

11.11 For the siphon shown in Fig. 11.7, obtain the discharge and absolute pressure at 2, if $l_{1,2} = 600$ m, $l_{2,3} = 400$, $D = 0.30$ m, $f = 0.02$, $h = 0.7$ m and $H = 12.0$ m. Take atmospheric pressure as 10.35 m of water.

Apply Bernoulli's equation between 1 and 3

$$\therefore \quad 12 = 0.5\frac{U^2}{2g} + \frac{flU^2}{2gD} + \frac{U^2}{2g}$$

$$\begin{pmatrix}\text{entrance}\\ \text{loss}\end{pmatrix} + \begin{pmatrix}\text{friction}\\ \text{loss}\end{pmatrix} + (\text{exit loss})$$

$$\therefore \quad 12 = \frac{U^2}{2g}\left(0.5 + \frac{0.02 \times 1000}{0.30} + 1.0\right) = 68.167\frac{U^2}{2g}$$

$$\therefore \quad U = \sqrt{\frac{2 \times 9.806 \times 12}{68.167}} = 1.858 \text{ m/s}$$

$$\therefore \quad Q = 0.785 \times 0.30^2 \times 1.858 = 0.1313 \text{ m}^3\text{/s.}$$

Now apply Bernoulli's equation between 1 and 2 using absolute pressures

$$10.35 + 0 + 0 = 0.70 + \frac{p_2}{\gamma} + \frac{U^2}{2g} + \frac{fl_{1,2}U^2}{2gD}$$

$$\frac{p_2}{\gamma} = 10.35 - 0.70 - \frac{1.858^2}{2 \times 9.806}\left(1 + \frac{0.02 \times 600}{0.30}\right)$$

$$= 10.35 - 0.70 - 7.217$$

$$= 2.433 \text{ m (abs)}$$

Pressure at 2 should not be equal to or less than vapour pressure of water; otherwise discontinuity in water column will develop due to vapourisation and siphon will not work. Lower limit for p_2/γ is normally taken as 2.0 m.

11.12 For the data given in Fig. 11.14, determine the energy to be supplied to pump 70 *l*/s of water from reservoir *A* to reservoir *B*.

$$U_1 = \frac{0.07}{0.785 \times 0.2^2} = 2.229 \text{ m/s}$$

$$U_2 = \frac{0.07}{0.185 \times 0.15^2} = 3.963 \text{ m/s}$$

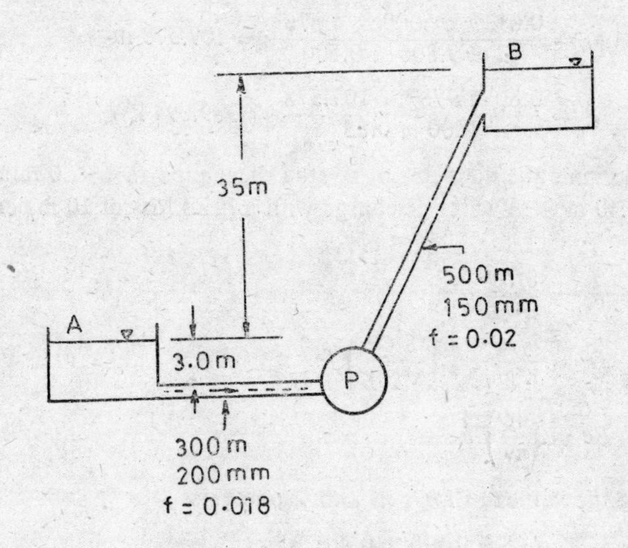

Fig. 11.14.

Applying Bernoulli's equation between *A* and *B*

$$3.00 + H = 38 + 0.5\frac{U_1^2}{2g} + \frac{0.018 \times 300}{0.20}\frac{U_1^2}{2g} + \frac{0.019 \times 500}{0.15}\frac{U_2^2}{2g} + \frac{U_2^2}{2g}$$

$\therefore \quad H = 38.0 - 3.0 + 0.2533\,(0.50 + 27.00) + 0.800\,(1 + 63.333)$

or $\quad H = 94.432$ m

$$\therefore \quad P = \frac{Q\gamma H}{1000} = \frac{0.07 \times 9787 \times 94.432}{1000} = 64.00 \text{ kW}$$

11.13 Determine the loss of head in 500 mm diameter steel pipe ($k = 0.045$ mm) carrying 0.80 m³/s of water over 5.0 km length. Also determine the power of the pump if its efficiency is 85 percent.

$$U = 0.80/0.785 \times 0.50^2 = 4.076 \text{ m/s}$$

Assuming water temperature to be 20°C, $\nu = 1 \times 10^{-6}$ m²/s

$$\therefore \quad Re = \frac{4.076 \times 0.50}{10^{-6}} = 2.038 \times 10^6 \text{ and } \frac{k}{D} = \frac{0.045}{500} = 0.00009$$

Using Jain's equation

$$\frac{1}{\sqrt{f}} = 1.14 - 2 \log\left(\frac{k}{D} + \frac{21.25}{Re^{0.90}}\right)$$

$$= 1.14 - 2 \log\,(0.00013457) = 8.88208$$

$$\therefore \quad f = 0.0127$$

(Moody's diagram can also be used to determine f for known Re and k/D).

$$h_f = \frac{0.0127 \times 5000 \times 4.076^2}{2 \times 9.806 \times 0.500} = 107.378 \text{ m}$$

$$\therefore \quad P = \frac{0.80 \times 9787 \times 107.378}{1000 \times 0.85} = 989.094 \text{ kW}$$

11.14 Determine the diameter of riveted steel pipe ($k = 2.0$ mm) which will carry 0.40 m³/s of water discharge with a head loss of 10 m per 1000 m length.

$$h_f = \frac{8flQ^2}{\pi^2 g D^5}$$

$$\therefore \quad D^5 = \frac{8flQ^2}{\pi^2 g h_f} = \left(\frac{8lQ^2}{\pi^2 g h_f}\right) f$$

$$Re = \left(\frac{4Q}{\pi\nu}\right)\frac{1}{D}$$

Substituting the values of l, Q, h_f and ν, one gets

$$D^5 = \left(\frac{8 \times 1000 \times 0.40^2}{3.142^2 \times 9.806 \times 10}\right) f = 1.232\, f$$

$$Re = \left(\frac{4 \times 0.40}{3.142 \times 10^{-6}}\right)\frac{1}{D} = (0.5092 \times 10^6)\,\frac{1}{D}$$

Assume $\quad f = 0.02 \qquad \therefore \; D^5 = 1.322 \times 0.02$

$$\therefore \quad D = 0.4835 \text{ m}$$

and $$Re = \frac{0.5092 \times 10^6}{0.4835} = 1.053 \times 10^6$$

Also $$k/D = 2/483.5 = 0.004136$$

Using Moody's diagram, for these values of Re and $k/D, f = 0.0285$

$$\therefore \quad D^5 = 1.322 \times 0.0285 = 0.03768$$

or $$D = 0.519 \text{ m}$$

Also $$k/D = 2/519 = 0.00385 \quad \text{and} \quad Re = 0.98 \times 10^6$$

For these values of k/D and Re, Moody's diagram gives $f = 0.028$. Since these two successive values of f are sufficiently close, correct value of D can be determined using $f = 0.028$.

$$D = (1.322 \times 0.028)^{1/5} = 0.5172 \text{ m}$$

One can also use the empirical equation to determine D.

$$\frac{k^3gs}{\nu^2} = \frac{(0.002)^3 \times 9.806 \times 0.01}{(10^{-6})^2} = 784.48$$

$$Q/k\nu = 0.40/0.002 \times 10^{-6} = 2 \times 10^8$$

$$\therefore \quad \left(\frac{k}{D}\right)^{8/3} = 2.3 \left(\frac{k^3gs}{\nu^2}\right)^{0.517} \Big/ (Q/k\nu)$$

$$= \frac{2.3 \times (784.48)^{0.517}}{2 \times 10^8} = (3.6074 \times 10^{-7})$$

$$\therefore \quad k/D = 0.003837 \qquad \therefore \quad D = 0.5212 \text{ m}$$

It can be seen that the two values are close to each other.

11.15 Find the discharge of natural gas carried by a 200 mm diameter smooth pipe at a pressure loss of 400 N/m² per km length if ρ for natural gas is 1.0 kg/m³ and μ is 8.55×10^{-6} Ns/m².

Assume $f = 0.01$.

Since $$\Delta p = \frac{8f\rho l Q^2}{\pi^2 D^5}$$

$$Q^2 = \frac{400 \times 3.142^2 \times 0.2^5}{8 \times 0.01 \times 1000 \times 1} = 0.015794$$

$$\therefore \quad Q = 0.1257 \text{ m}^3/\text{s}$$

$$Re = \frac{4Q\rho}{\pi D\mu} = \frac{4 \times 0.1257 \times 1.0}{3.142 \times 0.2 \times 8.55 \times 10^{-6}} = 9.36 \times 10^4$$

$$\therefore \quad f = 0.316/(9.36 \times 10^4)^{0.25} = 0.0181$$

$$\therefore \quad Q^2 = \frac{400 \times 3.142^2 \times 0.20^5}{8 \times 0.0181 \times 1000 \times 1} = 0.008\,727$$

$$\because \quad Q = 0.0934 \text{ m}^3/\text{s}$$

Hence $\quad Re = (4 \times 0.0934 \times 1.0)/(3.142 \times 0.20 \times 8.55 \times 10^{-6})$

$= 6.954 \times 10^4$

and $\quad f = 0.316/(6.954 \times 10^4)^{0.25} = 0.0195$

This gives $Q^2 = 0.0081$ and $Q = 0.090$ m³/s.

11.16 Two reservoirs with a difference in water levels of 180 m are connected by a 64 km long pipe of 500 mm diameter and f of 0.015. Determine the discharge through the pipe. In order to increase this discharge by fifty per cent, another pipe of the same diameter is to be laid from the lower reservoir for part of the length and connected to the first pipe (Fig. 11.15). Determine the length of additional pipe required.

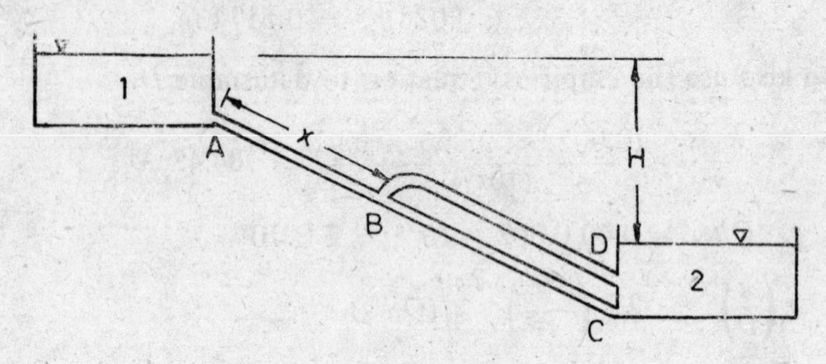

Fig. 11.15.

First determine the discharge through pipe *ABC*.

$$H = 0.5\frac{U^2}{2g} + \frac{flU^2}{2gD} + \frac{U^2}{2g}$$

$$\therefore \quad 180 = \frac{U^2}{2 \times 9.806}\left(0.5 + \frac{0.015 \times 64\,000}{0.60} + 1.0\right)$$

$$= \frac{1601.5U^2}{19.612}$$

$$\therefore \quad U^2 = (19.612 \times 180)/1601.5$$

$$= 2.204 \quad \text{or} \quad U = 1.485 \text{ m/s}$$

$$\therefore \quad Q = 0.785 \times 0.6^2 \times 1.485 = 0.420 \text{ m}^3\text{/s}$$

New discharge $= 1.5Q = 0.630$ m³/s

Velocity in length $AB = 0.63/0.785 \times 0.6^2 = 2.2275$ m/s

Pipes *BC* and *BD* will carry equal discharge.

Hence velocity in BC or $BD = \dfrac{0.315}{0.785 \times 0.6^2} = 1.115$ m/s

Since entrance and exit losses are very small compared to friction loss, they can be neglected.

$\therefore$ Bernoulli's equation along ABC or ABD gives

$$180 = \frac{0.015 \times x \times 2.2275^2}{2 \times 9.806 \times 0.60} + \frac{0.015 \times (64\,000 - x) \times 1.115^2}{2 \times 9.806 \times 0.60}$$

$$= 0.006\,327x + 101.347 - 0.001\,584x$$

$\therefore$ $78.653 = 0.004\,743x$

or $x = 16\,582.96$ or $16\,583$ m

Hence additional length of pipe $BD = 64\,000 - 16\,583$

$= 47\,417$ m or 47.417 km

11.17 In Fig. 11.15, $H = 100$ m, $ABC = 20\,000$ m long, 500 mm diameter pipe with $f = 0.015$ and $AB = 10000$ m.

$BD = 10\,000$ m long, 400 mm diameter pipe with $f = 0.02$

Neglecting the minor losses, determine the discharge through AB, BC and BD.

Assume head loss in length BC or $BD = h$

Hence head loss in length $AB = 100 - h$

Therefore

For BD, $h = \dfrac{0.02 \times 10000 U_{BD}^2}{2 \times 9.806 \times 0.40} = 25.495 U_{BD}^2$

For BC, $h = \dfrac{0.015 \times 10000 U_{BC}^2}{2 \times 9.806 \times 0.50} = 15.296 U_{BC}^2$

Therefore $25.495 U_{BD}^2 = 15.296 U_{BC}^2$

$\therefore$ $U_{BC} = 1.291 U_{BD}$ or $U_{BD} = 0.7746\, U_{BC}$

$Q_{BC} = 0.785 \times 0.5^2 U_{BC} = 0.1963\, U_{BC}$

$Q_{BD} = 0.785 \times 0.40^2 U_{BD} = 0.1256\, U_{BD} = 0.1256 \times 0.7746\, U_{BC}$

$= 0.097\,29 U_{BC}$

$\therefore$ $Q_{AB} = Q_{BC} + Q_{BD} = (0.1963 + 0.097\,29) U_{BC}$

$= 0.2936\, U_{BC}$

or $U_{AB} = \dfrac{Q_{AB}}{0.785 \times 0.5^2} = \left(\dfrac{0.2936}{0.785 \times 0.5^2}\right) U_{BC} = 1.4960 U_{BC}$

Further, apply Bernoulli's equation between A and C

$$\therefore \quad 100 = \frac{0.015 \times 100\,00 U_{AB}^2}{2 \times 9.806 \times 0.50} + \frac{0.015 \times 100\,00 U_{BC}^2}{2 \times 9.806 \times 0.50}$$

$$= 15.297\,[(1.4792 U_{BC})^2 + U_{BC}^2] = 48.767 U_{BC}^2$$

$\therefore$ $U_{BC} = 1.432$ m/s, $U_{AB} = 1.4960 \times 1.432 = 2.142$ m/s

and $U_{BD} = 0.7746 \times 1.432 = 1.109$ m/s

$$\therefore \quad Q_{AB} = 0.785 \times 0.5^2 \times 2.142 = 0.4203 \text{ m}^3/\text{s}$$

$$Q_{BC} = 0.785 \times 0.5^2 \times 1.432 = 0.2810 \text{ m}^3/\text{s}$$

$$Q_{BD} = 0.785 \times 0.4^2 \times 1.1092 = 0.1393 \text{ m}^3/\text{s}$$

11.18 For the pipes connected in parallel such as shown in Fig. 11.8, the following details are given:

Pipe	length m	diameter mm	f
1	200	200	0.020
2	300	250	0.018
3	150	300	0.015
4	100	200	0.020

If $Z_A = 150$ m and $Z_B = 144$ m, determine the discharge in each pipe.

For all the four pipes $h_f = Z_A - Z_B = (150 - 144) = 6.0$ m.

Hence

$$6 = \frac{8 \times 0.02 \times 200\, Q_1^2}{3.142^2 \times 9.806 \times 0.2^5} \quad \therefore \quad Q_1^2 = 0.046\,47$$

$$\text{or} \quad Q_1 = 0.076\,22 \text{ m}^3/\text{s}$$

$$6 = \frac{8 \times 0.018 \times 300\, Q_2^2}{3.142^6 \times 9.806 \times 0.25^5} \quad \therefore \quad Q_2^2 = 0.013\,13$$

$$\text{or} \quad Q_2 = 0.1146 \text{ m}^3/\text{s}$$

$$6 = \frac{8 \times 0.015 \times 150\, Q_3^2}{3.142^2 \times 9.806 \times 0.30^5} \quad \therefore \quad Q_3^2 = 0.078\,41$$

$$\text{or} \quad Q_3 = 0.2800 \text{ m}^3/\text{s}$$

$$6 = \frac{8 \times 0.02 \times 100\, Q_4^2}{3.142^2 \times 9.806 \times 0.20^5} \quad \therefore \quad Q_4^2 = 0.011\,62$$

$$\text{or} \quad Q_4 = 0.1078 \text{ m}^3/\text{s}$$

Total discharge $Q = Q_1 + Q_2 + Q_3 + Q_4$

or $Q = (0.076\,22 + 0.114\,60 + 0.28\,00 + 0.1078)$

$= 0.5786 \text{ m}^3/\text{s}$

11.19 Consider the data concerning f, l, and D given in Example 11.18. For a total discharge of 0.80 m³/s flowing through the system, determine the discharge distribution and $(Z_A - Z_B)$.

Assume a reasonable discharge in pipe 1, say

$Q_1' = 0.10$ m³/s. For this discharge $(Z_A - Z_B)$ can be calculated.

$$(Z_A - Z_B) = \frac{8 f_1 l_1 Q_1'^2}{\pi^2 g D_1^5} = \frac{8 \times 0.02 \times 200 \times 0.10^2}{3.142^2 \times 9.806 \times 0.2^5} = 10.33 \text{ m}$$

For this value of $(Z_A - Z_B)$, discharges Q_2' and Q_3' and Q_4' can be determined

$$10.33 = \frac{8 \times 0.018 \times 300\, Q_2'^2}{3.142^2 \times 9.806 \times 0.25^5}.$$

This gives $Q_2'^2 = 0.022\,61$ and $Q_2' = 0.150$ m³/s

$$10.33 = \frac{8 \times 0.015 \times 150 \times Q_3'^2}{3.142^2 \times 9.806 \times 0.3^5}.$$

Hence $Q_3'^2 = 0.135$ and $Q_3' = 0.367$ m³/s

$$10.33 = \frac{8 \times 0.02 \times 100 \times Q_4'^2}{3.142^2 \times 9.806 \times 0.2^5}.$$

Hence $Q_4'^2 = 0.02$ and $Q_4' = 0.141$ m³/s

$$\therefore \quad Q' = Q_1' + Q_2' + Q_3' + Q_4' = 0.1000 + 0.1500 + 0.3670 + 0.1410$$

$$= 0.758 \text{ m}^3/\text{s}$$

$$Q_1 = \frac{Q}{Q'} Q_1' = \frac{0.80 \times 0.1}{0.758} = 0.106 \text{ m}^3/\text{s}$$

$$Q_2 = \frac{Q}{Q'} Q_2' = \frac{0.8 \times 0.150}{0.758} = 0.158 \text{ m}^3/\text{s}$$

$$Q_3 = \frac{Q}{Q'} Q_3' = \frac{0.80 \times 0.367}{0.758} = 0.387 \text{ m}^3/\text{s}$$

$$Q_4 = \frac{Q}{Q'} Q_4' = \frac{.8 \times 0.141}{0.758} = 0.149 \text{ m}^3/\text{s}$$

For this value of Q_1, $(Z_A - Z_B)$ can be computed

$$(Z_A - Z_B) = \frac{8 \times 0.02 \times 200 \times 0.106^2}{3.142^2 \times 9.806 \times 0.20^5} = 11.61 \text{ m}$$

One can also solve this problem in the following manner

$$\text{Let } h_f = \frac{8 f_1 l_1 Q_1^2}{\pi^2 g D_1^5} = k_1^2 Q_1^2 \text{ where } k_1 = \sqrt{\frac{8 f_1 l_1}{\pi^2 g D_1^5}}$$

Then one can write $k_1 Q_1 = k_2 Q_2 = k_3 Q_3 = k_4 Q_4$

$$\therefore \quad Q = Q_1 + Q_2 + Q_3 + Q_4$$

$$= Q_1 \left(1 + \frac{k_1}{k_2} + \frac{k_1}{k_3} + \frac{k_1}{k_4}\right)$$

Knowing Q and $\frac{k_1}{k_2}$, $\frac{k_1}{k_3}$ and $\frac{k_1}{k_4}$, Q_1 can be determined. Then Q_2, Q_3 and Q_4 can be found out.

11.20 Three pipes of the same length l, diameter D and f value are connected in parallel. Determine the diameter of pipe of length l and friction factor f which will give the same total discharge for the same head loss.

Let h_f be the head loss across the parallel pipes. Hence Q in each pipe will be

$$h_f = \frac{8 f l Q_1^2}{\pi^2 g D^5} \quad \text{or} \quad Q_1 = \sqrt{\frac{\pi^2 g D^5 h_f}{8 f l}}$$

Hence total discharge through the three pipes will be $3Q_1$. Therefore

$$Q = 3\sqrt{\frac{\pi^2 g D^5 h_f}{8fl}}$$

For given h_f, f, l and diameter D_1, the discharge through the pipe is $\sqrt{\frac{\pi^2 g D_1^5 h_f}{8fl}}$. If the two discharges are to be equal

$$\sqrt{\frac{\pi^2 g D_1^5 h_f}{8fl}} = 3 \times \sqrt{\frac{\pi^2 g D^5 h_f}{8fl}}; \quad \text{or} \quad D_1^5 = 9D^5$$

$\therefore \qquad D_1 = 1.5518\ D.$

11.21 Power is transmitted through a pipe connected to the reservoir. Obtain expressions for power transmitted and the efficiency of power transmission. Also obtain the condition for maximum transmission of power for given head H in the reservoir (Fig. 11.16).

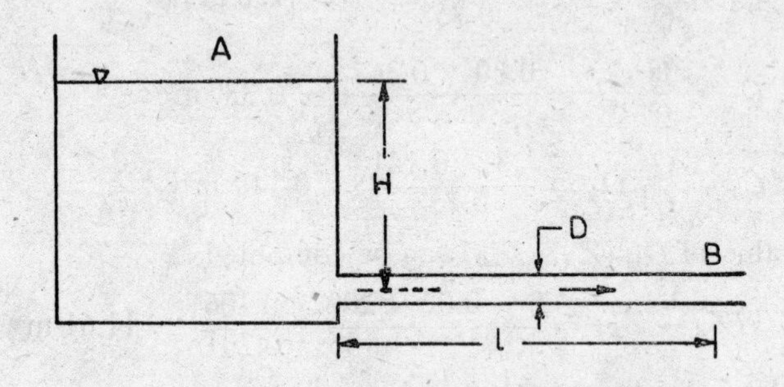

Fig. 11.16.

Apply Bernoulli's equation between A and B. Neglecting minor losses one gets

$$H + 0 + 0 = \left(\frac{p_B}{\gamma} + \frac{U_B^2}{2g}\right) + 0 + \frac{flU_B^2}{2gD}$$

Energy per unit weight of fluid at

$$B = \left(\frac{p_B}{\gamma} + \frac{U_B^2}{2g}\right) = H - \frac{flU_B^2}{2gD}$$

$$= H - \frac{8flQ^2}{\pi^2 g D^5}$$

$$= H - rQ^2$$

where $\qquad r = 8fl/\pi^2 g D^5$

$\therefore \qquad P = Q\gamma\,(H - rQ^2)$

$$\text{Transmission efficiency } \eta \text{ for power} = \frac{\text{Power available at outlet}}{\text{Power at inlet}}$$

$$\therefore \qquad \eta = \frac{Q\gamma\,(H - rQ^2)}{Q\gamma H}$$

$$= \left(1 - \frac{rQ^2}{H}\right)$$

If P is to be maximum for given H, $\frac{\partial P}{\partial Q} = 0$

$$\frac{\partial P}{\partial Q} = \gamma H - 3\gamma r Q^2 = 0$$

$$\therefore \qquad rQ^2 = \frac{H}{3}; \quad \text{but} \quad rQ^2 = h_f$$

Hence P will be maximum for given H when $h = \frac{H}{3}$.

$$P_{\max} = Q\gamma\left(H - \frac{H}{3}\right) = \frac{2}{3}\,Q\gamma H$$

and
$$\eta_{\max} = \frac{2}{3}\frac{Q\gamma H}{Q\gamma H} = 0.667 \quad \text{or} \quad 66.7 \text{ percent}$$

11.22 Obtain an expression for power transmitted through a nozzle at the end of a pipe connected to a reservoir and obtain the condition for power transmitted to be maximum for the given head H.

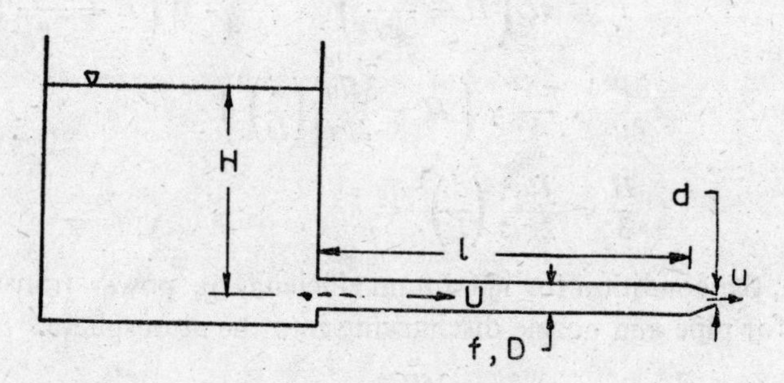

Fig. 11.17.

With respect to Fig. 11.17, head available at the nozzle can be found out.

$$H + 0 + 0 = \left(\frac{u^2}{2g} + 0 + 0\right) + \frac{flU^2}{2gD}$$

$$\therefore \quad \text{Head available at nozzl} \quad \frac{u^2}{2g} = \left(H - \frac{flU^2}{2gD}\right)$$

Here u is velocity at the nozzle. In writing the above equation, energy loss in the nozzle is neglected, hence $C_v = 1.0$; also $C_c = 1.0$.

$$\frac{\pi D^2}{4} U = \frac{\pi d^2}{4} u \quad \text{or} \quad U = \left(\frac{d}{D}\right)^2 u$$

$$\frac{u^2}{2g} = H - \frac{fl}{2gD}\left(\frac{d}{D}\right)^4 u^2$$

or
$$\frac{u^2}{2g}\left(1 + \frac{fl}{D}\left(\frac{d}{D}\right)^4\right) = H$$

and
$$u = \sqrt{\frac{2gH}{1 + \frac{fl}{D}\left(\frac{d}{D}\right)^4}}$$

Power transmitted $P = Q\gamma\left(H - \frac{flU^2}{2gD}\right) = \frac{Q\rho u^2}{2}$

or
$$P = \frac{\rho}{2}\frac{\pi d^2}{4}u^3$$

Power at inlet $= Q\gamma H$

$\therefore$ Transmission efficiency $\eta = \frac{\rho}{2}\frac{\pi d^2}{4}\frac{u^3}{Q\rho gH} = \frac{(\rho/2)(\pi d^2/4)u^3}{\rho gH(\pi d^2/4)u}$

$$= \frac{u^2}{2gH} = \frac{1}{1 + \frac{fl}{d}\left(\frac{d}{D}\right)^4}$$

Further for P to be maximum, one can write

$$P = \gamma Q\left(H - \frac{flU^2}{2gD}\right) = \gamma\frac{\pi d^2}{4}u\left(H - \frac{flu^2}{2gD}\left(\frac{d}{D}\right)^4\right)$$

$$\frac{\partial P}{\partial u} = \frac{\pi d^2}{4}\gamma\left(H - \frac{3flu^2}{2gD}\left(\frac{d}{D}\right)^4\right) = 0$$

$$\frac{H}{3} = \frac{flu^2}{2gD}\left(\frac{d}{D}\right)^4 = h_f$$

Therefore, the condition for maximum efficiency in power transmission is the same for pipe and nozzle discharging into the atmosphere.

$$H = 3h_f = \frac{3flU^2}{2gD}$$

and also
$$H = \frac{flU^2}{2gD} + \frac{u^2}{2g}$$

Equating the two values of H one gets

$$\frac{3flU^2}{2gD} = \frac{flU^2}{2gD} + \frac{u^2}{2g} \quad \text{or} \quad \frac{2flU^2}{D} = u^2$$

but $D^2U = d^2u$. Hence

$$U^2 = \left(\frac{d^2}{D^2}\right)^2 u^2 \quad \therefore \quad \frac{2fl}{D}u^2\left(\frac{d}{D}\right)^4 = u^2$$

which gives the condition for maximum efficiency for transmission of power as $\left(\frac{D}{d}\right)^2 = \sqrt{\frac{2fl}{D}}$.

Thus, for given pipe characteristics f, l, and D this gives the nozzle diameter.

11.23 Water is to be supplied to a turbine 200 m below the reservoir level through 1500 m length of penstock with $f = 0.025$. Determine the smallest diameter of penstock which will produce 530 kW of power with a turbine of efficiency of 80 percent.

$$\eta = \frac{\text{energy output}}{\text{energy input}} = 0.80$$

$$\therefore \quad \text{Input energy} = \frac{530}{0.80} = 662.5 \text{ kW}$$

$$p = \frac{Q\,(H - h_f)}{1000} \quad \text{and for maximum power transmission } h_f = H/3$$

$$\therefore \quad 662.5 = \frac{Q \times 9787 \times \frac{2}{3} \times 200}{1000}$$

$$Q = 0.508 \text{ m}^3/\text{s} \quad \text{and} \quad h_f = \frac{1}{3} H = 66.67 \text{ m}$$

But
$$h_f = \frac{8flQ^2}{\pi^2 g D^5}$$

$$\therefore \quad D^5 = \frac{8 \times 0.025 \times 1500 \times 0.508^2}{3.142^2 \times 9.806 \times 66.97} = 0.012$$

$$\therefore \quad D = 0.4128 \quad \text{or} \quad 412.8 \text{ mm}$$

11.24 A 75 mm diameter pipe of 500 m length operates under a head of 60 m at its inlet. If the nozzle is fitted at its outlet, determine its diameter and the power transmitted when the power transmitted is maximum. Take $f = 0.025$

$$\text{Under most efficient condition } h_f = \frac{H}{3} = \frac{60}{3} = 20.0 \text{ m}$$

$$\therefore \quad \text{Head at the nozzle} = 60 - 20 = 40.0 \text{ m}$$

Assuming C_v for nozzle to be unity, $u = \sqrt{2gH} = \sqrt{2g \times 40}$

$$u = \sqrt{2 \times 9.806 \times 40} = 28.01 \text{ m/s}$$

Also as seen is Ex. 11.22, the condition for maximum efficiency also gives

$$\therefore \quad \frac{A}{a} = \left(\frac{D}{d}\right)^2 = \sqrt{\frac{2fl}{D}} = \sqrt{\frac{2 \times 0.025 \times 500}{0.075}}$$

$$\therefore \quad \left(\frac{D}{d}\right)^2 = 18.257, \quad \text{or} \quad \frac{D}{d} = 4.2729$$

$$d = 0.075/4.2729 = 0.017\ 55 \text{ m} \quad \text{or} \quad 17.55 \text{ mm}$$

$$\therefore \quad Q = 0.785 \times 0.017\ 55^2 \times 22.01$$

$$= 0.006\ 772 \text{ m}^3/\text{s} \quad \text{or} \quad 6.772\ l/\text{s}$$

$$\text{Maximum power} = \frac{Q\gamma}{1000} \times \left(\frac{2H}{3}\right) = \frac{0.006\ 722 \times 9787 \times 40}{1000}$$

$$= 2.651 \text{ kW}$$

11.25 Water is pumped through a 600 mm diameter pipe with a head loss of 27 m. In order to reduce the power consumption it is proposed to lay another main of appropriate diameter (but with the same f and l as the first pipe) in parallel to the existing pipe and reduce the head loss to 9.6 m still carrying the same total discharge. Find the diameters of the two pipes.

For the single pipe $h_f = \left(\frac{8fl}{\pi^2 g}\right)\frac{Q_5^2}{D_1} = 27$

If $8fl/\pi^2 g = K$ which will have the same value for both the pipes

$$\frac{Q^2}{D_1^5} = \frac{27}{K}$$

When the same discharge is to be carried by two pipes of diameters D_1 and D_2 with head loss of 9.6 m,

$$\frac{KQ_1^2}{D_1^5} = 9.6 = \frac{KQ_2^2}{D_2^5}$$

$$\therefore \quad Q = \sqrt{\frac{27D_1^5}{K}}$$

$$Q_1 = \sqrt{\frac{9.6D_1^5}{K}}$$

and $$Q_2 = \sqrt{\frac{9.6D_2^5}{K}}$$

Also $Q = Q_1 + Q_2$ and $\frac{Q}{Q_1} = \sqrt{\frac{27}{9.6}} = 1.677$

Solution of these two equations yields $Q_1 = 0.5963Q$

$$Q_2 = 0.4037Q$$

Or, $\quad Q_2/Q_1 = 0.677$

and $\quad Q_2/Q_1 = (D_2/D_1)^{5/2}$

$\therefore \quad D_2/D_1 = 0.8552$

$\therefore \quad D_2 = 0.8552 \times 0.60 = 0.5131 \quad$ or $\quad 513.1$ mm

11.26 A 700 mm diameter water main for domestic water supply runs horizontally for 1600 m and then branches into two 450 mm diameter pipes of 3000 m length each. In the first branch, the entire discharge is withdrawn at the uniform rate along its length, while in the second branch half the discharge entering the pipe is withdrawn at a uniform rate along its length. Assume $f = 0.025$ for all the pipes, the flow entering 700 mm diameter pipe to be 0.30 m³/s and pressure at the outlet of two branches to be atmospheric and calculate the head loss between entrance and exit (Fig. 11.18).

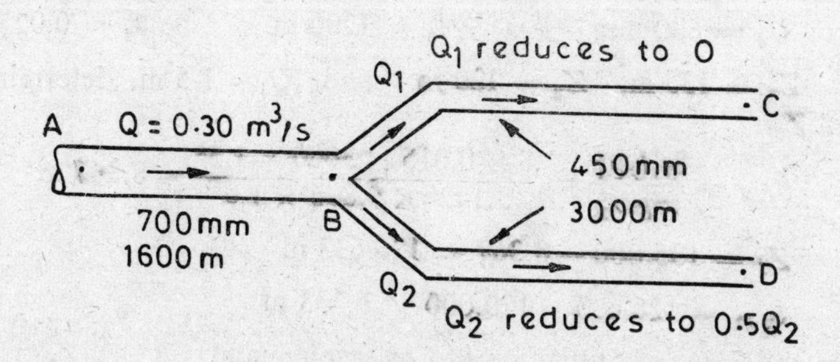

Fig. 11.18.

In order to solve this problem, make use of results obtained in Prob. 11.5.

$$\text{Head loss in } BC = \frac{8}{3}\frac{flQ_1^2}{\pi^2 g D^5}$$

$$\text{Head loss in } BD = \frac{14}{3}\frac{flQ_2^2}{\pi^2 g D^5}$$

$$\text{Head loss in } AB = \frac{8flQ^2}{\pi^2 g D_1^5} = \frac{8 \times 0.025 \times 1600 \times 0.30^2}{3.142^2 \times 9.806 \times 0.70^5}$$

$$= 1.77 \text{ m}$$

Since Head loss along BC = Head loss along BD

$$\frac{8}{3}\frac{flQ_1^2}{\pi^2 g D^5} = \frac{14}{3}\frac{flQ_2^2}{\pi^2 g D^5} \quad \therefore \quad 4Q_1^2 = 7Q_2^2$$

or $\quad Q_1 = 1.3229\, Q_2$

Also $\quad Q = Q_1 + Q_2$

$\therefore \quad 0.30 = Q_2 + 1.3229 Q_2 \qquad \therefore \quad Q_2 = \dfrac{0.30}{2.3229} = 0.129 \text{ m}^3/\text{s}$

$$Q_1 = 0.300 - 0.129 = 0.171 \text{ m}^3/\text{s}$$

$\therefore$ Head loss in $BC = \dfrac{8 \times 0.025 \times 3000 \times 0.171^2}{3 \times 3.142^2 \times 9.806 \times 0.45^5}$

$= 3.274$ m

$\therefore$ Total head loss $= 1.770 + 3.274$

$= 5.044$ m

11.27 In the case of 3 reservoirs which are interconnected (Fig. 11.9),

$D_1 = 1000$ mm $\quad l_1 = 3000$ m $\quad f_1 = 0.015$

$D_2 = 500$ mm $\quad l_2 = 600$ m $\quad f_2 = 0.024$

$D_3 = 600$ mm $\quad l_3 = 1200$ m $\quad f_3 = 0.02$

Further if $Z_1 = 135$ m, $Z_2 = 120$ m and $Q_1 = 1.5$ m, determine Q_2, Q_3, Z_j and Z_3.

$$h_{fj1} = \frac{8 f_1 l_1 Q_1^2}{\pi^2 g D_1^5} = \frac{8 \times 0.015 \times 3000 \times 1.5^2}{3.142^2 \times 9.806 \times 1.0^5} = 8.367 \text{ m}$$

$\therefore$ $Z_j = 135.000 - 8.367 = 126.633$ m

$h_{fj2} = 126.633 - 120.000 = 6.333$ m

$\therefore$ Water will flow from junction j to reservoir 2 and

$$Q_2^2 = \frac{6.333 \times 3.142^2 \times 9.806 \times 0.5^5}{8 \times 0.024 \times 600} = 0.1663;$$

or $Q_2 = 0.4078$ m³/s

$\therefore$ $Q_3 = Q_1 - Q_2 = 1.5000 - 0.4078 = 1.0922$ m³/s

$$h_{fj3} = \frac{8 \times 0.02 \times 1200 \times 1.0922^2}{3.142^2 \times 9.806 \times 0.60^5} = 30.426 \text{ m}$$

$\therefore$ $Z_3 = 126.633 - 30.426$

$= 96.207$ m

Hence $Q_2 = 0.4078$ m³/s, $\quad Q_3 = 1.0922$ m³/s

$Z_j = 126.533$ m and $Z_3 = 96.207$ m

11.28 Determine r and n in the equation $h_f = rQ^n$ for the 500 m long, 600 mm diameter concrete pipe ($k = 0.30$ mm) for a velocity range of 1.0 m/s to 5.0 m/s of water at 20°C.

$\therefore$ $$Re_1 = \frac{1.0 \times 0.60}{10^{-6}} = 6.0 \times 10^5$$

and $$Re_2 = \frac{5.0 \times 6.0}{10^{-6}} = 3.0 \times 10^6$$

Also $k/D = 0.30/600 = 0.0005$

Moody's diagram gives $f_1 = 0.0175$ and $f_2 = 0.0169$ for these two values of Re numbers and k/D value of 0.0005.

Also $Q_1 = 0.785 \times 0.6^2 \times 1.0 = 0.2826 \text{ m}^3/\text{s}$

$Q_2 = 0.785 \times 0.6^2 \times 5.0 = 1.4130 \text{ m}^3/\text{s}$

∴ On substitution of these values in the equation $f = aQ^b$ one gets

$$0.0175 = a\,(0.2826)^b$$

$$0.0169 = a\,(1.4130)^b$$

The solution of these equations gives $b = -0.0217$ and $a = 0.01702$.
Substitute the expression $f = 0.01702Q^{-0.0217}$ in $h_f = \dfrac{8flQ^2}{\pi^2 g D^5}$

$$\therefore \quad h_f = \frac{8 \times 0.01702 \times 500Q^2}{3.142^2 \times 9.806 \times 0.6^5 \times Q^{0.0217}}$$

$$= 9.044\, Q^{1.9783}$$

11.29 In the case of three reservoir problem (Fig. 11.9) take $r_1 = 10$, $r_2 = 40$ and $r_3 = 20$ for the three pipes, where $h_f = rQ^2$. Further $Z_1 = 100$ m, $Z_3 = 20$ and $Q_2 = 0.50$ m. Determine Q_1, Q_3 and Z_2.

One can assume different values of Z_j and obtain Q_1 and Q_3. Then determine $Q_2 - |Q_1 - Q_3|$. That value of Z_j is the correct value for which $Q_2 - |Q_1 - Q_3| = 0$. The calculations are summarised in the tabular form below.

Z_j (m)	Q_1 (m³/s)	Q_3 (m³/s)	$Q_2 - \lvert Q_1 - Q_3 \rvert$
50	2.236	1.225	− 0.510
70	1.732	1.581	0.349
60	2.000	1.414	− 0.086
62	1.949	1.449	0.000

The number of trials can be reduced by plotting Z_j versus $Q_2 - |Q_1 - Q_3|$ and reading the value of Z_j which gives $Q_2 - |Q_1 - Q_3| = 0$. Figure 11.19 shows plot of Z_j vs. $Q_2 - |Q_1 - Q_3|$. One can interpolate correct value of Z_j for which $Q_2 - |Q_1 - Q_3|$ is equal to zero.

11.30 For the three reservoir problem (Fig. 11.9) $Z_1 = 150$ m, $Z_2 = 100$ m and $Z_3 = 30$ m while r in the equation $h_f = rQ^2$ for the three pipes are $r_1 = 3.719$, $r_2 = 38.080$ and $r_3 = 15.176$. Determine Q_1, Q_2, Q_3.

As in Example 11.28, assume a certain value of Z_j and determine h_{fj1}, h_{fj2} and h_{fj3} and then Q_1, Q_2, Q_3. Determine $Q_2 - |Q_1 - Q_3|$ and plot against Z_j.

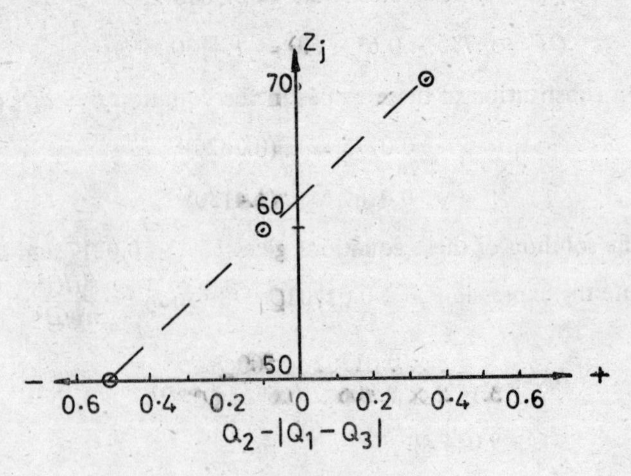

Fig. 11.19

Z_j (m)	Q_1 (m^3/s)	Q_2 (m^3/s)	Q_3 (m^3/s)	$Q_2 - \|Q_1 - Q_3\|$
120	2.841	0.725	2.435	0.315
110	3.280	0.512	2.296	−0.472
115	3.068	0.628	2.367	−0.073
116	3.024	0.648	2.380	0.004

Hence the correct elevation at junction is 116.0 m and $Q_1 = 3.024$ m³/s, $Q_2 = 0.648$ m³/s and $Q_3 = 2.380$ m³/s.

11.31 For the four reservoirs shown in Fig. 11.10, $Z_1 = 200$ m, $Z_2 = 100$ m, $Z_3 = 150$ m and $Z_4 = 80$ m. Further $r_1 = 20$, $r_2 = 100$, $r_3 = 10$ and $r_4 = 40$ and $r_5 = 30$, while $n = 2$ for all the pipes in the equation $h_f = rQ^n$. Determine the discharge distribution in various pipes.

Assume $Z_j = 170$ m.

$\therefore \quad (200 - 170) = 20\,Q_1^2;$ or $Q_1 = 1.225$ m³/s towards J.

$(170 - 100) = 100\,Q_2^2;$ or $Q_2 = 0.837$ m³/s towards reservoir 2.

$\therefore \quad Q_5 = (1.225 - 0.837) = 0.388$ m³/s from J to k.

Also $\quad h_{fjk} = 30 \times 0.388^2 = 4.52$ m³/s

$Z_k = 170.00 - 4.52 = 165.48$ m

Now one can determine Q_3 and Q_4.

$(165.48 - 150.00) = 10\,Q_3^2;$

or $\quad Q_3 = 1.224$ m³/s towards reservoir 3.

$(165.48 - 80.00) = 80\,Q_4^2;$

or $\qquad Q_4 = 1.034\ m^3/s$ towards reservoir 4.

$\therefore \qquad Q_3 + Q_4 = 2.258$ while $Q_5 = 0.338$

$$(Q_5 - Q_3 - Q_4) = -1.87\ m^3/s$$

if flow towards junction is assumed +ve and flow away from junction is taken as −ve. Similar calculations are made for other Z_j and shown below. It can be seen Z_j needs to be lowered so that Q_5 can increase.

Z_j (m)	Q_1 (m^3/s)	Q_2 (m^3/s)	Q_5 (m^3/s)	Q_3 (m^3/s)	Q_4 (m^3/s)	ΣQ_k
170	1.225	−0.837	−0.388	−1.228	−1.038	−1.870
150	1.580	−0.707	0.873	1.512	−0.768	1.617
162	1.378	−0.787	0.591	−0.390	−1.337	−1.136
158	1.449	−0.762	0.687	0.785	−1.263	0.209
161	1.396	−0.780	0.616	0.195	−1.319	−0.508
159	1.432	−0.768	0.664	0.650	−1.282	0.032
159.15	1.429	−0.769	0.660	0.626	−1.285	0.001

Hence junction elevation of 159.15 m at J satisfies the continuity equation at junction k reasonably well. Hence the discharges given in the last step can be taken as correct.

11.32 For the pipe network shown in Fig. 11.20 (i) determine the discharge distribution in the three pipes, if $n = 2.0$.

Assume the discharge distribution in various pipes such that continuity equation for flow is satisfied at the junctions (Fig. 11.20 (ii)).

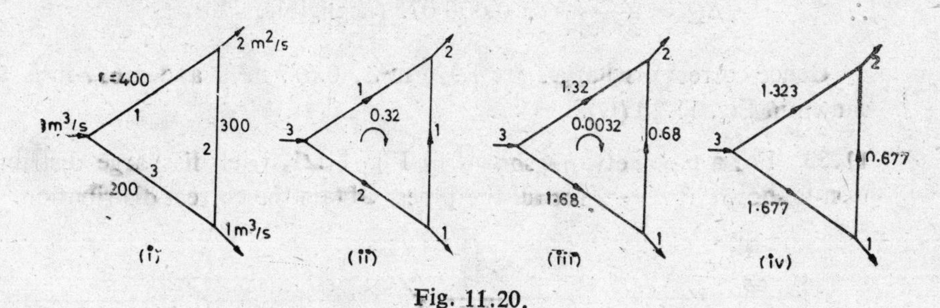

Fig. 11.20.

$\therefore$	Computation of ΣrQ^2	ΣnrQ^{n-1}
1	$400 \times 1^2 = 400$	$2 \times 400 \times 1 = 800$
2	$-300 \times 1^2 = -300$	$2 \times 300 \times 1 = 600$
3	$-200 \times 2^2 = -800$	$2 \times 200 \times 2 = 800$
	$\Sigma rQ^2 = -700$	$\Sigma nrQ^{n-1} = 2200$

$$\therefore \qquad \Delta Q = \frac{-(-700)}{2200} \approx +0.32 \text{ (clockwise)}$$

Corrected discharges are shown in Fig. 11.20 (iii).

ΣrQ^2		ΣnrQ^{n-1}	
$400 \times 1.32^2 =$	696.96	$2 \times 400 \times 1.32 =$	1056
$-300 \times 0.68^2 =$	-138.72	$2 \times 300 \times 0.68 =$	408
$-200 \times 1.68^2 =$	-564.48	$2 \times 200 \times 1.68 =$	672

$$\therefore \qquad \Sigma rQ^2 = -6.24 \qquad \Sigma nrQ^{n-1} = 2136$$

$$\therefore \qquad \Delta Q = \frac{-(-6.24)}{2136} = 0.003 \text{ (clockwise)}$$

Corrected discharges are again shown in Fig. 11.20 (iv).

ΣrQ^2		ΣnrQ^{n-1}
$400 \times 1.323^2 =$	700.13	1058.40
$-300 \times 0.677^2 =$	-137.50	406.20
$-200 \times 1.677^2 =$	-562.47	670.80
	0.16	2135.40

$$\therefore \qquad \Delta Q = \frac{-0.16}{2135.4} = 0.000\,075 \text{ (negligible)}$$

Hence correct discharges are 1.323 m^3/s, 0.677 m^3/s and 1.677 m^3/s as shown in Fig. 11.20 (iv).

11.33 For a pipe network shown in Fig. 11.21, trial discharge distribution is shown, if $n = 2$ for all the pipes. Obtain the correct distribution.

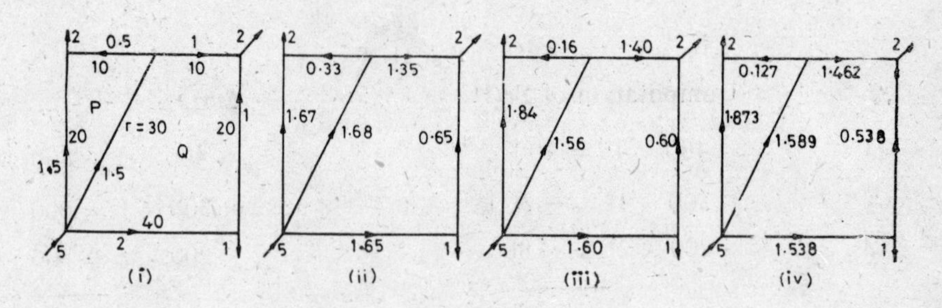

Fig. 11.21.

Loop A		Loop B	
$\Sigma rQ^2 =$ 45.0	$\Sigma 2rQ =$ 60.00	$\Sigma rQ^2 =$ 67.5	$\Sigma 2rQ =$ 75.00
− 2.5	10.00	10.0	20.00
− 67.5	75.00	− 20.0	40.00
		− 160.0	160.00
− 25.0	145.00	− 102.5	295.00

$$\therefore \quad \Delta Q_A = \frac{-(-25.0)}{145} \approx 0.17 \circlearrowright \qquad \Delta Q_B = \frac{-(-102.5)}{295} = 0.35 \circlearrowright$$

Corrected discharges are shown in Fig. 11.21 (ii). These are again corrected.

Loop A		Loop B	
$\Sigma rQ^2 =$ 55.778	$\Sigma 2rQ =$ 66.8	$\Sigma rQ^2 =$ 84.672	$\Sigma 2rQ =$ 100.8
− 1.090	6.6	18.225	27.0
− 84.672	100.8	− 8.450	26.0
		− 108.900	132.0
− 29.98	174.2	− 14.445	285.8

$$\therefore \quad \Delta Q_A = \frac{-(-29.98)}{174.2} = +0.17 \circlearrowright \qquad \Delta Q_B = \frac{-(-14.445)}{285.8} = 0.05 \circlearrowright$$

Corrected discharges are shown in Fig. 11.21 (iii) and are further corrected.

Loop A		Loop B	
$\Sigma rQ^2 =$ 67.712	$\Sigma 2rQ =$ 73.6	$\Sigma rQ^2 =$ 73.008	$\Sigma 2rQ =$ 93.60
− 0.260	3.2	19.600	28.00
− 73.008	93.6	− 7.200	24.00
		− 102.400	128.00
− 5.556	170.4	− 16.992	273.00

$$\Delta Q_A = \frac{-(-5.556)}{170.4} = 0.033 \circlearrowright \qquad \Delta Q_B = \frac{-(-16.992)}{273.0} = 0.062 \circlearrowright$$

Corrected flows are shown in Fig. 11.21 (iv).

With two further trials the following discharges are obtained:

$$Q_{AB} = 1.912 \text{ m}^3/\text{s}, \; Q_{CB} = 0.088 \text{ m}^3/\text{s}, \; Q_{AC} = 1.578 \text{ m}^3/\text{s},$$

$$Q_{AE} = 1.510 \text{ m}^3/\text{s}, \; Q_{ED} = 0.51 \text{ m}^3/\text{s and } Q_{CD} = 1.490 \text{ m}^3/\text{s}.$$

PROBLEMS

11.1 A 200 mm diameter pipe carrying 50 l/s of water discharge suddenly expands to such a diameter that it registers a rise of piezometric head of 0.0638 m across the sudden expansion. What is the diameter of the pipe after the expansion? (300 mm)

11.2 A horizontal pipe of 500 mm diameter suddenly contracts to 250 mm diameter. Assuming $C_c = 0.63$ and $K_1 = 0.015$, determine the pressure loss in the contraction if the pipe line carries a discharge 196.25 *l*/s. (15.669 kN/m²)

11.3 Part of the pipe area is symmetrically obstructed by the presence of the body, as shown in Fig. 11.22. If A is the area of pipe and a

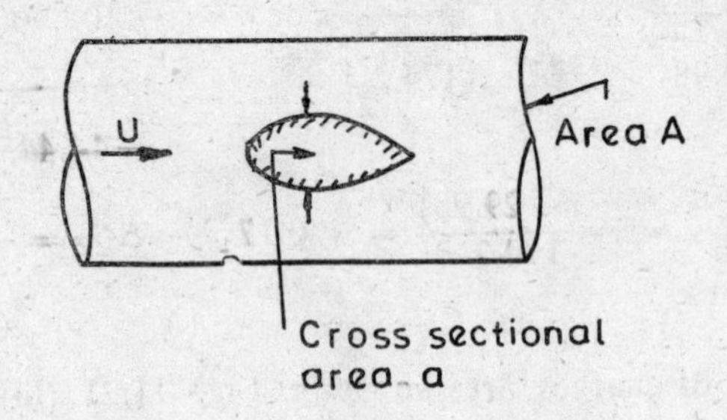

Fig. 11.22.

the cross-sectional area of the body, show that energy loss caused by the body is given by

$$h_L = \frac{Q^2}{2g}\left(\frac{A}{C_c(A-a)} - 1\right)^2.$$

11.4 A 300 mm diameter pipe gradually tapers to 150 mm diameter in a length of 10 m. If the discharge through the pipe is 0.15 m³/s, determine the loss of head due to friction if $f = 0.01$. (0.574 m)

11.5 Determine the total loss of head in a pipe of constant diameter in which discharge is uniformly withdrawn such that it reduces from Q to zero in length l. Also determine the head loss if discharge reduces from Q to $Q/2$ in the length l. Assume flow to be turbulent and f to be constant.

$$\left(h_f = \frac{8flQ^2}{3\pi^2gD^5},\quad h_f = \frac{14}{3}\,\frac{flQ^2}{\pi^2gD^5}\right)$$

11.6 Two reservoirs with a difference of water level of 60 m are connected by a 200 mm diameter, 500 m long pipe with the following nonuniformities:

Sharp entrance $K = 0.50$

Sudden contraction to 100 mm diameter pipe

Sudden expansion to 200 mm diameter pipe

A bend with loss coeff. $K = 0.80$

A globe valve fully open $K = 10$

Exit loss

If the pipe friction factor is 0.015, determine the pipe discharge.

(143.41 l/s)

11.7 Two reservoirs are connected by a 6000 m long and 1.0 m diameter cast iron pipe of surface roughness 0.30 mm. If the difference in water levels in the two reservoirs is 20 m, determine the percentage increase in discharge if the cast iron pipe is replaced by a steel pipe of surface roughness 0.10 mm (Hint: Assume the pipes to act as rough pipes and estimate f values. Later check your assumption and modify it if necessary).

(11.91 percent)

11.8 Determine Q and plot total energy line and hydraulic grade line if both pipes are of cast iron. Assume $f = 0.02$ for both the pipes (Fig. 11.23).

(0.0705 m³/s)

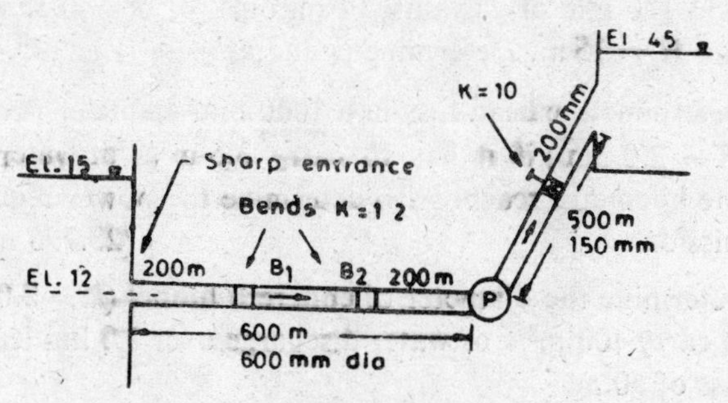

Fig. 11.23.

11.9 If 300 m length of 200 mm diameter pipe with $f = 0.018$ is to be replaced by 150 mm diameter pipe with $f = 0.02$ to carry the same discharge, what length will have to be provided?

(64.072 m)

11.10 A pipe line system consists of the following sources of energy loss

(i) Entrance loss in 300 mm diameter pipe

(ii) Friction loss in 300 m of 300 mm diameter pipe with $f = 0.019$

(iii) Sudden contraction from 300 mm to 200 mm diameter

(iv) Friction loss in 150 m of 200 mm diameter pipe with $f = 0.021$

(v) Sudden expansion from 200 mm to 250 mm diameter

(vi) Friction loss in 200 m of 250 mm diameter pipe with $f = 0.02$

(vii) Exit loss from 250 mm diameter pipe.

Obtain the equivalent length of 300 mm diameter pipe with $f = 0.019$. (2155.76 m)

11.11 A 200 mm diameter pipe of cast iron, which is 250 m long, has a globe valve with $K = 10.0$. Determine the equivalent length of the pipe line if it carries water discharge of 95.0 l/s at 20°C. (343.02 m)

11.12 A discharge of 60.70 l/s of water flows through a bend meter in 100 mm diameter pipe and gives 300 mm of differential mercury head across the bend. Determine the discharge coefficient of the bend. (0.60)

11.13 For the siphon shown in Fig. 11.7, $l_{1,3} = 400$ m, $D = 150$ mm of cast iron with $K = 0.26$ mm and $H = 8.0$ m. Assume water temperature to be 20°C and obtain the discharge flowing through the siphon.

(Note: Since f is not known, assume the pipe to act as rough pipe and then improve the solution). (0.0279 m³/s)

11.14 For the siphon shown in Fig. 11.7, $l_{1,2} = 100$ m, $l_{2,3} = 200$ m, $D = 150$ mm of galvanised iron pipe of $k = 0.15$ mm, $h = 3.0$ m and $H = 15$ m. Determine Q and p_2/γ. (45.41 l/s, 2.164 m)

11.15 Determine the head loss in a 1000 mm diameter revetted steel pipe ($K = 2.0$ mm) if it has to carry 3.5 m³/s of water discharge over one kilometre reach. Also determine the power required for transmission. (23.790 m, 814.915 kW)

11.16 Determine the diameter of concrete tunnel ($k = 2.0$ mm) required to carry 100 m³/s of water discharge over 2.0 km length with a head loss of 50 m. (3.548 m)

11.17 Natural gas is pumped through 600 mm diameter cast iron pipe with a pressure drop of 441 N/m² per 1000 m length of the pipe. If γ for the gas is 9.3 N/m³ and μ equals 8.55×10^{-6} Ns/m², determine the discharge flowing through the pipe. (1.60 m³/s)

11.18 Determine the diameter of cast iron ($K = 0.26$ mm) pipe required to carry 0.20 m³/s of kerosene with a hydraulic gradient of 0.00667. Take $\nu = 2.3 \times 10^{-6}$ m²/s. (0.394 m)

11.19 Determine the head loss in a 300 mm diameter reveted steel pipe ($K = 3.0$ mm) when it carries 0.124 m³/s of water discharge over a length of 600 m. Assume water temperature to be 20°C. (12.0 m)

11.20 In Fig. 11.15, $H = 50$ m, $ABC = 3000$ m of 200 mm diameter pipe with $f = 0.02$. If additional length of $BD = 1200$ m of 300 mm diameter pipe with $f = 0.02$ is added, determine what will be the increase in the discharge. (0.014 86 m³/s)

11.21 For a piezometric head difference $(Z_A - Z_B) = 7.5$ m (Fig. 11.8), determine the discharge in each of the three pipe lines and th total discharge for the following data.

Pipe	Length (m)	Diameter (mm)	f
1	300	200	0.020
2	200	250	0.018
3	300	300	0.015

(0.069 57 m³/s, 0.1569 m³/s, 0.2214 m³/s and 0.4479 m³/s)

11.22 Three pipes are connected in parallel (Fig. 11.8) the data for which are listed below.

Pipe	Length (m)	Diameter (mm)	Pipe material	k (mm)
1	400	300	Riveted steel	2.0
2	600	100	Galvanised iron	0.045
3	300	200	Concrete	1.0

For a total discharge of 0.50 m³/s through the three pipes, determine the distribution of discharge in each pipe and $(Z_A - Z_B)$. (Note that before hand f values are not known. Hence assume f, solve the problem and improve the f values and discharge distribution).

(0.372 m³/s, 0.0260 m³/s, 0.102 m³/s, 62.12 m)

11.23 Three pipes which are laid in parallel have the following details.

Pipes	Length (m)	Diameter (mm)	f
1	1500	1000	0.020
2	1200	800	0.020
3	1600	1200	0.024

If the total discharge through the system is 4.0 m^3/s, determine the discharge distribution in the three pipes and head loss.

(1.320 m^3/s, 0.894 m^3/s, 1.836 m^3/s, 4.32 m)

11.24 Two reservoirs with a level difference of 10 m are connected by two parallel pipes of 100 m length and diameters of 100 mm and 50 mm respectively. Assuming the friction factors for these two pipes to be 0.02 and 0.025 respectively, calculate the total discharge through them. Also determine the diameter of a single pipe of length 100 m $f = 0.02$ which will give the same discharge. (28.47 l/s, 0.106 m)

11.25 Water from a reservoir is discharged into the atmosphere from a 25 m length of 150 mm pipe followed by 15 m length of 300 m pipe, both pipes being horizontal and connected through a sudden expansion. If water level in the reservoir is 8.0 m above the outlet, determine the discharge assuming f to be 0.02 for both the pipes.

(104.77 l/s)

11.26 A pipe line of 1500 m length is used for the transmission of power. If 300 kW power is to be transmitted through this pipe under most efficient condition, determine the diameter of the pipe if $f = 0.03$ and the pressure at inlet is 5000 kN/m^2 (Fig 11.16). (185.7 mm)

11.27 A nozzle fitted to a pipe of 100 mm diameter and 300 m long pipe operates under a head of 90 m (Fig. 11.17). Find the area of the nozzle for maximum transmission of the power and the maximum power transmitted. Assume $f = 0.04$. (25.41 mm, 10.211 kW)

11.28 Water is supplied through a pipe of 50 mm diameter and 60 m length under a head of 16.5 m and is discharged through a nozzle having C_v value of 0.98. Calculate the diameter of the nozzle which will give a discharge of 3.8 l/s. Assume f to be 0.024. (12.32 mm)

11.29 Three new cast iron pipes of 200 mm, 300 mm and 400 mm diameters and 800 m, 1600 m and 2400 m in length respectively are connected in parallel. If a total discharge of 0.550 m^3/s is flowing through them, determine the pressure drop across them and the discharge distribution. (30.70 m, 85.0 l/s, 168.0 l/s and 297 l/s)

11.30 A tank is fitted with two horizontal galvanised iron pipes of 75 mm diameter and lengths 30 m and 90 m, H_1 and H_2 m below the water level in the tank. Entrance loss coefficient for both is 0.20 and both the pipes discharge into the atmosphere. Assume both the pipes to act as hydrodynamically rough (k = 0.15 mm) and determine the ratio of $\frac{H_1}{H_2}$ at which they will give the same discharge.

(0.361)

11.31 In Fig. 11.24, determine the air pressure p_A needed in the closed tank to maintain 15 l/s oil flow through the pipe. Assume kinematic viscosity of oil as 2.5×10^{-5} m^2/s. (73.680 kN/m^2)

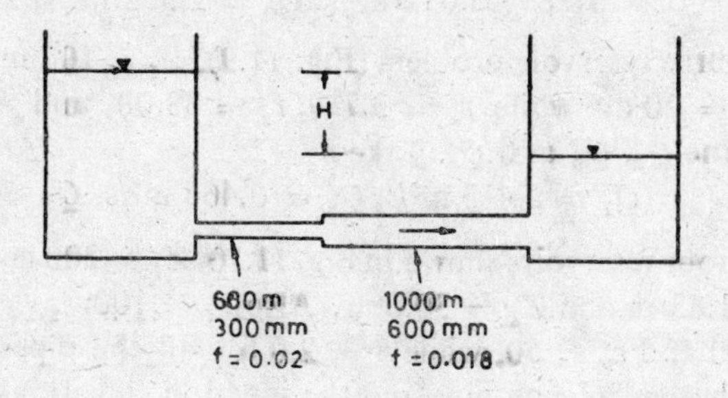

Fig. 11.24.

11.32 In Fig. 11.25, determine H if $Q = 0.135$ m^3/s. Take into account all types of losses. (7.968 m)

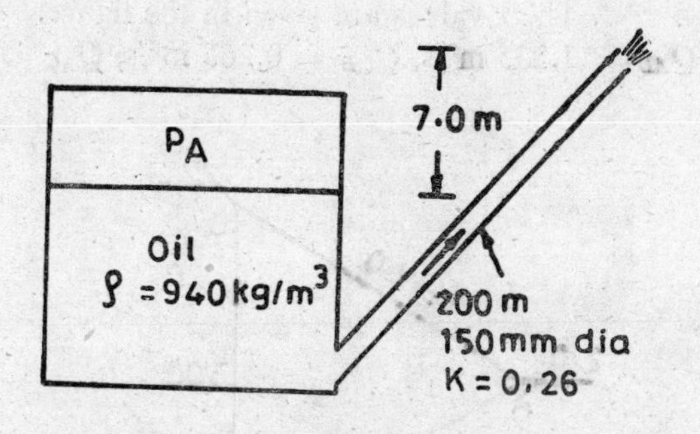

Fig. 11.25.

11.33 For a three reservoir problem (Fig. 11.9), the pipe characteristics are

$D_1 = 1000$ mm $\quad l_1 = 3000$ m $\quad f_1 = 0.015$

$D_2 = 500$ mm $\quad l_2 = 600$ m $\quad f_2 = 0.024$

$D_3 = 600$ mm $\quad l_3 = 1200$ m $\quad f_3 = 0.020$

Further if $Z_1 = 140$ m, $Z_2 = 120$ m and $Q_2 = 0.60$ m^3/s towards the reservoir, determine Q_1, Q_3 and Z_3.

($Q_1 = 1.30$ m^3/s, $Q_3 = 0.70$ m^3/s, $Z_3 = 121.209$ m)

11.34 For the cast iron pipe of 150 mm diameter and 80 m length the friction factor values for a velocity of 0.50 m/s and 3.0 m/s are 0.020 and 0.018 respectively. Determine r and n in the equation $h_f = rQ^n$. (1525.93, 1.9431)

11.35 For the three reservoir problem (Fig. 11.9), $Z_1 = 150$ m, $Z_3 = 30$ m, $Q_2 = 0.65$ m³/s and $r_1 = 4.0$, $r_2 = 40.0$ and $r_3 = 15.00$ for the three pipes while n in equation $h_f = rQ^n$ is 2.0. Determine Q_1, Q_3 and Z_2. ($Q_1 = 3.012$ m³/s, $Q_3 = 2.362$ m³/s, $Z_2 = 96.8$ m)

11.36 For the three reservoir problem (Fig. 11.9), $Z_1 = 100$ m, $Z_2 = 70$ m and $Z_3 = 20$ m while $r_1 = 3.719$, $r_2 = 38.08$ and $r_3 = 15.176$. Determine Q_1, Q_2 and Q_3. Take $n = 2$.
($Q_1 = 2.418$ m³/s, $Q_2 = 0.465$ m³/s, $Q_3 = 1.959$ m³/s)

11.37 For the four reservoirs shown in Fig. 11.10, $Z_1 = 200$ m, $Z_2 = 170$ m, $Z_3 = 91.83$ m and $Z_4 = 30.0$ m. Also $r_1 = 100$, $r_2 = 50$, $r_3 = 50$, $r_4 = 100$ and $r_5 = 50$, while $n = 2.0$ for all the pipes. Determine the discharges in each pipe and the junction elevations.

($Q_1 = 0.717$ m³/s, $Q_2 = 0.632$ m³/s, $Q_3 = 0.788$ m²/s

$Q_4 = -0.551$ m³/s, $Z_j = 150$, $Z_k = 60.35$ m)

11.38 For pipe network shown in Fig. 11.26, obtain the discharge distribution if $n = 2$. The r values are given in the figure.
($Q_{AB} = 1.235$ m³/s, $Q_{CB} = 0.265$ m³/s, $Q_{AC} = 0.765$ m³/s)

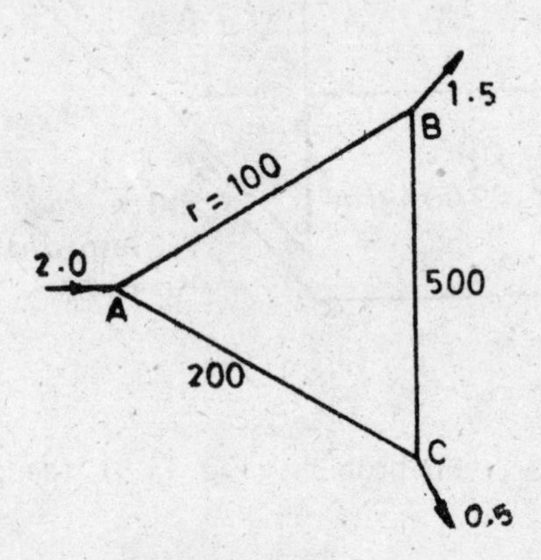

Fig. 11.26.

11.39 For pipe network shown in Fig 11.27, obtain the discharge distribution if $n = 2$.

($Q_{AB} = 1.098$ m³/s, $Q_{BC} = 0.598$ m³/s, $Q_{AC} = 1.010$ m³/s,

$Q_{AD} = 1.392$ m³/s, $Q_{DC} = 0.392$ m³/s)

11.40 For pipe network shown in Fig. 11.28, obtain the discharge distribution if $n = 2$.

($Q_{AB} = 0.507$ m³/s, $Q_{CB} = 1.993$ m³/s, $Q_{CD} = 1.514$ m³/s,

$Q_{CF} = 1.493$ m³/s, $Q_{ED} = 0.986$ m³/s, $Q_{EF} = 1.014$ m³/s)

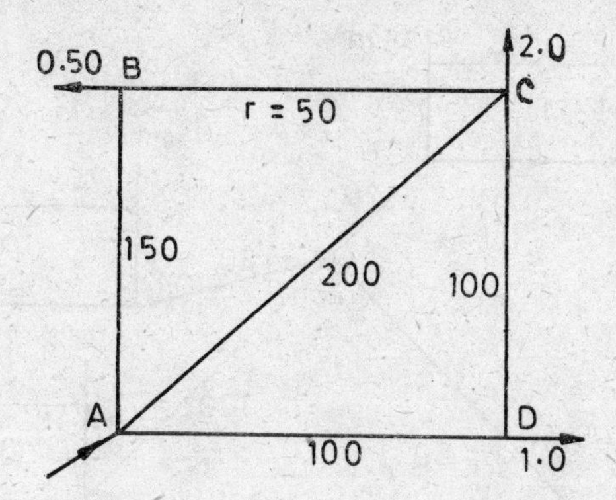

Fig. 11.27.

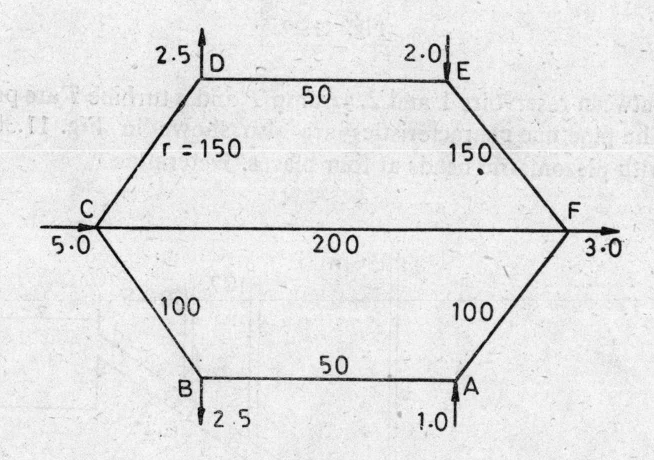

Fig. 11.28.

11.41 Two reservoirs, with a difference of elevation of 20 m between their water surfaces, are conected by a 4000 m length pipe of 600 mm diameter and 0.02 friction factor. If 0.10 m^3/s flow is withdrawn from the pipe line at a point 1000 m from higher reservoir, find the total discharge through the pipe and the discharge reaching the lower reservoir. (0.558 m^3/s, 0.458 m^3/s)

11.42 For the four reservoirs shown in Fig. 11.29, obtain Q_1, Q_2, Q_3, Q_4, and Z_j. Take $n = 2$.

($Q_1 = 0.758$ m^3/s, $Q_2 = 0.224$ m^3/s, $Q_3 = 0.363$ m^3/s, $Q_4 = 0.608$ m^3/s, $Z_j = 63.5$ m)

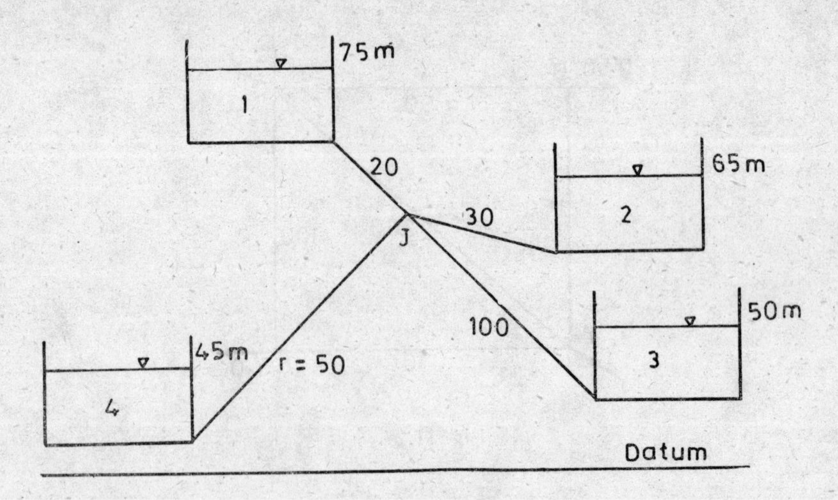

Fig. 11.29.

11.43 Between reservoirs 1 and 2, a pump P and a turbine T are provided. The pipe line characteristics are also shown in Fig. 11.30 along with piezometric heads at four places. Determine

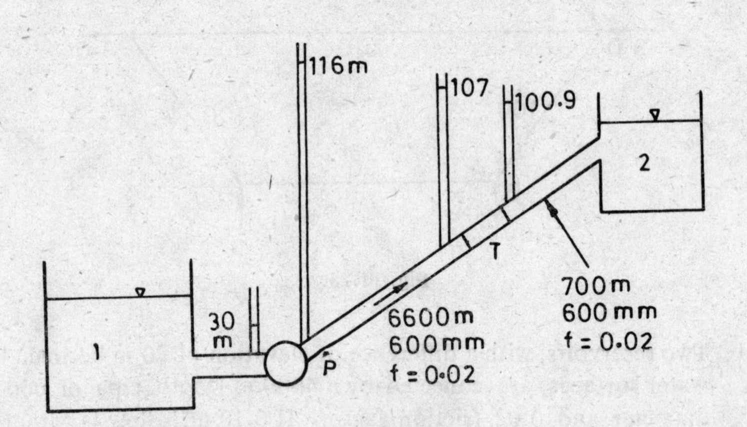

Fig. 11.30.

(i) the discharge and power supplied by pump;

(ii) power removed by turbine,

(iii) elevation of reservoir 2.

(0.253 m^3/s, Pump kW = 213.110, Turbine kW = 13.61

Z_2 = 99.947 m)

DESCRIPTIVE QUESTIONS

11.1 List the characteristics of flow in a wide angle diffuser.

11.2 Under what conditions will K for conical pipe diffuser depend on Reynolds number?

11.3 Identify the errors in Fig. 11.31 and correct the same.

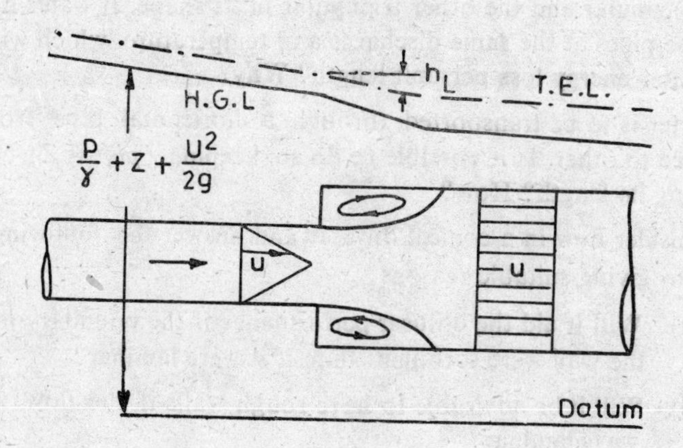

Fig. 11.31

11.4 Draw a neat sketch of a bend in a pipe indicating (i) pressure variation along the inside and outside of the bend, (ii) velocity distribution at the upstream and downstream sections, and (iii) secondary circulation pattern.

11.5 Draw the flow pattern in two consecutive 90° bends shown in Fig. 11.32. If $h_L = K\frac{U^2}{2g}$ represents the energy loss in one bend, can you say that total loss will be $h_L = 2k\frac{U^2}{2g}$? Justify your answer.

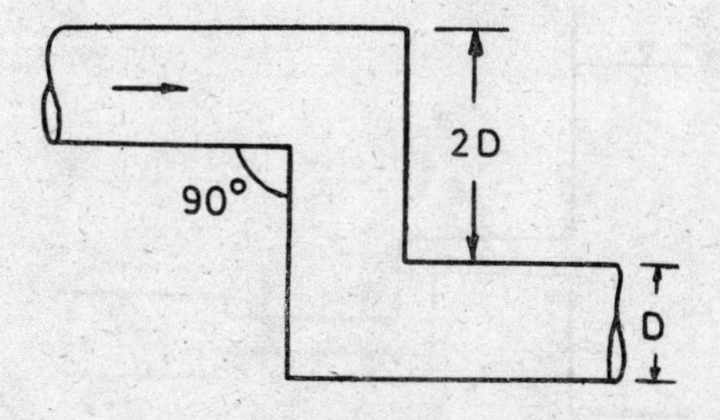

Fig. 11.32.

11.6 For turbulent flow in a smooth pipe with Re less than 10^5, obtain the value of n in $h_f = rQ^n$.

11.7 Two reserviors with a head differenee of H are connected by a pipe of length l, diameter D and friction factor 0.02. Determine the values of l/D at which computation of velocity, excluding entrance and exit losses, will cause 20 and 1 per cent error.

11.8 Consider two pipes of the same material and cross-sectional area—one circular and the other triangular in its shape. If water flows in these pipes at the same discharge and temperature, which will cause greater energy loss per unit length? Why?

11.9 Water is to be transported through a horizontal pipe from one place to other. Is it possible to do so keeping $(p/\gamma + Z)$ constant along its length? How?

11.10 Consider flow in a conical diffuser and answer the following questions giving suitable reasons:

(i) Will it aid the diffuser performance if the boundary layer on the wall were turbulent than if it were laminar?

(ii) Will it be advisable to have rough walls if the flow were always laminar?

(iii) Will the diffuser behave differently if the incoming flow were disturbed, e.g. due to fan etc. than when it is low turbulence undisturbed flow?

11.11 In Fig. 11.7 of the siphon, indicate the portion of the siphon where the pressure is subatmospheric.

11.12 What is priming of siphons? Why is the limit of minimum pressure imposed in the case of siphons?

11.13 Consider the pipe system shown in Fig. 11.33. What modifications in the system will you recommend to reduce losses and thereby to increase discharge for a given H?

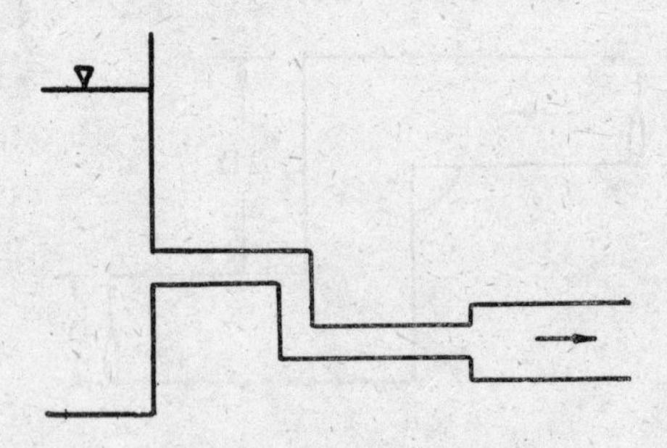

Fig. 11.33.

11.14 Hydraulic grade line for flow in a pipe of constant diameter

(i) always slopes downwards in the direction of flow

(ii) is always above the centreline of the pipe

(iii) is always parallel to the total energy line

(iv) none of the above.

11.15 One pipe is equivalent to the other when

(i) U and D are the same in both the pipes

(ii) h_f and Q are the same in both the pipes

(iii) l and Q are the same in both the pipes

(iv) none of the above.

11.16 Head loss in a sudden expansion in a pipe is given as

(i) $(U_1 + U_2)^2/2g$ (ii) $(U_1 - U_2)^2/2g$

(iii) $(U_1^2 - U_2^2)/2g$ (iv) $0.5\ U^2/2g$

11.17 Major loss in a sudden contraction is due to

(i) boundary friction

(ii) flow contraction

(iii) expansion of flow after sudden contraction

(iv) none of the above.

11.18 In $h_f = rQ^2$,

(i) f depends on R/k (ii) f is proportional to $\frac{1}{Re}$

(iii) f is proportional to $\frac{1}{Re^{0.25}}$ (iv) f is constant.

11.19 What are the undesirable effects of separation of flow in a conical diffuser?

11.20 If two pipes at right angle to each other are to be connected through a bend,

(i) what value of R_b/D will you recommend, and

(ii) what other method will you suggest to reduce the head loss?

11.21 For valves and gates, K in the equation $h_f = K\frac{U^2}{2g}$

(i) does not depend on Reynolds number

(ii) depends on Reynolds number only if oil flows through pipes

(iii) is independent of Reynolds number if it is large.

11.22 For maximum transmission of power through pipes for given H,

(i) $h_f = H$ (ii) $h_f = 0.50\,H$

(iii) $h_f = \frac{H}{3}$ (iv) none of the above.

11.23 If three pipes of different diameters, lengths and friction factors are connected in series,

(i) $f = f_1 + f_2 + f_3$, (ii) $U = U_1 + U_2 + U_3$

(iii) $h_f = h_{f1} + h_{f2} + h_{f3}$ (iv) $Q = Q_1 + Q_2 + Q_3$

(iv) $Q_1 = Q_2 = Q_3$.

11.24 If three pipes of different diameters, lengths and friction factors are connected in parallel,

(i) $U_1 = U_2 = U_3$ (ii) $Q_1 = Q_2 = Q_3$

(iii) $h_{f1} = h_{f2} = h_{f3}$ (iv) $f = f_1 + f_2 + f_3$

(iv) $Q = Q_1 + Q_2 + Q_3$

11.25 For pipe networks,

(i) $Q = Q_1 + Q_2 + Q_3 + \ldots\ldots$ irrespective of direction of Q

(ii) $Q = Q_1 + Q_2 + Q_3 + \ldots\ldots$ with due regard to direction

(iii) ΣnrQ^{n-1} has to be calculated with due regard to sign

(iv) ΣnrQ^{n-1} is to be calculated disregarding the sign.

CHAPTER XII

Forces on Immersed Bodies

12.1 INTRODUCTION

Whenever there is relative motion between a real fluid and a body, the fluid exerts a force on the body. The body exerts an equal and opposite force on the fluid. If the body is moving at a constant velocity in a stationary fluid, the fluid motion is unsteady, because, at a given point in space, the velocity changes with time. However, if the body is stationary and fluid flows at a constant velocity, it is steady motion. The magnitude of force in both cases is the same. The resultant force F acting on the body can be resolved into two components: F_D in the direction of flow, called drag, and F_L perpendicular to the direction of flow, called lift (Fig. 12.1). The fluid viscosity

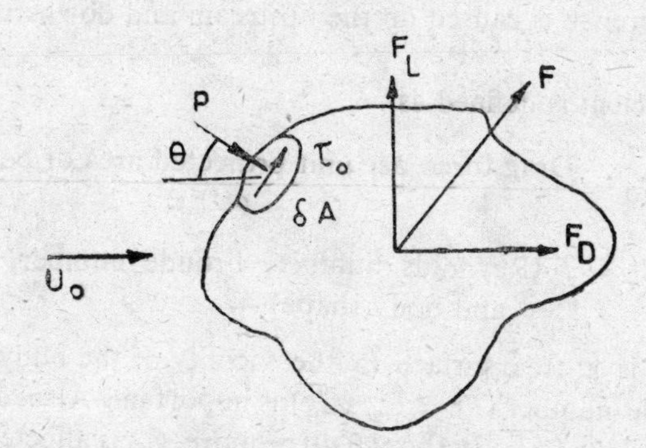

Fig. 12.1. Definition sketch.

affects the flow around the body in three ways to cause the force on the body:

1. At low Reyonlds number, the fluid is deformed in a very wide zone around the body causing pressure force and frictional force
2. As Reynolds number increases, viscous effects are confined to the boundary layer causing predominantly only frictional force on the boundary

3. For certain body shapes, the boundary layer can separate causing additional pressure force.

With reference to Fig. 12.1, one can write

$$\left.\begin{aligned} F_D &= \int_A \tau_0 dA \sin\theta + \int_A p dA \cos\theta \\ &\quad \text{(friction drag)} + \text{(pressure drag)} \\ F_L &= \int_A \tau_0 dA \cos\theta + \int_A p dA \sin\theta \end{aligned}\right] \quad 12.1$$

Here A is the surface area of the body. Usually the first part in expression for F_L is negligible.

12.2 DEFORMATION DRAG, FRICTION DRAG, FORM DRAG

Deformation drag is caused by widespread deformation of fluid around the body and it exists at low Reynolds numbers. It consists of partly skin friction drag and partly pressure drag. At high Reynolds numbers, deformation is limited to the boundary layer thickness and in such a case, deformation drag is exclusively friction drag.

Friction drag is due to the shear force exerted on the surface of the body due to the formation of the boundary layer. It is in fact the component of the shear force in the direction of motion.

Form drag is caused when separation of boundary layer takes place and a pressure difference is caused on the upstream and downstream side of the body.

Drag coefficient is defined as

$$C_D = \frac{\text{Drag force per unit projected area of body} \perp \text{to flow}}{\rho U_0^2/2}$$

In general $\quad C_D = f$ (Reynolds number, Froude number, Mach number and body shape)

Only if there is a free surface in the vicinity of the body and waves are formed, Froude number ($Fr = U_0/\sqrt{lg}$) is important. Also, only if the Mach number ($M = U_0/\sqrt{E/\rho}$) is about 0.40 or more, C_D is affected by M. Hence, for majority of Civil Engineering problems,

$$C_D = f\left(\frac{U_0 l \rho}{\mu}, \text{ body shape}\right)$$

Here l is the characteristic length of the body, e.g. diameter of sphere or width of rectangular plate.

12.3 VARIATION OF C_D WITH *Re*

12.3.1 Sphere

$$\left.\begin{aligned} F_D &= 3\pi D\mu U_0 \\ \text{and} \qquad C_D &= \frac{24}{Re} \end{aligned}\right\} \text{ if } Re < 0.20 \tag{12.2}$$

$$C_D = \frac{24}{Re}\left(1 + \frac{3}{16} Re\right) \text{ if } Re < 2.0 \tag{12.3}$$

$$C_D = \frac{24}{Re} + \frac{3}{\sqrt{Re}} + 0.34 \text{ if } 0.5 < Re < 10^4 \tag{12.4}$$

At Reynolds number of approximately 3×10^5, C_D for a smooth sphere decreases from 0.5 to 0.2, because, at this Reynolds number, the boundary layer changes from laminar to turbulent and the point of separation shifts from 80° to 110°. Figure 12.2 shows variation of C_D with *Re* for sphere while Fig. 12.3 shows pressure distribution around sphere when the fluid is ideal, when $Re = 7 \times 10^4$ and when $Re = 3 \times 10^5$.

Since $$C_D Re^2 = \frac{8F_D}{\pi D^2 \rho U_0^2} \times \frac{U_0^2 D^2 \rho^2}{\mu^2} = \frac{8}{\pi}\left(\frac{F_D \rho}{\mu^2}\right) \text{ or } \frac{8}{\pi}\left(\frac{F_D}{\rho \nu^2}\right)$$

Therefore C_D or *Re* can be related to $F_D/\rho\nu^2$ as given in Table 12.1.

Table 12.1 Variation C_D and *Re* with $F_D/\rho\nu^2$ for sphere

Re	0.10	1.0	10	100	1000	10,000	100,000
C_D	240	26	4.5	1.1	0.45	0.50	0.50
$F_D/\rho\nu^2$	0.943	10.21	176.7	392.8	1.77×10^5	2×10^7	2.36×10^9
Re	2000,000	1000,000					
C_D	0.42	0.20					
$F_D/\rho\nu^2$	6.6×10^9	7.8×10^{10}					

This facilitates direct solution for F_D, D and U_0 if other quantities are known. Increase in surface roughness or turbulence in free stream causes sudden reduction in C_D at lower Reynolds number and it slightly increases C_D.

At large Mach number values C_D for sphere is dependent on Mach number as can be seen from Table 12.2.

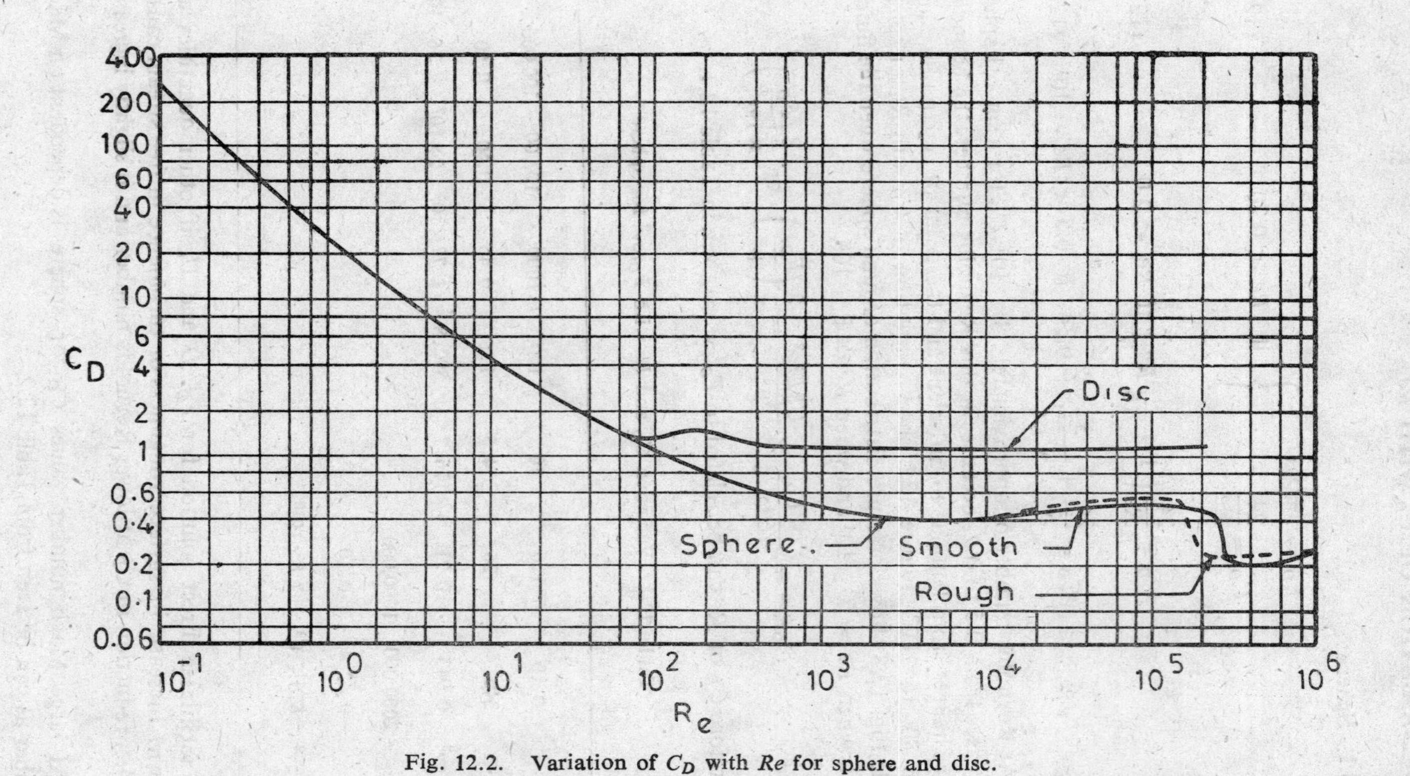

Fig. 12.2. Variation of C_D with *Re* for sphere and disc.

Table 12.2 Variation of C_D with M for sphere

M	0.50	1.0	1.5	1.7	2.0	2.1	3.7
C_D	0.54	0.81	0.98	1.00	0.97	0.95	0.95

12.3.2 Disc

$$C_D = \frac{20.37}{Re} \quad \text{if} \quad Re < 0.10 \tag{12.5}$$

C_D decreases with increase in Re upto about 10^3 and beyond this value of Re, C_D assumes a constant value of 1.10. Figure 12.2 shows the variation of C_D with Re for a disc.

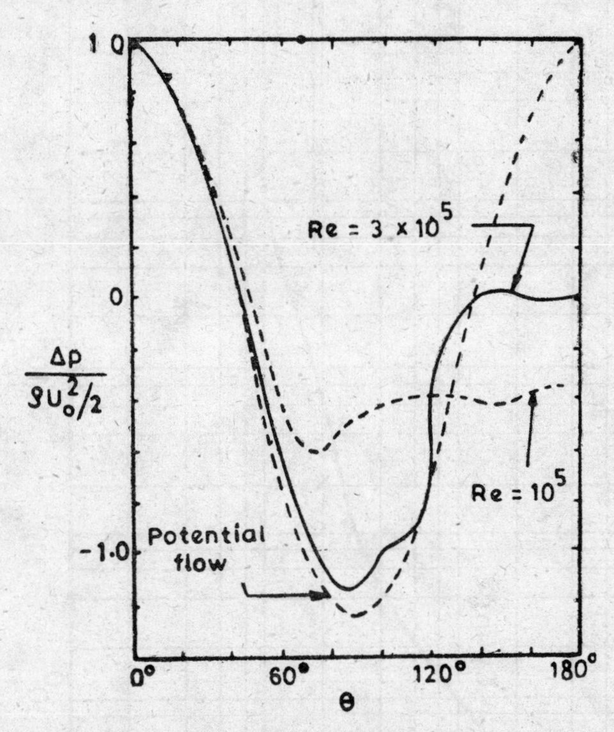

Fig. 12.3. Pressure distribution around a sphere.

12.3.3 Two-dimensional circular cylinder

$$C_D = \frac{10.9/Re}{0.87 - \log Re} \quad \text{if } Re < 0.20 \tag{12.6}$$

As Reynolds number increases, C_D decreases and reaches near constant value at 10^3. Between Re values of 10^3 and 10^5, C_D increases from 1.0 to 1.20 and then drops to 0.33 at 3×10^5 (Fig. 12.4). Figure 12.5 shows the pressure distribution around the cylinder for various Re values.

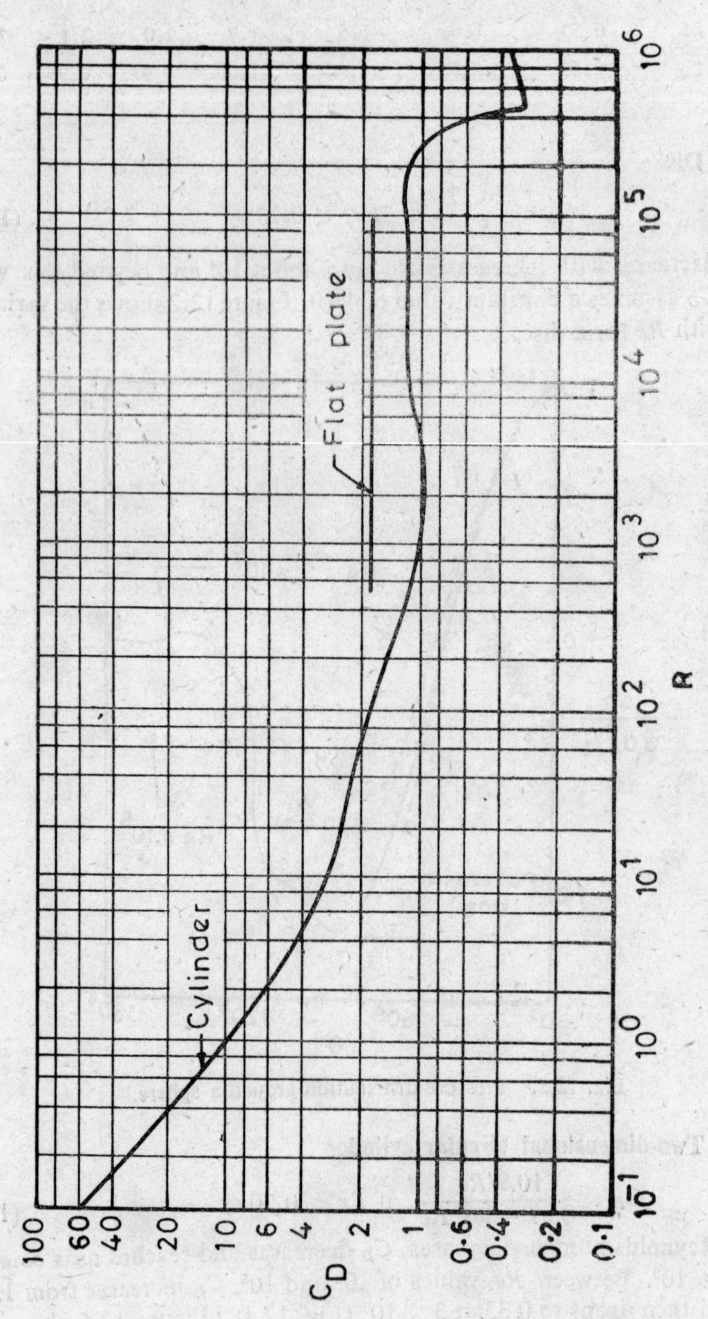

Fig. 12.4. Variation of C_D with R_e for cylinder and flat plate.

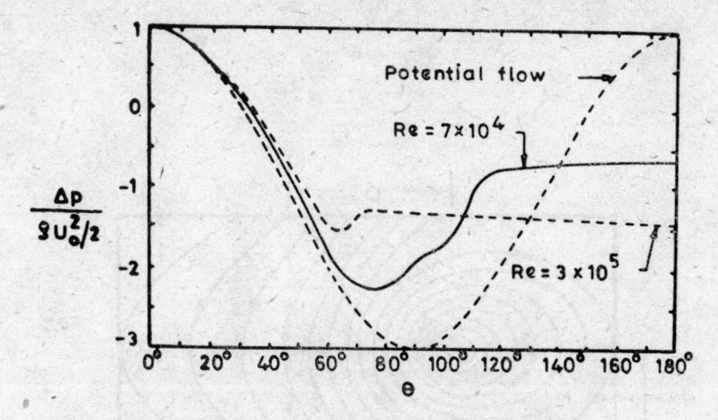

Fig. 12.5. **Pressure distribution around cylinder.**

Flow past two-dimensional bodies, in which separation takes place on the downstream side, are characterised by formation of Karman vortex trail (Fig. 12.6). This occurs at Reynolds numbers exceeding 30. For Karman vortex trail

$$\left.\begin{aligned} a/b &= \cosh^{-1}\sqrt{2} = 0.281 \\ S &= 0.198\left(1 - \frac{19.7}{Re}\right) \\ f &= \frac{U - u_r}{b} \\ u_r &= \Gamma/b\sqrt{8} \end{aligned}\right\} \tag{12.7}$$

Here f is the frequency of alternate shedding of vortices, u_r velocity of downstream movement of vortices, S is the Strouhal number which is equal to fD/U_0 and Γ is the circulation.

Strouhl number is related to C_D as

$$S = 0.21/C_D^{0.75} \tag{12.8}$$

This is valid for two-dimensional bodies of various shapes.

12.3.4 Flat plate held perpendicular to flow

For Re less than 0.20, C_D is given by

$$C_D = \frac{10.9/Re}{0.96 - \log Re} \tag{12.9}$$

The drag coefficient decreases as Re increases and beyond Re value of 10^3, C_D assumes a constant value of 1.9. For flat plates of finite length $C_D = f(b/l)$, here b is the width of plate and l its length. This relationship is given in Table 12.3.

Table 12.3 Variation of C_D with b/l for flat plate

b/l	1.0	0.50	0.2	0.1	0.04	0
C_D	1.10	1.15	1.20	1.25	1.50	1.9

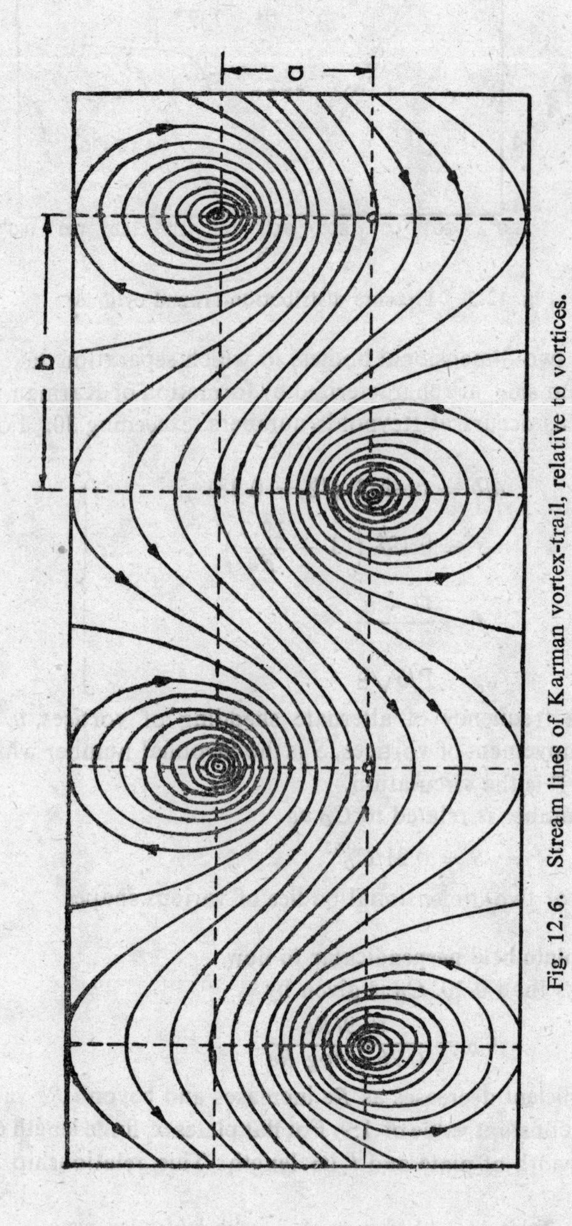

Fig. 12.6. **Stream lines of Karman vortex-trail, relative to vortices.**

12.3.5 Aerofoil

In the case of a symmetrical aerofoil held with its axis of symmetry parallel to flow direction, separation takes place only on a small length of its tail end. Hence almost all the drag is due to friction. Here C_D continuously decreases as *Re* increases. Pressure distribution over major portion agrees well with potential flow theory. When the aerofoil makes an angle with the direction of flow (known as the angle of attack), a lift force is also caused.

12.4 EFFECTS OF FREE SURFACE AND COMPRESSIBILITY ON DRAG

When a body is close to free surface or on the free surface, one can write

$$F_D = F_{\text{friction}} + F_{\text{residual}}$$

F_{residual} consists of partly wave drag and partly form drag. This part of the drag is considered to be a function of Froude number only, whereas F_{friction} is computed from boundary layer theory, F_{residual} is normally obtained from model studies.

In a compressible fluid the pressure wave travels at a speed of $\sqrt{E/\rho}$, where E is the volume modulus of elasticity of fluid. If Mach number is less than unity, U_0 is less than $\sqrt{E/\rho}$ (i.e. subsonic flow), and the pressure wave is always ahead of the moving body which produces it. In case of supersonic flow, U_0 is greater than $\sqrt{E/\rho}$; hence body always leaves behind the pressure wave it creates (Fig. 12.7).

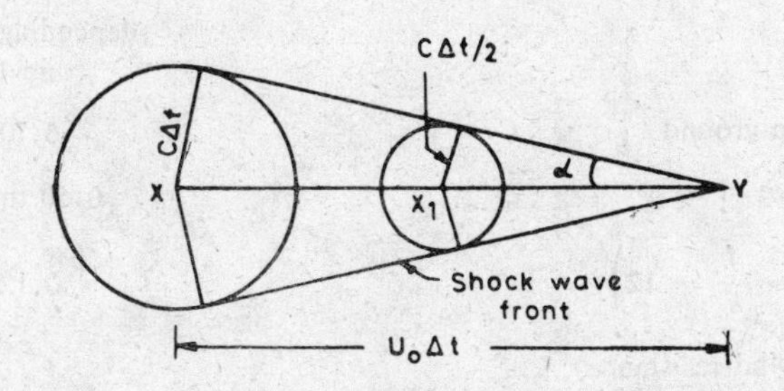

Fig. 12.7. Travel of pressure wave at supersonic speed.

$$\sin \alpha = \frac{C\Delta t}{U_0 \Delta t} = \frac{C}{U_0}$$

or

$$\alpha = \sin^{-1} (C/U_0)$$

α is known as Mach angle.

At high M number values,

$$C_D = f(M).$$

At supersonic flow, shape of tail end of projectile is not important but nose should be rounded or tapered. Table 12.2 shows variation of C_D with

M for a sphere. Table 12.4 gives average drag coefficients for some body shapes.

Table 12.4 Average drag coefficients of some body shapes

Body shape	C_D
Hollow hemisphere $\rightarrow \supset$	1.33
Hollow hemisphere $\rightarrow \subset$	0.34
Sphere ($10^4 - 10^5$)	0.50
Sphere (greater than 3×10^5)	0.20
Circular cylinder ($10^4 - 10^5$)	1.20
Circular cylinder (greater than 3×10^5)	0.33
Cylinder with long splitter plate ($10^4 - 10^5$) $\rightarrow$ o—	0.50
Square cylinder $\rightarrow \square$	2.0
Flat plate ($b/l = 0$) $\rightarrow \vert$	1.90
Disc	1.10
Empire state building, N.Y. (380 m high)	1.3 to 1.5 (depending on wind direction)
Sphere resting on ground	0.70
Rectangular barge	0.50 to 1.0
Submarine $\left(\frac{l}{b} = \frac{l}{h} = 12\right)$	0.12
Passenger ship wind resistance (projected frontal area above water line)	0.6 to 1.0
Passenger cars	0.40 to 0.60
Stream lined racing cars	0.20 to 0.30
Trucks and buses (box-like)	0.8 to 1.0
Railway engine with five bogies	1.90
Man facing wind from front	1.60 approx.
Man facing wind sideways	1.30 approx.
Man falling vertically	1.6 to 1.8

12.5 LIFT

The generation of lift can be explained on the basis of ideal fluid flow theory. Consider superposition of constant circulation flow around a cylinder and flow of an ideal fluid past the cylinder at velocity U_0. This results in a flow pattern as shown in Fig. 12.8 giving a pressure distribution which results in lift F_L

$$F_L = L\rho U_0 \Gamma \tag{12.10}$$

in which Γ is the circulation $= 2\pi R u_c$ and L is the length of cylinder.

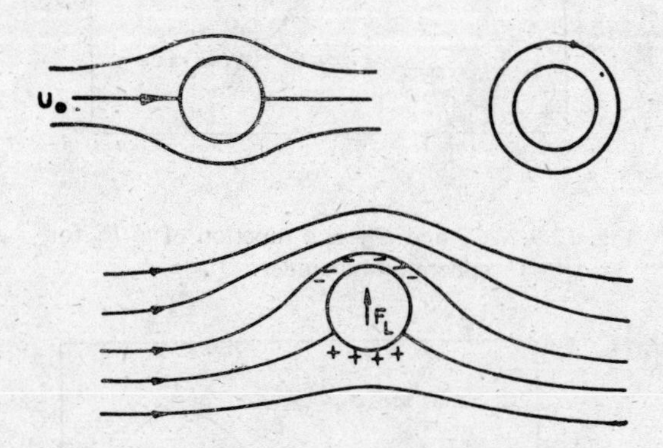

Fig. 12.8. Generation of lift around a cylinder.

$$\text{Lift coefficient } C_L = \frac{F_L/\text{area}}{\dfrac{\rho U_0^2}{2}} = 2\pi \frac{u_c}{U_0} \tag{12.11}$$

Here area $= D \times L$. The fact that lift force can be caused by superposition of rectilinear and circulatory flows around a cylinder was first discovered by Magnus in 1852. Hence it is known as Magnus effect. Similar phenomenon for sphere is known as Robin effect.

Figure 12.9 shows variation of C_L and C_D for a cylinder as a function of u_c/U_0. The large discrepancy (Fig. 12.9) between observed and theoretical lift is due to effects of viscosity. Figure 12.9 also shows experimental values of C_L and C_D as a function of u_c/U_0 for sphere.

In practice, unsymmetrical vanes kept at an inclination to flow are used to produce lift (Fig. 12.10). For lifting vanes, C_D and C_L are defined with respect to common area CL where C is the chord and L its length. Hence

$$C_D = \frac{F_D/LC}{\rho U_0^2/2}, \quad C_L = \frac{F_L/LC}{\rho U_0^2/2} \tag{12.12}$$

The circulation, which will develop a flow pattern such that streamline at trailing edge is tangential, is given by

$$\Gamma = \pi C U_0 \sin \theta_0 \tag{12.13}$$

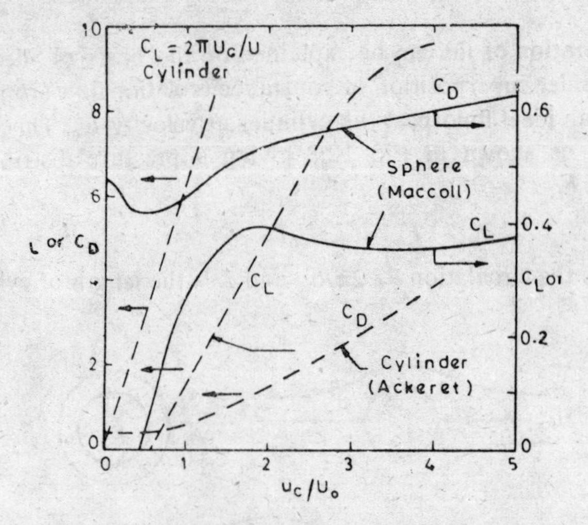

Fig. 12.9. C_L and C_D as a function of u_c/U_0 for sphere and cylinder.

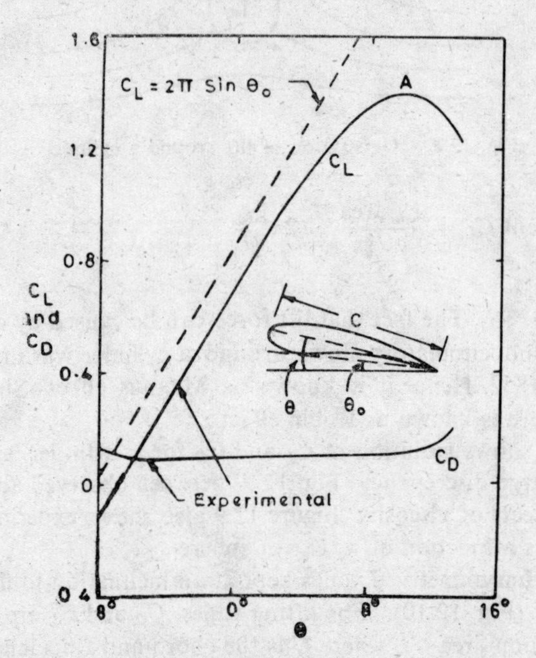

Fig. 12.10. Theoretical and experimental variation of C_L and C_D for lifting vane.

and hence as shown by Kutta and Joukowsky

$$F_L = L\rho U_0 \Gamma = \pi C L \rho U_0^2 \sin \theta_0$$

and

$$C_L = \frac{F_L/LC}{\rho U_0^2/2} = 2\pi \sin \theta_0 \tag{12.14}$$

The theoretical lift coefficient as given by the above equation and measured C_L and C_D values for a typical vane are shown in Fig. 12.10. At point A, the inclination of vane becomes so large that flow separates causing a sudden drop in C_L. This phenomenon is known as stall. Prandtl has given an explanation for generation of lift on a vane, on the basis of starting vortex.

The total drag coefficient C_D of a lifting vane is the sum of profile drag coefficient of two dimensional wing C_{D0} and induced drag coefficient C_{Di} due to finite value of $\frac{L}{C}$ i.e. due to three dimensionality of flow.

Induced drag coefficient is given by

$$C_{Di} = \frac{C_L^2}{\pi L/C} \tag{12.15}$$

Figure 12.11 shows variation of C_L with C_D for a typical vane as a function of angle of attack or inclination. Such a diagram is known as polar diagram.

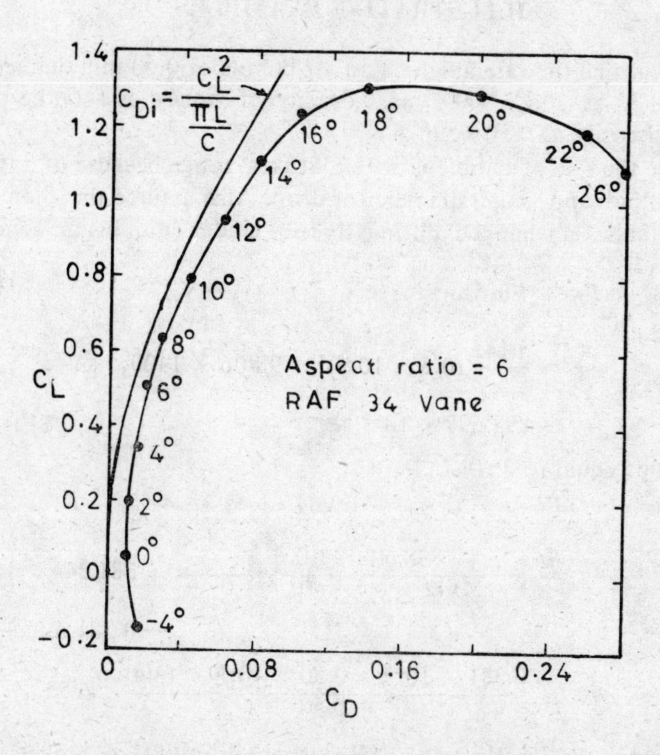

Fig. 12.11. **Polar diagram for a typical vane.**

For an aeroplane in flight, the weight of the plane W is balanced by the total lift force. Or

$$W = \frac{C_L A \rho U_0^2}{2} \tag{12.16}$$

where A is the total wing area. Total thrust required at a velocity U_0 is

$$F_D = (C_D + C_f) A \frac{\rho U_0^2}{2} \tag{12.17}$$

where C_D is the drag coefficient of wings and C_f which is known as parasite drag coefficient, is the drag coefficient of the rest of the plane. At the take-off stage of the plane, C_L should be maximum for a chosen wing section. This is obtained at an angle of attack slightly less than the stall angle. For this C_L and known W, the take-off speed will be

$$U_0 = \sqrt{2W/C_L A \rho} \tag{12.18}$$

In the level flight, the ratio of C_L/C_D should be maximum. This condition is obtained by drawing a tangent to polar diagram through the point of zero C_L. The point of tangency gives the corresponding angle of attack.

ILLUSTRATIVE EXAMPLES

12.1 Determine the rate at which an air bubble of 0.50 mm diameter will rise in honey at 20°C. Take mass density of honey as 1400 kg/m³ and dynamic viscosity as 0.50 kg/m s.

Assume the rise of bubble is within Stokes' range because of high viscosity of fluid and small diameter of drop. Also assume the diameter of bubble remains unchanged during its rise. Under equilibrium condition

$$F_D = \text{Buoyant force} = \frac{\pi}{6} D^3 (\gamma_h - \gamma_a)$$

$$= \frac{3.142}{6} (0.5 \times 10^{-3})^3 \times 9.806 \times 1400, \quad \text{since} \quad \gamma_a \ll \gamma_h.$$

$$= 898.698 \times 10^{-9} \text{ N}$$

This must be equal to $3\pi D \mu U_0$.

$$\therefore \qquad 3\pi D \mu U_0 = 898.698 \times 10^{-9}$$

or

$$U_0 = \frac{898.698 \times 10^{-9}}{3 \times 3.142 \times 0.5 \times 10^{-3} \times 0.50} = 381.344 \times 10^{-6} \text{ m/s}$$

$$= 0.381 \text{ mm/s.}$$

$$Re = \frac{0.381 \times 10^{-3} \times 0.50 \times 10^{-3} \times 1400}{0.50} = 533.4 \times 10^{-6}$$

which is much smaller than 0.20. Hence applicability of Stokes' law is justified.

12.2 A spherical sand particle of 0.10 mm diameter falls under the action of gravity in water at 20°C. Determine its terminal fall velocity.

Assume the particle motion to be within Stokes' range. With

$$\mu = 10^{-3} \text{ kg/m s},$$

$$U_0 = \frac{D^2}{18\mu}(\gamma_s - \gamma) = \frac{(0.1 \times 10^{-3})^2}{18 \times 10^{-3}}(2.65 - 1) \times 998 \times 9.806$$

$$= 8.971 \times 10^{-3} \text{ m/s}$$

Check the value of Reynolds number

$$Re = 8.971 \times 10^{-3} \times 0.1 \times 10^{-3} \times 998/10^{-3} = 0.895$$

Since Re is larger than 0.20, use Oseen's equation

$$C_D = \frac{24}{Re}\left(1 + \frac{3}{16}Re\right) = \frac{24}{Re} + 4.5$$

$$\therefore \quad C_D = \frac{24}{0.895} + 4.50 = 31.32$$

$$\therefore \quad C_D = \frac{\text{Submerged weight}}{(\text{area}) \times \rho U_0^2/2} = \frac{4}{3}\frac{D(\gamma s - \gamma)}{\rho U_0^2}$$

$$\therefore \quad U_0^2 = \frac{4 \times 0.1 \times 10^{-3} \times 1.65 \times 998 \times 9.806}{31.32 \times 3 \times 998}$$

$$= 0.689 \times 10^{-4} \quad \text{and} \quad U_0 = 0.823 \times 10^{-2} \text{ m/s}$$

$$Re = \frac{0.823 \times 10^{-2} \times 0.1 \times 10^{-3} \times 998}{10^{-3}} = 0.821$$

Compute refined value of C_D

$$C_D = \frac{24}{0.821} + 4.50 = 33.733$$

and $\quad U_0 = 0.793 \times 10^{-2}$ m/s

Again $\quad Re = 0.793 \times 10^{-2} \times 0.1 \times 10^{-3} \times 998/10^{-3} = 0.791$

$$C_D = \frac{24}{0.791} + 4.50 = 34.84$$

and $\quad U_0^2 = \frac{4}{3} \times \frac{4 \times 0.1 \times 10^{-3} \times 1.65 \times 998 \times 9.806}{3 \times 34.84 \times 998}$

$$U_0 = 7.869 \times 10^{-3} \text{ m}$$

or $\quad U_0 = 7.869$ mm/s.

This can be taken as reasonably correct since two consecutive values of U_0 are sufficiently close. Alternately Oseen's equation can be written as

$$U_0^2 + aU_0 - b = 0$$

where a and b are functions of μ, ρ, D and F_D. Solution of the above equation will give U_0.

12.3 What will be the diameter of a particle of relative density 2.65 which will have a fall velocity of 0.50 m/s in water of 20°C?

Assume $Re = 100$ for which C_D vs Re diagram can be referred and C_D determined.

$$C_D = \frac{24}{Re} + \frac{3}{\sqrt{Re}} + 0.34$$

$$= 0.24 + 0.30 + 0.34$$

$$= 0.88$$

$$\therefore \quad 0.88 = \frac{\pi D^3}{6} \frac{(\gamma_S - \gamma)}{(\pi D^2/4)\,(\rho U_0^2/2)} = \frac{4D\Delta\gamma_S}{3\rho U_0^2}$$

$$\therefore \quad D = \frac{3 \times 998 \times 0.50^2 \times 0.88}{4 \times 1.65 \times 998 \times 9.806} = 0.01 \text{ m}$$

and $$Re = 0.01 \times 0.50/10^{-6} = 5000$$

$$\therefore \quad C_D = \frac{24}{5000} + \frac{3}{\sqrt{5000}} + 0.34 = 0.387$$

$$D = \frac{3 \times 0.387 \times 998 \times 0.50^2}{4 \times 1.65 \times 998 \times 9.806} = 0.004\,48 \text{ m}$$

and $$Re = (0.004\,48 \times 0.50)/10^{-6} = 2240$$

$$C_D = \frac{24}{2240} + \frac{3}{\sqrt{2240}} + 0.34 = 0.414$$

and $$D = \frac{3 \times 998 \times 0.50^2 \times 0.414}{4 \times 1.65 \times 998 \times 9.806} = 0.0048 \text{ m}$$

Again Re can be calculated. $Re = \dfrac{0.0048 \times 0.50}{10^{-6}} = 2400$

and $$C_D = \frac{24}{2400} + \frac{3}{\sqrt{2400}} + 0.34 = 0.411$$

and $$D = \frac{3 \times 998 \times 0.50^2 \times 0.411}{4 \times 1.65 \times 998 \times 9.806} = 0.004\,76 \text{ m or } 4.76 \text{ mm}$$

This can be taken as correct diameter since two consecutive values are very close.

12.4 When the Reynolds number reaches its critical value, C_D for the sphere drops from 0.50 to a value of 0.20, due to onset of turbulence in the boundary layer. If the driving power remains the same, what will be corresponding increase in speed with such change?

Since power = force × velocity = $C_D A \dfrac{\rho U_0^3}{2}$

$$\therefore \quad C_{D_1} U_{01}^3 = C_{D_2} U_{02}^3$$

$$\therefore \qquad \left(\frac{U_{01}}{U_{02}}\right)^3 = \frac{C_{D2}}{C_{D1}} = \left(\frac{0.20}{0.50}\right)$$

or
$$\frac{U_{01}}{U_{02}} = \left(\frac{0.20}{0.50}\right)^{1/3} = 0.737$$

and
$$U_{02} = 1.357\, U_{01}$$

Thus there will be 35.7 percent increase in velocity.

12.5 A ball of 0.50 m diameter with a relative density 3.0 is held stationary in water and released to fall water gravity. Determine the acceleration of the ball (a) neglecting the added mass and (b) accounting for added mass.

Initially the velocity is zero and hence drag is zero. The driving force is the submerged weight of ball.

$$\therefore \qquad F = \frac{\pi D^3}{6}(\gamma s - \gamma) = \frac{3.142 \times 0.5^3}{6}(3 - 1) \times 9787$$

$$= 1281.281 \text{ N}$$

Since force = mass × acceleration,

$$a = \frac{F}{M}$$

(a) When added mass is not considered, the mass accelerated is the mass of ball; or $M = \frac{\pi D^3}{6} \times 998 \times 3$

$$= 195.982 \text{ kg}$$

$$\text{Acceleration} = \frac{1281.281}{195.982} = 6.538 \text{ m/s}^2$$

(b) Actually along with the ball, a certain mass of liquid is accelerated. Added mass coefficient is the ratio of the mass of fluid accelerated to the mass of fluid displaced by body. For sphere, added mass coefficient = 0.5.

$$\therefore \qquad \text{Mass of fluid accelerated} = \frac{\pi D^3}{6} \times 998 \times 0.50$$

$$= 32.664 \text{ kg}$$

$$\therefore \qquad M = 32.664 + 195.982$$

$$= 228.646 \text{ kg}$$

$$\therefore \qquad \text{Acceleration } a = \frac{1281.281}{228.646} = 5.604 \text{ m/s}$$

Added mass coefficients for other body shapes are given below:

Disc	0.636
Long plate ⊥ flow	1.05
Long cylinder ⊥ flow	1.0
Cube	0.67

12.6 A 10 mm ball of relative density 1.2 is suspended from a string, in air flowing at a velocity of 10 m/s and 20°C temperature. Determine the angle which the string will make with the vertircal (Fig. 12.12).

From the condition of equilibrium, one can write

$$F_D \cos\theta = W \sin\theta$$

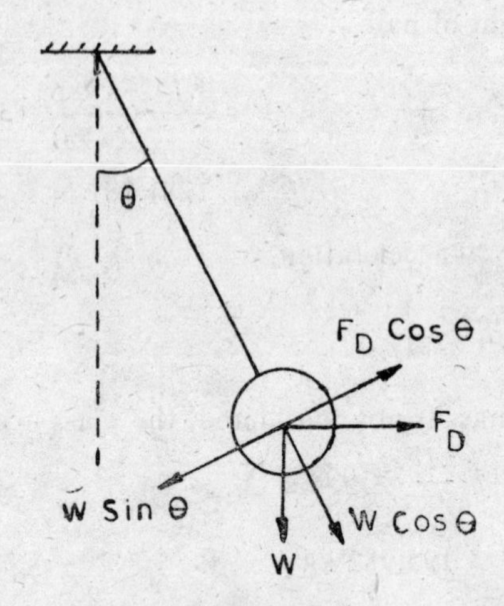

Fig. 12.12.

or $$C_D \frac{\pi D^2}{4} \frac{\rho U_0^2}{2} \cos\theta = \frac{\pi D^3}{6} \gamma_s \sin\theta$$

$$\therefore \quad \tan\theta = \frac{3}{4} C_D \frac{\rho U_0^2}{D\gamma_s} = \frac{3}{4} C_D \frac{U_0^2}{gD(\rho_s/\rho)}$$

$$Re = \frac{U_0 D \rho}{u} = (10.0 \times 0.01 \times 1.208)/1.85 \times 10^{-5}$$

$$= 6530$$

$$\therefore \quad C_D = \frac{24}{6530} + \frac{3}{\sqrt{6530}} + 0.34$$

$$= 0.381$$

$$\therefore \quad \tan\theta = \frac{3}{4} \times \frac{0.381 \times 10^2 \times 1.208}{9.806 \times 0.01 \times 1.2 \times 998}$$

$$= 0.2939$$

$\therefore \qquad \theta = 16.38°$

12.7 A bullet of mass M, cross-sectional area A and drag coefficient C_D is fired horizontally at an initial velocity U_0. Determine after how much distance S its horizontal velocity will reduce to half its initial value due to air resistance. Assume C_D to be constant. Find S if $M = 10$ gm, $A = 20$ mm^2 and $C_D = 0.40$.

Here added mass is negligible, since mass of displaced air is extremely small. Hence

$$MU \frac{dU}{dS} = - C_D A \rho \frac{U^2}{2}$$

Integration of this equation yields

$$S = - \frac{2M}{C_D A \rho} \ln U + C$$

The constant of integration C is determined from the initial condition that

when $\qquad S = 0, U = U_0$

$$\therefore \qquad C = \frac{2M}{C_D A \rho} \ln U_0$$

and

$$S = \frac{2M}{C_D A \rho} \ln \frac{U_0}{U}$$

When $\quad U = \frac{U_0}{2}, \quad S = \frac{2M}{C_D A \rho} \ln 2 = \frac{1.386M}{C_D A \rho}$

Substituting the values of M, C_D, A and ρ, one gets the distance S required to reduce the velocity to half its initial value as

$$S = \frac{1.386 \times 0.01}{0.40 \times 20 \times 10^{-6} \times 1.208} = 1434.19 \text{ m}$$

12.8 Determine the bending moment at the base of a chimney 25 m in height with its diameter gradually changing from 2.0 m at its base to 1.0 m at the top, if it is exposed to a wind of uniform velocity 150 km/hr at 20°C.

$$U_0 = \frac{150 \times 1000}{3600} = 41.67 \text{ m/s}$$

$$Re \text{ at top} = \frac{41.67 \times 1.0 \times 1.208}{1.85 \times 10^{-5}} = 2.721 \times 10^6$$

$$Re \text{ at bottom} = \frac{41.67 \times 2 \times 1.208}{1.85 \times 10^{-5}} = 5.442 \times 10^6$$

For this range of Re, an average value of C_D equal to 0.35 can be taken. With respect to Fig. 12.13,

$$y = \left(2 - \frac{x}{25}\right)$$

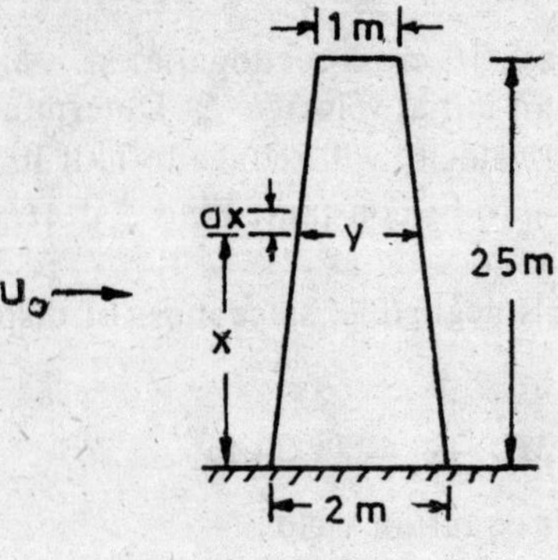

Fig. 12.13.

$$dF, \text{ Force on length } \delta x = C_D\left(2 - \frac{x}{25}\right)\frac{\rho U_0^2}{2}\, dx$$

δM, moment of force dF about base

$$= C_D \frac{U_0^2}{2}\left(2 - \frac{x}{25}\right) x dx$$

$$\therefore \qquad M = \int_0^{25} C_D \frac{U_0^2}{2}\left(2 - \frac{x}{25}\right) x dx$$

or

$$M = \frac{0.35 \times 1.208 \times 41.67^2}{2}\left[x^2 - \frac{x^3}{75}\right]_0^{25}$$

$$= 152\,948.31 \text{ Nm}$$

12.9 Estimate the drag force experienced by 1.0 m × 1.0 m trash rack made up of 5.0 mm deep and 1.0 m long strips kept 10 mm apart, when water flows at 2.0 m/s velocity through them.

Trash rack is used at the entrance to tunnels and in flows to remove floating matter and coarse material transported by the flow (Fig. 12.14).

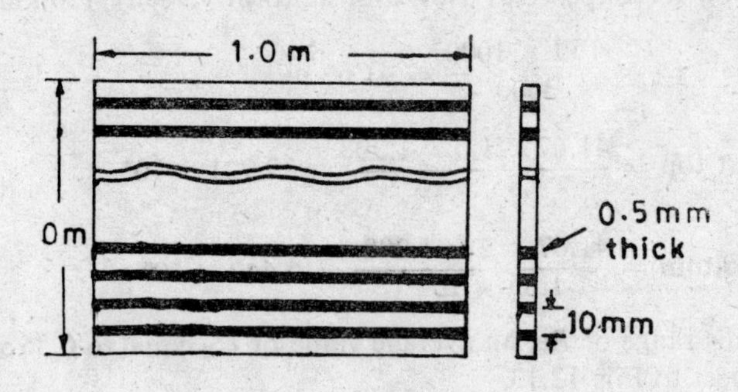

Fig. 12.14. Trash rack.

For strip of aspect ratio as large as $\frac{1000}{5} = 200$, the strip can be considered as two dimensional. Further

$$Re = \frac{2 \times 0.005}{10^{-6}} = 10 \times 10^3 \quad \text{or} \quad 10^4$$

Hence $\quad C_D = 1.9.$

$$\text{Force on each strip} = 1.9 \times (0.005 \times 1.0) \times \frac{998 \times 2^2}{2}$$

$$= 18.962 \text{ N}$$

Number of strips in 1.0 m height will be n

$$n \times 10 + (n - 1) \times 5 = 1000 \quad \text{or} \quad 15n = 1005$$

$$\text{or} \quad n \approx 67$$

$\therefore \quad$ Total force $= 18.962 \times 67 = 1020.454$ N

This calculation is based on the assumption that the flow past each strip is unaffected by the neighbouring strips. However, because of their close spacing, this assumption is not completely true. Hence in reality, trash rack will experience somewhat different drag.

12.10 Determine the frequency of vortex shedding for a 3.0 mm diameter transmission cable in wind of 72 km/hr speed at 20°C.

$$U_0 = \frac{72 \times 1000}{3600} = 20.0 \text{ m/s}$$

$$\therefore \quad Re = \frac{20 \times 0.003 \times 1.208}{1.85 \times 10^{-5}} = 3917$$

$$\therefore \quad \frac{fD}{U_0} = 0.198 \left(1 - \frac{19.7}{3917}\right)$$

$$= 0.197$$

$$\therefore \quad f = \frac{0.197 \times 20}{0.003} = 1313 \text{ Hz}$$

Human ear is capable of receiving sound in the frequency range of 20 to 20000 Hz. Hence vibrations of the cable at 1313 Hz will produce musical sound.

12.11 After bailing out from the aeroplane and before the parachute is released, the man accelerates to the terminal velocity. Taking typical values of body weight including accessories as 800 N, $\rho = 1.20$ kg/m³, area perpendicular to motion as 0.11 m² and C_D as 1.8, determine the fall velocity.

$$F = C_D A \frac{\rho U_0^2}{2}$$

$$\therefore \quad U_0^2 = \frac{800 \times 2}{1.8 \times 0.11 \times 1.2} = 6734.0 \text{ m}^2/\text{s}^2$$

$$\therefore \quad U_0 = 82.06 \text{ m/s}$$

12.12 The vertical component of the landing speed of a parachute is 6.0 m/s. Treat the parachute as an open hemisphere (Fig. 12.15) and determine its diameter if the total weight to be carried is 1500 N. Assume ρ to be 1.208 kg/m³.

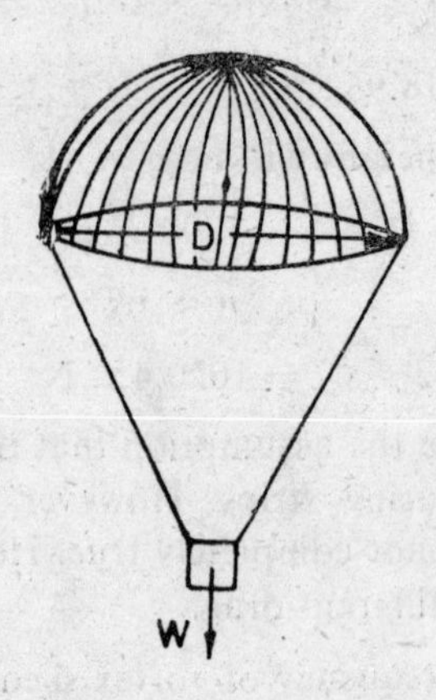

Fig. 12.15. Parachute.

Refer Table 22.4; $C_D = 1.33$

$$\therefore \quad 1500 = C_D \times \frac{\pi D^2}{4} \times \frac{\rho U_0^2}{2}$$

$$\therefore \quad D^2 = \frac{1500 \times 4 \times 2}{1.33 \times 3.142 \times 1.208 \times 6^2}$$

$$= 66.031 \text{ m}^2$$

$$\therefore \quad D = 8.126 \text{ m}$$

12.13 A rotary mixer is constructed from two circular discs (Fig. 12.16).

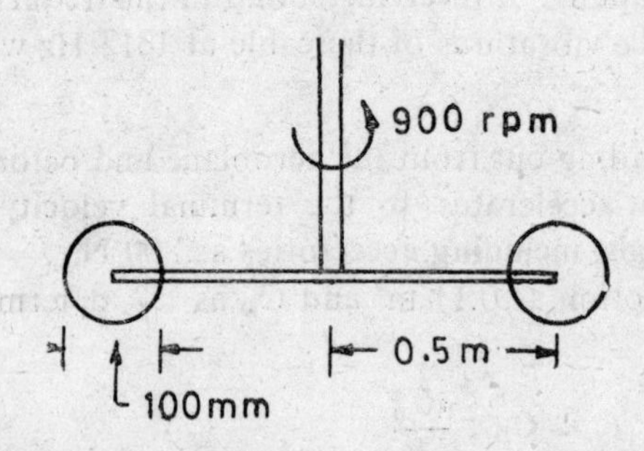

Fig. 12.16. Rotary mixer.

The mixer is rotated at 90 rpm in a large vessel containing liquid of relative density 1.1. Neglecting the drag on the rod and motion induced in the liquid, calculate the torque and energy required to drive the mixer. Take $\mu = 1 \times 10^{-3}$ kg/m s.

Speed at the centre of disc $= R\omega$

$$= \frac{2\pi N}{60} R = \frac{2 \times 3.142 \times 90}{60} \times 0.5$$

$$= 4.713 \text{ m/s}$$

$$Re = \frac{U_0 D \rho}{\mu} = \frac{4.713 \times 0.10 \times 1.1 \times 998}{10^{-3}} = 5.174 \times 10^5$$

At this Re value, C_D for disc is 1.10

Force on each disc, $$F = \frac{1.1 \times 3.142 \times 0.1^2}{4} \times (1.1 \times 998) \times \frac{4.713^2}{2}$$

$$= 105.348 \text{ N}$$

Torque $T = 2 \times 105.348 \times 0.5$

$= 105.348$ Nm

Work done per second $$= T\omega = \frac{105.348 \times 2 \times 3.142 \times 90}{60}$$

$$= 993.01 \text{ W} \quad \text{or} \quad 0.993 \text{ kW}$$

12.14 Assuming the cross-sectional area of a passenger car to be 2.25 m² with a drag coefficient of 0.60, estimate the energy requirement at a speed of 60 km/hr. Assume the weight of car to be 30 kN and coefficient of friction between the road and the tyres to be 0.01.

Total resisting force will be the sum of aerodynamic drag on the car and friction at the road surface.

$$\therefore \qquad F = \frac{C_D A \rho U_0^2}{2} + \mu' W$$

where μ' is the coefficient of friction

$$U_0 = 60 \text{ km/hr} = \frac{60 \times 1000}{3600} = 16.67 \text{ m/s}$$

$$F = \left(0.60 \times 2.25 \times 1.208 \times \frac{16.67^2}{2}\right) + (30000 \times 0.01)$$

$$= 526.541 \text{ N}$$

$$P = FU_0 = 526.591 \times 16.67$$

$$= 8778.27 \text{ W} \quad \text{or} \quad 8.778 \text{ kW}$$

This is the power required at the wheels.

12.15 A racing car of frontal area 1.2 m² is designed for a maximum speed of 200 km/hr and with inflation pressure of 210 kN/m² in the tyres. If the car weighs 15 kN and $C_D = 0.30$, determine the power of car at 90 per cent mechanical efficiency. Use the equation from Prob. 12.16 for μ'.

$$U_0 = 200 \text{ km/hr} = 200 \times 1000/3600$$
$$= 55.56 \text{ m/s}$$

$$\mu' = 0.005 + \frac{1048}{210\ 000} + \frac{0.0944 \times 200^2}{210\ 000}$$
$$= 0.005 + 0.00499 + 0.001380$$
$$= 0.01137$$

$$F = C_D A \frac{\rho U_0^2}{2} + \mu' W$$
$$= \left(0.30 \times 1.2 \times 1.208 \times \frac{55.56^2}{2}\right) + (15000 \times 0.01137)$$
$$= 671.21 + 170.50$$
$$= 841.76 \text{ N}$$

$$\text{Engine Power} = \frac{841.76 \times 55.56}{0.90 \times 1000} = 51.96 \text{ kW}$$

12.16 Idealised pressure distribution around a two-dimensional triangular rod of side a is shown in Fig. 12.17 in terms of $\Delta p / \frac{\rho U_0^2}{2}$. Determine the drag coefficient.

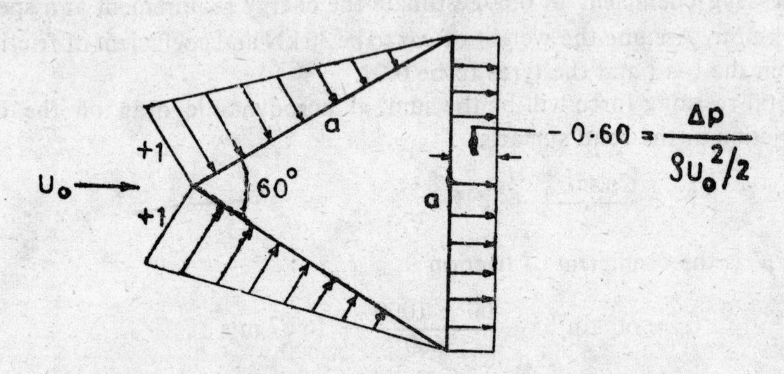

Fig. 12.17. Pressure distribution around triangular rod.

It can be seen that $\int \frac{\Delta p}{\frac{\rho U_0^2}{2}} dA \cos\theta$ on all the faces will give $\frac{F_D}{\rho U_0^2/2}$ and hence this divided by area perpendicular to flow will give C_D.

For unit length $\int \frac{\Delta p}{\frac{\rho U_0^2}{2}} dA \cos\theta = 0.5a \times \cos 60° + 0.5a \cos 60° - (-0.6a)$

$$= 1.1\ a$$

Area perpendicular to flow $= a \times 1$

$$\therefore \quad C_D = \frac{1.1a}{a \times 1} = 1.1$$

12.17 Total resistance and power requirement of a prototype ship are to be deduced from a 1 : 16 scale model tested under the following conditions:

area $= 0.40$ m², velocity $= 3.5$ m/s

length $= 3.0$ m, ρ for water $= 998$ kg/m³

ν at 20°C $= 1 \times 10^{-6}$ m²/s

Total resistance in model $= 10$ N

For the model

$$Re_L = \frac{3.0 \times 3.5}{10^{-6}} = 1.05 \times 10^7$$

$$\therefore \quad C_f = \frac{0.455}{[\log (1.05 \times 10^7)]^{2.58}} = 0.00298$$

$$\therefore \quad \text{Frictional force } F_{fm} = 0.00298 \times 0.40 \times \frac{998 \times 3.5^2}{2}$$

$$= 7.286 \text{ N}$$

$$\therefore \quad \text{Residual force } F_{rm} = 10.000 - 7.286$$

$$= 2.714 \text{ N}$$

For the prototype

$$U = 3.5 \times \sqrt{16} = 14.0 \text{ m/s}$$

$$L = 3 \times 16 = 48 \text{ m}, \qquad A = 0.4 \times 16^2 = 102.4 \text{ m}^2$$

$$R_{eL} = \frac{48 \times 14}{10^{-6}} = 6.72 \times 10^8$$

$$\therefore \quad C_f = \frac{0.455}{[\log (6.72 \times 10^8)]^{2.58}} = 0.00165$$

$$F_{fp} = (0.00165 \times 102.4 \times 998 \times 14.0^2)/2$$

$$= 16\,534.98 \text{ N} \quad \text{or} \quad 16.535 \text{ kN}$$

Force scale $= \rho_r l_r^3 = 16^3 = 4096$ as per Froude's law.

$$\therefore \quad F_{rp} = 2.714 \times 4096$$

$$= 11\,116.544 \text{ N} \quad \text{or} \quad 11.117 \text{ kN}$$

Total force $= 11.117 + 16.535$

$= 27.652$ kN

Power $= 27.652 \times 14$

$= 387.128$ kW

12.18 A circular cylinder of 2.0 m diameter and 12 m length is rotated at 300 rpm about its axis when it is kept in an air stream of 40 m/s velocity, with its axis perpendicular to the flow. Determine (i) circulation around the cylinder, (ii) theoretical lift, (iii) position of stagnation points and (iv) actual drag, lift and resultant force on the cylinder. Take $C_D = 0.52$, $C_L = 1.0$ (from Fig. 12.9) and $\rho = 1.208$ kg/m³.

$$u_c = \frac{\pi DN}{60} = \frac{3.142 \times 2.0 \times 300}{60} = 31.42 \text{ m/s}$$

$$\therefore \quad \frac{u_c}{U} = \frac{31.42}{40} = 0.7855$$

Circulation $\Gamma = 2\pi R u_c = 2 \times 3.142 \times 1.0 \times 31.42$

$$= 197.443 \text{ m}^2/\text{s}$$

Theoretical lift $F_L = \rho L U_0 \Gamma$

$$= 1.208 \times 12 \times 40 \times 197.443$$

$$= 114\,485.35 \text{ N} \quad \text{or} \quad 114.485 \text{ kN}$$

Points of stagnation

$$\frac{u}{U_0} = 2 \sin \theta + \frac{u_c}{U_0} \quad \text{and } u = 0$$

$$\therefore \quad \sin \theta = -\frac{1}{2}\frac{u_c}{U_0} = -\frac{0.7855}{2}$$

$$= -0.3928$$

$$\therefore \quad \theta = -23.13° \text{ and } 202.13°$$

Actual lift $F_L = C_L \times D \times L \dfrac{\rho U_0^2}{2}$

$$= 1.0 \times 2 \times 12 \times 1.208 \times \frac{40^2}{2}$$

$$= 23\,193.6 \text{ N}$$

$$F_D = 0.52 \times 2 \times 12 \times 1.208 \times \frac{40^2}{2}$$

$$= 12\,060.67 \text{ N}$$

$$F = \sqrt{F_L^2 + F_D^2} = 26141.97 \text{ N}$$

$$\tan \theta = \frac{F_L}{F_D} \quad \therefore \quad \theta = \tan^{-1} \frac{12\,060.67}{23\,193.60} = 27.47°$$

with the horizontal.

12.19 Wings of an aeroplane are replaced by two cylinders of 1.0 m diameter and 4.0 m length and it is proposed to make use of the lift caused by their rotation to lift the aeroplane. If the plane travels at 250 km/hr speed and weighs 80 000 N, determine the speed of rotation of the cylinders and the power required to overcome rotor friction.

$$U_0 = \frac{250 \times 1000}{3600} = 69.44 \text{ m/s}$$

$$W = F_L = C_L \times 2A \frac{\rho U_0^2}{2}$$

where A is area of each cylinder. Hence

$$80000 = C_L \times 2 \times 4 \times 1 \times 1.208 \times \frac{69.44^2}{2}$$

$\therefore$ $$C_L = 3.433$$

Figure 12.9 gives $\frac{u_c}{U_0} = 1.50$ and $C_D = 0.80$ for this value of C_L. Hence

$$u_c = 1.5 \times 69.44 = 104.16 \text{ m/s}$$

and $$104.16 = \frac{\pi \times 1.0 \times N}{60} \quad \therefore \quad N = \frac{104.16 \times 60}{3.142}$$

$\therefore$ $$N = 1989.17 \text{ rpm}$$

$$\text{Power } \frac{F_D U_0}{1000} = \frac{0.8 \times (2 \times 4 \times 1) \times 1.208 \times 69.44^3}{1000 \times 2}$$

$$= 1294.556 \text{ kW}$$

12.20 A kite has an effective area of 0.40 m² and weighs 2.0 N. In a wind of 40 km/hr, the drag on the kite is 11.9 N. Determine the tension in the cord if the cord makes an angle of 45° with the horizontal. Also determine the lift coefficient (Fig. 12.18).

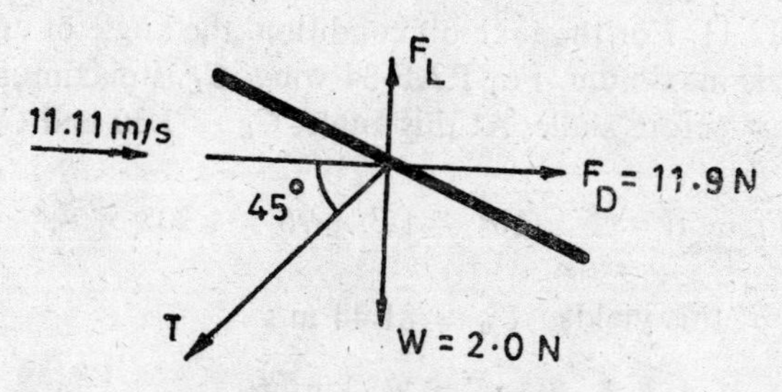

Fig. 12.18. Forces on a kite.

$$U_0 = 40 \times 1000/3600 = 11.11 \text{ m/s}$$

$$F_D = T \cos 45°$$

$\therefore$ $$T = 11.9 \times \sqrt{2} = 16.832 \text{ N}$$

Also $$F_L = W + T \cos 45°$$

$$= 2.00 + \frac{16.832}{\sqrt{2}}$$

$$= 13.9 \text{ N}$$

$$C_L = \frac{F_L}{A\rho U_0^2/2} = \frac{13.9 \times 2}{0.40 \times 1.208 \times 11.11^2} = 0.466$$

12.21 An aeroplane weighing 22 500 N has a wing area of 22.5 m² and span of 12.0 m. What is the lift coefficient if it travels at 320 km/hr in the horizontal direction? Also compute the theoretical value of circulation and angle of attack measured from zero lift axis.

$$U_0 = 320 \times 1000/3600 = 88.89 \text{ m/s}$$

$$\therefore \quad 22500 = 22.5 \times C_L \times 1.208 \times \frac{88.89^2}{2}$$

$$\therefore \quad C_L = 0.2095$$

But $\quad C_L = 2\pi \sin \theta_0 \quad \therefore \quad \theta_0 = \sin^{-1} \dfrac{0.2095}{2 \times 3.142}$

$$= 1.911°$$

$$\Gamma = \pi C U_0 \sin \theta_0 \quad \text{and} \quad C = \frac{22.5}{12} = 1.875 \text{ m}$$

$$= 3.142 \times 1.875 \times 88.89 \times \sin 1.911°$$

$$= 17.463 \text{ m}^2\text{/s}$$

12.22 An experimental plane is fitted with RAF-34 wing with total area of 96 m² and chord to length ratio of 1:6. If the plane weighs 5×10^5 N, determine the minimum take-off speed. How is the runway length related to this speed? Also determine the power required.

See Fig. 12.11. For the take-off condition, the angle of attack should be such that C_L is maximum. For RAF-34 wing, C_L is maximum at 19° angle of attack (just before stall). At this angle, $C_L = 1.30$ and $C_D = 0.16$.

$$\therefore \quad F_L = W = 5 \times 10^5 = 1.3 \times 96 \times 1.208 \times \frac{U_0^2}{2}$$

Solution of this yields $\quad U_0 = 81.44$ m/s

$$= 293 \text{ km/hr}$$

The power of the plane should be such that it accelerates the plane to achieve this velocity well within runway length S.

$$P = U_0 F_D = 0.16 \times 96 \times 1.208 \times \frac{81.94^3}{2} /1000 \text{ kW}$$

$$= 5011.20 \text{ kW}$$

If M is the mass of aeroplane whose velocity changes from 0 to U_0 in distance S, the constant acceleration

$$a = \frac{U_0^2}{2S} \quad \text{or} \quad S = \frac{U_0^2}{2a} \quad \text{but} \quad a = \frac{F_D}{M}$$

where M is the mass of plane.

$$\therefore \qquad S = \frac{U_0^2 M}{2F_D}$$

The actual distance provided will be larger in order to fulfill the condition that the plane should come to rest on the runway if in between some defect is developed and the pilot decides to stop the plane.

12.23 An aircraft has the following characteristics:

Weight = 13 500 N

Total wing area = 30 m²

Take-off speed = 30 m/s

The model tests have shown that C_D and C_L for the wing vary with the angle of attack θ_0 as

$$C_D = 0.008\,(1 + \theta_0)$$

$$C_L = 0.36(1 + 0.2\theta_0)$$

for small values of θ_0 measured in degrees. Determine the angle of attack that will ensure take-off at 30 m/s velocity and power required for take-off.

$$W = F_L = C_L A \frac{\rho U_0^2}{2}$$

$$\therefore \qquad C_L = (13500 \times 2)/1.208 \times 30 \times 30^2$$

$$= 0.828$$

$$\therefore \qquad 0.828 = 0.36(1 + 0.2\theta_0) \qquad \therefore \qquad \theta_0 = 6.5°$$

and $$C_D = 0.008(1 + 6.5) = 0.06$$

$$\therefore \qquad F_D = 0.06 \times 30 \times 1.208 \times \frac{30^2}{2} = 978.48 \text{ N}$$

$$P = \frac{978.48 \times 30}{1000} = 29.35 \text{ kW}$$

12.24 A supersonic plane travels at Mach two at a height of 18,000 m above the ground. How far ahead the plane will be when one hears the sonic boom on the ground? (Fig. 12.19).

$$\alpha = \sin^{-1} \tfrac{1}{2} \quad \text{or} \quad \alpha = 30°$$

If C is the observer's position and B is the position of plane when the sonic boom is heard, one must find AB

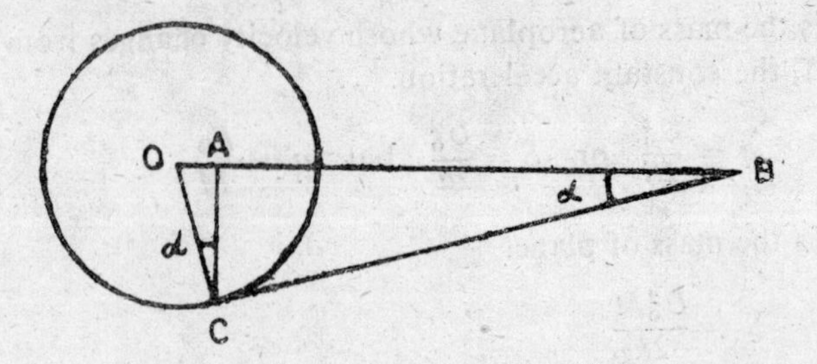

Fig. 12.19.

$$AB = \frac{AC}{\tan 30} = \frac{18}{0.5733} = 31.18 \text{ km}$$

12.25 A cup anemometer shown in Fig. 12.20 rotates freely without air friction. Calculate the speed of rotation against a wind speed of 15 km/hr.

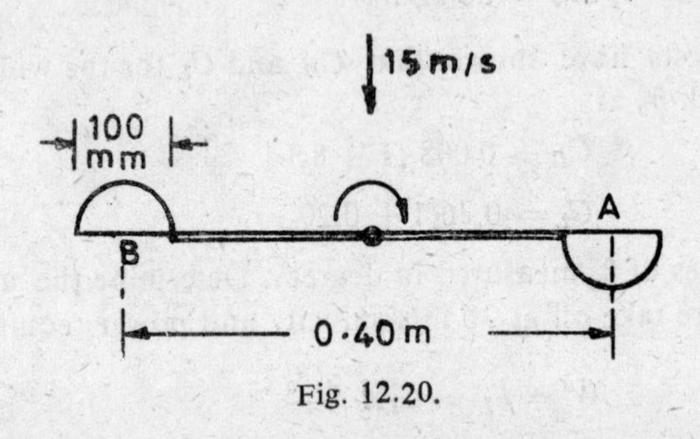

Fig. 12.20.

If the anemometer revolves at a uniform angular velocity ω, for steady rotation, net torque about the axis of rotation must be zero.

Fluid velocity relative to concave cup $A = 15 - 0.2\,\omega$

Fluid velocity relative to convex cup $B = 15 + 0.2\,\omega$

C_D **for cup** $A = 1.33$, C_D **for cup** $B = 0.34$

Corresponding drag forces on A **and** B **are**

$$F_{DA} = 1.33 \frac{A\rho}{2} (15 - 0.2\omega)^2$$

$$F_{DB} = 0.34 \frac{A\rho}{2} (15 + 0.2\omega)^2$$

Since net torque about 0 must be zero and since both the cups are at equal distance from 0

$$0.2\, F_{DA} = F_{DB} \times 0.2 \quad \text{or} \quad F_{DA} = F_{DB}$$

Hence $\frac{1.33}{1.34}(15-0.2\omega)^2 = (15+0.2\omega)^2$

$$29.667 - 0.3956\,\omega = 15 + 0.20\,\omega$$

$$14.667 = 0.5956\,\omega \quad \text{or} \quad \omega = 24.626 \text{ rad/s}$$

$$\therefore \qquad N = \frac{60\omega}{\pi D} = \frac{60 \times 24.626}{3.142 \times 0.4} = 1175.63 \text{ rpm}$$

12.26 A watercraft is fitted with a hydrofoil of area 0.70 m² to provide necessary lift to support the weight of the craft. If the drag and lift coefficients of the hydrofoil are 0.50 and 1.60 respectively, and weight of the craft is 18 000 N, determine the minimum speed when the water craft will be fully supported. What will be the power required to overcome hydrofoil resistance in water? If the craft is fitted with 90 kW engine, estimate the top speed.

The hydrofoil boats or water crafts are provided with one or two hydrofoils completely submerged in water (Fig. 12.21). As the boat speeds up,

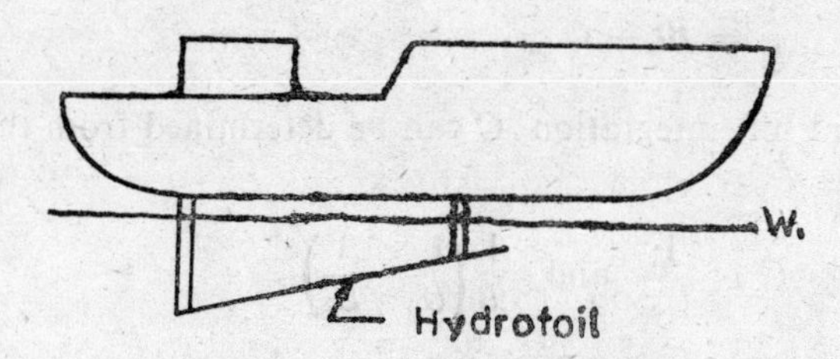

Fig. 12.21. Watercraft.

the lift on the hydrofoil increases and at a certain speed, the full weight of boat is supported by lift and the boat is out of water.

$$18\,000 = C_L A \frac{\rho U_0^2}{2}$$

$$\therefore \qquad U_0^2 = (2 \times 18\,000)/1.6 \times 0.7 \times 998 = 32.207$$

and $\qquad U_0 = 5.675$ m/s or 20.43 km/hr

$$F_D = \frac{C_D}{C_L} F_L = \frac{0.50 \times 18000}{1.60} = 5625 \text{ N}$$

$$P = F_D U_0 = \frac{5625 \times 5.675}{1000} = 31.921 \text{ kN}$$

At 90 kW power

$$90\,000 = F_D U_0 = C_D A \frac{\rho}{2} U_0^3$$

$$\therefore \quad U_0^3 = \frac{90\,000 \times 2}{0.5 \times 0.70 \times 998} = 515.316$$

$$\therefore \quad U_0 = 8.017 \text{ m/s} \quad \text{or} \quad 28.86 \text{ km/hr.}$$

12.27 An aircraft of mass M is slowed after landing by two parachutes employed from the rear. Each parachute is of diameter D and drag coefficient C_D. If the landing speed of the aircraft is U_0, estimate the time required for the aircraft to decelerate to a velocity $U_0/4$. Assume air resistance of aircraft to be negligible.

$$-M\frac{dU}{dt} = 2C_D \frac{\pi D^2}{2} \frac{\rho U^2}{2}$$

$$\therefore \quad \frac{dU}{dt} = -\left[\frac{\pi D^2 C_D \rho}{4M}\right] U^2 = -BU^2$$

where $B = \dfrac{\pi D^2 \rho C_D}{4M}$.

$$\therefore \quad \int \frac{dU}{U^2} = -\int B\, dt \quad \text{or} \quad -\frac{1}{U} = -Bt + C$$

or

$$\frac{1}{U} = Bt + C$$

The constant of integration C can be determined from the condition that $t = 0$, $U = U_0$

$$\therefore \quad C = \frac{1}{U_0} \quad \text{and} \quad \frac{1}{B}\left(\frac{1}{U} - \frac{1}{U_0}\right) = t$$

or

$$(U_0 - U) = BU\, U_0 t$$

or

$$U(1 + BU_0 t) = U_0$$

$$\therefore \quad \frac{U}{U_0} = \frac{1}{1 + BU_0 t}$$

Hence when

$$U/U_0 = 0.25, \quad 1 + BU_0 t = 4.0 \quad \text{and } t = \frac{3}{BU_0} = \frac{12M}{\pi D^2 C_D \rho U_0}$$

12.28 A light plane weighing 12.0 kN has a wing span of 9.0 m and chord length of 1.5 m. Assuming the lift characteristics of the wing are as shown in Fig. 12.11, determine the angle of attack for a take-off speed of 144 km/hr. Also determine the power required at take off speed if the parasite drag coefficient is 0.02. Further determine the stall speed.

$$U_0 = 144 \times 1000/3600 = 40.0 \text{ m/s}$$

$$\therefore \quad 12\,000 = C_L \times (9 \times 1.5) \times 1.203 \times \frac{40^2}{2}$$

$$\therefore \quad C_L = 0.920$$

From Fig. 12.11 for $C_L = 0.920$, angle of attack is 12°C.

Further for $C_L = 0.92$, $C_D = 0.07$ and $C_f = 0.02$

$$\therefore \quad \text{Power} = \frac{(0.07 + 0.02) \times (9 \times 1.5) \times 1.208}{1000} \times \frac{40^3}{2}$$

$$= 46.967 \text{ kW}$$

It can be seen that stall occurs at $C_L = 1.30$

$\therefore$ If U_0 is stall speed

$$12\,000 = 1.3 \times 9 \times 1.5 \times 1.208 \frac{U_0^2}{2}$$

$$\therefore \quad U_0^2 = 1132.054 \quad \text{and} \quad U_0 = 33.65 \text{ m/s}$$

PROBLEMS

12.1 A 3.0 mm diameter aluminium ball of relative density 2.8 falls under gravity in an oil of mass density 900 kg/m³ at a terminal fall velocity of 12 mm/s. Determine the dynamic viscosity of the oil. (0.774 kg/m s)

12.2 Obtain the relation between $F_D/\rho v^2$ and Re for a sphere within Stokes' range of Reynolds numbers. ($F_D/\rho v^2 = 9.426$ Re)

12.3 Determine the diameter of a sphere of relative density 2.65 which will have a fall velocity of 0.70 m/s in an oil of kinematic viscosity 1.9×10^{-4} m²/s and mass density of 900 kg/m³. (20 mm)

12.4 Determine the speed with which a 100 mm diamater sphere must travel in water at 20°C to experience a drag force of 5.0 N. Take $\mu = 10^{-3}$ kg/m s. (1.597 m/s)

12.5 Determine the terminal fall velocity of 50 mm diameter ball of relative density 3.50 dropped in oil of relative density 0.80 having dynamic viscosity of 10^{-1} kg/m s. (2.425 m/s)

12.6 A spherical ballon of 0.60 m diameter is filled with hydrogen. When held stationary in standard air, it exerts an upward force of 0.40 N. When it is anchored to ground with a weightless string, the string makes an angle of 60° with ground in a wind of 3.0 m/s velocity. Determine the drag coefficient of the balloon (Fig. 12.22). (0.52)

12.7 Determine the bending moment at the base of a 40 m high chimney of cylindrical shape of average diameter 2.5 m in a wind of uniform velocity 25 m/s. Assume air temperature to be 20°C. (264 250Nm)

12.8 Electrical transmission towers are loacted at 500 m spacing and 10 cables of 20 mm diamater are strung across them. Calculate the force on each tower when wind at 80 km/hr blows perpendicular

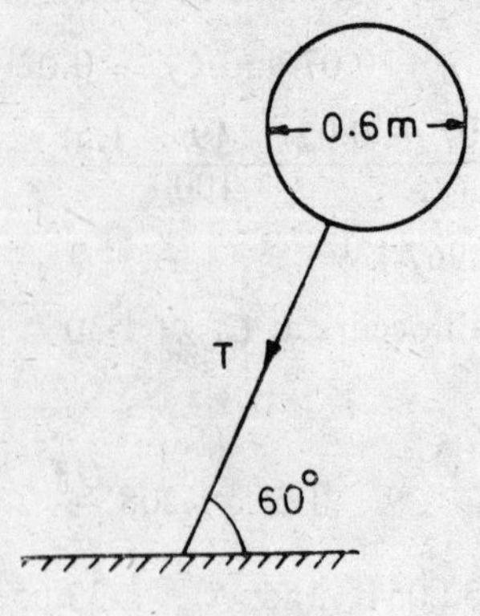

Fig. 12.22

to the cables. Assume $\rho = 1.20$ and $\mu = 1.7 \times 10^{-5}$ kg/m s. Also determine the frequency of lateral vibrations of cables.

(35.555 kN, 220 Hz)

12.9 A 15.0 m diameter spherical water tank rests on the top of 20 m high cylindrical vertical tower of 1.0 m diameter. Estimate the bending moment at its base for a wind velocity of 100 km/hr, if there is no interference between the tank and the tower. (Find C_D from the graph or use $C_D = 0.25$ for sphere and 0.35 for cylinder). (598.635 kN m)

12.10 The grill of the air cooler consists of 4.0 mm diameter light rods kept 10 mm apart centre to centre. What is the frequency of vortex shedding behind these rods at an air speed of 2.0 m/s. Assume air temperature to be 15°C. (96 Hz)

12.11 Determine the magnitude of wind force on a human being in standing position when he faces wind at 100 km/hr. Assume human body to be cylindrical in shape with frontal area of 0.80 m^2 and C_D as 1.6. (592. 68 N)

12.12 What will be the load carried by a parachute of diameter 6.0 m, if the velocity of descent is to be limited to 7.0 m/s? Take $\rho = 1.208$ kg/m^3. (1112.39 N)

12.13 Calculate the diameter of a parachute for dropping an object weighing 980 N so that it does not fall down at velocity greater than 5.0 m/s. Take $\rho = 1.208$ kg/m^3. (7.881 m)

12.14 Wind velocity is measeard by use of a cup anemometer which consists of two hollow hemispheres mounted in the opposite directions at the ends of a horizontal rod which turns freely about a vertical axis (Fig.12.20). What torque will be required to hold the rotating anemometer stationary at 60 km/hr wind speed. (Take C_D values from Table 12.4). (0.261 Nm)

12.15 A spherical balloon containing helium ascends through air at atmospheric pressure and 20°C. If the balloon and its payload weigh 15 N, determine the diameter of the balloon required so that the velocity of ascent is 1.0 m/s. (4.0 m)

12.16 Relation between the coefficient of rolling resistance of a motor vehicle, inflation pressure p of the tyres and vehicle speed U_0 is given by the empirical equation

$$\mu' = 0.005 + \frac{0.15}{p} + \frac{0.000\,035\,U_0^2}{p}$$

where μ' = coefficient of rolling resistance, and p and U_0 are in lb/in^2 and miles/hr respectively. Convert this formula so that p is expressed in N/m^2 and U_0 in km/hr.

$$\left(\mu' = 0.005 + \frac{1048}{p} + \frac{0.0944}{p} U_0^2\right)$$

12.17 A passenger car with frontal area of 2.4 m^2 has an inflation pressure of 200 000 N/m^2 in its tyres and travels at 80 km/hr. If the drag coefficiont of the car is 0.40, determine the power required. Assume mechanical efficiency of 90 percent and use formula in problem 12.16 for coefficient of rolling resistance. (11.735 kW)

12.18 A racing car with a frontal area of 1.40m^2, engine efficiency of 90 percent and with inflation pressure in tyres of 210 kN/m^2 needs 45 kW power at 200 km/hr design speed. If the car weighs 15 kN, determine the drag coefficient for the car. Use equation in Prob. 12.16 for μ'. (0.214)

12.19 A 750 mm diameter pipe of 300 m length and carrying gas was suspended on cables to carry it across the river. When subjected to 30 km/hr wind across it, it was found to sway up and down by about 1.5 m. Estimate its frequency of osciliation. What will be its natural frequency? (2.22 Hz, about 2.0 Hz)

12.20 Dimensionless pressure distribution on the three faces of triangular rod is shown in Fig. 12.23. Obtain C_D. (1.90)

12.21 A destroyer with 125 m length at water line and 3000 m^2 wetted area is cruising at 12.0 m/s velocity in sea water at 20°C (r.d = 1.025 and $\mu = 10^{-3}$ kg/ms). A 1:25 scale model designed using Froude's law gave a total drag of 50 N. Estimate the power required by the destroyer. (5831.98 kW)

12.22 A circular cylinder of 1.0 m diameter and 10 m length is rotated at 420 rpm about its axis when it is kept in an air stream with 11.0 m/s velocity perpendicular to its axis. Determine (i) circulation around the cylinder, (ii) theoretical lift and lift coefficient, (iii) positions of

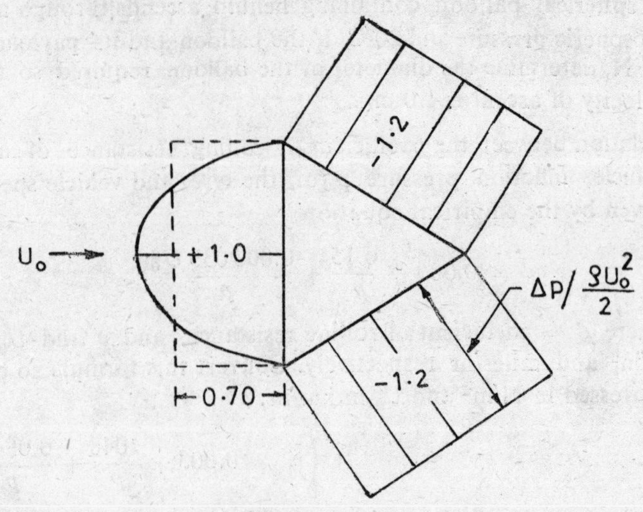

Fig. 12.23.

stagnation points, (iv) actual drag and lift force on the cylinder and (v) actual resultant force and its direction.

Take $\rho = 1.208$, and experimental values of C_D and C_L as 1.5 and 5.1 respectively.

(69.105 m²/s, 9182.672 N, 12.565, – 90°, 1096.26 N, 3727.28 N, 3885.155 N, 73.61°).

12.23 Wings of an aeroplane are fitted with two, 4 m long cylinders of 1.0 m diameter which rotate to cause lift on the plane. If 3711 kW power is required to overcome rotor friction when the plane travels at 288 km/hr speed, determine the speed of rotation of cylinders and the weight of the plane. (3055 rpm, 160.809 kN)

12.24 Calculate the ratio of lift force to drag force on a kite weighing 1.5 N if the tension in the cord is 6.0 N when the wind blows horizontally. The cord makes an angle of 35° with the horizontal. (1.616)

12.25 A square kite of area 0.64 m² weighs 3.922 N and makes an angle of 12° with the horizontal. The cord attached to the kite makes 45° angle with the horizontal. At a speed of 30 km/hr, the tension in the cord is 24.515 N: determine C_D and C_L if air weighs 12.257 N/m³. (0.624, 0.765)

12.26 For RAF-34 wing of aspect ratio six and chord length 2.0 m, determine the angle of attack of the wing at which lift to drag force ratio will be maximum. If the plane travels at 360 km/hr speed, determine the lift and total drag force if the parasite drag coefficient of the plane is 0.005. Also determine the induced drag coefficient.

(Note: Draw a tangent on polar diagram starting from zero lift to meet the polar diagram. This gives C_L/C_D maximum. For this $\theta° = 5°$, $C_D = 0.02$, $C_L = 0.45$). (5°, 65 232 N, 3654 N, 0.0107)

12.27 An experimental plane fitted with RAF-34 wing of aspect ratio 1:6 and area 96 m² has a take-off speed of 240 km/hr. What will be the maximum permissible weight of the plane? (335 kN)

12.28 A kite weighing 13 N has an area of 1.0 m². If it is flown in air at horizontal velocity of 10 m/s, the kite makes an angle of 8° with the horizontal. Assume C_L to be given by $C_L = 2\pi \sin \theta_0$. If the cord makes an angle of 45° with the horizontal, determine the tension T in the chord and C_D. (56.365 N, 0.66)

12.29 A supersonic plane travels at an elevation of 15 km above the ground surface. When the sonic bang was heard by the observer on the ground, the plane was 16.77 km ahead of him. Determine the speed of the plane if velocity of sound is taken as 290 m/s. (1566 km/hr)

12.30 A fighter plane passes by at a Mach number of 2.0 and height 10 km. How far ahead must one try to look for it when one hears the sonic boom? (17.32 km)

12.31 A ship is propelled by two cylindrical rotors of 1.2 m in diameter and 5.0 m in height with their axis vertical and rotating in air at 300 rpm (Fig. 12.24). If the ship is exposed to 40 km/hr wind flowing at 40° to the axis of ship, determine the longitudinal forces on the rotors and hence on the ship. (1592.952 N)

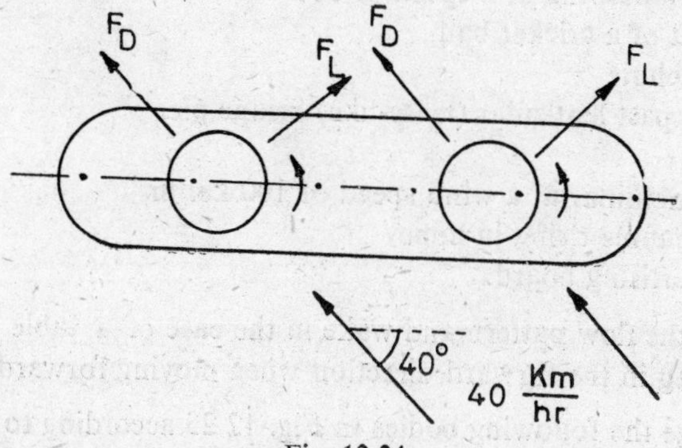

Fig. 12.24.

12.32 A watercraft is fitted with a hydrofoil of aspect ratio 4.0 and an angle of attack of 4° having C_L and C_D as 0.60 and 0.10 respectively. If the boat weighs 12 000 N, determine the hydrofoil dimensions and power required to overcome frictional resistance at 60 km/hr speed. (0.76 m, 0.19 m, 33.287 kW)

12.33 A wing of an aeroplane is rectangular in plan having a span of 10 m and chord of 1.20 m. In straight level flight at 300 km/hr, the total aerodynamic force on the wing is 31.25 kN. If C_L/C_D ratio for the wing is ten, calculate C_L. Assume $\rho = 1.208$ kg/m³. (0.618)

12.34 An experimental plane weighs 30 kN and has a wing of 1.826 m chord and 10.953 m span. When flying level at 216 km/hr, determine the power required to overcome friction if parasite drag coefficient is 0.015. Take $\rho = 1.1$ kg/m³. What is the angle of attack? (135.432 kW, 9.5°)

12.35 A small plane weighing 50 000 N has a wing area of 24 m² and span of 12.0 m. Determine the lift coefficient if the plane travels at 320 km/hr under level flight. Assume $\rho = 1.2$ and parasite drag coefficient as 0.01. Determine actual lift coefficient and corresponding angle of attack using Fig. 12.11. What is the power required? Also determine theoretical angle of attack and circulation. (0.44, 5°, 303.409 kW, 4.015°, 30.10 m²/s)

DESCRIPTIVE QUESTIONS

12.1 Why are cars not perfectly streamlined?

12.2 Differentiate between deformation drag, friction drag and form drag.

12.3 What type of drag predominates in the following cases:

—mist droplet falling in air
—air plane wing at a speed of 300 km/hr
—flight of a cricket ball
—parachute
—flow past lenticular (lense-like) bridge pier
—fish
—Kutubminar at a wind speed of 100 km/hr
—air bubble rising in honey
—advertising board.

12.4 Draw the flow pattern and wake in the case of a table tennis ball spinning in the forward direction when moving forward.

12.5 Arrange the following bodies in Fig. 12.25 according to decreasing order of magnitude of C_D, when kept in strong wind.

12.6 What do you conclude from the fact that at *Re* values less than 0.20, C_D for a sphere and a disc are nearly the same, viz.,

$$C_D = \frac{24}{Re} \text{ and } C_D = \frac{20.37}{Re} \text{ ?}$$

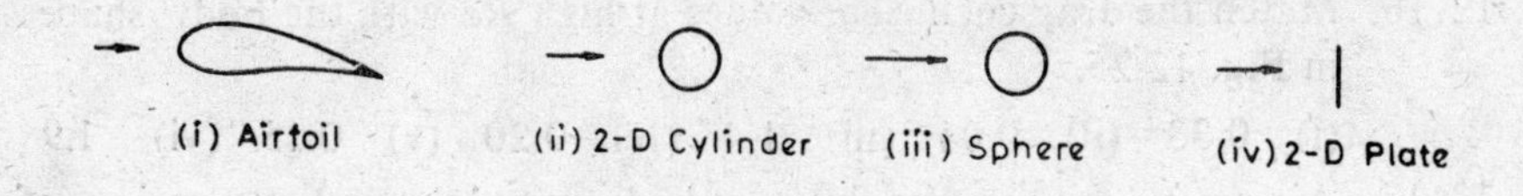

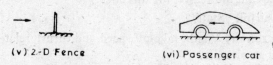

Fig. 12.25.

12.7 Give two examples where formation of Karman vortex trait produces undesirable effects.

12.8 Will it be easier to swim in fresh water or in sea water? Why? In which case will it be easier to float?

12.9 Why does a seem bowler hold the ball with the seam inclined to direction of flow?

12.10 A two dimensional cylinder is released along an inclined plane along which it rolls and drops down (Fig. 12.26). Trace its trajectory, as it leaves the plane.

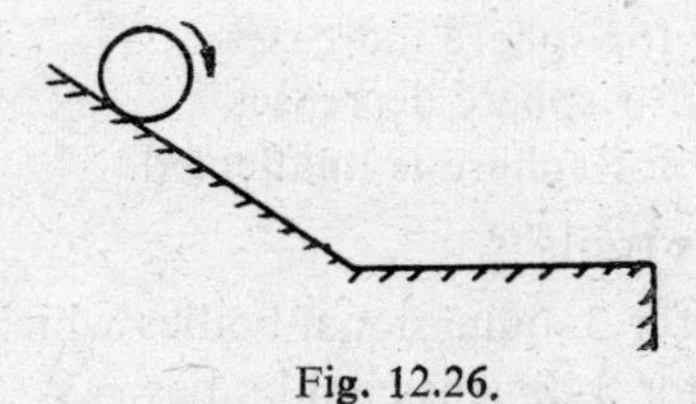

Fig. 12.26.

12.11 Sketch C_D vs *Re* diagram for a smooth and rough sphere in the *Re* range of 10^4 to 10^6 and explain the differences between the two.

12.12 What is the effect of finite value of L/C ratio of an airfoil on its drag and lift?

12.13 Does streamlining of the rear part of a body help in reduction of drag when (i) flow is subsonic (ii) flow is supersonic? Explain.

12.14 What is the origin of musical sound often heard when strong wind blows past electric cables?

12.15 Which is more advantageous for landing of planes:

(i) when wind is in the same direction?
(ii) against the wind?

Explain.

12.16 Match the drag coefficient values at high *Re* with the body shapes in Fig. 12.25.

(i) 0.33 (ii) 0.04 (iii) 1.4 (iv) 0.20 (v) 0.15 (vi) 1.9

12.17 Air crafts and missiles flying at high altitude may have region of laminar flow that becomes turbulent at lower altitude at the same speed. Explain the reason for this effect.

12.18 What is airfoil "Stall"?

12.19 How are runway length, take-off speed and thrust related to each other for an aircraft?

12.20 Deformation drag

(i) is independent of body length
(ii) occurs at high *Re*
(iii) depends mainly on cross-sectional shape
(iv) is primarily a friction drag
(v) occurs when *Re* is very small.

12.21 The drag coefficient at small *Re* values

(i) increases as *Re* increases
(ii) decreases as *Re* increases
(iii) remains constant
(iv) none of the above.

12.22 When turbulence level in the flow is increased,

(i) critical *Re* for sphere increases
(ii) critical *Re* for sphere decreases
(iii) critical *Re* for sphere is unaffected.

12.23 Karman vortex trail

(i) is formed for 2-dimensional bodies which are not streamlined
(ii) is formed for 3-dimensional bodies
(iii) induces longitudinal vibrations in the body
(iv) induces lateral vibrations in the body.

12.24 Suggest means of suppressing formation of Karman vortex trail behind a circular cylinder.

12.25 Draw pressure distribution around a sphere at *Re* values of 10^4, 10^6 and in ideal fluid.

12.26 Distinguish between residual drag and induced drag.

12.27 For a given body shape, at high *Re* values

(i) Form drag is proportional to U_0
(ii) Form drag is constant
(iii) Form drag is proportional to U_0^2.

12.28 If cricket ball is roughened on only one side, what will be its possible effect on the flight of the ball?

CHAPTER XIII

Open Channel Flow

13.1 SOME DEFINITIONS

A channel with constant bed slope and the same cross-section along its length is known as a prismatic channel.

When the depth or velocity of flow in a prismatic channel remains the same along its length at any given time, the flow is known as uniform flow; otherwise it is called nonuniform flow. Hence

$$\left(\frac{\partial y}{\partial x} \text{ or } \frac{\partial U}{\partial x}\right)_{t=t_0} = 0 \quad \text{for uniform flow}$$

$$\left(\frac{\partial y}{\partial x} \text{ or } \frac{\partial U}{\partial x}\right)_{t=t_0} \neq 0 \quad \text{for nonuniform flow}$$

In a channel, when any flow variable such as depth y or velocity U or discharge Q remains unchanged with respect to time at a given section, the flow is known as steady flow; otherwise it is called unsteady flow.

$$\left(\frac{\partial y}{\partial t}, \frac{\partial U}{\partial t} \text{ or } \frac{\partial Q}{\partial t}\right)_{x=x_0} = 0 \quad \text{for steady flow}$$

$$\left(\frac{\partial y}{\partial t}, \frac{\partial U}{\partial t} \text{ or } \frac{\partial Q}{\partial t}\right)_{x=x_0} \neq 0 \quad \text{for unsteady flow.}$$

$$\text{Hydraulic radius } R = \frac{\text{cross-sectional area of flow } A}{\text{wetted perimeter } P}$$

$$\text{Hydraulic mean depth } D = \frac{\text{cross-sectional area of flow } A}{\text{water surface width } T}$$

13.2 STEADY UNIFORM FLOW

In the case of steady uniform flow in a prismatic channel, channel slope S_0, water surface slope S_w and slope of total energy line S_f are equal.

For steady uniform flow, the average velocity U, hydraulic radius R and channel slope S_0 are related by the resistance laws of Manning, Chezy or Darcy-Weisbach.

Manning's equation: $$U = \frac{1}{n} R^{2/3} S_0^{1/2} \tag{13.1}$$

Chezy's equation: $$U = C\sqrt{RS_0} \tag{13.2}$$

Darcy-Weisbach equation: $$U = \sqrt{\frac{8g}{f}}\sqrt{RS_0} \tag{13.3}$$

Here n is Manning's roughness coefficient, C is Chezy's discharge coefficient and f is Darcy Weisbach friction factor. In nonuniform flow, the same equations are used for the determination of velocity at any section by replacing S_0 by the local friction slope, S_f. The most commonly used resistance equation is the Manning's equation.

The depth of flow in a prismatic channel, for steady uniform flow, for given values of Q, S_0 and n (or C or f) is known as the normal depth and is designated as y_0. Manning's n values vary from 0.011 for smooth plastered surface to about 0.08 for natural streams with boulder bed or with excessive weeds. Table 13.1 lists some typical values of n.

Table 13.1. Average values of Manning's n

Closed conduits flowing partly fully	0.013 – 0.014
Lined channels	0.011 – 0.027
Excavated earth channels	0.018 – 0.040
Natural streams	0.03 – 0.15

For hydrodynamically rough plane surface, Strickler's equation

$$n = d_{50}^{1/6}/21 \tag{13.4}$$

can be used to predict n; here d_{50} is the median size of bed material in *m*. This is valid for R/d_{50} varying from 10 to 700. Darcy—Weisbach friction factor f for channels is given by the following equations:

Hydrodynamically smooth surface

$$1/\sqrt{f} = 2 \log \frac{4UR}{\nu}\sqrt{f} - 0.80 \tag{13.5}$$

Hydrodynamically rough surface

$$1/\sqrt{f} = 2 \log \frac{R}{k_s} + 2.34 \tag{13.6}$$

The velocity distribution in smooth and rough channels is given by

$$\frac{u}{u_*} = 5.75 \log \frac{u_* Y}{\nu} + 5.50 \tag{13.7}$$

and

$$\frac{u}{u_*} = 5.75 \log \frac{Y}{k_s} + 8.50 \tag{13.8}$$

Following types of problems are encountered as regards computations for steady uniform flow:

1. Given y_0, n, S_0 and channel shape, find Q. First determine R and then U using Manning's equation. Finally $Q = UA$.
2. Given Q, y_0, S_0 and channel shape, find n and U. Determine A and then $U = Q/A$. Use Manning's equation and determine n.
3. Given Q, y_0, n and channel shape, find S_0 and U. Find A and then $U = Q/A$. Use Manning's equation and determine S_0.
4. Given Q, S_0 and n and channel shape, determine y_0 and U.

$$Q = \frac{1}{n} AR^{2/3} S_0^{1/2}$$

Hence,

$$\frac{Qn}{S_0^{1/2}} = AR^{2/3} = F(y_0) \tag{13.9}$$

The term $AR^{2/3}$ is known as the section factor Z_1. Since $Qn/S_0^{1/2}$ is known, plot $AR^{2/3}$ vs. y and choose $y = y_0$ when $AR^{2/3} = Qn/S_0^{1/2}$. For y_0 find A and $U = Q/A$. The term $\frac{AR^{2/3}}{n}$ is called conveyance and is denoted by K.

5. The design of lined channels can be carried by choosing a channel width which is found economical by field engineers. This is normally related to design discharge Q as follows:

Q (m³/s)	114	57	29	20	4.3	1.4
Bed width B (m)	9.0	4.6	3.4	2.75	1.83	1.10

This can be approximated by the equation

$$B = 0.70\sqrt{Q}$$

Side slope of the trapezoidal channel has to be adopted considering the nature of the soil and the freeboard as specified below is recommended.

Discharge less than 1.5 m³/s	0.5 m
Discharge between 1.5 and 75 m³/s	0.75 m
Discharge greater than 75 m³/s	1.0 m

13.3 MOST EFFICIENT CHANNEL SECTION

A channel section is known as most efficient when it carries maximum discharge for given area A, S_0 and n or C. This means that for given area A, R will be maximum. Under such condition $R = \frac{y_0}{2}$ for rectangular channel and trapezoidal channel with given side slope. For trapezoidal channel with given depth, the most efficient section is obtained with side slope of inclination of 60° to the horizontal.

13.4 SPECIFIC ENERGY AND SPECIFIC FORCE

Specific energy is defined as the energy per unit weight of liquid, using channel bottom as the datum. It is given by

$$H = y + \frac{U^2}{2g} = y + \frac{Q^2}{2gA^2} \tag{13.10}$$

which for rectangular channel becomes

$$H = y + \frac{q^2}{2gy^2} \tag{13.11}$$

where $q = Q/B$ is the discharge per unit width of channel.

Figure 13.1(a) shows variation of H with y for a given Q. It can be seen

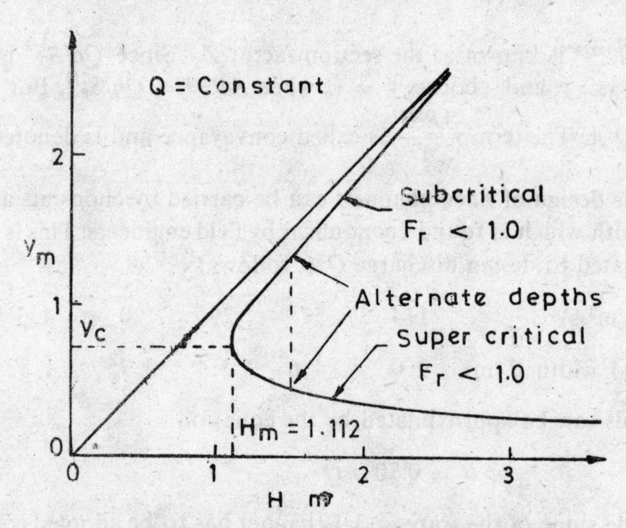

Fig. 13.1 (a). Variation of H with y for given Q.

that at $y = y_c$, H is minimum and equal to H_m. This depth at which for given Q, H is minimum is known as critical depth y_c. It can be shown that

$$H_m = \tfrac{3}{2}\, y_c$$

and at critical depth $Q^2T/gA^3 = 1$.

For rectangular channel this relation gives $y_c = \sqrt[3]{q^2/g}$, whereas for triangular channel $y_c = \sqrt[5]{2Q^2/(g\tan^2\theta)}$ where θ is half apex angle. Also the relation $Q^2T/gA^3 = 1$ can be written as

$$\frac{Q^2}{gA^2}\frac{T}{A} \text{ or } \frac{U^2}{gD} = 1$$

where D is hydraulic mean depth.

One can also study variation of Q or q with y for given H using Eq. 13.10 or 13.11 [Fig. 13.1(b)]. For depth y_c, Q is maximum and this condition is obtained when $Q^2T/gA^3 = 1.0$.

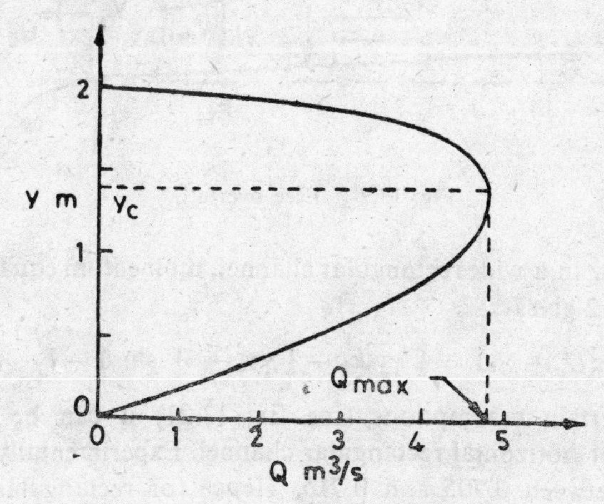

Fig. 13.1 (b). Variation of Q with y for given H.

When y is greater than y_c, flow is called subcritical or tranquil, while it is called supercritical or rapid when y is less than y_c. Two depths are possible for the same Q and same H—one subcritical and other supercritical. They are known as alternate depths.

Specific force, denoted by P, is defined as the force per unit weight of liquid and is given by

$$P = \frac{Q^2}{Ag} + A\bar{y} \tag{13.12}$$

where $\bar{y}$ is the depth of centroid of area A below the liquid surface. For given discharge, P is minimum when $Q^2T/gA^3 = 1$, i.e. at critical depth. Similarly for a given specific force P, Q is maximum at the critical depth. For given Q there are two depths for which P is same. They are known as conjugate depths. It may be noted that critical depths obtained from specific energy and specific force consideration are equal if momentum and energy correction factors are unity and pressure distribution in the vertical is hydrostatic.

13.5 BRINK DEPTH, BROAD CRESTED WEIRS, CONTRACTING TRANSITIONS

Brink depth, flow over broad crested weir, flow in transitions and in hydraulic jump are the examples of rapidly varied flow. A sudden drop in channel level is known as the free fall. Typical free overfall is shown in Fig. 13.2. The depth at the end of the free fall is known as brink depth y_b.

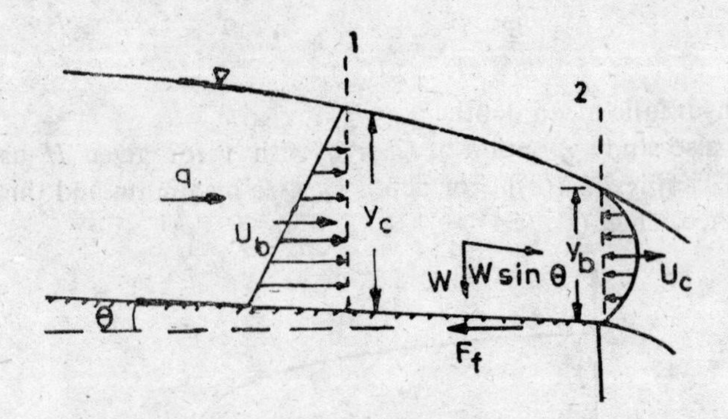

Fig. 13.2. Free overfall.

For steady flow in a wide rectangular channel, momentum equation between sections 1 and 2 gives

$$q\rho(U_c - U_b) = \tfrac{1}{2}\,\gamma y_b^2 k_b - \tfrac{1}{2}\,\gamma y_c^2 + W \sin\theta - F_f \qquad (13.13)$$

Making certain assumptions (see Ex. 13.21) it can be shown that $y_b/y_c = 2/3$, for horizontal rectangular channel. Experimentally this ratio is found to be between 0.705 and 0.715. Hence for rectangular channel of width B, one can write

$$Q = 1.654\, B g^{1/2} y_b^{3/2} \qquad (13.14)$$

Thus measurement of y_b gives an estimation of Q. For prismatic channels, the ratio of areas at brink depth and critical depth is given by

$$A_b/A_c = \frac{D_c}{D_c + \bar{y}_c}$$

where A_b and A_c are areas of cross-section at brink depth and critical depth, D_c is hydraulic mean depth at critical section and $\bar{y}_c$ is the depth of centroid of A_c below liquid surface.

Broad crested weir is an overflow structure with a horizontal crest of adequate length in the flow direction so that streamlines become parallel (Fig. 13.3). $2 \leqslant \dfrac{L}{H_1} \leqslant 12.5$.

Here the height W of weir is adequate so that critical depth occurs over the weir.

$$h_1 + \frac{U_1^2}{2g} = H_1 = y_c + \frac{U_c^2}{2g} \qquad (13.15)$$

and

$$y_c = \frac{2}{3}\, H_1$$

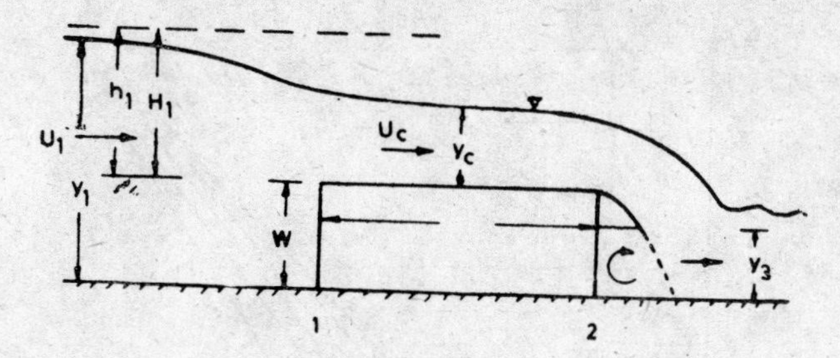

Fig. 13.3. Broad crested weir.

$$\therefore \qquad q = U_c y_c = \frac{2}{3}\sqrt{\frac{2}{3}g}\, H_1^{3/2} \tag{13.16}$$

$$Q = \frac{2}{3}\sqrt{\frac{2}{3}g}\, B\, H_1^{3/2}$$

where B is the width of the weir. If this equation is written in terms of h_1, one gets

$$Q = \frac{2}{3} C_d \sqrt{\frac{2}{3}g}\, B\, h_1^{3/2} \tag{13.17}$$

where

$$C_d = \left\{ 1 + \frac{\frac{4}{27} C_d^2}{\left(1 + \frac{W}{h_1}\right)^2} \right\}^{3/2} \tag{13.18}$$

The above equations are valid if y_3 is less than W, or y_3 is greater than W, but $(y_3 - W) < 0.80\, h_1$. Otherwise flow is submerged and for the same h_1 weir will carry less flow.

The contracting transitions considered here are of there types

1. Transition from wider to narrower channel
2. Gradual rise in the bed of prismatic channel, and
3. Combination of above two.

Transition from wider to narrower channel

Consider a contracting transition in a rectangular channel as shown in Fig. 13.4. Let y_1 be the subcritical depth of Section 1 corresponding specific energy curve is shown labelled as $q_1 = Q/B_1$. With reduction of B from B_1 to smaller widths at section 2, there will be corresponding increase in discharge per unit width; corresponding specific energy curves are shown as q_2, q_3, q_4, q_5. The vertical line AB cutting q_3 curve at B gives the corresponding depth y_2 at section 2; it is smaller than y_1. If width at section 2 is further reduced to B_4, one gets the specific energy curve q_4 and AB produced meets q_4 at B' giving critical depth $y_c = 3\sqrt{q_4^2/g}$ at section 2. Any further

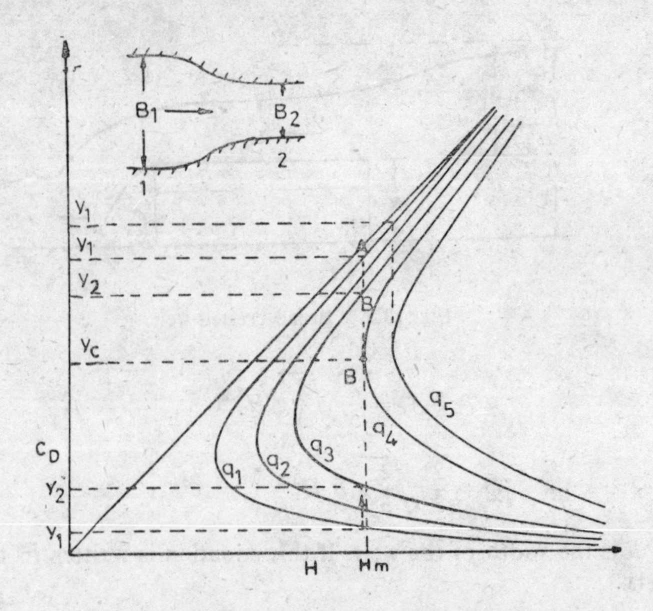

Fig. 13.4. Contracting transitions.

reduction in width at section 2 for given H will have $H < H_m$ for that discharge and flow cannot take place unless H is increased to H_m and y_1 to y_1' for that discharge q_5. For supercritical flow at section 1, flow at section 2 is supercritical and $y_2 > y_1$. When $B_2 = B_4$, $y_2 = y_c$, Any further reduction in B_2 causes radical changes in flow leading to hydraulic jump.

Transition with rise in bed level

One can write energy equation between sections 1 and 2 (Fig. 13.5) as

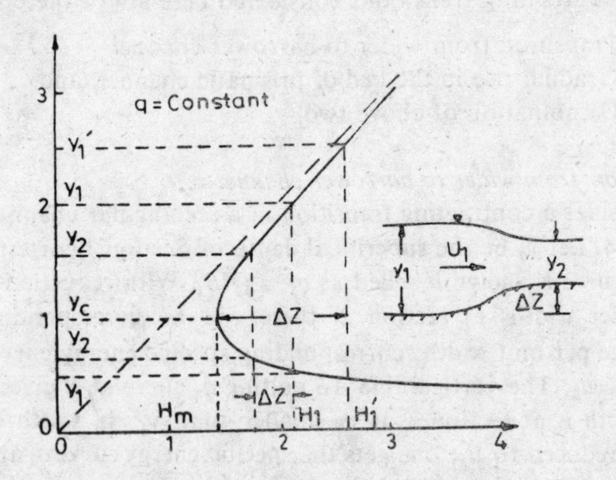

Fig. 13.5. Transition with rise in bed level.

$$\frac{q^2}{2gy_1^2} + y_1 = \frac{q^2}{2gy_2^2} + y_2 + \Delta Z \tag{13.19}$$

or $$(H_1 - \Delta Z) = H_2$$

For subcritical flow at section 1, as ΔZ increases, H_2 and hence y_2 decreases. When $H_1 - \Delta Z = H_m$, $y_2 = y_c$. Till such time y_1 is not affected by rise in bed level. If $(H_1 - \Delta Z) < H_m$, upstream depth increases to y_1' and H_1 to H_1' so that $(H_1' - \Delta Z) = H_{min}$.

When y_1 is supercritical, water surface will rise over the hump and y_2 will be greater than y_1. As ΔZ is increased, y_2 approaches y_c. When ΔZ is such that $(H_1 - \Delta Z)$ is less than H_{min}, radical changes take place in flow.

A transition where both contraction and rise in bed level are provided can be analysed by combining the two analysis presented above.

13.6 HYDRAULIC JUMP

If depth of flow in a channel is supercritical and there is adequate depth available in the downstream channel, the flow changes from supercritical to subscritical through a sudden rise in depth. This phenomenon is known as hydraulic jump which is analogous to shock wave in compressible fluids (Fig. 13.6). The ratio of $\frac{y_2}{y_1}$ and energy loss in a jump h_L can be determined

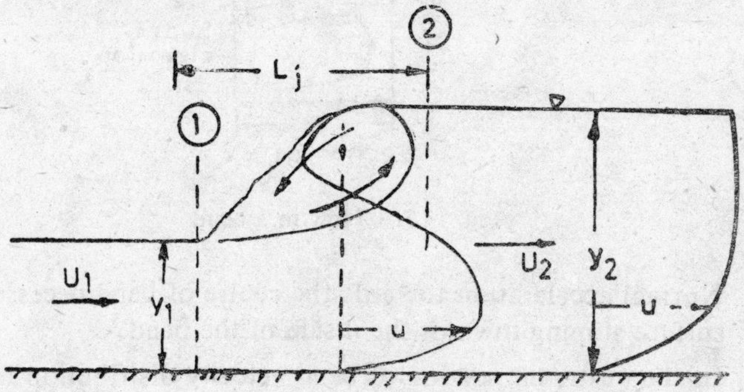

Fig. 13.6. Hydraulic jump.

using momentum, continuity and energy equations after making the following assumptions: (i) channel is horizontal, (ii) frictional force at the bottom is negligible, (iii) pressure distributions at sections 1 and 2 are hydrostatic and (iv) $\alpha = \beta = 1$ at both the sections. One can show that for rectangular channel,

$$\frac{y_2}{y_1} = \frac{1}{2}\left[\sqrt{1 + 8F_{r1}^2} - 1\right] \tag{13.20}$$

and $$h_L = \frac{(U_1 - U_2)^3}{2g(U_1 + U_2)} = \frac{(y_2 - y_1)^3}{4y_1y_2} \tag{13.21}$$

For jump in a nonrectangular channel or of irregular section, one can determine y_2 and h_L using specific force and specific energy diagrams. The dimensionless length of jump L_j/y_2 is a function of Froude number Fr_1. As an approximation, one can take

$$L_j = 6.9(y_2 - y_1) \tag{13.22}$$

13.7 FLOW IN A BEND

Subcritical flow in a rectangular channel bend is depicted in Fig. 13.7. The characteristics of flow in the bend are summarised below

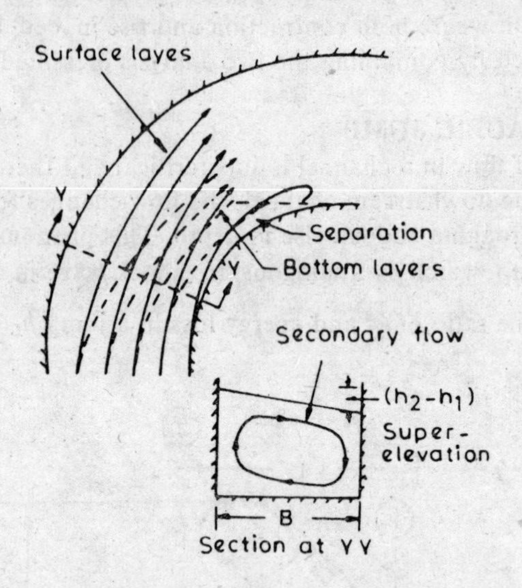

Fig. 13.7. Flow in a bend.

1. Normal acceleration towards the centre of bend necessitates water surface sloping towards the inside of the bend.
2. In the upstream portion of bend velocity distribution follows free vortex law, viz. $U \sim \frac{1}{r}$ while in the latter portion it is free vortex type $U \sim r$, where r is radius of streamline.
3. Secondary flow is generated which is towards the inside of bend near the bottom and towards outside near the surface.
4. Separation of flow on the inner side.
5. Greater energy loss due to secondary flow and increased turbulence level.

$$\text{Super-elevation } h_2 - h_1 = \frac{B}{r_c}\left[\frac{1}{1+\frac{B^2}{12Rr_c^2}}\right]\frac{U^2}{g} \tag{13.23}$$

Bend loss $h_L = K \frac{U^2}{2g}$ where K varies from 0.1 to 0.50; K is related to bend angle θ, $\frac{R_b}{B}$ and Re. Here r_c is the centreline bend radius.

13.8 GRADUALLY VARIED FLOW

Gradually varied flow is a steady nonuniform flow in which the depth of flow varies gradually along the length of the channel. Flow upstream of a weir, dam or sluice gate, flow upstream of an abrupt fall, flow at sudden changes in slope are but a few examples of gradually varied flow. If the depth of flow changes rapidly along the channel length, it is known as rapidly varied flow.

Starting from energy equation, one can develop the following differential equation (Fig. 13.8) for gradually varied flow.

$$\frac{dy}{dx} = \frac{S_0 - S_f}{1 - \frac{Q^2T}{gA^3}} = \frac{S_0 - S_f}{1 - F_r^2} \tag{13.24}$$

For wide rectangular channels, it takes the form

$$\frac{dy}{dx} = 1 - (y_0/y)^{0.3 \text{ or } 10/3} \Big/ 1 - \left(\frac{y_c}{y}\right)^3 \tag{13.25}$$

if one makes the following assumptions: (i) channel is prismatic; (ii) W.S. curvature is small so that pressure distribution in the vertical is hydrostatic, (iii) channel bottom slope is very small and (iv) Manning's or Chezy's equation can be used for gradually varied flow if S_0 is replaced by S_f and (v) $\alpha = 1.0$.

The exponent in Eq. 13.25 is 3.0 when Chezy's equation is used while it is 10/3 when Manning's equation is utilised.

Water surface profiles in gradually varied flow are first classified according to the slope on which they occur and then according to whether the depth lies between ∞ and y_0 and y_c or between y_0 and y_c, or below y_0 and y_c and bed :

C	Critical slope S_c	:	when $y_0 = y_c$
M	Mild slope	:	when $S_0 < S_c$
S	Steep slope	:	when $S_0 > S_c$
H	Horizontal	:	when $S_0 = 0$
A	Adverse		$S_0 < 0$
	Zone 1		$y > y_0$ and y_c
	Zone 2		y lies between y_0 and y_c
	Zone 3		$y < y_0$ and y_c

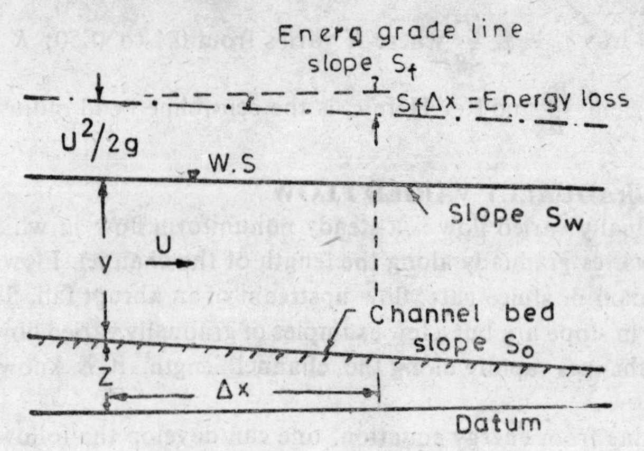

Fig. 13.8. (a) Definition sketch

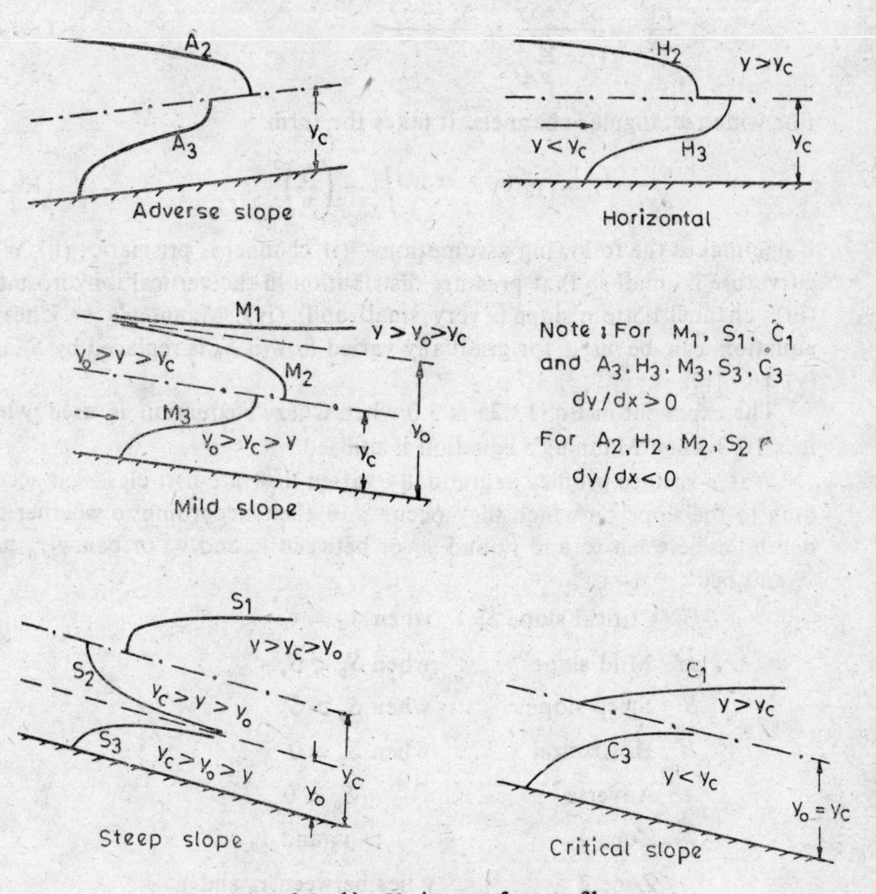

Fig. 13.8. (b) Various surface profiles.

Thus there are twelve possible profiles namely C_1, C_3, M_1, M_2, M_3, S_1, S_2, S_3, H_2, H_3, and A_2, A_3. These are shown in Fig. 13.8. Critical slope is that value of S_0 at which $y_0 = y_c$ for given Q, n or C and channel shape.

Control Section

Since a small disturbance in critical and supercritical flow cannot travel upstream, what happens in the downstream is controlled by the upstream condition in such flow. Hence rapid (supercritical) flow computations of gradually varied flow are started at upstream end. In subcritical (tranquil) flow, small changes at the downstream affect the condition upstream and hence are controlled by them. Therefore, subcritical flow computations in gradually varied flow must start at the downstream end. Control section is defined as one where there is a unique relationship beween discharge and depth. Sharp and broad crested weirs, gates, sudden drop, change in slope of channel from mild to steep, etc. act as control sections. In subcritical flow, downstream control prevails whereas in supercritical flow it is the upstream control that prevails.

Computation of surface profiles

The procedure for computations of surface profiles is as follows:

(i) Compute y_0, y_c, and determine whether the slope is mild, critical, steep, horizontal or adverse.

(ii) Locate control section and classify the profile and sketch it.

(iii) If in any reach, supercritical flow has to meet subcritical flow it will be through hydraulic jump.

Surface profile can now be computed by any one of the following methods.

1. Step method
2. Graphical integration method
3. Bresse's method
4. Chow's method.

These are explained through illustrative examples.

13.9 SURGES AND GRAVITY WAVES

A surge is a sudden change in depth of flow in a channel, which moves either upstream or downstream at velocity U_w. Moving hydraulic jump is a surge. Types of surges encountered in practice are shown in Fig. 13.9.

If U_w is constant these problems can be converted into steady state problems. Momentum equation and continuity equation applied together yield (for rectangular channel)

$$C = \sqrt{gy_1}\left\{1 + \frac{3}{2}\frac{h}{y_1}\right\}^{1/2} \tag{13.26}$$

where $h = y_2 - y_1$ = height of surge, and $C = U_w - U_1$ is known as celerily [Fig. 13.9 (a)].

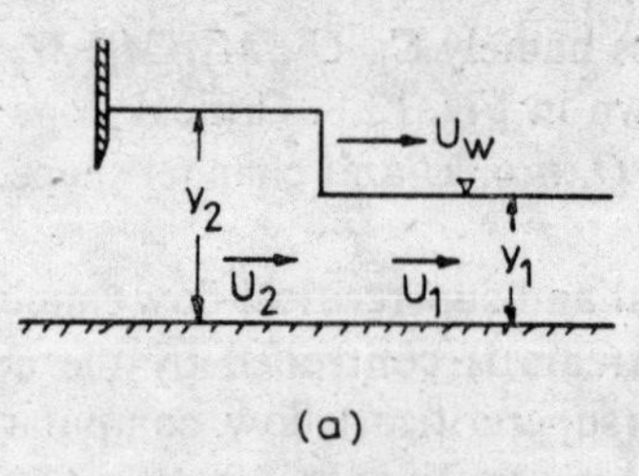

+ Ve surge moving d.s.

Occurrence:

(i) Sudden opening of gate at u.s. end

(ii) Sudden failure of dam

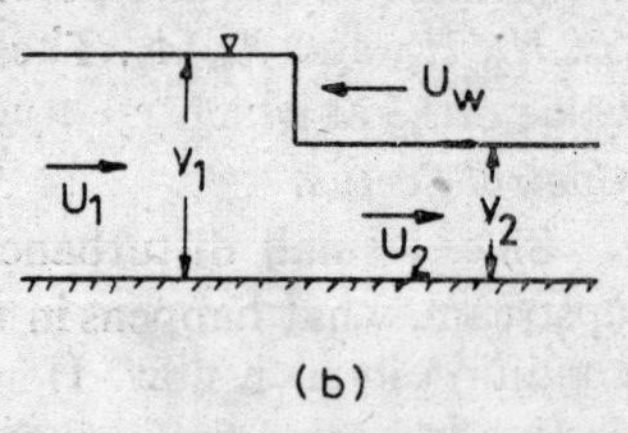

- Ve surge moving u.s.

Occurrence:

(i) Sudden opening of gate at d.s. end

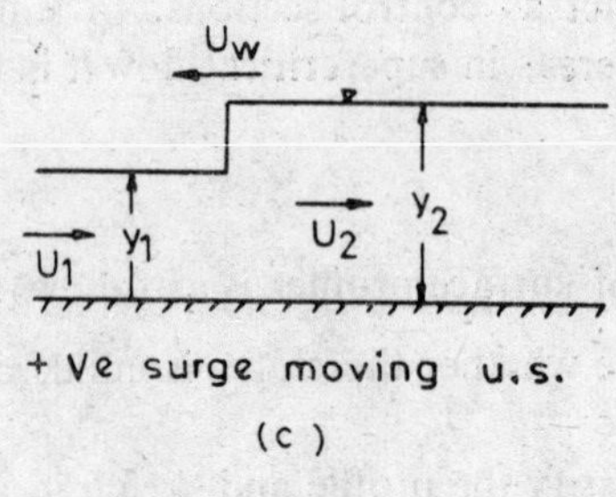

+ Ve surge moving u.s.

(c)

Occurence:

(i) Sudden closure of gate at d.s. end

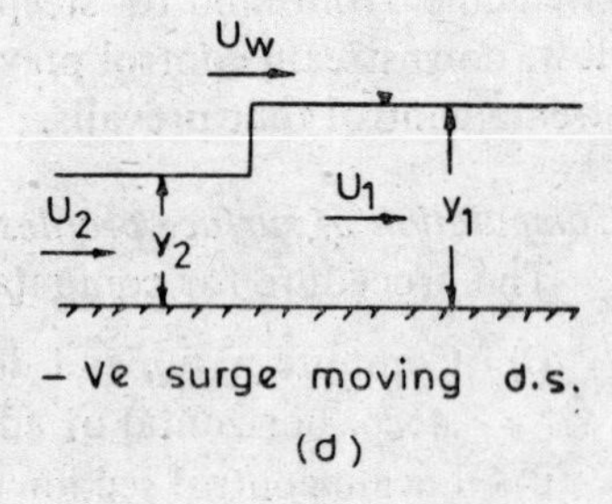

- Ve surge moving d.s.

(d)

Occurrence:

(i) Sudden closure of gate at u.s. end

Fig. 13.9. Types of surges.

For nonrectangular channels

$$C = (U_w - U_1) = \sqrt{\frac{(A_2\bar{y}_2 - A_1\bar{y}_1)\,g}{A_1\left(1 - \dfrac{A_1}{A_2}\right)}} \tag{13.27}$$

where $\bar{y}_1$ and $\bar{y}_2$ are positions of centroids of areas A_1 and A_2 below the water surfaces.

Wave is defined as a variation of velocity with time which travels through the fluid medium. Waves occurring on the water surface can be classified as gravity waves. These can be generated by the action of strong winds, motion of ships, or by geophysical activities such as earthquakes, volcanic eruptions or tides. Waves are classified as oscillatory waves and translatory waves.

Oscillatory wave is periodic in character which imparts to the liquid an undulating motion having both horizontal and vertical components. An oscillatory wave can be either progressive or standing. Waves are characterised by their wave length λ, amplitude a or height H (equal to $2a$) and velocity C. A translatory wave involves transport of fluid in the direction in which wave moves, e.g. moving hydraulic jump. In oscillatory wave, such permanent transport does not take place.

$$C^2 = \frac{g\lambda}{2\pi} \tanh \frac{2\pi h}{\lambda} \tag{13.28}$$

gives the relation between wave velocity C, wave length λ and depth of water h. If h/λ is less than 0.50, $\tanh \frac{2\pi h}{\lambda} \simeq \frac{2\pi h}{\lambda}$ and $C^2 = gh$. This is known as shallow water wave. If $\frac{h}{\lambda} > 0.50$, $\tanh \frac{2\pi h}{\lambda} \simeq 1.0$

$$\therefore \qquad C^2 = \frac{g\lambda}{2\pi} \quad \text{or} \quad C = \sqrt{g\lambda/2\pi}$$

This is known as deep water wave.

ILLUSTRATIVE EXAMPLES

13.1 For flow in open channels, derive continuity equation in differential form. Reduce it for case of a rectangular channel of constant width (Fig. 13.10).

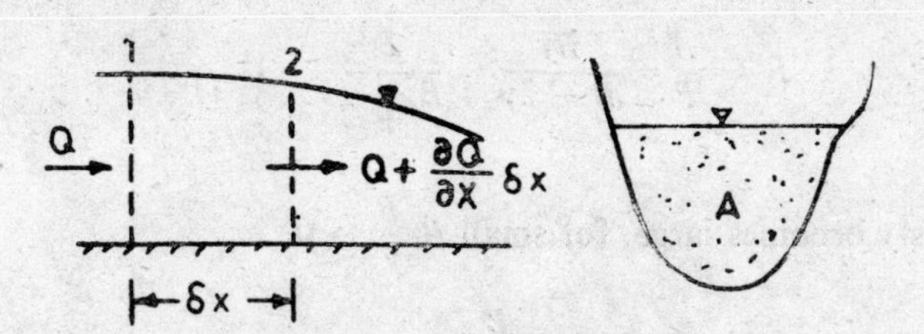

Fig. 13.10. Definition sketch.

Discharge coming in through section 1 $= Q$

Discharge coming out through section 2 $= Q + \frac{\partial Q}{\partial x}\delta x$

$$\therefore \qquad \text{Net rate of inflow} = Q - \left(Q + \frac{\partial Q}{\partial x}\delta x\right) = -\frac{\partial Q}{\partial x}\delta x$$

If there is net inflow, then liquid volume within sections 1 and 2 must increase. Liquid within the two sections $= A\delta x$.

$$\therefore \qquad \text{Its rate of increase} = \frac{\partial}{\partial t}(A\delta x) = \delta x \frac{\partial A}{\partial t}$$

$$\therefore \qquad \text{Equating the two, one gets } \delta x \frac{\partial A}{\partial t} + \frac{\partial Q}{\partial x}\delta x = 0$$

or

$$\frac{\partial A}{\partial t} + \frac{\partial Q}{\partial x} = 0 \tag{13.29}$$

For rectangular channel $A = By$ where B is the channel width. If B is constant, the above equation can be written as

$$\frac{\partial}{\partial t}(By) + \frac{\partial}{\partial x}(qB) = 0$$

or
$$\frac{\partial y}{\partial t} + \frac{\partial q}{\partial x} = 0$$

where q is the discharge per unit width. If $q = Uy$ is substituted in the above equation, one gets

$$\frac{\partial y}{\partial t} + U\frac{\partial y}{\partial x} + y\frac{\partial U}{\partial x} = 0 \qquad (13.30)$$

If flow is steady $\frac{\partial A}{\partial t} = 0$ and $\frac{\partial A}{\partial t} + \frac{\partial Q}{\partial x} = 0$ reduces to $\frac{\partial Q}{\partial x} = 0$ or

$$Q = \text{const.} \qquad \therefore \quad U_1A_1 = U_2A_2$$

13.2 Show that for narrow deep rectangular channels, hydraulic radius R is nearly equal to $\frac{B}{2}$.

$$A = By, \quad P = B + 2y$$

$$\therefore \qquad R = \frac{A}{P} = \frac{By}{B + 2y} = \frac{B}{\frac{B}{y} + 2}$$

Now as y becomes large, for small B, $\frac{B}{y} \to 0$.

Hence for large y values $R \to \frac{B}{2}$.

13.3 A rectangular channel of 2.0 m width and 0.0025 bottom slope carries 1.0 m³/s discharge of water at a depth of 0.45 m. Determine average velocity of flow, Manning's n and Chezy's C.

$$Q = By_0U \qquad \therefore \quad U = \frac{1.0}{2.0 \times 0.45} = 1.111 \text{ m/s}$$

$$U = \frac{1}{n} R^{2/3} S_0^{1/2} \quad \text{and} \quad R = \frac{2 \times 0.45}{(2 + 2 \times 0.45)} = 0.310 \text{ m}$$

$$\therefore \qquad n = \frac{(0.310)^{2/3}(0.0025)^{1/2}}{1.111} = 0.0206$$

Since $\qquad U = C\sqrt{RS_0}, \quad C = \frac{1.111}{\sqrt{0.310 \times 0.0015}} = 39.939 \text{ m}^{1/2}\text{/s}$

13.4 A trapezoidal channel with bottom width of 5.0 m and side slopes $1H: 1V$ has a bed slope of 0.0002. What discharge will it carry at a depth of 1.5 m, if its Manning's n is 0.018?

$$A = \frac{5.0 + (5.0 + 2 \times 1.5)}{2} \times 1.5 = 9.75 \text{ m}^2$$

$$P = 5.0 + 2 \times \sqrt{1.5^2 + 1.5^2} = 9.242 \text{ m}$$

$$R = 9.75/9.242 = 1.055 \text{ m}$$

$$U = \frac{1}{n} R^{2/3} S_0^{1/2} = \frac{1}{0.018} (1.055)^{2/3} (0.0002)^{1/2}$$

$$= 0.814 \text{ m/s}$$

$$\therefore \quad Q = AU = 9.75 \times 0.814 = 7.937 \text{ m}^3\text{/s}$$

13.5 A rectangular channel of 2.5 m width and 0.000 15 slope carries a discharge of 1.2 m³/s with Manning's n of 0.02. Determine the normal depth and average shear on the channel bed.

$$Q = By_0U$$

$$= \frac{(By_0)}{n} R^{2/3} S_0^{1/2}\ ; \quad \text{but } R = \frac{By_0}{(B + 2y_0)}$$

$$\therefore \quad \frac{(By_0)^{5/3}}{(B + 2y_0)^{2/3}} = \frac{Qn}{S_0^{1/2}} = \frac{1.2 \times 0.02}{(0.00015)^{1/2}} = 1.96$$

$$\therefore \quad 1.96 = (By_0)^{5/3}/(B + 2y_0)^{2/3} = F(y)$$

Assume various values of y and compute $F(y)$. Either by interpolation or by plotting, find that value of y say y_0, at which $F(y) = 1.96$.

Here $\quad y_0 = 1.115$ m

$$\therefore \quad R = \frac{2.5 \times 1.115}{(2.5 + 2 \times 1.115)} = 0.589 \text{ m}$$

$$\therefore \quad \tau_0 = \gamma R S_0 = 9787 \times 0.589 \times 0.00015$$

$$= 0.865 \text{ N/m}^2$$

Solution of problems involving computation of B or y_0 is facilitated by using section factor curves in dimensionless form namely $Z_1/B^{8/3}$ vs. y_0/B or $Z_1/y_0^{8/3}$ vs. y_0/B for rectangular and trapezoidal section (Figs 13.11 and 13.12). Here Z_1 is the section factor $AR^{2/3}$.

13.6 For a channel section shown in Fig. 13.13, $S_0 = 0.000169$ and $n =$ 0 02. Determine the discharge assuming

(i) channel section to be a single section,

(ii) section to be divided into three subsections as shown by dotted lines in Fig. 13.13.

$A_1 = 2 \times 1 = 2\text{m}^2 \qquad P_1 = 3.0 \text{ m} \qquad R_1 = 0.667 \text{ m}$

$A_2 = \frac{(3 + 5)}{2} \times 1 + (5 \times 1) = 9\text{m}^2 \qquad P_2 = 3 + 2\sqrt{2} = 5.828 \text{ m},$

$R_2 = 1.544 \text{ m}$

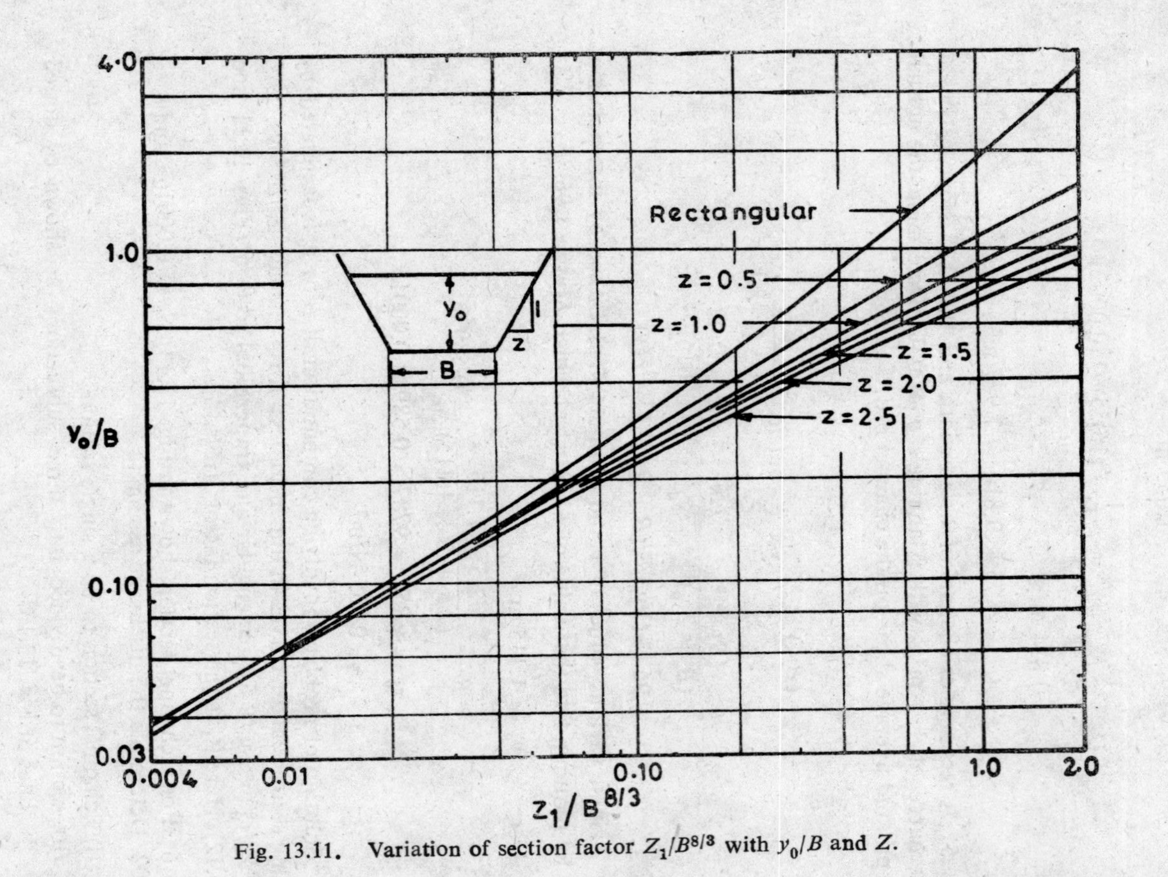

Fig. 13.11. Variation of section factor $Z_1/B^{8/3}$ with y_0/B and Z.

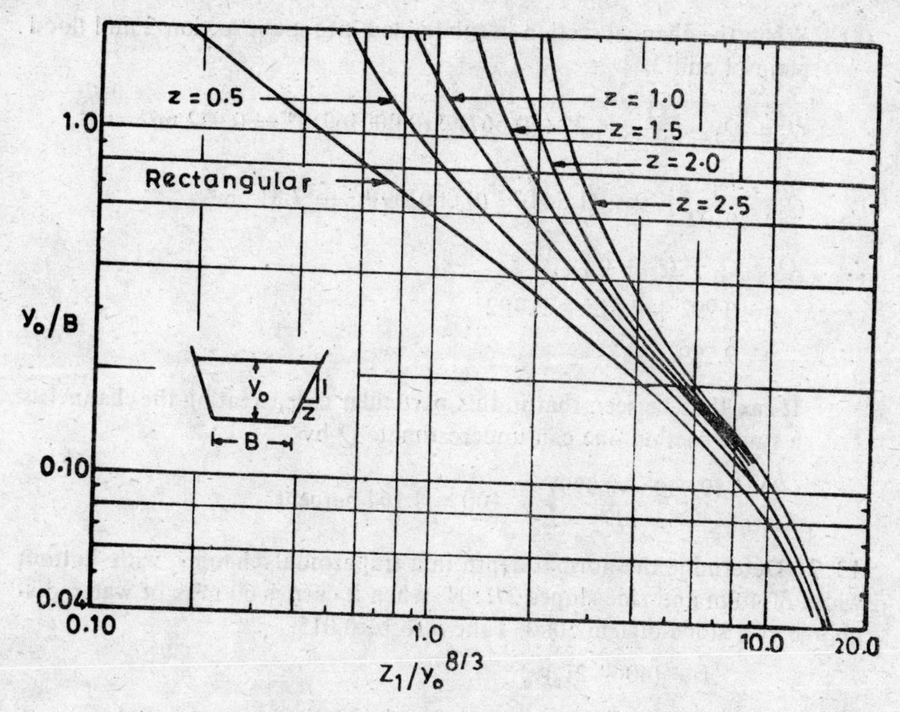

Fig. 13.12. Variation of section factor $Z_1/B^{8/3}$ with y_0/B and Z.

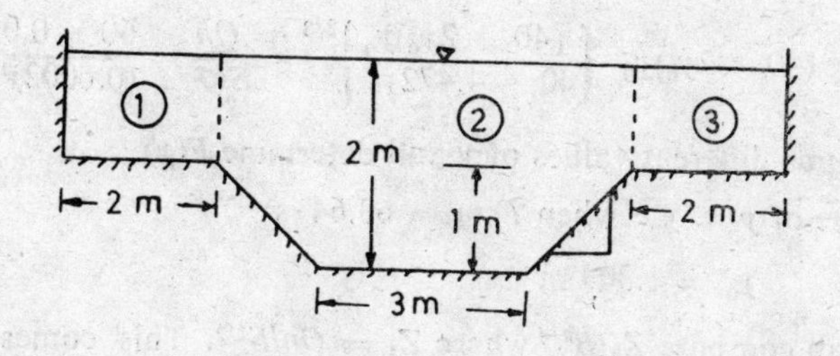

Fig. 13.13.

$A_3 = 2 \times 1 = 2 \text{ m}^2$ $\quad\quad P_3 = 3.0 \text{ m} \quad R_3 = 0.667 \text{ m}$

$A = A_1 + A_2 + A_3 = 2.0 + 9.0 + 2.0 = 13.0 \text{ m}^2$

$P = 1.0 + 2.0 + 1.414 + 3.0 + 1.414 + 2.0 + 1.0 = 11.828 \text{ m}$

$\therefore \; R = A/P = 13/11.828 = 1.099 \text{ m}$

(i) When the channel is considered as a single section

$$Q = \frac{1}{0.02} \times 13 \times (1.099)^{2/3} (0.000169)^{1/2}$$

$$= 8.999 \text{ m}^3$$

(2) When the channel section is subdivided into main section 2 and flood plains 1 and 3,

$$Q_1 = Q_3 = \frac{1}{0.02} \times 2 \times (0.667)^{2/3} (0.000169)^{1/2} = 0.992 \text{ m}^3/\text{s}$$

$$Q_2 = \frac{1}{0.02} \times 9 \times (1.544)^{2/3} (0.000169)^{1/2} = 7.815 \text{ m}^3/\text{s}$$

$$\therefore \quad Q = Q_1 + Q_2 + Q_3$$

$$= 0.992 + 7.815 + 0.992$$

$$= 9.799 \text{ m}^3/\text{s}$$

It can thus be seen that in this particular case, treating the channel as a single section one can underestimate Q by

$$\frac{(9.799 - 8.999)}{9.799} \times 100 = 8.164 \text{ percent.}$$

13.7 Determine the normal depth in a trapezoidal channel with bottom width of 40 m and side slopes $2H: 1V$ when it carries 60 m³/s of water discharge at a slope of 1 in 5000. Take n to be 0.015.

$$A = (40 + 2y_0)y_0$$

$$P = 40 + 2\sqrt{5}\, y_0 = (40 + 4.472\, y_0)$$

$$R = (40 + 2y_0)y_0/(40 + 4.472\, y_0)$$

$$\therefore \quad (40 + 2y_0)y_0 \left\{\frac{(40 + 2y_0)y_0}{40 + 4.472y_0}\right\}^{2/3} = \frac{Qn}{S_0^{1/2}} = \frac{60 \times 0.015}{(0.0002)^{1/2}} = 63.640$$

Assume different values of y and determine $F(y)$.

$\therefore$ Value of $y = y_0$ when $F(y) = 63.64$ is

$$y_0 = 1.309 \text{ m}$$

One can compute $Z_1/B^{8/3}$ where $Z_1 = Qn/S_0^{1/2}$. This comes out to be 0.0034. Fig. 13.11 now gives

$$y_0/B = 0.033 \quad \text{or} \quad y_0 = 1.32 \text{ m}$$

13.8 Design a concrete lined channel of trapezoidal shape to carry 200 m³/s discharge through a terrain where permissible slope is 0.0004. Take Manning's n as 0.014, and side slopes 1:1.

The bottom width can be calculated from the empirical formula

$$B = 0.70\sqrt{Q}$$

$$= 0.70\sqrt{200} = 9.9, \text{ say } 10 \text{ m}$$

Using side slopes as 1:1

$$A = (10 + y_0)y_0 = (10y_0 + y_0^2)$$

$$P = 10 + 2\sqrt{2}\, y_0$$

$$R = (10y_0 + y_0^2)/(10 + 2\sqrt{2}\, y_0)$$

$$\therefore \quad \frac{Qn}{S_0^{1/2}} = \frac{200 \times 0.014}{(0.0004)^{1/2}} = 140 = \frac{(10y_0 + y_0^2)^{5/3}}{(10 + 2.828y_0)^{2/3}}$$

Solution of this equation by trial-error procedure gives $y_0 = 4.653$ m.

Use a free board of 1.0 m. Hence channel depth will be

$$(4.653 + 1.00) = 5.653 \text{ m.}$$

13.9 The energy and momentum correction factors α and β for open channels are given by the equations

$$\alpha = 1 + 3\epsilon^2 - 2\epsilon^3$$

$$\beta = 1 + \epsilon^2$$

where $\epsilon = \left(\frac{u_m}{U} - 1\right)$, u_m being the maximum velocity and U the average velocity of flow. Determine α and β if velocity distribution is given by

$$\frac{u}{u_m} = \left(\frac{y}{y_0}\right)^{1/7}$$

where y_0 is the depth of flow.

$$y_0 U = q = \frac{u_m}{y_0^{1/7}} \int_0^{y_0} y^{1/7}\, dy = \frac{7}{8} u_m y_0$$

$$\therefore \quad \frac{u_m}{U} = \frac{8}{7} \quad \text{and} \quad \epsilon = \frac{8}{7} - 1 = \frac{1}{7}$$

$$\therefore \quad \alpha = 1 + 3 \times \left(\frac{1}{7}\right)^2 - 2 \times \left(\frac{1}{7}\right)^3$$

$$= 1 + 0.0612 - 0.0058$$

$$= 1.0554$$

$$\beta = 1 + \left(\frac{1}{7}\right)^2 = 1.0204$$

13.10 Obtain the condition for a rectangular channel of given area A to be most efficient.

$$A = By_0$$

$$P = B + 2y_0$$

$$\therefore \quad P = \frac{A}{y_0} + 2y_0$$

$\therefore$ For P to be minimum, $\dfrac{dP}{dy_0} = 0 = -\dfrac{A}{y_0^2} + 2$

$\therefore \quad A = 2y_0^2 = By_0 \quad \therefore \quad y_0 = \dfrac{B}{2}$ is the required condition.

13.11 Show that a trapezoidal channel with given area and side slopes Z:1 is most efficient when hydraulic radius is half the depth of flow.

The channel will be most efficient when P is minimum for given A, so that R is maximum and hence U and Q are maximum.

$$A = (B + Zy_0)y_0$$

$$P = B + 2y_0\sqrt{Z^2+1}$$

Eliminate B from these two equations

$$\therefore \quad P = \frac{A}{y_0} - Zy_0 + 2y_0\sqrt{Z^2+1}$$

For P to be minimum for constant A and Z, $\dfrac{dP}{dy_0} = 0$

$$\therefore \quad \frac{dP}{dy_0} = -\frac{A}{y_0^2} - Z + 2\sqrt{Z^2+1} = 0$$

$$\therefore \quad \frac{A}{y_0} = (2\sqrt{Z^2+1} - Z)\, y_0$$

ubstituting this value in the expression for P, one gets

$$P = (2\sqrt{Z^2+1} - Z)\, y_0 - Zy_0 + 2y_0\sqrt{Z^2+1}$$

$$= 4y_0\sqrt{Z^2+1} - 2Zy_0$$

$$\therefore \quad R = \frac{A}{P} = \frac{y_0^2\,(2\sqrt{Z^2+1} - Z)}{2y_0\,(2\sqrt{Z^2+1} - Z)} = \frac{y_0}{2}$$

$$\therefore \quad R = \frac{y_0}{2}$$

13.12 A rectangular channel is to carry 2.0 m³/s discharge of water at a slope of 0.0001 and Manning's n of 0.016. If it is designed as the most efficient section, determine its dimensions.

For the most efficient section, $\dfrac{B}{2} = y_0$

$$\therefore \quad A = By_0 = 2y_0^2 = \frac{B^2}{2}$$

$$P = B + 2y_0 = 2B$$

$$\therefore \quad R = R/P = B/4$$

$$\therefore \quad Q = \frac{1}{n} AR^{2/3} S_0^{1/2}$$

$$= \frac{1}{0.016}\left(\frac{B^2}{2}\right)\left(\frac{B}{4}\right)^{2/3} (0.0001)^{1/2}$$

$$= 2.0$$

$\therefore \quad B^{8.3} = 16.125 \quad \text{or} \quad B = 2.837 \text{ m}$

$$y_0 = 1.419 \text{ m}$$

13.13 Show that, for a circular conduit running part full, R is maximum for given A when $y = 0.813\ D$, where D is the conduit diameter (Fig. 13.14).

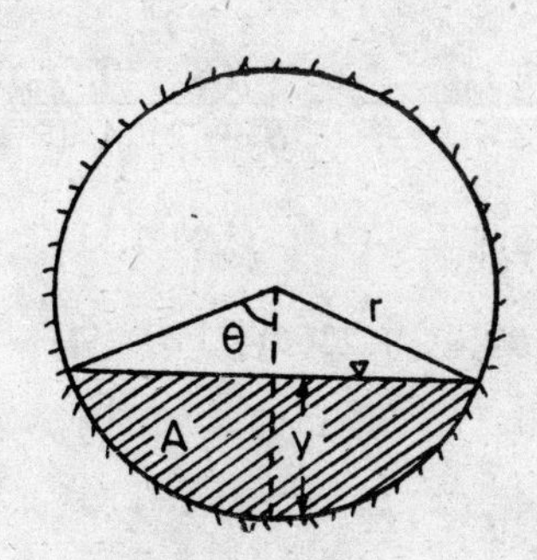

Fig. 13.14.

$$A = r^2\theta - r^2 \sin\theta \cos\theta$$

$$= r^2\theta - \frac{r^2}{2} \sin 2\theta$$

$$P = 2r\theta$$

For given A, R is to be maximised to get maximum velocity. Since

$$R = A/P$$

$$\therefore \quad \frac{d}{d\theta}\left(\frac{A}{P}\right) = \frac{P\dfrac{dA}{d\theta} - A\dfrac{dP}{d\theta}}{P^2} = 0$$

or

$$A\frac{dP}{d\theta} = P\frac{dA}{d\theta}$$

$$\left(r^2\theta - \frac{r^2}{2}\sin 2\theta\right) \times 2r = 2r\theta\,(r^2 - r^2 \cos 2\theta)$$

$$\therefore \quad 2\theta - \sin 2\theta = 2\theta - 2\theta \cos 2\theta$$

$$\therefore \quad \theta = \tfrac{1}{2} \tan 2\theta$$

The solution of this equation gives $2\theta = 257.5°$

$$\therefore \quad y = r - r\cos\theta$$

$$= r(1 + \cos 51.25°)$$

$$= 1.626r \quad \text{or} \quad y = 0.813D$$

where D is the conduit diameter.

13.14 A trapezoidal channel with side slopes of 0.50 $H : 1V$ is to be designed as the most efficient channel to carry 30 m³/s discharge at a slope of 0.000556. Using Chezy's C as 60, determine the bottom width and the depth of flow.

For most efficient channel $R = \frac{y_0}{2}$

$$\therefore \quad \frac{y_0}{2} = \frac{A}{P} = \frac{(B + 0.5y_0)y_0}{B + 2y_0\sqrt{1^2 + 0.5^2}}$$

$$\therefore \quad By_0 + 0.5y_0^2 = \frac{By_0}{2} + y_0^2 \times 1.118$$

$$\therefore \quad By_0 = 1.2366y_0^2 \quad \text{or} \quad B = 1.236y_0$$

$$\therefore \quad A = By_0 + 0.50y_0^2 = 1.236y_0^2 + 0.50y_0^2 = 1.736y_0^2$$

$$\text{and} \quad R = \frac{y_0}{2}$$

$$\therefore \quad Q = CA\sqrt{RS}$$

$$\therefore \quad 30 = 60 \times 1.736y_0^2 \sqrt{\frac{y_0 \times 0.000556}{2}}$$

$$\therefore \quad 30 = 1.737y_0^{5/2} \quad \text{and} \quad y_0 = 3.125 \text{ m}$$

$$B = 3.863 \text{ m}$$

13.15 A rectangular channel of 5.0 m width and 0.0004 bed slope has a water depth of 2.0 m at Manning's n of 0.011. If a bend of centre line radius 40 m is provided along its length, estimate the super-elevation.

$$Q = \frac{1}{0.011}(5 \times 2)\left(\frac{10}{9}\right)^{2/3}(0.0004)^{1/2}$$

$$= 19.504 \text{ m}^3/\text{s}$$

$$U = \frac{19.504}{5 \times 2} \text{ m}^3 = 1.950 \text{ m/s}$$

uper-elevation $\quad (h_2 - h_1) = \frac{B}{r_c}\left[\frac{1}{1 + \frac{B^2}{12r_c^2}}\right]\frac{U^2}{g}$

$$= \frac{5}{40}\left[\frac{1}{1 + \frac{5^2}{12 \times 40^2}}\right]\frac{1.95^2}{9.806}$$

$$= 0.0484 \text{ m} \quad \text{or} \quad 48.4 \text{ mm}$$

13.16 Show that for a given discharge, specific energy is minimum when $Q^2\ T/gA^3$ is unity. Here T is the water surface width. Obtain expressions for the critical depth and minimum specific energy for a rectangular channel (Fig. 13.15).

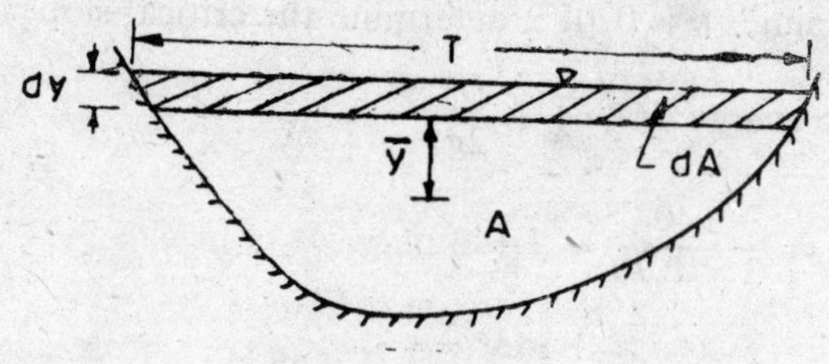

Fig. 13.15.

Specific energy $H = y + \dfrac{U^2}{2g} = y + \dfrac{Q^2}{2gA^2}$

For H to be minimum $\dfrac{\partial H}{\partial y} = 0 = 1 - \dfrac{Q^2}{gA^3}\dfrac{dA}{dy}$

but $\qquad Tdy = dA$ (Fig. 13.15).

$$\therefore \qquad \frac{dA}{dy} = T$$

$$\therefore \qquad 1 - \frac{Q^2T}{gA^3} = 0 \quad \text{or} \quad \frac{Q^2T}{gA^3} = 1$$

However one can define $A/T =$ hydraulic mean depth D and $Q^2/A^2 = U^2$

$\therefore$ Specific energy is minimum when $\dfrac{U^2}{gD} = 1$, i.e. the square of Froude number is unity.

If the channel is rectangular $A = By$ and $T = B$

$$\therefore \qquad \frac{Q^2T}{gA^3} = \frac{q^2}{gy_c^3} = 1$$

where q is the discharge per unit width and y_c is the depth at which H is minimum. y_c is known as the critical depth.

$$\therefore \qquad y_c = \sqrt[3]{\frac{q^2}{g}} \quad \text{and} \quad H_m = y_c + \tfrac{1}{2}\frac{q^2}{gy_c^2}$$

$$= \frac{3}{2}y_c \text{ since } \frac{q^2}{g} = y_c^3$$

$$\therefore \qquad H_m = \frac{3}{2}y_c \text{ and } y_c = \sqrt[3]{\frac{q^2}{g}} \text{ for rectangular channels.}$$

13.17 A 3.0 m wide rectangular channel carries 2.4 m^3/s discharge at a depth of 0.70 m.

(i) Determine specific energy at 0.70 m depth.
(ii) Determine the critical depth.
(iii) Is the flow subcritical or supercritical?
(iv) Determine the depth alternate to 0.70 m.
(v) If Manning's n is 0.015, determine the critical slope.

(i) Specific energy $H = y + \frac{U^2}{2g}$

$$U = \frac{2.40}{3 \times 0.70} = 1.143 \text{ m/s}$$

$$\therefore \quad H = 0.70 + \frac{1.143^2}{2 \times 9.806} = 0.7666 \text{ m}$$

(ii) Critical depth, $$y_c = \sqrt[3]{\frac{q^2}{g}} = \sqrt[3]{\left(\frac{2.4}{3}\right)^2 \frac{1}{9.806}}$$

$$= 0.403 \text{ m}$$

(iii) Since y is greater than 0.403 m, the flow is subcritical.

(iv) At $y = 0.70$ m, $H = 0.7666$ m

$$\therefore \quad 0.7666 = y + \frac{(0.8)^2}{2 \times 9.806 y^2} = y + \frac{0.0326}{y^2}$$

One of the solutions of this equation is $y = 0.70$ m. The other positive root which is less than 0.403 m will be the alternate depth. Solving this equation by trial error method one gets,

$$y_2 = 0.252 \text{ m}$$

(v) The channel slope which will make the critical depth normal will be the critical slope S_c.

$$R_c = \frac{3 \times 0.403}{3.806} = 0.318 \text{ m}$$

$$\therefore \quad q = \frac{1}{n} y_c R_c^{2/3} S_c^{1/2} \quad \therefore \quad S_c^{1/2} = \frac{0.8 \times 0.015}{0.403 \times 0.318^{2/3}} = 0.0733$$

$$\therefore \quad S_c = 0.00538$$

13.18 Obtain an expression for critical depth in a triangular channel with central angle of 2θ.

At critical condition $\frac{Q^2}{g} = \frac{A^3}{T}$ but $A = y_c^2 \tan\theta$

$$T = 2y_c \tan\theta$$

$$\therefore \quad \frac{Q^2}{g} = \frac{(y_c^2 \tan\theta)^3}{2y_c \tan\theta}$$

$$= \frac{y_c^5 \tan^2 \theta}{2}$$

$$\therefore \qquad y_c = \sqrt[5]{2Q^2/\tan^2 \theta}$$

13.19 A parabolic channel is defined by the equation $x^2 = 2y$. Determine the critical depth for a discharge of 2.0 m³/s (Fig. 13.16).

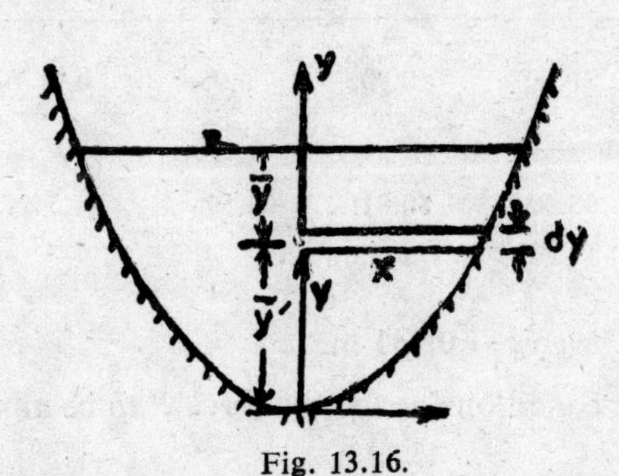

Fig. 13.16.

$$A = 2 \int_0^y x dy$$

$$= 2\sqrt{2} \int_0^y y^{1/2}\, dy$$

$$= \frac{4\sqrt{2}}{3} y^{3/2}$$

$$T = 2x = 2\sqrt{2y}$$

$$\therefore \qquad \text{At critical condition } \frac{Q^2}{g} = \frac{A^3}{T} = \frac{64 \times 2\sqrt{2}}{27} \frac{y_c^{9/2}}{2\sqrt{2}\, y_c^{1/2}} = \frac{64}{27} y_c^4$$

$$\therefore \qquad y_c = \left(\frac{2^2 \times 27}{9.806 \times 64}\right)^{1/4}$$

$$= 0.766 \text{ m}$$

13.20 Determine the critical depth in a trapezoidal channel of bed width 10 m and side slopes 1:1 at a discharge of 30 m³/s.

$$A = (10 + y)y$$

$$T = 10 + 2y$$

$\therefore$ At critical flow condition

$$\frac{Q^2}{g} = \frac{30^2}{9.806} = 91.781 = \frac{A^3}{T}$$

or
$$F(y) = \frac{\{(10 + y_c)\, y_c\}^3}{10 + 2y_c} = 91.781$$

Assume values of y and compute $F(y)$

y m	1	0.80	0.90	0.95	0.94	0.942	0.941
$F(y)$	110.917	55.60	80.01	94.595	91.542	92.147	91.843

$$\therefore \quad y_c = 0.941 \text{ m}$$

13.21 Obtain the condition for specific force P to be minimum for a constant discharge.

$$P = \frac{Q^2}{Ag} + A\bar{y}$$

where $\bar{y}$ is the depth of centroid of area A below the water surface (Fig. 13.15)

$$\therefore \quad \frac{dP}{dy} = -\frac{Q^2}{gA^2}\frac{dA}{dy} + dy\,(A\,\bar{y})$$

But
$$d\,(A\bar{y}) = A\,(\bar{y} + dy) + T\,dy.\frac{dy}{2} - A\,\bar{y}$$

$$= Ady + \frac{T\,(dy)^2}{2}$$

If dy is small, $\frac{T\,(dy)^2}{2}$ can be neglected.

$$\therefore \quad \frac{d}{dy}(A\bar{y}) = A$$

$$\therefore \quad \frac{dP}{dy} = -\frac{Q^2T}{gA^2} + A = 0 \quad \text{for } P \text{ to be minimum}$$

$$\therefore \quad \frac{Q^2T}{gA^3} = 1$$

Thus it can be seen that at critical depth, specific force is also minimum.

13.22 Consider a sudden drop in a wide rectangular horizontal channel. Show that the brink depth y_b is given by 0.667 y_c. What assumptions are made in the derivation?

Consider section 1 at critical depth and section 2 immediately downstream of sudden drop (Fig. 13.2). Assumptions to be made are:

(i) Channel is horizontal; so the component of gravity force in the direction of flow is zero, i.e., $W \sin \theta = 0$
(ii) Sections 1 and 2 are sufficiently close so that frictional force F_f can be neglected. Hence $F_f = 0$
(iii) Pressure distribution at section 1 is hydrostatic; pressure at section 2 is zero, i.e., $K_b = 0$
(iv) Velocity distribution at sections 1 and 2 is uniform; hence β is unity.

$\therefore$ Momentum equation between sections 1 and 2 yields

$$q \rho (U_b - U_c) = \frac{1}{2} \gamma y_c^2 - 0$$

But $$U_b = q/y_b,\ U_c = q/y_c \quad \text{and} \quad \frac{q^2}{g} = y_c^3$$

$$\therefore \quad \frac{q^2}{g}\left(\frac{1}{y_b} - \frac{1}{y_c}\right) = \frac{1}{2} y_c^2$$

$$\therefore \quad \frac{y_c}{y_b}\left(1 - \frac{y_b}{y_c}\right) = \frac{1}{2} \quad \text{or} \quad 1 - \frac{y_b}{y_c} = \frac{1}{2}\frac{y_b}{y_c}$$

$\therefore$ $y_b = \frac{2}{3} y_c$. Experimentally it is found that $y_b = 0.715\, y_c$

Hence a single measurement at the brink can give an estimation of discharge in the channel. Since for rectangular channel

$$y_c = \sqrt[3]{\frac{q^2}{g}}$$

$$y_b = 0.715 \frac{q^{2/3}}{g^{1/3}} \quad \text{or} \quad q = 1.654\, g^{1/2}\, y_b^{3/2}$$

and $$Q = 1.654\, B g^{1/2}\, y_b^{3/2}$$

13.23 If the channel shape is defined by $A = cy^n$, where c is constant, show that

$$\frac{y_b}{y_c} = \left(\frac{n+1}{2n+1}\right)^{1/n} \quad \text{and} \quad \frac{y_c}{H_m} = \left(\frac{2n}{2n+1}\right)$$

Since $$H = y + \frac{Q^2}{2gA^2} \quad \therefore \quad \frac{dH}{dy} = 1 - \frac{Q^2 T}{g\ A^3} = 0$$

However $$\frac{dA}{dy} = T = cny^{n-1}$$

When $$H = H_m, \quad \frac{Q^2}{g} = \frac{A^3}{T} = \frac{c^3 y^{3n}}{cny^{n-1}} = \frac{c^2}{n}\, y^{2n+1}$$

$$\therefore \quad H_m = y_c + \frac{Q^2}{g}\frac{1}{2A^2} = y_c + \frac{1}{2}\frac{c^2}{n}\frac{y_c^{2n+1}}{c^2 y_c^{2n}}$$

$$= y_c + \frac{y_c}{2n} = y_c\left(\frac{1+2n}{2n}\right)$$

$$\therefore \quad \frac{y_c}{Hm} = \frac{2n}{2n+1}$$

It can be shown that for nonrectangular channels

$$\frac{A_b}{A_c} = \frac{1}{1 + \dfrac{\bar{y}_c}{D_c}}$$

where A_b and A_c are the areas at brink and critical depths, $\bar{y}_c$ is the depth of centroid of A_c below water surface and $D_c = A_c/T_c$ where T_c is the water surface width of area A_c. If $\bar{y}'_c$ is distance of centroid of area above the bed (Fig. 13.16),

$$\bar{y}' = \frac{\int y dA}{\int dA}$$

but $$dA = cny^{n-1}\, dy$$

$$\bar{y}' = \frac{\int cny^n dy}{cy^n} = \frac{n}{n+1}\, y \quad \text{and} \quad \bar{y}_c = y_c - \bar{y}'_c$$

$$= \frac{y_c}{n+1}$$

$$D_c = \frac{A_c}{T_c} = \frac{cy_c^n}{ncy_c^{n-1}} \qquad \text{since } \frac{dA}{dy} = T$$

$$= \frac{y_c}{n}$$

Further $\dfrac{A_b}{A_e} = \left(\dfrac{y_b}{y_c}\right)^n$. Substituting these values in the above equation, one gets

$$\left(\frac{y_b}{y_c}\right)^n = \frac{1}{1 + \dfrac{y_c}{n+1}\dfrac{n}{y_c}} = \left(\frac{n+1}{2n+1}\right)$$

$$\therefore \quad \frac{y_b}{y_c} = \left(\frac{n+1}{2n+1}\right)^{1/n}$$

13.24 Water flows in a 4.0 m wide rectangular channel at the rate of 8.0 m³/s, at a depth of 1.0 m. Compare alternate and conjugate depths for 1.0 m.

$$q = 8/4 = 2.0 \text{ m}^3/\text{s m}$$

Specific energy $H = y + \dfrac{q^2}{2gy^2} = 1.0 + \dfrac{2^2}{2 \times 9.806 \times 1^2}$

$$= 1.204$$

Also at 1.0 m depth, $U = \dfrac{2.0}{1.0} = 2.0$ m/s and

$$U/\sqrt{gy} = \frac{2.0}{\sqrt{1 \times 9.806}} = 0.639;$$

Hence the flow is subcritical.

$$\therefore \qquad 1.204 = y_2 + \frac{2^2}{2 \times 9.806\, y_2^2} = y_2 + \frac{0.204}{y_2^2}$$

Solution of this equation gives $y_2 = 0.565$ m. This is the alternate depth.

$$P = \frac{q^2}{gy} + \frac{y^2}{2} = \frac{2^2}{9.806 \times 1.0} + \frac{1^2}{2}$$

$$= 0.408 + 0.50 = 0.908$$

$$\therefore \qquad 0.908 = \frac{0.408}{y'_2} + \frac{y_2'^2}{2}$$

Solution of this equation yields $y'_2 = 0.532$ m. This is the conjugate depth.

13.25 Obtain the equation for discharge over a broad crested weir and the expression for discharge coefficient C_d in terms of weir height to upstream depth ratio. What assumptions are made in the derivation?

See Fig. 13.3. Assumptions to be made are:

1. There are no surface tension and viscous effects on the flow
2. $2 < \dfrac{L}{H_1} < 12\ 5$; hence critical depth occurs on the weir
3. There is no energy loss between sections 1 and 2 (this follows from assumption 1)
4. Velocity distribution at sections 1 and 2 is uniform, hence $\alpha_1 = \alpha_2 = 1$
5. Pressure distribution at sections 1 and 2 is hydrostatic.

$\therefore$ Energy equation gives

$$y_1 + \frac{U_1^2}{2g} = W + y_c + \frac{U_c^2}{2g}$$

$$\therefore \quad (H_1 - y_c) = \frac{U_c^2}{2g} \quad \text{or} \quad U_c = \sqrt{2g(H_1 - y_c)}$$

$$\therefore \quad q = U_c y_c = y_c\sqrt{2g(H_1 - y_c)}, \quad \text{however } y_c = \frac{2}{3} H_1$$

$$\therefore \quad q = \frac{2}{3}\sqrt{\frac{2}{3} g}\, H_1^{3/2} \quad \text{or} \quad Q = \frac{2}{3}\sqrt{\frac{2}{3} g}\, B H_1^{3/2}$$

This equation can be written as

$$Q = C_d \frac{2}{3}\sqrt{\frac{2}{3} g} B\, h_1^{3/2}$$

where $C_d = \left(\frac{H_1}{h_1}\right)^{3/2}$. However $H_1 = h_1 + \frac{Q^2}{2gB^2(h_1 + W)^2}$

$$\therefore \quad \frac{H_1}{h_1} = 1 + \frac{Q^2}{2gB^2 h_1 (h_1 + W)^2}$$

$$= 1 + \frac{\frac{8}{27} C_d^2\, g\, B^2 h_1^3}{2gB^2 h_1 (h_1 + W)^2}$$

$$= 1 + \frac{4C_d^2/27}{\left(1 + \frac{W}{h_1}\right)^2}$$

$$\therefore \quad C_d = \left(\frac{H_1}{h_1}\right)^{3/2} = \left\{1 + \frac{4C_d^2/27}{\left(1 + \frac{W}{h_1}\right)^2}\right\}^{3/2}$$

13.26 Determine the discharge and upstream depth for flow over a broad crested weir of height 1.0 m in 10 m wide channel if the depth over the weir is 0.5 m. Weir is a free-flowing weir.

$$y_c = 0.50 = \sqrt[3]{\frac{q^2}{g}}$$

$$\therefore \quad q = \sqrt{0.50^3 \times 9.806} = 1.1071 \text{ m}^3\text{/s m}$$

and $\quad Q = qB = 1.1071 \times 10 = 11.071 \text{ m}^3$

Assume $\quad C_d = 1 \quad \therefore \quad Q = \frac{2}{3}\sqrt{\frac{2}{3}}\sqrt{g}\, B\, h_1^{3/2}$

Substituting the values of Q, g and B, one gets $h_1 = 0.75$ m.

$$\therefore \quad \frac{W}{h_1} = \frac{1.0}{0.75} = 1.333$$

and $$C_d^{2/3} = \left[1 + \frac{4C_d^2}{27(1 + 1.333)^2}\right]$$

$$= 1 + 0.0272\, C_d^2$$

Solution of this equation yields $C_d = 1.045$

$$11.07 = 1.045 \times \frac{2}{3}\sqrt{\frac{2}{3}}\sqrt{9.806}\, h_1^{3/2} \times 10$$

$$\therefore \quad h_1 = 0.728 \text{ m}$$

13.27 A 5.0 m wide rectangular channel carries a discharge of 20 m³/s at a depth of 2.0 m.

(i) Calculate the depth of flow over a hump of 0.25 m on the bed.

(ii) What will be the minimum rise in the bed level required to obtain critical depth over the rise?

(iii) What will be the water depths upstream and over the hump if bed level is raised by 0.6 m?

See Fig. 13.5.

$$q = 20/5 = 4\,0 \text{ m}^3/\text{s m} \quad \text{and} \quad U_1 = 4.0/2.0 = 2.0 \text{ m/s.}$$

$$\therefore \quad H_1 = 2.0 + \frac{2^2}{2 \times 9.806} = 2.204 \text{ m}$$

Critical depth $y_c = \sqrt[3]{4\,0^2/9.806} = 1.177$ m

Minimum specific energy $H_m = \frac{3}{2} \times 1.177 = 1.766$

Since $H_1 = \Delta Z_m + H_m$

$$\Delta Z_m = 2.204 - 1.766 = 0.438 \text{ m}$$

If the rise in bed is at least 0.438 m, depth over the hump will be critical.

Since $\Delta Z = 0.25$ m, water depth over the hump will be less than critical. If it is y_2

$$(H_1 - \Delta Z) = (2.204 - 0.250) = y_2 + \frac{q^2}{2gy_2^2}$$

or $$1.954 = y_2 + \frac{0.816}{y_2^2}$$

Solution of this equation will lie between y_1 and y_c. By trial-error method

$$y_2 = 1.657 \text{ m}$$

In the third case since $\Delta Z > \Delta Z_m$, the specific energy at section 1 is inadequate for water to flow at critical depth at section 2. Hence the depth at section 1 increases to y_1' and depth at section 2 is critical.

$$\therefore \quad H_1' - \Delta Z = H_m \quad \text{or} \quad H_1' = 1.766 + 0.600$$

$$= 2.366 \text{ m}$$

$$\therefore \qquad 2.366 = y_1' + \frac{0.816}{y_1'^2}$$

New y_1' will be greater than y_1, i.e. 2.0 m. By trial, one gets

$$y_1' = 2.197 \text{ m}$$

13.28 Consider a 5.0 m wide rectangular channel carrying 20 m³/s discharge at a depth of 2.0 m. (i) Determine the width to which the channel should be contracted so that the depth in contracted section is critical. (ii) What will be the depth at the contracted section if width there is 4.0 m? (iii) What will be the depths of flow in the upstream and in the contracted section if the width of the channel is reduced to 2.5 m?

$$\text{(i)} \qquad H_1 = 2.00 + \frac{4^2}{2 \times 9.806 \times 2^2} = 2.00 + 0.204$$

$$= 2.204 \text{ m}$$

With H_1 constant, q_{max} will occur when $y_c = \frac{2}{3} H_1 = 1.469$ m

$$\therefore \qquad y_c^3 = \frac{q_2^2}{g} \qquad \therefore \quad q_2 = \sqrt{1.469^3 \times 9.806}$$

$$= 5.578 \text{ m}^3\text{/s m}$$

$$\therefore \qquad B_2 = \frac{Q}{q_2} = \frac{20.0}{5.578} = 3.586 \text{ m}$$

(ii) Since $B_2 = 4.0$ m which is greater than 3.586 m, the flow in the contracted section will not be critical.

$$2.204 = y_2 + \frac{q_2^2}{2gy_2^2} \quad \text{and} \quad q_2 = \frac{20}{4} = 5 \text{ m}^3\text{/s m}$$

$$\therefore \qquad 2.204 = y_2 + \frac{1.275}{y_2^2}$$

The solution of this equation gives $y_2 = 1.82$ m which is greater than 1.469 m.

$$\text{(iii) When} \quad B_2 = 2.5 \text{ m} \qquad q_2 = \frac{20.0}{2.5} = 8.0 \text{ m}^3\text{/s m.}$$

$$\therefore \qquad y_c = \sqrt[3]{8^2/9.806} = 1.869 \text{ m} \quad \text{and} \quad H_2 = \frac{3}{2} y_c = 2.8035 \text{ m}$$

However since $H_1 = 2.204$, flow cannot occur unless $H_1' = 2.8035$. Thus

$$2.8035 = y_1' + \frac{4^2}{2 \times 9.806\, y_1'^2} = y_1' + \frac{0.816}{y_1'^2}$$

$$\therefore \qquad y_1' = 2.691 \text{ m}$$

The depth in the contracted section will be 1.869 m and the upstream depth will increase from 2.0 m to 2.691 m.

13.29 A 3.5 m wide rectangular channel carries a discharge of 10 m³/s at a depth of 1.75 m. If the width of the channel is reduced to 2.25 m and the bed level is lowered by 0.97 m, determine the difference in water level elevations between the upstream and contracted sections. Assume no energy loss (Fig. 13.17)

$$U_1 = \frac{10}{3.5 \times 1.75} = 1.633 \text{ m/s}$$

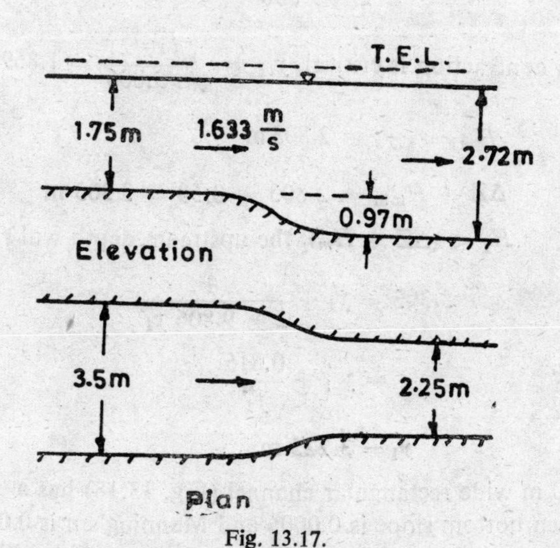

Fig. 13.17.

$$\therefore \qquad H_1 = 1.75 + \frac{1.633^2}{2 \times 9.806} = 1.866 \text{ m}$$

$$q_2 = \frac{10}{2.25} = 4.444 \text{ m}^3\text{/s m}$$

Since total energy at sections 1 and 2 is the same

$$0.97 + 1.75 + \frac{1.633^2}{2 \times 9.806} = y_2 + \frac{4.444^2}{2 \times 9.806\, y_2^2}$$

or

$$2.856 = y_2 + \frac{1.007}{y_2^2}$$

The solution of this equation in subcritical regime yields

$$y_2 = 2.72 \text{ m} \quad \text{whereas} \quad y_1 + \Delta Z = 1.75 + 0.97 = 2,72 \text{ m}$$

Hence water surface elevations at the two places will be the same. It may be mentioned that the assumption of no energy loss is not quite correct here because the flow is expanding.

13.30 Water flows in a 4.0 m wide rectangular channel at a depth of 2.0 m and velocity of 2.0 m/s. The channel is contracted to 2.0 m and bed raised by 0.5 m in a given reach. What will be the depth in the upstream reach?

$$Fr_1 = \frac{2}{\sqrt{2 \times 9.806}} = 0.452$$

Hence flow is subcritical and depth in transition will decrease

$$Q = 4 \times 2 \times 2 = 16 \text{ m}^3/\text{s}$$

$$H_1 = 2 + \frac{2^2}{2 \times 9.806} = 2.204 \text{ m}$$

If the flow in contraction is critical, $y_{c_2} = \sqrt[3]{\frac{(16/2)^2}{9.806}} = 1.869$ m

$$\therefore \quad H_m = \frac{3}{2} y_c = 2.803 \text{ m}$$

$$\Delta Z + H_{min} = 2.803 + 0.50 = 3.303 \text{ m}$$

Since $H_1 < (\Delta Z + H_m)$, the upstream depth will rise.

$$3.303 = y_1' + \frac{4^2}{2 \times 9.806\, y_1'^2}$$

$$= y_1' + \frac{0.816}{y_1'^2}$$

$$\therefore \quad y_1' = 3.225 \text{ m}$$

13.31 A 20 m wide rectangular channel (Fig. 13.18) has a normal depth of 2.5 m when bottom slope is 0.0006 and Manning's n is 0.015. Two piers of 2 m width each and 4 m long with rounded nose are constructed in the channel. What will be the water depth upstream of piers?

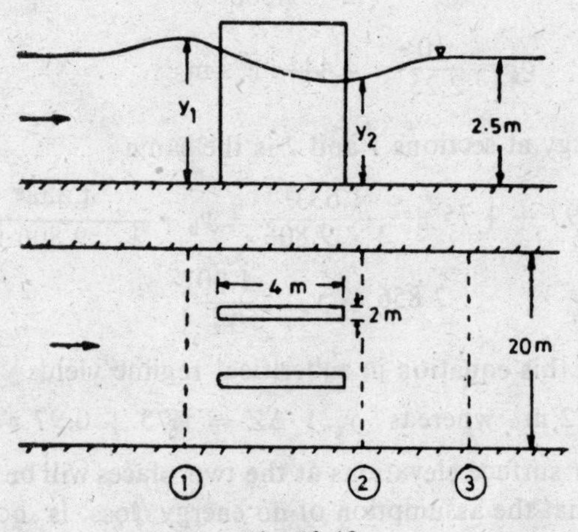

Fig. 13.18.

See Fig. 13.18. Under normal condition

$$R = \frac{20 \times 2.5}{25} = 2.0 \text{ m}$$

$$\therefore \qquad U_3 = \frac{1}{.015} \times 2^{2/3} \times (0.0006)^{1/2}$$

$$= 2.592 \text{ m/s}$$

and $$Q = 20 \times 2.5 \times 2.592$$

$$= 129.614 \text{ m}^3/\text{s}$$

At section 2 just downstream of piers

$$16\, U_2 y_2 = 129.614$$

At section 2, width of 16 m is taken because just downstream of piers is the wake region in which the flow has negligible velocity.

Or $$U_2 y_2 = 8.1$$

Application of momentum equation between sections 2 and 3 yields

$$Q\rho\,(U_3 - U_2) = \frac{1}{2}\,\gamma B\,(y_2^2 - y_3^2)$$

$$\frac{2 \times 129.614}{9.806 \times 20}\left(2.592 - \frac{8.1}{y_2}\right) = y_2^2 - (2.5)^2$$

which on simplification yields

$$y_2^3 - 9.677\, y_2 + 10.708 = 0$$

Its solution is $y_2 = 2.185$ m and $U_2 = \dfrac{8.1}{2.185} = 3.707$ m/s

Now apply energy equation between sections 1 and 2. Assume no energy loss since flow is contracting and sections 1 and 2 are sufficiently close.

$$y_1 + \frac{U_1^2}{2g} = y_2 + \frac{U_2^2}{2g}; \text{ but } \; U_1 = \frac{129.614}{20 \times y_1}$$

$$\therefore \qquad y_1 + \frac{2.142}{y_1^2} = 2.185 + 0.701$$

$$= 2.886$$

This gives $y_1 = 2.532$ m. Hence there is a rise in water level by

$$= 2.532 - 2.500 = 0.032 \text{ m.}$$

13.32 Show that for hydraulic jump in a rectangular channel

$$\frac{y_2}{y_1} = \tfrac{1}{2}\left[\sqrt{1 + 8F_{r1}^2} - 1\right]$$

and $$h_L = (y_2 - y_1)^3/4y_1 y_2$$

Hydraulic jump in a rectangular channel is shown in Fig. 13.6. The velocity distributions at sections 1 and 2 are uniform, the pressure distribution at these sections is hydrostatic, channel is horizontal, and boundary friction between sections 1 and 2 is negligible. Momentum equation between sections 1 and 2 gives

$$q\rho(U_2 - U_1) = \frac{\gamma}{2}(y_1^2 - y_2^2)$$

Continuity equation gives $U_1y_1 = U_2y_2 = q$

Combining these two equations one gets

$$y_1y_2(y_1 + y_2) = 2q^2/g$$

or

$$\frac{y_2}{y_1} + \left(\frac{y_2}{y_1}\right)^2 - 2F_{r1}^2 = 0$$

$$\therefore \quad \frac{y_2}{y_1} = \frac{-1 \pm \sqrt{1 + 8F_{r1}^2}}{2} = \tfrac{1}{2}[\sqrt{1 + 8F_{r1}^2} - 1]$$

since $\frac{y_2}{y_1}$ will be always +ve.

Also

$$h_L = H_1 - H_2 = y_1 + \frac{q^2}{2gy_1^2} - y_2 - \frac{q^2}{2gy_2^2}$$

$$= (y_1 - y_2) + \frac{q^2}{2g}\left(\frac{y_2^2 - y_1^2}{y_1^2y_2^2}\right)$$

Substituting from the equation $y_1y_2(y_1 + y_2) = \frac{2q^2}{g}$, the value of $\frac{q^2}{g}$ in the above equation and subsequent simplification yields

$$h_L = (y_2 - y_1)^3/4y_1y_2$$

13.33 Water flows over the spillway of a dam at a depth of 2.73 m over it (Fig. 13.19). The difference of elevation between spillway crest and downstream bed level is 30 m. If the discharge coefficient of spillway is 0.75,

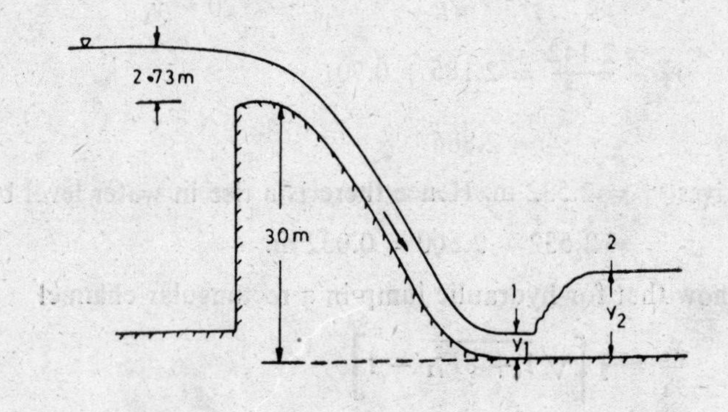

Fig. 13.19.

determine the water depth after the jump and head loss in the jump.

$$q = \frac{2}{3}\ C_d\ \sqrt{2g}H^{3/2}$$

$$= \frac{2}{3} \times 0.75 \times \sqrt{2 \times 9.806 \times 2.73^{3/2}}$$

$$= 9.99 \text{ m}^3/\text{s m}$$

Apply Bernoulli's equation between upstream of spillway and section 1.

$$\therefore \qquad 32.73 = y_1 + \frac{q^2}{2gy_1^2} = y_1 + \frac{5.089}{y_1^2}$$

Solving this equation by trial-error, for $y_1 = 0.396$, RHS $= 32.848$

$y_1 = 0.397$, RHS $= 32.686$

$\therefore$ $y_1 = 0.397$ m is taken to be correct depth.

$$U_1 = \frac{9.99}{.397} = 25.164 \text{ m/s and } \frac{U_1}{\sqrt{gy_1}} = F_{r1} = 12.754$$

$$\frac{y_2}{y_1} = \tfrac{1}{2}\,[\sqrt{1 + 8 \times 12.754^2} - 1]$$

$$= 17.543$$

$$\therefore \qquad y_2 = 17.543 \times 0.397 = 6.965 \text{ m}$$

$$h_L = \frac{(y_2 - y_1)^3}{4y_1y_2} = 25.612 \text{ m}$$

13.34 For discharge of 10 m³/m s and head loss of 20 m in the hydraulic jump, determine the depths of flow before and after the jump.

This problem needs to be solved by trial method. Following are the steps involved.

(i) Assume y_1

(ii) Determine $F_{r_1}^2 = q^2/gy_1^3$

(iii) Compute y_2 from the formula $\dfrac{y_2}{y_1} = \tfrac{1}{2}\,[\sqrt{1 + 8F_{r_1}^2} - 1]$

(iv) Compute $h_L = (y_2 - y_1)^3/4y_1y_2$

(v) If computed h_L is smaller than h_L given, decrease y_1; if computed h_L is greater than h_L given, increase y_1 and repeat computations till both are equal.

The computations are shown in tabular form.

y_1 (m)	$F_{r1}^2 = q^2/gy_1^3$	y_2/y_1	y_2 (m)	h_L (m)	
1.0	10.20	4.044	4.044	1.74	
0.50	81.58	12.28	6.140	14.63	
0.45	111.90	14.47	6.510	19.01	
0.44	119.71	14.98	6.591	20.06	Hence $y_1 = 0.44$ m
0.441	118.80	14.93	6.583	19.95	$y_2 = 6.591$ m

13.35 A negative step of height ΔZ is provided in a horizontal rectangular channel (Fig. 13.20). Determine expression for y_2/y_1 if the jump starts just at the step.

$$q = y_1 U_1 = y_2 U_2$$

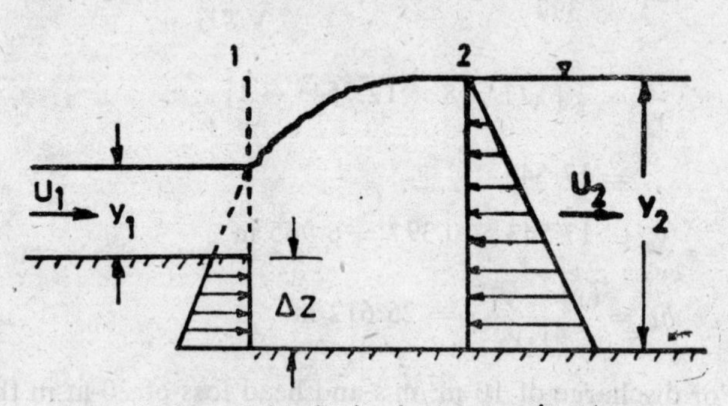

Fig. 13.20. Hydraulic jumb at a negative step.

Momentum equation must take into account additional force exerted by step on the fluid which is

$$\frac{[\gamma y_1 + \gamma(y_1 + \Delta Z)]}{2} \times \Delta Z$$

Hence

$$q\rho(U_2 - U_1) = \tfrac{1}{2}\gamma y_1^2 + \frac{\Delta Z}{2}[\gamma y_1 + \gamma(y_1 + \Delta Z)] - \tfrac{1}{2}\gamma y_2^2$$

Substituting, $q = U_1 y_1$ on LHS of the above equation, writing $U_2 = \dfrac{q}{y_2}$

and $U_1 = \dfrac{q}{y_1}$, one gets

$$\frac{U_1^2 y_1^2}{g}\left(\frac{y_1 - y_2}{y_1 y_2}\right) = \tfrac{1}{2}[y_1^2 - y_2^2 + 2y_1\Delta z + \Delta z^2]$$

This can be rewritten in the form

$$F_{r1}^2 = \frac{y_2}{2(y_2 - y_1)}\left[\left(\frac{y_2}{y_1}\right)^2 - \left(\frac{\Delta Z}{y_1} + 1\right)^2\right]$$

13.36 **Obtain the differential equation for water surface slope in the case of steady, gradually varied flow in a prismatic channel. List all the assumptions made.**

Assumptions made in the derivation are listed in section 13.8. See Fig. 13.8.

Total energy E at a section is given as

$$E = y + z + \frac{U^2}{2g}$$

$\therefore$ Differentiating w.r.t. x one gets

$$\frac{dE}{dx} = \frac{dy}{dx} + \frac{dz}{dx} + \frac{d}{dx}\left(\frac{U^2}{2g}\right)$$

However $\quad Q = AU \quad \therefore \quad \frac{d}{dx}\left(\frac{U^2}{2g}\right) = \frac{d}{dx}\left(\frac{Q^2}{2gA^2}\right) = \frac{d}{dy}\left(\frac{Q^2}{2gA^2}\right)\frac{dy}{dx}$

$$= -\frac{dy}{dx}\frac{Q^2}{gA^3}\frac{dA}{dy} = -\frac{dy}{dx}\left(\frac{Q^2T}{gA^3}\right)$$

$$\therefore \quad \frac{dE}{dx} = \frac{dy}{dx} + \frac{dz}{dx} - \frac{dy}{dx}\left(\frac{Q^2T}{gA^3}\right)$$

But $\frac{dE}{dx}$ = slope of total energy line $= -S_f$,

$\frac{dz}{dx}$ = channel bed slope $= -S_0$

$$\therefore \quad (S_0 - S_f) = \frac{dy}{dx}\left(1 - \frac{Q^2T}{gA^3}\right) \text{ or } \frac{dy}{dx} = \frac{S_0 - S_f}{1 - \dfrac{Q^2T}{gA^3}}$$

Since $\quad \frac{Q^2T}{gA^3} = \frac{1}{g}\frac{Q^2}{A^2}\left(\frac{T}{A}\right) = \frac{U^2}{gD} = F_r^2, \quad \frac{dy}{dx} = \frac{S_0 - S_f}{1 - F_r^2}$

If S_f is computed from Manning's or Chezy's equation assuming channel to be wide, the above equation can be expressed in the following two forms

$$\frac{dy}{dx} = \frac{1 - \left(\dfrac{y_0}{y}\right)^{10/3}}{1 - \left(\dfrac{y_c}{y}\right)^3} \quad \text{if Manning's equation is used,}$$

and

$$\frac{dy}{dx} = \frac{1 - \left(\dfrac{y_0}{y}\right)^3}{1 - \left(\dfrac{y_c}{y}\right)^3} \quad \text{if Chezy's equation is used.}$$

Other forms of the above equation can be obtained by introducing the definitions of section factor Z and conveyance K. Since section factor Z is defined as

$$Z = \sqrt{\frac{A^3}{T}}$$

$$Q^2T/gA^3 = \frac{Q^2}{gZ^2}$$

However for a given discharge Q, at the critical depth

$$\frac{Q^2}{g} = \left(\frac{A^3}{T}\right)_c = Z_c^2$$

$$\frac{Q^2T}{gA^3} = (Z_c/Z)^2$$

Further conveyance K is defined as $Q = K\sqrt{S}$ where $K = AR^{2/3}/n$ if Manning's equation is used.

For uniform flow,

$$S = S_0 \text{ and } Q = K_0\sqrt{S_0} \text{ or } S_0 = \frac{Q^2}{K_0^2}$$

where K_0 is value of K when depth is normal depth.

In gradually varied flow in Manning's equation, S is replaced by S_f

and
$$Q = K\sqrt{S_f} \quad \text{or} \quad \frac{Q^2}{K^2} = S_f$$

$$S_f/S_0 = (K_0/K)^2$$

Hence gradually varied flow equation can also be written as

$$\frac{dy}{dx} = \left(S_0 - S_f\right)\bigg/\left(1 - \frac{Q^2T}{gA^3}\right) = S_0 \frac{1 - \frac{S_f}{S_0}}{1 - \frac{Q^2T}{gA^3}}$$

or
$$\frac{dy}{dx} = S_0 \frac{1 - (K_0/K)^2}{1 - (Z_c/Z)^2}$$

13.37 Classify various surface profiles obtained in steady gradually varied flow in a prismatic channel.

The profiles occurring on mild slope ($y_0 > y_c$) are named M profiles, those occurring on critical slope ($y_0 = y_c$) are called C profiles, those occurring on steep slope ($y_0 < y_c$) are called S profiles, those occurring on horizontal bed are named as H profiles, and those occurring on adverse slope are named as A profiles. Further classification is done depending on the range of the depth of profile.

$M_1 : y > y_0 > y_c$ $\quad S_1 : y > y_c > y_0$ $\quad C_1 : y > y_c = y_0$

$M_2 : y_0 > y > y_c$ $\quad S_2 : y_c > y > y_0$ $\quad C_3 : y_c = y_0 > y$

$M_3 : y_0 > y_c > y$ $\quad S_3 : y_c > y_0 > y$

$H_2 : y > y_c$ $\quad A_2 : y > y_c$

$H_3 : y_c > y$ $\quad A_3 : y_c > y$

Further it can be shown from equation, $\dfrac{dy}{dx} = \dfrac{1 - \left(\dfrac{y_0}{y}\right)^3}{1 - \left(\dfrac{y_c}{y}\right)^3}$ that,

1. M_1 and S_1 curves meet y_0 line asymptotically and tend to be horizontal as $y \to \infty$.
2. M_2 and S_2 curves meet y_0 line asymptotically and y_c line normally.
3. M_3 and S_3 curves meet y_c line normally and also meet the channel bed normally.
4. C_1 and C_3 curves will be straight lines if Chezy's equation is used, otherwise they will be slightly curved.
5. A_2 tends to be horizontal as $y \to \infty$ and it intersects y_c line normally. A_3 curves tend to intersect y_c line and channel bed normally.
6. H_2 curve tends to be horizontal as $y \to \infty$ and it intersects y_c line normally. Similarly H_3 curve meets y_c line and channel bed normally.

Finally surface profiles can also be classified as backwater curves and drawdown curves. Backwater curves are those curves in which depth of flow increases in the direction of flow. (e.g. M_1, S_1, C_1, M_3, C_3, A_3, H_3). Drawdown curves are those curves in which depth of flow decreases in the direction of flow (e.g. H_2, S_2, A_2, H_2). These are shown in Fig. 13.8.

13.38 Consider a trapezoidal channel of 4.0 m bottom width with side slopes of 1:1 and bottom slope of 0.00015. If it carries a discharge of 2.485 m³/s with Manning's n of 0.02, determine the distance required to change the flow depth from 0.90 m to 0.50, using step method; classify the surface profile.

First determine normal depth y_0

$$2.485 = \frac{1}{0.02}(4y_0 + y_0^2)\left\{\frac{4y_0 + y_0^2}{4+2y_0\sqrt{2}}\right\}^{2/3}(0.00015)^{1/2}$$

Solving this, one gets $y_0 = 1.0$ m.

Also determine y_c

$$\frac{Q^2}{g} = \left(\frac{A^3}{T}\right) \text{or } \frac{(4y_c + y_c^2)^3}{4 + 2y_c} = \frac{2.485^3}{4.806}$$

Solution of this yields $y_c = 0.328$ m.
Since $y_c > y_0$ it is a mild slope and since 0.90 m and 0.50 m lie between y_0 and y_c it is type 2 profile namely M_2 (Fig. 13.21). Also it is a drawdown curve; hence depth will change from 0.90 m to 0.50 m in downstream direction.

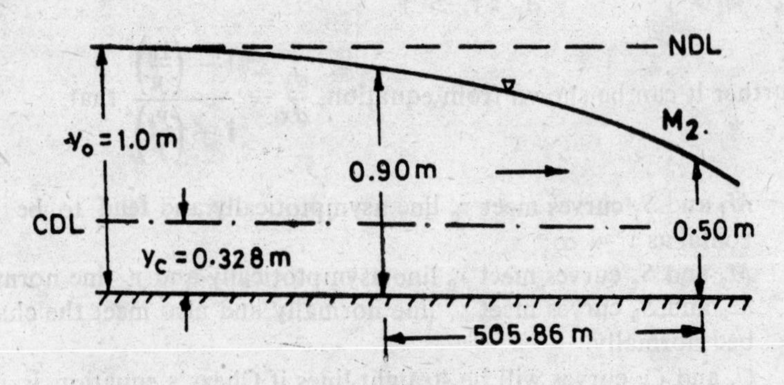

Fig. 13.21.

In a single step method $\Delta x = \dfrac{H_2 - H_1}{S_0 - \bar{S}_f}$

Here $\quad y_1 = 0.90 \text{ m} \quad U_1 = \dfrac{2.485}{(4 + 0.90) \times 0.90} = 0.563 \text{ m/s}$

$\quad y_2 = 0.50 \text{ m} \quad U_2 = \dfrac{2.285}{(4+ 0.50) \times 0.90} = 1.104 \text{ m/s}$

Further $\quad R_1 = \dfrac{(4 + 0.9) \times 0.90}{4 + 2\sqrt{2} \times 0.9} = 0.674 \text{ m}$

$$R_2 = \frac{(4 + 0.50) \times 0.50}{4 + 2\sqrt{2} \times 0.50} = 0.416 \text{ m}$$

$$\therefore \quad H_1 = 0.90 + \frac{0.563^2}{2 \times 9.806} = 0.9162 \text{ m}$$

$$H_2 = 0.50 + \frac{1.104^2}{2 \times 9.806} = 0.5621 \text{ m}$$

$$S_{f1} = \frac{U_1^2 n^2}{R_1^{4/3}} = \frac{0.563^2 \times 0.02^2}{(0.674)^{1.333}} = 0.000\,2145$$

$$S_{f2} = \frac{U_1^2 n^2}{R_2^{4/3}} = \frac{1.104^2 \times 0.02^2}{(0.416)^{1.333}} = 0.001\,569$$

$$\therefore \quad \bar{S}_f = \frac{0.0002145 + 0.001569}{2} = 0.000\,8918$$

$$\therefore \quad \Delta x = \frac{0.5621 - 0.9162}{0.00015 - 0.0008918} = 505.86 \text{ m}$$

In order to increase the accuracy of computations, one could have divided the depth range 0.50 to 0.90 m into small ranges such as 0.90 to 0.80, 0.80 to 0.70, 0.70 to 0.6 and 0.6 to 0.50 m and step method could have been used. In such a case, the computations are performed in the following tabular form

y m	A m²	U m/s	$\frac{U^2}{2g}$ m	$H = y + \frac{U^2}{2g}$ m	R m	$S_f = \frac{U^2 n^2}{R^{4/3}}$	$\bar{S_f}$	$S_0 - \bar{S_f}$	$\Delta H = H_2 - H_1$	$\Delta x = \frac{\Delta H}{S_0 - \bar{S_f}}$	$\Sigma \Delta x$
0.9											
0.8											

13.39 **Discuss graphical integration method for computing surface profiles.** Consider the differential equation for gradually varied flow

$$\frac{dy}{dx} = \frac{S_0 - S_f}{1 - \frac{Q^2 T}{g A^3}}$$

For a given discharge and channel shape, S_f, T and A are all functions of depth y. Hence the above equation can be written as

$$dx = f(y)dy$$

where $f(y) = \left(1 - \frac{Q^2 T}{g A^3}\right) \Big/ (S_0 - S_f).$

Integration of this equation yields

$$\int_{x_1}^{x_2} dx = \int_{y_1}^{y_2} f(y) dy$$

where y_1 and y_2 are depths at distances x_1 and x_2; or

$$x = (x_2 - x_1) = \frac{f(y_1) + f(y_2)}{2} (y_2 - y_1)$$

In the graphical method, one plots $f(y)$ vs y over the practical range of y values. The area under the curve between $y = y_1$ and $y = y_2$ gives the distance $x = (x_2 - x_1)$ required for the change in depth from y_1 to y_2. Alternately, one can make the area computations in the tabular form as given below, by dividing the depth range $(y_2 - y_1)$ into small ranges.

y m	T m	A m	P m	R m	$\frac{Q^2T}{gA^3}$	$1-\frac{Q^2T}{gA^3}$	$S_f=\frac{U^2n^2}{R^{4/3}}$	$S_0-\bar{S}_f$	$f(y)$	$\Delta x=\frac{f(y_1)+f(y_2)}{2}\times(y_1-y_2)$	$\Sigma\Delta x$

13.40 Value of Chezy's C for a wide rectangular channel carrying 5.3 m³/s m is 60. If the channel slope suddenly changes from 0.0009 to 0 0049, name the surface profiles occurring on the two slopes and determine their lengths from the control section upto $y/y_0 = 0.999$ on 0.0009 slope and upto $\frac{y}{y_0} = 1.001$ on the downstream slope using Bresse's method.

First determine normal depths on the two slopes of channel.

For $S_{01} = 0.0009$

$$y_{01} = \left(\frac{q^2}{C^2 S_{01}}\right)^{0.333} = \left(\frac{5.3^2}{60^2 \times 0.0009}\right)^{0.333} = 2.054 \text{ m}$$

For $S_{02} = 0.0049$

$$y_{02} = \left(\frac{5.3^2}{60^2 \times 0.0049}\right)^{0.333} = 1.168 \text{ m}$$

The critical depth

$$y_c = \sqrt[3]{\frac{5.3^2}{9.806}} = 1.420 \text{ m}$$

Since $y_{01} > y_c$, 0.0009 is a mild slope whereas since $y_c > y_{02}$, 0.0049 is a steep slope. Further, profile AB between y_{01} and y_c will be M_2 drawdown curve and profile BC between y_c and y_{02} will be S_2, also a drawdown curve (Fig. 13.22).

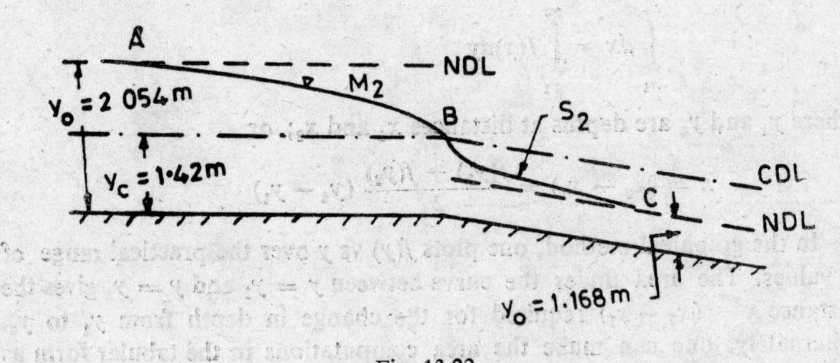

Fig. 13.22.

According to Bresse's method

$$\frac{dy}{dx} = \frac{1 - (y_0/y)^3}{1 - (y_c/y)^3}$$

can be integrated to yield

$$(x_2 - x_1) = \frac{y_0}{S_0}\left[(u_2 - u_1) - \left(1 - \left(\frac{y_c}{y_0}\right)^3\right)\int_{u_1}^{u_2}\frac{du}{1 - u^3}\right]$$

where $u = \frac{y}{y_0}$. For M_2 curve $u_1 = 0.999$ and $u_2 = \frac{y_c}{y_{01}} = 0.691$

Referring to Table 13.2, $\int_{0.909}^{0.691}\frac{du}{1 - u^3} = (0.763 - 2.788) = -2.025$

$$\therefore \quad (x_2 - x_1) = \frac{2.054}{0.0009}\{(0.691 - 0.999) - (1 - 0.691^3)(-2.025)\}$$

$$= 2282\{(-0.308) + 0.670 \times 2.025\}$$

$$= 2392.2 \text{ m}$$

For S_2 profile $u_1 = \frac{y_c}{y_{02}} = \frac{1.420}{1.168} = 1.216 \quad u_2 = 1.001$

From Table 13.2, $\int_{1.216}^{1.001}\frac{du}{1 - u^3} = 2.184 - 0\ 461$

$$(x_2 - x_1) = \frac{1.168}{0.0049}\left\{(1.001 - 1.216) - \left(1 - \left(\frac{1.420}{1.168}\right)^3\right)(2.184 - 0.461)\right\}$$

$$= 238.367\{-0.215 + 0.797 \times 1.723\} = 276.08 \text{ m}$$

Bresse has shown that the integral $\int\frac{du}{1 - u^3}$ is

$$\frac{1}{6}\ln\frac{u^2 + u + 1}{(u - 1)} - \frac{1}{\sqrt{3}}\cot^{-1}\frac{2u + 1}{\sqrt{3}} = \phi(u)$$

Hence in this particular case one can solve the problem without using Table 13.2.

13.41 Discuss Chow's method of integrating gradually varied flow equation.

In Example 13.36, it is shown that gradually varied flow equation can be written in terms of conveyance $K = \frac{Q}{\sqrt{S}}$ and section factor $Z = \sqrt{A^3/T}$ as

$$\frac{dy}{dx} = S_0\,\frac{1 - (K_0/K)^2}{1 - (Z_c/Z)^2}$$

For channels of a given shape K^2 and Z^2 can be expressed as $K^2 \sim y^N$ and

Table 13.2. The Varied-Flow Function

$$F(u, N) = \int_0^u \frac{du}{1 - u^N}$$ with constant of integration adjusted so that $F(0, N) = 0$ and $F(\infty, N) = 0$

u \ N	2.50	3.0	3.33	4.00	u \ N	2.50	3.00	3.33	4.00
0.00	0.000	0.000	0.000	0.000	0.20	0.201	0.200	0.200	0.200
0.02	0.020	0.020	0.020	0.020	0.22	0.222	0.221	0.220	0.220
0.04	0.040	0.040	0.040	0.040	0.24	0.243	0.241	0.240	0.240
0.06	0.060	0.060	0.060	0.060	0.26	0.263	0.261	0.261	0.260
0.08	0.080	0.080	0.080	0.080	0.28	0.284	0.282	0.281	0.280
0.10	0.100	0.100	0.100	0.100	0.30	0.305	0.302	0.301	0.300
0.12	0.120	0.120	0.120	0.120	0.32	0.326	0.323	0.322	0.321
0.14	0.140	0.140	0.140	0.140	0.34	0.347	0.343	0 342	0.341
0.16	0.161	0.160	0.160	0.160	0 36	0.368	0.364	0.363	0.361
0.18	0.181	0.180	0.180	0.180	0.38	0.391	0.385	0.383	0.381

(Contd.)

Table 13.2. *(Contd.)*

1	2	3	4	5	6	7	8	9	10
0.40	0.413	0.407	0.404	0.402	0.65	0.46	0.703	0.692	0.676
0.42	0.435	0.428	0.425	0.423	0.66	0.72	0.717	0.705	0.688
0.44	0.458	0.450	0.447	0.443	0.67	0.76	0.731	0.718	0.701
0.46	0.481	0.472	0.469	0.464	0.68	0.777	0.746	0.732	0.713
0.48	0.504	0.494	0.490	0.485	0.69	0.795	0.761	0.746	0.726
0.50	0.528	0.517	0.512	0.506	0.70	0.811	0.776	0.760	0.739
0.52	0.553	0.540	0.533	0.528	0.71	0.828	0.791	0.775	0.752
0.54	0.578	0.563	0.557	0.550	0.72	0.845	0.807	0.790	0.766
0.56	0.604	0.587	0.580	0.572	0.73	0.863	0.823	0.805	0.780
9.58	0 631	0.612	0.604	0.594	0.74	0.881	0.840	0.821	0.794
0.60	0.658	0.637	0.628	0.617	0.75	0.900	0.857	0.837	0.808
0.61	0.673	0.650	0.641	0.628	0.76	0.919	0.874	0.853	0.823
0.62	0.686	0.663	0.653	0.640	0.77	0.940	0.892	0.870	0.838
0.63	0.700	0.676	0.666	0.652	0.78	0.962	0.911	0.887	0.854
0.64	0.716	0.690	0.679	0.664	0.79	0.985	0.930	0.905	0 870

(Contd.)

Table 13.2. *(Contd.)*

1	2	3	4	5	6	7	8	9	10
0.80	1.008	0 950	0 924	0.887	0.950	1.605	1.467	1.398	1.296
0.81	1.032	0.971	0.943	0.904	0.960	1.703	1.545	1.468	1.355
0.82	1.057	0.993	0.396	0.922	0.970	1.823	1.644	1.559	1.431
0.83	1.083	1.016	0.985	0.940	0.975	1.899	1.707	1.615	1.478
0.84	1.110	1.040	1.007	0.960	0.980	1.996	1.783	1.684	1.536
0.85	1.139	1.065	1.030	0.980	0.985	2.111	1.880	1.772	1.610
0.86	1.171	1.092	1.055	1.002	0.990	2.273	2.017	1.895	1.714
0.87	1.205	1.120	1.081	1.025	0.995	2.550	2.250	2.106	1.889
0.88	1.241	1.151	1.109	1.049	0.999	3.195	2.788	2.590	2.293
0.89	1.279	1.183	1.139	1.075	1.000	∞	∞	∞	∞
0.90	1.319	1.218	1.172	1.103	1.001	2.786	2.184	1.907	1.508
0.91	1.362	1.257	1.206	1.133	1.005	2.144	1.649	1.425	1.107
0.92	1.400	1.300	1.246	1.166	1.010	1.867	1.419	1.218	0.936
0.93	1.455	1.348	1.290	1.204	1.015	1.705	1.286	1.099	0.836
0.94	1.520	1.403	1.340	1.246	1.020	1.602	1.191	1.014	0.766

(Contd.)

Table 13.2. (*Contd.*)

1	2	3	4	5	6	7	8	9	10
1.03	1.436	1.060	0.896	0.688	1.18	0.760	0.509	0.406	0.272
1.04	1.321	0.967	0.813	0.600	1.19	0.740	0.494	0.393	0.262
1.05	1.242	0.896	0.749	0.548	1.20	0.723	0.480	0.381	0.252
1.06	1.166	0.838	0.697	0.506	1.22	0.692	0.454	0.358	0.235
1.07	1.111	0.790	0.651	0.471	1.24	0.662	0.431	0.338	0.219
1.08	1.059	0.749	0.618	0.441	1.26	0.633	0.410	0.320	0.205
1.09	1.012	0.713	0.586	0.415	1.28	0.609	0.391	0.303	0.193
1.10	0.973	0.681	0.558	0.392	1.30	0.587	0.373	0.289	0.181
1.11	0.939	0.652	0.532	0.372	1.32	0.568	0.357	0.275	0.171
1.12	0.907	0.626	0.509	0.354	1.34	0.549	0.342	0.262	0.162
1.13	0.878	0.602	0.488	0.337	1.36	0.531	0.329	0.251	0.153
1.14	0.851	0.581	0.479	0.322	1.38	0.513	0.316	0.239	0.145
1.15	0.824	0.561	0.452	0.308	1.40	0.496	0.304	0.229	0.138
1.16	0.802	0.542	0.436	0.295	1.42	0.481	0.293	0.220	0.131
1.17	0.782	0.525	0.421	0.283	1.44	0.467	0.282	0.211	0.125

(*Contd.*)

Table 13.2. *(Contd.)*

1	2	3	4	5	6	7	8	9	10
1.46	0.455	0.272	0.203	0.119	2.3	0.204	0.098	0.064	0.028
1.48	0.444	0.263	0.196	0.113	2.4	0.190	0.089	0.057	0.024
1.50	0.432	0.255	0.188	0.108	2.5	0.179	0.082	0.052	0.022
1.55	0.405	0.235	0.172	0.097	2.6	0.169	0.076	0.048	0.019
1.60	0.380	0.218	0.158	0.087	2.7	0.160	0.070	0.043	0.017
1.65	0.359	0.203	0.145	0.079	2.8	0.150	0.065	0.040	0.015
1.70	0.340	0.189	0.135	0.072	2.9	0.142	0.060	0.037	0.014
1.75	0.322	0.177	0.125	0.065	3.0	0.135	0.056	0.034	0.012
1.80	0.308	0.166	0.116	0.060	3.5	0.106	0.041	0.024	0.008
1.85	0.293	0.156	0.108	0.055	4.0	0.087	0.031	0.017	0.005
1.90	0.279	0.147	0.102	0.050	4.5	0.072	0.025	0.013	0.004
1.95	0.268	0.139	0.095	0.046	5.0	0.062	0.019	0.010	0.003
2.00	0.257	0.132	0.089	0.043	6.0	0.048	0.014	0.007	0.002
2.10	0.238	0.119	0.079	0.037	7.0	0.038	0.010	0.005	0.001
2.20	0.220	0.107	0.071	0.032	8.0	0.031	0.008	0.004	0.001

(Contd.)

Table 13.2. *(Contd.)*

1	2	3	4	5	6	7	8	9	10
9.0	0.027	0.006	0.003	0.000	20.0	0.015	0.002	0.001	0.000
10.0	0.022	0.005	0.002	0.000					

$Z^2 \sim y^M$ where M and N are functions of channel shape. Hence the above equation can be written as

$$\frac{dy}{dx} = \frac{1-(y_0/y)^N}{1-(y_c/y)^M}$$

Values of N and M for different shapes of channel are listed below:

Shape	M	N
Wide rectangular	3.0	3.33
Narrow and deep rectangular	3.0	2.00
Triangular	5.0	5.33
Trapezoidal with different side slopes and y/B values	3.0 – 5.0	3.33 – 5.0

Actually for the range of depths to be studied, one can plot K^2 and Z^2 vs y on a log-log graph paper and obtain appropriate values of N and M for any channel. For channels of regular shape such as trapezoidal, M and N can be obtained as a function of side slope, bottom width to y_0 ratio and u. Then an average value of N and M has to be obtained for range of u_1 and u_2.

As in case of Bresse's method, assume $y/y_0 = u$. Then the above equation becomes

$$dx = \frac{y_0}{S_0}\left[1 - \frac{1}{(1-u^N)} + \left(\frac{y_c}{y_0}\right)^M \int \frac{u^{M-N}}{1-u^N}\right] du$$

which on integration yields

$$x = \frac{y_0}{S_0}\left[u - \int \frac{du}{1-u^N} + \left(\frac{y_c}{y_0}\right)^M \int \frac{u^{M-N}}{1-u^N}\, du\right] + C$$

where C is the constant of integration.

If one substitutes $V = u^{N/J}$ and $J = \dfrac{N}{(N-M+1)}$, the second integral on right hand side can be converted to the same form as the first one. Hence

$$(x_2 - x_1) = \frac{y_0}{S_0}\left[(u_2 - u_1) - \{\phi(u_2, N) - \phi(u_1, N)\} + \left(\frac{y_c}{y_0}\right)^M \frac{J}{N}\{\phi(V_2, J) - \phi(V_1, J)\}\right]$$

in which $\phi(u, N) = \displaystyle\int_0^u \frac{du}{1-u^N}$.

Function ϕ (u, N) or ϕ (V, J) is tabulated in Table 13.2 for N or J values of 2.5, 3 and 3.33. When $N = 3$, ϕ becomes Bresse's function.

13.42 A wide rectangular channel carries a discharge of 3.72 m³/m s on a slope of 0.001 having Manning's n of 0.025. Calculate the length required to change depth from 2.480 m to 3.05 m. Identify the surface profile and perform computations using Chow's method.

$$y_c = \sqrt[3]{\frac{3.72^2}{9.806}} = 1.122 \text{ m}$$

$$y_o = \left(\frac{3.72 \times 0.025}{0.001^{1/2}}\right)^{3/5} = 1.91 \text{ m}$$

Since y_1 and y_2 are greater than y_c and y_0, and y_c is less than y_0, it is M_1 profile. For wide rectangular channel $N = 3.33$ and $M = 3.0$.

$$J = \frac{N}{N - M + 1} = \frac{3.33}{1.33} = 2.504; \text{ also } N/J = 1.33.$$

Using Table 13.2, following values are computed

y	u	$V = u^{N/J}$	$\phi(u, N)$	$\phi(V, J)$
2.48	1.299	1.416	0.289	0.483
3.05	1.597	1.864	0.158	0.287

$$\therefore \quad (x_2 - x_1) = \frac{1.91}{0.001}\left\{(1.597 - 1.299) - (0.158 - 0.289) + \left(\frac{1.122}{1.910}\right)^3 \frac{2.504}{3.33}(0.287 - 0.483)\right\}$$

$$= 1910\,\{0.298 + 0.131 - .0299\}$$

$$= 762.28 \text{ m}$$

13.43 Sketch the possible gradually varied flow profiles when a steep slope is followed by a mild slope.

The relative positions of critical depth and normal depth lines on the slope combination are shown in Fig. 13.23. Depending on the magnitude of mild slope, the position of NDL on it will change and two such positions are shown. Far away on steep slope, normal depth prevails and it must rise to NDL_1.

One can check if the normal depth on steep slope is sequent to NDL_1 on mild slope. If so, hydraulic jump J_1 will take place at the toe. However, if NDL_2 is lower than the sequent depth to NDL on steep slope, M_3 profile

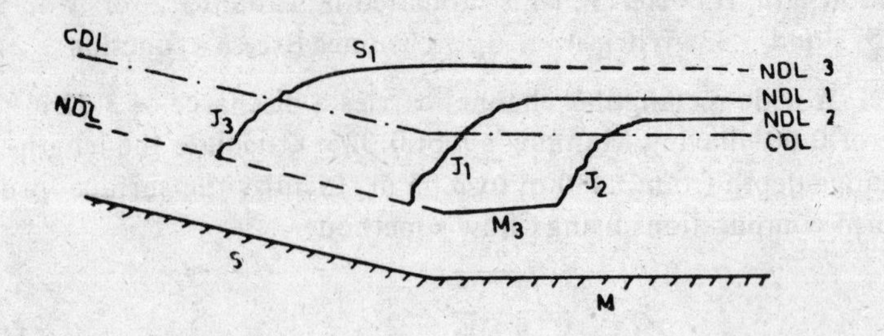

Fig. 13.23.

will prevail on mild slope and then hydraulic jump J_2 will occur. If NDL_3 on mild slope is larger than sequent depth to NOL on steep slope, hydraulic Jump J_3 followed by S_1 will occur.

13.43 A surge represents an abrupt change in the depth of flow caused by regulation of flow at a section due to gate control, etc. When friction is neglected, the surge moves upstream or downstream at a constant velocity. Obtain expression for celerity C.

Suppose a surge is created in prismatic open channel by opening a sluice gate at its upstream end [Fig. 13.9 (a)]. Let U_1 and U_2 be the velocities in the channel and under the surge and U_w be the velocity of surge. Make this problem a steady state problem by applying velocity U_w to the surge and to the flow to the left.

$$\therefore \quad U_2 \to U_2 - U_w$$

$$U_1 \to U_1 - U_w$$

$$U_w \to 0$$

Apply Momentum equation between sections 1 and 2.

$$\therefore \quad \gamma A_2 \bar{y}_2 - \gamma A_1 \bar{y}_1 = A_2 \rho \,(U_2 - U_w)\,[(U_1 - U_w) - (U_2 - U_w)]$$

$$= A_2 \rho \,(U_2 - U_w)\,(U_1 - U_2)$$

or
$$g\left(\bar{y}_2 - \frac{A_1}{A_2}\,\bar{y}_1\right) = (U_w - U_2)\,(U_2 - U_1) \tag{1}$$

Also, from continuity equation $A_1(U_1 - U_w) = A_2(U_2 - U_w)$

or
$$U_2 = \frac{A_1 U_1 + A_2 U_w - A_1 U_w}{A_2} \tag{2}$$

Substituting Eq. 2 in Eq. 1, one gets

$$g\left(\bar{y}_2 - \bar{y}_1\,\frac{A_1}{A_2}\right) = \left[\frac{A_1(U_w - U_1)}{A_2}\right]\left[\frac{-U_1(A_2 - A_1) + U_w(A_2 - A_1)}{A_2}\right]$$

$$= \frac{A_1}{A_2}\,(U_w - U_1).\ \frac{A_2 - A_1}{A_2}\,(U_w - U_1)$$

$$\therefore \qquad (U_w - U_1)^2 = \frac{g(A_2\bar{y}_2 - A_1\bar{y}_1)}{A_1\left(1 - \dfrac{A_1}{A_2}\right)}$$

or

$$U_w - U_1 = \sqrt{\frac{(A_2\bar{y}_2 - A_1\bar{y}_1)g}{A_1\left(1 - \dfrac{A_1}{A_2}\right)}}$$

$$\therefore \qquad U_w = U_1 + \sqrt{\frac{(A_2\bar{y}_2 - A_1\bar{y}_1)g}{A_1\left(1 - \dfrac{A_1}{A_2}\right)}} \tag{3}$$

This is the expression for the absolute velocity of the surge. Since surge is travelling in the direction of initial velocity, only +ve value of the square root is considered.

If $U_w - U_1 = C$, the celerity of wave w.r.t. velocity U_1,

$$C = \sqrt{\frac{(A_2\bar{y}_2 - A_1\bar{y}_1)g}{A_1\left(1 - \dfrac{A_1}{A_2}\right)}}$$

For rectangular channel, $A_1 = By_1$ and $A_2 = By_2$

$$\bar{y}_1 = \frac{y_1}{2} \qquad \bar{y}_2 = \frac{y_2}{2}$$

$$\therefore \qquad C = \sqrt{\frac{(y_2^2 - y_1^2)gy_2}{2y_1(y_2 - y_1)}} = \sqrt{\frac{gy_2}{2y_1}(y_1 + y_2)} \tag{4}$$

If one writes $\quad y_2 = y_1 + h$

$$C = \sqrt{\frac{g(y_1 + h)(2y_1 + h)}{2y_1}}$$

$$= \sqrt{gy_1}\sqrt{\tfrac{1}{2}\left(1 + \frac{h}{y_1}\right)\left(2 + \frac{h}{y_1}\right)}$$

$$= \sqrt{gy_1}\sqrt{1 + \frac{3}{2}\left(\frac{h}{y_1}\right) + \tfrac{1}{2}\left(\frac{h}{y_1}\right)^2}$$

If h is much smaller than y_1, $\frac{1}{2}\left(\dfrac{h}{y_1}\right)^2$ can be neglected in preference to 1 and $\dfrac{3h}{2y_1}$. Hence

$$C = \sqrt{gy_1}\left[1 + \frac{3}{2}\left(\frac{h}{y_2}\right)\right]^{1/2}$$

$$\therefore \qquad \frac{C}{\sqrt{gy_1}} \approx 1 + \frac{3}{4}\frac{h}{y_1} \tag{5}$$

Further if $\dfrac{h}{y_1} \ll 1$, $\dfrac{C}{\sqrt{gy_1}} = 1$ $\qquad (6)$

13.45 A gate is to be suddenly dropped into place closing a rectangular channel 2.0 m deep and 3.0 m wide in which 5.67 m^3/s discharge flows at a depth of 1·22 m. What will be the velocity of the travelling surge produced and its height (Fig. 13.24)?

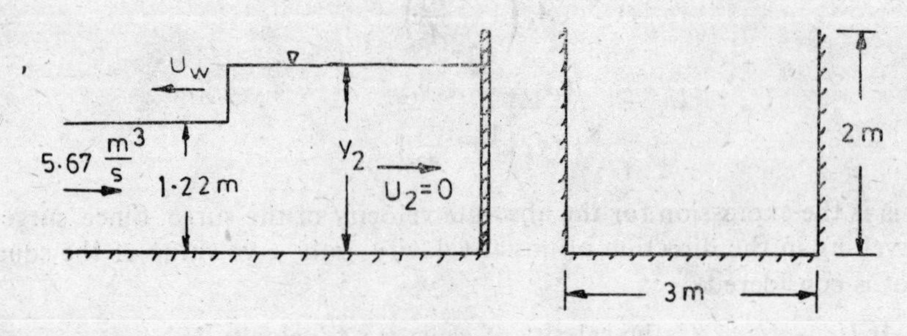

Fig. 13.24.

$U_1 = \dfrac{5.67}{3 \times 1.22} = 1.549$ m/s. Make the surge stationary by applying velocity U_w to the right.

∴ According to continuity equation $y_1(U_1 + U_w) = y_2 U_w$

Also momentum equation gives

$$y_1(U_1 + U_w)\,\rho\,[U_w - (U_1 + U_w)] = \tfrac{1}{2}\,\gamma(y_1^2 - y_2^2)$$

Substitute $y_1 = 1.22$ m, $U_1 = 1.549$ m/s

$$\therefore \qquad 1.22(1.549 + U_w) = y_2 U_w$$

$$\therefore \qquad \frac{1.89}{(y_2 - 1.22)} = U_w$$

Substitute this in momentum equation

$$1.22\left(1.549 + \frac{1.89}{y_2 - 1.22}\right) \times 1.549 = \frac{9.806}{2}\,(-\,1.22^2 + y_2^2)$$

This equation simplifies to

$$y_2^3 - 1.22y_2^2 - 5.409y_2 + 1.816 = 0$$

The solution of the above yields $y_2 = 1.82$ m

$$\therefore \qquad U_w = \frac{1.89}{(1.82 - 1.22)} = 3.15 \text{ m/s}$$

13.46 A rectangular channel carries water at a depth of 2.0 m and a velocity of 1.5 m/s. Sudden opening of a gate at its upstream end causes a surge of depth 3.5 m. Determine absolute velocity of surge and increased discharge (Fig. 13.25).

$$C = \sqrt{gy_1}\,\sqrt{\left(1 + \frac{3}{2}\left(\frac{h}{y_1}\right) + \frac{1}{2}\left(\frac{h}{y_1}\right)^2\right.}$$

Here $h = 3.5 - 2.0 = 1.50$ m

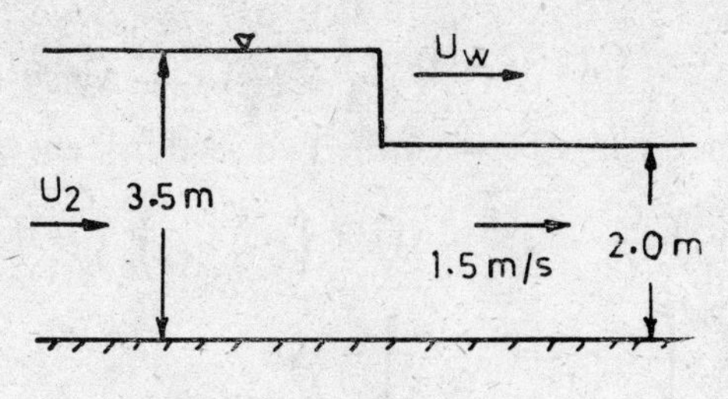

Fig. 13.25.

$\therefore \qquad h/y_1 = \dfrac{1.50}{2} = 0.75$

$\therefore \qquad C = \sqrt{9.806 \times 2.0}\sqrt{1 + \left(\frac{3}{2}\times 0.75\right) + \frac{1}{2}(0.75)^2}$

$= 6.869 \text{ m/s}$

But $\qquad C = U_w - U_1$

$\therefore \qquad U_w = 6.896 + 1.500 = 8.396 \text{ m/s}$

Also $\qquad (U_w - U_1)y_1 = (U_w - U_2)y_2$

$6.869 \times 2.0 = (8.369 - U_2) \times 3.50$

$\therefore \qquad U_2 = 4.444 \text{ m/s}$

$\therefore \qquad q = U_2 y_2 = 4.444 \times 3.5$

$= 15.554 \text{ m}^3\text{/s m}$

13.47 Consider propagation of a very small wave on the surface of water in a channel. The water is otherwise at rest. Apply energy equation and obtain the expression for celerity of the wave (Fig. 13.26).

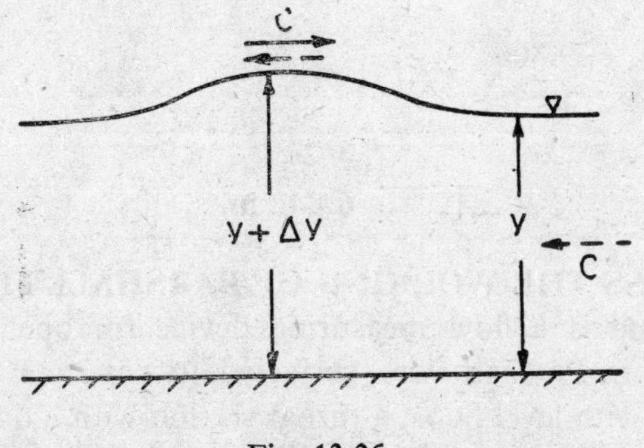

Fig. 13.26.

Make wave stationary by applying reverse velocity to wave and fluid. Apply continuity and momentum equations.

If C_1 is velocity of flow under wave under this condition

$$Cy = C_1(y + \Delta y) \quad \text{or} \quad C_1 = \frac{Cy}{y + \Delta y}$$

Assuming no energy loss between the two sections, energy equation gives

$$y + \frac{C^2}{2g} = (y + \Delta y) + \left(\frac{Cy}{y + \Delta y}\right)^2 \frac{1}{2g}$$

$$\therefore \quad \frac{C^2}{2g}\left[1 - \frac{y^2}{(y + \Delta y)^2}\right] = \Delta y$$

or

$$C^2 = \frac{2g\Delta y(y + \Delta y)^2}{\Delta y^2 + 2y\Delta y} = \frac{2g(y + \Delta y)^2}{2y + \Delta y}$$

$$\therefore \quad C = \sqrt{gy}\left\{\frac{2}{y}\frac{(y + \Delta y)^2}{(2y + \Delta y)}\right\}$$

This can be simplified by assuming that Δy is so small that Δy^2 can be neglected.

$$C = \sqrt{gy}\left(1 + \frac{3}{2}\frac{\Delta y}{y}\right)^{1/2}$$

$$= \sqrt{gy}\left(1 + \frac{3}{4}\frac{\Delta y}{y} + \ldots\right)$$

13.48 A floating object on the surface of deep sea is found to come on the crest 30 times per minute. Determine the wave length and velocity of propagation of waves.

$$\therefore \quad \text{Frequency} = \frac{C}{\lambda} = \frac{30}{60}\, s^{-1} \quad \therefore\ C = \frac{\lambda}{2}$$

Also for deep water waves $C^2 = g\lambda/2\pi$

$\therefore$ Substituting the value of λ from above, one gets

$$C^2 = \frac{2Cg}{2\pi}$$

$$\therefore \quad C = \frac{9.806}{3.142} = 3.121 \text{ m/s}$$

$$\therefore \quad \lambda = 2c$$

$$= 2 \times 3.121 = 6.242 \text{ m.}$$

13.49 DISCUSS THE WORKING OF PARSHALL FLUME

Parshall flume is a flow measuring device for open channels and is named after R.L. Parshall who perfected its design. It consists of a converging section with level floor, a throat section with a downstream sloping floor and a diverging section with an upward sloping floor, as shown in Fig. 13.27. The control section of the flume is not located in the throat but near the end of level 'crest' in the converging section. Typical dimensions of medium-sized Parshall flume are shown in Fig. 13.27. However, detailed dimensions of different sized flumes are available in the literature. The discharge equation for the flume is obtained by calibration and is of the form

$$Q = ky_a^n$$

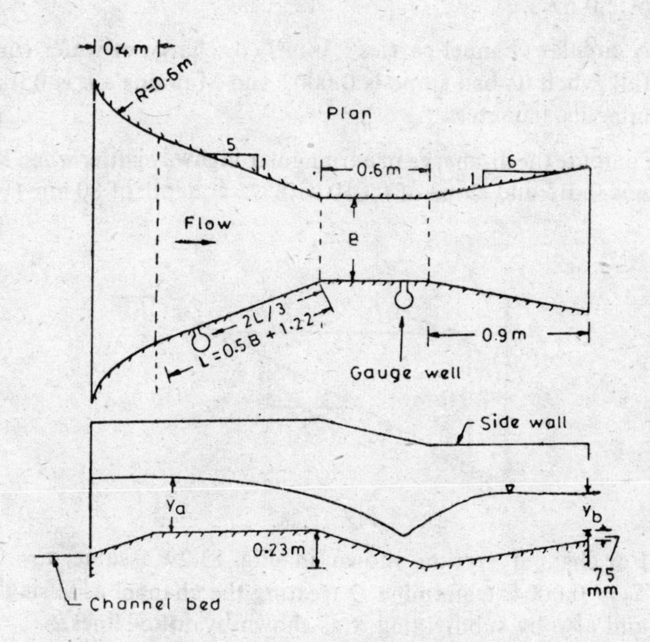

Fig. 13.27. Parshall Flume.

where k depends on throat width and n varies from 1.55 to 1.60 as throat width increases from 25 mm to 15 m. As long as y_b/y_a is less than 0.70, Q is uniquely determined by y_a. When y_b/y_a is larger, Q is somewhat reduced due to submergence. This decrease in Q is a function of y_a and y_b/y_a. Parshall flume should not be used as a discharge measuring device if y_b/y_a exceeds 0.95.

PROBLEMS

13.1 A Rectangular channel of 2.0 m width carries water at a depth of 0.30 m. If the channel slope is 0.0025 and Manning's n is 0.015, determine the mean velocity and discharge. (1.255 m/s, 0.753 m³/s)

13.2 A trapezoidal channel with bottom width of 4.0 m and side slopes $1H{:}1V$ has Manning's n of 0.02. If it carries a discharge of 2.485 m³/s at a depth of 1.0 m, calculate average velocity and bed slope.

(0.497 m/s, 0.00015)

13.3 Determine the normal depth in a 3.0 m wide rectangular channel carrying 1.5 m³/s discharge with channel slope of 0.0004 and Manning's n of 0.013. (0.581 m)

13.4 Determine the normal depth in a triangular channel with bed slope of 0.0004, $n = 0.015$ and central angle of 90° if it carries a discharge of 2.0 m³/s. (1.51 m)

13.5 A circular channel carries 2.0 m³/s discharge of water running half full when its bed slope is 0.0002 and Manning's n is 0.015. Determine its diameter. (2.662 m)

13.6 Compute the discharge in a triangular highway gutter when Manning's n is 0.015 and slope of 0.010 with water depth of 80 mm (Fig. 13.28) (54.93 l/s)

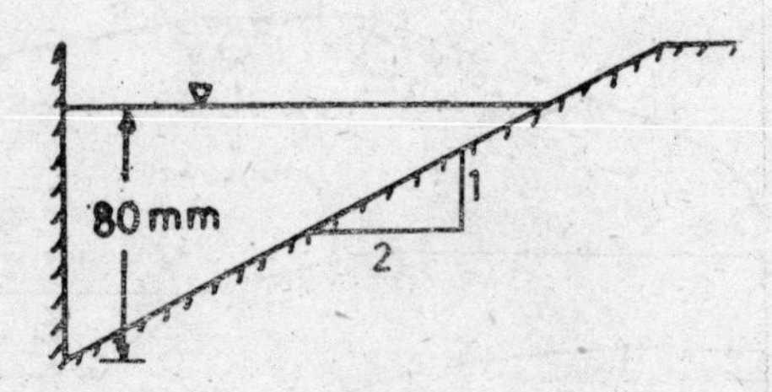

Fig. 13.28.

13.7 For channel section shown in Fig. 13.29 assume $n = 0.025$ and $S_0 = 0.0004$. Determine Q treating the channel as a single section and also by subdividing it as shown by dotted lines. (46.95 m³/s, 57.745 m³/s)

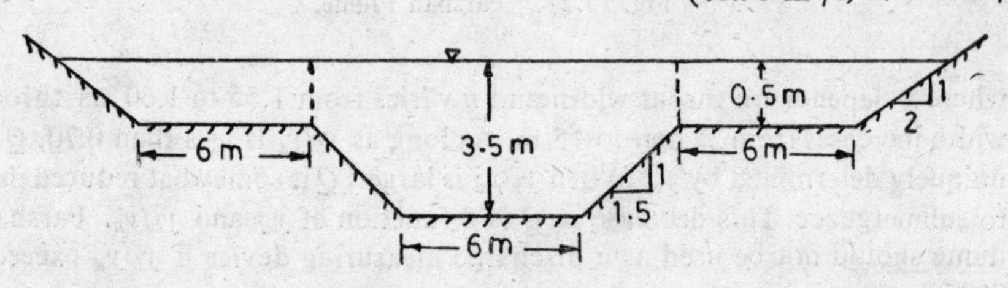

Fig. 13.29.

13.8 Determine the discharge flowing in the channel shown in Fig, 13.30. Take $n = 0.016$ and $S_0 = 1/2500$. (12.349 m³/s)

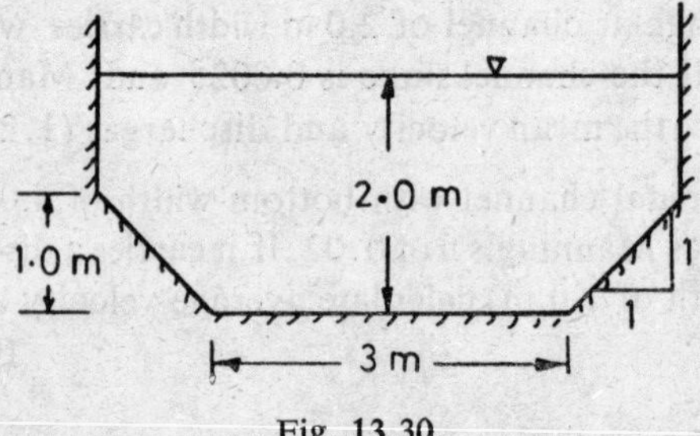

Fig. 13.30.

13.9 Determine the discharge through the channel shown in Fig. 13.31 with $S_0 = 0.0001$. Subdivide the section into 3 subsections as shown by dotted lines. (7.945 m³/s)

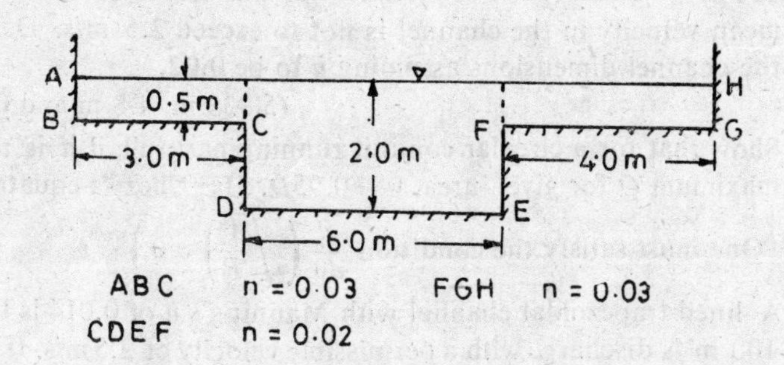

Fig. 13.31.

13.10 In Prob. 13.9, determine the equivalent Manning's n of the channel section.

(Hint: For each subsection, one can write $Q_i = \dfrac{A_i R_i^{2/3}}{n_i} S_0^{1/2}$

$\therefore \quad Q = \Sigma Q_i = S_0^{1/2}\, \Sigma A_i R_i^{2/3}/n_i$. Since $Q = \dfrac{1}{n} A R^{2/3} S_0^{1/2}$

$$\frac{1}{n} A R^{2/3} = \Sigma A_i R_i^{2/3}/n_i$$

Hence knowing A_i, R_i, n_i, and A and R for whole section determine n) (0.01835)

13.11 A trapezoidal channel with a bottom width of 10 m and side slopes 1 : 1 has a bed slope of 0.000 169 and Manning's n of 0.015. Determine the normal depth when it carries a discharge of 20 m³/s. (1.651 m)

13.12 Design a concrete lined trapezoidal channel to carry 40 m³/s discharge at a slope of 0.000 169. Assume side slopes as 1.3 : 1 and Manning's n of 0.013.
(4.5 m bottom width, 3.0 m depth using 0.75 m free board)

13.13 Velocity distribution in an open channel is given by

$$u = \left(1.0 + \frac{y}{y_0}\right).$$

Determine α and β. (1.259, 1.111)

13.14 Show that a trapezoidal channel section with given A and y is most efficient when its sides are inclined at 60° with the horizontal.

13.15 Show that a triangular section with half central angle θ is most efficient when $A = y^2$ and $\theta = 45°$.

13.16 A hydraulically efficient channel is to be excavated in rock to carry 100 m³/s discharge. The channel has side slopes of $0.5H : 1V$ and mean velocity in the channel is not to exceed 2.5 m/s. Determine the channel dimensions assuming n to be 0.02.

(5.933 m, 4.8 m and 0.00175)

13.17 Show that for a circular conduit running part full, if it is to carry maximum Q for given area, $y = 0.95D$. Use Chezy's equation

$$\left(\text{One must satisfy the condition } \frac{d}{d\theta}\left(\sqrt{\frac{A^3}{P}}\right) = 0\right)$$

13.18 A lined trapezoidal channel with Manning's n of 0.014 is to carry 100 m³/s discharge with a permissible velocity of 2.5 m/s. If the side slopes of the channel are $1H : 1V$, determine channel dimensions and slope if it is to be designed as the most efficient channel section. ($y_0 = 4.678$ m, $B = 3.873$ m, $S_0 = 0.000395$)

13.19 For the channel shape shown in Fig. 13.32, obtain the condition for most efficient section for given area A. ($y = a$)

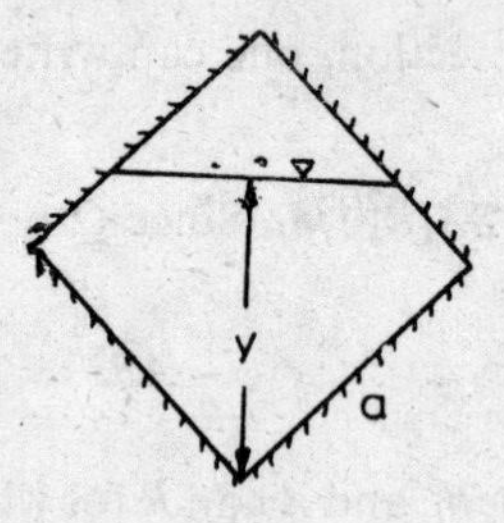

Fig. 13.32.

13.20 For an equilateral triangular flume of side a shown in Fig. 13.33, obtain the condition for R to be maximum. ($y = 0.536a$)

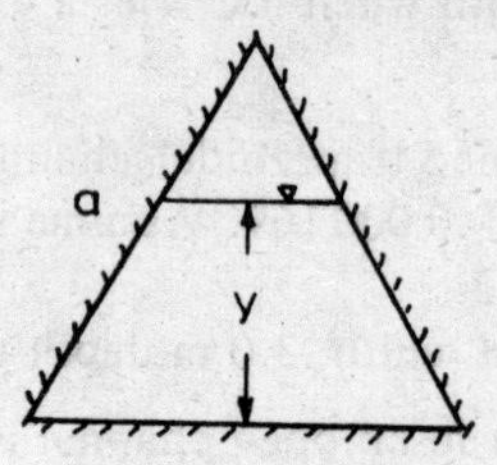

Fig. 13.33.

13.21 In a 5.0 m wide rectangular channel, a bend of centre line radius 30 m is provided. If the super-elevation observed is 80 mm estimate the discharge in the channel. Average depth of flow is 1.5 m.

(16.290 m³/s)

13.22 Show that for given specific energy, the discharge is maximum when $Q^2T/gA^3 = 1$.

13.23 A rectangular channel 4.0 m in width has normal depth of 0.50 m when the channel slope and Manning's n are 0.0049 and 0.013 respectively. Determine

(i) discharge per unit width
(ii) whether the flow is rapid (supercritical) or tranquil (subcritical)
(iii) specific energy and specific force at normal depth
(iv) critical depth and critical slope, and
(v) depth alternate to 0.50 m

(1.462 m³/s m, rapid, 0.936 m, 2.794 m³, 0.602 m, 0.002 78, 0.735 m)

13.24 A triangular channel with central angle of 120° has a critical depth of 0.50 m. What is the corresponding discharge? (0.678 m³/s)

13.25 For a triangular channel with the central angle of 90°, determine the critical depth at a discharge of 2.5 m³/s. (1.05 m)

13.26 What will be the discharge through a circular conduit of 0.841 m diameter, if it runs half full under critical flow condition?

(0.50 m³/s)

13.27 Determine the critical depth in a trapezoidal channel of bed width 5.0 m and side slopes 1 : 1 for a discharge of 20 m³/s. (1.090 m)

13.28 A trapezoidal channel of bottom width 5.0 m and side slopes 1:1 carries 30 m³/s discharge at the depth of 3.0 m. If Manning's n is 0.015, calculate channel slope S_0, critical depth and critical slope. What slope will make normal depth, the critical depth? What discharge will the channel carry?

(0.0015, 2.666 m, 0.0023, 0.0023, 111.02 m³/s)

13.29 A 3 m wide rectangular channel has a drop at its end. If the brink depth is 0.80 m, determine the channel discharge. (13.390 m³/s)

13.30 Show that for critical flow in a circular conduit of diameter D

$$\frac{Q^2}{gD^5} = \frac{\theta - \sin\theta\cos\theta}{64\sin\theta}$$

where 2θ is the angle subtended by w.s. at the centre.

13.31 For 2 m diameter circular conduit, water flows at critical condition making a central angle of 120°. Determine the discharge.

(1.864 m³/s)

13.32 Determine the critical depth in a circular conduit of 2.0 m diameter carrying 2.0 m³/s discharge. ($\theta = 64.5°$; 0.570 m)

13.33 Make assumptions as in Ex. 13.22 and show that for non-rectangular channel

$$\frac{A_b}{A_c} = 1 \bigg/ \left(1 + \frac{\bar{y}_c}{D_c}\right)$$

where A_b and A_c are cross-sectional areas at brink and critical depths, $\bar{y}_c$ is the depth of centroid of area A_c below water surface and $D_u = A_c/T_c$ where T_c is the water surface width of area A_c.

13.34 In Fig. 13.34 depicting a fall, determine X assuming $\theta = 90°$ and

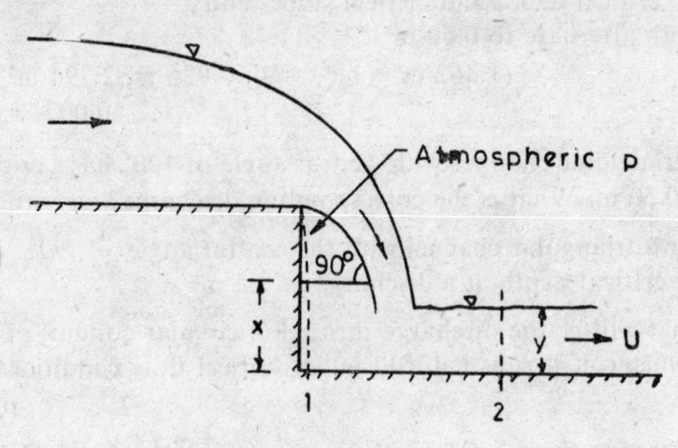

Fig. 13.34.

the pressure in the air pocket is zero. (Hint: apply momentum equation between sections 1 and 2).

$$\left(\frac{X}{y} = \sqrt{1 + 2F_r^2}\right)$$

13.35 Specific force per unit width of a rectangular channel is given by

$$P = \frac{q^2}{gy} + \frac{y^2}{2}.$$

Obtain the conditions for

(i) P to be minimum for constant q

(ii) q to be maximum for constant P

and compare these with the conditions for critical flow.

13.36 For a rectangular channel carrying 5.0 m³/s discharge at a width of 4.0 m and 0.80 m depth, determine the alternate and conjugate depths. (0.385 m, 0.347 m)

13.37 Determine the discharge over a broad crested weir of height 0.40 m fixed in a channel of 3.0 m width if the upstream depth is 1.0 m. (2.610 m³/s)

13.38 A 3.0 m wide rectangular channel carries a discharge of 15.0 m^3/s at a depth of 2.0 m. What will be the minimum height of hump at which the depth over the hump will be critical? Calculate the height of hump for which upstream water depth will be 2.5 m. What will be the depth of flow on the upstream and on the hump when its height is 0.20 m? (0.27 m, 0.655 m, 2.119 m, 1.366 m)

13.39 A 3.0 m wide rectangular channel carries 15.0 m^3/s discharge at a depth of 0.75 m. Determine the depth over the hump when its height is 0.50 m. (0.884 m)

13.40 A rectangular channel 8.0 m in width carries 40.0 m^3/s discharge at 3.0 m depth. If the channel is contracted to 6.0 m and 3.5 m, examine the flow conditions by computing the upstream depth and depth in the contracted section.
(upstream depth unaffected, 2.865 m, 3.449 m, 2.37 m)

13.41 A rectangular channel of 8.0 m width carries 40.0 m^3/s water discharge at a depth of 3.0 m. The channel is contracted to 3.5 m and the bed is raised by 0.5 m. Determine the upstream depth, depth in the contracted section and difference in water levels.
(3.975 m, 2.370 m)

13.42 A rectangular channel width B_1 is reduced to B_2 through a transition. By application of energy equation show that

$$\left(\frac{y_2}{y_1}\right)^3 - \left(1 + \frac{1}{2} Fr_1^2\right)\left(\frac{y_2}{y_1}\right)^2 + \frac{1}{2} Fr_1^2 \left(\frac{B_1}{B_2}\right)^2 = 0$$

What assumptions are made in the derivation?

13.43 A 6.0 m wide channel carrying 24.0 m^3/s water discharge at a depth of 2.5 m is provided with a hump of 0.4 m height. What will be the depths of flow in the upstream and over the hump if channel is contracted to 4.0 m width? (upstream depth unaffected, 1.934 m)

13.44 Flow in a trapezoidal channel at 5.0 m bottom width and 1 : 1 side slopes gives average velocity of 1.0 m/s at a depth of 1.5 m. Determine the height of the hump which will cause critical flow over the hump without affecting the upstream depth. Determine the depth over the hump when the hump height is 0.305 m.
(0.611 m, 1.23 m)

13.45 A pier of 3.0 m width and rounded nose is constructed in a 10 m wide channel carrying a discharge of 60 m^3/s when channel slope is 0.0009 and Manning's n is 0.015. Determine the rise in the water level upstream. (92 mm)

13.46 For the hydraulic jump in rectangular channel prove the following

(i) $\dfrac{y_1}{y_2} = \dfrac{1}{2}[\sqrt{1 + 8Fr_2^2} - 1]$

(ii) $h_L = (U_1 - U_2)^3/2g(U_1 + U_2)$

13.47 For a discharge of 20.0 m³/s in a 4.0 m wide channel, the depth of flow is 0.5 m. Determine the conjugate depth after the jump and energy loss in the jump. (2.953 m, 2.50 m)

13.48 In a 5.0 m wide rectangular channel, the depths before and after the hydraulic jump are 0.30 m and 1.8 m respectively. Determine the discharge in the channel and head loss in the hydraulic jump. (11.792 m³/s, 1.563 m)

13.49 For a discharge of 50.0 m³/s m and head loss of 59.5 m, determine y_1 and y_2 the depths before and after the jump. (1.278 m, 19.346 m)

13.50 Show that for hydraulic jump in a rectangular channel

$$Fr_2^2 = 8Fr_1^2/(\sqrt{1 + 8Fr_1^2} - 1)^3$$

13.51 For the case of hydraulic jump in a triangular channel, show that

$$Fr_1^2\left(1 - \frac{y_1^2}{y_2^2}\right) = \frac{1}{3}\left(\frac{y_2^3}{y_1^3} - 1\right)$$

where $Fr_1^2 = Q^2/A_1^2 g y_1$. What assumptions have been made?

13.52 Free hydraulic jump takes place immediately downstream of a sluice gate in a 4.0 m wide rectangular channel. If the channel carries a discharge of 5.3 m³/s at a depth of 1.47 m, determine depth before the jump, gate opening and water depth upstream of the gate. Take $C_c = 0.60$ for gate.
(Hint: First assume $C_c = C_d$ and solve and then improve the solution) (0.15 m, 0.25 m, 4.0 m)

13.53 Obtain relationship between Fr_1, $\dfrac{\Delta Z}{y_1}$ and $\dfrac{y_2}{y_1}$, for a hydraulic jump at a negative step shown in Fig. 13.35 when the jump is completed at the step.

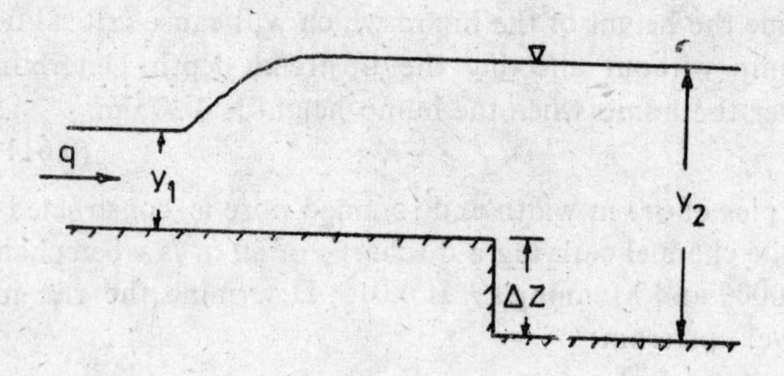

Fig. 13.35.

$$\left\{Fr_1^2 = \frac{y_2}{2(y_2 - y_1)}\left[\left(\frac{y_2}{y_1} - \frac{\Delta Z}{y_1}\right)^2 - 1\right]\right\}$$

13.54 Complete the following table for hydraulic jump in a rectangular channel.

q (m³/s m)	y_1 (m)	y_2 (m)	U_1 (m/s)	U_2 (m/s)	h_L (m)
	0.80		12.0		
10.0					19.01
	0.50	6.0			

Answers

9.6 m³/s m, 4.463 m, 2.251 m/s, 3.441 m

0.45 m, 6.51 m, 22.222 m/s, 1.536 m/s

9.778 m³/s m, 19.556 m/s, 1.63 m/s, 13.865 m

13.55 A sluice gate with 0.60 m opening has 6.0 m depth on the upstream side and it discharges into a wide horizontal channel with flow depth of 1.836 m. Determine how far downstream from vena contracta, the hydraulic jump will occur. Also classify the surface profile. Take $C_c = 0.603$ for gate and Manning's $n = 0.014$.
(Hints: Determine C_d for gate and q. Then determine depth at vena contracta. Also determine depth conjugate to 1.836 m. Then use step method). (H_3, 75.16 m)

13.56 A 10.0 m wide rectangular channel with bottom slope of 0.000 16 and Manning's n of 0.014 carries flow at a uniform depth of 2.0 m. If the dam is constructed at its downstream end at such a height that the depth immediately upstream of it is 12.0 m, determine how far upstream the flow depth will be 2.5. Use step method with following depths:

12.0, 10.0, 8.0, 6.0, 4.0, 2.5 m. Classify the surface profile.

(M_1, 65.828 km)

13.57 Solve problem 13.55 using graphical integration method.

13.58 Show that the differential equation for gradually varied flow in a rectangular channel of variable width B is

$$\frac{dy}{dx} = \frac{S_0 - S_f + \left(\frac{U^2}{gB}\right)\frac{dB}{dx}}{1 - F_r^2}$$

(Hint: Prove and use the relation $\frac{dA}{dy} = B + \frac{dB/dx}{dy/dx}$ in the derivation)

13.59 Show that for wide rectangular channel, gradually varied flow equation can be integrated to yield

$$(x_2 - x_1) = \frac{y_0}{S_0}\left[(u_2 - u_1) - \left(1 - \frac{y_c}{y_0}\right)^3 \int_{u_1}^{u_2} \frac{1}{1 - u^3}\, du\right]$$

where $u = \frac{y}{y_0}$ and Chezy's equation is used to express S_f. This is known as Bresse's method.

13.60 Solve Ex. 13.40 using direct single step method.

13.61 A 100 m wide natural stream having a bed slope of 0.0005 and Manning's n of 0.035 carries a discharge of 398.7 m³/s. Determine the distance required for a change in depth from 3.10 to 3.90 m using Chow's method. Identify the surface profile. (M_1, 4875 m)

13.62 Sketch the possible surface profiles in the slope combinations indicated below-

(i) A mild slope followed by a steep slope
(ii) A mild slope followed by a critical slope
(iii) A horizontal floor followed by a steep slope
(iv) A steep slope followed by a mild slope

[(i) M_2, S_2, (ii) M_2 (iii) H_2, S_2 (iv) J, S_1, or J or M_3, J]

13.63 Various surface profiles occurring on a sequence of slopes are shown in Fig. 13.36. Justify them.

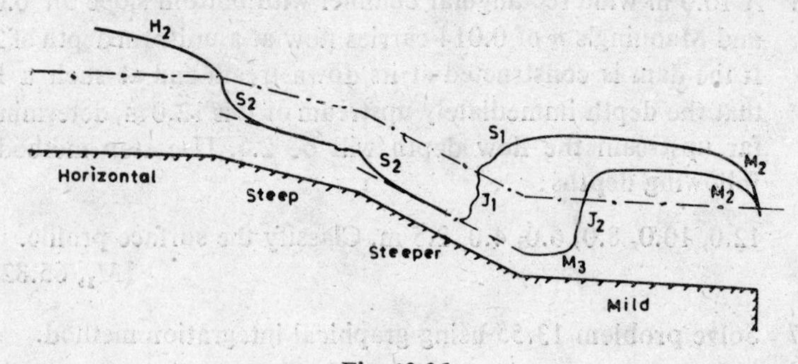

Fig. 13.36.

13.64 For the sequence of slopes shown in Fig. 13.37, assume that all the slopes are sufficiently long and draw the possible surface profiles.

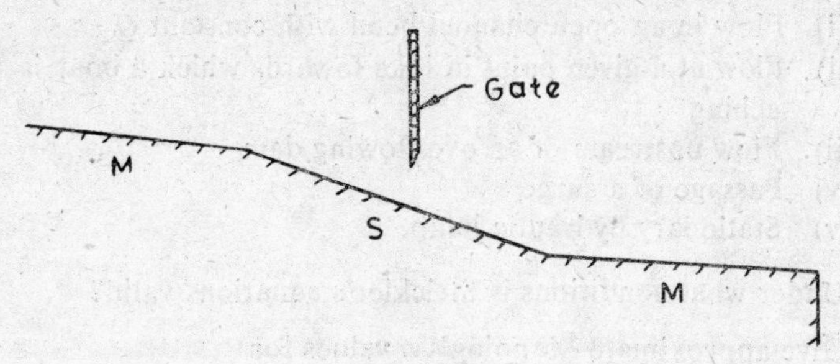

Fig. 13.37.

13.65 In a wide rectangular channel, the depth and velocity of flow are 1.0 m and 1.5 m/s respectively. If the inflow rate is doubled at the upstream end, determine the absolute velocity U_w of the surge and its height. (5.319 m/s, 0.282 m)

13.66 A rectangular channel carries water at a depth of 1.83 m and velocity of 3.05 m/s. By partial closure of the gate at the downstream end of channel, the discharge is reduced to 3.718 m³/s m. Calculate the surge height and absolute velocity of surge.

(0.75 m, 2.485 m/s)

13.67 A river joins the sea at a velocity of 2.0 m/s and a depth of 4.0 m. If the incoming tide from sea to the river has a depth of 6.0 m, estimate the absolute velocity of surge. What will be the velocity and direction of flow downstream of surge?

(−6.574 m/s, −0.858 m/s)

13.68 Consider a solitary wave as shown in Fig. 13.26 in Example 13.48. Apply momentum equation and show that

$$C = \sqrt{gy}\left(1 + \frac{3}{4}\frac{\Delta y}{y} + \ldots\right)$$

13.69 For deep water waves of 10 m wave length, determine the wave velocity and wave period. (3.95 m/s, 2.532 s)

13.70 In a 20 m deep large lake, waves of wave length 50 m are generated. What will be their velocity and frequency? (14.00 m/s, 0.285s⁻¹)

DESCRIPTIVE QUESTIONS

13.1 Give one example each of
(i) steady, uniform flow (ii) steady, nonuniform or gradually varied flow (iii) steady, rapidly varied flow and (iv) unsteady, nonuniform flow.

13.2 Classify the following flows as steady/unsteady, and uniform/nonuniform

(i) Flow in an open channel bend with constant Q
(ii) Flow at a given point in lake towards which a boat is approaching
(iii) Flow upstream of an overflowing dam
(iv) Passage of a surge
(v) Stationary hydraulic jump.

13.3 Under what conditions is Strickler's equations valid?

13.4 Give approximate Manning's *n* values for

(i) smooth plastered surface (ii) bed of gravel river with 50 mm diameter material, (iii) well maintained unlined irrigation channel (iv) badly maintained unlined channel.

13.5 Explain how secondary circulation is developed in a bend.

13.6 Why are canal head works preferably located on the outside of a bend?

13.7 Define critical depth in as many ways as you can.

13.8 Is it possible to know whether the flow in a channel is subcritical or supercritical by dropping a stone in it? Explain.

13.9 Draw pressure distribution in the vertical in case of (i) uniform flow (ii) flow over a spillway (iii) gradually varied flow (iv) flow in a curved bucket of an energy dissipator.

13.10 What assumptions are made in the derivation of expression for y_2/y_1 for hydraulic jump?

13.11 List the assumptions made in the derivation of differential equation for gradually varied flow.

13.12 A hydraulic jump takes place on the sloping floor of a channel. Write the momentum equation

$$\{q\rho(U_2 - U_1) = \frac{1}{2}\gamma \cos\theta\,(y_1^2 - y_2^2) + \frac{(y_1 + y_2)}{2} L\gamma K\}$$

where L is the length of jump and K is shape factor.

13.13 Discharge over a triangular notch is proportional to

(i) $h^{1/2}$ (ii) h^2 (iii) $h^{5/2}$ (iv) $h^{3/2}$

13.14 Depth at which specific energy is minimum is known as

(i) conjugate depth (ii) alternate depth (iii) critical depth (iv) normal depth.

13.15 Water surface profile computations in subcritical flow should always be carried out

(i) from downstream to upstream
(ii) from upstream to downstream
(iii) none of the above.

13.16 If dy/dx is positive, the surface profiles will be

(i) $M_1 S_2, H_3$ (ii) M_2, H_2, M_3 (iii) M_1, S_1, H_3
(iv) none of the above.

13.17 If dy/dx is negative, the surface profiles will be

(i) H_3, A_3, M_3 (ii) M_1, S_1, C_1 (iii) M_2, H_2, M_3
(iv) H_2, M_2, S_2.

13.18 In a channel with uniform flow

(i) $S_0 = S_w$ but S_f can be different
(ii) $S_0 = S_f$ but S_w can be different
(iii) $S_0 = S_w = S_f$
(iv) none of the above.

13.19 If y_1 and y_2 are the conjugate depths

(i) the total energies at these two depths are equal
(ii) specific energies at these two depths are equal
(iii) the specific force at these depths is the same
(iv) none of these.

13.20 If the channel slope changes from mild to steep, the gradually varied flow profiles are

(i) M_1, S_1 (ii) M_2, S_2 (iii) M_3, S_2 (iv) M_2, S_3

13.21 What is a control section? Explain how does it operate in subcritical and supercritical flows.

13.22 What conditions must be fulfilled before hydraulic jump takes place in a channel?

13.23 What factors govern the ratio of y_b/y_c at a drop?

13.24 Under what conditions will the specific force between two section in an open channel be the same?

13.25 What are the different ways in which waves can be generated on the water surface?

13.26 What are the basic principles on which the design of subcritical expanding transitions is based?

CHAPTER XIV

Compressible Flow

14.1 EFFECTS OF COMPRESSIBILITY

All real fluids are compressible to some extent and therefore their density will change with change in pressure or temperature. If the relative change in density $\Delta\rho/\rho$ is small, the fluid can be treated as incompressible. A compressible fluid, such as air, can be considered as incompressible with constant ρ if (i) changes in elevation are small, (ii) acceleration is small, and/or (iii) temperature changes are negligible. In other words, if Mach number U/C, where C is the sonic velocity, is small, compressible fluid can be treated as incompressible.

Compressibility affects the drag coefficients of bodies by formation of shock waves, discharge coefficients of measuring devices such as orifice meters, venturimeters and pitot tubes, stagnation pressure and flows in converging—diverging sections.

14.2 PERFECT GAS RELATIONSHIPS

A fluid can undergo compression or expansion under either the isothermal or the adiabatic process.

Isothermal process: (constant temperature) $$pv = \text{constant} \tag{14.1}$$

Adiabatic process: (constant heat content) $$pv^k = \text{constant} \tag{14.2}$$

Equation of state: $$pv = RT \tag{14.3}$$

Here p is the pressure in N/m², v is the specific volume in m³/N, T is the absolute temperature in K.

$$T = 273.16 + t°\text{C} \tag{14.4}$$

and R is universal gas constant in Nm/NK or m/K. For air, value of R can be taken as 29.57. k is the adiabatic constant and is related to R, specific heat at constant pressure c_p, and specific heat at constant volume c_v, as

$$c_p = \frac{k}{k-1} R \tag{14.5}$$

$$c_v = \frac{R}{k-1} \tag{14.6}$$

$$c_p/c_v = k \tag{14.7}$$

A perfect gas is defined as that gas which follows Eq. 14.3. Isentropic process is reversible adiabatic process. It approximates those processes in nature which occur with small or negligible friction, and with such rapidity that there is no opportunity for heat transfer. Appendix D gives properties of common gases at 0°C and atmospheric pressure.

14.3 INTERNAL ENERGY, ENTROPY AND ENTHALPY

Internal energy e is the energy possessed by 1 N of fluid due to state of molecular activity.

$$(e_2 - e_1) = c_v (T_2 - T_1) \tag{14.8}$$

in which T_1 and T_2 are the absolute temperatures in two states. Energy can be expressed in N m or J. In heat units, it will be in k Cal.

$$1 \text{ calorie} = 4.187 \text{ J}$$

According to first law of thermodynamics,

$$\begin{matrix}\text{Heat absorbed per} \\ \text{unit weight of fluid}\end{matrix} = \begin{matrix}\text{External work done} \\ \text{per unit weight}\end{matrix} + \begin{matrix}\text{Change in internal energy} \\ \text{per unit weight}\end{matrix}$$

or

$$H = W + (e_2 - e_1) \tag{14.9}$$

For isothermal process, T is constant; hence $(e_2 - e_1) = 0$

$$H = W = RT_1 \ln \frac{v_2}{v_1} = p_1 v_1 \ln \frac{v_2}{v_1} \tag{14.10}$$

For isentropic process $H = 0$

$$\therefore \quad W = (e_1 - e_2) = \frac{p_1 v_1 - p_2 v_2}{k-1} \tag{14.11}$$

Entropy is defined as the measure of availability of energy for conversion into work. If ΔH is the amount of heat absorbed per unit weight of fluid at temperature T K, the change in entropy $\Delta\phi$ is given by

$$\left.\begin{aligned} \Delta\phi &= \frac{\Delta H}{T} \\ \text{or} \quad (\phi_2 - \phi_1) &= \int \frac{dH}{T} \end{aligned}\right] \tag{14.12}$$

For isothermal process, from Eqs. 14.10 and 14.12 one gets

$$W = H = (\phi_2 - \phi_1)T = \frac{p_1 v_1}{J} \ln \frac{v_2}{v_1}$$

or $$\phi_2 - \phi_1 = \frac{p_1 v_1}{JT} \ln \frac{v_2}{v_1} \tag{14.13}$$

For isentropic process $\quad H = 0$, hence $\phi_2 - \phi_1 = 0 \quad$ (14.14)

The quantity $h = \left(\frac{p}{\gamma} + e\right)$ per unit weight is known as the *specific enthalpy*. Since

$$\frac{p}{\rho g} = RT \text{ and } (e_2 - e_1) = c_v(T_2 - T_1)$$

$$\Delta h = R(T_2 - T_1) + c_v(T_2 - T_1)$$

Further, since $R = c_p(k-1)/k$ and $c_v = c_p/k$

$$\Delta h = c_p(T_2 - T_1)\left(\frac{1}{k} + \frac{k-1}{k}\right)$$

or $$\Delta h = c_p(T_2 - T_1) \tag{14.15}$$

Using the 1st law of thermodynamics one can show that

$$\left.\begin{aligned}
\phi_2 - \phi_1 &= c_v \ln \left[\left(\frac{T_2}{T_1}\right)\left(\frac{\rho_1}{\rho_2}\right)^{k-1}\right] \\
\phi_2 - \phi_1 &= c_v \ln \left[\left(\frac{p_2}{p_1}\right)\left(\frac{\rho_1}{\rho_2}\right)^{k}\right] \\
\phi_2 - \phi_1 &= c_v \ln \left[\left(\frac{T_2}{T_1}\right)^{k}\left(\frac{p_1}{p_2}\right)^{k-1}\right]
\end{aligned}\right] \tag{14.16}$$

14.4 PROPAGATION OF ELASTIC WAVE AND FLOW CLASSIFICATION

A small pressure disturbance travels through a compressible medium at a velocity C, an expression for which can be obtained by applying continuity and momentum equations. One then gets

$$C^2 = \frac{dp}{d\rho}$$

For isothermal process, $\frac{dp}{d\rho} = \frac{p}{\rho}$ and for adiabatic process $\frac{dp}{d\rho} = \frac{kp}{\rho}$. Hence

$$\left.\begin{aligned}
&\text{Isothermal process} && C = \sqrt{p/\rho} \\
&\text{Isentropic process} && C = \sqrt{kp/\rho}
\end{aligned}\right] \tag{14.17}$$

It may be mentioned that since the propagation of elastic wave takes place rapidly, the process is considered adiabatic

$$\therefore \quad C = \sqrt{kp/\rho} \text{ or } \sqrt{kgRT} \tag{14.18}$$

According to the value of Mach number $M = \frac{U}{C}$ where U is the velocity of body in a compressible fluid, the flows are classified as follows

$0.40 < M < 1.0$ Subsonic

Slightly less than unity $< M <$ slightly greater than unity Transonic

$1.0 < M < 6.0$ Supersonic

$M > 6.0$ Hypersonic

Mach angle α is defined as

$$\alpha = \sin^{-1}\left(\frac{C}{U}\right) \tag{14.19}$$

14.5 EQUATIONS OF MOTION FOR ONE DIMENSIONAL STEADY COMPRESSIBLE FLOWS

14.5.1 Continuity Equation

$$\rho AU = \text{constant} \tag{14.20}$$

which after differentiation and rearrangement yields

$$\frac{dA}{A} + \frac{d\rho}{\rho} + \frac{dU}{U} = 0 \tag{14.21}$$

14.5.2 Momentum Equation

Simple form of momentum equation is

$$Adp = -A\rho UdU$$

or

$$dp = -\rho UdU \tag{14.22}$$

Combination of Eqs. 14.21 and 14.22 gives

$$\frac{dA}{A} = -\frac{dU}{U}(1 - M^2) \tag{14.23}$$

Significance of Eqs. 14.23 is illustrated in Table 14.1 which shows how

Table 14.1. Subsonic and supersonic flows in converging and diverging ducts

Geometry	Diverging		Converging	
Flow	$M < 1.0$	$M > 1.0$	$M < 1.0$	$M > 1.0$
Velocity	Decrease	Increase	Increase	Decrease
Pressure	Increase	Decrease	Decrease	Increase
Density	Increase	Decrease	Decrease	Increase
Temperature	Increase	Decrease	Decrease	Increase
Entropy	Constant	Constant	Constant	Constant

quantities such as velocity, pressure, density, temperature and entropy change in converging and diverging flows.

14.5.3 Energy Equation (Frictionless flow)

When energy equation is applied to frictionless flow of compressible fluids, one must take into account addition or subtraction of heat to the system and changes in internal energy. For isothermal and isentropic cases the energy equation between two sections 0 and 1 takes the form,

$$\text{Isothermal flow:}\quad (Z_0 - Z_1) + RT_0 \ln \frac{v_1}{v_0} = \frac{U_1^2 - U_0^2}{2g} \tag{14.24}$$

$$\text{Isentropic flow:}\quad (Z_0 - Z_1) + \frac{k}{k-1}(p_0 v_0 - p_1 v_1) = \frac{U_1^2 - U_0^2}{2g} \tag{14.25}$$

Further in most of the cases, effect of change in elevation is negligibly small; hence the term $(Z_0 - Z_1)$ is omitted from Eqs. (14.24) and (14.25).

14.6 NORMAL SHOCK WAVES

In a shock wave, which is analogous to hydraulic jump in an open channel, a supersonic flow changes to subsonic flow. This is accompanied by a rise in pressure, density and temperature. Normal shock wave elements can be obtained through use of continuity, momentum and energy equations. As a result, following equations can be derived for $\frac{p_2}{p_1}$ and $\frac{\rho_2}{\rho_1}$ (Fig. 14.1).

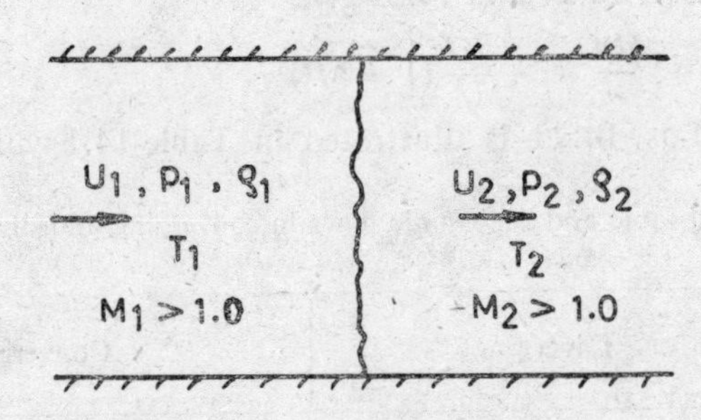

Fig. 14.1. Normal shock wave.

$$\left.\begin{aligned} \frac{p_2}{p_1} &= \frac{\left(\frac{k+1}{k-1}\right)\frac{\rho_2}{\rho_1} - 1}{\left(\frac{k+1}{k-1}\right) - \frac{\rho_2}{\rho_1}} \\ \frac{\rho_2}{\rho_1} &= \frac{1 + \left(\frac{k+1}{k-1}\right)\frac{p_2}{p_1}}{\left(\frac{k+1}{k-1}\right) + \frac{p_2}{p_1}} \end{aligned}\right] \tag{14.26}$$

Equations (14.26) are known as Rankine-Hugoniot equations. One can also express $\frac{p_2}{p_1}$, $\frac{U_2}{U_1}$, $\frac{\rho_2}{\rho_1}$ and $\frac{T_2}{T_1}$ in terms of Mach number M_1 as follows:

$$\left.\begin{aligned}
\frac{p_2}{p_1} &= \frac{2kM_1^2 - (k-1)}{(k+1)} \\
\frac{U_1}{U_2} &= \frac{(k+1)\,M_1^2}{(k-1)\,M_1^2 + 2.0} \\
\frac{\rho_2}{\rho_1} &= \frac{(k+1)\,M_1^2}{(k-1)\,M_1^2 + 2.0} \\
\frac{T_2}{T_1} &= \frac{[(k-1)\,M_1^2 + 2.0]\,[2kM_1^2 - (k-1)]}{(k+1)^2\,M_1^2}
\end{aligned}\right] \tag{14.27}$$

By algebraic manipulation the following equation between M_1 and M_2 can be obtained.

$$M_2^2 = \frac{(k-1)\,M_1^2 + 2}{2kM_1^2 - (k-1)} \tag{14.28}$$

14.7 STAGNATION PRESSURE IN COMPRESSIBLE FLUIDS

When frictionless adiabatic (i.e. isentropic) flow comes to rest, the pressure p_s, mass density, ρ_s, and temperature T_s at the stagnation point can be shown to be related to ambient flow conditions, p_0, ρ_0, U_0 and T_0 by the equations

$$\frac{p_s}{p_0} = \left(1 + \frac{k-1}{2}\,M_0^2\right)^{k/k-1} \tag{14.29}$$

$$\frac{\rho_s}{\rho_0} = \left(1 + \frac{k-1}{2}\,M_0^2\right)^{1/k-1} \tag{14.30}$$

$$\frac{T_s}{T_0} = \left(1 + \frac{k-1}{2}\,M_0^2\right) \tag{14.31}$$

If Mach number M_0 of the ambient flow is much less than unity, Eq. (14.29) can be approximated by Eq. (14.32).

$$\frac{p_s}{p_0} = 1 + \frac{1}{4}\,M_0^2 + \frac{2-k}{24}\,M_0^4 + \ldots \tag{14.32}$$

14.8 FLOW THROUGH NOZZLES, ORIFICES AND VENTURIMETERS

Consider flow through a nozzle or an orifice of area a connected to a large tank (Fig. 14.2). Energy equation yields

$$\frac{T_0}{T_1} = \left(1 + \frac{k-1}{2}\,M_1^2\right) \tag{14.33}$$

$$\frac{\rho_0}{\rho_1} = \left(1 + \frac{k-1}{2}\,M_1^2\right)^{1/k-1} \tag{14.34}$$

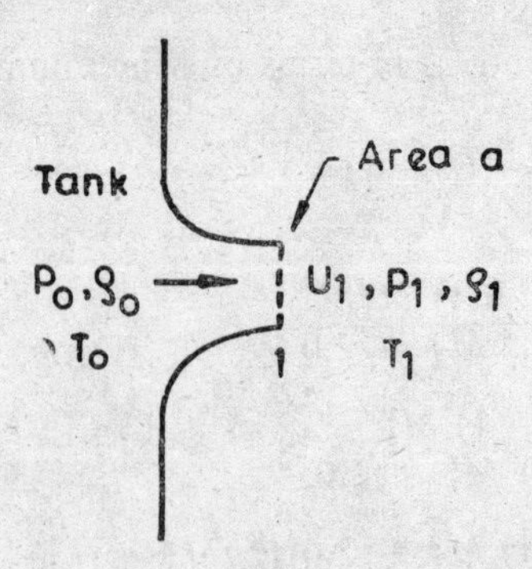

Fig. 14.2. Flow through a nozzle.

$$\frac{p_0}{p_1} = \left(1 + \frac{k-1}{2} M_1^2\right)^{k/k-1} \tag{14.35}$$

Also
$$U_1 = \sqrt{\frac{2k}{k-1}\frac{p_0}{\rho_0}\left(1 - n^{(k-1)/k}\right)} \tag{14.36}$$

and
$$G = a\sqrt{\frac{2kg}{k-1}\frac{p_0}{v_0} n^{2/k}\left(1 - n^{(k-1)/k}\right)} \tag{14.37}$$

where $n = p_1/p_0$ and G is the weight rate of flow of gas in N/s. For given p_0, ρ_0, T_0 and k, G is maximum when

$$n = \left(\frac{2}{k+1}\right)^{k/(k-1)}$$

At this value of n, $U_1 = \sqrt{\frac{kp_1}{\rho_1}}$ or $M_1 = 1.0$, i.e. flow is critical at section 1. Under this condition, critical values of T, ρ and p are (from Eqs. 14.33-14.35).

$$\frac{T_*}{T_0} = \frac{2}{k+1} = 0.833 \text{ for } k = 1.4$$

$$\frac{\rho_*}{\rho_0} = \left(\frac{2}{k+1}\right)^{1/(k-1)} = 0.634 \text{ for } k = 1.4$$

$$\frac{p_*}{p_0} = \left(\frac{2}{k+1}\right)^{k/(k-1)} = 0.528 \text{ for } k = 1.4$$

If $\frac{p_1}{p_0}$ is greater than 0.528, the weight rate of flow can be calculated using Eq. 14.37. If $p_1/p_0 < 0.528$, the weight rate of flow will be maximum which occurs at $M_* = 1$.

$$G_* = a\rho_* g\sqrt{kgRT_*}$$

Flow through venturimeter can be studied using Eq. 14.25. One gets

$$G = \frac{A}{\sqrt{1 - n^{2/k}\left(\frac{a}{A}\right)^2}} \exp\left(\sqrt{\frac{2kg}{k-1}\frac{p_1}{v_0}} n^{2/k}(1-n)^{(k-1)/n}\right) \tag{14.38}$$

It can be seen that when $A \gg a$, Eq. 14.38 reduces to Eq. 14.37.

ILLUSTRATIVE EXAMPLES

14.1 A closed container holding 5.0 kg of helium at 80 kN/m² pressure and 30°C temperature is so heated that its pressure rises to 200 kN/m². What will be its final temperature, work done, heat added and change in internal energy? Take $c_v = 315$ m/K.

Since volume of gas is constant $\frac{p_1}{p_2} = \frac{T_1}{T_2}$ and $T_1 = 273.16 + 30$

$$= 303.16 \text{ K}$$

$$\therefore \quad T_2 = \frac{303.16 \times 200}{80} = 759.90 \text{ K}$$

or

$$t_2 = (759.90 - 273.16) = 486.74°\text{C}$$

$$\text{Heat added} = H = 5.0 \times 9.806\,(T_2 - T_1)\,C_v$$

$$= 5.0 \times 9.806 \times 456.74 \times 315 \text{ Nm or J}$$

$$= 7054,098.09 \text{ J}$$

$$= 1684.762 \text{ kCal}$$

Since gas is heated at constant volume, no work is done on the gas or by the gas. Therefore

Heat added = change in internal energy

$$\therefore \quad (e_2 - e_1) = 1684.762 \text{ kCal}$$

14.2 If 2.5 kg of oxygen gas is compressed adiabaticall from 2.0 m³ volume and 20°C temperature to 300 kN/m² pressure, determine the initial pressure, final volume, final temperature, work done, heat added or removed, and change in internal energy. Take

$$c_p = 92.75 \text{ m/K}, \; c_v = 66.25 \text{ m/K, and } R = 26.50 \text{ m/K.}$$

$$k = c_p/c_v = \frac{92.75}{66.25} = 1.40$$

$$p_1 v_1 = RT_1 \quad \text{and} \quad T_1 = 273.16 + 20.00 = 293.16 \text{ K, and}$$

$$v_1 = \frac{2.0}{2.5 \times 9.806} = 0.0816 \text{ m}^3/\text{N}$$

$$p_1 = \frac{26.5 \times 293.16}{0.0816} = 88\,710.05 \text{ N/m}^2$$

Also $$p_1 v_1^k = p_2 v_2^k$$

$\therefore$ $$v_2 = \left(\frac{p_1}{p_2}\right)^{1/k} v_1 = \left(\frac{88.710}{300}\right)^{1/1.4} \times 0.0816$$

or $$v_2 = 0.034\ 18\ \text{m}^3/\text{N}$$

$\therefore$ $$\text{volume } V_2 = 0.03418 \times 2.5 \times 9.806 = 0.8378\ \text{m}^3$$

$$T_2 = \frac{p_2 v_2}{R} = \frac{300,\ 000 \times 0.03418}{26.5} = 386.94\ \text{K}$$

$\therefore$ $$t_2 = (386.94 - 273.16) = 113.78°\text{C}$$

$$\text{Work done per } N = \frac{p_1 v_1 - p_2 v_2}{k-1} = \frac{(88\ 710.05 \times 0.0816 - 300,\ 000 \times 0.034\ 18)}{0.40}$$

$$= -7538.15\ \text{N m/N}$$

$\therefore$ $$\text{Work done} = -7538.15 \times 2.5 \times 9.806$$

$$= -184,\ 797.74\ \text{N m}$$

Since the process is adiabatic, heat added or removed is zero.

Also $$(e_2 - e_1) = c_v(T_2 - T_1)$$

$$= 66.25\ (386.94 - 293.16)$$

$$= 6212.93\ \text{N m/N}$$

$$\text{Total change in internal energy} = 6212.93 \times 2.5 \times 9.806$$

$$= 152,\ 309.86\ \text{N m}$$

$$= (152,\ 309.86)/4.187 \times 1000)\ \text{k Cal}$$

$$= 36.377\ \text{kCal}$$

14.3 Show that the work done in compressing the gas under isothermal condition is given by

$$W = RT_1 \ln \frac{v_2}{v_1}$$

For isothermal process (Fig. 14.3).

$$\text{Work done} = \int_{v_1}^{v_2} p\,dv$$

but $$p_1 v_1 = pv$$

$\therefore$ $$W = \int_{v_1}^{v_2} \frac{p_1 v_1}{v}\,dv$$

$$= p_1 v_1 \ln \frac{v_2}{v_1} \text{ or } RT_1 \ln \frac{v_2}{v_1}$$

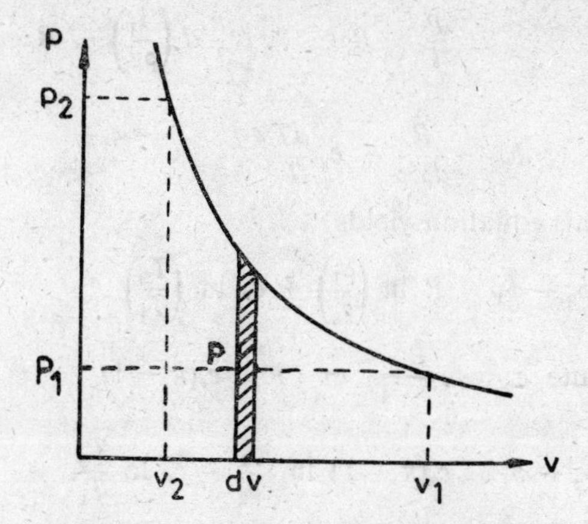

Fig. 14.3. Isothermal process.

14.4 Compute change in enthalpy of 3.5 kg of oxygen when its initial and final conditions are

$$p_1 = 150 \text{ kN/m}^2 \qquad t_1 = 4.0°\text{C}$$

$$p_2 = 600 \text{ kN/m}^2 \qquad t_2 = 90.0°\text{C}$$

$$(h_2 - h_1) = c_p(T_2 - T_1)$$

From Appendix D, $c_p = 91.3$ m/K

$$\therefore \qquad (h_2 - h_1) = 91.3\ (363.16 - 277.16)$$

$$= 91.3 \times 86 = 7851.8 \text{ J/N}$$

$\therefore$ Change in enthalpy for 3.5 kg of oxygen

$$= 7851.8 \times 3.5 \times 9.806$$

$$= 269\ 481.63 \text{ J or } 269.482 \text{ kJ}$$

14.5 According to first law of thermodynamics

heat added = work done + change in internal energy

From this show that change in entropy is given by

$$\phi_2 - \phi_1 = c_v \ln \left(\frac{T_2}{T_1}\right)\left(\frac{\rho_1}{\rho_2}\right)^{k-1}$$

First law of thermodynamics can be expressed as (see Eqs. 14.8 and 14.12).

$$Td\phi = pdv + c_v dT$$

$$= pd\left(\frac{1}{\rho g}\right) + c_v dT$$

$$\therefore \qquad d\phi = \frac{p}{T}\, d\left(\frac{1}{\rho g}\right) + c_v \frac{dT}{T}$$

But $$\frac{p}{T} = R\rho g \quad \therefore \quad \frac{p}{T} d\left(\frac{1}{\rho g}\right) = R\rho g d\left(\frac{1}{\rho g}\right) \text{ or } \frac{R}{d\rho/\rho}$$

$$\therefore \quad d\phi = \frac{R}{d\rho/\rho} + c_v \frac{dT}{T}$$

Integration of this equation yields

$$\phi_2 - \phi_1 = R \ln \left(\frac{\rho_1}{\rho_2}\right) + c_v \ln \left(\frac{T_2}{T_1}\right)$$

One can substitute $c_v = \dfrac{R}{k-1}$ or $R = c_v(k-1)$

$$\therefore \quad \phi_2 - \phi_1 = c_v(k-1) \ln \frac{\rho_1}{\rho_2} + c_v \ln \frac{T_2}{T_1}$$

$$\therefore \quad \phi_2 - \phi_1 = c_v \ln \left(\frac{T_2}{T_1}\right)\left(\frac{\rho_1}{\rho_2}\right)^{k-1}$$

14.6 Determine the velocity of sound wave in international standard atmosphere (see Appendix E) at altitudes 1.0 km, 10.0 km and 30.0 km above sea level. If the aeroplane travels at 1400 km/hr, what will be the corresponding values of Mach numbers? Take $k = 1.40$ and $R = 29.57$ m/K.

$$U = \frac{1400 \times 1000}{3600} = 388.89 \text{ m/s}$$

Referring to Appendix E, absolute temperatures at these elevations are as recorded below:

Height	T K	$C = \sqrt{kgRT}$	$M = \dfrac{U}{C}$
1.0 km	281.7	338.17 m/s	1.150
10.0 km	222.3	300.40 m/s	1.295
30.0 km	226.5	303.23 m/s	1.282

It can thus be seen that for the same velocity of aeroplane, its Mach number will change with change in elevation at which it travels.

14.7 What will be the difference in Mach numbers for an aeroplane travelling at 1300 km/hr at sea level at 20°C and in stratosphere at − 40°C?

$$T_1 = 273.16 + 20.00 = 293.16 \text{ K}$$

$$T_2 = 273.16 - 40.00 = 233.16 \text{ K}$$

$$\therefore \quad C_1 = \sqrt{kgRT_1} = \sqrt{1.40 \times 9.806 \times 29.57 \times 293.16} = 344.96 \text{ m/s}$$

$$C_2 = \sqrt{kgRT_2} = \sqrt{1.40 \times 9.806 \times 29.57 \times 233.16} = 307.65 \text{ m/s}$$

$$U = \frac{1300 \times 1000}{3600} = 361.11 \text{ m/s}$$

$\therefore \quad M_1 = 361.11/344.96 = 1.047$

$M_2 = 361.11/307.65 = 1.174$

14.8 **A supersonic plane flying at Mach number of 1.50 is observed directly overhead at a height of 15.0 km. How far ahead will it be when one hears the sonic boom? (Fig. 14.4).**

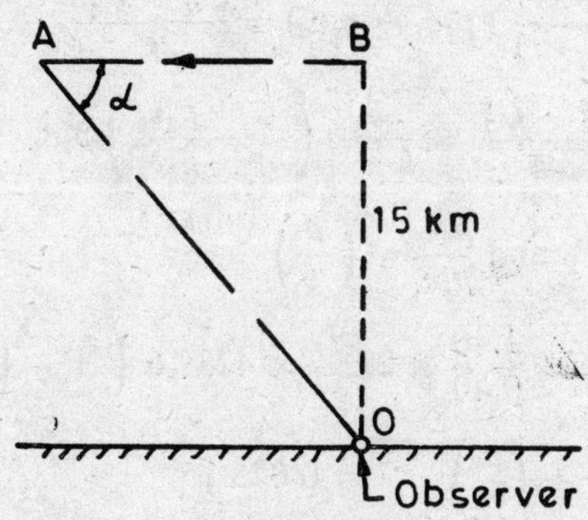

Fig. 14.4.

$$\sin \alpha = \frac{1}{M} = \frac{1}{1.50}$$

$$= 0.6667$$

$$\therefore \quad \alpha = 41.81°$$

$$AB = \frac{OB}{\tan \alpha} = \frac{15.0}{0.8944} = 16.77 \text{ km}$$

14.9 **In a given conduit flow, Mach number is 1.50. If the velocity undergoes 25 percent increase, what percent of the original area is needed for this? What would be the area if $M = 0.50$?**

Since
$$\frac{dA}{A} = -\frac{dU}{U}(1 - M^2)$$

$$= -0.25(1 - 1.5^2)$$

$$= 0.3125$$

$\therefore$ Area will be 31.25 percent larger than the original area or the area required will be 1.3125 times the original area.

When $M = 0.50$

$$\frac{dA}{A} = -0.25(1 - 0.50^2)$$

$$= -0.1875$$

$\therefore$ Area will be 18.75 percent smaller than the original area. Or the area will be 0.8125 times the original area.

14.10 In the case of air flow in a conduit transition, the pressure, velocity and temperature at the upstream section are 35.0 kN/m², 30.0 m/s and 150°C. If at the downstream section the velocity is 150.0 m/s, determine the temperature and pressure if the process followed is isentropic. Take $k = 1.40$ and $R = 29.57$ m/K.

According to energy equation

$$\frac{k}{k-1}(p_1 v_1 - p_2 v_2) = \frac{U_2^2 - U_1^2}{2g}$$

or

$$\frac{U_2^2 - U_1^2}{2g} = \frac{k}{k-1}\left(1 - \frac{p_2 v_2}{p_1 v_1}\right) p_1 v_1$$

However $p_1 v_1 = RT_1$ and $\frac{p_2 v_2}{p_1 v_1} = \left(\frac{p_2}{p_1}\right)^{(k-1)/k}$

$$\therefore \quad \frac{150^2 - 30^2}{2 \times 9.806} = \frac{1.40}{0.40} \times 29.57 \times 423.16 \left(1 - \left(\frac{p_2}{p_1}\right)^{0.4/1.4}\right)$$

This gives

$$1 - \left(\frac{p_2}{p_1}\right)^{0.2857} = 0.025\,15$$

$$\therefore \quad \left(\frac{p_2}{p_1}\right) = 0.9147 \quad \therefore \quad p_2 = 32.015 \text{ kN/m}^2$$

$$\frac{T_2}{T_1} = \left(\frac{p_2}{p_1}\right)^{(k-1)/k} = (0.9147)^{0.2857} = 0.9748$$

$$\therefore \quad T_2 = 423.16 \times 0.9748$$

$$= 412.50 \text{ K}$$

or

$$t_2 = 412.50 - 273.16 = 139.34°\text{C}$$

14.11 A 100 mm diameter pipe reduces to 50 mm diameter through a sudden contraction. When it carries air at 20°C under isothermal condition, the absolute pressures observed in the two pipes just before and after the contraction are 400 kN/m² and 320 kN/m² respectively. Determine the mass rate of flow through the pipe as well as the density and flow velocity at two sections.

For isothermal condition $\frac{p_1}{\rho_1} = \frac{p_2}{\rho_2}$ or $\frac{p_1}{p_2} = \frac{\rho_1}{\rho_2}$

$$\therefore \quad 400/320 = \rho_1/\rho_2 = 1.25$$

$$T_1 = 273.16 + 20.00 = 293.16 \text{ K}$$

$$\frac{p_1}{\rho_1 g} = RT_1 \quad \therefore \quad \rho_1 = \frac{400\,000}{9.806 \times 29.57 \times 293.16} = 4.706 \text{ kg/m}^3$$

$$\therefore \quad \rho_2 = \rho_1/1.25 = \frac{4.706}{1.25} = 3.765 \text{ kg/m}^3$$

According to continuity equation $A_1 \rho_1 U_1 = A_2 \rho_2 U_2$

$$\therefore \qquad \frac{U_2}{U_1} = \frac{A_1\rho_1}{A_2\rho_2} = \left(\frac{100}{50}\right)^2 \times 1.25 = 5.0$$

Assuming $Z_1 = Z_2$, $(U_2^2 - U_1^2)/2g = RT_1 \ln \dfrac{\rho_1}{\rho_2}$

Substituting $U_2/U_1 = 5.0$, one gets

$$\frac{24\, U_1^2}{2 \times 9.806} = 29.57 \times 293.16 \ln 5$$

or $$U_1^2 = 1580.40 \quad \text{and} \quad U_1 = 39.754 \text{ m/s}$$

$$U_2 = 5 \times 39.754$$

$$= 198.771 \text{ m/s}$$

14.12 **Show that the mass rate of flow of a compressible fluid through a venturimeter for isentropic conditions of flow is given by**

$$\dot{m} = \frac{a\rho_1 (n)^{1/k}}{\sqrt{1 - \left(\dfrac{a}{A}\right)^2 n^{2/k}}} \sqrt{\frac{2k}{k-1}\frac{p_1}{\rho_1}(1 - n^{(k-1)/k})}$$

where $n = p_2/p_1$ the ratio of absolute pressures at the throat and in the pipe.

For isentropic condition of flow, energy equation between sections 1 and 2 in Fig. 14.5 yields

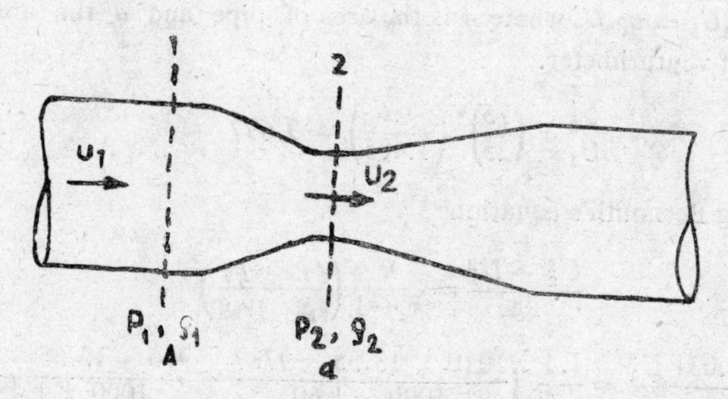

Fig. 14.5. Venturimeter.

$$\frac{k}{k-1}(p_1 v_1 - p_2 v_2) = \frac{U_2^2 - U_1^2}{2g}$$

while continuity equation gives $U_1 A\rho_1 = U_2 a\rho_2$, and $v = \dfrac{1}{\rho g}$.

Hence energy equation reduces to

$$\frac{k}{k-1}\frac{p_1}{\rho_1}\left(1 - \frac{p_2}{p_1}\frac{\rho_1}{\rho_2}\right) = \frac{U_2^2}{2}\left[1 - \left(\frac{a}{A}\right)^2\left(\frac{\rho_2}{\rho_1}\right)^2\right]$$

$$U_2 = \sqrt{\frac{2k}{k-1}\frac{p_1}{\rho_1}\left(1-\frac{p_2}{p_1}\frac{\rho_1}{\rho_2}\right)\Big/\left[1-\left(\frac{a}{A}\right)^2\left(\frac{\rho_2}{\rho_1}\right)^2\right]}$$

However

$$p_1/\rho_1^k = p_2/\rho_2^k \quad \therefore \quad \left(\frac{\rho_1}{\rho_2}\right) = \left(\frac{p_1}{p_2}\right)^{1/k}$$

Substitution of this value in the expression for U_2 yields

$$U_2 = \sqrt{\frac{2k}{k-1}\frac{p_1}{\rho_1}\left(1-\left(\frac{p_2}{p_1}\right)^{(k-1)/k}\right)\Big/\left[1-\left(\frac{a}{A}\right)^2\left(\frac{p_2}{p_1}\right)^{2/k}\right]}$$

Substituting $p_2/p_1 = n$ and $\dot{m} = aU_2\rho_2 = aU_2\rho_1 n^{1/k}$, $\dot{m}$ is given by

$$\dot{m} = a\rho_1 n^{1/k} \sqrt{\frac{2k}{k-1}\frac{p_1}{\rho_1}(1-n^{(k-1)/k})} \Big/ \sqrt{1-\left(\frac{a}{A}\right)^2 n^{2/k}}$$

14.13 A 25 mm diameter venturimeter is fixed in a 75 mm diameter pipe to measure the rate of flow of gas. If the absolute pressures at the inlet and the throat of venturimeter are equivalent to 1010 mm and 910 mm of mercury, determine the volumetric flow rate of gas. Assume the flow to be isentropic, $k = 1.40$ and $\rho_1 = 1.60$ kg/m³.

$$\left(\frac{\rho_2}{\rho_1}\right)^k = \frac{p_2}{p_1} \quad \therefore \quad \rho_2 = \left(\frac{p_2}{p_1}\right)^{1/k}\rho_1 = \left(\frac{910}{1010}\right)^{0.7143} \times 1.60$$

$$= 1.485 \text{ kg/m}^3$$

Also $A\rho_1 U_1 = a\rho_2 U_2$ where A is the area of pipe and a the area of the throat of venturimeter.

$$\therefore \quad \frac{U_2}{U_1} = \left(\frac{75}{25}\right)^2\left(\frac{1.6}{1.485}\right) = 9.697$$

Applying Bernoulli's equation

$$\frac{U_2^2 - U_1^2}{2g} = \frac{k}{k-1}\left(\frac{p_1}{\rho_1 g} - \frac{p_2}{\rho_2 g}\right)$$

$$\frac{93.031\, U_1^2}{2} = \frac{1.4}{0.40}\left(\frac{1010 \times 13.55 \times 9787}{1000 \times 1.60} - \frac{910 \times 13.55 \times 9787}{1000 \times 1.485}\right)$$

$$= 464.184 \times 18.46$$

or $\quad U_1 = 13.573$ m/s

$$\therefore \quad \text{Volumetric flow rate} = 0.785 \times \left(\frac{75}{1000}\right)^2 \times 13.573$$

$$= 0.0599 \text{ m}^3/\text{s}$$

14.14 For a normal shock wave in air, $M_1 = 2.0$. If the atmospheric pressure and air density are 0.265×10^5 N/m² and 0.413 kg/m³, determine the flow conditions before and after the shock wave. Take $k = 1.4$.

$$M_2^2 = \frac{(k-1)\,M_1^2 + 2.0}{2kM_1^2 - (k-1)} = \frac{0.4 \times 2.0^2 + 2.0}{2 \times 1.4 \times 2.0^2 - 0.40}$$

$$= 0.333 \qquad \therefore \quad M_2 = 0.577$$

$$\frac{p_2}{p_1} = \frac{2kM_1^2 - (k-1)}{(k+1)} = \frac{2 \times 1.4 \times 2.0^2 - 0.40}{2.4} = 4.5$$

$$\therefore \quad p_2 = 4.5 \times 0.265 \times 10^5 = 1.193 \times 10^5 \text{ N/m}^2 \text{ absolute.}$$

$$\frac{\rho_2}{\rho_1} = \frac{(k+1)\,M_1^2}{(k-1)\,M_1^2 + 2.0} = \frac{2.4 \times 2.0^2}{0.4 \times 2.0^2 + 2.0} = 2.667$$

$$\therefore \quad \rho_2 = 2.667 \times 0.413 = 1.101 \text{ kg/m}^3$$

Since $\dfrac{p_1}{\rho_1 g} = RT_1 \qquad T_1 = \dfrac{0.265 \times 10^5}{0.413 \times 29.57 \times 9.806} = 221.29\ K$

$$\therefore \quad t_1 = 221.29 - 273.16 = -51.87°\text{C}$$

Further $\dfrac{T_2}{T_1} = \dfrac{[(k-1)\,M_1^2 + 2.0]\,[2kM_1^2 - (k-1)]}{(k+1)^2\,M_1^2}$

$$= \frac{(0.4 \times 2.0^2 + 2.0)\,(2 \times 1.4 \times 2.0^2 - 0.40)}{2.4^2 \times 2.0^2}$$

$$= 1.6275$$

$$\therefore \quad T_2 = 221.29 \times 1.6875 = 373.43K$$

$$t_2 = 373.43 - 273.16 = 100.27°C$$

$$C_1 = \sqrt{kgRT_1} = \sqrt{1.4 \times 9.806 \times 29.57 \times 221.29}$$

$$= 299.72 \text{ m/s}$$

Since $\dfrac{U_1}{C_1} = M_1 = 2.0 \quad \therefore \quad U_1 = 299.72 \times 2.0 = 599.44 \text{ m/s}$

$$C_2 = \sqrt{kgRT_2} = \sqrt{1.4 \times 9.806 \times 29.57 \times 373.43}$$

$$= 389.35 \text{ m/s}$$

Since $\dfrac{U_2}{C_2} = M_2 = 0.577 \qquad U_2 = 0.577 \times 389.35$

$$= 224.66 \text{ m/s}$$

14.15 Determine the velocity of a bullet fired in the atmosphere at 15.0°C, if the Mach angle is 30°. Take $k = 1.40$, $R = 29.57$ m/K

$$T = 273.16 + 15.00 = 288.16\ K$$

$$C = \sqrt{kgRT} = \sqrt{1.40 \times 9.806 \times 29.57 \times 288.16}$$

$$= 342.02 \text{ m/s}$$

$$\sin 30° = \frac{C}{U} = 0.50 \quad \therefore \quad U = \frac{342.02}{0.50} = 684.04 \text{ m/s}$$

$$= 2462.54 \text{ km/hr.}$$

14.16 A normal shock moves through still air (101.0 kN/m^2 pressure and 298 K temperature) with a constant velocity of 1500 m/s. Calculate the velocity of air behind the wave, static pressure and temperature there. Take $R = 29.27$ m/K.

In order to apply the equations of shock wave, the shock wave must be made stationary by applying a velocity of 1500 m/s in the opposite direction, so that the observer on the wave sees it stationary (Fig. 14.6).

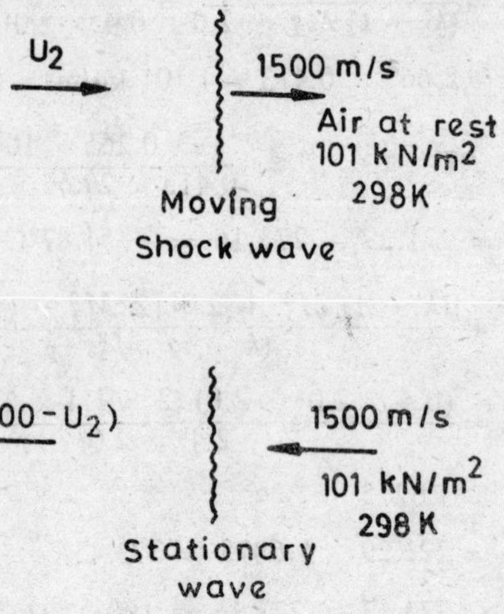

Fig. 14.6.

$$C_1 = \sqrt{1.4 \times 9.806 \times 29.27 \times 298}$$

$$= 346.04 \text{ m/s}$$

$$M_1 = 1500/346.04 = 4.335$$

Use of Eq. (14.27) with $k = 1.4$ yields

$$\frac{p_2}{p_1} = 21.76, \quad \frac{T_2}{T_1} = 4.591$$

and

$$\frac{\rho_2}{\rho_1} = \frac{U_1}{U_2'} = 4.739$$

∴ Conditions after or behind the shock wave will be

$$p_2 = 21.76 \times 101 = 2197.76 \text{ kN/m}^2$$

$$U_2' = \frac{1500}{4.739} = 316.52 \text{ m/s} \quad \therefore \quad 1500 - U_2 = 316.52 \text{ m/s}$$

or

$$U_2 = 1183.48 \text{ m/s}$$

$$T_2 = 4.591 \times 298 = 1368.15 \text{ K}$$

14.17 Air at a velocity of 1400 km/hr has a pressure of 10 kN/m² vacuum and temperature of 50°C. Calculate local Mach number and stagnation pressure, density and temperature. Take $k = 1.40$ and $R = 28.7$ m/K and barometric pressure of 101.325 kN/m².

$$T_0 = 273.16 + 50.00 = 323.16\text{K}$$

$$C_0 = \sqrt{1.4 \times 9.806 \times 28.7 \times 323.16} = 357.02 \text{ m/s}$$

$$M_0 = U_0/C_0 = \frac{1400 \times 1000}{3600 \times 357.02} = 1.089$$

$$p_0 \text{ abs} = (101\,325 - 10\,000) = 91\,325 \text{ N/m}^2$$

$$\frac{p_s}{p_0} = \left(1 + \frac{k-1}{2} M_0^2\right)^{k/(k-1)} = (1 + 0.2 \times 1.089^2)^{3.5}$$

$$p_s = 91\,325 \times 2.1063 = 192\,357.9 \text{ N/m}^2$$

or $$192.358 \text{ kN/m}^2$$

$$\left(\frac{\rho_s}{\rho_0}\right) = \left(\frac{p_s}{p_0}\right)^{\frac{1}{k}} = \left(\frac{192.358}{91.325}\right)^{0.7143} = 1.703$$

However $$\rho_0 = \frac{p_0}{gRT_0} = \frac{91\,325}{9.806 \times 28.7 \times 323.16} = 1.003 \text{ kg/m}^3$$

$\therefore$ $$\rho_s = 1.703 \times 1.003 = 1.708 \text{ kg/m}^3$$

$$T_s/T_0 = \left(1 + \frac{k-1}{2} M_0^2\right)$$

$\therefore$ $$T_s = 323.16\,(1 + 0.2 \times 1.089^2)$$

$$= 399.81 \text{ K}$$

$\therefore$ $$t_2 = (399.81 - 273.16) = 126.65°\text{C}$$

14.18 Determine the pressure on the nose of a bullet travelling through standard air at 400.0 m/s speed, assuming compressibility effects to be negligible and also considering the fluid to be compressible.

For standard atmosphere $t_0 = 15°\text{C}$, $p_0 = 101.325 \text{ kN/m}^2$

$$\rho_0 = 1.225 \text{ kg/m}^3$$

When fluid is considered incompressible

$$p_s - p_0 = \rho_0 U_0^2/2$$

$$= \frac{1.225 \times 400^2}{2} = 98\,000 \text{ N/m}^2$$

$\therefore$ $$p_s = 101325 + 98000 = 199,325 \text{ N/m}^2$$

or $$199.325 \text{ kN/m}^2$$

When compressibility is considered

$$C = \sqrt{kgRT_0} = \sqrt{1.4 \times 9.806 \times 28.7 \times 288.16}$$

$$= 336.95 \text{ m/s}$$

$$M_0 = 400.00/336.95 = 1.187$$

$$\frac{p_s}{p_0} = (1 + 0.2 \times 1.187^2)^{3.5}$$

$$= 2.384$$

$$\therefore \quad p_s = 101.325 \times 2.384 = 241.591 \text{ kN/m}^2$$

$$\frac{(p_s)_{com} - (p_s)_{incom}}{(p_s)_{com}} = \frac{241.591 - 199.325}{241.591} = 0.1749 \text{ or } 17.49 \text{ percent.}$$

Hence assuming air to be incompressible underestimates the stagnation pressure by 17.49 percent.

14.19 Oxygen flows in a conduit at an absolute pressure of 170 kN/m². If the absolute pressure and temperature at the nose of small object in the stream are 200 kN/m² and 70°C respectively, determine the velocity in the conduit. Take $k = 1.4$ and $R = 28.7$ m/K.

$$p_0 = 170 \text{ kN/m}^2 \quad p_s = 200 \text{ kN/m}^2$$

$$T_s = 273.16 + 70.00 = 343.16 \text{ K}$$

Since

$$\frac{p_s}{p_0} = (1 + 0.2\, M_0^2)^{3.5}$$

$$(1 + 0.2\, M_0^2) = \left(\frac{200}{170}\right)^{1/3.5}.$$ Solution of this gives $M_0 = 0.488$.

Also

$$\frac{T_s}{T_0} = (1 + 0.2\, M_0^2) = 1 + 0.2 \times 0.488^2$$

$$= 1.0478$$

$$\therefore \quad T_0 = \frac{343.16}{1.0478} = 327.51 \text{ K}$$

$$C_0 = \sqrt{kgRT_0} = \sqrt{1.4 \times 9.806 \times 28.7 \times 327.51}$$

$$= 359.22 \text{ m/s}$$

$$\therefore \quad \frac{U_0}{C_0} = M_0 \quad \therefore \quad U_0 = 359.22 \times 0.488$$

$$= 175.30 \text{ m/s}$$

14.20 Determine the mass rate of flow through a nozzle of 25 mm diameter connected to a large tank if the air in the tank is at 140 kN/m² pressure absolute, and barometric pressure is 100 kN/m². Assume air temperature to be 20°C, $k = 1.40$ and $R = 28.57$ m/K. What will be the corresponding value of $\dot{m}$ if pressure in the tank is 300 kN/m²?

$$T_0 = 273.16 + 20.00 = 293.16$$

$$\frac{p_0}{\rho_0 g} = RT_0 \quad \therefore \quad \rho_0 = \frac{140\,000}{28.57 \times 9.806 \times 293.16}$$

$$= 1.705 \text{ kg/m}^3$$

$$\frac{p_1}{p_0} = \frac{100}{140} = 0.7143.$$

Since this is more than the critical value of 0.528, flow in nozzle will be subsonic.

$$U_1 = \sqrt{\frac{2k}{k-1}\frac{p_0}{\rho_0}(1 - n^{(k-1)/k})}$$

$$= \sqrt{\frac{2 \times 1.4}{0.40} \times \frac{140\,000}{1.705}(1 - 0.7143^{.2857})}$$

$$= 229.52 \text{ m/s}$$

$$\frac{\rho_0}{\rho_1} = \left(\frac{p_0}{p_1}\right)^{1/k} \quad \therefore \quad \rho_1 = \rho_0\left(\frac{p_1}{p_0}\right)^{1/k} = 1.341$$

$$\therefore \quad \dot{m} = 1.341 \times 229.52 \times 0.785 \times 0.025^2$$

$$= 0.151 \text{ kg/s}$$

When the pressure in the tank is 300 kN/m²

$\frac{p_1}{p_0} = \frac{100}{300} = 0.333$. This being less than the critical ratio of 0.528, the flow in the nozzle will be sonic i.e. $M_1 = 1.0$. At critical condition

$$T_*/T_0 = 0.833 \quad \therefore \quad T_* = 0.833 \times 293.16$$

$$= 244.30 \text{ K}$$

Sonic velocity $\quad C_* = \sqrt{kgRT_*} = \sqrt{1.4 \times 9.806 \times 28.57 \times 244.30}$

$$= 309.49 \text{ m/s}$$

$$\therefore \quad \frac{U_*}{C_*} = 1 \quad \text{or} \quad U_* = 309.49 \text{ m/s}$$

$$\frac{\rho_*}{\rho_0} = 0.634 \quad \therefore \quad \rho_* = 1.705 \times 0.634$$

$$= 1.081 \text{ kg/m}^3$$

$$\therefore \quad \dot{m} = a\rho_* U_*$$

$$= 0.785 \times .025^2 \times 1.081 \times 309.49$$

$$= 0.164 \text{ kg/s}$$

14.21 What is Laval nozzle? Describe the flow through such a nozzle. (Fig. 14.7)

Laval nozzle is a convergent—divergent nozzle named after de Laval (1845-1913), the Swedish scientist who invented it. In such a nozzle, subsonic flow prevails in the converging section, critical or transonic conditions in the throat and supersonic flow in the diverging section.

Let p_* be the value of pressure in the throat when the flow is sonic for given pressure p_0. When the receiver pressure p_3 equals the pressure in the tank p_0, there will be no flow through the nozzle (line *a* in Fig. 14.7).

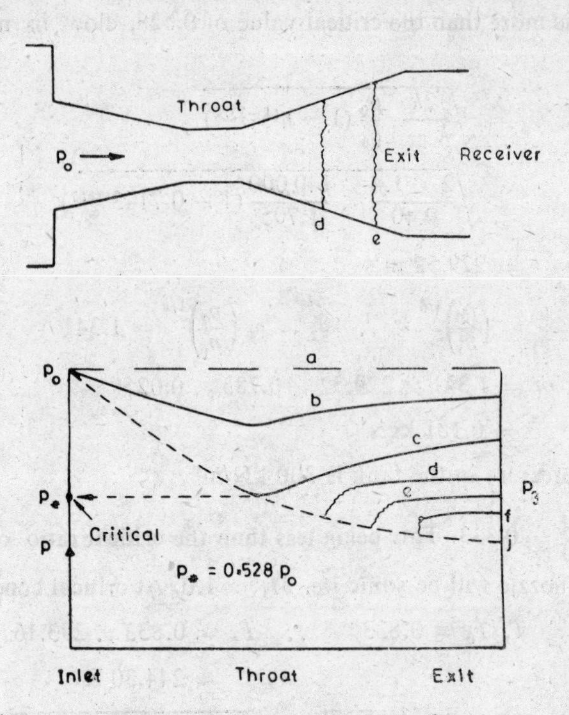

Fig. 14.7. Flow in Laval Nozzle.

When the receiver pressure is reduced, flow will occur through the nozzle. As long as p_3 value is such that throat pressure p_t is greater than the critical pressure 0.528 p_0, the flow in the converging and diverging sections will be subsonic. This condition is depicted by line *b*. With further reduction in p_3, a stage is reached when p_t is equal to critical pressure $p_* = 0.528\ p_0$, at this time $M = 1$ in the throat. This condition is shown by line *C*. Flow is subsonic on the upstream as well as the downstream of the throat. The flow is also isentropic. If p_3 is further reduced, it does not affect the flow in the convergent section. The flow in throat is sonic, downstream it is supersonic. Somewhere in the diverging section a shock wave occurs and flow changes to subsonic (curve *d*). The flow across the shock is nonisentropic. Downstream of the shock wave the flow is subsonic and decelerates. If the value of p_3 is further reduced, the shock wave forms somewhat downstream (curve *e*). For p_3 value equal to p_j, the shock wave will occur just at the exit of divergent section. If p_3 value lies between p_f and p_j oblique waves are formed at the exit.

PROBLEMS

14.1 Four kg of a gas is compressed adiabatically from 100 kN/m² pressure and 17°C temperature to a final pressure of 400 kN/m². Determine initial volume, final volume, final temperature, work done, heat added or removed and change in internal energy. Take $c_p = 99.98$ m/K, $c_v = 71.29$ m/K, $R = 28.7$ m/K.

(3.266 m³, 1.215 m³, −396, 625.3 Nm, 158.9°C, 0, 326, 625.3 N m)

14.2 What will be the work done on 1.5 kg of carbon dioxide at 25°C if it is compressed under isothermal condition from 200 kN/m² to 500 kN/m²? Also determine the initial and final volumes of gas and the density. Take $R = 19.3$ m/K.

(0.424 m³, 0.1696 m³, −77 481.33 Nm, 3.541 kg/m³, 8.868 kg/m³)

14.3 Show that the work done in compressing a gas under isentropic conditions is given by

$$W = (p_1 v_1 - p_2 v_2)/(k-1)$$

14.4 When 0.75 kg of oxygen gas was compressed isothermally to 0.10 m³ volume at 30°C, 30 000 N m work was done on the gas. Determine its initial volume and pressure of the gas. Take $R = 26.8$ m/K.

(0.165 m³, 361.4 kN/m²)

14.5 Two kg of a gas at 20°C is compressed isentropically to 50 percent of its initial volume. If the initial pressure is 100 kN/m², determine the final pressure, final temperature, work done, and changes in internal energy and enthalpy. Take c_p=99.98 m/K, c_v=71.29 m/K and $R = 28.7$ m/K.

(263.90 kN/m², 113.66°C, − 52 956.13 J, 31.275 k Cal, 43.862 k Cal)

14.6 In continuation of Example 14.5, prove that change in entropy can also be expressed as

$$\phi_2 - \phi_1 = c_v \ln \left(\frac{p_2}{p_1}\right)\left(\frac{\rho_1}{\rho_2}\right)^k$$

and
$$\phi_2 - \phi_1 = c_v \ln \left(\frac{T_2}{T_1}\right)^k \left(\frac{p_1}{p_2}\right)^{k-1}$$

14.7 Determine the velocity of sound in air at 20°C, 200°C and 500°C. (Take $k = 1.40$). (344.98 m/s, 438.27m/s, 560.24 m/s)

14.8 A projectile travels through water at 20°C at a speed of 3240 km/hr. What will be its Mach number? (0.624)

14.9 An aeroplane is observed directly overhead. It is first heard 15 seconds later at an angle of 30° with respect to the vertical. For a velocity of sound of 330 m/s, calculate Mach number, altitude of the plane and its speed. (1.155, 4.95 km, 1372.14 km/hr)

14.10 In the case of a compressible fluid with flow velocity of 600 m/s and velocity of pressure wave of 300 m/s, the flow area is reduced from 0.50 m^2 to 0.40 m^2. Determine the corresponding change in velocity. (Reduction by 40.0 m/s)

14.11 What has to be the Mach number of the flow if 30 percent increase in velocity is produced by 30 percent increase in area? (1.414)

14.12 Determine the change in temperaturc of carbon dioxide when its velocity changes from 10.0 m/s to 50.0 m/s as it passes through the nozzle. Assume the process to be isentropic. Take R=19.5 m/K and $k = 1.30$. (1.448°C)

14.13 Show that for isentropic flow

$$\frac{T_1}{T_2} = \left(\frac{\rho_1}{\rho_2}\right)^{k-1}$$

14.14 Oxygen gas flows in a 50 mm diameter pipe at 10°C at a mass rate of 0.20 kg/s at a pressure of 350 kN/m^2 absolute. If the pipe diameter is suddenly reduced to 35 mm diameter, determine the velocity, pressure and density in 35 mm section. Take $k = 1.29$ and $R = 26.8$ m/K and assume the process to be isentropic.
(47.658 m/s, 324.745 kN/m^2. 4.364 kg/m^3)

14.15 A 50 mm diameter pipe carries oxygen at the gauge pressure of 100 mm of mercury and 20°C. The pipe is fitted with a venturimeter of throat diameter 25 mm. If the mass rate of flow in the pipe is 0.1319 kg/s, determine the density of oxygen at the throat and the throat pressure. Assume barometric pressure to be 760 mm of mercury. Take $k = 1.40$ and $R = 26.5$ m/K.
(1.160 kg/m^3, −158 mm of mercury)

14.16 Compare the mass rate of flow of air through 100 mm × 50 mm venturimeter with inlet and throat pressures of 160 kN/m^2 and 130 kN/m^2 absolute if

(i) air is considered to be incompressible
(ii) air is treated as compressible fluid.

Take $\rho_1 = 1.953$ kg/m^3 and $k = 1.40$.

(0.6939 kg/s, 0.6143 kg/s)

14.17 An aeroplane is travelling at an altitude of 18,000 m above sea level where the temperature is −56.50°C. Assuming k and R to be 1.40 and 29.57 m/K respectively, determine the speed of aeroplane a Mach two. (2135.3 km/hr)

14.18 A missile travels in the atmosphere where the pressure and temperature are 0.70 kN/m² abs. and −4.0°C respectively. For Mach angle of 40°, determine the speed of the missile. Take $R = 29.57$ m/K, $k = 1.40$. (1851.24 km/hr)

14.19 Just downstream of a normal shockwave in air, the pressure, speed and temperatare are 360.0 kN/m², 110.0 m/s and 50.0°C. Compute M_1 and pressure and speed prior to shock wave. (3.787, 21.72 kN/m² abs., 489.285 m/s)

14.20 Upstream of a normal shock wave in air flow, the sonic velocity, flow velocity and absolute pressure are 265.0 m/s, 525.8 m/s and 101.3 kN/m². Determine the corresponding quantities downstream of shock wave. (343.8 m/s, 199.4 m/s, 446.0 kN/m²)

14.21 If in a compression shock, p_1=42.13 kN/m², t_1 = 0°C and U_1=1000 m/s, determine the conditions just downstream of the shock. (435.64 kN/m², 259.32°C, 259.0 m/s)

14.22 For stagnation flow of a compressible fluid show that

$$\frac{T_s}{T_0} = \left(1 + \frac{k-1}{k} M_0^2\right)$$

and

$$\frac{\rho_s}{\rho_0} = \left(1 + \frac{k-1}{k} M_0^2\right)^{1/(k-1)}$$

14.23 An aeroplane travels at an altitude of 14,000 m above sea level at a speed of 2000 km/hr. If the ambient pressure and temperature at this elevation are 14.170 kN/m² and 216.7 K, determine the stagnation pressure, temperature and air density. (95.095 kN/m² abs., 100.21°C, 0.963 kg/m³)

14.24 A flying aeroplane records the ambient pressure and stagnation pressure and temperature on the nose of its fuselage as 90.00 kN/m², 110.0 kN/m² absolute and −4.0°C respectively. Determine the speed of aeroplane and the temperature and density of air through which the plane travels. Also determine the density at stagnation point. (618.59 km/hr, 19.0°C, 1.258 kg/m³, 1.452 kg/m³)

14.25 A pitot tube at the nose of the fuselage of the aeroplane records the stagnation pressure of 96.00 kN/m² while the ambient flow conditions are 90.00 kN/m² pressure abs. and 263 K. Determine the speed of the plane. (600 km/hr)

14.26 A missile travels in the earth's atmosphere at Mach number of 2.5 relative to the undisturbed air. Assuming $k = 1.40$, determine the temperature at the nose of missile when the air temperature at the flight altitude is −60°C. (206.45°C)

14.27 Air is discharged from a large tank in which the gauge pressure is 700 kN/m^2 and temperature is 26.84°C, through a convergent nozzle of 25 mm diameter. Calculate the mass rate of flow when the outside pressure is (i) 200 kN/m^2 and (ii) 500 kN/m^2, and barometric pressure is 100 kN/m^2. Also calculate the pressure, temperature, velocity and sonic velocity at the nozzle in both the cases. Take $R = 28.57$ m/K.

(0.927 kg/s, 322.4 kN/m^2 gauge, −23.26°C, 313.05 m/s, 0.819 kg/s, 3.18°C, 215.461 m/s, 329.22 m/s)

14.28 In a normal shock wave, the conditions before the shock are $p_1 = 45.0$ kN/m^2, $U_1 = 1000$ m/s and $t_1 = 26.84$°C. Determine the flow conditions after the shock. Take $R = 29.57$.

(412.18 kN/m^2, 268.0 m/s, 1.929 kg/m^3, 483.87°C)

14.29 If the density ratio $\frac{\rho_2}{\rho_1}$ across a normal shock is 3.71 determine M_1, M_2, p_2/p_1 and U_2/U_1. Take $k = 1.3$.

(2.50, 0.423, 6.935, 0.2695)

DESCRIPTIVE QUESTIONS

14.1 Give two examples where liquid is treated as a compressible fluid.

14.2 When is the compressibility of fluid important?

14.3 Give two examples when air can be treated as an incompressible fluid.

14.4 Why is change in potential energy considered negligible in the treatment of compressible fluids?

14.5 What is the difference between isentropic and adiabatic flows?

14.6 When can a thermodynamic process be considered as isothermal or adiabatic?

14.7 Will the velocity of pressure wave increase or decrease as the altitude increases? Explain.

14.8 What is the qualitative dependence of velocity of pressure wave with molecular weight of gas? (Refer Appendix D)

14.9 A pitot tube is calibrated in a jet of air issuing from a large tank, at a velocity range of 10–30 m/s, and the equation obtained is $U = 0.98\sqrt{\Delta p/\rho}$. Can this equation be used in the velocity range of 100 – 400 m/s? Explain. Can Eq. 14.32 be used for supersonic flows?

14.10 An aeroplane is travelling while you are observing it from the ground. How will you know whether it is subsonic or supersonic?

14.11 Compare shock wave in compressible fluid with hydraulic jump in an open channel flow.

14.12 A supersonic aerofoil has a sharp leading edge while the subsonic aerofoil has rounded lead edge. Why?

14.13 Compare the structures of Mach number, Weber number and Froude number.

14.14 Using Eq. 14.23, give arguments to show that if throat velocity is not sonic, it will be maximum in subsonic flow and minimum in supersonic flow.

14.15 If the pressure at the nozzle attached to a large reservoir is critical, what can you say about the magnitude of pressure in the reservoir?

14.16 Do pressure waves always travel at sonic speed in compressible fluid? Give an example to suit your answer.

14.17 Give three examples where shock waves occur.

14.18 C_p and C_v are related as

(i) $c_p\, c_v = k$ (ii) $c_v/c_p = k$ (iii) $c_p/c_v = k$ (iv) $c_p = c_v$

14.19 k for atmosphere is

(i) 1.0 (ii) 1.2 (iii) 1.4 (iv) 1.6

14.20 An isentropic process is

(i) adiabatic and irreversible
(ii) isothermal and reversible
(iii) irreversible and isothermal
(iv) adiabatic and reversible.

14.21 Speed of sound in atmosphere is given by

(i) $\sqrt{\rho/k}$ (ii) $\sqrt{p/\rho}$ (iii) $\sqrt{kp/\rho}$ (iv) none of the above.

14.22 A normal shock wave is analogous to

(i) sinusoidal wave in open channels
(ii) elementary wave in still water
(iii) hydraulic jump
(iv) a moving hydraulic jump.

14.23 Across a normal shock wave

(i) entropy remains constant
(ii) entropy decreases
(iii) entropy increases.

14.24 Continuity equation for one dimensional compressible flows is

(i) $\frac{A}{dA} + \frac{\rho}{d\rho} + \frac{U}{dU} = 0$ (ii) $\frac{dA}{A} + \frac{d\rho}{\rho} + \frac{dU}{U} = 0$

(iii) $\frac{dA}{A} + (\rho U)dU = 0$

14.25 Complete the following table by writing increases or decreares

Geometry Flow	Diverging		Converging	
	$M < 1.0$	$M > 1.0$	$M < 1.0$	$M > 1.0$
V			increases	
P			decreases	
ρ			decreases	
T			decreases	
ϕ			constant	

CHAPTER XV

Pumps

15.1 INTRODUCTION

A pump is a device which gives energy to the fluid it transports, thereby increasing its pressure head, kinetic energy or both. Fluids transported may be a gas or a liquid. Transport of gases involves a certain degree of compression and hence volume change. Such volume change does not take place in the case of liquids. Here only liquid transport is considered. Either pure liquids such as water, oils, liquid chemicals can be transported, or two phase mixtures such as water and ores or coal, sewage or sludge can be transported. Pumps find use in water supply, sewage treatment, agricultural engineering, chemical and mechanical engineering, hydraulic control systems and many more specialisations in engineering.

15.2 PUMP CLASSIFICATION

Pumps can be broadly classified into positive displacement pumps and rotodynamic or dynamic pressure pumps. Reciprocating pump is a positive displacement pump in which fluid is drawn or forced into a finite space and is then sealed in it by mechanical means. The fluid is then forced out to flow and the cycle is repeated. Reciprocating pump can be single acting or double acting (Fig. 15.1). Among the positive displacement pumps, there are pumps in which there is rotatory action instead of reciprocating action. Under this category fall gear pumps, lobe pumps and vane pumps, etc. They can be called special types of pumps; there are also shown in Fig. 15.1. In rotodynamic pumps, there is a free passage of fluid between the inlet and outlet of machine without any intermittent sealing. Rotodynamic pumps have a rotating part called impeller, rotor or runner which rotates continuously in fluid and transfers energy from rotor to fluid. Hence there is a rate of change of angular momentum. Centrifugal pump is a rotodynamic pump. Rotodynamic pumps are further classified according to direction of fluid flow in relation to the plane of impeller rotation. If flow is perpendicular to this plane, it is called axial flow pump; if flow enters axially and leaves radially, it is called radial flow pump (or conventional centrifugal

pump). These are shown in Fig. 15.5. In mixed flow pump, the flow enters axially and flow through their impellers is partly radial and partly axial.

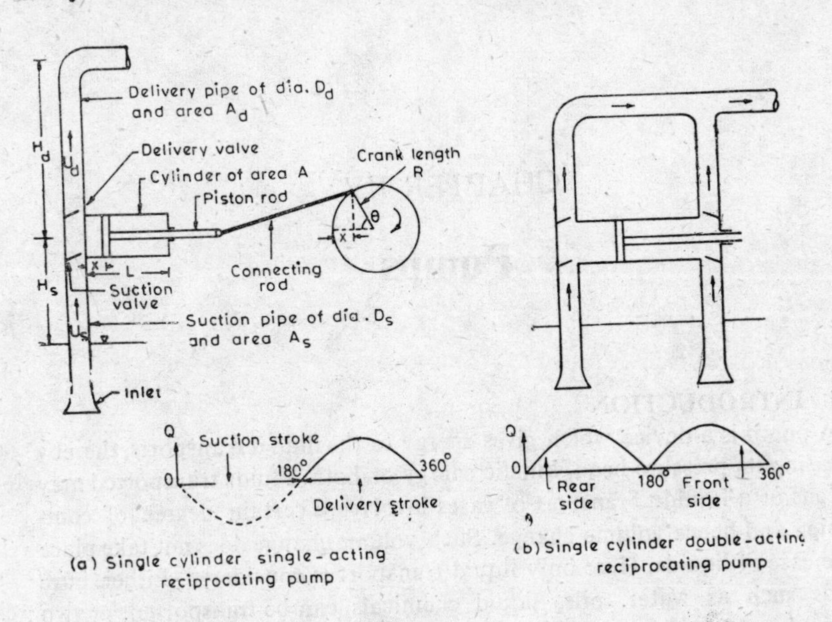

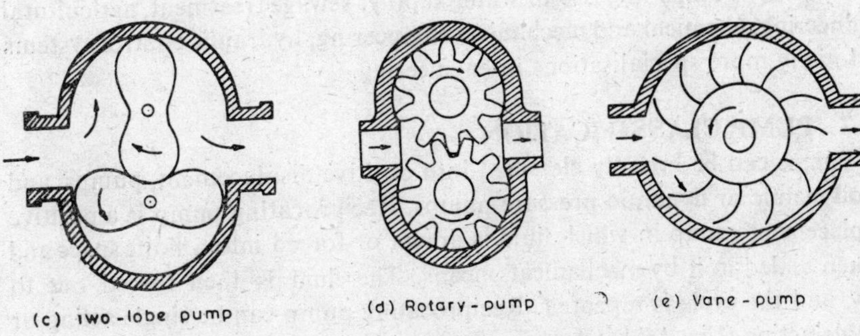

Fig. 15.1. Positive-displacement pumps.

15.3 RECIPROCATING PUMPS

Figure 15.1 (a) shows a single cylinder single acting reciprocating pump giving notation details.

Theoretical discharge of single acting reciprocating pumps $$Q_t = \frac{ALN}{60} \tag{15.1}$$

Theoretical discharge of double acting reciprocating pump $$Q_t = \frac{2ALN}{60} \tag{15.2}$$

Volumetric efficiency $$\eta = \frac{Q_a}{Q_t} \times 100 \tag{15.3}$$

where N is the rotational speed of pump in rpm and Q_a is the actual discharge of the pump. The slip of pump is defined as

$$\text{Slip} = \left(\frac{Q_t - Q_a}{Q_t}\right) \times 100 \tag{15.4}$$

The slip of the pump is usually positive and one to two percent in its magnitude. However slip can be negative if suction pipe is long, delivery pipe is short and speed rotation N is high. The speed varies from 20 to 150 rpm, stroke length L from 0.15 to 1.0 m and generally the higher the speed, the smaller the length.

Following expression can be derived from the geometry and working of the pump.

Displacement of piston $x = R(1 - \cos \omega t) = R \cos \theta$

where R is the crank length or radius, $\omega = 2\pi N/60$ and $\theta = \omega t$.

Velocity of piston $U_p = \dfrac{dx}{dt} = R\omega \sin \omega t$

Continuity equation: $$AU_p = A_s U_s = A_d U_d \tag{15.5}$$

where A, A_s and A_d are the areas of piston, suction pipe and delivery pipe respectively, and U_s and U_d are the velocities in suction and delivery pipe respectively.

Acceleration head represents the energy required to accelerate the water column in suction or delivery pipe.

$$\left.\begin{aligned}
&\text{Acceleration} \quad = \frac{dU_p}{dt} = R\omega^2 \cos \omega t \\
&\text{Acceleration head in suction pipe} \\
&\qquad H_{as} = \frac{l_s}{g}\left(\frac{A}{A_s}\right) R\omega^2 \cos \omega t \\
&\text{Acceleration head in delivery pipe} \\
&\qquad H_{ad} = \frac{l_d}{g}\left(\frac{A}{A_d}\right) R\omega^2 \cos \omega t
\end{aligned}\right] \tag{15.6}$$

where l_s and l_d are the lengths of suction and delivery pipes.

Friction head in suction and delivery pipes, h_{fs} and h_{fd} are given by

$$\left.\begin{aligned}
h_{fs} &= \frac{fl_s}{2gD_s}\left(\frac{A}{A_s} R\omega \sin \omega t\right)^2 \\
h_{fd} &= \frac{fl_s}{2gD_d}\left(\frac{A}{A_d} R\omega \sin \omega t\right)^2
\end{aligned}\right] \tag{15.7}$$

Since h_{fs} and h_{fd} vary with the position of piston (or ωt), their averages during the suction and delivery strokes are $\frac{2}{3}h_{fsm}$ and $\frac{2}{3}h_{fdm}$ where h_{fsm} and h_{fdm} are the maximum values of h_{fs} and h_{fd}. Hence average head losses are

$$\frac{fl_s}{3gD_s}\left(\frac{A}{A_s} R\omega\right)^2 \quad \text{and} \quad \frac{fl_s}{3gD_d}\left(\frac{A}{A_d} R\omega\right)^2$$

Figure 15.2 shows the indicator diagram which indicates how pressure head varies during the suction and delivery strokes.

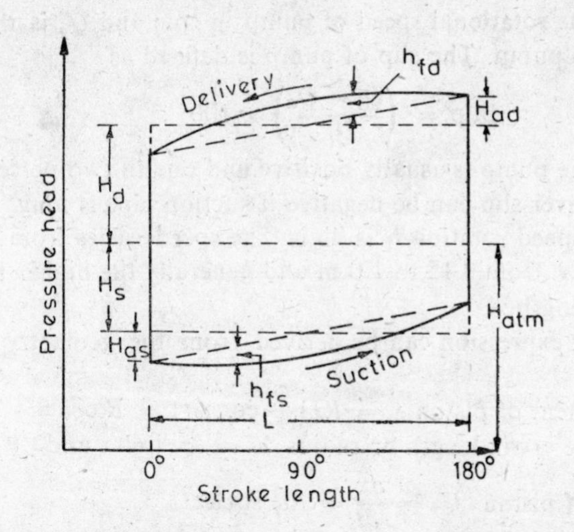

Fig. 15.2 Indicator diagram for reciprocating pump.

$$\text{Work done per second in kW} = \frac{Q\gamma}{1000}\left[H_s + H_d + \frac{fl_s}{3gD_s}\left(\frac{A}{A_s}R\omega\right)^2 + \frac{fl_d}{3gD_d}\left(\frac{A}{A_s}R\omega\right)^2\right] \tag{15.8}$$

Safe speed of the pump is decided from the consideration that at the beginning of the suction stroke and the end of the delivery stroke, the net head is more than the minimum head at which separation of flow can occur.

15.4 AIR VESSELS

Air vessels can be fitted on one or both sides of the pump, i.e. on suction and delivery side. They are located close to the cylinder (Fig. 15.3). Volume of air vessel on the delivery side is about 6 to 9 times the volume of cylinder and that on the suction side 3 to 4 times the cylinder volume. Air vessels serve the following functions:

1. Fluctuations in discharge are decreased.
2. Pump can be run at higher speed because there is less danger of separation of flow.
3. There is less friction loss and hence saving in work done.

When air vessels are provided, the flow before air vessel on the delivery side and the flow between air vessel and cylinder on the suction side are varying. Hence acceleration heads are calculated only in lengths l_{sa} and l_{sd}.

Velocity beyond air vessel on delivery side $\overline{U}_d = \dfrac{ALN}{60A_d} = \dfrac{A}{A_d}\dfrac{R\omega}{\pi}$

Discharge beyond air vessel $\qquad Q = \dfrac{AR\omega}{\pi}$

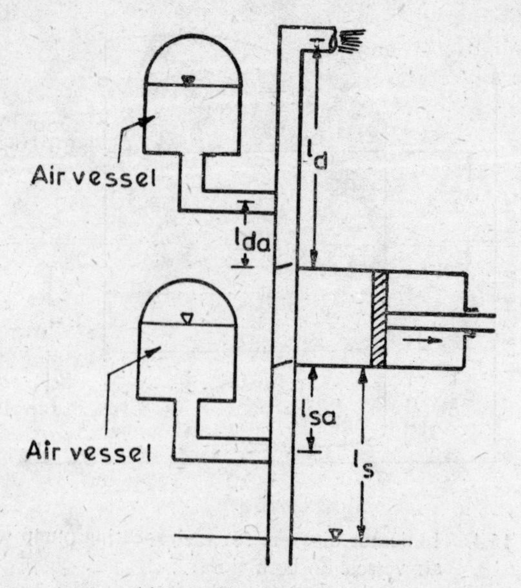

Fig. 15.3. Air vessels in a reciprocating pump.

Discharge in air vessel $Q_a = Q_d - Q$

$$= AR\omega\left(\sin\omega t - \frac{1}{\pi}\right) \tag{15.9}$$

If Q_a is positive, water flows into air vessel and if it is negative, it flows out of it. When air vessel is fitted on the suction side, the same expression is valid for Q_a but signs get reversed. For air vessel on the suction side, water flow out of air vessel if Q_a is +ve and into it, if it is −ve. For double acting cylinder,

$$Q_a = AR\omega\left(\sin\omega t - \frac{2}{\pi}\right) \tag{15.10}$$

Work done by a single acting reciprocating pump with air vessels on both sides is given by Eq. (15.11).

Work done per second, P is given as

$$P = \frac{Q\gamma}{1000}\left[H_s + \frac{f(l_s - l_{sa})}{2gD_s}\left(\frac{Q}{A_s}\right)^2 + \frac{1}{3}\frac{fl_{sa}}{gD_s}\left(\frac{A}{A_s}R\omega\right)^2 + H_d + \frac{f(l_d - l_{da})}{2gD_d}\left(\frac{Q}{A_d}\right)^2 + \frac{1}{3}\frac{fl_{sa}}{gD_d}\left(\frac{A}{A_d}R\omega\right)^2\right] \tag{15.11}$$

Similar expressions can be written for double acting reciprocating pump with air vessels. Figure 15.4 shows the indicator diagram for single acting reciprocating pump in the presence of air vessels on both sides.

15.5 ROTODYNAMIC PUMPS

Rotodynamic pumps, also known as centrifugal pumps, can be broadly

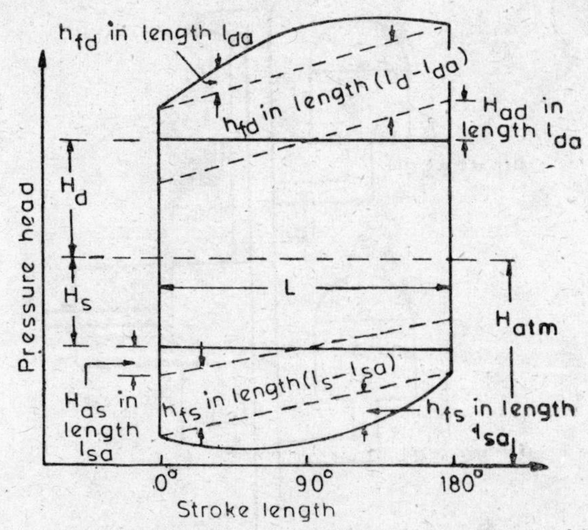

Fig. 15.4. Indicator diagram for reciprocating pump with air vessels on both sides.

classified into three categories, namely radial, axial and mixed. Figure 15.5 shows the three types of pumps; radial pumps are more often used. Parts

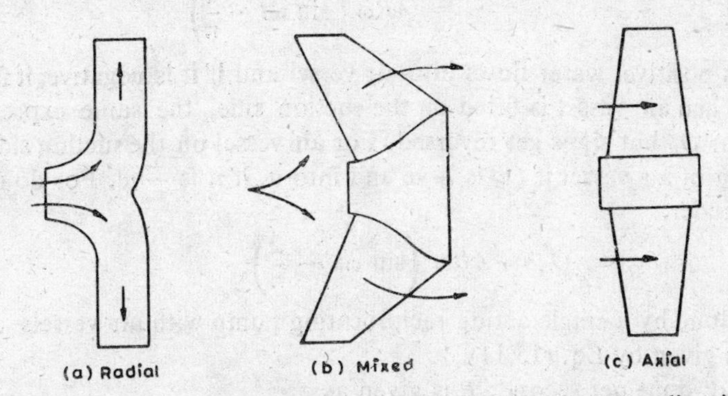

Fig. 15.5. Classification of centrifugal pumps based on flow direction through impeller.

of pumps are shown in Fig. 15.6. Blades of vanes can be straight, radial, bending backwards or bending forwards. Their number varies from six to twelve.

Vanes curving backward present convex surface to the liquid which is impelled by them and they are most efficient. For them α and ϕ are less than 90° (Fig. 15.8). Radial vanes have got their outlet tips in radial direction and for them α is less than 90° and ϕ equal to 90°. For vanes curving forward, α is less than 90° and ϕ greater than 90°. The impeller consisting

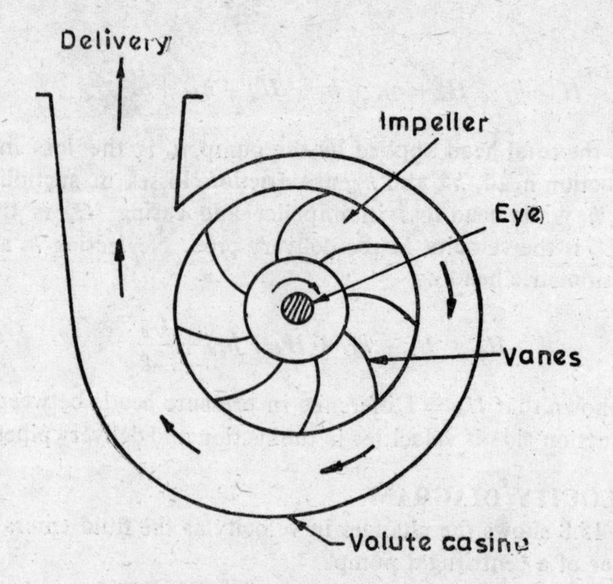

Fig. 15.6. Parts of centrifugal pump.

of vanes is usually enclosed in a casing which can be of volute of turbine type.

Typical installation of a pump along with different definitions are shown in Fig. 15.7.

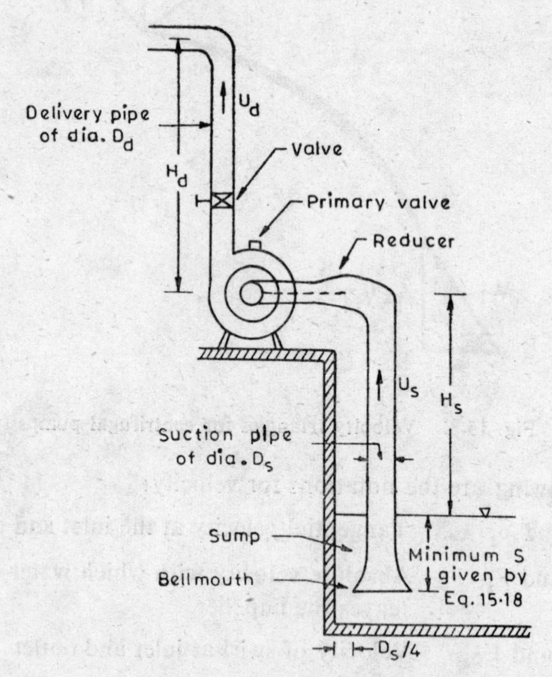

Fig. 15.7. Typical pump installation.

Here

$$H = h_e + H_s + h_{fs} + h_1 + H_d + h_{fd} + \frac{U_d^2}{2g} \tag{15.12}$$

where H is the total head applied by the pump, h_e is the loss in entrance, H_s is the suction head, h_{fs} and h_{fd} are friction losses in suction and delivery pipes, h_1 is the head loss in impeller and casing, H_d is the delivery head and U_d is the velocity in the delivery pipe. Neglecting h_e and h_1, one defines manometric head as

$$H_m = H_s + h_{fs} + H_d + h_{fd} + \frac{U_d^2}{2g} \tag{15.13}$$

It can be shown that H_m = Difference in pressure heads between the delivery and suction side if velocities in the suction and delivery pipes are equal.

15.6 VELOCITY DIAGRAM

Figure 15.8 shows the changes in velocity as the fluid enters and leaves the impeller of a centrifugal pump.

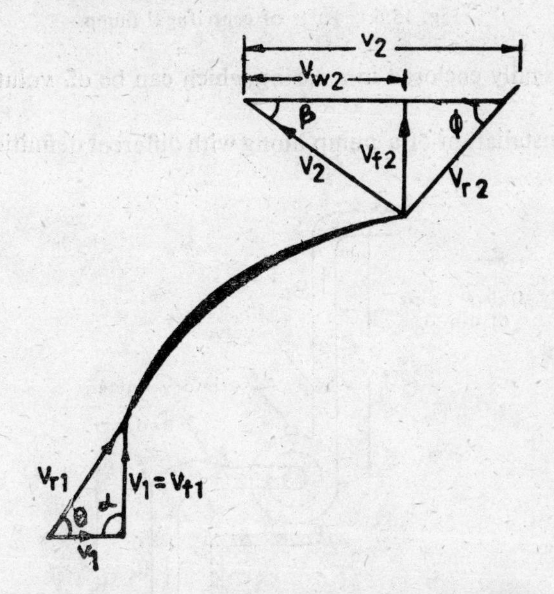

Fig. 15.8. Velocity triangles for centrifugal pumps.

The following are the notations for velocity:

v_1 and v_2	Tangential velocity at the inlet and outlet
V_1 and V_2	Absolute velocity with which water enters and leaves the impeller
V_{w1} and V_{w2}	Velocity of swirl at inlet and outlet
D_1 and D_2	Inner and outer diameter of the impeller

V_{r1} and V_{r2}	Relative velocity of flow at inlet and outlet
V_{f1} and V_{f2}	Velocity of flow at inlet and outlet
α	Angle at which water enters the impeller
β	Angle at which water leaves the impeller
θ	Vane angle at inlet
ϕ	Vane angle at outlet

Work done per second $= T\omega$

$$= Q\rho(V_{w2}v_2 - V_{w1}v_1) \qquad (15.14)$$

where T is the torque developed and ω is angular velocity.

$$\text{Theoretical head developed} = \frac{1}{g}(V_{w2}v_2 - V_{w1}v_1) \qquad (15.15)$$

For radial entry $\alpha = 90°$ and $V_{w1} = 0$

$$\text{Theoretical head developed} = \frac{V_{w2}v_2}{g} \qquad (15.16)$$

Different efficiencies for the pump are defined as follows

$$\text{Overall efficiency } \eta = \eta_{mo}\eta_m\eta_v$$

where

$$\text{Manometric efficiency } \eta_{mo} = \frac{\text{Manometric head}}{\text{Theoretical head}} \times 100$$

$$= \frac{gH_m}{V_{w2}v_2} \times 100$$

$$\text{Mechanical efficiency } \eta_m = \frac{\text{Available head at impeller}}{\text{Energy given by prime mover}} \times 100$$

$$\text{Volumetric efficiency } \eta_v = \frac{Q}{Q+q} \times 100$$

where Q is the actual discharge and q is leakage discharge. η_v varies from 95 to 98 percent, while η varies from 70 to 85 percent.

15.7 SPECIFIED SPEED

Specific speed N_s is defined as

$$N_s = NQ^{1/2}/H^{3/4} \qquad (15.17)$$

where N is the rotational speed in rpm, Q is in m^3/s and H is in m. The

type of pump chosen depends on the range of specific speed as shown below.

N_s	Type
Less than 10	Low speed radial flow
10–20	Medium speed radial flow
20–80	High speed radial flow
80–100	Mixed flow
Greater than 100	Axial flow

15.8 PUMP INTAKES AND SUMPS

Water should enter the suction pipe without much disturbance; hence a bellmouth entrance is provided. The bellmouth portion is usually a quarter of an ellipse. Also there must be adequate water depth over the entrance so that air entraining vortices do not form. Formation of such vortices can reduce pump efficiency and cause vibration. A very approximate criterion for minimum depths to avoid vortex formation is

$$\frac{S}{D_s} = 2.2 \frac{U_s}{\sqrt{gD_s}} \tag{15.18}$$

For single pump sump, certain basic requirements of a hydraulically good sump can be mentioned. Such a sump should have adequate volume, there should be no stagnation zones and no abrupt change in flow direction and there should be adequate clearance from sump bottom and sump sides. Pump manufacturers normally specify these dimensions.

15.9 MISCELLANEOUS

The characteristic curves of a pump show graphically its performance under various operating curves. The data are plotted as $\frac{\eta}{\eta_0}$, $\frac{H_m}{H_{m0}}$, $\frac{P}{P_0}$, vs $\frac{Q}{Q_0}$ where η, H_m, P and Q are values of overall efficiency, manometric head, and power at any discharge Q while those with subscript 0 refer to corresponding quantities at design discharge Q_0, all at constant speed. One can also prepare H_m vs Q curve with constant η and N lines. Such curves enable one to determine optimum working conditions for given efficiency.

When pressure at any point in the pump falls below the vapour pressure of liquid, cavitation will occur. In a centrifugal pump, the pressure is lowest on the underside of vanes at the entrance where cavitation would first start. Factors contributing towards the onset of cavitation are higher rotational speeds, restricted suctions, higher temperature of liquid, and higher specific speed.

Cavitation parameter

$$\sigma = \frac{H_{atm} - H_s - h_v}{H}$$

where h_v is the equivalent vapour pressu e head. If σ is less than σ_c, cavitation occurs. σ_c is given as

$$\sigma_c = (1.042 \times 10^{-3})\, N_s^{4/3} \tag{15.19}$$

ILLUSTRATIVE EXAMPLES

15.1 For a single acting reciprocating pump, piston diameter D is 150mm, stroke length L is 300 mm, rotational speed N is 50 rpm and the water is to be raised through 18m. Determine theoretical discharge Q_t. If the actual discharge is 4.0 l/s, determine volumetric efficiency, slip and actual power required; take the mechanical efficiency as 80 percent.

$$Q_t = \frac{ALN}{60} = \frac{0.785 \times 0.150^2 \times 0.300 \times 50}{60} = 4.416 \times 10^{-3} \text{ m}^3\text{/s}$$

or 4.416 l/s

$$\text{Volumetric efficiency } \eta = \frac{4.0}{4.416} = 0.9058 \text{ or } 90.58 \text{ percent}$$

$$\text{Slip} = \frac{Q_t - Q_a}{Q_t} = \frac{4.416 - 4.0}{4.416} = 0.0942 \text{ or } 9.42 \text{ percent}$$

$$\text{Power} = \frac{Q_a \gamma H}{1000} = \frac{0.004 \times 9787 \times 18}{0.80 \times 1000} = 0.881 \text{ kW}$$

$$\text{Theoretical power} = \frac{Q_t \gamma H}{1000} = \frac{0.004416 \times 9787 \times 18}{1000} = 0.778 \text{ kW}$$

15.2 A single acting reciprocating pump has a plunger diameter of 125 mm and stroke of 300 mm. The length of suction pipe is 10.0 m and diameter 75 mm. Find acceleration head at the beginning, middle and end of suction stroke. If the suction head is 3.0 m, determine the pressure head in the cylinder at the beginning of stroke when the pump runs at 30 rpm. Take atmospheric head as 10.23 m.

$$R = L/2 = 300/2 \text{ or } 150 \text{ mm}$$

Acceleration head on suction side H_{as}

$$= \frac{l_s}{g}\left(\frac{A}{A_s}\right) R\omega^2 \cos \omega t$$

$$= \frac{10.0}{9.806}\left(\frac{125}{75}\right)^2 \times 0.150 \left(\frac{2\pi \times 30}{60}\right)^2 \cos \omega t$$

$$= 4.195 \cos \omega t$$

At the beginning of stroke, $\omega t = 0$ and $\cos \omega t = 1$

$\therefore$ H_{as} at the beginning $= 4.195$ m

$\therefore$ Pressure head in cylinder at the beginning of stroke (Fig. 15.2)

$$= H_{atm} - H_s - H_{as}$$

$$= (10.23 - 3.0 - 4.195)$$

$$= 3.035 \text{ m abs}$$

At the middle of stroke $\omega t = \pi/2$ and $\cos \omega t = 0$; hence $H_{as} = 0$

$\therefore$ Pressure head at the middle of stroke $= H_{atm} - H_s$

$$= 10.23 - 3.00 = 7.23 \text{ m abs}$$

At the end of suction stroke $\omega t = 180°$ and $\cos \omega t = -1$

Pressure head at the end of suction stroke

$$= 10.23 - 3.0 + 4.195$$

$$= 11.425 \text{ m abs}$$

15.3 A single acting reciprocating pump has the following data

Stroke $L = 300$ mm, Piston diameter $D = 125$ mm

Suction pipe length $l_s = 5.0$ m,

Diameter of suction pipe $D_s = 75$ mm

Suction head $H_s = 3.0$ m, $H_{atm} = 10.23$ m abs

Safe minimum pressure head $= 2.0$ m (abs)

What is the maximum speed at which it can be run without causing separation during suction stroke?

Pressure head during the beginning of suction stroke will be minimum.

H_{as} at the beginning of suction stroke $= \frac{l_s}{g}\left(\frac{A}{A_s}\right) R\omega^2$

$$= \frac{5.0}{9.806}\left(\frac{125^2}{75}\right) \times 0.150 \times \left(\frac{2 \times 3.142 \times N}{60}\right)^2$$

$$= 0.002\,33\ N^2$$

$$\therefore \quad 2.0 = 10.23 - 3.00 - 0.002\,33\ N^2$$

$$0.002\,33\ N^2 = 5.23$$

or $$N = 47.377 \text{ rpm}$$

15.4 A single acting reciprocating pump has the followig details:

$L = 150$ mm $\quad N = 75$ rpm $\quad D = 1.333$ times D_s

$l_s = 7.5$ m $\quad H_s = 2.5$ m $\quad D_s = 75$ mm

$f = 0.025$

Determine the pressure head in the cylinder in the beginning, middle and end of stroke.

$$H_{as} = \frac{7.5}{9.806}\left(\frac{1.333D_s}{D_s}\right)^2 \times 0.075 \times \left(\frac{2 \times 3.142 \times 75}{60}\right)^2$$
$$= 6.292 \text{ m}$$

Friction head at the middle of suction stroke will be given by substituting

$$\omega t = 90° \text{ in } h_{fs} = \frac{fl_s}{2gD_s}\left(\frac{A}{A_s} R\omega \sin \omega t\right)^2$$

$$\therefore \qquad h_{fsm} = \frac{0.025 \times 7.5}{2 \times 9.806 \times 0.075}\left(\frac{16}{9} \times 0.075 \times \frac{2 \times 3.142 \times 75}{60}\right)^2$$
$$= 0.140 \text{ m}$$

$\therefore$ Pressure head at the beginning of suction stroke

$$= H_{atm} - (6.292 + 2.5) \text{ m (gauge)}$$

or $$= 8.792 \text{ m (vacuum)}$$

Pressure head at the middle of the stroke

$$= H_{atm} - (H_s + h_{fs})$$
$$= H_{at} - (2.50 + 0.140) \text{ m (gauge)}$$

or $$= 2.642 \text{ m (vacuum)}$$

Pressure head at the end of the suction stroke

$$= H_{atm} - H_s + H_{as}$$
$$= H_{atm} - 2.5 + 6.292 \text{ m (gauge)}$$

or $$= 3.792 \text{ m (vacuum)}$$

15.5 A single cylinder double acting reciprocating pump has a piston diameter of 300 mm and stroke length of 400 mm. When the pump runs at 45 rpm, it discharges 0.039 m³/s under a total head of 15 m. What will be the volumetric efficiency, work done per second and power required if the mechanical efficiency of the pump is 75 percent?

$$Q_t = \frac{2ALN}{60} = \frac{2.0 \times 0.785 \times 0.3^2 \times 0.40 \times 45}{60}$$
$$= 0.042\ 39 \text{ m}^3/\text{s}$$

$$\text{Volumetric efficiency} = \frac{0.039}{0.042\,39} = 0.92 \quad \text{or} \quad 92.00 \text{ percent}$$

$$\text{Work done per second} = \frac{0.039 \times 15 \times 9787}{1000} = 5.725 \text{ kW}$$

$$\text{Power required} = \frac{5.725}{0.75} = 7.633 \text{ kW}$$

15.6 For a single acting single cylinder reciprocating pump, following data are given:

$L = 400$ mm	$N = 30$ rpm	$D = 200$ mm
$H_s = 3.0$ m	$l_s = 5.0$ m	$D_s = 100$ mm
$H_d = 20$ m	$l_d = 20$ m	$D_d = 100$ mm
$f = 0.015$ for both the pipes.		$H_{atm} = 10.0$ m

Determine the pressure heads at the beginning, middle and end of delivery stroke, work done during the suction and the delivery strokes, and the power.

$$\omega = \frac{2\pi N}{60} = \frac{2 \times \pi \times 30}{60} = \pi$$

Maximum acceleration head during delivery stroke

$$H_{ad} = \frac{l_d}{g}\left(\frac{A}{A_d}\right) R\omega^2$$

or

$$H_{ad} = \frac{20}{9.806}\left(\frac{200}{100}\right)^2 \times 0.20 \times 3.142^2$$

$$= 16.108 \text{ m}$$

$$h_{fd} = \frac{fl_d}{2gD_d}\left(\frac{A}{A_d} R\omega\right)^2$$

$$= \frac{0.015 \times 20}{2 \times 9.806 \times 0.100}\left(\frac{200^2}{100^2} \times 0.20 \times 3.142\right)^2$$

$$= 0.967 \text{ m}$$

$\therefore$ Pressure head at the beginning of delivery stroke

$$= H_{atm} + H_d + H_{ad}$$

$$= 10.00 + 20.00 + 16.108$$

$$= 46.108 \text{ m abs}$$

Pressure head at the middle of delivery stroke

$$= H_{atm} + H_d + h_{fdm}$$

$$= 10.00 + 20.00 + 0.957$$

$$= 30.957 \text{ m abs}$$

Pressure head at the end of delivery stroke

$$= H_{atm} + H_d - H_{ad}$$
$$= 10.00 + 20.00 - 16.108$$
$$= 13.892 \text{ m abs}$$

$$h_{fsm} = \frac{fl_s}{2gD_s}\left(\frac{A}{A_s} R\omega\right)^2$$
$$= \frac{0.015 \times 5}{2 \times 9.806 \times 0.10} \times \left[\left(\frac{200}{100}\right)^2 \times 0.20 \times \left(\frac{2\pi \times 30}{60}\right)\right]^2$$
$$= 0.242 \text{ m}$$

Work done during suction stroke $= \frac{\pi D^2}{4} L\gamma \times \left(H_s + \frac{2}{3} h_{fsm}\right)$

$$= (0.785 \times 0.20^2 \times 0.400 \times 9.787)(3.0 + \tfrac{2}{3} \times 0.242)$$
$$= 388.602 \text{ Nm}$$

Work done during delivery stroke $= \frac{\pi D^2}{4} L\gamma \left(H_d + \frac{2}{3} h_{fd}\right)$

$$= (0.785 \times 0.20^2 \times 0.400 \times 9787)(20 + \tfrac{2}{3} \times 0.967)$$
$$= 2537.781 \text{ N m}$$

$\therefore$ Work done per stroke $= 388.602 + 2537.781$

$$= 2926.383 \text{ N m}$$

$$\text{Power} = \frac{2926.383 \times 30}{60 \times 1000} = 1.269 \text{ kW}$$

15.7 Following details of a single acting, single cylinder, reciprocating pump are given:

$L = 500$ mm	$D = 125$ mm	$H_{atm} = 10.2$ m
$l_s = 5.0$ m	$D_s = 100$ mm	$H_s = 3.0$ m
$l_d = 15$ m	$D_d = 100$ mm	$H_d = 10.0$ m

$f = 0.02$ for both suction and delivery pipes.

Safe minimum head $= 2.4$ m

Neglect slip and calculate (i) maximum permissible speed, and (ii) energy required to drive the pump if an air vessel is provided on the delivery side very close to the cylinder.

$$H_{as} = \frac{l_s}{g}\left(\frac{A}{A_s}\right) R\omega^2$$
$$= \frac{5.0}{9.806}\left(\frac{125}{100}\right)^2 \times 0.250 \left(\frac{2\pi N}{60}\right)^2$$
$$= 0.002\,186\, N^2$$

$$\therefore \quad 2.40 = H - H_s - H_{as}$$

$$= 10.20 - 3.00 - 0.002\,186\, N^2$$

$$\therefore \quad N^2 = 4.80/0.002\,186 \quad \text{or} \quad N = 46.86 \text{ rpm}$$

$$Q = \frac{ALN}{60} = \frac{0.785 \times 0.125^2 \times 0.500 \times 46.86}{60}$$

$$= 0.004\,791 \text{ m}^3/\text{s} \quad \text{or} \quad 4.791 \text{ l/s}$$

Maximum head loss on suction side $h_{fsm} = \dfrac{fl_s}{2gD_s}\left(\dfrac{A}{A_s}\, R\omega\right)^2$

$$= \frac{0.02 \times 5.0}{2 \times 9.806 \times 0.100}\left[\left(\frac{125}{100}\right)^2 \times 0.25\left(\frac{2 \times 3.142 \times 46.86}{60}\right)\right]^2$$

$$= 0.188 \text{ m}$$

Average frictional head loss on suction side

$$= \tfrac{2}{3}\, h_{fsm} = \tfrac{2}{3} \times 0.188 = 0.125 \text{ m}$$

Because of the air vessel that is provided on the delivery side is close to the cylinder, velocity in delivery pipe will be constant.

$$\bar{U}_d = \frac{Q}{A_d} = \frac{0.004\,791}{0.785 \times 0.10^2} = 0.61 \text{ m/s}$$

$$h_{fd} = \frac{fl_d \bar{U}_d^2}{2gD_d} = \frac{0.02 \times 15 \times 0.61^2}{2 \times 9.806 \times 0.10} = 0.0569 \text{ m}$$

Further

$$\text{Work done per second} = \frac{Q\gamma}{1000}\left(H_s + H_d + \tfrac{2}{3} h_{fs} + h_{fd}\right)$$

$$= \frac{0.004\,791 \times 978}{1000}\,(3.0 + 10.0 + 0.125 + 0\,0569)$$

$$= 0.621 \text{ kW}$$

15.8 Determine the percentage of work saved in one cycle when an air vessel is provided on the delivery side of a single cylinder single acting reciprocating pump.

Because of the air vessel very near the cylinder on the delivery side, velocity in delivery pipe will be uniform

$$\bar{U}_d = \frac{ALN}{60A_d}$$

Loss of head due to friction in delivery pipe

$$= \frac{fl_d \bar{U}_d^2}{2gD_d} = \frac{fl_d}{2gD_d}\left(\frac{ALN}{60A_d}\right)^2$$

Work done per stroke per unit weight of liquid $= h_{fd}$

When there is no air vessel, $h_{fdm} = \dfrac{fl_d}{2gD_d}\left(\dfrac{A}{A_d}\,R\omega\right)^2$

$$= \frac{fl_d}{2gD_d}\left(\frac{A}{A_d}\frac{L}{2}\frac{2\pi N}{60}\right)^2$$

Average head loss $= \frac{2}{3}\, h_{fdm}$

$$= \frac{fl_d}{3gD_d}\left(\frac{A}{A_d}\right)^2 \frac{L^2}{4}\frac{4\pi^2 N^2}{60^2}$$

Percentage of saving in work
$$= \frac{\dfrac{fl_d}{3gD_d}\left(\dfrac{A}{A_d}\right)^2 \dfrac{L^2}{4}\dfrac{4\pi^2N^2}{60^2} - \dfrac{fl_d}{2gD_d}\dfrac{A^2L^2N^2}{60^2A_d^2}}{\dfrac{fl_d}{3gD_d}\left(\dfrac{A}{A_d}\right)^2\dfrac{L^2}{4}\dfrac{4\pi^2N^2}{60^2}} \times 100$$

$$= \frac{\dfrac{\pi^2}{3} - \dfrac{1}{2}}{\pi^2/3} \times 100 = \frac{1}{2}\left(2 - \frac{3}{\pi^2}\right) \times 100$$

$$= 84.8 \text{ percent}$$

15.9 A single cylinder double acting reciprocating pump has the following details:

$D = 200$ mm $N = 120$ rpm

$L = 400$ mm $D_s = 150$ mm

If an air vessel is fitted on the suction side, determine the crank angle at which there is no flow to the air vessel or from the air vessel.

$$A = 0.785 \times 0.20^2 = 0.0314 \text{ m}^2$$

$$R = 0.50\, L = 0.20 \text{ m}$$

$$\omega = \frac{2\pi N}{60} = \frac{2 \times \pi \times 120}{60} = 4\pi \text{ rad/s}$$

Discharge from the sump to the air vessel $= Q = \dfrac{ALN}{60}$

$$= \frac{2AR\omega}{\pi}$$

$$= \frac{2 \times 0.0314 \times 0.20 \times 4\pi}{4} = 0.0512 \text{ m}^3/\text{s}$$

Discharge beyond air vessel $= AR\omega \sin\theta$

$$= 0.0314 \times 0.20 \times 4\pi \sin\theta$$

If there is no flow to or from the air vessel, these two discharges will be equal

$$\therefore \quad 0.0512 = 0.0314 \times 0.20 \times 4\pi \sin\theta$$

$\therefore \qquad \sin\theta = 0.6356$

or $\qquad \theta = 39.47°$

15.10 Determine the maximum permissible speed which will not cause separation on the suction or delivery stroke for the reciprocating pump with following data

$D = 100$ mm $\qquad L = 200$ mm $\qquad H_{atm} = 10.20$ m

$D_s = 40$ mm $\qquad l_s = 5.0$ m $\qquad H_s = 3.5$ m

$D_d = 30$ mm $\qquad l_d = 25$ m $\qquad H_d = 7.0$ m

Safe minimum pressure head = 2.40 m

$$H_{as} = \frac{l_s}{g}\left(\frac{A}{A_s}\right) R\omega^2$$

$$= \frac{5.0}{9.806}\left(\frac{100}{40}\right)^2 \times 0.10 \times \left(\frac{2 \times 3.142\, N}{60}\right)^2$$

$$= 0.0035\ N^2$$

$$H_{ad} = \frac{25}{9.806}\left(\frac{100}{30}\right)^2 \times 0.10\left(\frac{2 \times 3.142 \times N}{60}\right)^2$$

$$= 0.0155\ N^2$$

$\therefore$ Safe speed to avoid separation on the suction stroke is given by

$$2.4 = 10.2 - 3.5 - 0.0035 N_1^2$$

or $\qquad N_1^2 = \dfrac{4.3}{0.0035} \qquad \therefore N_1 = 35\ 05$ rpm

Safe speed to avoid separation on the delivery stroke is given by

$$2.4 = 10.2 + 7.0 - 0.0155\, N_2^2$$

$\therefore \qquad N_2^2 = 14.8/0.0155;\quad$ or $\quad N_2 = 30.9$ rpm

Hence pump should be run at 30.9 rpm speed.

15.11 Consider a single acting reciprocating pump having the following details:

$D = 125$ mm $\qquad L = 225$ mm $\qquad H_{atm} = 10.2$ m

$D_s = 75$ mm $\qquad l_s = 3.5$ mm $\qquad H_s = 2.5$ m; $f = 0.04$

$D_d = 75$ mm $\qquad l_d = 13.5$ m $\qquad H_d = 9.0$ m

Safe minimum pressure head = 2.4 m

If an air vessel is provided on the delivery side very close to the cylinder, determine the power required and the maximum speed at which the pump can be run.

Since the air vessel is provided on the delivery side, separation needs to be avoided on the suction side only.

$$H_{as} = \frac{3.5}{9.806} \times \left(\frac{125}{75}\right)^2 \times 0.1125 \times \left(\frac{2 \times 3.142\, N}{60}\right)^2$$

$$= 0.001\ 223\ N^2$$

$$\therefore \quad 2.4 = 10.2 - 2.5 - 0.001\,223\,N^2$$

$$N^2 = 5.3/0.001\,223 \quad \text{or} \quad N = 65.83 \text{ rpm}$$

$$Q = \frac{ALN}{60} = \frac{0.785 \times 0.125^2 \times 0.225 \times 65.83}{60}$$

$$= 0.003\,03 \text{ m}^3/\text{s}$$

$$\bar{U}_d = \frac{0.003\,03}{0.785 \times 0.075^2} = 0.686 \text{ m/s and it will be constant.}$$

$$\therefore \quad h_{fd} = \frac{0.04 \times 13.5 \times 0.686^2}{2 \times 9.806 \times 0.075} = 0.173 \text{ m}$$

Maximum head loss on suction side $h_{fsm} = \dfrac{fl_s}{2gD_s}\left(\dfrac{AR\omega}{A_s}\right)^2$

$$= \frac{0.04 \times 3.5}{2 \times 9.806 \times 0.075}\left[\left(\frac{125}{75}\right)^2 \times 0.1125 \times \frac{2 \times 3.142 \times 65.83}{60}\right]^2$$

$$= 0.4418 \text{ m}$$

$$P = \frac{Q\gamma}{1000}\,[H_s + H_d + h_{fd} + \tfrac{2}{3}\,h_{fs}]$$

$$= \frac{0.003\,03 \times 9787}{1000}\,(2.5 + 9.0 + 0.173 + 0.295)$$

$$= 0.355 \text{ kW}$$

15.12 A single cylinder double acting reciprocating pump has the following details—

$D = 190$ mm	$L = 380$ mm	$H_{atm} = 10.2$ m	$N = 36$ rpm
$D_s = 100$ mm	$l_s = 9.0$ m	$H_s = 3.6$ m	$f = 0.03$
$D_d = 100$ mm	$l_d = 60$ m	$H_d = 30$ m	

Air vessels are provided 3.0 m away from the pump on the suction side and 6.0 m away from the pump on the delivery side. Estimate, for the beginning of the stroke (i) the heads in the two ends of the cylinder, and (ii) the load on the piston rod neglecting its size.

$$Q = \frac{2ALN}{60} = \frac{2 \times 0.785 \times 0.19^2 \times 0.38 \times 36}{60} = 0.0129 \text{ m}^3/\text{s}$$

$$\bar{U}_s = \bar{U}_d = \frac{0.0129}{0.785 \times 0.10^2} = 1.643 \text{ m/s}$$

Head on piston on suction side $= H_{atm} - H_s - H_{as} - h_{fs} - \dfrac{\bar{U}_s^2}{2g}$

$$H_{as} = \frac{3.0}{9.806} \times \left(\frac{190}{100}\right)^2 \times 0.190 \times \left(\frac{2 \times 3.142 \times 36}{60}\right)^2$$

$$= 2.983 \text{ m}$$

$$h_{fs} = (0.03 \times 6.0 \times 1.643^2)/2 \times 9.806 \times 0.100$$
$$= 0.248 \text{ m}$$

$\therefore$ Head on the piston on suction side

$$= 10.20 - 3.60 - 2.983 - 0.248 - \frac{1.643^2}{2 \times 9.806}$$
$$= 3.231 \text{ m}$$

Head on the discharge side $= H_{atm} + H_d + H_{ad} + h_{fd} + \dfrac{U_d^2}{2g}$

$\therefore$ Head on the discharge side

$$= 10.2 + 30.0 + 5.966 + \frac{1.643^2}{2 \times 9.806} + \frac{0.03 \times 54 \times 1.643^2}{2 \times 9.806 \times 0.10}$$
$$= 10.2 + 30.6 + 5.966 + 0.138 + 2.230$$
$$= 48.534 \text{ m abs}$$

Difference in head on two sides of piston $= 48.534 - 3.231$

$= 45.303$ m

or $= 45.303 \times 9.787$ kN/m²

$= 443.38$ kN/m²

Net head on piston $= 0.785 \times 0.19^2 \times 443.38$

$= 12.565$ kN

15.13 A double acting reciprocating pump has the following details:

$D = 200$ mm $\qquad L = 350$ mm $\qquad N = 30$ rpm

If an air vessel is fitted on the suction side, determine the crank angle at which discharge to or from the air vessel is zero. Also determine the discharge to or from the air vessel on a crank angle of 45°.

$$\omega = \frac{2\pi N}{60} = \frac{2 \times 3.142 \times 30}{60} = 3.142$$

$$Q_a = AR\omega\left(\sin\theta - \frac{2}{\pi}\right)$$

$\therefore$ At 45°,

$$Q_a = 0.785 \times 0.20^2 \times 0.175 \times 3.142\left(0.7071 - \frac{2}{3.142}\right)$$
$$= 0.001\,22 \text{ m}^3/\text{s}$$

Since this is +ve, Q_a is from the air vessel.

Also Q_a will be zero when $\sin\theta = \dfrac{2}{\pi} = 0.6356$

$\therefore \quad \theta = 39.53° \quad$ or $\quad 140.47°$

15.14 Consider a single acting reciprocating pump with following data:

$D = 150$ mm $\quad L = 200$ mm $\quad N = 60$ rpm

$D_s = 75$ mm $\quad l_s = 10$ m $\quad H_s = 3.0$ m

$D_d = 50$ mm $\quad l_d = 40$ m $\quad H_d = 30$ m

$f = 0.03$ for both pipes

Air vessels are provided 3.0 m and 5.0 m from the suction and delivery side. Determine the power required if pump efficiency is 85 percent.

Here $l_{sa} = 3.0$ m and $l_{da} = 5.0$ m, $R = 100$ mm

$$P = \frac{Q\gamma}{1000\eta}\left[H_s + \frac{f(l_s - l_{sa})}{2gD_s}\left(\frac{Q}{A_s}\right)^2 + \frac{1}{3}\frac{fl_{sa}}{gD_s}\left(\frac{A}{A_s}R\omega\right)^2 + H_d + \frac{f(l_d - l_{da})}{2gD_a}\left(\frac{Q}{A_d}\right)^2 + \frac{1}{3}\frac{fl_{da}}{gD_a}\left(\frac{A}{A_d}R\omega\right)^2\right]$$

$$Q = \frac{ALN}{60} = \frac{0.785 \times 0.15^2 \times 0.2 \times 60}{60} = \mathbf{0.003\ 53\ m^3/s}$$

$$\omega = \frac{2\pi N}{60} = 6.284 \text{ rad/s}$$

$$\therefore \quad \bar{U}_s = \frac{Q}{A_s} = \frac{0.003\ 53}{0.785 \times 0.075^2} = 0.799 \text{ m/s}$$

$$\bar{U}_d = \frac{Q}{A_d} = \frac{0.003\ 53}{0.785 \times 0.05^2} = 1.799 \text{ m/s}$$

$$\frac{A}{A_s} R\omega = \left(\frac{150}{75}\right)^2 \times 0.10 \times 6.284 = 2.5136 \text{ m/s}$$

$$\frac{A}{A_d} R\omega = \left(\frac{150}{50}\right)^2 \times 0.1 \times 6.284 = 5.6556 \text{ m/s}$$

$$\therefore \quad P = \frac{0\ 003\ 53 \times 9787}{1000 \times 0.85}\left[3.000 + \frac{0.03 \times 7 \times (0.799)^2}{2 \times 9.806 \times 0.075} + \frac{0.03 \times 3 \times (2.5136)^2}{3 \times 9.806 \times 0.075} + 30.000 + \frac{0.03 \times 35 \times 1.799^2}{2 \times 9.806 \times 0.050} + \frac{0.03 \times 5 \times 5.6556^2}{3 \times 9.806 \times 0.05}\right]$$

$$= 0.040\ 64(3.000 + 0.0911 + 0.0258 + 30.000 + 3.4654 + 3.2619)$$

$$= 1.619 \text{ kW}$$

15.15 A centrifugal pump with outer impeller diameter of 400 mm and inner impeller diameter of 200 mm runs at 1000 rpm. Radial velocity of water through the pump is constant at 2.5 m/s. If the vanes are set back at

an angle of 30° to the outer rim, determine the angle of vane at inlet and velocity and direction of water at outlet, (Fig. 15.8).

$$\text{Tangential velocity at inlet } v_1 = \frac{\pi D_1 N}{60}$$

$$= \frac{3.142 \times 0.20 \times 1000}{60} = 10.473 \text{ m/s}$$

$$\tan\theta = V_{f1}/v_1 = 2.50/10.473 = 0.2387$$

$$\therefore \quad \theta = \tan^{-1} 0.2387 = 13.426°$$

$$v_2 = \frac{\pi D_2 N}{60} = \frac{3.142 \times 0.40 \times 1000}{60} = 20.946 \text{ m/s}$$

$$\text{Velocity of swirl } V_{w2} = v_2 - \frac{V_{f2}}{\tan\phi}$$

$$\therefore \quad V_{w2} = 20.946 - \frac{2.50}{\tan 30} = 20.946 - 4.330$$

$$= 16.616 \text{ m/s}$$

$$V_2 = \sqrt{V_{w2}^2 + V_{f2}^2} = \sqrt{16.616^2 + 2.50^2} = 16.800 \text{ m/s}$$

$$\tan\beta = \frac{V_{f2}}{V_{w2}} = \frac{2.50}{16.616} = 0.1505. \text{ Hence } \beta = 8.556°.$$

15.16 A centrifugal pump runs at 1000 rpm against a head of 16 m and carries 145 l/s of water discharge. The impeller diameter at the outlet is 300 mm and the width there is 60 mm. If the vane angle ϕ at the outlet is 40°, determine the manometric efficiency.

$$v_2 = \frac{\pi D_2 N}{60} = \frac{3.142 \times 0.30 \times 1000}{60} = 15.71 \text{ m/s}$$

Since $\quad Q = \pi D_2 b_2 V_{f2}$

$$V_{f2} = \frac{0.145}{3.142 \times 0.30 \times 0.06} = 2.564 \text{ m/s}$$

$$\tan\phi = \frac{V_{f2}}{(v_2 - V_{w2})} \quad \therefore \quad v_2 - V_{w2} = \frac{2.564}{\tan 40°}$$

$$\therefore \quad v_2 - V_{w2} = \frac{2.564}{0.839} = 3.056 \text{ m/s}$$

$$\therefore \quad V_{w2} = 15.710 - 3.056 = 12.674 \text{ m/s}$$

$$\text{Manometric efficiency } \eta_{mo} = \frac{gH_m}{V_{w2} v_2} = \frac{9.806 \times 16}{15.71 \times 12.674}$$

$$= 0.7892 \quad \text{or} \quad 78.92 \text{ percent}$$

15.17 Show that the pressure rise in the impeller of a centrifugal pump is given by

$$\frac{(p_2 - p_1)}{\gamma} = \frac{1}{2g}[V_{f1}^2 + v_2^2 - V_{f2}^2 \operatorname{cosec}^2 \phi]$$

Apply Bernoulli's equation between the entrance and the exit of the impeller assuming that (i) their elevations are same, and (ii) there is no energy loss between the two sections.

$$\therefore \quad \frac{p_1}{\gamma} + \frac{V_1^2}{2g} = \frac{p_2}{\gamma} + \frac{V_2^2}{2g} - \text{(work done by impeller per unit weight of fluid)}$$

$$\therefore \quad \frac{p_2 - p_1}{\gamma} = \frac{V_1^2}{2g} - \frac{V_2^2}{2g} + \frac{V_{w2} v_2}{g}$$

Since the entry of the fluid is radial

$$V_1 = V_{f1} \text{ and } V_{w2} = v_2 - V_{f2} \cot \phi$$

Further

$$V_2^2 = V_{f2}^2 + V_{w2}^2$$

$$= V_{f2}^2 + (v_2 - V_{f2} \cot \phi)^2$$

$$= V_{f2}^2 (1 + \cot^2 \phi) + v_2^2 - 2v_2 V_{f2} \cot \phi$$

$$= V_{f2}^2 \operatorname{cosec}^2 \phi + v_2^2 - 2v_2 V_{f2} \cot \phi$$

Substituting the values of V_1^2, V_2^2 and $V_{\omega 2}$ in the expression for $(p_2 - p_1)/\gamma$, one gets

$$\frac{(p_2 - p_1)}{\gamma} = \frac{V_{f1}^2}{2g} - \frac{V_{f2}^2 \operatorname{cosec}^2 \phi + v_2^2 - 2v_2 V_{f2} \cot \phi)}{2g} + \frac{2v_2(v_2 - V_{f2} \cot \phi)}{2g}$$

$$= \frac{1}{2g}[V_{f1}^2 + v_2^2 - V_{f2}^2 \operatorname{cosec}^2 \phi]$$

15.18 A centrifugal pump discharges 2000 l/s of water per second developing a head of 20 m when running at 300 rpm. The impeller diameter at the outlet and outflow velocity there, are 1.5 m and 2.5 m/s respectively. If vanes are set back at angle of 30° at the outlet, determine

(i) manometric efficiency, and
(ii) power required by the pump.

If inner diameter is 750 mm, find the minimum speed to start the pump.

Here $H_m = 20$ m $\qquad N = 300$ rpm

$D_2 = 1.5$ m $\qquad V_{f2} = 3.0$ m/s $\qquad \phi = 30°$

$$v_2 = \frac{\pi D_2 N}{60} = \frac{3.142 \times 1.50 \times 300}{60} = 23.565 \text{ m/s}$$

Considering the outlet velocity triangle,

$$V_{w2} = v_2 - \frac{V_{f2}}{\tan \phi}$$

$$= 23.565 - \frac{3.0}{\tan 30} = 18.369 \text{ m/s}$$

$$\eta_{mo} = \frac{gH_m}{V_{w2}v_2} = \frac{9.806 \times 20}{18.369 \times 23.565} = 0.453 \quad \text{or} \quad 45.3 \text{ percent}$$

$$\text{Power required to drive the pump} = \frac{Q\gamma V_{w2}v_2}{1000g}$$

$$= \frac{2.0 \times 9787 \times 18.369 \times 23.565}{9.806 \times 1000}$$

$$= 864.053 \text{ kW}$$

Minimum starting speed:

$$\frac{v_1}{D_1} = \frac{v_2}{D_2}$$

$$\therefore \quad v_1 = v_2 \frac{D_1}{D_2}$$

or

$$v_1 = 0.5\, v_2$$

Also

$$\frac{v_2^2}{2g} - \frac{v_1^2}{2g} = H_m$$

$$\frac{v_2^2}{2g} - \frac{0.25 v_2^2}{2g} = 20$$

or

$$0.75\, v_2^2 = 20 \times 2 \times 9.806$$

$$v_2 = \sqrt{\frac{20 \times 2 \times 9.806}{0.75}} = 22.86 \text{ m/s}$$

But

$$\frac{\pi D_2 N}{60} = v_2 = 22.86$$

$$\therefore \quad N = \frac{60 \times 22.86}{3.142 \times 1.5} = 291.3 \text{ rpm}$$

15.19 A centrifugal pump has the following characteristics

$$Q = 0.005 \text{ m}^3\text{/s} \qquad D_1 = 0.10 \text{ m}$$

$$N = 1500 \text{ rpm} \qquad D_2 = 0.20 \text{ m}$$

Total head = 22.0 m. Width of impeller at inlet and outlet: 10 mm and 5 mm. Vanes are curved back at an angle of 30° to the tangent at the outlet. Neglect the losses and determine the increase in pressure head.

$$Q = \pi D_2 V_{f2} b_2$$

$$\therefore \quad V_{f2} = \frac{0.005}{3.142 \times 0.20 \times 0.005} = 1.591 \text{ m/s}$$

$$V_{f1} = \frac{0.005}{3.142 \times 0.10 \times 0.01} = 1.591 \text{ m/s}$$

$$v_2 = \frac{\pi D_2 N}{60} = \frac{3.142 \times 0.2 \times 1500}{60} = 15.71 \text{ m/s}$$

$$\left(\frac{p_2 - p_1}{\gamma}\right) = \frac{1}{2g}\left[V_{f1}^2 + v_2^2 - V_{f2}^2 \operatorname{cosec}^2 \phi\right]$$

$$= \frac{1}{2 \times 9.806}(1.591^2 + 15.71^2 - 1.591^2 \operatorname{cosec}^2 30)$$

$$= 12.23 \text{ m}$$

15.20 A centrifugal pump lifts water under a static head of 36 m of which 4.0 m is the suction head. Suction and delivery pipes are both of 150 mm diameter having their lengths such that they cause headloss of 1.5 m and 7.0 m in suction and delivery pipes respectively. Other data are: impeller diameter $D_2 = 0.380$ m, Width at outlet $b_2 = 25$ mm, $N = 1200$ rpm and exit blade angle $\beta = 38°$. Manometric efficiency of pump $\eta_{mo} = 0.80$.

Determine the discharge and pressure at suction and delivery sides of pump.

$$v_2 = \frac{\pi D_2 N}{60} = \frac{3.142 \times 0.38 \times 1200}{60} = 23.879 \text{ m/s}$$

If flow is assumed radial, $\alpha = 90°$

Work done per unit weight of water $= \dfrac{v_2 V_{\omega 2}}{g}$

Total head to be supplied by pump $= 36.0 + 7.0 + 1.5 = 44.5$ m and manometric efficiency $\eta_{mo} = \dfrac{H_m}{v_2 V_{\omega 2}/g}$.

$$\therefore \quad 0.80 = \frac{44.5 \times 9.806}{V_{\omega 2} \times 23.879} \qquad \therefore \quad V_{\omega 2} = 22.843 \text{ m/s}$$

$$\tan\beta = \frac{V_{f2}}{v_2 - V_{\omega 2}} \quad \text{or} \quad 0.7813 = \frac{V_{f2}}{23.879 - 22.843}$$

$$\therefore \quad V_{f2} = 0\ 809 \text{ m/s}$$

$$Q = \pi D_2 b_2 V_2 = 3.142 \times 0.380 \times 0.025 \times 0.809$$

$$= 0.0241 \text{ m}^3\text{/s} \quad \text{or} \quad 24.1 \ l\text{/s}$$

Velocity in suction or delivery pipe $= \dfrac{0.0241}{0.785 \times 0\ 130^2} = 1.817$ m/s and

velocity head $= \dfrac{1.817^2}{2 \times 9.806} = 0.168$ m.

$\therefore$ Pressure on suction side $= 4.0 + 1.50 + 0.168$

$= 5.668$ m

$= 5.668 \times 9.787$ i.e. 55.473 kN/m^2 (vacuum)

Pressure on delivery side of pump $= 32.0 + 7.0 + 0.168$

$= 39.168$ m or 383.337 kN/m^2

15.21 The diameter of an impeller of a centrifugal pump is 350 mm with outer width of 50 mm. Pump runs at 1000 rpm against a total head of 15 m. If vane angle at the outlet is 35° and manometric efficiency is 80 percent, determine (i) velocity of flow at the outlet, (ii) velocity of water leaving the vane, (iii) angle made by the absolute velocity at outlet with direction of motion there and (iv) discharge.

$$v_2 = \frac{\pi D_2 N}{60} = \frac{3.142 \times 0.350 \times 1000}{60} = 18.328 \text{ m/s}$$

$$\eta_{mo} = \frac{gH_m}{v_2 V_{\omega 2}} \quad \therefore \quad 0.80 = \frac{9.806 \times 15}{18.328 \times V_{\omega 2}}$$

$$\therefore \quad V_{\omega 2} = 10.032 \text{ m/s}$$

$$\tan \phi = \frac{V_{f2}}{v_2 - V_{\omega 2}} \quad \therefore \quad \tan 35 = \frac{V_{f2}}{18.328 - 10.032}$$

$$\therefore \quad V_{f2} = 0.7002 \times 8.296 = 5.809 \text{ m/s}$$

Velocity of water leaving the vane, V_2

$$= \sqrt{V_{f2}^2 + V_{\omega 2}^2} = \sqrt{5.809^2 + 10.032^2} = 11.592 \text{ m/s}$$

Angle made by the absolute velocity at outlet, β is given by

$$\tan \beta = \frac{V_{f2}}{V_{\omega 2}}$$

$$\therefore \quad \tan \beta = \frac{5.809}{10.032} = 0.5790$$

$$\therefore \quad \beta = 30.073°$$

$$Q = \pi D_2 b_2 V_{f_2} = 3.142 \times 0.35 \times 0.05 \times 5.809$$

$$= 0.3194 \text{ m}^3\text{/s} \quad \text{or} \quad 319.4 \ l\text{/s}$$

15.22 Obtain the similarity conditions for geometrically similar pumps.

One can assume that for geometrically similar pumps, the independent variables are discharge (Q), head (gH), characteristic or impeller diameter (D), angular velocity (ω) and mass density of fluid (ρ).

Hence $\quad \eta$ or $P = \phi(Q, gH, D, \omega, \rho)$

Choose ρ, ω and D as the repeating variables; dimensional analysis, then gives,

$$\eta \text{ or } \frac{P}{\rho \omega^3 D^5} = \phi\left(\frac{Q}{\omega D^3}, \frac{gH}{\omega^2 D^2}\right)$$

Thus if $\frac{Q}{\omega D^3}$ and $\frac{gH}{\omega^2 D^2}$ are maintained the same in the two pumps, their

efficiency η or parameter $\frac{P}{\rho\omega^3 D^5}$ will be the same. Replacing N by ω, one can then write

$$\frac{ND}{H^{1/2}} = C_1$$

$$\frac{Q^2}{N^2D^6}\,\frac{N^2D^2}{gH} = \frac{Q^2}{D^4Hg} = \text{const. or } \frac{Q}{H^{1/2}D^2} = C_2$$

Similarly one can show $\frac{P}{D^2H^{3/2}} = C_3$ and $\frac{NQ^{1/2}}{H^{3/4}} = C_4$

where C_1, C_2, C_3, C_4 are constants. When N is in rpm. Q in m³/s and H in m, $NQ^{1/2}/H^{3/4}$ is known as the specific speed N_s.

15.23 A combination of centrifugal pumps of specific speed 20 and overall efficiency 75 percent, running at 1500 rpm is to be used to pump 0.03 m³/s water to a height of 65 m. Determine their number and arrangement.

$$20 = \frac{NQ^{1/2}}{(H/\eta)^{3/4}} \quad \therefore \quad H^{3/4} = \frac{0.03^{1/2} \times 0.75^{3/4} \times 1500}{20}$$

or $H = 22.88$ m. Hence $\frac{65}{22.88} = 3.0$. Three pumps will be required and they must be connected in series to develop a head of 65 m.

15.24 A discharge of 1.0 m³/s is to be pumped to 40 m height using centri fugal pumps of specific speed 50, running at 1500 rpm at 75 percent overall efficiency. Determine the number and their arrangement.

$$20 = \frac{NQ^{1/2}}{(H/\eta)^{3/4}} \quad \therefore \quad Q^{1/2} = \frac{50 \times \left(\frac{40}{0.75}\right)^{3/4}}{1500} = 0.6579$$

$$\therefore \qquad Q = (0.6579)^2 = 0.433 \text{ m}^3/\text{s}.$$

Hence, to deliver 1.0 m³/s, 3 pumps will be required and these will be connected in parallel.

15.25 A centrifugal pump discharges water at the rate of 0.167 m³/s at 2000 rpm under a head of 100 m, and consumes 300 kW power. A 1 : 4 scale model is to run at 1500 rpm. Determine its power, discharge and head.

$D_m/D_p = \frac{1}{4}$. According to the similarity criteria established in Ex. 15.22,

$$\frac{Q_m}{N_mD_m^3} = \frac{Q_p}{N_pD_p^3} \quad \therefore \quad \frac{Q_m}{Q_p} = \left(\frac{N_m}{N_p}\right)\left(\frac{D_m}{D_p}\right)^3 = \left(\frac{1500}{2000}\right)\left(\tfrac{1}{4}\right)^3$$

$$= 0.011\,72$$

$$\therefore \qquad Q_m = 0.011\,72 \times 0.167 = 1.957 \times 10^{-3} \text{ m}^3/\text{s}$$

Also $\dfrac{H_m}{N_m^2 D_m^2} = \dfrac{H_p}{N_p^2 D_p^2}$ or $H_m = H_p \left(\dfrac{N_m}{N_p}\right)^2 \left(\dfrac{D_m}{D_p}\right)^2$

$$= 100 \times \left(\frac{1500}{2000}\right)^2 \times (\tfrac{1}{4})^2$$

$$= 3.516 \text{ m}$$

In the same manner $\dfrac{P_m}{N_m^3 D_m^5} = \dfrac{P_p}{N_p^3 D_p^5}$ or $P_m = P_p \left(\dfrac{N_m}{N_p}\right)^3 \left(\dfrac{D_m}{D_p}\right)^5$

$$\therefore \quad P_m = 300 \left(\frac{1500}{2000}\right)^3 (\tfrac{1}{4})^5 = 0.1236 \text{ kW}$$

15.26 A geometrically similar model to scale 1 : 6 of a large centrifugal pump is tested. The prototype parameters are:

$$N_p = 400 \text{ rpm}, \; Q_p = 1.7 \text{ m}^3\text{/s}, \; H_p = 36.5 \text{ m and } p_p = 720 \text{ kW}.$$

If the model is tested under a head of 9 m, determine the speed and discharge at which it should be run and the power required to drive it.

$$\frac{H_p}{N_p^2 D_p^2} = \frac{H_m}{N_m^2 D_m^2};$$

or $$N_m^2 = N_p^2 \left(\frac{H_m}{H_p}\right) \left(\frac{D_p}{D_m}\right)^2 = 400^2 \left(\frac{9}{36{\cdot}5}\right) (6)^2$$

$$\therefore \quad N_m = 1191.75 \text{ rpm}$$

$$\frac{Q_p}{N_p D_p^3} = \frac{Q_m}{N_m D_m^3};$$

or $$Q_m = Q_p \left(\frac{N_m}{N_p}\right) \left(\frac{D_m}{D_p}\right)^3 = 1.70 \times \left(\frac{1191.75}{400}\right) \times (\tfrac{1}{6})^3$$

$$= 0.0235 \text{ m}^3\text{/s}$$

$$\frac{P_m}{N_m^3 D_m^5} = \frac{P_p}{N_p^3 D_p^5};$$

or $$P_m = P_p \left(\frac{N_m}{N_p}\right)^3 \left(\frac{D_m}{D_p}\right)^5 = 720 \times \left(\frac{1191.75}{400}\right)^3 \times (\tfrac{1}{6})^5$$

$$= 2.449 \text{ kW}$$

15.27 Impellers of a three stage centrifugal pump have their external diameter and width as 380 mm and 20 mm respectively. The pump delivers 0.06 m³/s at 900 rpm. If the vanes are curved back at 45° to the tangent at the outlet and 8 percent of area is blocked by vanes, determine the manometric head developed when $\eta_{mo} = 84$ percent.

$$v_2 = \frac{\pi D_2 N}{60} = \frac{3.142 \times 0.380 \times 900}{60} = 17.909 \text{ m/s}$$

$$V_{f2} = \frac{0.06}{3.142 \times 0.380 \times 0.02} = 2.513 \text{ m/s}$$

$$V_{w2} = v_2 - \frac{V_{f2}}{\tan \phi} = 17.909 - \frac{2.513}{1.0} = 15.396 \text{ m/s}$$

$$\therefore \quad \eta_0 = \frac{gH_m}{V_{w2}v_2} \quad \text{or} \quad H_m = \frac{0.84 \times 15.396 \times 17.909}{9.806}$$

$$= 23.619 \text{ m}$$

Total manometric head developed $= 3 \times 23.619$

$= 70.857$ m

PROBLEMS

15.1 A single acting reciprocating pump has plunger diameter of 500 mm, stroke of 300 mm and it runs at 40 rpm. If the total lift is 20.0 m, actual discharge 0.036 m³/s, determine theoretical discharge, theoretical power required, volumetric efficiency and the slip.

(0.03925 m³/s, 7.683 kW, 91.72 percent and 8.28 percent)

15.2 A single acting reciprocating pump has the following data: $L = 300$ mm, Plunger diameter $D = 125$ mm.

Suction pipe diameter $D_s = 75$ mm, Suction length $l_s = 5.0$ m
$N = 40$ rpm, Minimum permissible pressure head $= 2.4$ m abs.
Determine the maximum suction head which will not cause separation at the beginning of suction stroke. Take $H_{atm} = 10.23$ m abs.

(4.101 m)

15.3 A single cylinder single acting reciprocating pump has the following details:

$N = 30$ rpm $\quad L = 400$ mm $\quad D = 200$ mm

$H_s = 3.0$ m $\quad l_s = 5.0$ m $\quad D_s = 100$ mm

Friction factor for suction pipe $f = 0.015$

Determine absolute pressure on the piston at the beginning, middle and end of suction stroke. Take $H_{atm} = 10.20$ m absolute.

(3.173 m, 6.958 m, 11.227 m abs.)

15.4 A single cylinder single acting reciprocating pump without air vessels has the following details:

$L = 600$ mm $\quad D = 150$ mm $\quad H_s = 3.50$ m $\quad l_s = 6.0$ m

$D_s = 100$ mm $\quad H_{atm} = 10.2$ m $\quad$ Safe minimum pressure head in cylinder $= 2.4$ m. Determine the safe maximum speed.

(30.81 rpm)

15.5 Determine the water delivered and power required to drive a double acting single cylinder reciprocating pump if following data are given:

Total head ($H = H_s + H_d$) $= 15$ m
Frictional loss in pipes $= 1.5$ m
Stroke length L $= 0.50$ m
Piston diameter D $= 0.40$ m
Speed N $= 30$ rpm
Positive slip $= 5.0$ percent
Mechanical efficiency $= 80$ percent

(0.059 66 m³/s, 12.043 kW)

15.6 Consider a single cylinder single acting reciprocating pump with the following details:

$L = 500$ mm $D = 300$ mm $D_s = 200$ mm $l_s = 10$ m

$H_s = 3.5$ m $f = 0.04$ $H_{atm} = 10.3$ m

Find the speed at which separation of flow will take place on the suction stroke. Also determine the speed if an air vessel is provided on the suction side at 3.0 m before the cylinder. Take minimum safe pressure as 2.5 m absolute. (26.14 rpm, 49.4 rpm)

15.7 A single cylinder single acting reciprocating pump has the following details:

$D = 250$ mm $L = 45$ mm $D_s = 125$ mm

$l_s = 12$ m $H_s = 3.0$ m $H_{atm} = 10.0$ m

Safe minimum presssure head $= 2.4$ m, $f = 0.03$

Determine the maximum permissible speed which will avoid separation, when an air vessel is fitted on the suction side at a distance of 1.5 m from the cylinder. (54.40 rpm)

15.8 Prove that the saving in the work done by fitting an air vessel on the suction side of a single cylinder double acting reciprocating pump is 39.20 percent.

15.9 Determine the maximum permissible speed for a single cylinder single acting reciprocating pump which will ensure no separation of flow either during the suction or during the delivery stroke. The data for the pump are:

$D = 100$ mm $L = 200$ mm $H_{atm} = 10.3$ m

$D_s = 40$ mm $l_s = 6.0$ m $H_s = 4.0$ m

$D_d = 30$ mm $\quad l_d = 18.0$ m $\quad H_d = 14.0$ m

Minimum permissible pressure head = 2.3 m. (30.89 rpm)

15.10 For a single acting reciprocating pump fitted with an air vessel on the suction side, the following data are given:

$D = 0.250$ m $\quad L = 450$ mm $\quad N = 60.0$ rpm

Determine the discharge into or out of air vessel when the crank angle is 45°. Also determine the crank angle when flow into or out of air vessel is zero. (0.0270 m³/s from air vessel, 18.56°)

15.11 For a double acting reciprocating pump the following data are given: $D = 200$ mm, $L = 400$ mm, $D_s = 100$ mm, $N = 120$ rpm. If an air vessel is fitted on the suction side, determine the flow in or out of the air vessel when the crank angle is 30° and 120°. (0.010 78 m³/s into air vessel, and 0.0181 m³/s from the air vessel)

15.12 For a single acting reciprocating pump fitted with air vessels on both sides, determine the power required, if the following data are given:

$D = 300$ mm $\quad L = 500$ mm $\quad N = 45$ rpm $\quad D_s = 200$ mm

$l_s = 10.0$ m $\quad H_s = 3.5$ m $\quad D_d = 150$ mm $\quad l_d = 25$ m

$H_d = 15$ m

$f = 0.03$ for both pipes, $l_{sa} = 2.0$ m, $l_{da} = 4.0$ m

Mechanical efficiency of pump = 80 percent. (6.229 kW)

15.13 A centrifugal pump having an overall efficiency of 75 percent discharges 30 l/s of water 20 m above the sump level through 10 m long suction pipe and 80 m long delivery pipe both of 100 mm diameter having $f = 0.035$. Calculate the power required to run it. (17.015 kW)

15.14 Determine the head imparted to the fluid as it passes through the impeller of inlet and outlet diameters 100 mm and 250 mm respectively rotating at 1440 rpm. Outlet vane angle is set back at 20° to the tangent. Assume velocity of flow as 4.0 m/s and radial entrance. (15.115 m)

15.15 A centrifugal pump running at an overall efficiency of 80 percent delivers 25 l/s at a height of 20 m through 100 m lenght of 100 mm diameter pipe having a friction factor of 0.04. Calculate the energy required to drive the pump. (12.603 kW)

15.16 A centrifugal pump will start developing the flow when the liquid head developed is equal to manometric head. A centrifugal pump rotating at 300 rpm and delivering 1800 l/s against a head of 16 m has 25 m/s velocity at outlet. If the inlet and outlet diameters of the impeller are 0.60 m and 1.20 m respectively, determine the manometric efficiency, power required to drive the pump and the minimum starting speed. Vanes are set back at 30° at the outlet. (57.3 percent, 991.83 kW, 325.5 rpm)

15.17 A centrifugal pump discharges water against a head of 14.5 m when running at a design speed of 1000 rpm. The impeller diameter is 300 mm, outlet width 60 mm and the vanes are set back to an angle of 30° with the periphery. If the manometric efficiency is 90 percent, determine the pump discharge. (184.6 l/s)

15.18 A centrifugal pump running at 1400 rpm at a manometric efficiency of 85 percent carries 180 l/s of water discharge. The outlet vane angle of the impeller ϕ is 45° and velocity of flow at the outlet is 3.0 m/s. If the pump works aganist a total head of 20 m, determine the diameter of the impeller and its width at the outlet. (0.23 m, 83.0 mm)

15.19 The following data are given for a centrifugal pump:

$N = 1000$ rpm $\quad D_1 = 0.22$ m $\quad D_2 = 0.40$ m

$Q = 0.045$ m³/s $\quad D_s = 150$ mm $\quad D_d = 100$ mm

Constant velocity of flow $V_{f1} = V_{f2} = 2.20$ m/s

$H_s = 6.0$ m $\quad H_d = 30.0$ m $\quad \beta = 45°$

Power = 18.6 kW

Determine the vane angle θ of the impeller at inlet and overall and manometric efficiencies. (11.86°, 60 percent, 63.3 percent)

15.20 A set of centrifugal pumps with specific speed of 50 and running at 1500 rpm delivers 1.2 m³/s of water at 40 m head, 80 percent overall efficiency. Determine their number and the arrangement. (three pumps in parallel)

15.21 0.40 m³/s of water is to be pumped against a head of 160 m. If pump units of 80 percent overall efficiency and running at 1500 rpm and having a specific speed of 50 are to be used, determine their number and arrangement. (four pumps in series)

15.22 A centrifugal pump with impeller of 150 mm diameter discharges 0.038 m³/s water when running at 1500 rpm aganist a head of 10 m. Determine the corresponding speed and the head of a geometrically similar pump with impeller of 375 mm diameter delivering 0.750 m³/s. (1894.74 rpm, 99.72 m)

15.23 Two geometrically similar pumps are running at the same speed of 1000 rpm. One pump has an impeller diameter of 300 mm and lifts water at the rate of 0.02 m³/s against a head of 15 m. Determine the head and impeller of the other pump to deliver a discharge of 0.01 m³/s. (9.44 m, 238 mm)

15.24 A centrifugal pump running at 750 rpm discharges 0.056 m^3/s water against a head of 20 m, consuming 20 kW power. Find the discharge, the head and power if the same pump is run at 1500 rpm.
(0.112 m^3/s, 80m, 160 kW)

15.25 A centrifugal pump discharges 1560 *l*/s against a lift of 6.0 m when the impeller rotates at 200 rpm. The impeller diameter is 1.20 m and ratio of outer to inner diameter of impeller is 2.0. If the area at the outlet periphery is 0.645 m^2 and if the vanes are set back at an angle of 26° at the outlet, determine (i) the hydraulic efficiency, (ii) power required to drive the pump and (iii) the minimum starting speed. (61.57 percent, 91.606 kW, 199.33 rpm)

15.26 A multistage centrifugal pump discharge 0.75 m^3/s against a manometric head of 60 m. There are four pumps connected to the same shaft running at 350 rpm. If $\phi = 60°$, $V_{f_2} = 0.27v_2$ and hydraulic losses are 0.33 times the velocity head at outlet of impeller, determine impeller diameter and manometric efficiency.
(783 mm, 84.5 percent)

DESCRIPTIVE QUESTIONS

15.1 What is the slip of a reciprocating pump? Why is it sometimes negative?

15.2 What factors govern the speed of reciprocating pumps?

15.3 Why is the suction height of pumps limited?

15.4 What are the functions of air vessels in reciprocating pumps?

15.5 As L increases, should N for reciprocating pumps increase or decrease?

15 6 Draw the indicator diagram for a single cylinder single acting reciprocating pump when air vessel is provided only on the suction side.

15.7 What is the function of nonreturn valve in a reciprocating pump?

15.8 State whether N for reciprocating pump should increase or decrease as (i) fluid viscosity decreases, (ii) area A increases, (iii) mass density of fluid increases.

15.9 Compare the conditions under which reciprocating and centrifugal pumps can be used.

15.10 Draw sketches of a double cylinder and triple cylinder reciprocating pumps and their discharge delivery diagrams.

15.11 What decides the maximum speed of a reciprocating pump?

15.12 How are rotodynamic pumps classified? Draw their sketches.

15.13 What is priming of a pump? How is it done?

15.14 Explain the conditions under which centrifugal pumps are connected in series and in parellel.

15.15 Which part of a centrifugal pump is most susceptible to cavitation? What are the undesirable effects of cavitation ?

15.16 Define the different efficiencies of a centrifugal pump and give their practical ranges.

15.17 What are the characteristic curves of a pump? How are they prepared and what is their utility?

15.18 What hydraulic considerations govern the design of intake and sump for a pump?

CHAPTER XVI

Turbines

16.1 INTRODUCTION

A turbine is a machine which converts the fluid energy into mechanical energy which is then utilised to run the electric generator of a power plant. Fluid used can be water or steam. The first water wheel was probably designed and constructed by Leonardo da Vinci in 15th century, even though they were known earlier. Water wheels are simple, rugged, and have high efficiency; however, they are relatively heavy and hence costly. Different types of water wheels are shown in Fig. 16.1.

Depending on the action of water on their blades, turbines are classified into impulse turbines and reaction turbines. Pelton wheel is an impulse turbine in which a high velocity jet strikes the buckets under atmospheric condition. The buckets change the magnitude and direction of velocity of water, and due to torque developed in this process, the wheel rotates. In a reaction turbine the water enters the turbine under pressure; water therefore has both the kinetic and pressure energy. Reaction turbine always runs full. As in the case of an impulse turbine, in the case of a reaction turbine also, change in momentum produces a force and hence a torque which causes rotation. Francis, Kaplan and propeller turbines fall under the category of reaction turbines. Turbines can also be classified on the basis of direction of flow through the runner into radial flow turbines, axial flow turbine (e.g. Kaplan), tangential flow turbine (e.g. Pelton wheel), and mixed flow turbines (i.e. radial and axial).

16.2 IMPULSE TURBINES: PELTON WHEEL

Pelton wheel setup is shown in Fig. 16.2 in which a series of buckets receive a jet of water having a velocity V_1. The Pelton wheel rotates at a peripheral velocity v at the centre of the buckets. It can be shown that power developed by the Pelton wheel, for a given Q, is maximum when $v = V_1/2$ and

$$P_{max} = Q\rho(1 - \cos\phi)\frac{V_1^2}{4} \tag{16.1}$$

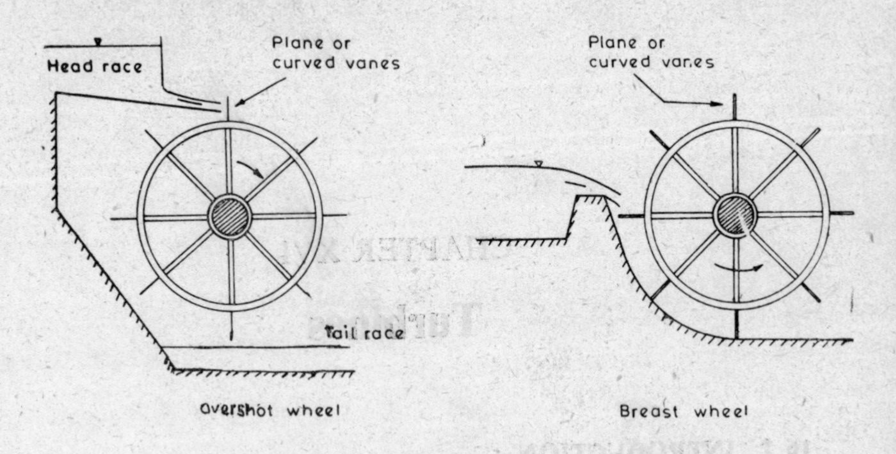

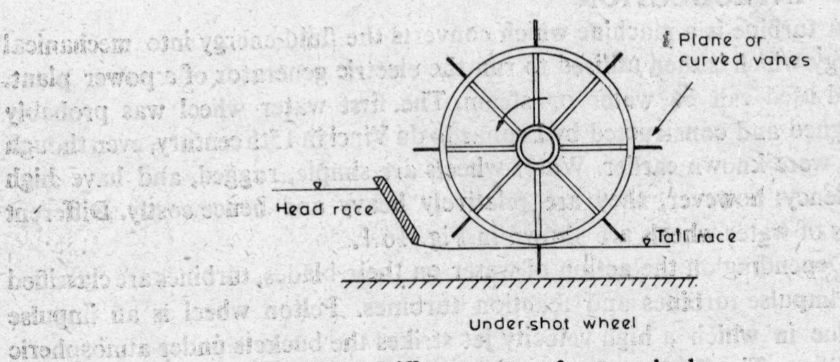

Fig. 16.1. Different types of water wheels.

where ϕ is the angle between direction of impinging jet and the direction of relative velocity at the exit, (Fig. 16.2). Tests have shown that the best operating conditions prevail when $\phi_1 = v/V_1 = 0.43$ to 0.48. Overall efficiency η is related to power developed in kW by the relation

$$P = \frac{QrH}{1000}\eta \tag{16.2}$$

and $$\eta = \eta_h \eta_m \tag{16.3}$$

where η_h is the hydraulic efficiency,

$$\eta_h = 2\phi_1(1-\phi_1)(1-\cos\phi) \tag{16.4}$$

and mechanical efficiency η_m is

$$\eta_m = \frac{\text{Power available at shaft}}{Q\rho(1-\cos\phi)(V_1-v)v} \tag{16.5}$$

For Pelton wheel, η_h is about 84–90 percent while η_m varies from 94–98 percent. The design parameters (Fig. 16.2) of the **Pelton** wheel are listed below.

$\phi = 10° \text{ to } 20°$ so that the jet deflects through $170° - 160°$

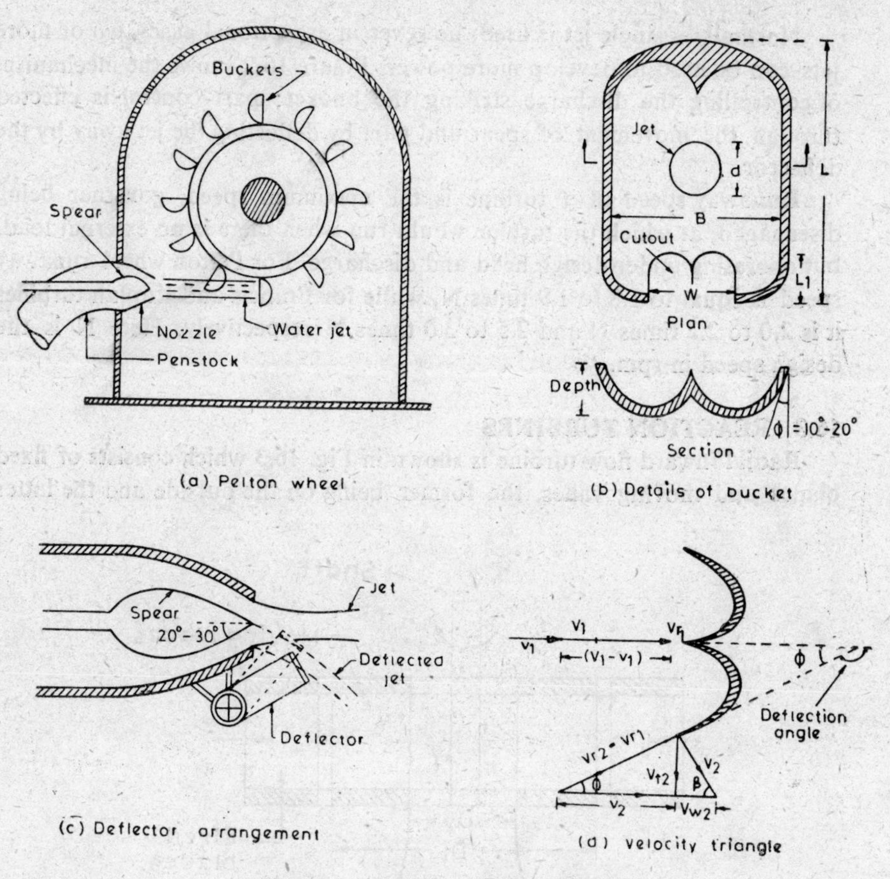

Fig. 16.2. **Details of Pelton wheel.**

$\frac{B}{d} = (3 \text{ to } 4)$ where d is the diameter of jet and B is the width of bucket.

$\frac{L}{d} = (2.5 \text{ to } 3.0)$ where L is radial length.

$Z = \frac{1}{2}\left(\frac{D}{d}\right) + 15$ where Z is the number of buckets and D is the diameter of wheel.

The ratio $\frac{D}{d}$ varies from six to twenty as the specific speed

$$N_s = N\sqrt{P}/H^{5/4}$$

reduces from 35 to 10. Here P is in kW. This dependence is shown in Table 16.1.

Table 16.1. Variation of D/d with N_s for Pelton wheels

N_s	35	32	24	10
D/d	6.5	7.5	10	20

Normally a single jet is used; however in exceptional cases two or more jets can be used to develop more power. Figure 16.2 shows the mechanism of controlling the discharge striking the bucket; part control is effected through the movement of spear and part by deflecting the jet away by the deflector.

Runaway speed of a turbine is the maximum speed, governer being disengaged, at which the turbine would run when there is no external load, but operating under design head and discharge. For Pelton wheel runaway speed is equal to 1.8 to 1.9 times N, while for Francis and Kaplan turbines it is 2.0 to 2.2 times N and 2.5 to 3.0 times **N** respectively. Here N is the design speed in rpm.

16.3 REACTION TURBINES

Radial inward flow turbine is shown in Fig. 16.3 which consists of fixed blades and moving vanes, the former being on the outside and the latter on the inside. Figure 16.4 also shows the velocity triangles for inward flow turbine.

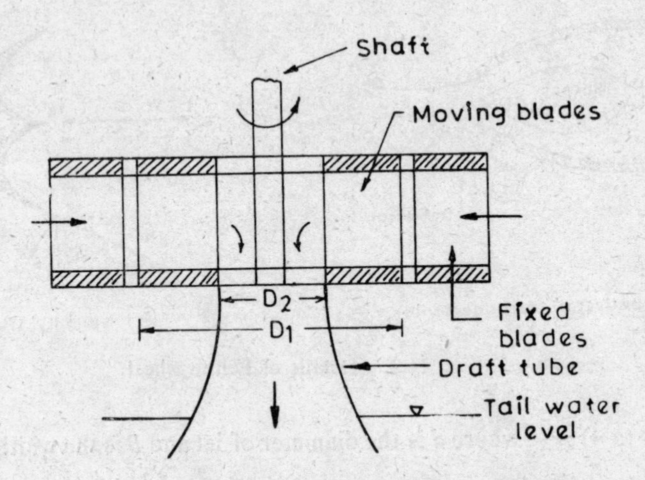

Fig. 16.3. Radially inward flow turbine.

$$Q = \pi D_1 b_1 V_{f1} = \pi D_2 b_2 V_{f2} \tag{16.6}$$

The notation of velocity triangles is as follows:

v_1, v_2 — Tangential velocity (peripheral velocity) of the runner at inlet and outlet respectively $\left(= \frac{\pi D_1 N}{60} \text{ and } \frac{\pi D_2 N}{60}\right)$

V_1, V_2 — Absolute velocity of water entering and leaving the vane

V_{f1}, V_{f2} — Velocity of flow at inlet and outlet

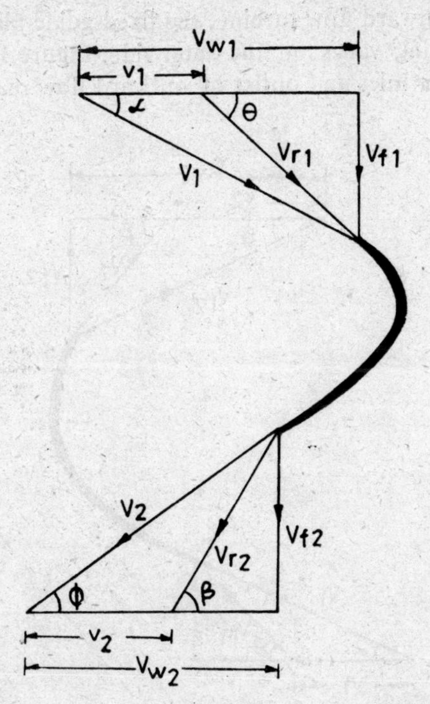

Fig. 16.4. Velocity triangles for inward flow turbine.

$V_{\omega 1}$, $V_{\omega 2}$	Velocity of swirl at inlet and outlet (i.e. velocity component parallel to the direction of vane)
V_{r1}, V_{r2}	Relative velocity of water with respect to vane at inlet and outlet
α	Guide blade angle at inlet i.e. angle at which water enters the vane
θ	Vane angle at inlet
β	Angle at which water leaves the runner
ϕ	Vane angle at outlet
D_1, D_2	Outer and inner diameter of the runner
b_1, b_2	Flow widths at outer and inner end

$$P = Q\rho\,(r_1\omega V_1 \cos\alpha - r_2\omega V_2 \cos\beta)$$

$$= Q\rho\,(v_1 V_{\omega 1} - v_2 V_{\omega 2}) \qquad (16.7)$$

If flow at exit is radial, $\phi = 90°$ and $V_2 = V_{f2}$; also $V_{\omega 2} = 0$. Then $P = Q\rho\, v_1 V_{\omega 1}$

$$\eta_m = \frac{P}{Q\gamma H} = \frac{1}{gH}(v_1 V_{\omega 1} - v_2 V_{\omega 2}) \qquad (16.8)$$

For radially outward flow turbine, the fixed guide blades are on the inner side and rotating vanes on the outer side. Figure 16.5 shows the velocity triangles for the inlet and outlet of outward flow reaction turbine.

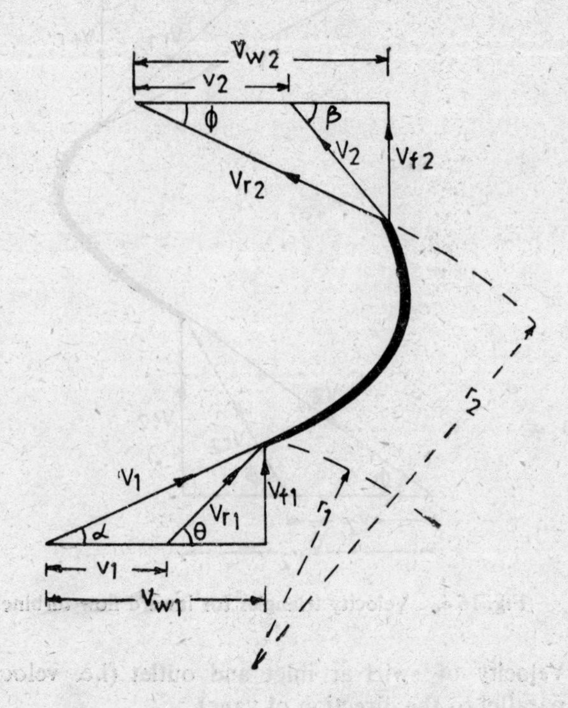

Fig. 16.5. Velocity triangles for outward flow turbine.

16.4 FRANCIS TURBINE

Figure 16.6 shows the sketch of a Francis turbine which is an inward mixed flow turbine where the inflow is radial and outflow is axial. The following steps can be followed in the design of the essential elements of Francis runner. For known P, H and generator speed N

(i) Assume η and η_h (η_h varies from 85 to 95 percent and η from 80 to 90 percent),

(ii) $P = \frac{Q\gamma H}{1000}$; hence find Q.

(iii) $n = \frac{b_1}{D_1}$ where n varies from 0.1 to 0.3; here b_1 is the width of opening at inlet.

(iv) $V_{f1} = \psi_1 \sqrt{2gH}$ where ψ_1 varies from 0.15 to 0.35.

(v) $\frac{Q}{\pi D_1 b_1 k} = V_{f1}$ where $k = 0.95$ and accounts for the thickness of blades.

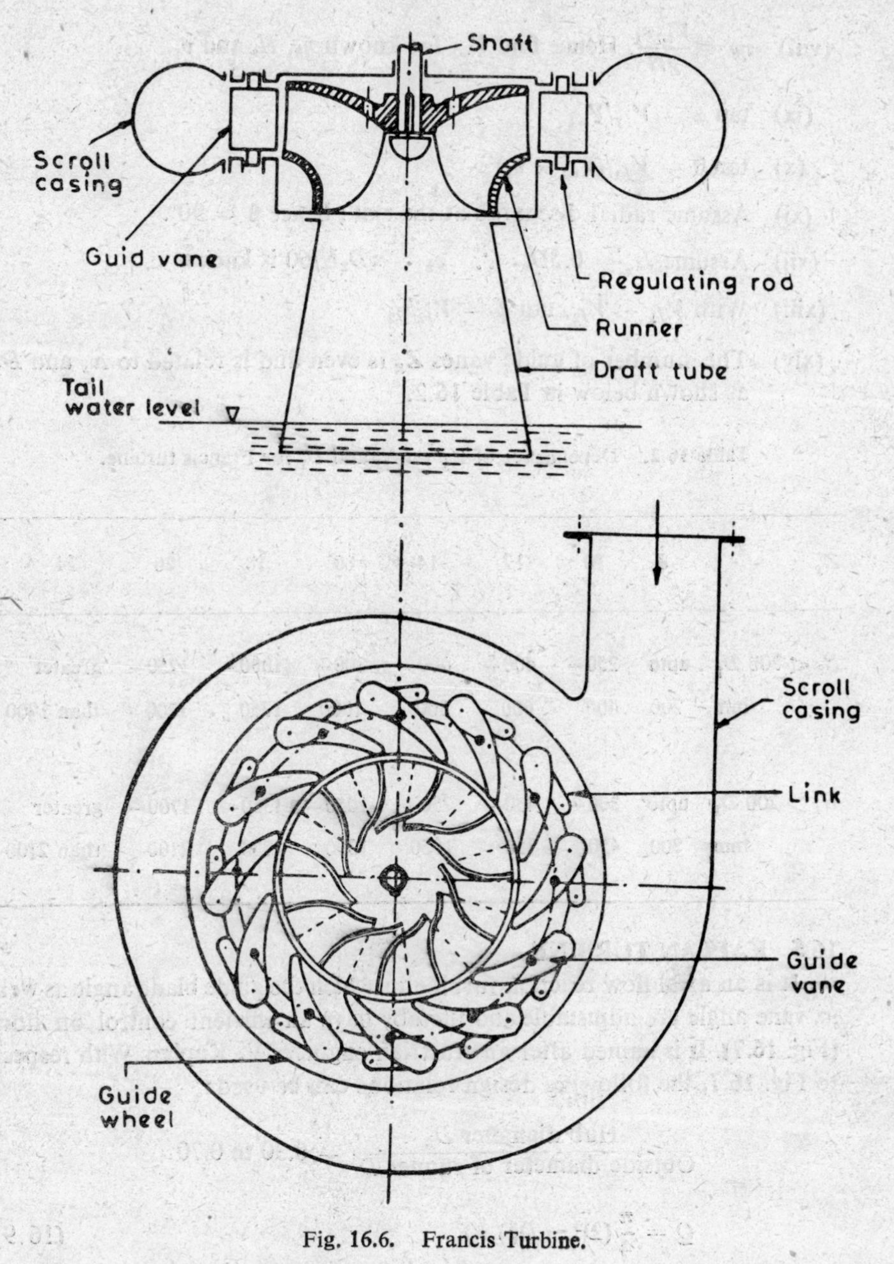

Fig. 16.6. Francis Turbine.

(vi) From the above determine b_1 and D_1.

(vii) For known N determine v_1 from the relation $v_1 = \frac{\pi D_1 N}{60}$

Otherwise $v_1 = \phi_1 \sqrt{2gH}$, where $\phi_1 = 0.6$ to 0.8

(viii) $\eta_h = \dfrac{V_{\omega1}v_1}{gH}$. Hence find $V_{\omega1}$ for known η_h, H, and v_1.

(ix) $\tan\alpha = V_{f1}/V_{\omega1}$

(x) $\tan\theta = V_{f1}/(V_{\omega1} - v_1)$

(xi) Assume radial discharge at the exit; hence $\beta = 90°$.

(xii) Assume $D_2 = 0.5D$, $\therefore \; v_2 = \pi D_2 N/60$ is known.

(xiii) With $V_{f1} = V_{f2}$, $\tan\phi = V_{f2}/v_2$

(xiv) The number of guide vanes Z_g is even and is related to N_s and D_1 as shown below in Table 16.2.

Table 16.2. Dependence of Z_g on N_s and D_1 for Francis turbine.

Z_g	8	10	12	14	16	18	20	24
$N_s < 200$ D_1 mm	upto 200	250–400	400–600	600–800	800–1000	1000–1250	1250–1700	greater than 1700
$N_s > 200$ D_1 mm	upto 300	300–450	450–750	750–1050	1050–1350	1350–1700	1700–2100	greater than 2100

16.5 KAPLAN TURBINE

It is an axial flow reaction turbine in which the guide blade angle as well as vane angle are adjustable and thereby have an efficient control on flow (Fig. 16.7). It is named after an Austrian engineer V. Kaplan. With respect to Fig. 16.7, the following design relations can be used:

$$\frac{\text{Hub diameter } D_0}{\text{Outside diameter of runner } D} = 0.30 \text{ to } 0.70$$

$$Q = \frac{\pi}{4}(D^2 - D_0^2)\, V_{f1} \tag{16.9}$$

$$V_{f1} = \psi_1\sqrt{2gH} \quad \text{where } \psi_1 = 0.70 \text{ to } 0.75$$

$$v_1 = v_2 = \frac{\pi DN}{60}$$

$$V_{f1} = V_{f2}$$

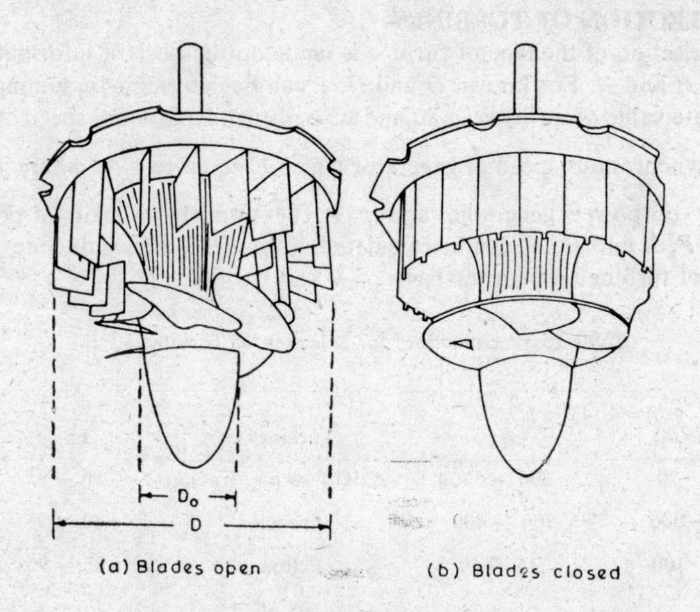

Fig. 16.7. Kaplan turbine runner.

Francis and Kaplan turbines are provided with an expanding closed passage called draft tube which carries water from exit of runner to the tail water. Draft tube serves the following purposes: (i) It allows the setting of turbine runner above the tail race without losing available head. (ii) Because of gradual reduction in kinetic energy of flow as it passes through the draft tube, it reduces the final head loss and increases turbine efficiency. Various shapes of draft tubes are used depending on the site conditions.

Parts of turbine which are susceptible to cavitataion are the guide vanes and vanes of runner. As a result of cavitation, the vanes are pitted, greater energy loss takes place and turbine efficiency is drastically reduced. Cavitation characteristic of a turbine is defined in terms of Thomas number σ

$$\sigma = \frac{\frac{p_a}{\gamma} - \frac{p_v}{\gamma} - h_s}{H} \tag{16.10}$$

where h_s is the vertical distance between centre of turbine runner and tail water, H is the working head on the turbine, p_a is the atmospheric pressure and p_v is the vapour pressure of water. The cavitation occurs when σ reduces to σ_c. This minimum value σ_c is given by the following equation

$$\left.\begin{aligned} \sigma_c &= 3.17 \times 10^{-6}\, N_s^2 \quad \text{for Francis turbine} \\ \sigma_c &= 0.30 + 0.0024 \left(\frac{N_s}{100}\right)^{2.73} \quad \text{for Propeller and Keplan turbines} \end{aligned}\right\} \tag{16.11}$$

16.6 SELECTION OF TURBINES

The selection of the type of turbine is made on the basis of information about Q, H and N. For known Q and H, P can be determined assuming an appropriate value of η (between 80 and 95 percent). Rotational speed N is equal to synchronous speed of generator which is equal to $\frac{60f}{p}$ where f = frequency of power generated and p is the number of pairs of poles. Knowing P, N and H, N_s can be calculated. Table 16.3 gives guidelines for selection of turbine type on the basis of H and N_s.

Table 15.3 Guidelines for Selection of Turbines

H (m)	N_s	Turbine	η
7 — 70	300 — 1100	Axial flow e.g. Kaplan	90 — 95
30 — 300	100 — 400	Francis	80 — 95
Above 100	15 — 60	Pelton	85 — 90

ILLUSTRATIVE EXAMPLES

16.1 An overshot wheel, which is driven by the weight of water, is supplied water from a sluice gate 2.0 m in width, 0.2 m in depth and with water velocity of 2.5 m/s. If the total drop available is 9.0 m and wheel efficiency is 75 percent, estimate the power developed.

An overshot wheel is shown in Fig. 16.1.

$$Q = 2.0 \times 0.20 \times 2.5 = 1.00 \text{ m}^3\text{/s}$$

$$P = \frac{Q\gamma H\eta}{1000}$$

$$= 1.0 \times 9787 \times 9.0 \times 0.75/1000$$

$$= 66.062 \text{ kW}$$

16.2 A Pelton wheel has a mean bucket speed of 10.0 m/s and the jet discharges 0.70 m³/s under a head of 30.0 m. If the deflection angle β is 160° and C_v for the nozzle is 0.98, estimate the power developed and the wheel efficiency.

For Pelton wheel $v_1 = v_2$ and it is calculated at the centre of bucket.

$$v = 10.0 \text{ m/s}$$

$$v_1 = C_v\sqrt{2gH} = 0.98 \times \sqrt{2 \times 9.806 \times 30}$$

$$= 23.771 \text{ m/s}$$

$$\therefore \quad V_r = (V_1 - v) = (23.771 - 10.000) = 13.771 \text{ m/s}$$

$$P = Q\rho (V_1 - v)(1 - \cos\beta)\, v/1000$$
$$= (0.70 \times 998) \times 13.771 \times (1 + 0.9373) \times 10/1000$$
$$= 186.607 \text{ kW}$$

$$\eta_h = 2\frac{v}{V_1}\left(1 - \frac{v}{V_1}\right)(1 - \cos\beta) \times 100 \quad \text{and} \quad \frac{v}{V_1} = \frac{10}{23.771} = 0.4207$$
$$= 2 \times 0.4207 \times 0.5793 \times 1.9373 \times 100$$
$$= 94.541 \text{ percent}$$

16.3 Tests carried out on a Pelton wheel gave the following results:

Head at the base of nozzle	= 27.0 m
Discharge in the nozzle	= 270 l/s
Diameter of jet	= 123 mm
Power developed	= 60.00 kW
Power absorbed in mechanical resistances, etc.	= 3.5 kW

Determine the power lost in the nozzle and due to hydraulic resistance.

$$\text{Power available at nozzle} = \frac{Q\gamma H}{1000} = \frac{0.27 \times 9787 \times 27}{1000} = 71.347 \text{ kW}$$

But power available at nozzle,

P = Power available at shaft + Power lost in nozzle and runner + Power lost in mechanical resistances

$\therefore$ Power lost in nozzle and runner = (71.347) – (60.00) – (3.50)
= 7.847 kW

$$V_1 = Q\bigg/\frac{\pi d^2}{4} = \frac{0.27}{0.785 \times 0.123^2} = 22.734 \text{ m/s}$$

$$\text{Velocity head of jet} = V_1^2/2g = \frac{22.734^2}{2 \times 9.806} = 26.354 \text{ m}$$

$\therefore$ Head lost in nozzle = (27.000 – 26.354) = 0.646 m

$$\therefore \text{Power lost in nozzle} = \frac{0.270 \times 0.646 \times 9787}{1000}$$
$$= 1.707 \text{ kW}$$

$\therefore$ Power lost in runner = 7.847 – 1.707
= 6.140 kW

16.4 A total discharge of 4.0 m³/s under a head upto nozzle of 250 m is available for power generation. A Pelton turbine with two runners, with two jets each and having same diameter, is provided. The penstock length is 3000 m and efficiency of power transmission through the pipe line and through nozzle is 91 percent. The efficiency of each runner is 90 percent. Assume C_v for nozzle to be 0.97 and friction factor f for penstock as 0.013 and determine (i) power developed by the turbine, (ii) diameter of jet, and (iii) diameter of penstock.

Head lost in friction in penstock $= H(1 - 0.91)$

$$h_f = 0.09 \times 250 = 22.5 \text{ m}$$

$\therefore$ Head available for power generation $= 250.0 - 22.5$

$$= 227.5 \text{ m}$$

Velocity of jet $V_1 = C_v\sqrt{2gH}$

$$= 0.97 \times \sqrt{2 \times 9.806 \times 227.5}$$

$$= 64.792 \text{ m/s}$$

Power at inlet to turbine $= \frac{1}{2}MV_1^2$

$$= \frac{0.5 \times 4.0 \times 998 \times 64.792^2}{1000} \text{ kW}$$

$$= 8379.215 \text{ kW}$$

Power developed by turbine $= 0.90 \times 8379.215$

$$= 7541.293 \text{ kW}$$

Discharge per jet $= \frac{4.0}{4.0} = 1.0 \text{ m}^3$

$$\therefore \quad 1.0 = 0.785d^2 \times V_1 \quad \therefore \quad d^2 = \frac{1.0}{0.785 \times 64.792} = 0.019\,66$$

$$\therefore \quad d = 0.140 \text{ m}$$

$$h_f = \frac{8flQ^2}{\pi^2 g D^5}$$

$$\therefore \quad D^5 = \frac{8 \times 0.013 \times 3000 \times 4.0^2}{3.142^2 \times 9.806 \times 22.5} = 2.292$$

$$\therefore \quad D = \sqrt[5]{2.292} = 1.18 \text{ m}$$

16.5 Pelton wheels at Khopoli power house near Bombay work under a head of 525 m and produee 11 170 kW power each while running at 300 rpm. If the overall efficiency of the wheels is 84 percent and C_v for nozzle is 0.98, determine (i) diameter of jet, (ii) diameter of nozzle, (iii) mean diameter of wheel, and (iv) number of buckets.

$$11170 = \frac{Q\gamma H\eta}{1000}$$

or $$Q = \frac{11\,170 \times 1000}{9787 \times 525 \times 0\,84} = 2.588 \text{ m}^3/\text{s}$$

$$Q = \frac{\pi d^2}{4} C_v \sqrt{2gH} \quad \therefore \quad d^2 = \frac{4 \times 2.588}{3.142 \times 0.98 \times \sqrt{2 \times 9.806 \times 525}}$$

$$\therefore \quad d^2 = 0.033\,12 \quad \text{or} \quad d = 0.182 \text{ m}$$

$v_1 = \phi_1 \sqrt{2gH}$. For maximum efficiency assume $\phi_1 = 0.45$

$$\therefore \quad v_1 = 0.45 \times \sqrt{2 \times 9.806 \times 525} = \frac{\pi DN}{60}$$

$$\therefore \quad D = \frac{60 \times 0.45\sqrt{2 \times 9.806 \times 525}}{3.142 \times 300} = 2.907\text{m}$$

$$\frac{\text{Diameter of jet}}{\text{Diameter of nozzle}} = 0.8.$$

Hence diameter of nozzle $= \frac{0.182}{0.80} = 0.228$ m

Further $\frac{D}{d} = \frac{2.907}{0.182} = 15.97$ or 16

$$\therefore \quad Z = 0.5 \times 16 + 15 = 23$$

16.6 **The net head on the Pelton wheel is 55 m and discharge through the jet is 0.03 m³/s. If the Pelton wheel has a tangential velocity of 22.0 m/s, determine the power developed by the Pelton wheel. Take side clearance as 15° and C_v for the jet to be 0.98.**

$$v = 22.0 \text{ m/s}, \; H = 55 \text{ m}, \; Q = 0.03 \text{ m}^3/\text{s and } C_v = 0.98$$

$$V_1 = C_v\sqrt{2gH} = 0.98 \times \sqrt{2 \times 9.806 \times 55} = 32.186 \text{ m/s}$$

$$V_{\omega 1} = V_1 = 32.186 \text{ m/s}$$

$$V_{r1} = (V_1 - v) = (32.186 - 22.000) = 10.186 \text{ m/s}$$

From outlet velocity triangle (Fig. 16.2),

$$V_{r2} = V_{r1} = 10.186 \text{ m/s}$$

$$V_{r2} \cos\phi = 10.186 \cos 15° = 9.839 \text{ m/s}$$

$$V_{\omega 2} = v - V_{r2}\cos\phi = (22.000 - 9.839) = 12.161 \text{ m/s}$$

Work done per second or power will be

$$P = Q\rho\,(V_{\omega 1} - V_{\omega 2})\,v/1000$$

$$= \frac{0.03 \times 998\,(32.186 - 12.161) \times 22}{1000}$$

$$= 13.190 \text{ kW}$$

16.7 Calculate the number of nozzles required for a Pelton wheel producing 13 000 kW power under a head of 250 m when running 430 rpm. Also determine the size of jets, diameter of wheel and the specific speed if overall efficiency is 84 per cent.

$$13\,000 = \frac{Q\gamma H}{1000}\eta \quad \therefore \quad Q = \frac{13\,000 \times 1000}{9787 \times 250 \times 0.84}$$

$$\therefore \qquad Q = 6.325 \text{ m}^3/\text{s}$$

$$V_1 = 0.98\sqrt{2gH} = 0.98 \times \sqrt{2 \times 9.806 \times 250}$$

$$= 68.621 \text{ m/s}$$

$$\frac{v}{V_1} = 0.46 \quad \therefore \quad v = 68.621 \times 0.46 = 31.566 \text{ m/s}$$

But $$v = \frac{\pi DN}{60} \quad \therefore \quad D = \frac{31.566 \times 60}{3.142 \times 430} = 1.402 \text{ m}$$

Specific speed $$N_s = \frac{N\sqrt{P}}{H^{5/4}} = \frac{430\sqrt{13\,000}}{250^{5/4}} = 49.32$$

Looking at the dependance of $\frac{D}{d}$ on N_s, $\frac{D}{d}$ is assumed to be 6.0.

$$\therefore \qquad d = \frac{1.402}{6} = 0.234 \text{ m}$$

$$\text{Total jet area} = \frac{Q}{V_1} = \frac{6.325}{68.621} = 0.0909 \text{ m}^2$$

$$\text{Area of one jet} = 0.785 \times (0.234)^2 = 0.043 \text{ m}^2$$

$$\therefore \qquad \text{Number of jets} = \frac{0.0909}{0.043} \approx 2.0$$

$$\text{Number of buckets} = 0.5 \times 6 + 15 = 18.$$

16.8 When running at full load, a Pelton wheel develops 250 kW power under a head of 330 m with an overall efficiency of 86 per cent. Estimate the power that the Pelton wheel will develop at the same speed if discharge is reduced by 35 percent (i) by use of spear in the nozzle, and (ii) by use of sluice in the supply pipe, nozzle remaining fully open. In the second case, take overall efficiency of 78 per cent.

Case I: When the regulation is with spear, the speed ratio remains the same and hence efficiency is not materially affected.

$$\therefore \qquad \text{Power} \sim Q$$

$$P_2 = 0.65\, P_1 = 0.65 \times 250 = 162.50 \text{ kW}$$

Case II: With the use of sluice valve in the supply pipe $Q \sim \sqrt{H}$

$$\therefore \quad \sqrt{\frac{H_1}{H_2}} = \frac{Q_1}{Q_2} \quad \text{or} \quad \sqrt{H_2} = \sqrt{H_1} \times \frac{Q_2}{Q_1} = 0.65\sqrt{H_1}$$

$$\therefore \quad H_2 = 0.4225\, H_1$$

$$\frac{P_2}{P_1} = \frac{Q_2 H_2 \eta_2}{Q_1 H_1 \eta_1} = \frac{0.65\, Q_1 \times 0.4225\, H_1 \times 0.78}{Q_1 H_1 \times 0.86} = 0.249$$

$$\therefore \quad P_2 = 0.249 \times 250 = 62.27 \text{ kW}$$

16.9 **A Pelton wheel has a tangential velocity of 18.0 m/s when discharge and head are 0.25 m³/s and 40 m respectively. If the bucket deflects the jet through 160°, determine the power produced by the turbine. Take $C_v = 0.98$**

In Fig. 16.2,

$$v = 18.0 \text{ m/s}, \quad \phi = 180 - 160 = 20°$$

$$V_1 = C_v\sqrt{2gH}$$

$$= 0.98 \times \sqrt{2 \times 9.806 \times 40} = 27.448 \text{ m/s}$$

For inlet condition $V_{\omega 1} = V_1 = 27.448$ m/s

$$V_{r1} = (V_1 - v) = (27.448 - 18.000) = 9.448 \text{ m/s}$$

For outlet triangle $\quad V_{r2} = V_{r1} = 9.448$ m/s

$$v_2 = 18.000 \text{ m/s}$$

$$V_{\omega 2} = 18.00 - V_{r2}\cos 20° = 18.000 - 9.448 \times 0.9397$$

$$= 9.122 \text{ m/s}$$

Work done per unit weight of fluid $= \dfrac{v\,(V_{\omega 1} + V_{\omega 2})}{g}$

$$= \frac{18.000}{9.806}(27.448 + 9.122)$$

$$= 67.128 \text{ Nm/N}$$

$$\therefore \quad P = \frac{Q \times \gamma \times 67.128}{1000} = \frac{0.25 \times 9787 \times 67.128}{1000}$$

$$= 164.25 \text{ kW}$$

16.10 A double overhung Pelton wheel is to operate at 25 760 kW under a head of 280 m at the base of the nozzle. Find the size of the jet, mean diameter of the runner, synchronous speed and specific speed of each wheel. Assume generator efficiency = 92 percent, Pelton wheel efficiency = 87 percent, $C_v = 0.97$, Speed ratio = 0.46 and jet ratio = 12.

In the case of a double overhung Pelton wheel, two identical Pelton wheels are keyed to the same shaft as shown in Fig. 16.8.

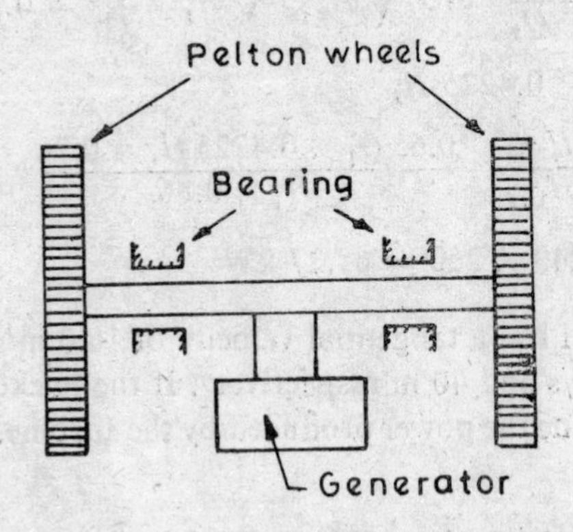

Fig. 16.8. Double overhung Pelton wheel.

$$\text{Power of each turbine} = \frac{25\,760}{0.92 \times 2} = 14\,000 \text{ kW}$$

$$\text{Available power for each turbine} = \frac{14\,000}{0.87} = 16\,091.95 \text{ kW}$$

$$\therefore \quad \frac{QrH}{1000} = 16\,091.95 \quad \text{or} \quad Q = \frac{16\,091.95 \times 1000}{9787 \times 280}$$

$$= 5.872 \text{ m}^3/\text{s}$$

$$V_1 = C_v\sqrt{2gH} = 0.97 \times \sqrt{2 \times 9.806 \times 280}$$

$$= 71.88 \text{ m/s}$$

$$\text{Area of jet, } d = \frac{5.872}{71.88} = 0.0817 \text{ m}^2$$

$$\therefore \quad \text{Diameter of jet, } d = \sqrt{\frac{0.0817}{0.785}} = 0.3226 \text{ m}$$

$$\frac{D}{d} = 12$$

$$\therefore \quad \text{Diameter of wheel} = 12 \times 0.3226 = 3.871 \text{ m}$$

$$\frac{v}{V_1} = 0.46 \quad \therefore \quad v = 0.46 \times 71.88 = 33.065 \text{ m/s}$$

But $\quad v = \dfrac{\pi DN}{60}$;

hence $\quad N = \dfrac{33.065 \times 60}{3.142 \times 3.871} = 163.11 \quad \text{or} \quad 163 \text{ rpm}$

$$f = \frac{pN}{60}$$

where f is frequency in H_z of power generation and p is number of pairs of poles. Hence for $f = 50$,

$$N_{syn} = \frac{50 \times 60}{p}$$

If $\quad p = 18$, Synchronous speed $= \frac{3000}{18} = 166.67$ rpm

Specific speed $\quad N_s = N\sqrt{P}/H^{5/4}$

$$= \frac{163\sqrt{14\,000}}{280^{5/4}} = 16.838$$

16.11 Discuss the dimensional analysis of geometrically similar turbines. What are unit speed, unit discharge and unit power?

Geometrically similar turbines can be characterised by characteristic diameter D, angular velocity ω or speed N, head gH, discharge Q, mass density ρ and viscocity μ of fluid. The dependent variables can be either P or η. Hence one can write

$$\eta \text{ or } P = f(D, \omega, gH, Q, \rho, \mu)$$

Choosing ω, D and ρ as the repeating variables, one gets

$$\eta \text{ or } \frac{P}{\rho\omega^3 D^5} = f\left(\frac{gH}{\omega^2 D^2}, \frac{Q}{\omega D^3}, \frac{\omega D^2 \rho}{\mu}\right)$$

If the fluid has small viscosity, Reynolds number $\omega D^2\rho/\mu$ will be large and can be omitted from the analysis. Therefore, if for two geometrically similar pumps, $Q/\omega D^3$ and $gH/\omega^2 D^2$ are maintained the same, their efficiency η or dimensionless power $P/\rho\omega^3 D^5$ will be equal. One can eliminate D from $P/\rho\omega^3 D^5$ and $gH/\omega^2 D^2$ to get a parameter $\omega\sqrt{P}/(gH)^{5/4}\rho^{1/2}$ which can be called dimensionless specific speed $\bar{N}_s$. Usually specific speed is defined replacing ω by N, eliminating $\rho^{1/2}g^{5/4}$ and expressing P and H in kW and m respectively. Thus specific speed $N_s = N\sqrt{P}/H^{5/4}$.

From the three dimensionless parameters involving P, Q, and ω or N, one can eliminate D, and express P, Q and N in terms of H, which yields

$$\frac{P}{H^{3/2}} = \text{const. } C_1, \quad \frac{Q}{\sqrt{H}} = \text{const. } C_2, \quad \frac{N}{\sqrt{H}} = \text{const. } C_3$$

These are known as unit power, unit discharge and unit speed respectively. For a given machine, these unit quantities can tell how P, Q and N will change if H is changed.

16.12 An impulse turbine of 2.75 m diameter is rated at 11 000 kW at 300 rpm under a head of 490 m. It uses 2.7 m³/s discharge. (i) If this turbine is operated under a head of 390 m, what will be the speed, power and discharge? (ii) Determine the size of the wheel to develop 7000 kW power under a head of 300 m. What will be its speed and discharge?

(i) $P_2/P_1 = (H_2/H_1)^{3/2} \quad \therefore \quad P_2 = 11\,000 \times (390/490)^{3/2}$
$= 7810.80$ kW

$Q_2/Q_1 = (H_2/H_1)^{1/2} \quad \therefore \quad Q_2 = 2.7 \times (390/490)^{1/2}$
$= 2.409$ m³/s

$N_2/N_1 = (H_2/H_1)^{1/2} \quad \therefore \quad N_1 = 300 \times (390/490)^{1/2}$
$= 267.64$ rpm

(ii) Since $N \sim \dfrac{1}{D}, \dfrac{N_2}{N_1} = \dfrac{D_1}{D_2} = \left(\dfrac{H_2}{H_1}\right)^{1/2} = \left(\dfrac{300}{490}\right)^{1/2} = 0.7825$

$$\therefore D_2 = \frac{D_1}{0.7825} = \frac{2.75}{0.7825} = 3.514 \text{ m}$$

$$\frac{Q_2}{Q_1} = \left(\frac{H_2}{H_1}\right)^{1/2} = 0.7825 \quad \therefore \quad Q_2 = 2.7 \times 0.7825 = 2.112 \text{ m}^3\text{/s}$$

$$\frac{N_2}{N_1} = \left(\frac{H_2}{H_1}\right)^{1/2} = 0.7825 \quad \therefore \quad N_2 = 300 \times 0{,}7825 = 234.75 \text{ rpm}$$

16.13 An inward flow reaction turbine has outer and inner diameters of wheel as 1.0 m and 0.50 m respectively. Vanes are radial at inlet and discharge is radial at outlet. Water enters the vanes at an angle of 10°. Assuming velocity of flow as constant equal to 3.0 m/s, determine the speed of the turbine and vane angle at the outlet.

Inward flow turbine is shown in Fig. 16.4. Here basically the flow is guided between near parallel circular passage by guide vanes. Then the flow encounters runner blades which move the fluid radially inward and then outwardly in the axial direction. Figure 16.4 also shows the velocity triangles at inlet and outlet. Here

$D_1 = 1.0$ m, $D_2 = 0.5$ m, $\alpha = 10°$. Since the vane is radial at inlet $\theta = 90°$.

$$v_1 = \frac{\pi D_1 N}{60} \quad \therefore \quad N = \frac{60 v_1}{\pi D_1}$$

Further $\dfrac{V_{f_1}}{v_1} = \tan\alpha = \tan 10° = 0.1763$

$$\therefore \quad v_1 = \frac{3.0}{0.1763} = 17.016 \text{ m/s}$$

$$\therefore \quad N = \frac{60 \times 17.016}{3.142 \times 1.0} = 324.95 \text{ rpm} \approx 325 \text{ rpm}$$

$$v_2 = \frac{\pi D_2 N}{60} = \frac{3.142 \times 0.5 \times 325}{60} = 8.508 \text{ m/s}$$

Since discharge is radial, $\tan\phi = \dfrac{3.0}{8.508} = 0.3526 \quad \therefore \quad \phi = 19.42°$

16.14 An inward flow turbine has the following data.

$$Q = 0.60 \text{ m}^3/\text{s},\ H = 18 \text{ m, power developed} = 89 \text{ kW}$$

$$N = 375 \text{ rpm},\ D_1 = 0.75 \text{ m},\ D_2 = 0.50 \text{ m, exit velocity} = 3.5 \text{ m/s}$$

and is radial, $b_1 = b_2$.

Determine overall efficiency η, hydraulic efficiency η_h and inlet angles of guide and runner blades.

$$v_1 = \frac{\pi D_1 N}{60} = \frac{3.142 \times 0.75 \times 375}{60} = 14.728 \text{ m/s}$$

$$\text{Work done/wt} = \frac{1}{g}(r_1\omega_1 V_1 \cos\alpha - r_2\omega_2 V_2 \cos\beta)$$

Since exit is radial $\beta = 90°$, $\therefore$ $V_2 r_2 \omega_2 \cos\beta = 0$

$$\therefore \quad \text{Work done/wt} = \frac{r_1\omega_1 V_1 \cos\alpha}{g}$$

Bernoulli's equation gives

$$\frac{V_{\omega 1} v_1}{g} = 18 - \frac{3.5^2}{2g} = 18 - 0.625 = 17.375$$

$$\therefore \quad V_{w_1} = \frac{17.375 \times 9.806}{14.728} = 11.568 \text{ m}$$

$$P = \frac{Q\gamma v_1 V_{\omega 1}}{1000\, g} = \frac{0.60 \times 9787 \times 11.568 \times 14.728}{1000 \times 9.806} = 102.02 \text{ kW}$$

$$\text{Power available} = \frac{Q\gamma H}{1000} = \frac{0.60 \times 18 \times 9787}{1000} = 105.70 \text{ kW}$$

$$\eta = \frac{89}{105.70} = 84.2 \text{ percent}$$

$$\eta_h = \frac{102.02}{105.70} = 96.51 \text{ percent}$$

$$Q = \pi D_1 V_{f1} b_1 = \pi D_2 V_{f2} b_2 \quad \text{and} \quad b_1 = b_2$$

$$V_{f1} = \frac{V_{f2} D_2}{D_1} = \frac{3.5 \times 0.5}{0.75} = 2.333 \text{ m/s}$$

Since

$$\frac{V_{f1}}{V_1} = \sin\alpha$$

$$\text{and} \quad V_1 = \sqrt{V_{f1}^2 + V_{\omega 1}^2} = \sqrt{2.333^2 + 11.568^2}$$

$$= 11.801 \text{ m/s}$$

$$\sin\alpha = \frac{2.333}{11.808} = 0.1980 \quad \text{or} \quad \alpha = 11.40°$$

$$\tan\theta = \frac{V_{f1}}{(V_{\omega 1} - v_1)} = \frac{2.333}{11.568 - 14.728} = -0.738$$

$$\therefore \quad \theta = 143.56°$$

16.15 An outward flow reaction turbine has the following details:

Inner diameter $D_1 = 0.50$ m;

Velocity of flow $= V_{f1} = V_{f2} = 4.5$ m/s; $H = 12$ m

Outer diameter $D_2 = 1.00$ m;

Guide blade angle $\alpha = 18°$; $\quad N = 300$ rpm

If discharge at the outlet is radial, determine (i) runner vane angles at inlet and outlet, (ii) work done/wt. on the runner and (iii) hydraulic efficiency. Here $V_2 = V_{f2}$ and $V_{\omega 2}$ is zero since discharge at the outlet is radial.

$$v_1 = \frac{\pi D_1 N}{60} = \frac{3.142 \times 0.50 \times 300}{60} = 7.855 \text{ m/s}$$

$$v_2 = \frac{\pi D_2 N}{60} = \frac{3.142 \times 1.0 \times 300}{60} = 15.710 \text{ m/s}$$

$$\tan\alpha = \frac{V_{f1}}{V_{\omega 1}} \text{ and } \alpha = 18° \quad \therefore \quad V_{\omega 1} = \frac{4.5}{0.3249} = 13.854 \text{ m/s}$$

$$\tan\theta = \frac{V_{f1}}{(V_{\omega 1} - v_1)} = \frac{4.5}{(13.854 - 7.855)} = \frac{4.5}{5.999} = 0.7501$$

$$\therefore \quad \theta = 36.87°$$

$$\tan\phi = V_{f2}/v_2 = \frac{4.5}{15.710} = 0.2684 \quad \text{or} \quad \phi = 15.98°$$

$$\text{Work done/wt} = \frac{1}{g} V_{\omega 1} v_1 = \frac{13.854 \times 15.71}{9.806} = 11.098 \text{ Nm/N}$$

$$\text{Hydraulic efficiency} \quad \eta_h = \frac{V_{\omega 1} v_1}{gH} = \frac{11.098}{12} = 0.9248 \text{ or } 92.48 \text{ percent.}$$

16.16 For an inward flow reaction turbine which discharges radially with constant velocity of flow equal to velocity of flow from suction tube, show that

$$H = \frac{1}{\dfrac{1 + \tan^2 \alpha/2}{1 - \tan\alpha/\tan\theta}}$$

From the inlet velocity triangle; see Fig. 16.4,

$$V_{\omega 1} = V_{f1}/\tan\alpha \quad \text{and} \quad V_{f1}/(V_{\omega 1} - v_1) = \tan\theta$$

Solving these two equations, one gets

$$v_1 = V_{f1}(\tan\theta - \tan\alpha)/\tan\alpha \tan\theta$$

As the turbine discharges radially, $V_{\omega 2} = 0$, $V_2 = V_{f2}$, and $V_{f1} = V_{f2}$ as given.

$$H = \frac{v_1 V_{\omega 1}}{g} + \frac{V_2^2}{2g}$$

Substituting the values of $V_{\omega 1}$, v_1 and v_2, one gets

$$H = \frac{V_{f1}^2}{g} \frac{\tan\theta - \tan\alpha}{\tan^2\alpha \tan\theta} + \frac{V_{f1}^2}{2g}$$

$$= \frac{V_{f1}^2}{g}\left(\frac{1}{2} + \frac{\tan\theta - \tan\alpha}{\tan^2\alpha \tan\theta}\right)$$

$$\eta_h = \frac{V_{\omega 1} v_1}{gH} = \frac{V_{f1}^2}{g}\left\{\frac{\tan\theta - \tan\alpha}{\tan^2\alpha \tan\theta}\right\} \Big/ \frac{V_{f1}^2}{g}\left\{\frac{1}{2} + \frac{\tan\theta - \tan\alpha}{\tan^2\alpha \tan\theta}\right\}$$

which on simplification yields

$$\eta_h = \frac{1}{1 + \dfrac{1}{2}\dfrac{\tan^2\alpha}{1 - \dfrac{\tan\alpha}{\tan\theta}}}$$

16.17 An inward flow reaction turbine has the following data:

$N = 240$ rpm, velocity of flow which is constant $= 3.5$ m/s

$D_1 = 1.0$ m, $b_1 = 0.135$ m

Assume radial discharge at inlet and outlet, and determine, (i) work done per unit weight of water, (ii) power developed by turbine, (iii) water head on the machine and (iv) hydraulic efficiency.

See Fig. 16.4. $V_{f1} = V_{f2} = 3.5$ m/s

$$v_1 = \frac{\pi D_1 N}{60} = \frac{3.142 \times 1.0 \times 240}{60} = 12.568 \text{ m/s}$$

Since discharge is radial at inlet and outlet, θ and β are equal to 90°. Also $V_{\omega 1} = v_1$.

$$\therefore \quad \text{Work done/wt} = \frac{1}{g} V_{\omega 1} v_1 = \frac{12.568^2}{9.806} = 16.108 \text{ Nm/N}$$

$$Q = \pi D_1 b_1 V_{f1} = 3.142 \times 1.0 \times 0.135 \times 3.5$$
$$= 1.485 \text{ m/s}$$

$$\therefore \quad P = \frac{1.485 \times 9787 \times 16.108}{1000} = 234.045 \text{ kW}$$

According to Bernoulli's equation

$$\frac{V_{w1} v_1}{2} = H - \frac{V_2^2}{2g} \quad \text{and} \quad V_1 = V_{f1}$$

$$\therefore \quad 16.108 = H - \frac{3.5^2}{2 \times 9.806} \qquad \therefore \quad H = 16.108 + 0.625 = 16.733 \text{ m}$$

$$\eta_h = \frac{V_{\omega 1} v_1}{gH} = \frac{16.108}{16.733} = 0.9626 \quad \text{or} \quad 96.26 \text{ percent}$$

16.18 An outward flow turbine running at 200 rpm works on a discharge of 5.0 m^3/s under a head of 40 m. Internal and external diameters of the wheel are 2.0 m and 2.5 m respectively while the width at the inlet and outlet is 200 mm. Assuming the discharge to be radial at outlet, determine angles of the turbine at the inlet and outlet.

See Fig. 16.5. $v_1 = \frac{\pi D_1 N}{60} = \frac{3.142 \times 2.0 \times 200}{60}$

$$= 20.947 \text{ m/s}$$

$$v_2 = \frac{\pi D_2 N}{60} = \frac{3.142 \times 2.5 \times 200}{60} = 26.183 \text{ m/s}$$

$$Q = \pi D_1 b_1 V_{f1} = \pi D_2 b_2 V_{f2}$$

$$\therefore \quad V_{f1} = 5.0/(3.142 \times 2.0 \times 0.20) = 3.978 \text{ m/s}$$

$$V_{f2} = 5.0/(3.142 \times 2.5 \times 0.20) = 3.182 \text{ m/s}$$

Since the discharge is radial at outlet, $V_{\omega 2} = 0$ and $V_2 = V_{f2}$.

$$\frac{V_{\omega 1} v_1}{g} = H - \frac{V_2^2}{2g}$$

$$\therefore \quad \frac{V_{\omega 1} \times 20.947}{9.806} = 40 - \frac{3.182^2}{2 \times 9.806} = 39.483 \text{ m}$$

$$\therefore \quad V_{\omega 1} = \frac{39.483 \times 9.806}{20.947} = 18.484 \text{ m/s}$$

Since $V_{\omega 1}$ is less than v_1, θ is greater than 90°

$$\tan(180 - \theta) = \frac{V_{f1}}{v_1 - V_{\omega 1}} = \frac{3.978}{(20.947 - 18.484)} = 1.6151$$

$$\therefore \quad 180 - \theta = 58.24°$$

Vane angle $\theta = 121.76°$

$$\tan \alpha = V_{f1}/V_{\omega 1} = \frac{3.978}{18.484} = 0.2152$$

$\therefore$ Guide blade angle = 12.15°

16.19 In the case of a Francis turbine

$$P = 130 \text{ kW}, \ H = 9.0 \text{ m}, \ N = 120 \text{ rpm}, \quad \eta = 75 \text{ percent}$$

$$\eta_h = 80 \text{ percent}, \quad v_1 = 3.45\sqrt{H}, \quad V_{f1} = 1.15\sqrt{H}$$

Determine the guide blade angle α at the inlet, vane angle θ at the inlet and the diameter of the wheel (Fig. 16.9).

$$v_1 = 3.45\sqrt{9} = 10.35 \text{ m/s}$$

$$V_{f1} = 1.15\sqrt{9} = 3.45 \text{ m/s}$$

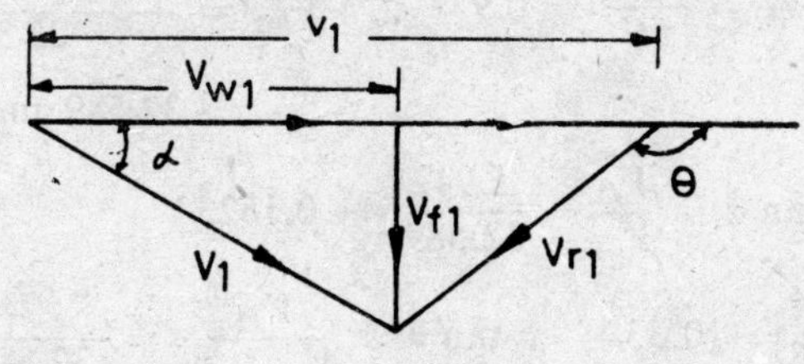

Fig. 16.9.

$$\eta_h = \frac{V_{\omega 1} v_1}{gH} = 0.80 \qquad \therefore \quad V_{\omega 1} = \frac{0.8 \times 9.806 \times 9}{10.35}$$

$$= 6.822 \text{ m/s}$$

$$\tan \alpha = \frac{V_{f1}}{V_{\omega 1}} = \frac{3.45}{6.822} = 0.5057$$

$$\therefore \qquad \alpha = 26.83°$$

$$\tan (180 - \theta) = \frac{V_{f1}}{(v_1 - V_{\omega 1})} = \frac{3.45}{(10.35 - 6.822)} = \frac{3.45}{3.528} = 0.9779$$

$$\therefore \qquad 180 - \theta = 44.36° \quad \text{and} \quad \theta = 135.64°$$

Further, since $v_1 = \dfrac{\pi D_1 N}{60}$, $D_1 = \dfrac{60 \times 10.35}{3.142 \times 120}$

$$= 1.647 \text{ m}$$

16.20 Design a Francis turbine runner for the following data:

$H = 80$ m, $N = 750$ m, $P = 375$ kW, $\eta_h = 93$ percent, $\eta = 84$ percent, $n = 0.10$, $\psi_1 = 0.15$, $k = 0.95$. Assume radial flow at the exit.

Since $P = \dfrac{Q\gamma H}{1000}\eta$, $\quad Q = \dfrac{375 \times 1000}{9787 \times 80 \times 0.84} = 0.57 \text{ m}^3/\text{s}$

$$V_1 = \psi_1 \sqrt{2gH} = 0.15 \times \sqrt{2 \times 9.806 \times 80} = 5.942 \text{ m/s}$$

$$V_{f1} = \frac{Q}{\pi D_1 b_1 k} \text{ but } \frac{b_1}{D_1} = n = 0.10 \text{ and } V_1 = V_{f1}$$

$$\therefore \qquad V_{f1} = \frac{0.57}{3.142 \times 0.10 D_1^2 \times 0.95} = 5.942 \text{ m/s}$$

$$\therefore \qquad D_1^2 = 0.321 \text{ and } D_1 = 0.567 \text{ m} \quad \text{or} \quad 0.57 \text{ m}$$

$$b_1 = 0.10 \times 0.57 = 0.057 \text{ m} \quad \text{or} \quad 57 \text{ mm}$$

$$v_1 = \frac{\pi D_1 N}{60} = \frac{3.142 \times 0.570 \times 750}{60} = 22.387 \text{ m/s}$$

$$\eta_h = \frac{V_{\omega 1} v_1}{gH} = 0.93 \quad \therefore \quad V_{\omega 1} = \frac{0.93 \times 9.806 \times 80}{22.387}$$

$$= 32.589 \text{ m/s}$$

$$\tan \alpha = \frac{V_{f1}}{V_{\omega 1}} = \frac{5.942}{32.589} = 0.1823$$

$$\therefore \quad \alpha = 10.33° \qquad \tan \theta = \frac{V_{f1}}{V_{\omega 1} - v_1} = \frac{5.942}{(32.589 - 22.387)}$$

$$= 0.5824$$

$$\therefore \quad \theta = 30.2°$$

Take $D_2/D_1 = 0.5$; hence $D_2 = 0.5 \times 0.57$

$$= 0.285 \text{ m}$$

$$\therefore \quad v_2 = 0.5\, v_1 = 11.194 \text{ m/s}$$

$$V_{f1} = V_{f2} = 5.942 \text{ m/s}$$

$$\therefore \quad \tan \phi = \frac{V_{f2}}{v_2} = \frac{5.942}{11.194} = 0.5308 \text{ and } \phi = 27.96°$$

Since N_s comes out to be 60, use 12 guide blades and 13 runner blades.

16.21 A Francis turbine produces 375 kW power when running at 750 rpm under a head of 80 m. A model of this turbine to a scale of 1: 6 is to be tetted under a head of 6.0 m. Determine the speed and power generated by the model.

For dynamic similarity, $\frac{Q}{ND^3}$, $\frac{gH}{N^2D^2}$ and $\frac{P}{N^3D^5}$ will remain the same in the model and the prototype.

$$\therefore \quad \frac{H_m}{N_m^2 D_m^2} = \frac{H_p}{N_p^2 D_p^2} \qquad \therefore \; N_m^2 = N_p^2 \left(\frac{D_p}{D_m}\right)^2 \left(\frac{H_m}{H_p}\right)$$

$$\text{but} \quad \frac{D_p}{D_m} = 6.0 \qquad \text{or} \quad N_m = N_p \left(\frac{D_p}{D_m}\right) \left(\frac{H_m}{H_p}\right)^{1/2}$$

$$= 750 \times 6 \times (6/80)^{1/2}$$

$$= 1232.38 \text{ rpm}$$

$$\frac{P_m}{P_p} = \left(\frac{N_m}{N_p}\right)^3 \left(\frac{D_m}{D_p}\right)^5 \qquad \therefore \; P_m = 375 \left(\frac{1232.38}{750}\right)^3 \left(\frac{1}{6}\right)^5$$

$$= 0.214 \text{ kW}$$

16.22 Francis turbine working under a head of 30 m produces 13 500 kW power when running at 120 rpm. If at the site, the atmospheric pressure is equivalent to 10.2 m of water head and vapour pressure is 0.22 m of water, calculate the safe height of the discharge end of the runner above the tail water.

$$N_s = \frac{120 \times \sqrt{13\,500}}{30^{5/4}} = 194.97$$

$$\sigma_c = 0.0317 \left(\frac{N_s}{100}\right)^2 = 0.1205$$

But
$$\sigma_c = \frac{\frac{p_a}{\gamma} - \frac{p_v}{\gamma} - h_t}{H}$$

$\therefore$
$$h_s = \frac{p_a}{\gamma} - \frac{p_v}{\gamma} - H\,\sigma_c$$

$$= 10.2 - 0.22 - 30 \times 0.1205$$

$$= 6.362 \text{ m}$$

16.23 A Kaplan turbine develops 20 000 kW power under a head of 35 m. Assume overall efficiency of 85 percent, velocity ratio $v/\sqrt{2gH}$ of 2.0, flow ratio of 0.65 and D_0/D ratio of 0.35. Calculate the diameter, speed and specific speed of the turbine.

$$v = 2.0 \times \sqrt{2gH} = 2.0 \times \sqrt{2 \times 9.806 \times 35} = 52.4 \text{ m/s}$$

$$V_{f_1} = 0.65\sqrt{2gH} = 0.65 \times \sqrt{2 \times 9.806 \times 35} = 17.03 \text{ m/s}$$

Diameter of hub $D_0 = 0.35D$

$$\eta = 0.85 = \frac{20\,000 \times 1000}{Q \times 9787 \times 35} \qquad \therefore\ Q = 68.69 \text{ m}^3/\text{s}$$

but
$$Q = \frac{\pi}{4}(D^2 - D_0^2)\,V_{f1}$$

$\therefore$
$$68.69 = 0.785 \times (D^2 - 0.35^2 D^2) \times 17.03$$

$\therefore$
$$D^2 = 5.855 \quad \text{and} \quad D = 2.42 \text{ m}$$

$$D_0 = 2.42 \times 0.35 = 0.847 \text{ m}$$

$$v = \frac{\pi D N}{60} \qquad \therefore\ N = \frac{52.4 \times 60}{3.142 \times 2.42} = 413.49 \text{ rpm}$$

$$N_s = \frac{N\sqrt{P}}{H^{5/4}} = \frac{413.49\sqrt{20\,000}}{35^{5/4}} = 686.9$$

16.24 A Kaplan turbine has the hub diameter of 2.0 m and runner diameter of 5.0 m. If it develops 25 000 kW when running at 150 rpm under a head of 25 m, with η_h and η of 90 and 85 percent, determine the discharge through the turbine and the guide blade angle α at the inlet.

Since
$$\eta = \frac{P \times 1000}{Q\gamma H} \qquad Q = \frac{25\,000 \times 1000}{0.85 \times 9787 \times 25}$$

$$= 120.207 \text{ m}^3/\text{s}$$

$$\eta_h = \frac{V_{\omega 1} v_1}{gH} \text{ and } v_1 = \frac{3.142 \times 5.0 \times 150}{60} = 39.275 \text{ m/s}$$

$$\therefore \quad V_{\omega 1} = \frac{0.9 \times 9.806 \times 25}{39.275} = 5.618 \text{ m/s}$$

Since $\quad Q = \frac{\pi}{4}(D^2 - D_0^2)\, V_{f1},$

$$V_{f1} = \frac{120.207}{0.785\,(5.0^2 - 2.0^2)} = 7.292 \text{ m/s}$$

Since $V_{\omega 1}$ is less than v_1, outlet triangle is as shown in Fig. 16.9.

$$\tan \alpha = \frac{V_{f1}}{V_{\omega 1}} = \frac{7.292}{5.618} = 1.298 \qquad \therefore \quad \alpha = 52.39°$$

16.25 A turbine is to operate under a head of 30 m while running at 120 rpm using 12.0 m³/s discharge. Assuming overall efficiency of 90 percent, determine the type of turbine.

$$P = \frac{12.0 \times 9787 \times 30}{1000} \times 0.90$$

$$= 3171 \text{ kW}$$

$$\therefore \quad N_s = N\sqrt{P}/H^{5/4} = \frac{120 \times \sqrt{3171}}{(30)^{5/4}} = 96.24$$

From Table 16.3 it can be seen that for this specific speed, Francis runner can be used.

PROBLEMS

16.1 An overshot wheel is supplied water from a rectangular channel at the end of which is a sluice gate. If the gate width is 2.0 m, gate opening 0.30 m and velocity through the sluice is 3.0 m/s, determine the power developed if operating head is 7.0 m and wheel efficiency is 70 percent. (86.321 kW)

16.2 A Pelton wheel of 0.60 m diameter and with semicircular buckets runs at 600 rpm when working under a head of 150 m. If the discharge through the nozzle is 0.05 m³/s, determine the power available at the nozzle and the hydraulic efficiency. Assume C_v for nozzle as 0.97. (73.403 kW, 91.968 percent)

16.3 Pykara power house on the Pykara river is equipped with Pelton wheels each of which develops a maximum of 14 180 kW power while working under a head of 855 m and when running at 600 rpm. Determine the smallest diameter of the jet and mean diameter of the wheel, Also determine the approximate diameter of the nozzle

tip. Assume overall efficiency of 89 percent and specify the number of buckets for the wheel. Take $C_v = 0.98$ and $\phi_1 = 0.45$.
(Note: C_c for nozzle can be taken as about 0.8)
(138.2 m. 1.855 m, 172 mm, 22)

16.4 A Pelton wheel running at 500 rpm produces 10 000 kW power under a head of 320 m. Assuming the jet diameter to be one tenth of the wheel diameter, $C_v = 0.98$ and overall efficiency η to be 84 percent, determine the diameter and number of jets required. What assumptions have you made? (0.1364 m, 4)

16.5 A double jet Pelton wheel with a specific speed of 23 develops 5360 kW power under a head of 350 m. Water is supplied to the wheels through 800 m long pipe line. Allow five percent friction loss in pipe and calculate (i) speed in rpm (ii) diameter of jets (iii) mean diameter of wheel and (iv) diameter of supply pipe. Assume $\eta = 85$ percent, $C_v = 0.98$, $\phi_1 = 0.46$ and f for pipe = 0.025.
(475.59 rpm, 120 mm, 1.548 m, 0.6035 m)

15.6 A Pelton wheel running at 500 rpm produces 4660 kW power under a head of 300 m. Assuming the jet diameter to be one tenth of the wheel diameter, $C_v = 0.98$, and overall efficiency of 85 percent. determine the diameter and number of jets required. Also determine the diameter of wheel. (132 mm, 2, 1.321 m)

16.7 A Pelton wheel works under the following conditions:

$$H = 45 \text{ m}, \; Q = 0.80 \text{ m}, \; v = 14.00 \text{ m/s}, \; \phi = 15°, \; C_v = 0.985$$

Find overall efficiency and power produced.
(95.19 percent, 335.382 kW)

16.8 Following data are given for a Pelton wheel:

$H = 172$ m, $\phi = 20°$, $Q = 190$ l/s, $N = 500$ rpm, $C_v = 0.98$, $v/V_1 = 0.45$

Find the overall efficiency, power developed and diameter of wheel.
(92.22 percent, 294.98 kW, 0.978 m)

16.9 A double overhung Pelton wheel working under a head of 350 m develops 24 000 kW power. Determine the size of the jet, diameter of the wheel, synchronous speed and specific speed of each wheel. Take $\eta = 85$ percent, $C_v = 0.98$, $v/V_1 = 0.46$, jet ratio = 0.11.
(0.254 m, 2.309 m, 300 rpm, 22.35)

16.10 A Pelton wheel develops 5620 kW power under a head of 245 m at an overall efficiency of 85 percent and speed of 200 rpm. Determine discharge, speed and power if the head is decreased to 175 m.
(2.33 m³/s, 169 rpm, 3392.69 kW)

16.11 A turbine model 400 mm in diameter develops 195 kW power under a head of 3.0 m, discharge of 0.40 m^3/s and speed of 180 rpm. If geometrically similar turbine of 1.2 m diameter is to operate under a head of 9.0 m, what will be its speed, power and discharge?

(103.93 rpm, 343.73 kW, 5.31 m^3/s)

16.12 A turbine develops 7300 kW power under a head of 25 m and at 135 rpm speed. What is its specific speed? What would be the normal speed and output under a head of 29 m?

(206.33, 120.75 rpm, 5223.45 kW)

16.13 An inward flow reaction turbine running at 300 rpm has inner and outer diameters of 0.50 m and 1.0 m respectively while the breadth at inlet and outlet of 0.20 m and 0.30 m. If water enters radially at 4.0 m/s, determine the discharge, and velocity of flow at the outlet,

(2.516 m^3/s, 5.333 m/s)

16.14 An inward flow reaction turbine has the following data: $D_1 = 1.0$ m, $b_1 = 0.25$, velocity flow V_{f_1} at inlet $= 2.0$ m/s, $N = 210$ rpm. Ten percent of area of flow is blocked by blade thickness. Further, guide blades make an angle of 10° to the wheel tangent. Determine (i) weight of water passing through the turbine (ii) runner vane angle of inlet (iii) velocity of wheel at inlet (iv) absolute velocity of water leaving the guide vanes and (v) relative velocity of water entering the runner blades.

(13.839 kN, 80.129°, 10.997 m/s, 11.519 m/s, 2.03 m/s)

16.15 Inner and outer diameters of an outward flow turbine running at 250 rpm are 2.5 m and 3.5 m respectively. If total head on the turbine is 45 m and the width of the wheel is constant at 0.30 m, determine the inlet and exit velocities, and velocity of swirl at inlet if the radial discharge is 6.2 m^3/s. (2.631 m/s, 1.879 m/s, 13.429 m/s)

16.16 For a Francis turbine the vanes are radial at inlet and velocity of flow is constant. Show that

$$\eta_h = \frac{2}{2 + \tan^2\alpha}$$

(Hint: Refer to Example 16.16)

16.17 A radially inward flow reaction turbine working under a head of 9.0 m develops 165 kW power when running at 225 rpm. The peripheral velocity at inlet v_1 is $0.90\sqrt{2gH}$ and velocity of flow at inlet V_{f_1} is $0.30\sqrt{2gH}$. If the overall and hydraulic efficiencies are 78 and 88 percent respectively, determine D_1, b_1, guide blade angle at the inlet α and vane angle at inlet θ.

(1.015 m, 0.189 m, 31.54°, 143.89°)

16.18 A Francis turbine working under a head of 15 m has guide blade angle of 20° and vanes are radial at inlet. The inlet diameter to outlet diameter ratio is 1.5, wheel velocity of flow at exit is 4.5 m/s. Assuming velocity of flow to be constant, determine the peripheral velocity at inlet and the vane angle ϕ at the outlet.

(12.363 m/s, 28.63°)

16.19 The following data are given for a Francis turbine:

$H = 65$ m, $N = 700$ rpm, shaft power $= 315$ kW, $\eta = 85$ percent,

$\eta_h = 90$ percent, Flow ratio $= 0.20$, $n = 0.10$, $D_1/D_2 = 2.0$, $K = 0.95$.

Further, velocity of flow is constant at inlet and outlet and discharge is radial at the outlet. Determine guide blade angle α at the inlet and runner blade angles at inlet and outlet θ and ϕ.

(13.42°, 33.59°, 36.69°)

16.20 Design essential elements of a Francis turbine for the following data:

$P = 100\,000$ kW, $H = 120$ m, $N = 170$ rpm, $\eta = 84$ percent,

$\eta_h = 96$ percent, $D_2/D_1 = 0.50$, $n = 0.10$, $\psi = 0.15$, $k = 0.95$.

Flow is radial at exit.

($D_1 = 6.831$ m, $b = 0.683$ m, $\alpha = 21.39°$, $\theta = 170.22°$
$\phi = 13.46°$, guide vanes 24, runner vanes 25)

16.21 A model of a Francis turbine built to a scale of 1 : 8 gives 10 kW power at 600 rpm under a head of 2.0 m. Calculate the speed and power of the prototype working under a head of 30 m. What will be the specific speed of the runner?

(290.47 rpm, 3717.92 kW, 252.26)

16.22 A Francis turbine develops 130 kW power when running at 120 rpm under a head of 9.0 m. The diameter of runner is 1.65 m. A 1:4 scale model of this turbine is tested under a head of 2.0 m. Determine the speed and power of the model. (226.27 rpm, 0.851 kW)

16 23 The following data are given for a Kaplan turbine:

$P = 15\,000$ kW, $H = 8.0$ m, $\eta = 90$ percent, Diameter ratio $= 0$ 30, speed ratio $= 2.0$, Flow ratio $= 0.60$. Determine the operating speed unit power, unit speed and unit discharge.

(6.30 m, 75.97 rpm, 662.57, 26.86, 75.26)

16.24 The following data are given for a Kaplan turbine:

$P = 11000$ kW, $H = 4.5$ m, $\eta = 90$ percent, $D_0 = 3.0$ m,

$v\sqrt{2gH} = 2.0$ (Speed ratio), $V_{f1}/\sqrt{2gH} = 0.65$ (flow ratio)

Determine the operating speed and the runner diameter.

(48 rpm, 7.476 m)

16.25 Determine the relation between specific speeds computed using P in kW (N_s), P in metric horse power (N_{sH}) and dimensionless specific speed $\overline{N}_s$. ($N_{sH} = 1.166\ N_s = 193.85\ \overline{N}_s$)

16.26 Suggest a suitable type of turbine to operate under a head of 300 m with 9.0 m^3/s when running at 200 rpm and 85 percent overall efficiency. Also determine its approximate diameter. (Pelton wheel, 3.30 m)

16.27 At a certain hydroelectric project in India, 63.5 m^3/s discharge is available under a head of 55 m. If overall efficiency of the turbine is 80 percent and speed of runner 187 rpm, determine the type of runner and its approximate diameter. (Francis, 1.51 m)

DESCRIPTIVE QUESTIONS

16.1 Enumerate the differences between reaction turbines and impulse turbines.

16.2 Draw a sketch of a hydropower plant and label its various components.

16.3 Compare methods of controlling flow in the case of Pelton wheel, Francis turbine and Kaplan turbine.

16.4 Why is the lower portion of the bucket of a Pelton wheel cut off?

16.5 What factors govern the speed of a Pelton turbine?

16.6 The specific speed is defined as

(i) $H^{5/4}/N\sqrt{P}$ (ii) $NP^{5/4}/\sqrt{H}$ (iii) $N\sqrt{P}/H^{5/4}$ (iv) $N^{5/4}P/\sqrt{H}$

16.7 What are unit speed, unit power and unit discharge? What is their utility?

16.8 What is governing of a turbine? How is it done for a Pelton wheel?

16.9 Match the range of specific speed with the type of turbine:

(i) Pelton wheel (a) 100 – 400
(ii) Francis (b) 300 – 1100
(iii) Kaplan (c) 15 – 60

16.10 Match the range of head with the type of turbine

(i) Pelton wheel (a) 7 m – 70 m
(ii) Francis (b) above 100 m
(iii) Kaplan (c) 30 m – 300 m

16.11 Specify the type of turbine which can be recommended for 20 m, 60 m, 200 m head.

16.12 What are the functions of a draft tube? What are the different shapes of draft tube that are used commonly?

16.13 For a given turbine if P is the power generated at head H, and head is halved, the power will be (i) $0.707P$ (ii) $0.354P$ (iii) P.

16.14 What preliminary data are needed for selection of the type of turbine in a hydropower project?

16.15 What are the characteristic curves of a turbine? How are they obtained? What is their utility?

16.16 If the head is halved for a given turbine, the speed will be

(i) $0.354\,N$ (ii) N (iii) $0.707\,N$

CHAPTER XVII

Unsteady Flows

17.1 TYPES OF UNSTEADY FLOWS

As defined earlier, in the case of unsteady flow, a fluid or flow characteristic changes with respect to time at a given section. The following types of unsteady flows are normally encountered in practice:

1. *Flows with random and rapid fluctuations*

Turbulent flow falls under this category, in which local velocity or pressure fluctuates rapidly. Normally such flow is analysed using time-averaged quantities and their standard deviations.

2. *Flows caused by motion of a body at constant velocity in a fluid*

Such a flow can be converted into equivalent steady flow by applying equal and opposite velocity to the fluid and the body, e.g. moving hydraulic jump.

3. *Flows involving slow or rapid periodic or near periodic variations*

Oscillations in a U-tube, surge tanks and water hammer problems fall in this category.

4. *Flows involving nonperiodic changes*

Discharge under falling head, flood routing and similar problems fall under this category.

17.2 EQUATION OF MOTION

For small acceleration or deceleration, when frictional effect is neglected, one dimensional equation of motion is

$$\frac{1}{g}\frac{\partial U}{\partial t} = -\frac{\partial H}{\partial s} \tag{17.1}$$

where H is the total head. For a pipe of constant cross-section, this equation reduces to

$$\frac{1}{g}\frac{\partial U}{\partial t} = -\frac{\partial h}{\partial s} \tag{17.2}$$

where h is the piezometric head. This equation shows that when flow is decelaratcd, i.e. $\frac{\partial U}{\partial t}$ is $-$ve, piezometric head will increase in the direction of flow. Similarly when flow is accelerated at a section, i.e. $\frac{\partial U}{\partial t}$ is $+$ve, h will decrease in the direction of flow.

If the valve at the end of a pipe of length l is suddenly opened, it takes some time for velocity to attain its equilibrium value U_0 corresponding to the operating head h. When friction and compressibility effects are neglected, variation of U with t is given by the equation

$$t = \frac{l}{\sqrt{2gh}} \ln \left(\frac{1 + \frac{U}{U_0}}{1 - \frac{U}{U_0}} \right) \tag{17.3}$$

17.3 PROPAGATION OF PRESSURE WAVE, WATER HAMMER

A pressure wave will travel in an infinite compressible fluid at a velocity C given by

$$C = \sqrt{E/\rho} \tag{17.4}$$

where E is the volume modulus of elasticity of the fluid. For water at 20°C, $E = 2.075 \times 10^9$ N/m². Hence $C = \sqrt{\frac{2.075 \times 10^9}{998}} = 1441.93$ m/s. When the fluid is bounded by an elastic boundary, e.g. pipe, $C = \sqrt{K/\rho}$ where K is the effective volume modulus of elasticity of fluid and is given by

$$K = \frac{E}{1 + \frac{D}{t}\left(\frac{E}{E_m}\right)} \tag{17.5}$$

where D and t are the diameter and thickness of pipe and E_m is Young's modulus of elasticity of pipe material. Average values of E_m for steel, cast iron and wood are 2.47×10^{11}, 1.04×10^{11} and 1.92×10^{10} N/m² respectively.

A sudden decrease in flow in a pipe by partial or complete closure of valve at the downstream end causes a sudden rise in pressure and a hammering sound. This phenomenon known as water hammer.

Pressure rise due to sudden closure of valve

$$\Delta p_m = -\rho C \Delta U = -\rho C U \tag{17.6}$$

The closure is considered sudden if closure time t_c is less than $\frac{2l}{C}$ where l is length of pipe. If t_c is slightly greater than $2l/C$, the pressure rise Δp is given by

$$\frac{\Delta p}{\Delta p_m} = \frac{2l}{C t_c} \tag{17.7}$$

Allievi's charts give information about pressure rise or fall when valve in a pipe line is closed or opened such that the opening area varies linearly with respect to time. The three parameters used in Allievi charts are

$$\alpha = \frac{UC}{2gH}, \ \beta = \frac{Ct_c}{2l} \quad \text{and} \quad Z^2 = \frac{H+h}{H}$$

where h is the rise in pressure head and H is operating head. Figures 17.1 and 17.2 give diagrams for computation of pressure rise or fall. Figure 17.1 also gives the approximate time when the maximum pressure rise occurs. If a point falls between S_n and S_{n+1}, the maximum rise will occur between $n\frac{2l}{C}$ and $(n+1)\frac{2l}{C}$ from the beginning of closure.

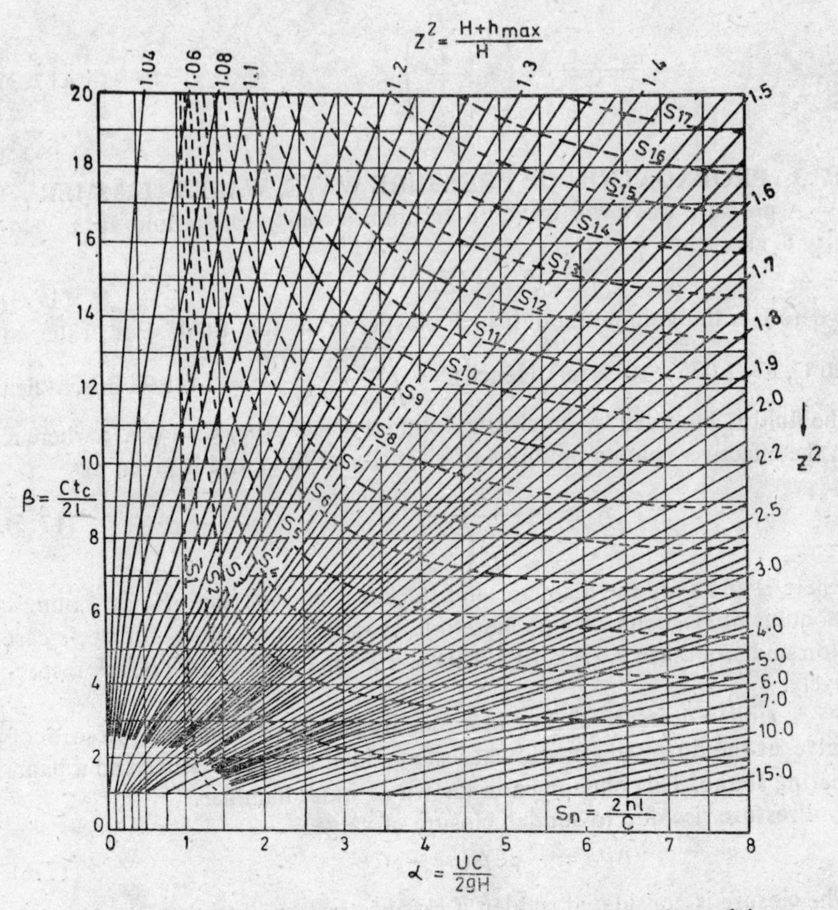

Fig. 17.1. Allievi diagram: Maximum pressure rise in simple conduits.

As a first approximation, Allievi's charts can be used to compute water hammer pressures in pipes of different diameters connected in series, if equivalent velocity of wave propagation is determined.

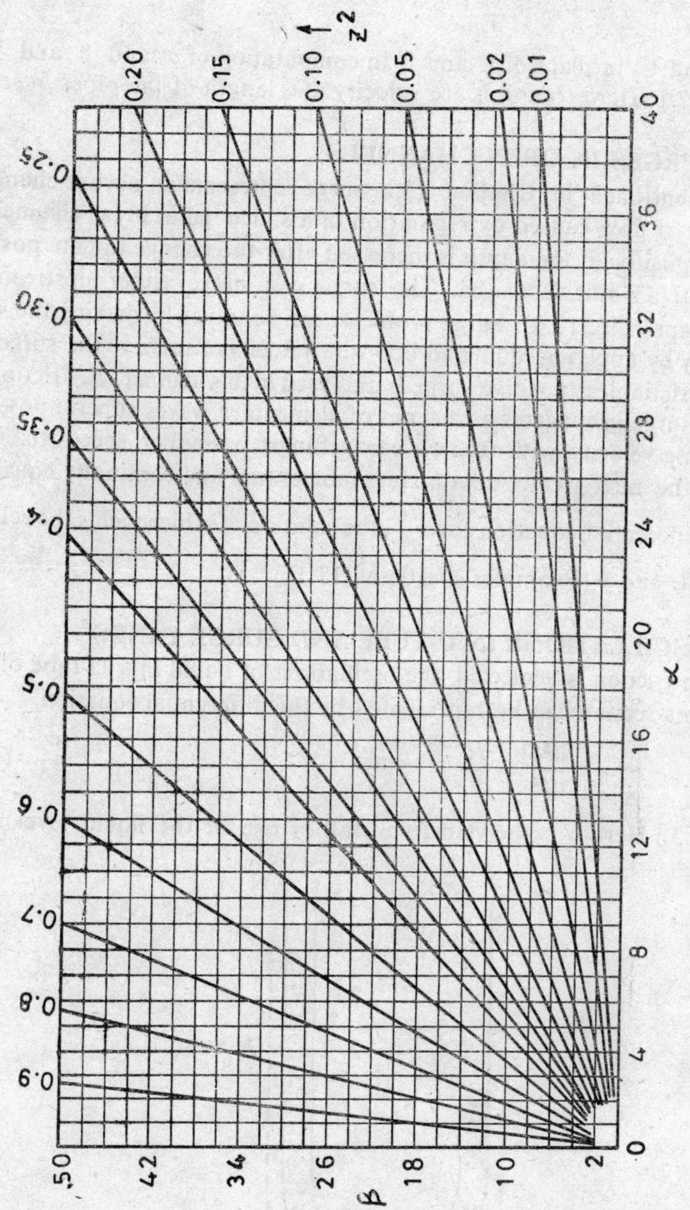

Fig. 17.2. Allievi diagram: Maximum fall in pressure in simple conduits.

$$\frac{l}{C_e} = \sum_1^n l_i/C_i$$

and
$$U_e = \sum_1^n U_i l_i/l$$

Use C_e and U_e in place of C and U in computation of α and β and then use Fig. 17.1. Here U_i and l_i are velocity and length of the pipes in series.

17.4 SURGES IN OPEN CHANNELS

As mentioned in Chapter 13, a surge represents an abrupt change in the depth of flow caused by regulation of a sluice gate in a channel. A surge is positive if the depth is increased after the change in gate position and negative if it is decreased. The surge can move either upstream or downstream (Fig. 13.9). Surge problems can be solved by making the surge stationary by applying equal and opposite velocity of surge to the surge and flow. A frictionless positive surge is analysed in this manner. A frictionless negative surge propagates as a series of elementary waves superimposed on the existing velocity, with each wave travelling at a smaller speed than the one at the next greater depth. Here momentum and continuity equations yield a differential equation for $\frac{dU}{dy}$, the solution of which yields the relation between U and y (Examples 17.10 and 17.11).

17.5 OSCILLATIONS IN U-TUBE AND SURGE TANKS

When friction is neglected, the oscillations of liquid in a U-tube of constant cross-section can be represented by the differential equation

$$\frac{d^2S}{dt^2} = \frac{dU}{dt} = -\frac{2gS}{l} \tag{17.8}$$

(Fig. 17.3). Here U is the velocity of fall or rise of the liquid level. With

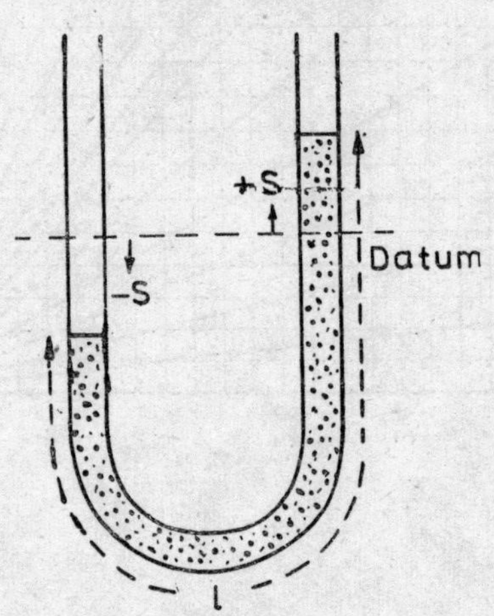

Fig. 17.3. Oscillations in a *U*-tube.

the boundary condition $S = S_{max}$ and $\frac{dS}{dt} = 0$ at $t = 0$, solution of Eq. 17.8 is

$$S = S_{max} \cos\left(\sqrt{\frac{2g}{l}}\, t\right) \tag{17.9}$$

Thus it executes a simple harmonic motion with period of oscillation T,

$$T = 2\pi \sqrt{\frac{l}{2g}} \tag{17.10}$$

Equations (17.9) and (17.10) are applicable to two interconnected reservoirs of area A_1 and A_2 shown in Fig. 17.4 if some simplifying assumptions are made, viz. acceleration of large masses of fluid at either ends is neglected, and also kinetic energy at the two surfaces is neglected. Then Eqs. 17.9 and 17.10 can be used if one replaces $2g/l$ by $\frac{g}{l}\left\{\frac{A\,(A_1 + A_2)}{A_1 A_2}\right\}$ where A is the cross-sectional area of connecting tube.

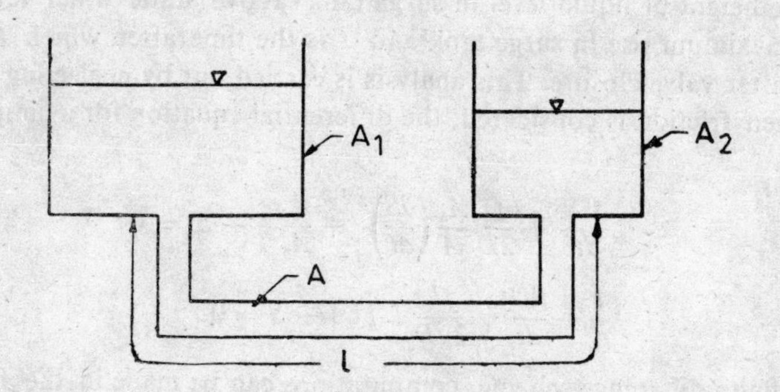

Fig. 17.4. Oscillation in two-reservoirs.

A simple surge tank is shown in Fig. 17.5. The surge tank serves the following purposes

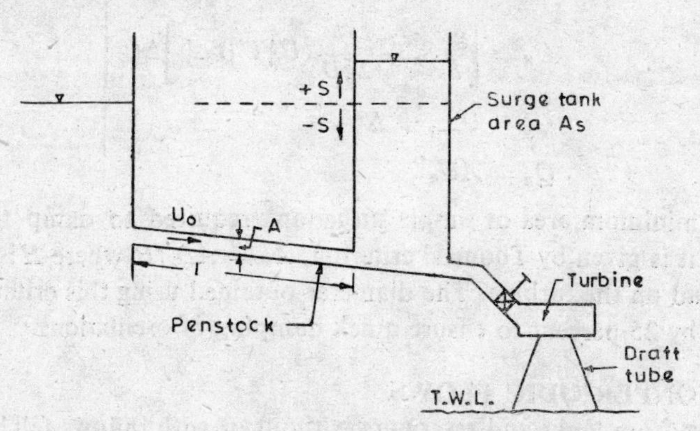

Fig. 17.5. Simple surge tank.

1. It supplies additional water to turbine when the demand suddenly increases, until the water in the pipe is accelerated. Similarly, when the demand is decreased, it stores excess flow.
2. The length of pipe, in which effects of excess pressure due to the sudden closure are felt, is reduced.

Oscillations in a surge tank are relatively slow and hence are analysed assuming liquid to be incompressible. If the valve is closed suddenly,

$$S_{\max} = U_0 \sqrt{\frac{A}{A_s} \frac{l}{g}}$$

$$t = \sqrt{\frac{A_s}{A} \frac{l}{g}} \sin^{-1}\left(\frac{S}{S_{\max}}\right) \tag{17.11}$$

and

$$T = \frac{\pi}{2} \sqrt{\frac{A_s}{A} \frac{l}{g}}$$

Here A and A_s are the areas of pipe and surge tank, l is the length of pipe, S is the height of liquid level in surge tank above static water level, $S_{\max}$ is the maximum rise in surge tank and T is the time after which $S_{\max}$ will occur after valve closure. This analysis is carried out by neglecting friction.

When friction is considered, the differential equation for a simple surge tank is

$$\frac{d^2S}{dt^2} + \frac{f}{2D} \frac{A_s}{A} \left(\frac{dS}{dt}\right)^2 + \frac{A}{A_s} \frac{g}{l} - S = 0 \tag{17.12}$$

or

$$\frac{l}{g} - \frac{dU}{dt} + \frac{fl}{2gD} U|U| + S = 0$$

Using finite difference scheme, computations can be made in the following manner

$$\left.\begin{aligned}
\Delta S_n &= \frac{A}{A_s} S_{n-1} \, \Delta t \\
S_n &= S_{n-1} - \Delta S_n \\
U_n &= \left[\frac{g}{l} S_n - \frac{fl}{2gD} (U|U|)_{n-1}\right] \Delta t \\
U_n &= U_{n-1} + \Delta U_n \\
Q_n &= AU_n
\end{aligned}\right] \tag{17.13}$$

The minimum area of simple surge tank required to damp the oscillations in it is given by Thomas' criterion, $A_s = AD/fH$ where H is the steady state head on the turbine. The diameter obtained using this criterion is increased by 25 percent to ensure quick damping of oscillations.

17.6 NONPERIODIC FLOWS

Flows from tanks and reservoirs without or with inflows fall in this category. Consider a tank of variable area A from which water comes out

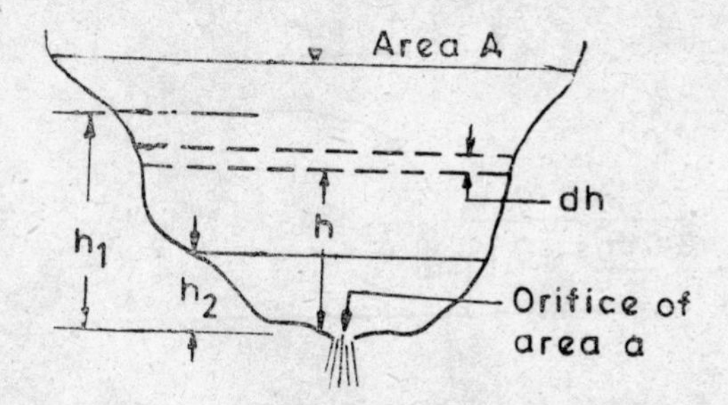

Fig. 17.6. Definition sketch.

through an orifice of area a and discharge coefficient C_d (See Fig. 17.6). Continuity equation gives,

$$-A dh = C_d a\sqrt{2gh}\, dt$$

and $$(t_2 - t_1) = \frac{-1}{C_d a\sqrt{2g}} \int_{h_1}^{h_2} A h^{-1/2}\, dh$$

If A can be expressed as a mathematical function of h, the above integration can be performed and the time required for liquid level to fall from h_1 and h_2 determined. Otherwise integration has to be done graphically.

ILLUSTRATIVE EXAMPLES

17.1 When a body moves through still water at a constant speed of 5.0 m/s, the velocity of water 0.5 m in front of the nose of the body is 3.0 m/s. What will be the difference of pressure between the nose and a point 0.50 m ahead of it?

This is a problem of unsteady flow. However, since the body is moving at a constant speed of 5.0 m/s, apply a velocity of 5.0 m/s in reverse direction to water and body and change the problem to a problem of steady motion.

Hence velocity at nose $= (5.0 - 5.0) = 0$ m/s

Velocity at 0.5 m away from nose $= (3.0 - 5.0)$

$= 2.0$ m/s towards nose.

$$\Delta p = \rho U^2/2 = \frac{998 \times 2^2}{2} = 1996 \text{ N/m}^2$$

17.2 A hydraulic jump formed in a 5.0 m wide rectangular channel carries a discharge of 30 m³/s with upstream depth of 1.0 m. If the hydraulic jump is found to travel downstream at a velocity of 1.0 m/s, determine the depth and velocity after the jump (Fig. 17.7).

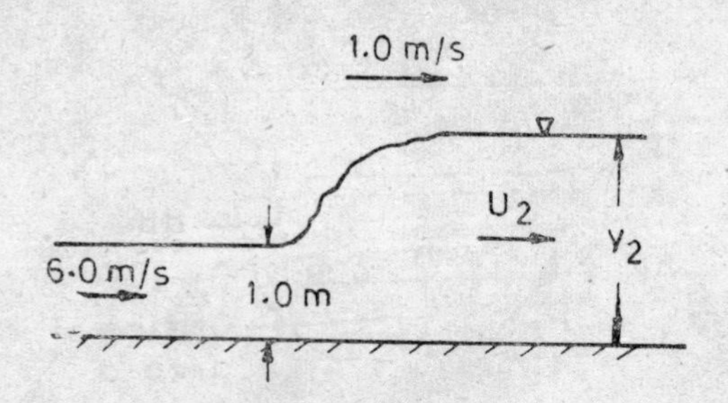

Fig. 17.7. Moving hydraulic jump.

$$U_1 = 30/(5 \times 1) = 6.0 \text{ m/s}$$

The hydraulic jump is made stationary by applying a velocity of U_w towards left of the water and jump. The resulting velocities at sections 1 and 2 are $(U_1 - U_w)$ and $(U_2 - U_w)$.

Writing in general terms, the continuity equation becomes

$$y_1(U_1 - U_w) = y_2(U_2 - U_w)$$

The momentum equation should consider unit discharge under steady conditions, i.e. $y_2(U_2 - U_w)$. The equation will be

$$\rho y_2(U_2 - U_w)\,[(U_1 - U_w) - (U_2 - U_w)] = \tfrac{1}{2}\gamma(y_2^2 - y_1^2)$$

Substituting the values of y_1, U_1 and U_0 one gets

$$1(6-1) = y_2(U_2 - 1) \quad \text{or} \quad U_2 = \left(\frac{5}{y_2} + 1\right)$$

and $$y_2(U_2 - 1)(6 - U_2) = \frac{9.806}{2}(y_2^2 - 1^2)$$

Substituting the value of U_2 in this equation, one gets

$$5\left(5 - \frac{5}{y_2}\right) = \frac{9.806}{2}(y_2^2 - 1)$$

or $$y_2^3 - 6.099y_2 + 1 = 0$$

Solution of this equation yields $\qquad y_2 = 2.384$ m

$$\therefore \quad U_2 = \left(\frac{5}{2.384} + 1\right) = 3.097 \text{ m/s}$$

17.3 A tank of volume 57 *l* contains air at 700 kN/m², density 6.0 kg/m³, and temperature of 15°C. If air escapes through an opening of 60 mm² area at a velocity of 400 m/s, determine the instantaneous rate of change of density in the tank at $t = 0$, assuming that properties in the rest of the tank remain uniform.

Applying continuity equation to the control volume shown in Fig. 17.8,

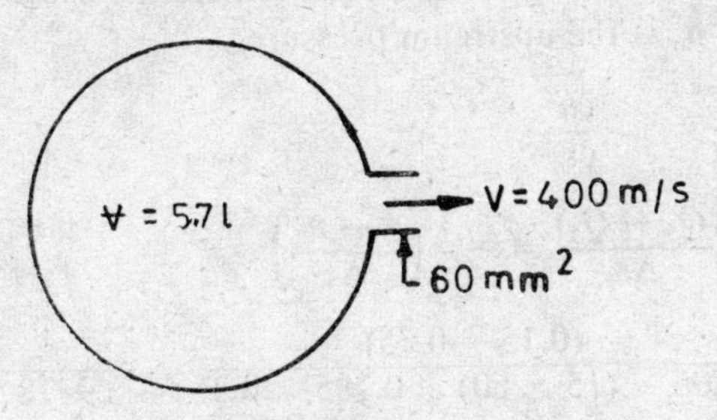

Fig. 17.8.

$$(a\rho V) + \frac{\partial}{\partial t}(\forall \rho) = 0$$

or
$$\frac{\partial \rho}{\partial t} = -\frac{a\rho V}{\forall}$$

$$\therefore \quad \left(\frac{\partial \rho}{\partial t}\right)_{t=0} = -\frac{(60 \times 10^{-6}) \times 6.0 \times 400}{57 \times 10^{-3}}$$

$$= -2.526 \; \frac{\text{kg/m}^3}{\text{s}}$$

Hence the density will decrease at this rate.

17.4 During a flood in a stream of width 500 m, the depth of flow at the gauging station was found to increase at the rate of 0.10 m per hour. What will be the difference in discharges at two sections 0.70 km apart at that time?

Continuity equation for open channels of any cross section is

$$\frac{\partial A}{\partial t} + \frac{\partial Q}{\partial x} = 0$$

If the channel is assumed to be rectangular $A = Bh$.

Hence the above equation can be written as

$$B\frac{\Delta h}{\Delta t} = -\frac{\Delta Q}{\Delta x} \quad \text{or} \quad \Delta Q = -\frac{B \times \Delta h \times \Delta x}{\Delta t}$$

$$\therefore \quad \Delta Q = -\frac{500 \times 0.10}{3600} \times 0.7 \times 1000$$

$$= -9.722 \text{ m}^3/\text{s}$$

The discharge will be less at the downstream section by 9.722 m³/s.

17.5 A pipe of 0.30 m diameter carries water at a discharge of 0.25 m³/s. The valve at the down stream end is closed uniformly at such a rate that discharge is changed to 0.15 m³/s in 15 minutes. Neglecting the effects of compressibility and viscosity, determine the difference of pressure between gate and a section 50 m upstream of it.

For such a slow closure, Euler's equation can be used. If p is the pressure at gate and p_0 is the upstream pressure,

$$\frac{1}{g}\frac{\partial U}{\partial t} = -\frac{\partial h}{\partial s}$$

or

$$\frac{1}{gA}\frac{(Q_2 - Q_1)}{\Delta t} = -\frac{1}{\gamma}\left(\frac{p - p_0}{\Delta s}\right)$$

$$\frac{(0.15 - 0.25)}{9.806 \times (15 \times 60) \times 0.785 \times 0.3^2} = -\frac{1}{9.787}\frac{(p - p_0)}{50}$$

$\therefore$ $$(p - p_0) = 0.078\ 48\ \text{kN/m}^2$$
$$= 78.48\ \text{N/m}^2$$

Thus the pressure at the gate will be 78.45 N/m² higher than the pressure at a section 50 m upstream. Note that this analysis will not be valid if the valve were closed very rapidly. In such a case, effect of compressibility will have to be taken into account.

17.6 A 0.25 m diameter pipe of 50 m length is connected to a reservoir which provides a constant head of 3.0m. The other end of the pipe has a quick acting valve. If the valve is suddenly opened fully, determine the time required for the water in the pipe to attain 90 percent of ultimate velocity. Neglect friction and compressibility effects.

If U_0 is the ultimate velocity, $U_0 = \sqrt{2gh}$ or $U_0 = \sqrt{2 \times 9.806 \times 3}$
$= 7.67$ m/s

$$t = \frac{l}{\sqrt{2gh}} \ln\left(\frac{1 + \dfrac{U}{U_0}}{1 - \dfrac{U}{U_0}}\right) \quad \text{and} \quad \frac{U}{U_0} = 0.9$$

$$= \frac{50}{7.67} \log_e \frac{1.9}{0.1}$$

$$= 19.195\ \text{s.}$$

17.7 Compare the velocity of propagation of a pressure wave in 300 mm diameter pipe of 10 mm thickness if it is made of mild steel and if it carries (i) water, and (ii) ethyl alcohol. What will be the corresponding velocities if the pipe is considered to be rigid?

Take $E_m = 2.47 \times 10^{11}$ N/m² for steel while E for water and ethyl alcohol are 2.075×10^9 N/m² and 1.118×10^9 respectively Take mass density of alcohol as 789 kg/m³.

When pipe acts as rigid $C = \sqrt{E/\rho}$. Hence

For water $$C = \sqrt{\frac{2.075 \times 10^9}{998}} = 1441.93\ \text{m/s}$$

For alcohol $\quad C = \sqrt{1.118 \times 10^9/789} = 1190.37 \text{ m/s}$

When pipe elasticity is to be taken into account,
For water:

$$K = \frac{E}{1+\dfrac{D}{t}\dfrac{E}{E_m}} = \frac{2.075 \times 10^9}{1+\dfrac{300}{10}\times\dfrac{2.075\times 10^9}{2.47\times 10^{11}}} = 1.657 \times 10^9 \text{ N/m}^2$$

Hence $\quad C = \sqrt{1.657 \times 10^9/998} = 1285.53 \text{ m/s}$

For alcohol:

$$K = \frac{1.118 \times 10^9}{1+\dfrac{300}{10}\times\dfrac{1.118\times 10^9}{2.47\times 10^{11}}} = 0.9843 \times 10^9 \text{ N/m}^2$$

Hence $\quad C = \sqrt{0.9843 \times 10^9/789} = 1116.93 \text{ m/s}.$

17.8 A 300 mm diameter pipe of mild steel having 6 mm thickness carries water at the rate of 200 l/s. What will be the rise in pressure if the valve at the downstream end is closed instantaneously? Compare results assuming the pipe to be rigid as well as elastic. What should be the maximum closing time for the computed results to be valid? Take pipe length as 8.0 km.

If pipe is rigid $\quad C = \sqrt{E/\rho} = \sqrt{2.075 \times 10^9/998}$

$$= 1441.93 \text{ m/s}$$

Average velocity of flow in pipe, $U = 0.200/(0.785 \times 0.3^2)$

$$= 2.831 \text{ m/s}$$

$$\therefore \quad \Delta p = -\rho C U = 998 \times 1441.93 \times 2.831$$

$$= 4073\,939.62 \text{ N/m}^2 \quad \text{or} \quad 4073.94 \text{ kN/m}^2$$

If the pipe is elastic

$$K = \frac{2.075 \times 10^9}{1+\dfrac{300}{6}\times\dfrac{2.075\times 10^9}{2.47\times 10^{11}}} = 1.461 \times 10^9 \text{ N/m}^2$$

$$\therefore \quad C = \sqrt{1.461 \times 10^9/998} = 1209.93 \text{ m/s}$$

$$\therefore \quad \Delta p = -\rho C U = 998 \times 1209.93 \times 2.631$$

$$= 3418\,458.0 \text{ N/m}^2 \quad \text{or} \quad 3418.46 \text{ kN/m}^2$$

If the valve is considered to be closed suddenly, the closure time should be less than the time required for the wave to travel to the other end and to return the valve. Hence $t_c \leqslant \dfrac{2l}{C}$.

$\therefore$ If pipe is rigid $t_c = \dfrac{2 \times 8 \times 1000}{1441.93} = 11.096$ s

If pipe is elastic $t_c = \dfrac{2 \times 8 \times 1000}{1209.93} = 13.224$ s

17.9 A 2.0 m diameter steel penstock of 15 mm thickness carries water under a head of 50 m. When the penstock carries a water discharge of 18.0 m³/s, the valve at the downstream end of 1.0 km long penstock is closed completely in 35 s. Determine the maximum rise in pressure due to water hammer using Allievi's charts.

Since $E_m = 2.47 \times 10^{11}$ N/m² and $E = 2.075 \times 10^9$ N/m²

$$K = \frac{2.075 \times 10^9}{1 + \dfrac{2000}{15} \times \dfrac{2.075 \times 10^9}{2.47 \times 10^{11}}} = 0.9787 \times 10^9 \text{ N/m}^2$$

and $C = \sqrt{K/\rho} = \sqrt{0.9787 \times 10^9/998} = 990.28$ m/s

$$U = \frac{4Q}{\pi D^2} = \frac{18.00 \times 4}{3.142 \times 2^2} = 5.732 \text{ m/s}$$

$$\therefore \quad \alpha = \frac{UC}{2gH} = \frac{5.732 \times 990.28}{2 \times 9.806 \times 50} = 5.788$$

$$\beta = \frac{Ct_c}{2l} = \frac{990.29 \times 35}{2 \times 1000} = 17.330$$

From Fig. 17.1, for $\alpha = 5.788$ and $\beta = 17.330$

$$Z^2 = 1.39 = \frac{H + h}{H} \text{ or } \left(1 + \frac{h}{H}\right)$$

$\therefore$ Rise in pressure head $= 50\,(1.39 - 1) = 19.5$ m

Since the point (17.33, 5.788) falls close to S_{15}, maximum pressure will be reached $15 \times \dfrac{2l}{C} = \dfrac{15 \times 2 \times 1000}{990.28}$ i.e. 30.29 seconds after closure begins.

17.10 Analyse the frictionless negative surge in the downstream channel caused by the partial closure of the gate at the upstream.

Let U_w be the velocity of surge, dy its height, and U and y the initial velocity and depth in wide rectangular channel. Further assume $(U - dU)$ be velocity under the surge. Make the surge stationary (Fig. 17.9). Hence continuity equation gives

$$(U - dU - U_w)\,(y - dy) = y\,(U - U_w)$$

or

$$(U_w - U)\,dy = y dU$$

or

$$dU/dy = (U_w - U)/y \qquad (1)$$

after neglecting the second order term of $(dy.dU)$. Momentum equation yields

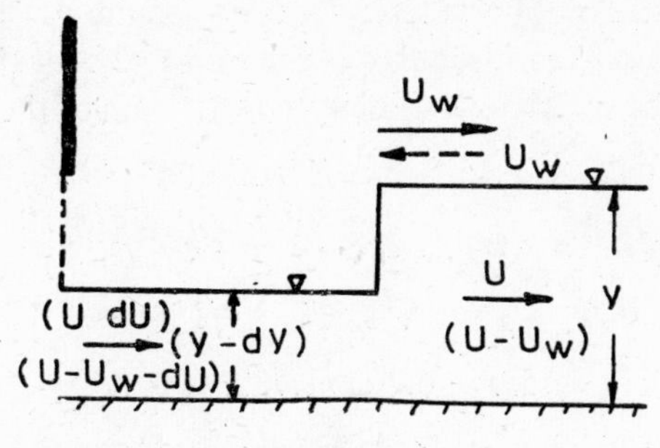

Fig. 17.9.

$$\frac{1}{2}\gamma (y - dy)^2 - \frac{1}{2}\gamma y^2 = \rho (U - U_w)\, y\, [(U - U_w) - (U - U_w - dU)]$$

which gives on simplification

$$dy = \frac{U_w - U}{g}\, dU$$

or $$dU/dy = g/(U_w - U) \qquad (2)$$

Equating values of dU/dy from (1) and (2), one gets

$$\frac{(U_w - U)}{y} = \frac{g}{(U_w - U)} \quad \text{or} \quad U_w = U \pm \sqrt{gy} \qquad (3)$$

Equating the value of U_w from Eqs. 2 and 3, one gets

$$U \pm \sqrt{gy} = g\frac{dy}{dU} + U \quad \text{or} \quad \frac{dU}{dy} = \pm\sqrt{g/y}$$

which on integration yields $U = \pm 2\sqrt{gy} + C_1$

where C_1 is the constant of integration. It can be evaluated from the condition that at the time of partial instantaneous closure of gate,

$$U = U_0 \quad \text{and} \quad y = y_0$$

$$\therefore \qquad C_1 = U_0 + 2\sqrt{gy_0}$$

and $$U = U_0 - 2\sqrt{g}\,(y_0^{1/2} - y^{1/2})$$

Since the wave is moving in downstream direction

$$U_w = U + \sqrt{gy} = U_0 - 2\sqrt{gy_0} + 3\sqrt{gy} \qquad (4)$$

Position of the wave from the gate is given by $x = U_w t$

$$\therefore \qquad x = [U_0 - \sqrt{2gy_0} + 3\sqrt{gy}]\, t \qquad (5)$$

If y is eliminated from Eq. 4 and 5, one gets an expression for U as

$$U = \frac{U_0}{3} + \frac{2}{3}\frac{x}{t} - \frac{2}{3}\sqrt{gy_0} \qquad (6)$$

17.11 Downstream of a gate $U_0 = 5.0$ m/s and $y_0 = 3.0$ m/s. By partial closure of the gate the discharge is reduced to 2/3 of the original discharge. What will be the depth and velocity downstream of the gate. Also find U_w.

Original discharge $= 5 \times 3 = 15$ m³/s m

New discharge $= U_1 y_1 = \frac{2}{3} \times 15 = 10$ m³/s m

Also, since $\quad U_1 = U_0 - 2\sqrt{g}\,(y_0^{1/2} - y_1^{1/2})$

$$= 5.0 - 6.263\,(1.732 - \sqrt{y_1})$$

Substituting the value of y_1 from $U_1 y_1 = 10$ in the second equation, one gets

$$U_1 = 5.0 - 6.263\left(1.732 - \frac{3.162}{\sqrt{U_1}}\right)$$

Solving this equation by trial and error one obtains

$$U_1 = 4.025 \text{ m} \qquad y_1 = 2.484 \text{ m}$$

$$U_w = U_1 + \sqrt{gy_1} = 4.025 + \sqrt{9.806 \times 2.484}$$

$$= 8.960 \text{ m/s}$$

17.12 A simple surge tank of 20 m diameter is connected to a reservoir through penstock of 3 m diameter and 5.0 km length. If the turbine valve is suddenly closed when the penstock was carrying 70 m³/s discharge, determine the maximum rise in surge tank and its period of oscillation.

$$A = 0.785 \times 3^2 = 7.065 \text{ m}^2$$

$$A_s = 0.785 \times 20^2 = 314.00 \text{ m}^2$$

$$L = 5000 \text{ m}$$

$$U_0 = 70/7.065 = 9.908 \text{ m/s}$$

$$\therefore \qquad S_{max} = U_0 \sqrt{\frac{AL}{A_s g}}$$

$$= 9.908\sqrt{(7.065 \times 5000)/(314 \times 9.806)}$$

$$= 33.56 \text{ m}$$

$$t_{max} = \frac{\pi}{2}\sqrt{A_s L/Ag} = \frac{3.142}{2} \times \sqrt{\frac{314 \times 5000}{7.075 \times 9.806}}$$

$$= 236.5 \text{ s}$$

$$\therefore \qquad T = 4\, t_{max} = 946.0 \text{ s.}$$

17.13 For a simple surge tank, $f = 0.02$, $D = 5.0$ m, $L = 10$ km, surge tank diameter $= 35$ m, $Q = 100$ m³/s. Obtain variation of s and Q with time when the valve is suddenly closed. Take $\Delta t = 50$ s.

$$A = 0.785 \times 5^2 = 19.625 \text{ m}^2, \quad A_s = 0.785 \times 35^2 = 961.625 \text{ m}^2$$

$$U_0 = 100/19.625 = 5.096 \text{ m/s}$$

$$\text{Head loss} = \frac{fLU^2}{2gD} = \frac{0.02 \times 1000 \times 5.096^2}{2 \times 9.806 \times 5} = 52.96 \text{ m} = S_0$$

Measure S from reservoir level as +ve downwards. Computations can be performed in tabular form using steps indicated in Section 17.5.

Time Step	t (s)	U (m/s)	Δs (m)	S (m)	$\frac{flU^2}{2gD}$ (m)	ΔU (m/s)	Q (m³/s)
0	0	5.096		52.96			100
1	50	4.816	5.20	47.76	47.32	−0.250	94.51
2	100	4.596	4.91	42.85	43.09	−0.220	90.20
3	150	4.356	4.69	38.16	38.71	−0.240	85.49
4	200	4.096	4.44	33.72	34.22	−0.240	80.38
5	250	3.866	4.18	29.54	30.49	−0.230	75.88
6	300	3.426	3.94	25.60	23.94	−0.240	67.24
7	350	3.336	3.49	22.11	22.70	−0.090	65.47
8	400	3.091	3.40	17.71	19.49	−0.245	60.66
9	450	2.849	3.15	14.56	16.56	−0.242	55.91

Similar calculations can be performed for larger time intervals. It can then be seen that S and Q will reach zero and then become negative. These oscillations will continue for some time and then both will tend to zero.

17.14 A tank in the form of paraboloid of revolution of radius R at the top and height H is completely filled with water and drained through an orifice of area a and discharge coefficient C_d at its bottom (Fig. 17.10). Determine the time required to drain the tank completely.

From geometry it is known that $r^2/R^2 = h/H$

If h is the depth of water at any time t and it falls through dh in time dt, then according to continuity equation

$$-\pi r^2 dh = C_d a\sqrt{2gh}\, dt$$

The negative sign indicates that as t increases h decreases.

$$\therefore \quad dt = \frac{-\pi\left(\frac{R}{H}\right)^2 h^{1/2}\, dh}{C_d\, a\sqrt{2g}}$$

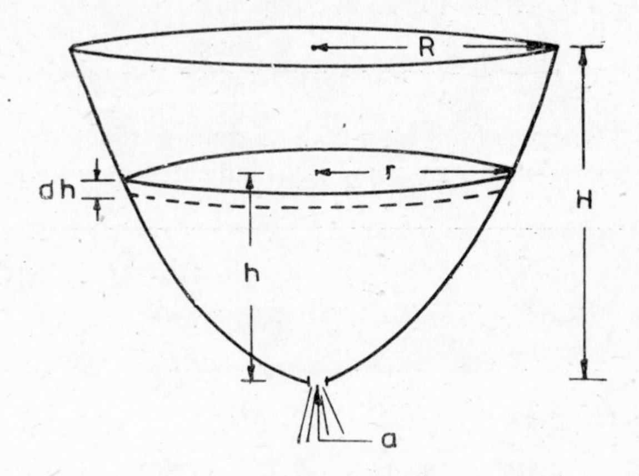

Fig. 17.10.

$$\therefore \quad \text{On integration} \quad t = \frac{-\pi R^2}{H\, C_d a \sqrt{2g}} \int_H^0 h^{1/2}\, dh$$

$$\therefore \quad t = \frac{2\pi R^2 H^{1/2}}{3C_d a\, H\sqrt{2g}} = \frac{2\pi R^2 H^{1/2}}{3C_d a\sqrt{2g}}$$

17.15 Two tanks of cross-sectional areas A_1 and A_2 are interconnected by means of an opening of area a and discharge coefficient C_d. If h_1 is the difference in the height of water levels at the beginning, water level in reservoir 1 being higher, determine the time required to reduce the difference in water levels to h_2 (Fig. 17.11).

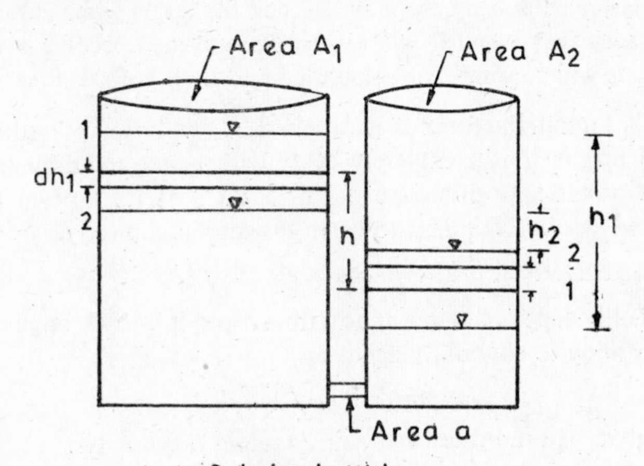

Fig. 17.11. Interconnected reservoirs.

Let h be the water level difference in two tanks at time t, and it falls through dh_1 in time dt in tank 1. Corresponding rise in tank 2 will be A_1dh_1/A_2 and net difference in water levels in two tanks will be

$$\left[h - dh_1\left(1 + \frac{A_1}{A_2}\right)\right]$$

$$\therefore \qquad dh = h - \left[h - dh_1\left(1 + \frac{A_1}{A_2}\right)\right] = dh_1\left(1 + \frac{A_1}{A_2}\right)$$

Also according to continuity equation

$$C_d\, a\sqrt{2gh}\, dt = -A_1 dh_1$$

$$\therefore \qquad dt = \frac{-A_1\, dh}{\left(1 + \frac{A_1}{A_2}\right) C_d a\sqrt{2g}\, h^{1/2}}$$

which on integration yields

$$T = (t_2 - t_1) = \frac{2A_1\,(h_1^{1/2} - h_2^{1/2})}{\left(1 + \frac{A_1}{A_2}\right) C_d a\sqrt{2g}}$$

17.16 A cylindrical tank of area A is supplied water through a pipe of diameter D and length l having friction factor f. The pump at the other end of the pipe maintains a constant pressure equivalent to head H at its upstream end. Determine the time required to fill the tank from depth h_1 to depth h_2 (Fig. 17.12).

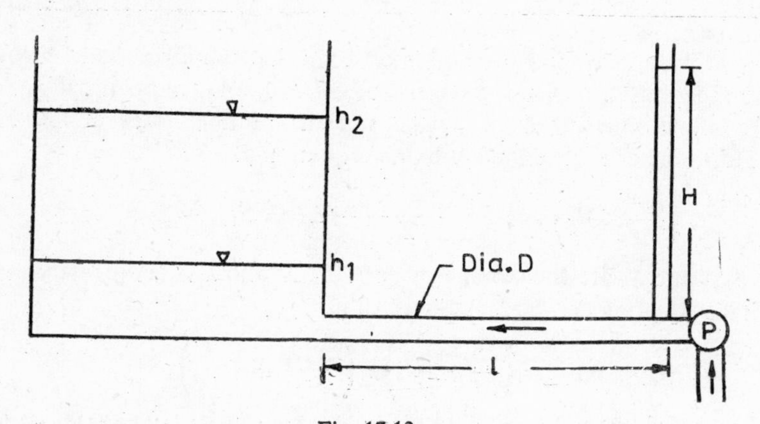

Fig. 17.12.

When head in the tank is h

$$H = h + \frac{flU^2}{2gD}$$

$$= h + \frac{8flQ^2}{\pi^2 gD^5}$$

$$\therefore \quad Q = \sqrt{\frac{\pi^2 gD^5}{8fl}(H-h)} = \sqrt{(KH - Kh)}$$

where $\quad K = \frac{\pi^2 gD^5}{8fl}.$

Also according to continuity equation

$Qdt = Adh$ where dh is the rise in water level in the tank in time dt.

$$\therefore \quad dt = \frac{A}{\sqrt{K}}\frac{dh}{\sqrt{H-h}} \quad \text{or} \quad t_2 - t_1 = \frac{2A}{\sqrt{K}}(\sqrt{h-h})_{h_1}^{h_2}$$

$$\therefore \quad T = (t_2 - t_1) = \frac{2A}{\sqrt{K}}[(H-h_2)^{1/2} - (H-h_1)^{1/2}]$$

17.17 A cylindrical tank of cross-sectional area 10 m² has an orifice at its bottom whose discharge equation is $Q = 0.60\ h^{1/2}$ where h is the head above tank bottom in m and Q is in m³/s. Under equilibrium condition, it carried a discharge of 0.60 m³/s under a head of 1.0 m and equal discharge was flowing in the tank. If the inflow I as a function of time (known as inflow hydrograph) is as given below, determine the variation of outflow with time (known as outflow hydrograph) and variation of head in the tank as a function of time.

t (s)	0	5	10	15	20	25	30	35	40	45
$I\left(\frac{m^3}{s}\right)$	0.60	1.20	1.60	1.40	1.00	0.60	0.30	0	0	0

This problem is similar to routing of flood through a reservoir having uncontrolled outlet. Hence a general method is developed below. If I_n and I_{n+1} are inflow rates and O_n and O_{n+1} are the outflow rates at $n\Delta t$ and $(n+1)\,\Delta t$, one can write the continuity equation as

$$\left(\frac{I_n + I_{n+1}}{2}\right)\Delta t - \frac{(O_n + O_{n+1})}{2}\Delta t = S_{n+1} - S_n$$

where S_{n+1} and S_n are the storages at $(n+1)\,\Delta t$ and $n\Delta t$ in the reservoir or tank. This equation can be rearranged to give

$$(I_n + I_{n+1}) + \left(\frac{2S_n}{\Delta t} - O_n\right) = \left(\frac{2S_{n+1}}{\Delta t} + O_{n+1}\right)$$

At the beginning of time step, one knows I_n, I_{n+1}, S_n, O_n and Δt. Hence left hand side of above equation is known and therefore $\left(\frac{2S_{n+1}}{\Delta t} + O_{n+1}\right)$. Realising that storage S and outflow O are unique functions of head h, one can beforehand prepare a graph of $\left(\frac{2S}{\Delta t} + 0\right)$ as a function of h. Hence knowing $\left(\frac{2S_{n+1}}{\Delta t} + O_{n+1}\right)$, one can determine h_{n+1}. The computations can now be done for the next time step. For a reservoir with odd shape, one needs to know $I(t)$, $O(h)$, $S(h)$ and the time step Δt along with the graph of $\left(\frac{2S}{\Delta t}+O\right)$ vs h. The computations are performed in tabular form as shown in Table 17.1.

For cylindrical reservoir of surface area 10 m², $\Delta t = 5.0$ s and outflow $O = 0.6h^{1/2}$, one can compute $\left(\frac{2S}{\Delta t}+O\right)$ for various values of h. These are plotted in Fig. 17.13.

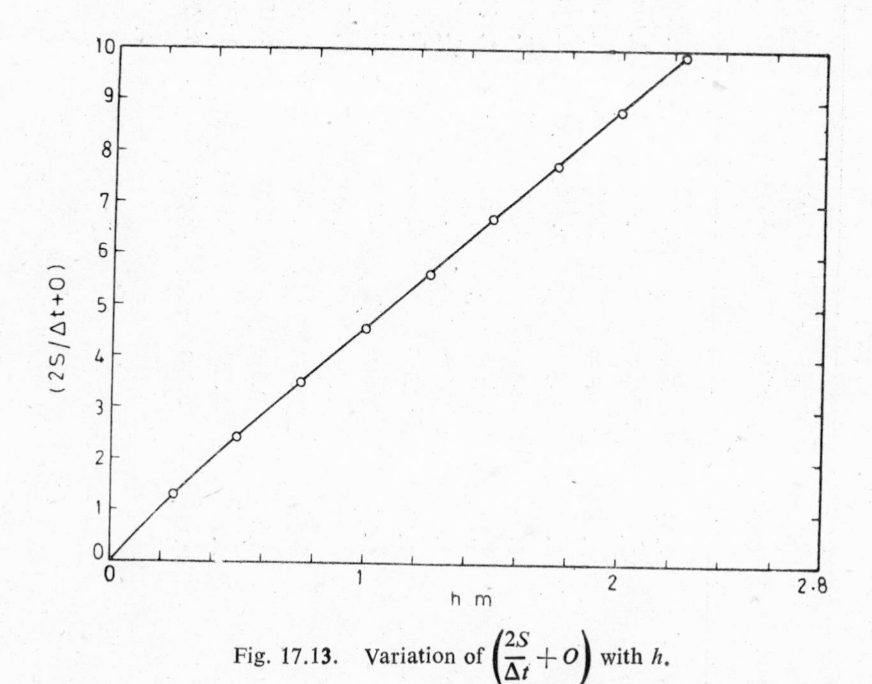

Fig. 17.13. Variation of $\left(\frac{2S}{\Delta t}+O\right)$ with h.

Figure 17.14 shows the plot of I and O as a function of time. It can be seen that the peak of the outflow hydrograph is considerably reduced; thus the downstream channel will be subject to a less severe flood. The duration of outflow, however, is much larger. This is basically the function of a flood control reservoir.

17.18 A broad crested weir is provided at the downstream end of a 10 m wide and 1.0 km long rectangular channel. At a certain instant, the water level upstream is 3.0 m above the weir crest. Determine the time required to lower it to 1.0 m if the discharge over the broad crested weir is given by $Q = 17h^{3/2}$. Assume there is no inflow.

$$\text{Area of channel} = 1000 \times 10 = 10\,000 \text{ m}^2$$

$$-10{,}000\, dh = 17\, h^{3/2}\, dt$$

Table 17.1. Computations for reservoir routing in Example 17.17

n	t (s)	I_n (m^3/s)	$(I_n + I_{n+1})$	h_n (m)	O_n (m^3/s)	$2S_n/\Delta t$ (m^3/s)	$\frac{2S_n}{\Delta t} - O_n$	$\frac{2S_{n+1}}{\Delta t} + O_{n+1}$
1	0	0.60	1.80	1.00	0.600	4.000	3.400	5.200
2	5	1.20	2.80	1.14	0.641	4.560	3.919	6.719
3	10	1.60	3.00	1.48	0.730	5.920	5.190	8.190
4	15	1.40	2.40	1.85	0.816	7.400	6.550	8.950
5	20	1.00	1.60	2.02	0.853	8.080	7.227	8.827
6	25	0.60	0.90	2.00	0.849	8.000	7.151	8.051
7	30	0.30	0.30	1.82	0.809	7.280	6.471	6.771
8	35	0	0	1.50	0.735	6.000	5.265	5.265
9	40	0	0	1.15	0.643	4.600	3.957	3.957
10	45	0	0	0.85	0.553	3.400	2.847	2.847
11	50	0	0	0.59	0.461	2.360	1.899	1.899
12	55	0	0	0.38	0.370	1.520	1.150	1.150
13	60	0	0	0.22	0.281	0.880	0.599	0.599
14	65	0	0	0.12	0.209	0.480	0.271	0.271

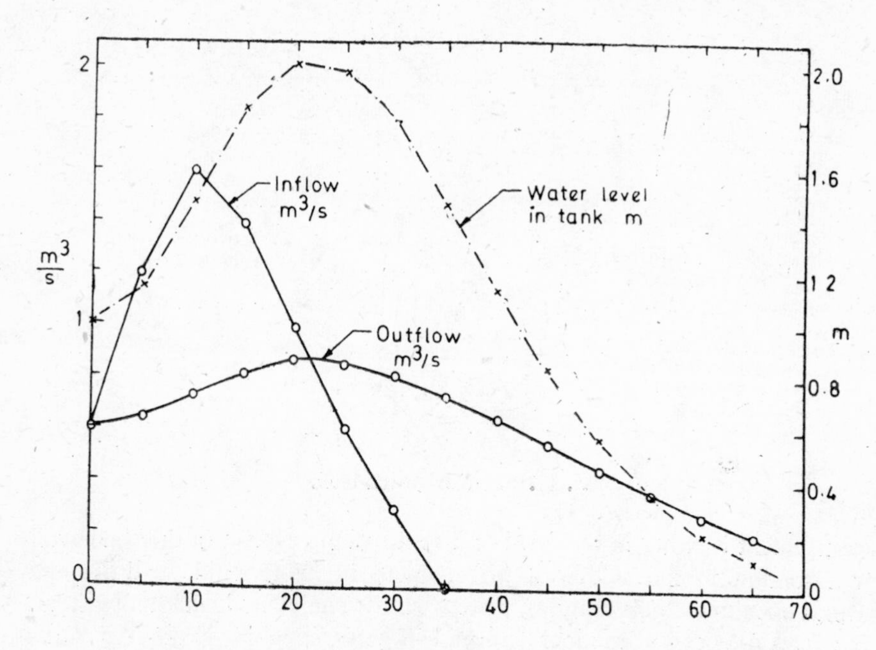

Fig. 17.14. Variation of inflow, outflow and water level with time.

or $$dt = -\frac{10\,000}{17}\,h^{-3/2}\,dh$$

$\therefore$ $$t_2 - t_1 = \frac{10\,000 \times 2}{17}\left(\frac{1}{1^{1/2}} - \frac{1}{3^{1/2}}\right)$$

$$= \frac{10{,}000 \times 2}{17} \times (1 - 0{\cdot}577) = 497.23 \text{ s}$$

17·19 Describe navigation lock and its functioning.

When a river is obstructed by construction of a dam or barrage and a provision has to be made for the passage of ships and barges from upstream to downstream or vice-versa, a navigation lock is provided. Its essential features are shown in Fig. 17.15.

When a ship has to travel upstream, gates *A* are closed and water in the chamber is at downstream water level. Gates *B* are opened and the ship is allowed inside after which gates *B* are also closed. Water is allowed in the chamber through lock filling arrangement and water level in the chamber is raised to upstream water level. Gates *A* are now opened and the ship is allowed to travel upstream. When a ship has to travel downstream, the chamber is first filled up to upstream water level and the ship is allowed to enter the chamber by opening gates *A*. Now gates *A* are closed and water in the chamber is gradually drained out through lock draining arrangement until the water level inside is the same as the downstream water level. The gates *B* are now opened so that the ship can sail downstream. Lock filling

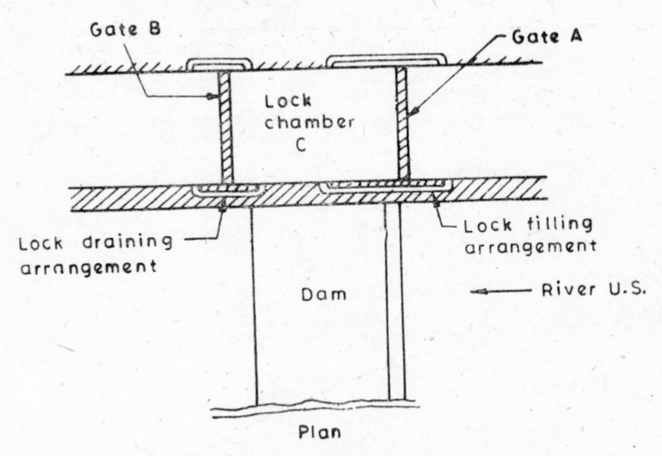

Fig. 17.15. Navigation lock.

arrangement is either in the form of a venturi loop system in the side wall or a conduit in the side wall which opens in the lock side wall through several openings. Lock draining system can be similar to lock filling system. The gates provided at the lock can be leaf gates capable of moving vertically, or tainter or twin leaf gates opening horizontally. The design of filling and emptying system of chamber ensures that water is let in or let out at uniform rate and gradually so that the ship is not subjected to undesirable stresses. With this constraint, the filling and emptying should be rapid enough so that ships are allowed to go through lock without much waiting.

17.20 A flood detention tank of area A is provided with a waste weir of length L whose discharge equation is $Q = KLh^{3/2}$. When the tank is full, a constant discharge Q_0 flows into the tank. Determine the time after which weir will pass the entire discharge downstream after the tank is just full.

Estimate the time when $A = 4.00$ km^2, $Q_0 = 200$ m^3/s and $Q = 17Lh^{3/2}$, L being 100 m.

Discharge coming into the tank in time $dt = Q_0 dt$

Discharge going out over the weir in time $dt = Klh^{3/2}\, dt$

Increase in storage in time $dt = Adh$

$$\therefore \qquad Adh = Q_0 dt - Klh^{3/2}\, dt$$

$$= (Q_0 - Klh^{3/2})\, dt$$

$$\therefore \qquad dt = \frac{Adh}{(Q_0 - Klh^{3/2})}$$

$$\text{and} \qquad t_2 - t_1 = \int_{h_1}^{h_2} \frac{Adh}{(Q_0 - Klh^{3/2})}$$

When all the discharge passes over the waste weir

$$Q_0 = Klh_2^{3/2} \quad \therefore \quad h_2 = \left(\frac{Q_0}{Kl}\right)^{2/3} = \left(\frac{200}{1700}\right)^{2/3} = 0.24 \text{ m}$$

$\therefore$ Time required for water level to rise from 0 to 0.24 m is

$$T = \int_0^{0.24} \frac{4.00 \times 10^6 \, dh}{(200 - 1700h^{3/2})} = \int_0^{0.24} F(h) \, dh$$

h	0	0.05	0.10	0.15	0.20	0.239
$F(h)$	20 000	22 100	27 352	39 510	83 425	2920 615

$\Sigma F(h)\, dh = 1052.5 + 1236.3 + 1671.55 + 3104.3 + 58578.8$

or $\quad = 65643.5$ s 18.23 hrs approx.

PROBLEMS

17.1 The model of a barge was towed in a towing tank at 8.0 m/s velocity. If the velocity at a point 2 m in front of it was 4.0 m/s, determine the pressure difference between the front of barge and the point 2.0 m in front of it. (7984 N/m²)

17.2 Draw stream line pattern for steady irrotational flow past a two dimensional circular cylinder when fluid flows at a constant velocity U. Now apply a velocity U in opposite direction to the fluid and the cylinder. The cylinder will now move at constant speed U. By vector additions of velocities at few points in the flow field, sketch streamline pattern for this unsteady case.

17.3 A hydraulic jump moving in the downstream direction has upstream depth of 2.50 m, upstream velocity of 11.20 m/s and downstream depth of 6.0 m. Determine U_w and U_2. (1.50 m/s, 5.54 m/s)

17.4 For a moving hydraulic jump the following data are given: $y_1 = 1.0$ m, $y_2 = 6.0$ m and $U_w = 1.0$ m/s (in upstream direction). Determine U_1 and U_2. (10 35 m/s, 1.392 m/s)

17.5 Determine the instantaneous rate of change of density of air in a 0.5 m³ tank if air at 1000 kN/m² pressure at 15°C and 12 kg/m³ density comes out through an opening of area 1×10^{-4} m² at a velocity of 350 m/s.

$$\left(-0.84 \frac{\text{kg/m}^3}{\text{s}}\right)$$

17.6 In a rectangular channel of width 1000 m, simultaneous measurements of discharge at two sections one km apart indicated the discharge to be higher at downstram section by 30 m³/s. At what rate is the water level changing there? (0.108 m/hr falling)

16.7 The pressure gradient developed upstream of a valve in a 200 mm

diameter pipe carrying water due to its gradual closure is 2.5 N/m^2 per metre. Determine the rate of decrease of discharge.
(7.866×10^{-5} m^3/s^2)

17.8 A 300 mm diameter pipe of 35 m length is connected to a reservoir which provides a head of 3.5 m. If the valve at the downstream end of the pipe is suddenly opened fully, what will be the velocity in the pipe 10 seconds after the valve is opened? Neglect friction and compressibility effects. (6.865 m/s)

17.9 How much faster will a pressure wave travel in water in a 2 m diameter, 15 mm thick steel pipe if the pipe were rigid than when it was elastic? (450.91 m/s)

17.10 A sonar transmitter used to measure depth of water in sea or a lake gives out 2 impulses per second near the water surface. If 0.30 seconds elapse between the initial impulse and its reflection, what is the depth of flow in the lake? (216.18 m)

17.11 A sonar transmitter is to be used to detect submarines which are within a maximum radius of 10 km. Determine the minimum time interval between two sound impulses if all returning signals are to be received within the first half of the singal period. Take E for water as 2.075×10^9 N/m^2 and $\rho = 1025$ kg/m^3. (14.056 s)

17.12 Water is supplied to the turbines through 3.0 m diameter and 1.5 km long steel penstock of 20 mm thickness. The turbine takes 50 m^3/s discharge under a head of 100 m. What will be the maximum rise in pressure at its dowsntream end if the valve there is closed in 15 seconds? Use Allievi's charts. After what time will it occur after the closure begins? (105 m, 11.6 s)

17.13 A one metre diameter pipe carries water at 5.0 m^3/s over a length 2.5 km. If the valve at its downstream end is closed in 10 seconds, determine the pressure rise at the valve using approximate relationship. Assume pipe to be rigid. (3188.75 kN/m^2)

17.14 In a wide rectangular channel, the depth and velocity downstream of a sluice gate are 5.0 m and 3.0 m/s respectively. If the gate is instantaneously closed partially, determine the depth immediately downstream of the gate and velocity there if the new discharge is 5.0 m^3/s ?

17.15 A U-tube has a water column of length 1.50 m. From one side the liquid is sucked up to a height of 0.30 m above the equilibrium level and released. Determine the period of oscillation and speed when $t = 0.50$ s and the maximum speed.
(1.738 s, 1.054 m/s, 1.084 m/s)

17.16 A simple surge tank of 30 m diameter is connected to a reservoir through 10 km length of 6.0 m diameter tunnel. When the tunnel carries a discharge of 90 m^3/s, determine the maximum rise in water level in the surge tank and time required to attain it if the flow is suddenly stopped. (20.34 m, 250.84 s)

17.17 A circular tank of 2.0 m has a sharp edged orifice of 10 mm diameter at its bottom. Determine the time required to lower the water level from 3.0 m to 1.5 m. Derive the formula you will use. Take $C_d = 0.60$.

$$\left(t = \frac{2A}{C_d a \sqrt{2g}} (h_2^{1/2} - h_1^{1/2}),\ 4.242 \text{ hrs}\right)$$

17.18 In Problem 17.17, show that the time t required to lower the water level from h_2 to h_1 is equal to the volume of water discharged in time t divided by the average of discharges at heads h_2 and h_1.
[Hint: multiply the numerator and denominator of expression for t by $(h_2^{1/2} + h_1^{1/2})$]

17.19 A tank of length L has a semicircular cross-section with radius R. This is filled to a depth h_2. If it discharges through an orifice of area a and discharge coefficient C_d, located at its bottom, determine the time required to lower the water level from h_2 to h_1.

$$\left(t = \frac{4L}{3\, C_d a \sqrt{2g}} [(2R - h_1)^{3/2} - (2R - h_2)^{3/2}]\right)$$

17.20 What will be the time required to lower the water level in a hemispherical tank of radius R from h_2 to h_1 if it is drained through an orifice of area a and discharge coefficient C_d located at its bottom?

$$t = \frac{2\pi}{C_d a \sqrt{2g}} \left[\left(\frac{2R}{3} h_2^{3/2} - \frac{1}{5} h_2^{5/2}\right) - \left(\frac{2R}{3} h_1^{3/2} - \frac{1}{5} h_1^{5/2}\right)\right]$$

17.21 A canal lock with area of 290 m^2 has a water level difference of 3.0 m between upstream and downstream. It is filled with two upstream sluices of 1 m^2 area each whose centres are located 2.0 m below upstream water level and is emptied through two identical sluices of 1 m^2 area each. If discharge coefficient of the sluices is 0.60 determine the time required to empty and fill the lock.
(Hint: It will be emptied at varying head of 3 to 0 m while it will be filled for a depth of 1 m at constant head of 2 m and then at varying head from 2 to 0 m) (189.02 s and 192.9 s)

17.22 A cylindrical tank of constant area A discharges through an orifice at its bottom, the outflow being given by $Q_o = kh^{1/2}$. If there is a constant inflow into the tank at the rate Q_i, show that time T required for water level to rise from h_1 to h_2 is given by

$$T = \frac{2A}{k^2}\left[k(h_1^{1/2} - h_2^{1/2}) - Q_i \ln\left(\frac{Q_i - kh_2^{1/2}}{Q_i - kh_1^{1/2}}\right)\right]$$

17.23 A rectangular tank of 2.25 km^2 surface area is provided with a weir whose discharge equation is $Q = 90\, h^{3/2}$ where h is the head over the weir. When a constant discharge of 20 m^3/s flows into the tank, determine approximately the time required for water level on the weir to rise from 0.20 m to 0.30 m. (Hint: Use graphical integration technique). (8.219 hrs when $dh = 0.005$ m)

17.24 Two rectangular tanks which are connected through an orifice 10 mm diameter have their areas in the ratio of 4:1. If the total area of two tanks is 10 m^2 and C_d for orifice is 0.60, determine the time required for the water levels in the two tanks to equalise when initially the water levels in larger and smaller tank are 4 m and 1 m above the orifice. (5.739 hrs)

17.25 A navigation lock chamber has area A. A sluice of maximum area a_o and discharge coefficient C_d is provided at the base of the lock. To fulfill the condition of uniform filling rate, it is proposed to open the sluice according to the law $a = a_0 t/T$ where t is the time and T is the filling time of the chamber. If h_1 is the constant head under which sluice works, determine the time T_1 required to fill it.

$$\left(T_1 = \frac{4Ah_1^{1/2}}{C_d a_0 \sqrt{2g}}\right)$$

DESCRIPTIVE QUESTIONS

17.1 What category of unsteady flows can be converted into equivalent steady state case? Give two examples.

17.2 One dimensional Euler's equation of motion can be used to predict the pressure changes in a pipe if

(i) friction is neglected
(ii) friction is neglected and acceleration/deceleration is small
(iii) acceleration/deceleration is small
(iv) friction is neglected and acceleration/deceleration is large.

17.3 If $\frac{\partial U}{\partial t}$, for frictionless one dimensional flow in a pipe, is positive,

(i) total energy will increase in the direction of flow
(ii) total energy remains constant
(iii) total energy decreases in the direction of flow.

17.4 Will the pressure wave in a pipe travel faster when the pipe is elastic or when it is rigid? How much change is expected in the velocity of pressure wave?

17.5 Draw pressure vs time diagram at a section along the pipe length in which water hammer occurs.

17.6 What happens to a pressure wave in a pipe line
(i) when it reaches dead end? (ii) when it reaches reservoir?

17.7 What are the functions of a surge tank?

17.8 What requirements should a well-designed surge tank fulfill?

17.9 What will be the effect of friction in penstock on the oscillations in a surge tank?

17.10 Flow is unsteady if

(i) magnitude of velocity changes along the direction of flow,
(ii) magnitude of velocity changes with time at a given point,
(iii) velocity changes in magnitude and/or direction with time at a given point,
(iv) none of the above.

17.11 What functions does a flood control reservoir fulfill?

17.12 Consider a long rectangular channel in which steady uniform flow prevails. If at the upstream end the inflow is in the form of + ve half sine wave superimposed on the existing flow, describe qualitatively the flow conditions in the channel.

17.13 What hydraulic considerations are involved in the design of navigation locks?

Appendix A

Properties of some common fluids at 20°C and atmospheric pressure

Fluid	Mass density ρ kg/m³	Specific weight kN/m³	Dynamic Viscosity μ Poise p	Dynamic Viscosity μ kg/m s	Kinematic Viscosity ν Stoke St	Kinematic Viscosity ν m²/s	Modulus of elasticity E (N/m²)	Surface tension in contact with air σ (N/m)	Vapour pressure (N/m²)
Water	998	9.787	0.010	1.00×10^{-3}	1.00×10^{-2}	1.00×10^{-6}	2.075×10^{9}	0.0736	0.239×10^{4}
Glycerine	1260	12.356	8.350	8.35×10^{-1}	6.63	6.63×10^{-4}	4.354×10^{9}	0.0637	1.373×10^{-2}
Carbon Tetrachloride	1594	15.632	0.010	1.00×10^{-3}	6.04×10^{-3}	6.04×10^{-7}	1.104×10^{9}	0.0265	1.275×10^{4}
Kerosene	800	7.845	0.020	2.00×10^{-3}	2.50×10^{-2}	2.50×10^{-6}	—	0.0235	—
Benzene	860	8.434	0.007	7.00×10^{-4}	8.14×10^{-3}	8.14×10^{-7}	1.0356×10^{9}	0.0255	1.000×10^{4}
Castor Oil	960	9.414	9.800	9.80×10^{-1}	1.00×10^{1}	1.00×10^{-3}	1.441×10^{9}	0.0392	—
Ethyl Alcohol	789	7.737	0.012	1.20×10^{-3}	1.52×10^{-2}	1.52×10^{-6}	1.118×10^{9}	0.0216	5.786×10^{3}
Mercury	13550	132.880	0.016	1.60×10^{-3}	1.18×10^{-3}	1.18×10^{-7}	2.431×10^{10}	0.510	1.726×10^{-1}
Air	1.208	0.01185	1.85×10^{-4}	1.85×10^{-5}	1.53×10^{-1}	1.53×10^{-5}	—	—	—

Appendix B

Properties of water at different temperatures

Temperature (°C)	Density, ρ (kg/m^3)	Viscosity, μ (kg/m s)	Kinematic viscosity, ν (m^2/s)	Surface tension, σ (N/m)	Vapour-pressure head, p_v/γ (m)
0	999.9	1.792 ($\times 10^{-3}$)	1.792 ($\times 10^{-6}$)	7.62 ($\times 10^{-2}$)	0.06
5	1000.0	1.519	1.519	7.54	0.09
10	999.7	1.308	1.308	7.48	0.12
15	999.1	1.140	1.141	7.41	0.17
20	998.0	1.000	1.000	7.36	0.24
25	997.1	0.894	0.897	7.26	0.33
30	995.7	0.801	0.804	7.18	0.44

(Contd.)

35	994.1	0.723	0.727	7.10	0.58
40	992.2	0.656	0.661	7.01	0.76
45	990.2	0.599	0.605	6.92	0.98
50	998.1	0.549	0.556	6.82	1.26
55	985.7	0.506	0.513	6.74	1.61
60	983.2	0.469	0.477	6.68	2.03
65	980.6	0.436	0.444	6.58	2.56
70	977.8	0.406	0.415	6.50	3.20
75	974.9	0.380	0.390	6.40	3.96
80	971.8	0.357	0.367	6.30	4.86
85	968.6	0.336	0.347	6.20	5.93
90	965.3	0.317	0.328	6.12	7.18
95	961.9	0.299	0.311	6.02	8.62
100	958.4	0.284	0.296	5.94	10.33

Appendix C

Properties of air at different temperatures and at atmospheric pressure

Temperature, t (°C)	Density, ρ (kg/m^3)	Dynamic viscosity, μ (kg/m s)	Kinematic viscosity, ν (m/s) $\times 10^{-5}$
– 40	1.52	14.94 ($\times 10^{-6}$)	0.983
– 20	1.40	15.92	1.137
0	1.29	17.05	1.321
20	1.21	18.50	1.530
40	1.12	19.05	1.701
60	1.06	19.82	1.870
80	0.99	20.65	2.086
100	0.94	21.85	2.324
120	0.90	23.20	2.578

Appendix D

Properties of common gases at 273.16 K and atmospheric pressure

Gas	Molecular weight	Density ρ (kg/m³)	Gas constant R m/K	Specific heats c_p m/K	c_v m/K	Specific heat ratio, γ $(= c_p/c_v)$	Dynamic viscosity μ (kg/m s)
Air		1.293	29.3	101.3	72.2	1.402	17.05 $(\times 10^{-6})$
Carbon monoxide	28.0	1.250	30.3	107.1	76.3	1.404	16.60
Carbon dioxide	44.0	1.977	19.3	85.0	65.3	1.303	14 00
Helium	4.0	0.179	211.8	534.4	321.9	1.660	18.60
Hydrogen	2.02	0.090	420.3	145.8	1034 1	1.410	8.35
Methane	16.04	0.717	—	224.4	170 9	1.313	10.30
Nitrogen	28.0	1.250	30.2	106.1	75.6	1.404	16.70
Oxygen	32.0	1.429	26.5	93.1	66.5	1.400	19.20
Water vapour	18.0	0.800	47.1	206.0 (373 K)	154.9	1.338	8.70

Appendix E

Properties of Standard atmosphere

Altitude above sea level (m)	Absolute pressure (bar)	Absolute temerature (K)	Mass density ρ (kg/m^3)	Kinematic viscosity ν (m^2/s)
0	1.0132	288.16	1·2250	1.461 ($\times 10^{-5}$)
1000	0.8988	281.70	1.1117	1.581
2000	0.7950	275.20	1.0066	1.715
4000	0.6166	262.20	0.8194	2.028
6000	0.4722	249.20	0.6602	2.416
8000	0.3565	236.20	0.5258	2.904
10000	0.2650	223.30	0.4134	3.525
11500	0.2098	216.70	0.3375	4.213
14000	0.1417	216.70	0.2279	6.239
16000	0.1035	216.70	0.1665	8.540
18000	0.07565	216.70	0.1216	11.686
20000	0.05529	216.70	0.08892	15.989
22000	0.04047	218.60	0.06451	22.201
24000	0.02972	220.60	0.04694	30.743
26000	0.02188	222.50	0.03426	42.439
28000	0.01616	224.50	0.02508	58.405
30000	0.01197	226.50	0.01841	80.134
32000	0.00889	228.50	0.01356	109.62

Appendix F

Properties of areas

Shape	*Sketch*	*Area*	*Location of centroid*	*I* or I_{gg}
1	2	3	4	5
Rectangle	b, h, h_c	bh	$h_c = \frac{h}{2}$	$I_{gg} = \frac{bh^3}{12}$
Trapezium	a, b, h, h_c	$\left(\frac{a+b}{2}\right)^h$	$h_c = \frac{(2a+b)h}{3(a+b)}$	$I_{gg} = \frac{h^3}{24}(3b-a)$
Triangle	b, h, h_c	$\frac{bh}{2}$	$h_e = \frac{h}{3}$	$I_{gg} = \frac{bh^3}{36}$

(*Contd.*)

1	2	3	4	5
Circle	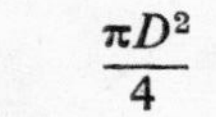 	$\frac{\pi D^2}{4}$	$h_c = \frac{D}{2}$	$I_{gg} = \frac{\pi D^4}{64}$
Semicircle	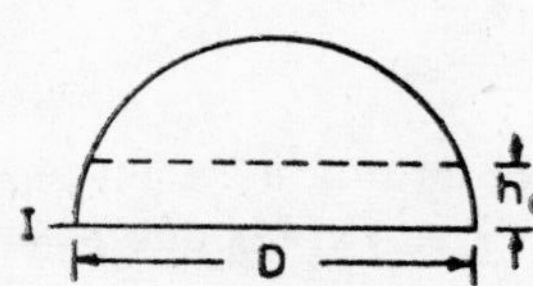	$\frac{\pi D^2}{8}$	$h_c = \frac{4R}{3\pi}$	$I = \frac{\pi D^4}{128}$
Parabola	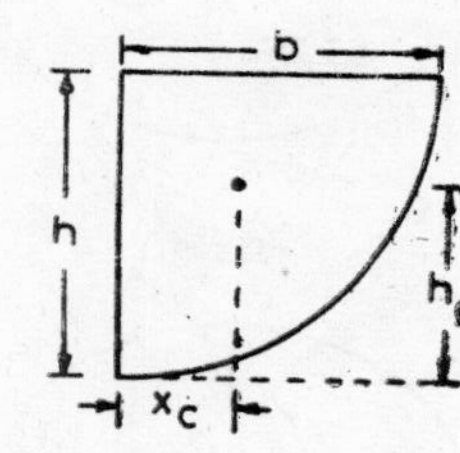	$\frac{2bh}{3}$	$x_c = \frac{3b}{8},\ h_c = \frac{3h}{5}$	$I = \frac{2bh^3}{7}$

(Contd.)

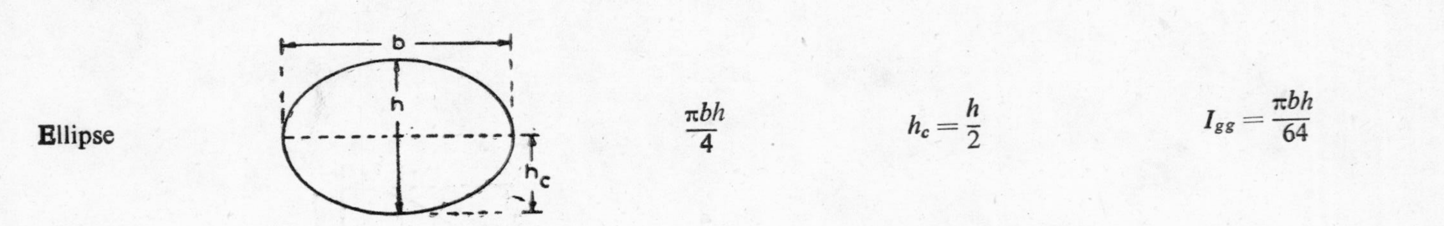

Ellipse		$\frac{\pi bh}{4}$	$h_c = \frac{h}{2}$	$I_{gg} = \frac{\pi bh}{64}$
Semiellipse		$\frac{\pi bh}{4}$	$h_c = \frac{4h}{3\pi}$	$I_{gg} = \frac{\pi bh^3}{16}$

The relation between I and I_{gg} is $I = I_{gg} + A\bar{h}^2$ where $\bar{h}$ is the depth of centroid from the axis about which I is calculated. I_{gg} is moment of inertia about an axis passing through centroid.

Notations

Notation	*Quantity*	*Dimensions*
a	Acceleration	LT^{-2}
a	Area	L^2
A	Area of flow, pipe, cylinder	L^2
A_s	Area of suction pipe, surge tank	L^2
A_d	Area of delivery pipe	L^2
b	Width	L
B	Bed width of rectangular or trapezoidal channel	L
c_f	Local drag coefficient	$M^0L^0T^0$
c_p	Specific heat at constant pressure	$L\theta^{-1}$
c_v	Specifie heat at constant volume	$L\theta^{-1}$
C	Coefficient	$M^0L^0T^0$
C	Chezy's discharge coefficient	$L^{1/2}T^{-1}$
C	Celerity of a pressure wave	LT^{-1}
C_c	Coefficient of contraction	$M^0L^0T^0$
C_d	Discharge coefficient of weirs, orifice plates	$M^0L^0T^0$
C_D	Drag coefficient	$M^0L^0T^0$
C_f	Average drag coefficient	$M^0L^0T^0$
C_v	Coefficient of velocity	$M^0L^0T^0$
d	Diameter of orifice plate, pipe, particle	L
D	Diameter of pipe, hydraulic mean depth	L

Symbol	Description	Dimensions
T	Absolute temperature in Kelvins	θ
T	Torque	ML^2T^{-2}
T	Water surface width	L
u	Instantaneous velocity at a point in x direction	LT^{-1}
u'	Turbulent fluctuations in u	LT^{-1}
u_*	Shear velocity ($=\sqrt{\tau_0/\rho}$)	LT^{-1}
u_m	Maximum velocity in x direction	LT^{-1}
u_r, u_θ, u_x	Velocity components in r, θ and x directions	LT^{-1}
$\bar{u}$	Time averaged velocity at a point	LT^{-1}
U, U_0	Average velocity of flow in a pipe or channel, Free stream velocity	LT^{-1}
U_d	Velocity of flow in delivery pipe	LT^{-1}
U_p	Velocity of flow in the cylinder	LT^{-1}
U_s	Velocity of flow in suction pipe	LT^{-1}
v	Instantaneous velocity at a point in y direction	LT^{-1}
v	Specific volume	$L^2T^2M^{-1}$
v'	Turbulent fluctuation in v	LT^{-1}
$\bar{v}$	Time averaged velocity at a point in y direction, absolute velocity	LT^{-1}
V_r	Relative velocity	LT^{-1}
V_f	Velocity of flow (in turbines and pumps)	LT^{-1}
V_w	Velocity of swirl (in turbines and pumps)	LT^{-1}
V_t	Tangential velocity	LT^{-1}
V_n	Normal component of velocity	LT^{-1}
$\forall$	Volume	L^3
w	Instantaneous velocity at a point, in z direction	LT^{-1}
w'	Turbulent fluctuation in w	LT^{-1}
$\bar{w}$	Time averaged velocity at a point, in z direction	LT^{-1}
W	Weir height	L
W	Weight of fluid	MLT^{-2}
W	Work done	ML^2T^{-2}

x	Distance in x direction	L
X	Component of body force per unit volume in x direction	$ML^{-2}T^{-2}$
y	Distance in y direction, depth of flow	L
y_b	Brink depth	L
y_c	Critical depth	L
y_0	Normal depth	L
$\bar{y}$	Depth of centroid of area below water surface	L
Y	Component of body force per unit volume in y direction	$ML^{-2}T^{-2}$
Z	Distance in Z direction, elevation	L
Z	Component of body force per unit volume in Z direction	$ML^{-2}T^{-2}$
Z_1	Section factor in open channels	$L^{8/3}$

Greek Notations

α	Energy correction factor, Mach angle, angle	$M^0L^0T^0$
β	Momentum correction factor, angle	$M^0L^0T^0$
γ	Specific weight	$ML^{-2}T^{-2}$
γ_s	Specific weight of sediment	$ML^{-2}T^{-2}$
δ	Boundary layer thickness	L
δ'	Laminar sub-layer thickness	L
δ_*	Displacement thickness of boundary layer	L
λ	Wave length	L
$\Delta\phi$	Change in entropy	$ML^2T^{-2}\theta^{-1}$
ϵ	Eddy kinematic viscosity	L^2T^{-1}
η	Efficiency, dimensionless distance (y/δ)	$M^0L^0T^0$
η	Eddy dynamic viscosity	$ML^{-1}T^{-1}$
θ	Angle	$M^0L^0T^0$
θ	Momentum thickness of boundary layer	L
μ	Coefficient of dynamic viscosity	$ML^{-1}T^{-1}$
ν	Kinematic viscosity	L^2T^{-1}
ρ	Mass density of fluid	ML^{-3}
σ	Coefficient of surface tension	MT^{-2}